MINERALOGY

CONCEPTS AND PRINCIPLES

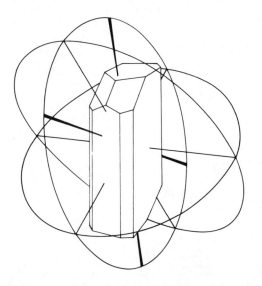

TIBOR ZOLTAI and JAMES H. STOUT

University of Minnesota
Minneapolis

Burgess Publishing Company
Minneapolis, Minnesota

Acquisitions editor: Wayne Schotanus
Assistant editor: Sharon Harrington
Developmental editor: Anne Heller
Copy editor: Betsey I. Rhame
Art director: Judy Vicars
Illustrations: Dennis Tasa
Composition: Computype, Inc.
Cover design: Dennis Tasa

Library of Congress Cataloging in Publication Data

Zoltai, Tibor.
    Mineralogy: concepts and principles.

    Bibliography: p.
    Includes index.
    1. Mineralogy.   I. Stout, James H.   II. Title.
QE363.2.Z65   1985   549   83-20992
ISBN 0-8087-2606-4

Burgess Publishing Company
7108 Ohms Lane
Minneapolis, Minnesota 55435

J I H G F E D C B A

To Olga and Ann

CONTENTS

PREFACE

Education in mineralogy has undergone rapid change in the last few years with the development of new instrumentation and the input of current research. For an undergraduate student, the demands are great. Since the turn of the century, mineralogy as a science has developed more along the lines of classical chemistry and physics. Today's student of mineralogy must be well versed in a scientific community in which *crystal chemistry* and *crystal physics* are common terms. At the same time, every mineralogist should be exposed to and aware of the large body of observational data on which the explanations of physics and chemistry are based. Going to the extreme in either direction has disadvantages. Too much emphasis on the descriptive aspects of mineralogy leaves the student with an unmanageable body of observational data, with few underpinnings as a basis for understanding. At the other extreme, a crystal chemist who cannot distinguish quartz from feldspar cannot be considered a mineralogist. Students are somehow expected to learn all of descriptive mineralogy that has gone before, and then to learn much of what modern research considers important today.

For the instructor, the demands are also great. Decisions must be made as to what of the old can be sacrificed for the new. This seems unavoidable in light of current research and the realization that the needs of the modern student of mineralogy are changing. The contents of this book have evolved from more than 20 years of collective teaching in response to those needs. It is designed for an introductory course at the university level, and is prerequisite to a first course in petrology or geochemistry. We emphasize the importance of fundamental principles in the belief that students who are well grounded in basic concepts can better assimilate and systematize in their own mind the vast body of descriptive mineralogy. Much of the conventional descriptive mineral data has been excluded. In its place in Part II of this text are nearly 100 stereographic representations of mineral structures. These, we believe, are essential to the understanding of how mineral chemistry and physical properties are related to mineral structure. Students are expected to refer continually to these representations where appropriate to the content of Part I. Throughout the text, the mineral names are followed by a representative mineral formula designed to convey information about both structure and chemistry. We hope that by repeated exposure, students will be better able to relate a mineral name to the crystalline substance it represents.

Through the years we have experimented with the organization of the subject matter of mineralogy. In Part I, the sequence of topics finally adopted begins with a review of descriptive mineralogy. The purpose of the review is in part historical, for this descriptive information led early mineralogists to formalize the concepts of symmetry that are so important to all that follows in the text. The purpose of our review is also practical, for it starts students off with a three-dimensional conception of real minerals that is an immense aid when crystal structures and their relation to physical properties are introduced. In Chapters 2 and 3 we explain the language of symmetry and crystallography. That language combined with concepts of chemical bonding in Chapter 4 prepares the student for a discussion of crystal structures in Chapter 5. Chapters 6 and 7 focus on crystal physics and crystal growth, and Chapters 8 and 9 explore the relationships between mineral chemistry, mineral structure, the stability of minerals, and mineral associations. Chapters 10 and 11 describe the diagnostic techniques of x-ray mineralogy and optical mineralogy. We introduce the methods of x-ray diffraction (Chapter 10) and optical mineralogy (Chapter 11) in our mineralogy laboratories as routine tools to be used and refined in later courses. The bulk of descriptive mineralogy is included in Part II of the text. Part II is intended as a supplement to Part I and for laboratory use.

Some chapters may have more depth of coverage than necessary for the average student. In our experience, most students gain from knowing that mineralogy involves more than just mineral identification, and the more conscientious students are provided the stimulation and challenge they want and need.

This is an exciting time to be in the field of mineralogy. We are now learning for the first time by direct experiment how single crystals behave at high temperature and pressure—the very conditions under which most minerals form. We are also on the verge of being able to predict the structure and properties of minerals that lie deep in the earth's crust and mantle where no human will probably ever tread. With modern microscopy, we are also beginning to see actual atoms in minerals—a goal mineralogists have long dreamed of but have not realized until now.

Throughout the text, we have adopted the international units of measurement (SI). Putting to rest the revered calorie and Ångström units may seem an unnecessarily harsh step for U.S. readers, but to resist is to postpone the inevitable. Our experience has been that resistance to change is more from educators than from students to whom the units of Ångström and nanometer (nm) are equally unfamiliar. A list of conversion factors is included in Appendix 1.

We sincerely thank the many colleagues and students who through the years have contributed to the final content of the text. Without their patience and cooperation we would have fallen short of our goal. Professors Charles W. Burnham, Robert M. Gates, Guy L. Hovis, Rodney C. Ewing, Edward P. Meagher, and Deane K. Smith read the manuscript in its entirety, and to them we extend our sincere thanks. Lastly, we thank Anne E. Heller, Betsey I. Rhame, and Wayne E. Schotanus, of Burgess Publishing Company, for their assistance in bringing the manuscript to its present form.

PART I

PRINCIPLES AND CONCEPTS

The first part of this text is devoted to the basic principles and concepts on which the science of mineralogy is based. The subject matter is organized in 11 chapters, each of which relies to some extent on material presented earlier. We begin in Chapter 1 with some historical background followed by an overview of the common physical properties of minerals. Chapters 2 and 3 introduce the language of symmetry and crystallography. In Chapter 4 these important relationships are combined with the concepts of chemical bonding to set the stage for crystal structures, the subject of Chapter 5. The following chapters, 6 and 7, on crystal physics and crystal growth, rely heavily on symmetry and structure. Chapters 8 and 9 deal with the important relationships between mineral chemistry and crystal structure, and the stability of minerals and their associations. The final two chapters deal with the theory and techniques of x-ray mineralogy and optical mineralogy.

CHAPTER ONE

MINERALS AND THEIR PROPERTIES

Mineralogy as a science dates back to the time when humans first wondered at the beauty and order of natural crystals. The use of minerals is as old as civilization itself. The first civilized humans survived because they recognized that certain minerals would provide sustenance and protection when fashioned as tools or weapons. As people came to depend on minerals, they became more selective of their various properties. Minerals for adornment and even for presumed magical powers became an important part of their life. Human intelligence to seek out certain minerals for their desirable properties was the basis for an early form of descriptive mineralogy.

Since then, nearly 3000 distinct minerals have been systematically studied and described in the mineralogical literature. This great accumulation of printed information has made impractical, if not impossible, the task of reading, analyzing, and digesting all of the information on one's own. Data collection is a necessary function of every science, and when gathered and assimilated, new concepts always emerge that guide further studies. In view of this, the approach employed in the following chapters is based on the belief that once students are armed with a relatively small number of fundamental concepts, they are able to discover and understand additional examples and applications far too numerous to be covered in a college course in mineralogy.

Before embarking on the subject, let us consider a few of the reasons why scientists study minerals. An important reason is that minerals occur in one form or another in nearly all inorganic materials of everyday life. Most familiar are the natural metals, which are derived from the earth and are used in everything from space-age alloys to amalgam to fill cavities in our teeth. A glance around any classroom quickly reveals that nearly every item has a mineral basis. The window glass is made primarily from feldspar and quartz, the chairs are metal, and even the wood and synthetic plastic items can be traced back ultimately to minerals in the earth.

Much of our space-age technology stems directly or indirectly from mineralogical studies. The quest to find and synthesize exceedingly pure crystals made possible the development of lasers and of oscillators, semiconductors, and transistors that form the essence of modern electronics. The Space Program and the first Apollo 12 samples from the moon marked a new era for mineralogical research in which numerous important advances have been made. These studies indicate that exploration of our neighboring planets is likely to reveal mineralogical problems similar to those encountered on earth.

Another reason to study minerals is that they are the constituents of nearly all rocks in the earth's crust and mantle. Research in the last 50 years has shown that some minerals can exist in a stable form over only a narrow range of pressure and temperature. Experimental confirmation of this range enables the scientist to make a correlation between the occurrence of a mineral in a rock and the conditions under which the rock was formed. This approach, perhaps more properly considered in the field of petrology, has helped scientists to unravel the earth's complicated history.

Finally, the study of minerals satisfies our natural curiosity about how things happen in nature. The most fundamental principles of science are reflected in one form or another in all natural phenomena. By understanding how and why minerals behave in their unique manner, we gain insight into other processes. Those insights can then be applied to predict the properties and behavior of different minerals or substances.

SOME EARLY HISTORY

The emergence of mineralogy as a separate field of study was signaled in 1556 by the publication of *De re metallica libris XII* by the German physician Georgius Agricola, known also as Georg Bauer. Agricola summarized for the first time a collection of observations that had accumulated for centuries in the midst of folklore and legend. Aided by his own factual observations, he outlined several categories of mineral properties that included color, transparency, luster, hardness, flexibility, and cleavage.

Based on observations of this kind, attempts were made as early as the 17th century to relate the physical appearance and properties of minerals to their internal structure. The internal structure of exactly *what* was still a fundamental question that would not be answered for another three centuries. Nonetheless, early scientists and philosophers conceived of tiny building blocks arranged in an internally periodic array long before the existence of atoms was demonstrated. The physicist and astronomer Johannes Kepler speculated in 1611 on the planar arrangement of spheres as a possible explanation of the symmetry of snowflakes. He recognized the two unique geometrical arrangements that today form the basis for the study of closest-packed structures (Chapter 5). Other 17th century scholars, such as René Descartes, Robert Hooke, and Christian Huygens, subsequently made contributions to the concept that mineral properties were more fundamentally related to internal structure.

Niels Stensen, whose name was latinized to Nicolaus Steno, studied further the external forms of crystals. Stensen was a physician known in his time for his contributions to anatomy. Drawing on his knowledge of how plants and animals grow and are nourished by internal fluids, he asserted that crystals grow by the addition of particles from an external fluid. He concluded, correctly, that the process of crystal growth is directional in nature, and that the eventual shape of a crystal depends on growth rates in different directions. Based on his study of the morphology of natural crystals, in 1669 he formulated the general relationship that the angles between the same two faces of a crystal are always constant, regardless of the size or shape of the crystal faces. This relationship later became known as Steno's law.

An important stage in the development of mineralogy came with the rising industrialism of the 18th century. There was a growing need to develop methods for identifying minerals and for systematizing mineral properties. In 1768, Carolus Linnaeus put forth one of the first systems of mineral classification. He was convinced that he had discovered in the mineral kingdom the same principles that had played such an important role in his botanical work. Based on common external properties, his system was divided into orders, classes, and species. The external form of the crystal was central to his classification, and his descriptions of crystals were so carefully done that later researchers referred to him as the founder of the science of crystallography.

In 1774, Abraham Werner, a professor of mineralogy at Freiburg, Germany, proposed the first comprehensive and consistent system of mineralogy. In his system of about 300 species, external form was only one of several properties that required evaluation before identification could be made. Werner's students spread his teachings over most of the civilized world, but his work was later attacked as being unscientific because it was not based on any single established theory of that day.

A milestone came in 1784 when René Just Haüy, in his *Traité de cristallographie*, conceived of crystal structures made up of identical *molecules integrantes*, or integral molecules. Figure 1.1, from Haüy's treatise, illustrates how similar his thoughts were to the modern concept of the space lattice and how clearly he visualized the relationship of the lattice to external crystal form. Haüy's successful synthesis of the current ideas of his time was coupled with his own studies of calcite ($CaCO_3$) and its distinctive rhombohedral cleavage. Because the rhombohedral form could be observed down to the eye's limit of resolution, Haüy believed that form to be characteristic of one of several basic building blocks from which all minerals were derived. Implicit to Haüy's theory was that each integral molecule had a specified chemistry, although he relied much more on form than chemistry in his classification. In his

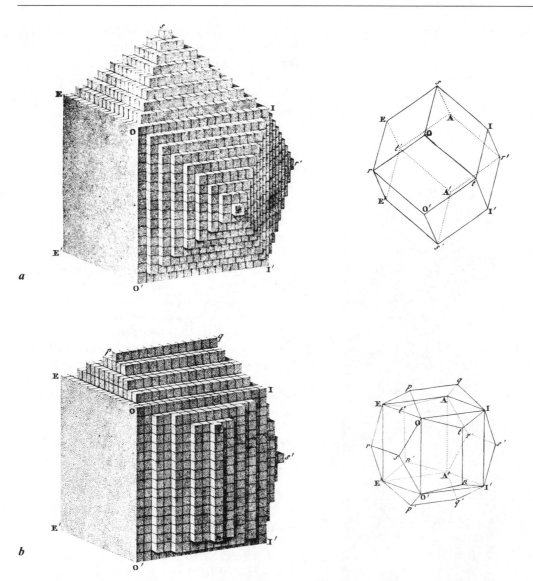

FIGURE 1.1 **(a)** René Haüy's illustration of how the dodecahedron (12 faces) crystal form with rhombic faces is derived from a cubic nucleus or integral molecule. **(b)**

Haüy's derivation of the dodecahedron with pentagonal faces from the cubic nucleus. (From Haüy, R. J., 1801. *Traité de minéralogie*. vol. 5, plate II.)

view, the integral molecule was the basic substance itself, and any further division would destroy its identity as such.

The priority of Haüy's concept of the relationship between internal structure and external morphology is not entirely clear. Eight years earlier in 1773, the Swedish chemist and mineralogist Torbern O. Bergman demonstrated how the scalenohedron of calcite ($CaCO_3$) could be formed from the rhombohedral nucleus (Figure 1.2) with angles of 101.5 and 78.5 degrees. Haüy was apparently aware of Bergman's results, but he discusses them only in the context of Bergman's erroneous attempt to relate the dodecahedral form of garnet to the rhombohedral nucleus of calcite. Bergman's most significant contribution was in the chemical

analysis of minerals in the laboratory. In 1779, Bergman gave a full account of the use of the blowpipe for chemical tests on minerals. It was the chemical approach to mineralogy more than anything else that eventually led to the downfall of Haüy's theory.

As chemistry began to contribute more to the science of mineralogy, the role of crystal forms became less dominant. After the turn of the century, Francois Beudant, a former student of Haüy, discovered that aqueous solutions containing differing proportions of dissolved ferrous sulfate ($FeSO_4$) and zinc sulfate ($ZnSO_4$) always crystallized the same distinctive rhombohedral crystals on evaporation. Haüy's theory held that $FeSO_4$ and $ZnSO_4$ had different forms, and thus Beudant's observation led to controversy. Beudant concluded that chemically composite compounds were not mechanical mixtures of two different, integral molecules as Haüy believed, but rather were chemical mixtures that exhibit a continuous range of physical properties. About the same time, William H. Wollaston completed a similar study

of calcite ($CaCO_3$), magnesite ($MgCO_3$), and siderite ($FeCO_3$), concluding like Beudant that those minerals should be regarded as homogeneous chemical mixtures having the same crystal form. With these results, the modern concept of *solid solution* or *isomorphism* was established.

Similar problems with Haüy's theories arose with the recognition that even though calcite ($CaCO_3$) and aragonite ($CaCO_3$) have entirely different physical properties, they have identical chemistries. In 1821, Eilhard Mitscherlich, a student of the Swedish chemist and mineralogist Jons J. Berzelius, proposed the modern concept of *polymorphism* by demonstrating that an identical number of the same elements can arrange themselves in such different ways that resulting external forms and physical properties are different.

With evidence mounting that the internal arrangement of chemical elements in minerals is highly symmetric, Christian Samuel Weiss set the science of crystallography on a new course. A brilliant scholar who received his doctorate at the age of 20, Weiss was a proponent of the polar theory of matter, which held that elementary particles were drawn together by attractive forces and held apart by repulsive forces. His formulation of the concept of crystallographic axes and their relationship to symmetry axes in a three-dimensional space was published in 1815. By considering mutually perpendicular axes, he identified the isometric, tetragonal, and orthorhombic crystal systems. He also recognized the natural divisions of minerals into those with sixfold rotational symmetry and those with threefold rotational symmetry. Hence hexagonal and trigonal crystal systems came into being. Then in 1825, Friedrich Mohs, the successor to Werner at Freiburg and inventor of the hardness scale that bears his name, demonstrated that both monoclinic and triclinic systems exist by considering nonorthogonal crystallographic axes.

The science of crystallography now developed rapidly. Johann Hessel, the German physician and mineralogist, derived in 1830 the fact that there are exactly 32 crystal classes, and that only twofold, threefold, fourfold, and sixfold axes of rotational symmetry are compatible with translation. Hessel's findings were based on his exhaustive study of the possible types of symmetry any geometrical form might have. In 1840, Gabriel Delafosse wrote correctly that the integral molecule of Haüy was the outline of nodes in the crystal lattice—that is, a geometrical element distinct from the physical and chemical entity we now call the unit cell. This observation was a major turning point conceptually, for it stripped

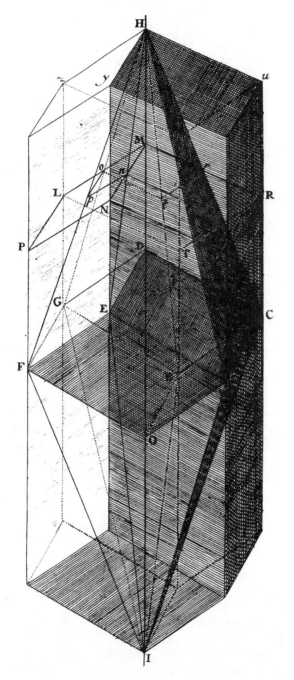

FIGURE 1.2   Torbern Bergman's illustration of how the calcite prism (HURCIFP) and scalenohedron (HCIF) could be derived from a rhombic nucleus (COFD). (From Bergman, T. O., 1773. Variae crystallorium formae, e spatho ortae. *Nova Acta Regiae Societatis Scientiarum Upsaliensis*. vol. 2, plate II.)

Haüy's form of any chemical significance. In 1848, Auguste Bravais proposed independently the 32 crystal classes that Hessel derived earlier. More importantly, Bravais proposed the 14 space lattices that were the forerunner of space group theory. Bravais perceived that the 14 lattices consisted of seven different lattice symmetries, which corresponded to the previously recognized seven crystal systems. His work established the fact that external symmetry is grounded on the concept of the space lattice.

The last major question that remained was exactly how atoms were arranged within unit cells. It became a geometrical problem for scientists to determine the number of symmetrical ways in which points could be arranged in space such that the environment around each point was identical. In 1879, Leonard Sohncke provided part of the answer by recognizing two new symmetry elements called a screw axis and a glide plane. Armed with these concepts, the Russian crystallographer E. S. Federov derived the 230 space groups and published his results in 1881. Artur Schoenflies, a German mathematician who was unaware of Federov's work, used mathematical group theory and published the same results in 1890. About the same time, William Barlow, a self-educated genius who had the unusual advantage of independent wealth, also concluded after studying the symmetrical arrangement of spheres that there must be 230 space groups.

With Röntgen's discovery of x-rays in 1895, the stage was set to test the earlier structural models. In 1911, Max von Laue, professor of physics at the University of Munich, and his two assistants Walter Friedrich and Paul Knipping, passed a narrow beam of x-rays through a crystal of copper sulfate, and history was made. Diffraction spots appeared on a photographic plate placed behind the crystal. The group concluded that such a response could only occur if (1) x-rays were electromagnetic in nature and had short wavelengths and (2) those wavelengths were diffracted by a regular arrangement of atoms with spacing comparable to those wavelengths.

This result launched the entire field of crystal structure analysis. Two years later in 1913, the first crystal structure was determined by the father-and-son team of William H. Bragg and William L. Bragg. For their discovery, the Braggs jointly received the Nobel Prize for physics in 1915. The fields of solid-state physics and crystal chemistry were born. The precise positions of atoms in structures and the distances between atoms were determined for the first time. Inferences on the actual sizes of atoms, and the types and strengths of chemical bonds between atoms followed.

In the 70 years since the first crystal structure was determined, mineralogy as a science has benefited from important technological advances. The advent of sophisticated x-ray precession cameras in the 1940s enabled crystallographers to collect better data for crystal structure refinements. Today, with the aid of automated diffractometers, the basic structures of all but a few minerals are known. With the development of the electron probe microanalyzer (microprobe) in the 1960s, determining the chemical composition of a single crystal over areas no greater than a few hundredths of a millimeter in diameter is now a routine procedure. The scanning electron microscope (SEM) is another instrument that has opened an entirely new world of observation, providing magnification and resolution far beyond that obtainable with the optical microscope. A related instrument, the transmission electron microscope (TEM), has signaled a new, exciting era of mineralogical research. With the TEM, atoms and their arrangements are directly imaged to yield a view of minerals never before possible.

MINERAL CLASSIFICATION

Every beginning student of mineralogy soon discovers an incredible variety of minerals. They occur in all colors, shapes, and sizes, and in all rocks from the most beautiful museum specimens to common roadside gravel. All minerals have one common property—they are crystalline. That is, their internal structure is characterized by a periodic and predictable array of atoms, ions, or molecules. This property alone distinguishes minerals from chemically equivalent liquids and gases, and imparts to minerals their unique chemical and physical characteristics. We therefore define a *mineral* as *any naturally occurring crystalline substance*. The condition that a mineral must be naturally occurring is mostly for convenience. There is an entire field of study that involves the manufacture of exceedingly pure crystals for electronic and other industrial purposes. Some of these crystals are equivalent in every respect to naturally occurring minerals. The condition that all minerals be crystalline is generally valid, although in a few instances the degree of crystallinity might be questioned. The term *mineraloid* applies to naturally occurring substances such as obsidian and opal, the structures of which may be only partially crystalline or even noncrystalline.

An important consequence of the periodic arrangement of atoms in crystalline structures is that certain groups of minerals have common properties that clearly set them apart from all other mineral groups. Crystalline structure imposes definite limits on the range of chemical composition in each group, and no intermediate states of symmetry, chemistry, or other physical properties will exist within certain limits.

One objective of mineralogical studies is to provide a meaningful scheme by which a large body of mineral properties can be organized. In creating such a scheme, the best place to begin is to recognize that all properties of minerals are fundamentally related to their *chemistry* and *structure*. Any attempt to classify minerals in terms of chemistry at the exclusion of structure, or vice versa, shows immediately how closely these two factors are related. An examination of the elemental composition of the earth's crust (all of the rocks on earth down to a depth of about 40 km beneath the continents and to a depth of about 10 km beneath the oceans) reveals that oxygen is unquestionably the most abundant element (Table 1.1). For every 100 atoms in the earth's crust, 63 are oxygen and 21 are silicon. By volume, oxygen constitutes well over 90% of most crustal rocks and over 90% of most minerals that occur within crustal rocks (Table 1.1).

Because of the abundance and relatively large size of the oxygen anion (O^{2-}), the possible geometric arrangements of O^{2-} determine to a great extent the structural framework of most of the silicate and oxide minerals. As a first approximation, the structure of these minerals may be considered in terms of a symmetrical framework of oxygen anions, with the interstitial voids that remain being occupied by the smaller, less abundant cations. The other common anions are sulfur (S^{2-}), chlorine (Cl^{1-}), and fluorine (F^{1-}). Although much less abundant than O^{2-}, they behave in much the same way by dominating the structural framework of the sulfide and halide minerals. In addition, a number of cations combine with oxygen to form *anionic groups*, which, like oxygen or sulfur alone in other minerals, dominate the volume and structural framework of mineral groups such as the carbonates, sulfates, and nitrates.

The generally recognized chemical classification of minerals based on the predominant anion or anionic group is listed here in the same order as described in Part II. Of these categories, the silicate minerals are easily the most abundant in most crustal rocks.

1. Silicate minerals: Oxygen anions combined with silicon cations to form either SiO_2 or various SiO_x anionic groups. The latter groups combine with various cations. Example: wollastonite ($CaSiO_3$).
2. Native elements: naturally occurring metallic and nonmetallic elements. Example: gold (Au).
3. Sulfide minerals: Sulfur anions combined with various cations. Example: pyrite (FeS_2). The anions As, Se, and Te are included in this category.
4. Halide minerals: F, Cl, and I anions combined with various cations. Example: halite (NaCl).
5. Oxide minerals: Oxygen anions combined with various cations other than Si. Example: periclase (MgO).
6. Hydroxide minerals: Hydroxyl (OH) groups combined with various cations. Example: brucite ($Mg(OH)_2$).

TABLE 1.1. *Atomic Abundances of the Common Elements in the Earth's Crust and in Some Common Silicate Minerals*

| Element | Ionic Radius* | Crustal Composition | | | Quartz (SiO_2) Vol.% | Albite ($NaSi_3AlO_8$) Vol.% | Anorthite ($CaSi_2Al_2O_8$) Vol.% | Diopside ($CaMgSi_2O_6$) Vol.% |
		Weight Percent	Atom Percent	Volume Percent				
O	0.134	46.6	62.6	93.8	98.7	93.3	91.6	88.6
Si	0.034	27.7	21.2	0.9	1.3	0.9	0.6	0.8
Al	0.047	8.1	6.5	0.5	—	1.0	1.9	—
Fe	0.069	5.0	1.9	0.4	—	—	—	—
Mg	0.080	2.1	1.8	0.3	—	—	—	3.0
Ca	0.108	3.6	1.9	1.0	—	—	5.9	7.6
Na	0.124	2.8	2.6	1.3	—	4.8	—	—
K	0.168	2.6	1.4	1.8	—	—	—	—

*Ionic radii from Appendix 5 are for the most common coordinations: O^{2-} (8), Si^{4+} (4), Al^{3+} (4), Fe^{2+} (6), Mg^{2+} (6), Ca^{2+} (6), Na^{1+} (8), and K^{1+} (12).

7. Carbonate and nitrate minerals: CO_3 and NO_3 anionic groups combined with various cations. Example: calcite ($CaCO_3$).
8. Borate minerals: BO_x anionic groups combined with various cations. Example: sinhalite ($MgAlBO_4$).
9. Sulfate minerals: SO_x anionic groups combined with various cations. Example: anhydrite ($CaSO_4$).
10. Chromate, tungstate, and molybdate minerals: CrO_x, WO_x, and MoO_x anionic groups combined with various cations. Example: scheelite ($CaWO_4$).
11. Phosphate, arsenate, and vanadate minerals: PO_x, AsO_x, and VO_x anionic groups combined with various cations. Example: triphylite ($LiFePO_4$).

An important, useful feature of this classification is that most of the minerals in each group have certain properties in common. All of the carbonates, for example, have much closer resemblances to each other than exist between all calcium-bearing minerals. All carbonate minerals have similar structures, whereas the structures of the calcium-bearing silicates, oxides, and phosphates, for example, are quite different. This raises an important question: Is the observed range of chemistry in the carbonates due to their similar structures, or are the structures the result of their similar chemistries?

Every classification scheme must provide for categorizing observations. The problem with the scheme just described is that it does only that. It is much like an encyclopedia with an index that grows ever larger as the number of entries increases. A more useful scheme would place greater emphasis on the fundamental principles that govern how atoms or ions are arranged. In addition to categorizing, the ideal classification system would also provide a basis for predicting mineral structures, even if those structures were not previously described. When we consider that 75% of the earth's volume is at a depth where no human can ever explore directly, we realize that we must have tools to understand minerals we cannot see. To accomplish this goal, the fundamental structural relationships that exist between all minerals must be built into the classification scheme.

The systematics of crystal structures are based principally on the *symmetry* of ion and atom arrangements. In the discussions that follow here and in later chapters, keep in mind that the symmetry of physical properties is a consequence of more fundamental principles that are ultimately related to crystal structure. With some historical perspective, we can now understand how these principles developed from the observation of mineral properties.

HAND SPECIMEN MINERALOGY

Every mineral has a distinctive chemistry and crystalline structure, and the complete description of a mineral requires that both the chemistry and structure be known. These determinations, however, require special instruments that are frequently not available for routine mineral identification. Fortunately, the crystal structure and the chemistry of most minerals are sufficiently distinctive to yield diagnostic physical properties. Determination of these properties is generally sufficient for the identification of most minerals, or at least to distinguish between groups of minerals that are structurally unrelated. The physical properties that are most commonly and readily observed in minerals are:

1. Crystal form and habit
2. Luster and transparency
3. Color and streak
4. Cleavage, fracture, and parting
5. Tenacity
6. Density or specific gravity
7. Hardness
8. Unique properties such as taste, magnetism, fluorescence, and radioactivity

Our purpose in this chapter is simply to define these properties and to draw attention to their usefulness in describing minerals. The fundamental explanations of these properties are the subjects of later chapters.

CRYSTAL FORM AND HABIT

One of the most striking, attractive features of well-developed single crystals is their *form*. The term is used in everyday language in reference to shape, but in mineralogy it has an additional meaning. *Crystal form* is defined by symmetrically equivalent crystal faces. The external faces of a single crystal are an expression of the symmetrical arrangement of the atoms within the structure. If the crystal symmetry is known, the symmetrical equivalents of a particular crystal face are predicted easily, and these faces collectively constitute a single form. Conversely, a study of crystal forms can lead to the determination of crystal symmetry. Most minerals can crystallize in any of several different forms or combinations of forms (Figure 1.3) depending on the growth rate and environment of formation. Although possibly not apparent at this stage, all of the forms of calcite

($CaCO_3$) in Figure 1.3 have a common property. Their symmetry is the same, or put in the parlance of crystallography, they belong to the same crystal class.

The term *habit* is used in mineralogy to describe the general shape of a single crystal and aggregates of crystals. In the case of single crystals, the habit may correspond to a particular form or combinations of forms, but in general, the habit differs from the form. Figure 1.4 illustrates some of the common habits of single crystals.

Minerals such as natrolite ($Na_2Si_3Al_2O_{10} \cdot 2H_2O$) and scolecite ($CaSi_3Al_2O_{10} \cdot 3H_2O$), which have long, needlelike shapes, are said to be *acicular*. Minerals such as beryl ($Al_2Si_6Be_3O_{18}$), which are thicker than acicular minerals and have a parallel arrangement of crystal faces much like the faces of a column, are said to have a *columnar* or *prismatic* habit. Stubby, equidimensional crystals such as garnet ($Fe_3Al_2Si_3O_{12}$) and zircon ($ZrSiO_4$) have an *equant* habit. Minerals may also be *tabular* if they occur as flattened plates (e.g., barite,

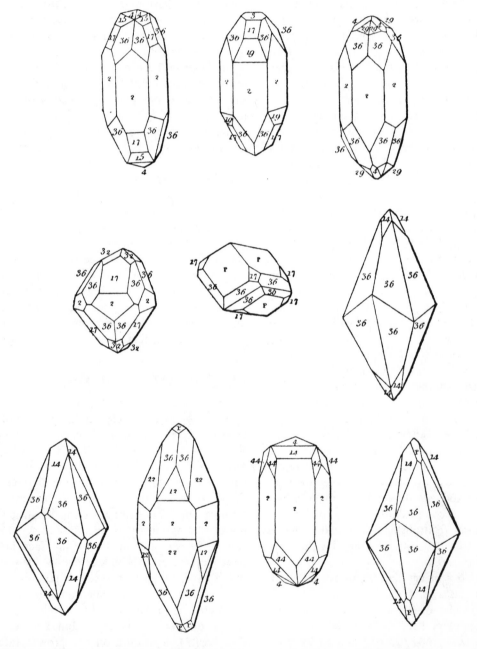

FIGURE 1.3   Several crystal forms of calcite ($CaCO_3$) from Goldschmidt, V. M., 1913. *Atlas der Kristalformen.* vol. 2, table 25. (Reprinted courtesy of Carl Winter, Universitätsverlag, Heidelberg.)

FIGURE 1.4   The common habits of single crystals.

a   Acicular

b   Columnar or prismatic

c   Tabular

d   Foliated

$BaSO_4$), or *foliated* or *micaceous* if they occur as thin, flexible sheets. The term *bladed* refers to flat, elongated crystals with a shape resembling that of a knife blade. Kyanite (Al_2OSiO_4) commonly has a bladed habit. The terms *capillary* and *filiform* are used interchangeably to describe delicate, hairlike shapes such as those commonly observed in millerite (NiS). *Blocky* is a term used to describe the habit of equant crystals with square or rectangular cross sections. Typical forms of all of the common rock-forming minerals are illustrated in Part II of this book directly above the appropriate column of the mineral properties tables.

A different category of terms is used to describe the habits of mineral aggregates. Several of the more common habits in this category are illustrated in Figure 1.5. *Dendritic* habit refers to mosslike patterns commonly developed by native copper (Cu), pyrolusite (MnO_2), and other minerals on joint surfaces in rocks. Such patterns when developed in chalcedony give rise to the varietal name moss agate. *Arborescent* is a term used to describe treelike growth patterns such as those developed by ice (H_2O), native silver (Ag), copper (Cu), and gold (Au). *Reticulated* refers to a pattern of crisscrossing crystals, whereas *stellate* describes a radiating, starlike shape. The term *fibrous* is generally applied to minerals that appear to be composed of individual crystals that resemble organic fibers. Serpentine ($Mg_6(OH)_8Si_4O_{10}$), gypsum ($CaSO_4 \cdot 2H_2O$), and many other minerals may develop fibrous habit.

Additional descriptive terms for mineral aggregates include *botryoidal* (grapelike), *reniform* (kidneylike), and *mammillary* (breastlike). *Lamellar* describes the growth pattern of thin, parallel plates. The term *oolitic* is used to describe fine-grained, spherical aggregates. *Pisolitic* refers to the same pattern, but the spheres are larger than the approximate size of a pea. The terms *massive*, *granular*, *radiating*, *plumose* (featherlike), and *stalactitic* are self-explanatory and may also be used when appropriate.

Unfortunately, minerals do not always develop as good individual crystals or crystalline aggregates. More often than not they occur as massive aggregates composed of many small, irregularly shaped crystals. In these cases, the usefulness of crystallographic principles in the hand-specimen identification of minerals is clearly limited. However, in spite of the submicroscopic size of these tiny crystals, they still possess the same basic symmetry as the macroscopic crystals, and this symmetry is readily revealed by x-ray diffraction. Indeed, x-ray diffraction is one of the most powerful tools in mineralogical research today.

FIGURE 1.5   The common habits of crystal aggregates.

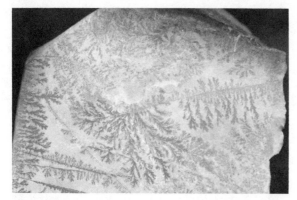

a   Dendritic

b   Reticulated

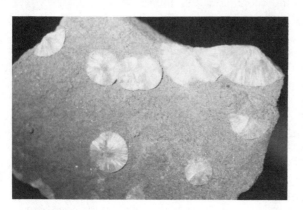

c   Stellate

d   Fibrous

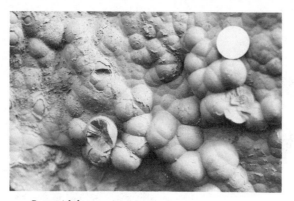

e   Botryoidal

LUSTER AND TRANSPARENCY

Every student who has looked over a suite of minerals can pick out those that look like metals. All minerals have the property of *luster*, which is a measure of light reflectivity. Metals and many metal sulfides and oxides reflect nearly all of the visible light that impinges on them. They are therefore considered opaque to visible light. Their high reflectivity imparts the property of *metallic luster*. Minerals that transmit visible light are said to be *transparent*, or if the quality of transmission is poor, they may be *translucent*. Both transparent and translucent minerals have *nonmetallic luster*. The actual mechanism by which absorption and reflection take place requires some knowledge of the differences in energy states of electrons in atoms and ions and how those differences compare to the energy of light. This topic is covered in more detail in Chapters 4 and 11.

Highly transparent minerals like diamond (C) appear brilliant to the eye and are said to have *adamantine* luster. Minerals such as quartz (SiO_2) that appear bright and glassy on a freshly broken surface are said to have *vitreous* luster. The terms *resinous*, *pearly*, *silky*, *dull*, *earthy*, and *greasy* are self-explanatory and are used to describe the luster of certain minerals.

The property of luster is one of the quickest and easiest to determine. The mineral hardness and identification table in Appendix 2 is divided into two major groups on the basis of luster.

COLOR AND STREAK

Another striking feature that imparts much of the natural beauty to minerals is their color. The brilliant red color of rubies and the deep blue color of sapphires have always been prized for their aesthetic value. Color in this case is not a diagnostic property, because both ruby and sapphire are varieties of the same mineral, corundum (Al_2O_3). Crystals of fluorite (CaF_2) may be purple, brown, yellow, black, or colorless, depending on the absorptive properties of small amounts of impurity ions.

The normal light to which we are accustomed consists of a range of energies that the eye records as white. Everyone has seen how this range is dispersed by refraction in the form of a rainbow. Each color is associated with a particular wavelength, frequency, and energy. The constituent ions and atoms of minerals have electrons in various energy states. When some of these electron energy states are identical with certain component energies of white light, those components are absorbed. Ferrous (Fe^{2+}) iron is a good energy absorber at the red end of the visible spectrum. The remaining part of the spectrum is transmitted, and in the case of minerals like fayalite (Fe_2SiO_4) and actinolite ($Ca_2Fe_5(OH)_2Si_8O_{22}$), a green color results. Chromium behaves in a similar manner, imparting the green color to emerald, a variety of beryl ($Al_2Si_6Be_3O_{18}$), and to certain varieties of garnet (uvarovite, $Ca_3Cr_2Si_3O_{12}$), and to mica ($K(Al, Cr)_2(OH)_2Si_3AlO_{10}$).

Color is diagnostic for only a few minerals. Azurite ($Cu_3(CO_3)_2(OH)$) is always blue, and malachite ($Cu_2CO_3(OH)_2$) is always green. Most minerals, however, exhibit a range of color depending on their specific chemistry and structure.

The presence of disseminated inclusions of other substances may impart color to any mineral. Milky quartz, for example, is filled with tiny fluid and gas inclusions that scatter visible light. Contrast this with the color of rose quartz, which is due to small amounts of manganese in its structure, and with the color of smoky quartz, which is probably due to radiation damage. In arid regions when the same phenomenon occurs in the ultraviolet part of the spectrum, clear bottle glass is converted to the pink or violet varieties so sought after by bottle collectors.

The color of a powdered mineral, referred to as its *streak*, may be different from its color as a single crystal. For example, the color of hematite (Fe_2O_3) is commonly dark red to black, but it gives a reddish brown streak when drawn across an unglazed piece of porcelain. The color of the mineral's streak is much more diagnostic than the color of the unfragmented mineral. The streak is one of the simplest criterion used to distinguish between most gray, metallic minerals. Indeed, streak is such an easily determined property that it also provides a useful criterion for distinguishing categories of minerals with nonmetallic luster.

CLEAVAGE, FRACTURE, AND PARTING

Like nearly all objects, minerals break along surfaces of least resistance. In most minerals, the chemical bonding of atoms is not uniformly strong in all directions throughout the structure. The chemical bonding along certain planes may be weaker, and consequently, a mineral may *cleave* along that interface rather than break randomly elsewhere. Some outstanding and important examples of this are the perfect basal cleavages of graphite (C), which make it useful as a lubricant, and of talc ($Mg_3(OH)_2Si_4O_{10}$), which is the primary ingredient of baby powders. Scientists made these observations of cleavage long before the nature of atoms was understood, and their observations provided an important clue that minerals were made in a regular, symmetric way.

A crystal such as sillimanite ($AlSiAlO_5$) that cleaves along one set of parallel planes is said to have only one cleavage plane. Other minerals may have two or more distinct, nonparallel sets of cleavage planes. The symmetrical arrangement of the planes is usually expressed in the nomenclature of crystal forms even though crystal faces and cleavage develop by quite different processes. A single cleavage plane is called *basal* or pinacoidal, whereas two cleavage planes intersect to produce *prismatic* fragments, as in tremolite ($Ca_2Mg_5(OH)_2Si_8O_{22}$) (Figure 1.6a). Three cleavage planes commonly produce different closed forms, as in halite (NaCl) and galena (PbS). These minerals develop *cubic* cleavage fragments (Figure 1.6b). Calcite ($CaCO_3$) develops *rhombohedral* cleavage

FIGURE 1.6   Some common cleavage forms of minerals.

a   Prismatic cleavage of tremolite

b   Cubic cleavage of galena

c   Rhombohedral cleavage of calcite

d   Octahedral cleavage of fluorite

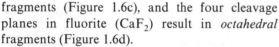

fragments (Figure 1.6c), and the four cleavage planes in fluorite (CaF_2) result in *octahedral* fragments (Figure 1.6d).

The quality of the cleavage planes and the relative tendency of the mineral to develop cleavage is usually expressed in terms such as *perfect* (e.g., mica), *good* (e.g., realgar), and *poor* or *imperfect* (e.g., apatite).

If there are no preferred planes of weakness that are controlled by crystal structure, a mineral breaks along a random *fracture*. These surfaces are not planar, but frequently result in distinctive patterns. Quartz (SiO_2) provides an outstanding example of *conchoidal* fracture in which the surface resembles that of broken bottle glass. Small, irregular steps characterize the fracture surface of native metals and many sulfide minerals, in which case the term *hackly* fracture may be used. The important point is that fractures have little or no symmetry, and therefore do not express any fundamental behavior of the internal arrangement of atoms.

Occasionally a mineral breaks along a surface that is too regular to be called a fracture but not sufficiently planar or repeating to be termed cleavage. The property of *parting* results. Parting frequently develops along planes of weakness between twins or along isolated planes of deformation. Minerals that cool slowly from a high temperature or that come to the earth's surface from a great depth may develop parting due to a change in volume. Unlike cleavage, parting is not repeated throughout the crystal on a fine scale, and is distinguished from cleavage on this basis.

TENACITY

The term *tenacity* refers to the mechanical properties of minerals. If a mineral breaks easily when struck by a hammer, it is said to be *brittle*. Cleavage may or may not develop, depending on the mineral. Some minerals resist breaking and are literally *tough*. Jade, a submicroscopic aggregate of either the pyroxene jadeite ($NaAlSi_2O_6$) or the amphibole actinolite ($Ca_2Fe_5(OH)_2Si_8O_{22}$), is the toughest natural substance known. Many minerals, particularly the softer metals, are said to be *malleable* because of their ability to be pounded into a sheet. *Ductile* minerals are those that can be drawn out mechanically into a wire, and *sectile* minerals can be cut with a knife.

DENSITY AND SPECIFIC GRAVITY

The density of a mineral is one of its most important properties, because it is an expression of both chemistry and crystal structure. *Density* is defined as mass per unit volume, and conventionally has been expressed in grams per cubic centimeter (g/cm^3). In the International System of Units (SI) adopted herein, density is expressed in units of megagrams per cubic meter (Mg/m^3). Quartz (SiO_2), for example, has a density of 2.65 g/cm^3, which we express as 2.65 Mg/m^3 so that the number 2.65 does not lose its usual physical impression. Minerals composed of heavy elements (e.g., Fe, Cu, Cr) have a high density compared with minerals composed of lighter elements. A simple example is aragonite ($CaCO_3$) with a density of 2.95 Mg/m^3 compared with witherite ($BaCO_3$) with a density of 4.30 Mg/m^3. Both have the same structure.

The efficiency with which atoms are packed together in a crystal structure also affects mineral density. Two or more minerals may have the same chemistry but different structures. In general, a mineral that forms in a high-pressure environment is denser than a different mineral with the same chemistry that forms in a low-pressure environment. The mineral formed in a high-pressure environment has more atoms of the same kind put into the same volume. A good example of a structure formed under high pressure is diamond (C), with a density of 3.50 Mg/m^3. Graphite (C), with a density of 2.30 Mg/m^3, is formed in a low-pressure environment. Both have the chemistry of elemental carbon.

Specific gravity is closely related to density but easier to measure. It gives the relative density of a mineral with respect to water, and is defined as the weight of the mineral divided by the weight of an equal volume of water at a temperature of 4 °C. In practice, the mineral is first weighed in air. Then the mineral is totally submerged in a beaker of water and weighed again. Given the relationship

$$specific\ gravity = \frac{weight\ in\ air}{weight\ in\ air - weight\ in\ water}$$

we see that specific gravity, unlike density, has no units. Its numerical value differs from density by a factor of 0.0001, which is considerably less than any error of measurement.

The disadvantage of measuring density directly

is that the volume of an irregularly shaped mineral specimen is extremely difficult to determine. Specific gravity eliminates this difficulty with no loss of information. It is an important property that is useful for identifying minerals.

HARDNESS

One of the more important and most easily determined physical properties is *hardness*. Early mineralogists must have wondered why diamond (C) was so hard, and talc ($Mg_3(OH)_2Si_4O_{10}$), for example, so soft. Long before the answer was known, Friedrich Mohs in 1812 proposed a qualitative order of hardness based on a series of ten standard minerals. Diamond was the hardest (10), and talc (1) the softest. In between were common minerals that defined a scale of "scratchability" (Figure 1.7). That is, any mineral on the Mohs scale can be scratched by all minerals above it. If an unknown mineral could scratch orthoclase (hardness 6) but could be scratched by quartz (hardness 7), the hardness of the unknown mineral was determined to be 6.5.

Little did Mohs know that he was measuring, albeit in a qualitative way, a property directly related to the strength of chemical bonds, for at that time no one had proved that minerals consisted of atoms held together by different kinds of forces. We now know that the "scratchability" of a mineral depends on how strong the bonds are that hold atoms and ions together and on the density of the bonds in the crystal structure. Hard minerals have strong bonds that are symmetrically distributed, whereas soft minerals generally have weak bonds that are constrained to certain specific directions in a structure. The foremost example is diamond (C) and graphite (C). Some of the

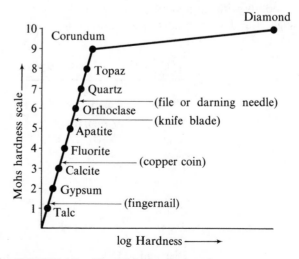

FIGURE 1.7   The Mohs hardness scale.

carbon-carbon bonds in graphite are actually stronger than those in diamond, but some of the bonds that hold the graphite sheets together are weak. These weak bonds are the ones that are important as far as hardness is concerned, because once these are broken by a scratch test, the mineral is deemed soft.

Another fact unknown to Mohs was that diamond (C) has a hardness far greater than its nearest competitor, corundum (Al_2O_3). Based on actual bond energy (Chapter 4), diamond is nearly five times harder than corundum. To eliminate the nonlinearity in the Mohs scale, in 1935 Charles Woodel introduced a modified hardness scale (Table 1.2) based on the standardized method of abrasion. Corundum remained at a hardness of 9, but diamond was assigned a hardness of 42.5.

The National Bureau of Standards devised yet another scale to measure hardness by making an indentation in the material of interest. Known as the Knoop scale, it measured the force per unit area required to produce a permanent deformation in a polished surface. This method is more sensitive than the Mohs scale to small differences in hardness and can be used to determine differences in hardness from face to face on the same crystal. Diamond cutters have long known that every diamond has "easy" and "hard" directions, trade expressions for small differences in hardness between faces. A skillful cutter uses these differences in the preparation of quality gemstones.

Hardness on the Mohs scale is determined easily. The only special materials required are a small piece of each standard mineral, excluding diamond. Care should be taken to choose pointed or sharp-cornered standards to facilitate an unambiguous scratch test. In drawing a particular standard mineral across the smooth surface of an unknown mineral, the standard either slides smoothly with no noticeable friction or it catches slightly as a scratch is generated. Only light pressure is required to make the test, and examining the results with a hand lens or magnifying glass is always helpful before drawing definite conclusions.

In the absence of the Mohs set of minerals, one can make do with a fingernail, a copper coin or wire, a knife blade, and either a darning needle or a broken file (Figure 1.7). Once determined, relative hardness is an extremely useful property and provides one of the most reliable criteria for mineral identification (Appendix 2).

UNIQUE PROPERTIES

Numerous other properties are present only in a few minerals and can be of aid in identifying them. The property of *taste* is diagnostic for halite (NaCl) and sylvite (KCl). Some sulfide minerals give off a distinctive *odor* when crushed. Magnetite (Fe_3O_4) is strongly *magnetic*, whereas most minerals are not. Tourmaline has a *pyroelectric effect*, the property of generating an electric current by the application of heat, and quartz (SiO_2) has a *piezoelectric effect*, the property of generating an electric current by the application of stress.

The optical behavior of minerals in transmitted light is another category of important properties. We need a polarizing microscope for the quick determination of properties such as the index of refraction, pleochroism, and birefringence. Chapter 11 is devoted to the subject of optical mineralogy.

An optical property that does not require a microscope for observation is *luminescence*. Certain minerals emit light in response to an external influence, or absorb light of one wavelength and reemit it at a different wavelength. The latter quality is called *fluorescence* or *phosphorescence*, depending on whether the emission of light ceases immediately (fluorescence) or continues on (phosphorescence) after the exciting light is turned

TABLE 1.2. *Comparison of the Various Hardness Scales*

Mineral	Composition	Mohs Hardness	Knoop Hardness	Mohs-Woodel Hardness
Talc	$Mg_3(OH)_2Si_4O_{10}$	1	—	1
Gypsum	$CaSO_4 \cdot 2H_2O$	2	32	2
Calcite	$CaCO_3$	3	135	3
Fluorite	CaF_2	4	163	4
Apatite	$Ca_5F(PO_4)_3$	5	430	5
Orthoclase	KSi_3AlO_8	6	560	6
Quartz	SiO_2	7	820	7
Topaz	$Al_2(OH, F)_2SiO_4$	8	1340	8
Corundum	Al_2O_3	9	2100	9
Diamond	C	10	7000	42.5

off. Several minerals are fluorescent, and when exposed to ultraviolet light, emit visible light of different colors (e.g., scheelite ($CaWO_4$), fluorite (CaF_2), and calcite ($CaCO_3$)).

Radioactivity is a unique property associated with minerals that contain the unstable isotopes of uranium, namely, ^{238}U and ^{235}U, or thorium, ^{232}Th. These isotopes disintegrate spontaneously to produce lighter elements along with a variety of radiations that can be detected with a Geiger counter. Uraninite (UO_2), thorianite (ThO_2), and carnotite ($K_2(UO_2)_2(VO_4)_2 \cdot 3H_2O$) are examples of radioactive minerals.

SUMMARY OF PROPERTIES

We can readily appreciate how rapidly the descriptive data on minerals can accumulate when all of their properties are described. One wonders what must have gone through the minds of earlier investigators who tried to synthesize these observations into a few basic principles. Which of the many properties did they consider important, and why? One investigative approach was to identify a more fundamental property that all the other properties had in common. This fundamental property turns out to be *symmetry*. It is the only feature common to such diverse properties as cleavage, hardness, and crystal form. Its most obvious expression is in crystal forms and cleavage because these are immediately apparent to the unaided eye. A closer examination of the optical, magnetic, or electrical properties of minerals, and indeed all other properties that are fundamentally related to structure, reveals the presence of symmetry. This topic is of such importance to all that follows that we will treat it next. As we proceed, keep in mind that symmetry, more than any other concept, draws together the many and diverse observations of descriptive mineralogy.

THE LITERATURE OF MINERALOGY

Numerous professional societies around the world publish journals on a regular basis in which recent advances in mineralogy are reported. In the United States, the *American Mineralogist* is published monthly by the Mineralogical Society of America. In Canada, the *Canadian Mineralogist* serves the same function. Other scientific journals that are devoted to mineralogy are *Mineralogical Magazine*, which is published by the Mineralogical Society of Great Britain, and *Physics and Chemistry of Minerals*, which was first published in 1977.

The above journals include many contributions to the field of crystallography. In addition, the journals *Acta Crystallographica*, *Zeitschrift für Kristallographie*, the *Journal of Applied Crystallography*, and publications of the American Crystallographic Association are devoted to reporting research on the structures and properties of materials in general, including many minerals.

The *Mineralogical Record*, first published in 1970, is dedicated to the amateur mineralogist. High-quality photography, most in color, and practical information on famous localities around the world make this a valuable supplement.

A standard reference work for the common rock-forming minerals is the five volume set *Rock-Forming Minerals* by Deer, Howie, and Zussman. A condensed version is also available in paperback. Another standard reference work on minerals in general is *Dana's System of Mineralogy*, consisting of three volumes devoted to specific mineral groups. Lastly, a useful German reference is *Mineralogische Tabellen* by H. Strunz.

ADDITIONAL READINGS

Battey, M. H., 1981. *Mineralogy for students*. 2nd ed. London and New York: Longman.

Berry, L. G., and Mason, B., 1959. *Mineralogy*. San Francisco: Freeman.

Burke, John G., 1966. *Origins of the science of crystals*. Berkeley: Univ. of California Press.

Deer, W. A., Howie, R. A., and Zussman, J., 1962. *Rock-forming minerals*. 5 vols. New York: John Wiley and Sons.

Fleischer, M., 1980. *A glossary of mineral species 1980*. Tucson, Arizona: Mineralogical Record.

Ford, W. E., 1918. The growth of mineralogy from 1818 to 1918. *American Journal of Science* 46:240–254.

Hurlbut, C. S., 1968. *Minerals and man*. New York: Random House.

Hurlbut, C. S., and Klein, C., 1977. *Manual of mineralogy* (after James D. Dana). 19th ed. New York: John Wiley and Sons.

Palache, C., Berman, H., and Frondel, C., 1944. *Dana's system of mineralogy*, vol. 1; 1951, vol. 2; 1962, vol. 3. New York: John Wiley and Sons.

Strunz, H., and Tennyson, C., 1970. *Mineralogische tabellen*. 5th ed. Leipzig: Akademische Verlag.

Wenk, H.-R., ed., 1976. *Electron microscopy in mineralogy*. New York: Springer-Verlag.

CHAPTER TWO

SYMMETRY AND CRYSTALLOGRAPHY

One objective of all scientific endeavor is to state in the simplest terms possible those natural laws and concepts that explain natural phenomena. Isaac Newton's discovery of the universal law of gravitational attraction and Einstein's recognition of the relationship between energy and mass are two examples of scientific "laws" or concepts that enable us to understand a bewildering assortment of observations. We recognize such laws as fundamental because of their universal application, their simple mathematical form, and the common link they provide between apparently unrelated phenomena. In 1933, Einstein summarized these points in the following statement:

> Nature is the realization of the simplest mathematical ideas. I am convinced that we can discover, by means of purely mathematical constructions, those concepts and those lawful connections between them which furnish the key to the understanding of natural phenomena.

The concept of symmetry must rank among the fundamental relationships in science. Its universal application can hardly be questioned. We observe symmetry at every scale in nature, from the orbits of celestial bodies to the behavior of electrons in atoms. The concept of symmetry is also broad in its application, providing a fundamental similarity between such diverse and apparently unrelated phenomena as the formation of snowflakes, the rhythms of music, and the behavior patterns of bees. More important for our purposes is the relationship of physical properties to the arrangement of atoms and ions on an atomic scale. Once the symmetry of a crystal is known, we can use that knowledge to predict many of the crystal's physical properties before these are measured or even observed. Thus, being a fundamental concept, symmetry provides science with a set of relationships that require no deeper explanation.

DEFINING SYMMETRY

Unlike the classical laws of mechanics, the concept of symmetry is not defined easily. The term has a variety of connotations, depending on the context in which it is used. For example, in everyday language we use "symmetry" to describe the regular arrangement of patterns in space and time. In this context, symmetry is usually a periodic and systematic repetition such as one observes in patterned wallpaper. The term symmetry is also used to describe the intrinsic beauty of the entire object in relation to its parts. This usage refers to a certain order that can be more easily sensed and appreciated than explained.

Aristotle formulated one of the earliest and broadest definitions of symmetry:

> In symmetrical concepts the parts build up the whole, not as a pile of components but as a coherent entity.

The parts referred to in this definition are purposely left unspecified so that the definition retains its generality. The parts might be material objects, patterns of design, sound, color, or any quality that can follow a pattern of repetition. These repeated qualities or features we will hereafter refer to as *motifs*. Some motifs occur in space. We see examples in the symmetry of crystals, of plants and animals, and in architecture. Other motifs are repeated in time, such as the regularity of night and day or the rhythm of music. The repetitive nature of the medium, whether in space or time, makes the collection of parts a "coherent entity" according to Aristotle's early definition.

We define symmetry in a much narrower sense as it applies to mineralogy. *Symmetry is the order in arrangement and orientation of atoms in minerals, and the order in the consequent distribution of mineral properties.* This is a modern interpretation of crystallographic symmetry in the sense that our definition presupposes a knowledge of how individual atoms are arranged in the crystal structure. The atomic structure of most minerals, and certainly all of those found in common rocks, is known and is related directly to the symmetry of mineral properties.

A useful and popular definition of symmetry that was formulated by Georges Friedel in 1904 is based solely on the symmetry of properties. That is, *the symmetry of a crystal is the symmetry common to all of its properties.* The only problem with Friedel's definition is that the apparent true symmetry of a mineral is likely to change as new properties are discovered or as the symmetry of existing properties becomes better understood. If we accept the crystal structure as a property, our definition and Friedel's are equivalent.

CRYSTALLOGRAPHY

The systematic study of the symmetries of crystals is called crystallography. The crystals of galena (PbS) shown in Figure 2.1 possess symmetry by any definition. A mineralogist who is well versed in the language and methods of crystallography can determine that each of these crystals possesses $4/m\overline{3}2/m$ point group symmetry. The nomenclature may seem like a foreign language at this juncture, but to a mineralogist the symbols convey a concise description of the external symmetry of galena. This information is not sufficient in itself to distinguish galena from most other minerals. Elimination of the other possibilities (i.e., those other minerals with the same external symmetry) requires some additional criteria such as chemistry, or the value of some other physical property.

The use of crystallographic symmetry is not

FIGURE 2.1   Galena (PbS) crystals all have the common property of identical internal symmetry.

only valuable for mineral identification but also indispensable in modern experimental mineralogy. One of the ultimate objectives of crystallographic research is to know in detail the crystal structures of all important minerals. Symmetry operates on all parts of a crystal and controls the arrangement of atoms and the chemical bonds between them. Once we determine the relative position of a few atoms, we can use the crystal's symmetry to reconstruct the positions of the several trillion other atoms in a small crystal.

SYMMETRY OPERATIONS

The basic components of crystallographic symmetry are the symmetry operations. Only seven unique symmetry operations are necessary for the description of crystal symmetry. These are divided into the basic and the compound symmetry operations:

1. Basic symmetry operations
   a. Translation
   b. Rotation
   c. Reflection
   d. Inversion
2. Compound symmetry operations
   a. Screw rotation
   b. Glide reflection
   c. Rotoinversion or rotoreflection

Various sets of these operations are present in real minerals. Therefore, the operations singly or in permissible sets have a three-dimensional representation. All but three of the operations (inversion, screw rotation, and rotoinversion) can be represented in two dimensions. The ease with which symmetry operations can be described in a plane is convenient and realistic, and the arguments can be extended to three dimensions with little modification.

BASIC SYMMETRY OPERATIONS

Translation

Basic symmetry operations are those operations that cannot be broken down into more fundamental components. *Translation* is the basic symmetry operation that expresses the systematic repetition of points by displacement. The operation may be performed in one-dimensional, two-dimensional, or three-dimensional space. Figure 2.2 is a two-dimensional electron image of a mixed chain silicate called chesterite $(Mg_{17}(OH)_6Si_{20}O_{54})$. The magnification is approximately four million times,

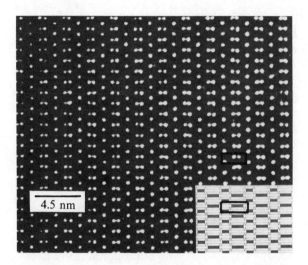

FIGURE 2.2 Electron microscope image of electron distribution in the mineral chesterite, an alternating double and triple chain silicate. (Courtesy of Dr. D. Veblen.)

and is produced by a technique known as transmission electron microscopy. The light regions are areas of high electron density and correspond to the positions of actual atoms along chains in the crystal structure. A striking feature of this image is its translational symmetry. The light-colored "motif" is repeated in two dimensions by the operation of translation.

A similar but more familiar example of translational symmetry is the repeated patterns of commercial wallpaper (Figure 2.3). Several other symmetry operations are present as well in this particular pattern, and we will discuss each of these as we proceed. Collectively, these operations constitute a *symmetry group*.

The periodicity, or translational symmetry, in Figure 2.3 can be described by two unit translations. These are vectors and hence have both direction and magnitude. Any pair of several possible unit translations could be chosen to describe the repetition of motifs in the wallpaper. For example, the star motif *A* at the origin of the heavily outlined vector pair is translated by the vertical vector to *B*, and again (dashed line) to *D*. The diagonal vector of the unit pair translates the motif at *A* to *C*, then to *F*, and so on. Note that the entire pattern around each of these points is the same, so when a single point (or motif) is translated, all space around it is also translated.

Unit translations *AB* and *AG*, or *AH* and *AI*, or any other combination could be chosen with the motif at point *A* as the origin. The hexagonal motif at *J* or the hexagon apex at *K* could have been chosen as easily. In each case, the arbitrarily chosen origin and all two-dimensional space

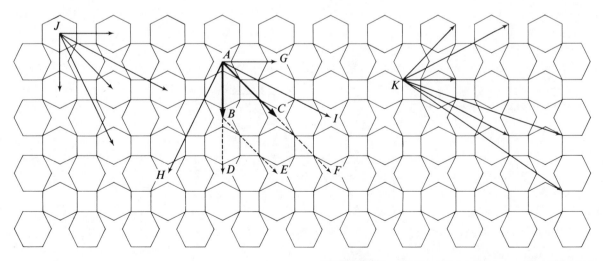

FIGURE 2.3   Commercial wallpaper pattern illustrating alternate choices of unit vectors to describe lattice translations.

around it is repeated periodically according to the choice of unit translations.

Of the possible choices of unit translations in Figure 2.3, the two shortest vectors are unique. As Figure 2.4 illustrates, their origin in the pattern is arbitrary. The translational symmetry of any pattern can be described from any point in the pattern. Whatever motif or point is chosen as the origin, that motif or point is repeated by the unit translations. A collection of translationally equivalent points in two-dimensional space defines a *mesh*, and in three-dimensional space defines a *lattice*. Stated another way, the lattice is the translational symmetry of the mineral structure. The translationally equivalent points in the lattice are referred to as *lattice points*.

The area defined by two unit translations is called a *unit mesh*. The volume defined by three unit translations in three dimensions is called the *unit cell*. By that definition, the unit mesh is always a parallelogram, and the unit cell is a parallelepiped. The unit mesh in the upper left part of Figure 2.5 is a square, as the two unit translations chosen have equal length and the angle between them is 90 degrees. Each of the four lattice points at the corners of the unit mesh is also a corner for three other adjoining unit meshes. Only one fourth of each lattice point belongs to a given mesh, and consequently, each unit mesh contains only one lattice point ($1/4 \times 4 = 1$). Meshes of this kind are referred to as *primitive* because no additional lattice points are internal to the mesh. In three dimensions, a unit cell with only one lattice point is called a *primitive unit cell*.

If the unit mesh is defined by the two shortest

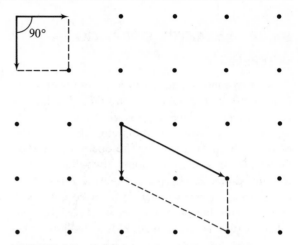

FIGURE 2.5   Primitive and multiple unit meshes.

unit translations, the unit mesh is said to be *reduced*. The primitive unit cell, defined by the three shortest noncoplanar unit translations, is called a *reduced unit cell*. The square unit mesh in Figure 2.5 is both primitive and reduced. For some purposes that we will consider later, unit translations that yield a *multiple* unit mesh or unit cell are a convenient choice. Such a mesh or cell contains more than one lattice point. In Figure 2.5, the parallelogram in the lower right is a multiple unit mesh and contains two lattice points.

A unique unit cell can be defined for every mineral, and the dimensions and shape of the cell are diagnostic properties for identification. Methods by which these properties are actually determined are discussed in Chapter 10. Actual unit cell dimensions range from about 0.5 nm[1] to over

[1]One nanometer = 10^{-9} meters = 10 Ångstrom units.

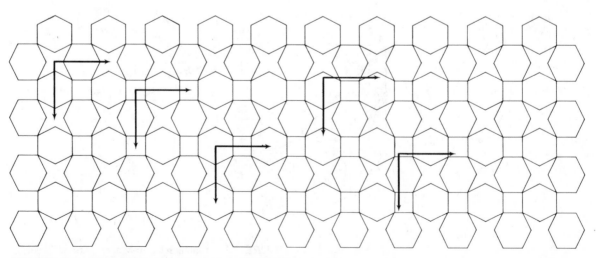

FIGURE 2.4   Wallpaper pattern illustrating that origin of the unit mesh is arbitrary.

2 nm for the prevalent rock-forming minerals. The common atoms and ions in the earth's crust have radii that range from about 0.03 nm to over 0.3 nm. Thus, unit cells contain many atoms, at least as many as are represented in a standard chemical formula. The unit mesh of the projection of the chesterite structure (Figure 2.2), for example, is a unit mesh with an angle of 90 degrees between the unit translations. This unit mesh is actually the projection of the unit cell along the third unit translation, which is perpendicular to the plane of the paper.

The symbol for a general unit translation is t. Symbols for the specific unit translations that define the unit cell of a lattice are a, b, and c. (These parameters for all of the common minerals are tabulated in Part II.) By international convention, the labeling of a, b, and c unit translations should always be made as in the right-handed coordinate system. If a and b unit translations are equal in magnitude because of the total symmetry of the crystal, they are labeled a_1 and a_2. If c is equal to a_1 and a_2, it is labeled a_3.

Rotation

The symmetry operation of *rotation* is the repetition of a point or motif about an axis. Repetition through an angular interval of 360 degrees brings the motif back exactly to its original position. Repetition at intervals of 180 degrees requires two steps to achieve the original position, and hence is referred to as a twofold axis of rotation. Repetition at intervals of 120 degrees, 90 degrees, and 60 degrees corresponds to threefold, fourfold, and sixfold axes of rotation, respectively. That is, an angular interval of α corresponds to an n-fold rotation of $n = 360/\alpha$. The value of n is frequently referred to as the *rank* of the rotation axis. If one imagines a crystal positioned on a threefold axis of rotation, it will appear exactly the same to a viewer at every 120 degrees of rotation around that axis. If the viewer's eyes are closed to the actual rotation, and then opened, no evidence appears from the morphology of the crystal itself that the rotation has taken place.

Rotational symmetry is extremely common in nature. In the symmetry of starfish, flowers, and other organisms, we find fivefold rotation and angular intervals other than those mentioned above. The only permissible axes of rotation in minerals, however, are those that have angular intervals of 360, 180, 120, 90, and 60 degrees. These are the angular intervals that are compatible with the concept of translational symmetry. Figure 2.6 illustrates a simple proof of this fact. Let the origin be at A, and the unit translations

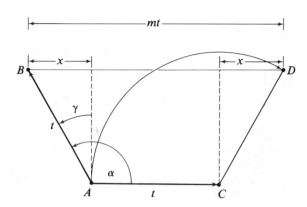

FIGURE 2.6 **Graphical representation of how only twofold, threefold, fourfold, and sixfold axes of rotation are compatible with translational symmetry.**

have length t with an angle of α between them. Translation requires the rotation axis at A (vertical to the page) also to be at points B, C, D, and so on. Other values of α are limited to those values that permit the rotation of one lattice point into translationally equivalent lattice points. In Figure 2.6, the rotation axis at point A must also exist at point C, because these points are translationally equivalent. Lattice points A and D must then be equivalent and related through the angle α as measured from C. Because B and D must also be equivalent, the distance between them must be an integer multiple m of the unit translation t. That is,

$$BD = mt = t + 2x$$

where $x = t \sin \gamma$. It follows that

$$mt = t + 2t \sin \gamma$$
$$= t + 2t \sin(\alpha - 90)$$
$$= t - 2t \cos \alpha$$

or

$$m = 1 - 2 \cos \alpha$$

or

$$\cos \alpha = \frac{1 - m}{2}$$

The only values of m in this equation that yield values of α consistent with a lattice are -1, 0, 1, 2, and 3 (Table 2.1). The permissible values of α,

TABLE 2.1 *Permissible Values of Rotation α Consistent With Translation*

m	$\sin(\alpha - 90°)$	α	Symbol	Axis Name
0	$-1/2$	60°	⬡	Sixfold
1	0	90°	■	Fourfold
2	$1/2$	120°	▲	Threefold
3	1	180°	◆	Twofold

in that order, are 0 or 360, 60, 90, 120, and 180 degrees. The reason we find objects in nature with rotation axes other than these is because many objects are not constrained by translational symmetry.

In a similar fashion, the locations of rotation axes (except for the onefold) are restricted by the lattice. Rotation axes must be oriented parallel to unit translations and perpendicular to planes of lattice points. This is because motifs repeated by rotation must remain in a plane perpendicular to the axis. Furthermore, rotation axes either must coincide in position to such translations or must be located halfway between them. This is simply because motifs repeated by rotation are at equal distances from the rotation axis.

Returning to our wallpaper, Figure 2.7a shows the presence of fourfold axes at each of the star motifs. All of the translationally equivalent fourfold axes are also shown. Note that another fourfold axis, which is symbolized with a tilted square, is not equivalent to the first because it cannot be generated by translation from the first. Both operate on each other to generate the overall symmetry pattern.

In addition to the two fourfold axes, Figure 2.7 illustrates single twofold axes. All of the twofold axes are equivalent by the operations of translation and fourfold rotation. Thus, only one twofold axis is unique. The total symmetry, both translational and rotational, discussed in the pattern up to this point is summarized in Figure 2.7b. The origin of the unit translations, and consequently of the unit mesh, has been taken arbitrarily at one of the fourfold axes. In this representation, the shapes of the motifs are immaterial—only the symmetry relationships between them are important.

Reflection

A point or motif can be repeated in space by the operation of *reflection*. The motif is repeated across a reflection plane by a parallel, equidistant projection, just as an image is reflected by a mirror. Thus, reflection planes are usually referred to as *mirror planes*. Like rotation axes, the positions of mirror planes in a structure must be compatible with the lattice. This means that mirror planes can be present only in those lattice planes and halfway between those lattice planes that have lattice rows in the third dimension perpendicular to them.

A mirror plane in Figure 2.8 that is denoted by the heavy vertical line and the symbol *m* is projected from the third dimension perpendicular to the page. All of the two-dimensional space on one side of the mirror is reflected to the other side in such a way that motifs *A* and *B* or *C* and *D*, for example, are equivalent. Translation requires a family of parallel planes, each of which passes vertically through the star motifs. The presence of the fourfold axis requires an equivalent mirror *m'* at 90 degrees to the first mirror *m*. These are

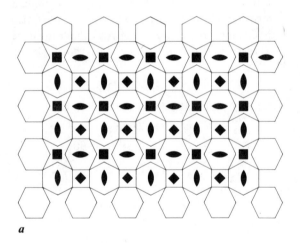

a

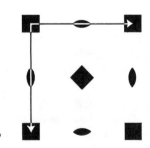

b

FIGURE 2.7   (a) Wallpaper pattern with locations of fourfold and twofold axes of rotation. Note how each symmetry element operates on all others. (b) Summary of both translational and rotational symmetries in the same pattern.

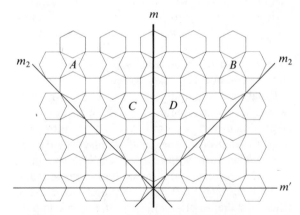

FIGURE 2.8   Symmetrically equivalent mirror planes *m* and *m'* are related by fourfold axes on the star motifs. Mirror planes m_2 are symmetrically unrelated to *m*.

considered crystallographically as a single mirror, because they are all symmetrically equivalent by virtue of the fourfold operation. A second set of mirror planes, m_2, runs diagonally through both the star and square motifs. The previously identified operations must operate on m_2 to generate the entire family of symmetrically equivalent planes.

Inversion

Inversion is a symmetry operation in which a point or motif is repeated by equidistant projection through a point. The operation can be visualized as reflection across a point, rather than across a plane as in the preceding mirror example. The point of inversion is called a *center of inversion* (or center of symmetry), and is denoted by the symbol *i*. Figure 2.9a illustrates the operation whereby every point of the motif is projected through *i*.

In a crystal, the permissible locations of centers of inversion are restricted to lattice points and points halfway between them. These are the only positions in a lattice across which all pairs of lattice points are equivalent by inversion. In the three-dimensional unit cell shown in Figure 2.9b, centers of symmetry must exist at all corners and

halfway between them (i.e., at points halfway along all edges, at the centers of all faces, and at the center of the cell).

Unlike the operations of translation and rotation, the operations of a mirror plane and a center of inversion reverse the "sense" of the motif. That is, the first two operations, translation and rotation, repeat a left-handed motif as a left-handed motif. Such operations are referred to either as *congruent*, or as operations of the *first sort*. In contrast, mirror planes and centers of inversion repeat left-handed motifs as right-handed motifs, as Figure 2.9a illustrates. These operations are referred to as *enantiomorphous*, or as operations of the *second sort*.

COMPOUND SYMMETRY OPERATIONS

Compound symmetry operations are combinations of basic symmetry operations. When two operations are combined into one, and a motif is repeated only when both operations are completed, the resulting pattern of motifs may be different from the pattern of the two basic operations acting on the motif in sequence. That is, certain combinations of basic symmetry operations yield a new type of operation. Different possible combinations of symmetry operations are numerous, but only three give unique symmetry patterns that cannot be duplicated as sets of basic operations acting in sequence.

Screw Rotation

Screw rotation consists of two basic operations —rotation and translation—combined in a single operation. The axis of rotation is referred to as the *screw axis*. The location of screw axes in a lattice is limited to permissible sites of rotation axes. To be compatible with a lattice, the magnitude of the translational component (τ) must be a submultiple of the unit translation parallel to the axis. For every *n*-fold rotation axis, the possible values of τ can be derived from

$$\tau = \frac{mt}{n}$$

where *m* is an integer and *t* is the unit translation along the rotation axis. For example, the values of τ and *m* consistent with a fourfold axis are $\tau = t/4$ ($m = 1$), $\tau = t/2$ ($m = 2$), and $\tau = 3t/4$ ($m = 3$). The conventional symbol of a screw axis is n_m, and for this example the values are 4_1, 4_2, and 4_3. The first symbol (Figure 2.10a) can be read to mean that the motif is rotated 90 degrees, and then translated by $1/4t$. The motif is then

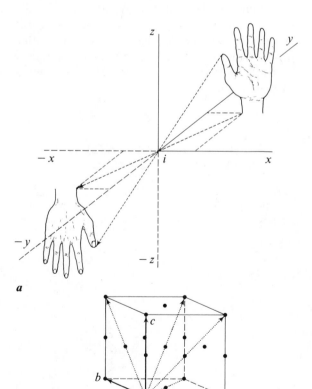

FIGURE 2.9 **(a) Operation of an inversion center. (b) Inversion centers must be present in a lattice at centers of all faces and halfway along cell edges.**

NOTE: $\tau = \frac{mt}{n}$　$n = 4$

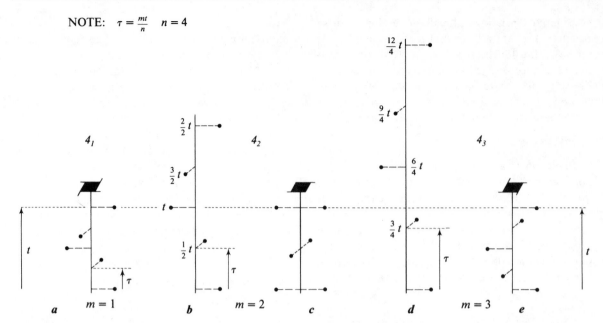

FIGURE 2.10 Compound symmetry operation of four-fold screw axes. (a) 4_1, (b) 4_2 showing $2t$ total translation after four 90° rotations, (c) 4_2 after restoring transla-tional equivalence of lattice points, and (d) and (e) 4_3 with $3t$ total translation and restoration of translational equivalence.

rotated another 90 degrees and again translated by $1/4t$. This operation is repeated two more times until the collective translation, after 360 degrees of rotation, is $4/4t$ or a single unit transla-tion. This cycle of operations returns the original motif to a translationally equivalent position. The second symbol (Figure 2.10b) can be read to mean that the motif is rotated 90 degrees, and then translated by $\tau = 2/4t$, and so on until 360 degrees of rotation have been achieved and the total translation parallel to the screw axis is $8/4t$ or $2t$. Normal translation requires that if a motif is at $2t$, it must also be at t. Similarly, if the motif is at $5/4t$, it must also be at $1/4t$ as Figure 2.10c illustrates. In the case of a 4_3 screw axis (Figure 2.10d), the first 90-degree rotation is accompanied by a translation of $3/4t$. The operation is repeated three more times until the total rotation is 360 degrees and the total translation is $12/4t$ or $3t$. Lattice translation requires that the motif at $3t$ be repeated at $2t$ and t, and that the motif at $9/4t$ be repeated at $5/4t$ and $1/4t$ (Figure 2.10e). The other permissible screw axes and their sym-bols are illustrated in Figure 2.11.

Glide Reflection

The compound symmetry operation called *glide reflection* consists of a translation and a reflection. The plane of reflection is called a *glide plane*. To be consistent with a lattice, glide planes are re-stricted to the same locations as mirror planes. The translational component τ of the glide must be parallel to a unit translation t in the plane and have a magnitude of $t/2$ (Figure 2.12). If τ is parallel to a cell edge, we use the term *axial glide*. If τ points to the center of a unit cell or to the center of any of its faces, we use the term *diagonal glide*. The *diamond glide* has one half the τ value of the diagonal glide and is restricted to centered unit cells. The various types of glides are summa-rized in Table 2.2.

TABLE 2.2. *Types of Glide Planes, Their Translation Directions, and Translational Components*

Type of Glide	Translation Direction	τ	Symbol for Glide
Axial	a	$\dfrac{a}{2}$	a
	b	$\dfrac{b}{2}$	b
	c	$\dfrac{c}{2}$	c
Diagonal	n	$\dfrac{a}{2} + \dfrac{b}{2}$	n
		$\dfrac{a}{2} + \dfrac{c}{2}$	n
		$\dfrac{b}{2} + \dfrac{c}{2}$	n
Diamond	d	$\dfrac{a}{4} + \dfrac{b}{4}$	d
		$\dfrac{a}{4} + \dfrac{c}{4}$	d
		$\dfrac{b}{4} + \dfrac{c}{4}$	d

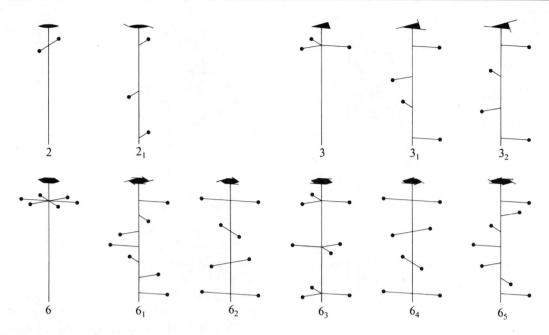

FIGURE 2.11 Illustration of twofold, threefold, and sixfold screw axes (after restoration of translational equivalence).

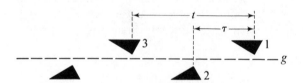

FIGURE 2.12 Operation of a glide plane *g*.

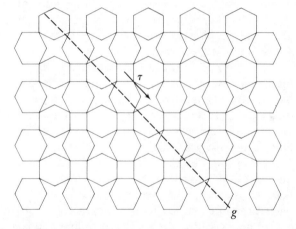

FIGURE 2.13 Wallpaper pattern with dashed line indicating location of one of a set of glide planes, *g*.

Returning once more to our wallpaper, we see a diagonal glide plane, the final symmetry element in the pattern (Figure 2.13). It is symbolized by a dashed line to indicate that the plane is perpendicular to the page, and that τ is in the plane of the page. A dotted line would denote a perpendicular glide plane but with τ vertical to the page. An alternating dashed and dotted line would indicate a vertical glide plane with τ at some angle to the page other than 90 degrees. In the example here, the glide plane is operated on by all of the other symmetry elements to generate a family of parallel glides with a symmetrically equivalent family perpendicular to it.

Note in Figure 2.8 a family of mirror planes parallel to *m* but passing through the square motifs. These are symmetrically equivalent to *m* by virtue of the glide plane operation.

Rotoinversion

The final compound symmetry operation of *rotoinversion* combines a rotation axis and a center of inversion. The axis of rotation is called a *rotoinversion axis*. The four types of rotoinversion axes are illustrated in Figure 2.14 and are symbolized by $\bar{2}$, $\bar{3}$, $\bar{4}$, and $\bar{6}$ to distinguish them from normal rotation axes. The sequence of operations *n* and *i* is immaterial. The motif may be rotated first and then inverted, or inverted first and then rotated. As we will see in the next section, only the fourfold rotoinversion represents a unique symmetry operation.

Other possible compound symmetry operations would include the combined operations of rotation and reflection, reflection and inversion, and inversion and translation. None of these is unique

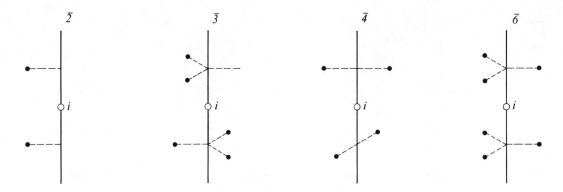

FIGURE 2.14   Operation of rotoinversion.

because they have simpler representations through the sequence of basic symmetry operations.

SYMMETRY GROUPS

All crystal structures can be analyzed in terms of the four basic symmetry operations and the three compound operations. This is also true of crystals, because the external faces that develop during growth are an expression of the internal structure. In all symmetry patterns, only certain allowable sets of symmetry operations are present. These are called *symmetry groups*.

For example, Figure 2.15a summarizes all of the symmetry operations in the same wallpaper pattern we examined earlier. This pattern consists of (1) two unit translations, (2) two fourfold axes and one twofold axis of rotation, (3) two mirror planes, and (4) one glide plane. Every symmetry element operates on all of the others in such a way that any motif can always be returned to its initial position. An equivalent statement is that any sequence of operations within a symmetry group constitutes a *closed cycle*. The two-dimensional symmetry group corresponding to our wallpaper pattern is shown in Figure 2.15b. This pattern is an example of one of 17 different *plane groups*. This means that only 17 different symmetry patterns are possible for the two-dimensional repetition of a set of wallpaper motifs. This does not mean, however, that there are no other possible wallpaper patterns. The choice of motifs is unlimited, and the magnitude of the unit translations can be varied.

The three-dimensional analogues of the two-dimensional plane groups are called *space groups*. There are 230 of these, and as Table 2.3 indicates, they represent permissible patterns of all the symmetry operations. If we remove the translational symmetry operations, the 230 space groups reduce to 32 three-dimensional space point groups, and

the 17 plane groups reduce to 10 two-dimensional plane point groups.

We can derive the various symmetry groups mathematically through the application of *group theory*. Although the mathematical derivation of symmetry groups is important and is used extensively in advanced crystallography, it has the shortcoming of being less illustrative than the geometric derivation and representation. The latter is preferable in an elementary discussion of crystallography, because it is easier to follow and the fundamental concepts are easier to under-

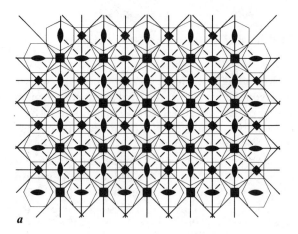

a

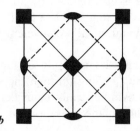

b

FIGURE 2.15   (a) Complete collection of symmetry elements present in wallpaper pattern. (b) Representation of the plane point group that fully describes the two-dimensional symmetry of the wallpaper pattern.

TABLE 2.3. *Summary of the Two- and Three-Dimensional Symmetry Groups and the Symmetry Operations of Which They Consist*

Symmetry Groups	Applicable Symmetry Operations							No. of Groups
	t	n	m	i	n_m	g	$\bar{n}$	
Space groups (3-d)	√	√	√	√	√	√	√	230
Plane groups (2-d)	√	√	√			√		17
Space point groups (3-d)		√	√	√			√	32
Plane point groups (2-d)		√	√					10
Space lattice symmetry (3-d)	√	√	√	√			√	7
Plane lattice symmetry (2-d)	√	√	√					5

NOTE: t = translation, n = rotation, m = reflection, i = inversion, n_m = screw rotation, g = glide reflection, and $\bar{n}$ = rotoinversion.

stand. The visual conception of symmetry groups offers additional advantages in mineralogy, because in most cases we study observable crystals.

PLANE GROUPS

Plane groups represent the symmetry of crystal structures projected in two dimensions. This is a useful representation because it can be shown on a page and illustrates many important symmetry relationships. Figure 2.16 is the projection of the crystal structure of the mineral marcasite (FeS$_2$). This two-dimensional pattern is defined fully by giving:

1. The dimensions of its unit mesh (a and b)
2. The plane group symmetry
3. The location of iron (Fe) and sulfur (S) atoms in the asymmetric unit of its mesh

The unit mesh is defined by giving the magnitude of the unit translations in nanometers (nm) and the angle between them in degrees. An *asymmetric unit* is that portion of the unit mesh within which no atoms are repeated by symmetry operations, that is, the content of an asymmetric unit is not controlled by symmetry. In the example of marcasite, the asymmetric unit is any quarter of the unit mesh. The content of an asymmetric unit is then repeated by the symmetry of the plane

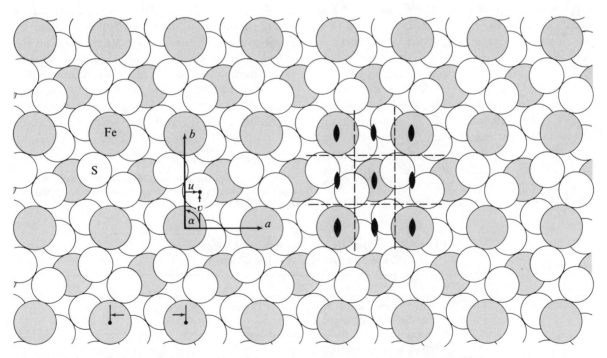

FIGURE 2.16 Two-dimensional projection of the structure of marcasite (FeS$_2$). Unit mesh dimensions are $a = 0.444$ nm, $b = 0.539$ nm, and $\alpha = 90°$. Plane group is $p2gg$. Atomic positions are Fe($x = 0$, $y = 0$), S($x = 0.203$, $y = 0.375$). (After Buerger, Martin J. *Contemporary crystallography* [New York, 1970], Figure 5, p. 9. Copyright © 1970 and used with kind permission of McGraw-Hill Book Company.)

group. The coordinates of the atoms in an asymmetric unit are usually given in terms of fractions of unit cell translations. For example, in a three-dimensional structure, $x = u/a$, $y = v/b$, and $z = w/c$, where u, v, and w are the atomic positions along the a, b, and c crystallographic axes, respectively. All the measurable units (a, b, c, u, v, w) are given in nanometers and x, y, and z are decimal fractions.

The plane group symmetry of marcasite in this projection is symbolized by *p2gg*. For each of the 17 plane groups, the first letter refers to cell type. The next entry in the symbol is a number (2, 3, 4, or 6) indicating the rotation axis, which is perpendicular to the plane. The next two (or one) letters refer to the reflection symmetry perpendicular to the unit translations. There are only two cell types for unit meshes. The symbol p in this example denotes a primitive unit mesh. The symbol c in other plane groups refers to a multiple unit mesh. The geometric representation of plane group *p2gg* is shown with the other 16 plane groups in Figure 2.17.

In the example of marcasite, the plane group symmetry represented by *p2gg* in Figure 2.17 is superimposed on the right side of the projected structure (Figure 2.16) in order to show how the atoms in the asymmetric unit are repeated. Note in Figure 2.16 that although an Fe atom occupies the center of the unit mesh, that atom is not translationally equivalent to the Fe atoms at the corners of the mesh. The S atom is said to be in a *general position*, meaning that the position occupied by S is not coincident with a symmetry operation. According to the symmetry operations in the plane group *p2gg*, the S atom is repeated four times. This number (4) is called the *equipoint number*. On the other hand, Fe is located right on the twofold axis and halfway between the two glide planes. Fe is, therefore, in a *special position*. Because of the speciality of this position, the Fe atom is not repeated by the twofold axis and is repeated to only one other position in the unit cell

by the glide planes, so the equipoint number of this special position is 2. The symmetry of the general and special positions is defined by the *point symmetry* of the position. Point symmetry refers to the presence of symmetry operations at the site where the atom or ion is located. The only symmetry element present at a general position is the trivial onefold axis of rotation. Hence, the point symmetry of S in marcasite is 1, and of Fe is 2, because Fe is on the twofold axis. Note that the symmetry operations repeat S twice as many times as Fe, consistent with the chemical formula of marcasite—FeS_2.

In marcasite, a total of 2 Fe and 4 S is in a unit cell, which makes two formula units ($2(FeS_2)$). The number of formula units per unit cell is denoted by the letter Z.

SPACE GROUPS

Space groups define the complete symmetry relationships between atoms or ions systematically distributed in a crystal structure. Because atoms are systematically arranged in minerals, the symmetry of their distribution can be described precisely with the appropriate space group. An infinite number of different crystal structures exists, but the number of their possible structural symmetries is limited to 230.

When considering space groups, the addition of the third dimension requires a more complicated nomenclature than that used for description of plane groups. Symmetry operations may be inclined or perpendicular to each other, and may be at different levels in the third dimension. Table 2.4 gives the appropriate symbols and explanations. If the elevation of a particular symmetry operation is not at zero,[2] the fraction of the vertical translation (perpendicular to the page) at

[2] Note that an operation at zero elevation is repeated at 1/2, and that operations at 1/4 elevation are repeated by translation at 3/4 elevation.

TABLE 2.4.   *Symbols and Nomenclature for Horizontal Symmetry Operations in Space Groups*

Symbol	Explanation
⌐	Mirror plane, located at 0 (and 1/2) elevation, parallel with the paper
⌐→	Axial glide, with the translational component pointing to the right, at elevation 0 (and 1/2)
↖	Diagonal glide, elevation 0 (and 1/2)
→	Horizontal twofold axis, elevation 0 (and 1/2)
⇀	Horizontal twofold screw axis, elevation 0 (and 1/2)

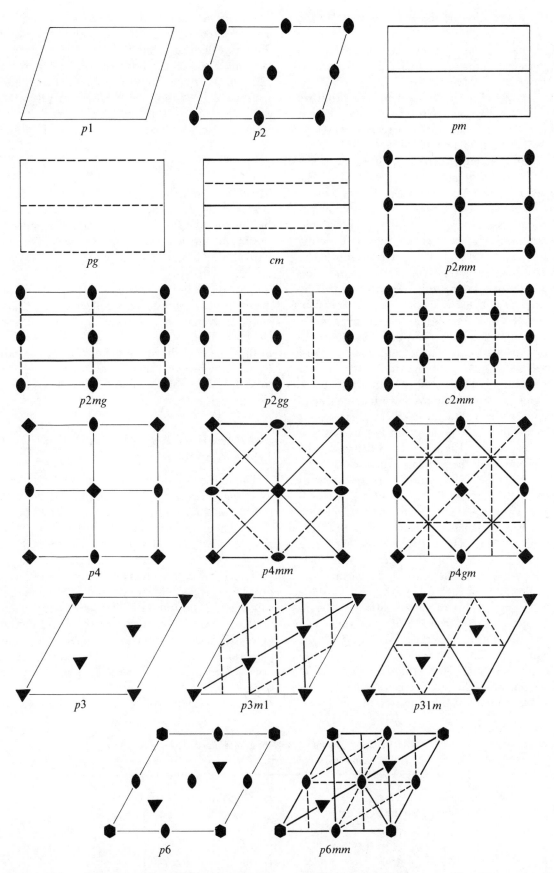

FIGURE 2.17 Representation of the 17 plane groups and their symbols. Solid and dashed lines denote mirror and glide planes, respectively, oriented perpendicular to the page.

which the operation occurs is written next to the symbol.

The space group symbols are similar to the plane group symbols. The first digit is an upper-case letter that represents the cell type (*P*, *A*, *B*, *C*, *I*, *F*, or *R*). The symbol *P* denotes a primitive unit cell. The other symbols refer to multiple unit cells and are discussed in a later section. The next one, two, or three digits refer to the major symmetry operations along or across the three unit translations, or symmetrically distinct directions. The monoclinic system has only one symmetry axis (2, $1/m$, or $2/m$). The orthorhombic system has three similar symmetry axes. In the tetragonal and hexagonal systems, the first two digits refer to the symmetry of the *c* and the a_1 axes, respectively. The third entry in the symbol is the direction between the a_1 and a_2 unit translations. In the isometric system, the first digit denotes the symmetry of the unit translation, the second denotes the symmetry of the direction given by the resultant of the three unit translations, and the third denotes the symmetry of the direction given by the resultant of the a_1 and a_2 unit translations. The significance of these symbols and their symmetrically distinct directions will be clearer after our discussion of the point group symmetries.

For example, Figure 2.18 illustrates the space group and the crystal structure of beryl ($Al_2Si_6Be_3O_{18}$). The space group symbol $P6/m2/c2/c$ reveals that the unit cell is primitive, the symmetry along the *c* axis is sixfold with a mirror perpendicular to it, and there are twofold axes with perpendicular *c*-glide planes, one along the a_1 unit translation and the other perpendicular to a_2 (Figure 2.18a). The unit cell dimensions and atomic positions of each atom are given in Table 2.5, including, in this case, two different sites in the structure for oxygen. Each of these atoms can then be positioned and repeated by *all* of the symmetry operations. Oxygen (2) is in a general position (point symmetry 1) and is repeated 24 times within the unit cell. Oxygen (1) and silicon atoms are both located on mirror planes (point

symmetry *m*), and so they are in special positions. They are repeated only half as many times within the unit cell. The atoms Be and Al are located at even more specialized positions (point symmetries 222 and 32, respectively) involving various intersections of rotation axes. Consequently, their equipoint numbers are smaller. If we total the repetitions on each atom, the chemical formula per unit cell is:

$$Al_4Si_{12}Be_6O_{12}^{(1)}O_{24}^{(2)}$$

The common denominator of the subscripts in this formula is 2. The formula can be simplified by dividing the subscripts by 2 and by combining the two symmetrically distinct oxygens. The standard formula unit ($Al_2Si_6Be_3O_{18}$) results where $Z = 2$.

The relationship between atom position and symmetry can be visualized better in three dimensions with the stereoscopic structural diagram shown in Figure 2.18b. Although the entire depth of the cell is not shown, note how the Be atom, for example, is repeated throughout the structure and how the environment around that atom is always exactly the same.

SYMMETRY GROUPS WITHOUT TRANSLATION

The unit cells of minerals are exceedingly small by visual standards. Unit translations and the translational components of screw axes and glide planes have dimensions on the order of 0.1 to 1.0 nm. Modern electron microscope techniques can now resolve such small differences, but in the normal observation of crystals, the operation of translation is not directly observed. Mineralogists can observe the translation-free symmetry operations, however, from the symmetrical positioning of crystal faces or from the symmetry of certain physical properties. These translation-free operations are *rotation*, *reflection*, *inversion*, and *rotoinversion*.

The translation-free symmetry groups can be

TABLE 2.5.   *Atomic Coordinates and Site Symmetry in the Asymmetric Unit of Beryl*

Equipoint Number	Point Symmetry	Atom or Ion				Coordinates in Asymmetric Unit
4	(32)	Al	1/3	2/3	3/4	
6	(222)	Be	1/2	0	3/4	
12	(*m*)	Si	*x*	*y*	0	$x = 0.39, y = 0.12$
12	(*m*)	O(1)	*x*	*y*	0	$x = 0.30, y = 0.24$
24	(1)	O(2)	*x*	*y*	*z*	$x = 0.48, y = 0.15, z = 0.85$

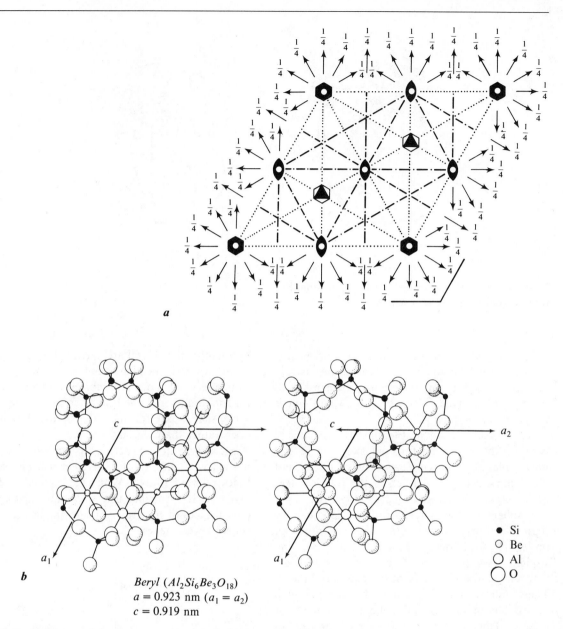

Beryl $(Al_2Si_6Be_3O_{18})$
$a = 0.923$ nm $(a_1 = a_2)$
$c = 0.919$ nm

- Si
- Be
- Al
- O

FIGURE 2.18 **(a) Space group diagram of** $P6/m2/c2/c$. **(b) Stereoscopic diagram of beryl shows three-dimensional positions of atoms.**

derived from space groups and plane groups by removing their translational components. In so doing, the remaining symmetry elements are condensed to a common point. The resulting symmetry groups express the symmetry of space about that point, and hence are referred to as *point groups*. They are more useful in the primary study of crystal symmetry than space groups and plane groups.

To discuss point groups and to apply the concept to real minerals, we must develop a method to portray the three-dimensional nature of symmetry operations. This method is called *stereographic projection*.

STEREOGRAPHIC PROJECTION

In stereographic projection, the components of a three-dimensional object are projected into a two-dimensional plane, called the *plane of projection*. To retain the spatial and angular relationships between the components of the object in the projection plane, the object is placed in the center of a sphere, called the *sphere of projection* (Figure 2.19). The equatorial plane of the sphere is the plane of projection. The various components of the object are first projected from the center to the surface of the sphere of projection. These points are then connected by straight lines to the south

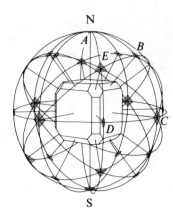
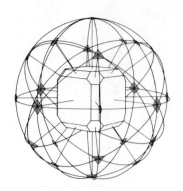

FIGURE 2.19 Projection of symmetry axes from center of crystal to surface of a sphere.

pole of the sphere of projection. The connecting lines intersect the equatorial plane as points that are the projections of the original elements onto the equatorial plane. In this case, the south pole is the *point of projection*.

The crystal shown in Figure 2.19 has several twofold, threefold, and fourfold axes and several mirror planes in its symmetry. For simplicity, not all of these are shown. Each of the rotation axes of the crystal, when extended to pierce the upper hemisphere of the sphere of projection, can be represented by unique points on the sphere of projection (e.g., points B, C, D, and E). The straight lines that connect each of these points with the south pole intersect the equatorial plane. These intersection points are the stereographic projections of the original symmetry elements of interest. Thus, in Figure 2.20, point B projects to B', A projects to A', and points on the equatorial plane (horizontal elements of the original objects) such as point C remain unchanged.

The stereographic projections of each of the 32 point groups are shown in Figure 2.21. The example of the cube discussed here is indicated by the symbol $4/m\bar{3}2/m$. The columns in the figure represent *isogonal* groups, that is, each point group in a column has a common rotation axis that is oriented normal to the page.

Mirror planes in projection are shown as heavy lines. If no horizontal mirror is in the equatorial plane, a dotted line is used. The standard symbols for rotation axes in the upper hemisphere are as given in Figure 2.6. The crystal also has a bottom half, and unless there is a horizontal mirror, its symmetry is different from the top half. For projection purposes, symmetry elements on the bot-

tom half of the crystal are extended to the surface of the lower hemisphere, and are then projected from the north pole. The corresponding symbols for rotation axes in the lower hemisphere have the same shape, but for the sake of distinction are open rather than closed.

Stereographic projection is normally accomplished with the use of a *stereographic net*, sometimes referred to as a Wulff net, named after its inventor. Its principal use is for representing the crystal faces of minerals and the angular relationships between faces.

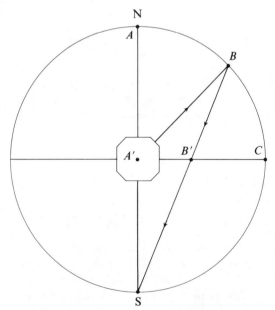

FIGURE 2.20 Cross section (plane NBCS) of sphere in Figure 2.18 illustrating principle of stereographic projection. Point B is projected to point B' from point of projection S.

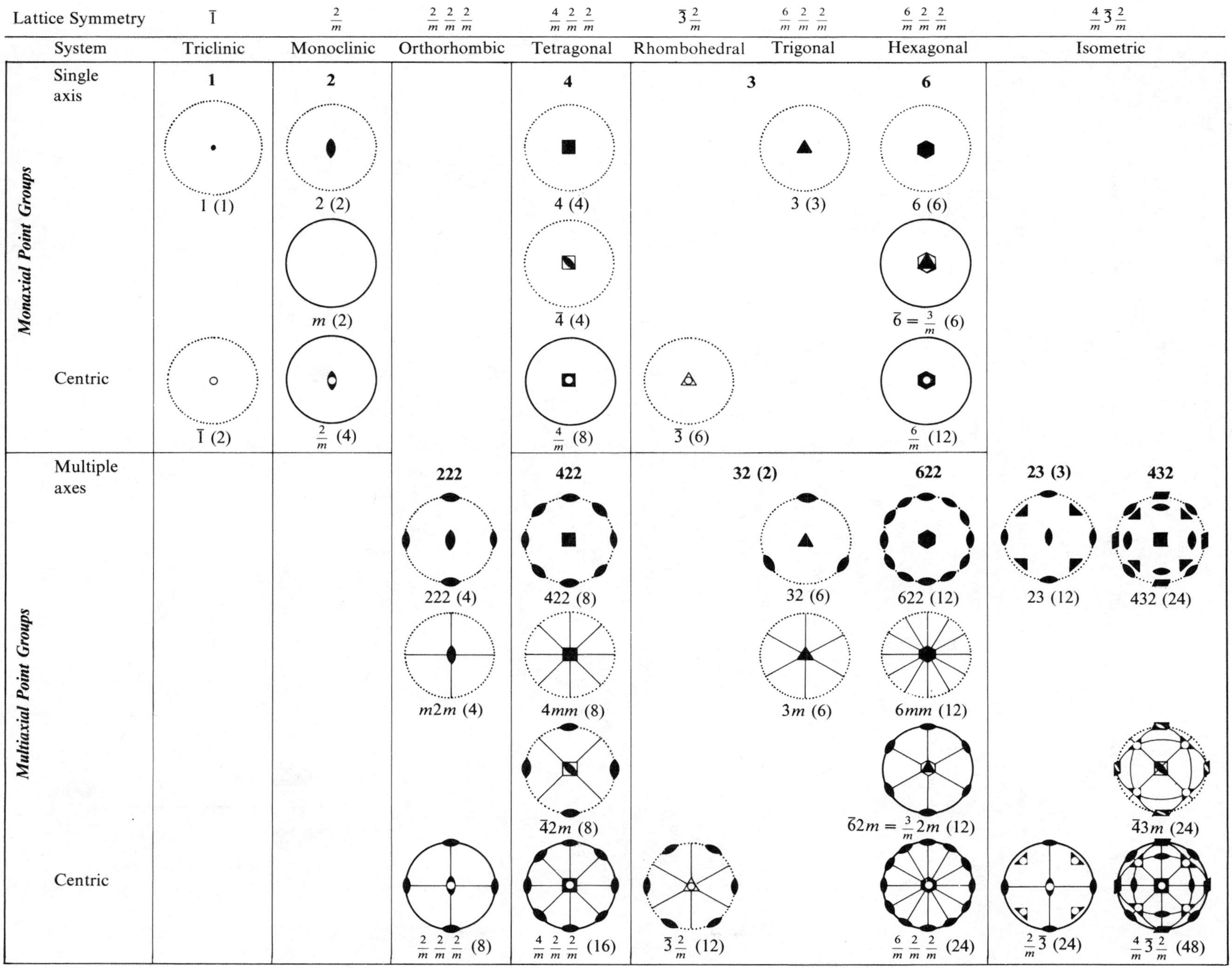

FIGURE 2.21 Summary of the 32 point groups, their seven lattice symmetries (top row), and the corresponding seven crystal systems (second row). Point groups within each system are organized in columns with the 11 centric point groups at the bottom of each isogonal group. Numbers following point group symbols are the equipoint numbers.

PLANE POINT GROUPS

Table 2.3 shows that when translation is removed from each of the 17 two-dimensional plane groups, only 10 plane point groups remain. The only translation-free symmetry operations in two dimensions are rotation and reflection. Each of the five rotation axes ($n = 1, 2, 3, 4,$ and 6) constitutes by itself a plane point group. A single mirror plane yields a sixth group. By combining rotation axes with mirrors, the following relationships emerge: (1) The mirror planes must be parallel to the rotation axes. (2) The mirror planes must be repeated at the same angular interval, α, as the angular interval of the rotation to be compatible with that rotation. (3) A second mirror plane is automatically generated at one half the angular interval ($\alpha/2$) of the rotation operation from the first mirror plane. Figure 2.22 illustrates why this must be. The motif at point A is reflected to A' by the mirror plane m_1 normal to the page. The sixfold axis then operates on m_1 and the motif to repeat them at an angular interval of 60 degrees. The new positions are denoted by B and B'. Now, however, the motifs at B and A' are mirror images, as are the motifs at A and B' and the two mirrors. A new mirror plane (m_2) must exist between the first two.

The full symmetry of this group can be obtained by any two of the following pairs of operations:[3]

$$R_\alpha \cdot m_1 \cdot (R_\alpha) \cdot m_2 = R_\alpha m_1 m_2$$

[3] The dot symbol between operations means that one symmetry operation follows the other.

The permissible $R_\alpha mm$ groups are then $2mm$, $3m(m)$, $4mm$, and $6mm$. Note that combining a onefold rotation axis with a mirror yields only a mirror. In the symbol $3m(m)$, the second mirror may be omitted because it is automatically implied by the argument given for Figure 2.22. The 10 plane point groups are illustrated in Figure 2.23. They are truly two dimensional in that they have no dimension normal to the page. Thus, the mirrors are mirror "lines," and the axes are points about which rotation takes place. When the vertical dimension is considered, the mirror lines become vertical mirror planes, the rotation points become vertical rotation axes, and 10 of the 32 space point groups (Figure 2.21) are generated.

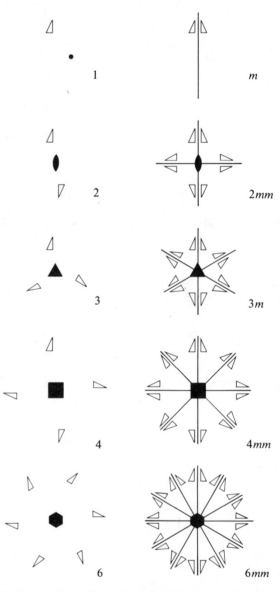

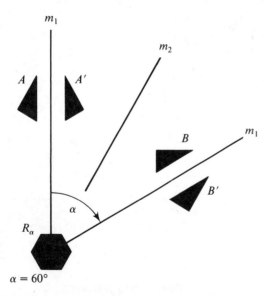

FIGURE 2.22   Graphical proof that combined rotation axis R_α and mirror plane m_1 generate second mirror m_2.

FIGURE 2.23   Operation of the 10 plane point groups. Each axis is perpendicular to the page. Small triangle represents an arbitrary motif.

The notation of a rotation followed directly by a reflection, *nm* (e.g., 3*m*), in both plane and space point groups implies that the rotation axis (or point in two-dimensional space) is in the mirror plane (or line). The symbol 2*mm* then means that the two mirrors intersect, and that the line of their intersection is the twofold rotation axis. If a rotation is followed by a mirror perpendicular to the rotation axis, the space point group symbol is given as *n/m* (e.g., 4/*m*).

Each of the plane two-dimensional groups in Figure 2.17 reduces to a plane point group. Without translation, glide planes become mirrors, and screw axes become simple rotation axes.

SPACE POINT GROUPS

The 32 three-dimensional point groups, or space point groups, can be derived directly from the 230 space groups by removing the translation components. An alternative derivation that better illustrates the mutual relationships between symmetry operations is to consider the allowable sets of such operations. For point groups, we need consider only the basic and compound symmetry operations that are translation-free, namely, rotation, reflection, inversion, and rotoinversion. Space groups could be derived in the same way if the operations of translation, glide planes, and screw axes were included. Our objective here is to demonstrate, with the example of point groups, how symmetry groups are derived in general. In so doing, a more thorough understanding of point groups is achieved.

Our procedure is first to demonstrate that all four of the translation-free operations, including reflection, can be considered as either rotation or rotoinversion axes. The operation of inversion *i* is equivalent to a onefold rotoinversion axis, $\bar{1}$. That is, a rotation of 360 degrees followed by inversion yields the same result as inversion alone. Table 2.6 shows that the operation of reflection *m* by itself, or in combination with perpendicular rotation axes of ranks 2, 3, 4, and 6, can be represented by either rotoinversion axes or pairs of equal ranking and coincident rotation and rotoinversion axes, $n \cdot \bar{n}$.

Equivalent operations of the kind in Table 2.6 can be proved easily with the aid of stereographic projection. For example, Figure 2.24 illustrates the

TABLE 2.6.   *Equivalent Symmetry Operations for Rotoinversion*

Rotoinversion Axes	Pairs of Rotation and Rotoinversion Axes
$\bar{1} = i$	$1 \cdot 1 = i = \bar{1}$
$\bar{2} = \dfrac{1}{m}$	$2 \cdot 2 = \dfrac{2}{m}$
$\bar{3} = 3 \cdot i$	$3 \cdot \bar{3} = 3 \cdot i = \bar{3}$
$\bar{4} = \bar{4}$	$4 \cdot \bar{4} = \dfrac{4}{m}$
$\bar{6} = \dfrac{3}{m}$	$6 \cdot \bar{6} = \dfrac{6}{m}$

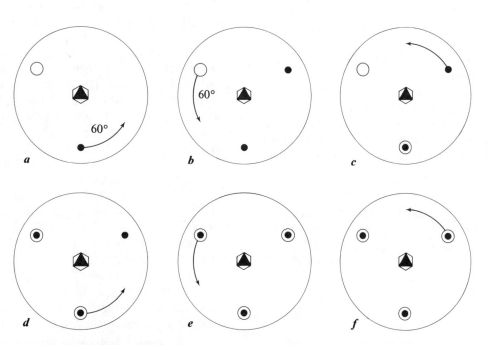

FIGURE 2.24   Stereographic projection illustrating sequence of steps required by a sixfold rotoinversion axis.   Solid dots are in upper hemisphere, open circles in lower hemisphere.

operation of a vertical sixfold axis of rotoinversion on an arbitrary point or motif. The solid dot denotes a position in the upper hemisphere; an open circle denotes a position in the lower hemisphere. The projection starts with a solid dot (Figure 2.24a), rotates it 60 degrees, and then inverts it through the center of the projection to the open dot. The open dot is then rotated 60 degrees (Figures 2.24a–f) and inverted through the center to give a new solid dot. This combination of operations ($6 \cdot i = \bar{6}$) is repeated six times, and only then is the original solid dot restored. The result (Figure 2.24f) could also be obtained by a threefold axis of rotation to generate the three solid dots separated by an angle of 120 degrees, and then operating with a perpendicular mirror to get the open dots directly below. The combination $3/m$ is therefore equivalent to $\bar{6}$.

We can now derive the 32 point groups by considering the following combinations:

1. All rotation axes, n
2. All rotoinversion axes, $\bar{n}$
3. All pairs of equal ranking and coincident rotation and rotoinversion axes, $n \cdot \bar{n}$

From these combinations, 13 point groups having a single symmetry axis are possible. They are called the *monaxial point groups* (Table 2.7). The two table entries in parentheses are redundant.

The remaining 19 point groups are all multiaxial. Their derivation is more complex because of the additional axes. Before we proceed with the derivation, three questions need to be resolved:

1. How many axes can or must be present in a multiaxial point group?
2. What are the ranks of rotation axes in the permissible groups?
3. What are the permissible combinations of rotation and rotoinversion axes?

The answer to the first question is that if more than one axis is in the point group, three must be present in a multiaxial point group. The situation is analogous to the generation of a second mirror plane in the plane point groups when a rotation

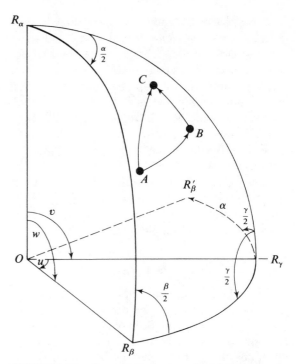

FIGURE 2.25 **Steps in derivation of permissible ranks and angular combinations of three rotation axes.**

axis is operated on by a mirror. Figure 2.25 illustrates how two intersecting rotation axes generate a third. Axis R_α rotates a point from position A to position B. Axis R_β rotates the point at B to a new position at C. Positions C and A, and axes R_α and R_β, however, are symmetrically equivalent through an angle γ generated by R_γ.

The answer to the second question is that the permissible ranks are those of the following six groups of rotation axes. They are 222, 422, 322, 622, 233, and 432. The permissible groups of rotation axes and the angles between them are determined as follows. The positions of the three rotation axes in Figure 2.25 define a spherical triangle, and for them to operate in a closed cycle, the surface angles of the spherical triangle between R_α, R_β, and R_γ must be $\alpha/2$, $\beta/2$, and $\gamma/2$. This is because a rotation by α around R_α repeats R_β to R'_β, and a rotation by γ around R_γ returns R'_β to R_β. The plane $OR_\alpha R_\gamma$ is then a mirror plane that in turn requires the angle between the arc $R_\alpha R_\gamma$ and the arc $R_\gamma R_\beta$ to be $\gamma/2$. By the same procedure, the other two surface angles can be shown to be $\alpha/2$ and $\beta/2$.

The possible values of $\alpha/2$, $\beta/2$, and $\gamma/2$ are related to the angles u, v, and w between the rotation angles. Taking the angle w between R_α and R_β as an example, the law of cosines states:

$$\cos w = \frac{\cos(\gamma/2) + \cos(\alpha/2)\cos(\beta/2)}{\sin(\alpha/2)\sin(\beta/2)}$$

TABLE 2.7. *The 13 Monaxial Point Groups*

n	$\bar{n}$	$n \cdot \bar{n}$
1	$\bar{1}$	$(1 \cdot \bar{1} = \bar{1})$
2	$\bar{2} = m$	$2 \cdot \bar{2} = \dfrac{2}{m}$
3	$\bar{3}$	$(3 \cdot \bar{3} = \bar{3})$
4	$\bar{4}$	$4 \cdot \bar{4} = \dfrac{4}{m}$
6	$\bar{6} = \dfrac{3}{m}$	$6 \cdot \bar{6} = \dfrac{6}{m}$

The equations for $\cos u$ and $\cos v$ are similar. When these equations are solved by substituting the angular values of permissible rotation axes (60, 90, 120, 180, and 360 degrees), only six axial groups yield real and nonzero values for the cosines of u, v, and w. These multiaxial point groups and the angles between their component axes are illustrated in Figure 2.26.

The third question, concerning the permissible combinations of rotation and rotoinversion axes, really addresses the issue of how rotation axes, n, and rotoinversion axes, $\bar{n}$, can be mixed three at a time. Recall that a rotation axis always preserves the "handedness" of the motif, whereas a rotoinversion axis does not. Only those combinations of three axes that have the same sense of the motif at the beginning and at the end of the operation are permissible. If we let the symbol L represent a left-handed motif and R its enantiomorph, only four (*) of the following sequences yield closed cycles.

$$L \cdot n \to L \cdot n \to L \cdot n \to L \ (*)$$
$$L \cdot n \to L \cdot n \to L \cdot \bar{n} \to R$$
$$L \cdot n \to L \cdot \bar{n} \to R \cdot \bar{n} \to L \ (*)$$

$$L \cdot n \to L \cdot \bar{n} \to R \cdot n \to R$$
$$L \cdot \bar{n} \to R \cdot \bar{n} \to L \cdot \bar{n} \to R$$
$$L \cdot \bar{n} \to R \cdot \bar{n} \to L \cdot n \to L \ (*)$$

$$L \cdot \bar{n} \to R \cdot n \to R \cdot n \to R$$
$$L \cdot \bar{n} \to R \cdot n \to R \cdot \bar{n} \to L \ (*)$$

The first sequence should be read to mean that an initial left-handed motif is operated on by an n-fold rotation axis to yield a left-handed motif. It is then operated on by a second n-fold axis to yield a left-handed motif, which in turn is operated on by a third n-fold axis to yield a left-handed motif identical to the original. This sequence, like the others denoted by an asterisk, constitutes a closed cycle.

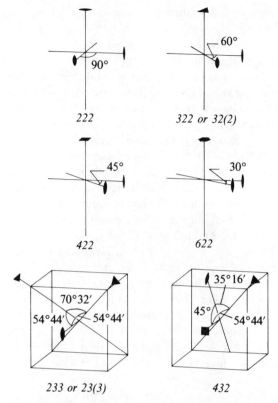

FIGURE 2.26 **Permissible axial combinations and angles in multiaxial point groups. (After Buerger, Martin J.** *Elementary crystallography*, **rev. printing [New York, 1963], Fig. 13, p. 43. By kind permission of John Wiley & Sons, Inc., copyright © 1963.)**

The permissible combinations can be represented as nnn, $n\bar{n}\bar{n}$, $\bar{n}n\bar{n}$, and $\bar{n}n\bar{n}$. The coaxial combinations of nnn and $\bar{n}\bar{n}\bar{n}$, denoted by $nnn \cdot \bar{n}\bar{n}\bar{n}$, are also possible. These are shown as the top row of Table 2.8. The permissible axial combinations derived earlier are listed in the first column. The rest of the table then yields the 19 multiaxial point groups. The point group symbols are derived for each of the conditions by using the basic

TABLE 2.8. *The 19 Multiaxial Point Groups*

nnn	$n\bar{n}\bar{n}$	$\bar{n}n\bar{n}$	$\bar{n}\bar{n}n$	$nnn \cdot \bar{n}\bar{n}\bar{n}$
222	$2\bar{2}\bar{2} = 2mm$	$(\bar{2}2\bar{2} = 2mm)$	$(\bar{2}\bar{2}2 = 2mm)$	$222 \cdot \bar{2}\bar{2}\bar{2} = \dfrac{2}{m}\dfrac{2}{m}\dfrac{2}{m}$
32(2)	$3\bar{2}\bar{2} = 3m(m)$	$\bar{3}2\bar{2} = \bar{3}\dfrac{2}{m}(\dfrac{2}{m})$	$(\bar{3}\bar{2}2 = \bar{3}\dfrac{2}{m}\dfrac{2}{m})$	$(322 \cdot \bar{3}\bar{2}\bar{2} = \bar{3}\dfrac{2}{m}\dfrac{2}{m})$
422	$4\bar{2}\bar{2} = 4mm$	$\bar{4}2\bar{2} = \bar{4}2m$	$(\bar{4}\bar{2}2 = \bar{4}2m)$	$422 \cdot \bar{4}\bar{2}\bar{2} = \dfrac{4}{m}\dfrac{2}{m}\dfrac{2}{m}$
622	$6\bar{2}\bar{2} = 6mm$	$\bar{6}2\bar{2} = \bar{6}2m$	$(\bar{6}\bar{2}2 = \bar{6}2m)$	$622 \cdot \bar{6}\bar{2}\bar{2} = \dfrac{6}{m}\dfrac{2}{m}\dfrac{2}{m}$
23(3)	$2\bar{3}\bar{3} = \dfrac{2}{m}\bar{3}(3)$	$(2\bar{3}3 = \dfrac{2}{m}\bar{3}3)$	$(\bar{2}3\bar{3} = \dfrac{2}{m}\bar{3}3)$	$(233 \cdot \bar{2}\bar{3}\bar{3} = \dfrac{2}{m}\bar{3}3)$
432	$4\bar{3}\bar{2} = \dfrac{4}{m}\bar{3}\dfrac{2}{m}$	$\bar{4}\bar{3}2 = \bar{4}3m$	$(\bar{4}3\bar{2} = \dfrac{4}{m}\bar{3}\dfrac{2}{m})$	$(432 \cdot \bar{4}\bar{3}\bar{2} = \dfrac{4}{m}\bar{3}\dfrac{2}{m})$

operational identities in Table 2.6 or by tracing the steps of the operations in stereographic projection. Combined with the 13 monaxial point groups, 32 unique point groups result.

The point group symbols used here are the internationally accepted Hermann-Mauguin symbols. They are very descriptive and the complete symmetry of a point group can be reconstructed easily from them. One need remember only the angular relationships between axes in Figure 2.21 and the fact that a mirror symbol immediately following an axis symbol is parallel to that axis, whereas a mirror in the denominator is perpendicular to the immediately preceding axis in the numerator. For example, point group $\bar{4}2m$ has a fourfold axis of rotoinversion, a twofold axis perpendicular to it (Figure 2.21), and a single mirror that passes through both axes. Point group $4/m\,2/m\,2/m$ has a fourfold axis with a perpendicular mirror, a twofold axis perpendicular to the fourfold with its own perpendicular mirror, and finally, another twofold axis halfway between the first two and with a mirror normal to it (in the plane of the first two axes).

Recall that in the plane point group $3m(m)$, the second mirror in parentheses is identical with the first mirror and therefore can be deleted. The same is true for point groups $32(2)$ and $23(3)$. The final axis shown in parentheses will be omitted hereafter.

Each of the 32 point groups illustrated in stereographic projection in Figure 2.21 belongs to one of 11 isogonal groups. Two or more point groups are said to be *isogonal* if the derivation of their symmetries is based on identical axial groups. The point groups 4, $\bar{4}$, and $4/m$ are isogonal, for example. The point group at the bottom of each isogonal column always has the highest symmetry and always has a center of inversion. Point groups with an inversion center are called *centric*. As a rule of thumb, the 11 centric point groups can be recognized if rotation axes of even rank ($n = 2$, 4, and 6) are perpendicular to mirror planes, or if axes of odd rank ($n = 1$ or 3) are rotoinversion axes. This is proved easily by performing the operations in stereographic projection as outlined in Figure 2.24. In that example of the operation $\bar{6}$, n is even, so a center of inversion ought not to exist, and indeed it does not. The operation $\bar{6}$ is equivalent to $3/m$ where n is odd. If n were even, an inversion center would appear.

Throughout the text we frequently refer to crystals with relatively *high symmetry* or with relatively *low symmetry*. A measure of this distinction is the general equipoint number, which is the number of symmetrically equivalent positions a given motif in a general position would occupy after operation by the point group symmetry. The motif must be in a general position, rather than coincident with a position occupied by a symmetry element. Low equipoint numbers for point groups at the top of Figure 2.21 represent low symmetry. As we look down a column, the equipoint number remains the same until a centric point group is encountered. A centric point group will have a higher equipoint number, and consequently higher symmetry. Several of the isogonal groups have the same lattice symmetry (e.g., 4 and 422, 23 and 432) and are shown in the same full-page column. The equipoint number of point groups in the lower half of those columns is the same as the centric point group above it. Again, the number increases for the centric point groups at the bottom of the figure.

The 32 point groups were derived in 1830 by Johann Hessel, a German physician and mineralogist. He was also the first to show that only the onefold, twofold, threefold, fourfold, and sixfold axes are compatible with a lattice. His derivation of the 32 point groups was based on an exhaustive study of the possible point symmetries any geometrical form might have. Every mineral has a unique point group, or *crystal class*. In Hessel's time and for the following century, the geometrical name of the crystal form that expressed the symmetry of a particular point group was used rather than a symbol. Thus, the point group $4/m\bar{3}2/m$ was referred to as the hexoctahedral or isometric-holohedral class, $\bar{4}$ was referred to as the tetartohedral or tetragonal-disphenoidal class, $32(2)$ was referred to as the trapezohedral or trigonal-trapezohedral class, and so on. These names are still used in some textbooks.

LATTICE SYMMETRIES

As the lattice is a three-dimensional array of equivalent points, the symmetry about each of those points must be the same. By describing the point group symmetry about a lattice point, we have described the *lattice symmetry*. As lattices and unit cells are defined by three unit translations, all unit cells must be parallelepipeds, and that requires opposite faces to be parallel and equivalent by inversion. It follows that all lattices are *centric* (i.e., have a center of inversion) and that their possible symmetries are limited to centric point groups.

The requirement that unit cells be centric reduces the possible number of lattice symmetries to 11. One additional unique property of parallelepipeds further reduces the number to 7. If two opposite faces of the unit cell are considered independently and the faces are extended to infinity, they will be related to each other by a mirror plane located halfway between them. The faces also will be related by a twofold axis perpendicular to the mirror. In lattices in which the $2/m$ symmetry is possible (i.e., having lattice rows perpendicular to lattice planes), this relationship exists for all faces. Thus, we see why crystals with point groups 4, $\bar{4}$, $4/m$, 422, $4mm$, and $\bar{4}2m$ must have the same lattice symmetry of $4/m\,2/m\,2/m$. Table 2.9 demonstrates the relationship between the 11 centric point groups and the 7 lattice symmetries. Each of the lattice symmetries has a special name that will be used hereafter.

We can summarize the relationships between the three major symmetry groups as follows:[4]

Space groups: symmetry of crystal structures (i.e., the regular arrangement of atoms in space)

Point groups: translation-free or "visible" symmetry of the crystal structure as illustrated by the faces of crystals

Lattices: symmetry of the translational pattern of the crystal structure

Of these three symmetry groups, the concept of the lattice is the most difficult to understand initially. The lattice is not fixed to a definite point in the crystal structure, but can originate at any arbitrary point. That point, wherever we choose it to be, is defined as a lattice point, and all translationally equivalent points will be identical with it. In other words, the lattice can be viewed as a three-dimensional "overlay" that has its own intrinsic symmetry. As an example, the lattice symmetry of $2/m$ is common to the three point groups $2/m$, 2, and m shown at the top of Figure 2.27. When the translational symmetry operations of this lattice are added to each of the point groups, three columns of permissible space groups are generated.

CRYSTALLOGRAPHIC COORDINATE SYSTEMS

A simple Cartesian coordinate system with nanometer divisions on the ordinates is adequate to describe all crystal lattices. This, however, would require complex mathematical expressions to fully describe most crystal lattices. Instead, crystallographers prefer to use tailor-made coordinate systems, one for each of the seven symmetrically distinct lattices. In these seven coordinate systems, the angles between the three ordinates are the same as the angles between the three unit translations, and the unit division on the ordinates is equal to the length of the corresponding unit translation.

For example, a crystal with lattice symmetry $2/m\,2/m\,2/m$ has an orthogonal coordinate system, because its symmetry requires that the three unit translations be perpendicular to each other. The unit divisions on the x, y, and z ordinates are equal to the magnitudes of the a, b, and c unit translations of the crystal lattice. The magnitudes of a, b, and c need not be the same, as that condition is not required by the $2/m\,2/m\,2/m$ symmetry. The magnitudes can be the same, however, because that condition is not excluded by the $2/m\,2/m\,2/m$ symmetry.

To assure that all crystallographers describe a given crystal lattice in the same way, certain conventions have been developed through international agreement. One convention is that all lattices are described in a *right-handed coordinate system*. That is, if the thumb of the right hand points up and represents the c crystallographic axis, the extended index finger points in the direction of the a axis and the partially closed other

---

[4] A complete list of the 230 space groups and the related point groups and lattices is given in Table A3.1, Appendix 3.

TABLE 2.9. *Relationships Between the 11 Centric Point Groups and the 7 Lattice Symmetries*

Centric Point Groups	Possible Lattice Symmetries	Names of Lattice Groups
$\bar{1}$	$\bar{1}$	Triclinic
$\dfrac{2}{m}$	$\dfrac{2}{m}$	Monoclinic
$\dfrac{2}{m}\dfrac{2}{m}\dfrac{2}{m}$	$\dfrac{2}{m}\dfrac{2}{m}\dfrac{2}{m}$	Orthorhombic
$\dfrac{4}{m}$; $\dfrac{4}{m}\dfrac{2}{m}\dfrac{2}{m}$	$\dfrac{4}{m}\dfrac{2}{m}\dfrac{2}{m}$	Tetragonal
$\bar{3}$; $\bar{3}\dfrac{2}{m}\left(\dfrac{2}{m}\right)$	$\bar{3}\dfrac{2}{m}\left(\dfrac{2}{m}\right)$	Trigonal
$\dfrac{6}{m}$; $\dfrac{6}{m}\dfrac{2}{m}\dfrac{2}{m}$	$\dfrac{6}{m}\dfrac{2}{m}\dfrac{2}{m}$	Hexagonal
$\dfrac{2}{m}\bar{3}(\bar{3})$; $\dfrac{4}{m}\bar{3}\dfrac{2}{m}$	$\dfrac{4}{m}\bar{3}\dfrac{2}{m}$	Isometric

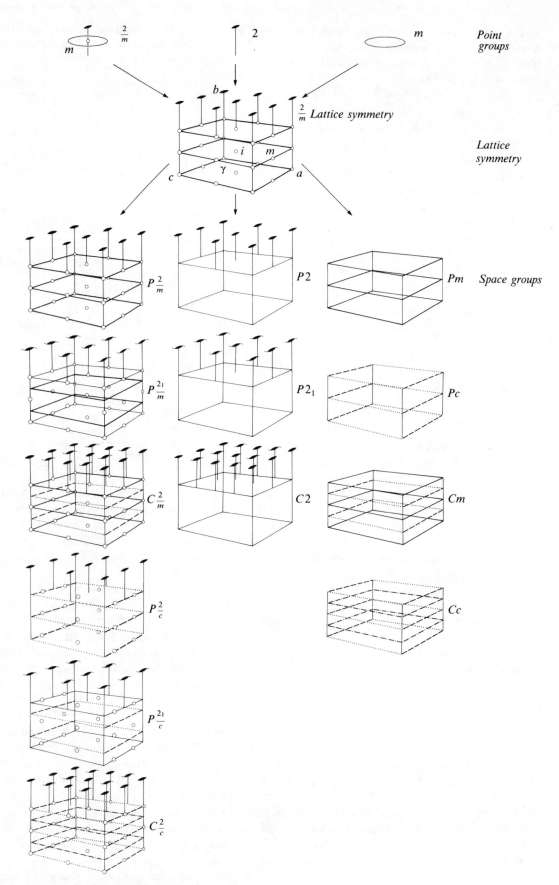

FIGURE 2.27 Relationships between monoclinic point groups, lattice symmetry, and space groups.

fingers point in the direction of the b axis. The other important agreements concern the choice of labeling the three unit translations a, b, and c. These unit translations are referred to as *crystal axes*, because they coincide with the coordinate axes. The unit translation that coincides with the highest ranking rotation axis of the lattice symmetry, if $n > 2$, is labeled c. The other two translations are labeled a and b, or if they have equal magnitude, a_1 and a_2. In crystals with lattice symmetry of $4/m\bar{3}2/m$, all three unit translations have equal magnitude and therefore are labeled a_1, a_2, and a_3.

In the $2/m$ point group, the twofold axis has the highest rank, but most mineralogists prefer the so-called *second setting* in which the translation parallel to the twofold axis is labeled b. In the first setting, the shorter of the other two unit translations is labeled a, and in the second setting it is labeled c. Confusion is possible in either of the two settings, for the directions of the two axes a and b in the first setting, and a and c in the second setting, could be given so that the angle between them is either acute or obtuse. By convention, the choice with the obtuse angle is preferred.

Lattices with symmetries of $2/m2/m2/m$ do not have a unique rotation axis. Consequently, a different convention is needed. In the first setting, the unit translations are labeled a, b, and c in order of their increasing magnitude. In the second setting, which mineralogists prefer, the order is $c < a < b$.

In crystal lattices having $\bar{1}$ symmetry, the choice of labeling can be rather confusing. The simple procedure of labeling the axes a, b, and c in accordance with their increasing magnitude, without paying attention to the angles between them, has been used in the past. More recently, however, crystallographers prefer a labeling system that results in the angles between the axes (α, β, and γ) being either *all acute* or *all obtuse*. In any coordinate system where all three axes are inclined to each other, there are always two octants in which all the angles are either acute or obtuse.[5]

A summary is given in Table 2.10 of labeling conventions and of the relationships between the lattice symmetries and the various lattice parameters[6] used to describe them. The seven symmetrically distinct lattice symmetries are called *crystal systems*. Keep clearly in mind, however, that in spite of the usual meaning of "system," crystal systems are not higher ranking criteria than crystal classes. To the contrary, the crystal system is determined by the point group symmetry, and not by the relationship between the unit translations. An example is coesite (SiO_2), a mineral that forms only in high-pressure environments. Its lattice parameters are: $a = c = 0.723$ nm, $b = 1.252$ nm, and $\beta = 120$ degrees. The shape and the dimensions of this lattice conform to all of the requirements of the hexagonal system. Studies with x-ray diffraction show that the point group of coesite is $2/m$, and therefore the mineral is monoclinic. The observations that $a = c$ and $\beta = 120$ degrees are not contrary to the restrictions imposed on the lattice by the $2/m$ symmetry. Note that in Table 2.10 the equals sign ($=$) means that equality is a requirement, and the symbol NR means that the value of the parameter has no restriction. In the monoclinic system, the values of a, b, c, and β have no restriction. The only restriction is that the angles α and γ be equal to 90 degrees.

Table 2.10 lists the seven distinct lattice symmetries and the seven coordinate systems that go with them. Current usage recognizes only six coordinate systems and considers trigonal, rhombohedral, and hexagonal under the heading of hexagonal. Elimination of the trigonal system is justified by the fact that trigonal and rhombohedral lattices are either identical with or can be transformed into hexagonal coordinate systems by using a multiple unit cell. An International Committee on Crystallography agreed in 1946 that all point groups with a single threefold or sixfold axis be considered hexagonal. This restriction excludes the threefold axes in isometric point groups. Table 2.10 is consistent with this usage. The terms "trigonal" or "rhombohedral" subdivisions are still popular in both foreign and American literature.

A unique entry in Table 2.10 is the modification of the point group symbols to include alternative choices for the positions of unit translations relative to the symmetry elements of the point group. For example, only one point group has the symbol $\bar{4}2m$. The a_1 and a_2 crystallographic axes are coincident with the twofold axes. The symbol can be written as $\bar{4}m2$ to designate that a_1 and a_2 are perpendicular to the mirror planes. This distinction is not important when only macroscopic symmetries are considered. It is important, however, in crystal structures where $P\bar{4}m2$ and $P\bar{4}2m$ are in fact different space groups. The same distinction applies to $32(1)$ and 312, $3m(1)$ and $31m$, and $\bar{6}2m$ and $\bar{6}m2$ point groups.

---

[5] Of these two octants, only one will yield a right-handed coordinate system for the appropriate a, b, c labeling.

[6] The term "lattice parameters" refers collectively to the magnitude of the unit cell translations and the angles between them.

TABLE 2.10. *The Seven Crystal Systems and Conditions for Assignment to Them*

Crystal System and Lattice Symmetry	a b c α β γ	Point Group	Conventions of Labeling
Triclinic $\bar{1}$	a(NR), b(NR), c(NR) α(NR), β(NR), γ(NR)	$\boxed{\bar{1}}$	Either α, β, γ are all $> 90°$ or are $< 90°$ 1st setting: $a < b < c$ 2nd setting: $c < a < b$
		1	
Monoclinic $\frac{2}{m}$	a(NR), b(NR), c(NR) $\alpha = \beta = 90°$, γ(NR) $\alpha = \gamma = 90°$, β(NR)	$\boxed{\frac{2}{m}}$	1st setting: $c = $ 2-fold, $a < b$ and $\gamma > 90°$ 2nd setting: $b = $ 2-fold, $c < a$ and $\beta > 90°$
		2	1st setting: $c = $ 2-fold 2nd setting: $b = $ 2-fold
		m	1st setting: $c = \frac{1}{m}$ 2nd setting: $b = \frac{1}{m}$
Orthorhombic $\frac{2}{m}\frac{2}{m}\frac{2}{m}$	a(NR), b(NR), c(NR) $\alpha = \beta = \gamma = 90°$	$\boxed{\frac{2}{m}\frac{2}{m}\frac{2}{m}}$	1st setting: $a < b < c$ 2nd setting: $c < a < b$
		222	
		$mm2$	$c = $ 2-fold, $a < b$ 1st setting
		$(m2m)$	$b = $ 2-fold, $c < a$ 2nd setting
Tetragonal $\frac{4}{m}\frac{2}{m}\frac{2}{m}$	$a = b$, c(NR) $\alpha = \beta = \gamma = 90°$	$\boxed{\frac{4}{m}\frac{2}{m}\frac{2}{m}}$	$c = $ 4-fold, $a_1, a_2 = $ 2-fold
		4	$c = $ 4-fold, $a_1, a_2 = \frac{1}{4}$
		$\bar{4}$	$c = \bar{4}$-fold, $a_1, a_2 = \frac{1}{4}$
		$\frac{4}{m}$	$c = $ 4-fold, $a_1, a_2 = \frac{m}{4}$
		422	$c = $ 4-fold, $a_1, a_2 = $ 2-fold
		$4mm$	$c = $ 4-fold, $a_1, a_2 = \frac{1}{m}$
		$\bar{4}2m$	$c = \bar{4}$-fold, $a_1, a_2 = $ 2-fold
		$(\bar{4}m2)$	$c = \bar{4}$-fold, $a_1, a_2 = \frac{1}{m}$
(Rhombohedral) $\bar{3}\frac{2}{m}$	$a = b$, c(NR) $\alpha = \beta = 90°$, $\gamma = 120°$	$\boxed{\bar{3}\frac{2}{m}}$	$c = \bar{3}$-fold, $a_1, a_2 = $ 2-fold
		$(\bar{3}1\frac{2}{m})$	$c = \bar{3}$-fold, $a_1, a_2 = \frac{m}{2}$
		$\bar{3}$	$c = \bar{3}$-fold, $a_1, a_2 = \frac{1}{3}$
(Trigonal) $\frac{6}{m}\frac{2}{m}\frac{2}{m}$	$a = b$, c(NR) $\alpha = \beta = 90°$, $\gamma = 120°$	3	$c = \bar{3}$-fold, $a_1, a_2 = \frac{1}{3}$
		32	$c = $ 3-fold, $a_1, a_2 = $ 2-fold
		(312)	$c = $ 3-fold, $a_1, a_2 = \frac{1}{2}$

TABLE 2.10 (*continued*)

Crystal System and Lattice Symmetry	a b c α β γ	Point Group	Conventions of Labeling
		$3m$	$c = 3\text{-fold}, a_1, a_2 = \dfrac{1}{m}$
		$(31m)$	$c = 3\text{-fold}, a_1, a_2 = \dfrac{m}{3}$
Hexagonal $\dfrac{6}{m}\dfrac{2}{m}\dfrac{2}{m}$	$a = b, \quad c(\text{NR})$ $\alpha = \beta = 90°, \quad \gamma = 120°$	$\boxed{\dfrac{6}{m}\dfrac{2}{m}\dfrac{2}{m}}$	$c = 6\text{-fold}, a_1, a_2 = 2\text{-fold}$
		6	$c = 6\text{-fold}, a_1, a_2 = \dfrac{1}{6}$
		$\bar{6}$ or $\dfrac{3}{m}$	$c = \bar{6}\text{-fold}, a_1, a_2 = \dfrac{m}{\bar{6}}$
		$\dfrac{6}{m}$	$c = 6\text{-fold}, a_1, a_2 = \dfrac{m}{6}$
		622	$c = 6\text{-fold}, a_1, a_2 = 2\text{-fold}$
		$6mm$	$c = 6\text{-fold}, a_1, a_2 = \dfrac{1}{m}$
		$\bar{6}2m$	$c = \bar{6}\text{-fold}, a_1, a_2 = 2\text{-fold}$
		$(\bar{6}m2)$	$c = \bar{6}\text{-fold}, a_1, a_2 = \dfrac{1}{m}$
Isometric $\dfrac{4}{m}\bar{3}\dfrac{2}{m}$	$a = b = c$ $\alpha = \beta = \gamma = 90°$	$\boxed{\dfrac{4}{m}\bar{3}\dfrac{2}{m}}$	$a_1, a_2, a_3 = 4\text{-fold}$
		432	$a_1, a_2, a_3 = 4\text{-fold}$
		$\bar{4}3m$	$a_1, a_2, a_3 = 4\text{-fold}$
		23	$a_1, a_2, a_3 = 2\text{-fold}$
		$\dfrac{2}{m}\bar{3}$	$a_1, a_2, a_3 = 2\text{-fold}$

NOTE: Symbol (NR) following a lattice parameter denotes no restriction on the value of that parameter. Although both settings are tabulated for each point group, the *second setting* is used exclusively in mineralogy.

MULTIPLE UNIT CELLS

Up to now we have assumed that the unit translations of the reduced cell are parallel to rotation axes whenever they are present. This is not always true. In some lattices, the rotation axes may coincide with a face-diagonal or body-diagonal of the reduced cell. Instead of defining a new coordinate system for such a cell, the common procedure is to enlarge the cell to a *multiple unit cell* in which the unit translations are made to coincide with rotation axes or to be perpendicular to mirrors. Use of a multiple cell in this example achieves the objective of providing coincidence of unit translations and rotation axes. In so doing, a coincidence is provided also between lattice orientation and external morphology.

As we learn in the next chapter, a close connection exists between crystal form and crystal symmetry. Historically, the possible shapes and dimensions of unit cells are conceived by observation of a large number of crystal forms. In 1848, Auguste Bravais derived the seven possible lattice symmetries and, in addition, seven multiple cells that he referred to as "modes" of the first seven. He clearly stated that his multiple cells possessed no additional symmetry over the primitive cells. Unfortunately, Bravais's arguments were misinterpreted by others who assumed that the lattices requiring multiple cells had different symmetry from the primitive cells.

The 14 space lattices are illustrated in Figure 2.28 and listed in Table 2.11. The reduced, primitive triclinic cell is almost always retained as it has

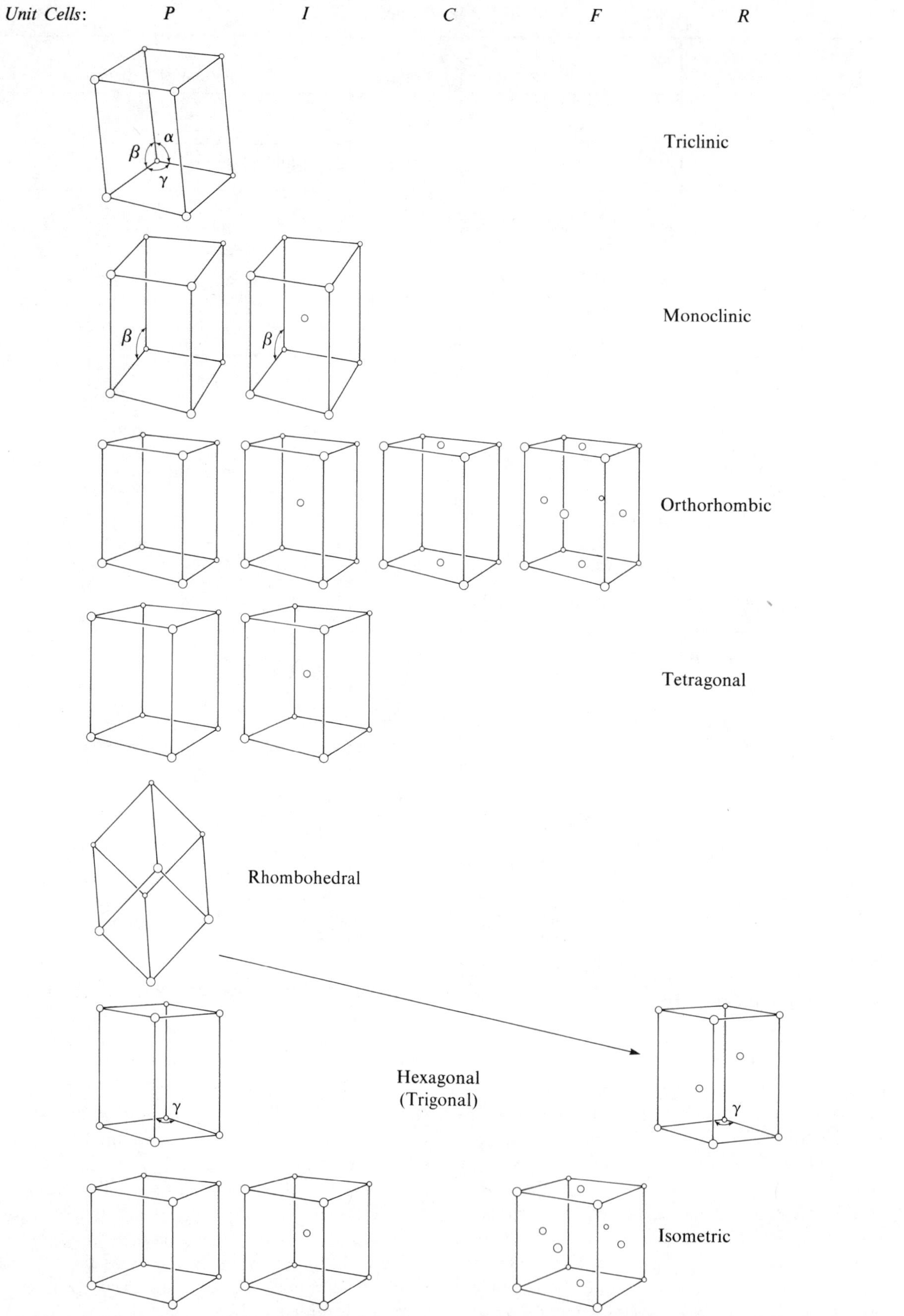

Unit Cells: P   I   C   F   R

Triclinic

Monoclinic

Orthorhombic

Tetragonal

Rhombohedral

Hexagonal
(Trigonal)

Isometric

FIGURE 2.28   Summary of the seven primitive space lattices and permissible multiple cells derived from them. (After Buerger, Martin J. *Elementary crystallography*, rev. printing [New York, 1963], Fig. 1, p. 101. By kind permission of John Wiley & Sons, Inc., copyright © 1963.)

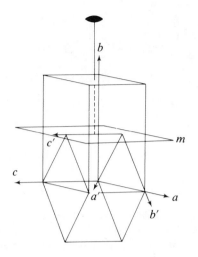

FIGURE 2.29   Primitive and multiple unit cell (*C*-side centered) in microcline.

no rotation axes to justify a multiple cell. Occasionally, a triclinic multiple cell is chosen for the purpose of comparison. In Figure 2.29, microcline (KSi_3AlO_8) has triclinic symmetry but is commonly considered to have a *C*-centered unit cell. The reason for this choice is that the microcline structure is similar to that of other feldspars, which have monoclinic symmetry. The reduced cell of triclinic microcline differs in shape and in dimensions from the monoclinic unit cells. A multiple cell, namely, the *C*-side centered cell of microcline, is more similar to the monoclinic cell, and therefore provides a better comparison of properties such as cell volume, cell content, and structural orientation.

Only two uniquely different unit cells are among the monoclinic minerals. They are the primitive (*P*) unit cell and the body-centered (*I*) cell, which is equivalent to the *A*-side, *B*-side, or *C*-side centered cells. The relationship between the primitive and multiple unit cells is shown in Figure 2.30. The twofold axis of the primitive unit cell (Figure 2.30a) is the diagonal of the *B* side, that is, the side that includes the *a* and *c* axes. The mirror plane includes the *b* axis and the other diagonal of the *B* side. This means that the twofold axis is not parallel with a unit translation. To derive a multiple cell in which the twofold axis does coincide with a unit translation, the primitive cell is tilted (Figure 2.30b) such that the points marked with a dot and circle are at elevations 0 and 1 along the twofold axis. The lattice points marked as dots are at $1/2$ elevation. A body-centered *I* cell is outlined with a dashed line. A

TABLE 2.11.   *The 14 Space Lattices That Define the Possible Unit Cells in Each Crystal System*

Lattice Symmetry	Crystal System	Possible Unit Cells
$\bar{1}$	Triclinic	P
$\dfrac{2}{m}$	Monoclinic	P, I (or A, B, C)
$\dfrac{2}{m}\dfrac{2}{m}\dfrac{2}{m}$	Orthorhombic	P, I, A (or B, C) F
$\dfrac{4}{m}\dfrac{2}{m}\dfrac{2}{m}$	Tetragonal	P, I
$\bar{3}\dfrac{2}{m}$	Trigonal	R
$\dfrac{6}{m}\dfrac{2}{m}\dfrac{2}{m}$	Hexagonal	P
$\dfrac{4}{m}\bar{3}\dfrac{2}{m}$	Isometric	P, I, F

C-side centered cell is outlined with a dotted line. Both are multiple unit cells. The choice between them depends on which gives the shorter *a* and *b* unit translations. In this case it is the *C*-side centered cell. *A*-side and *B*-side centered cells are derived in the same way by using a different axis setting. If the first setting is used, the *c* axis is the twofold axis, and an *A*-side centered cell results.

ANALYTICAL REPRESENTATION OF SYMMETRY OPERATIONS

The symmetry operations considered up to now have been treated in a visual manner. This is appropriate, because most of the work of a miner-

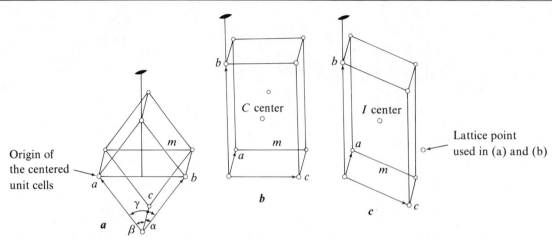

FIGURE 2.30   **Alternative unit cells for a lattice having 2/m symmetry. (a) Reduced primitive cell. (b) C-centered multiple unit cell in which the b axis coincides with the twofold axis of the lattice. (c) I-centered multi-** **ple unit cell in which the b axis coincides with the twofold axis of the lattice. The choice between (b) and (c) depends on which yields the shorter a and c unit translations.**

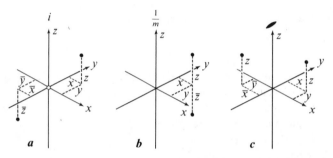

FIGURE 2.31   **Operations on general coordinates x, y, and z of (a) inversion: xyz, $\bar{x}\bar{y}\bar{z}$, (b) reflection: xyz, xy$\bar{z}$, (c) twofold rotation: xyz, $\bar{x}\bar{y}z$.**

alogist deals with real three-dimensional minerals. Some applications, however, require a rapid, analytical formulation of the effects of symmetry. Examples are the determination and refinement of crystal structures, the computer simulation of a crystal during growth, or calculations involving symmetry-dependent physical properties.

Given an arbitrary point with coordinates x, y, and z (Figure 2.31a), the position of the point changes after a symmetry operation. A center of inversion reverses the sign of each of the original coordinates, but maintains their magnitude. A bar symbol over the coordinate symbolizes the negative sign. In Figure 2.31b, the effect of a mirror plane in the xy plane is to reverse the sign of the z coordinate only. A mirror in the xz plane changes the sign of the y coordinate. A rotation axis coincident with z (Figure 2.31c) does not affect the sign or the magnitude of the z coordinate. The signs of both the x and y coordinates change if the rotation axis is twofold. Higher ranking axes parallel to z can change both the sign and magnitude of x and y.

The x, y, and z coordinates of *symmetrically*

equivalent points generated by all of the basic and compound symmetry operations are given in Table A3.2, Appendix 3. The location or orientation of the symmetry operations, or both, and the appropriate coordinate systems are identified for each operation.

An analytical representation is convenient when we know the initial coordinates of an atom and wish to repeat that atom to symmetrically equivalent positions in the structure. Several symmetry operations may apply. Successive coordinate changes must always return the atom to the original position, that is, to the original coordinate set. Equivalent atomic sites for space groups in the International Table for Crystallography are given in terms of equivalent coordinates.

ADDITIONAL READINGS

Buerger, M. J., 1963. *Elementary crystallography*. rev. ed. New York: John Wiley and Sons.

Buerger, M. J., 1975. *Contemporary crystallography*. New York: John Wiley and Sons.

Phillips, F. C., 1963. *An introduction to crystallography*. 3rd ed. New York: John Wiley and Sons.

CHAPTER THREE

SYMMETRY
OF CRYSTALS

Most minerals occur in rocks where their external form is largely controlled by the minerals around them. The development of well-formed crystal faces is common only in environments where crystals precipitate from a gas or liquid. A distinctive feature of crystals so formed is the symmetrical arrangement of crystal faces. A casual inspection of the photographs in this chapter emphasizes an important property of crystals— they all have symmetry. One objective of this chapter is to learn how crystal faces are described, and how symmetry can be deduced from that information.

As we proceed, keep in mind that many variables control the extent to which crystal faces develop during growth. As a prelude to Chapter 7 on crystal growth, note that external morphology does not depend exclusively on internal structure. The symmetry of the structure always controls the symmetry of the faces because the faces are the planar terminations of the structure, but other factors such as the mechanism or the environment of growth are commonly important. Crystal forms cannot be solely dependent on growth mechanisms either. The forms of freely grown crystals of a given chemistry do not appear with the same frequency. The six-sided prisms of beryl ($Al_2Si_6Be_3O_{18}$), for example, are much more common and better developed than are other faces. This observation indicates a close relationship between crystal structure and form.

The study of crystal morphology has its practical side. When the point group symmetry can be determined, mineral identification often becomes an easy matter. In some instances, crystal faces reveal an absence of a center of symmetry that can be difficult to determine by other methods. This is essential to distinguishing between centric and noncentric space groups, and useful for predicting the existence of certain polar properties such as pyroelectricity and piezoelectricity (Chapter 6). A better understanding of the interdependence of external form and internal structure is also useful in formulating crystal growth models, since in detail, all crystal faces must develop through various growth mechanisms.

INDEXING OF RATIONAL FEATURES IN THE LATTICE

Points, lines, and planes that contain lattice points are said to be *rational*. Crystal faces are parallel to lattice planes and can be described in terms of lattice parameters. The generally accepted notation is as follows:

1. A *lattice point* is located in the lattice by giving the number of unit translations, along each of the three translation directions, by which it is removed from the origin. The three unit translations are the unit cell edges, and their directions are the crystallographic axes *a*, *b*, and *c*. Figure 3.1 illustrates a two-dimensional lattice in which the vector **T**

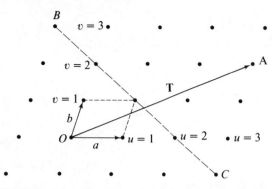

FIGURE 3.1 Two-dimensional lattice and primitive cell with unit translations of magnitudes *a* and *b*. Vector **T** is a rational line and describes the direction [320] to the lattice point *A* (·320·). Dashed line *BC* is a two-dimensional lattice plane described by the intercept method with *u* = 2 and *v* = 2.

from the origin to lattice point *A* is **T** = *u***a** + *v***b** where the coefficients *u* and *v* are the number of unit translations along the component axes *a* and *b*, respectively. In this example, *u* = 3 and *v* = 2. The appropriate symbol to describe a lattice point in three dimensions is ·*uvw*· where *w* is the number of unit translations along the *c* axis. In this example, the lattice point is denoted by ·320·. The same coefficients placed between double dots (i.e., ··320··) denote not only the lattice point ·320· but all points that are symmetrically equivalent to it.

2. *Rational lines* or rational directions are defined in the same way as lattice points. The direction of vector **T** in Figure 3.1 is symbolized by [320]. If placed between angular brackets ⟨320⟩, all symmetrically equivalent lines are included.

3. *Rational planes* in a lattice can be described in two ways. One way, no longer in use, is called the *intercept notation*, and like the notation for lattice points and lines, is based on the coefficients *u*, *v*, and *w*. The method, illustrated in Figure 3.1, was first developed by Samuel Weiss in 1808. The intercepts of the lattice plane (*BC* dashed line) with the crystallographic axes are *u* = 2, *v* = 2, and *w* = ∞. The Weiss symbol for a general plane is (*uvw*) and for this plane in particular (22∞). Weiss's contribution was to reiterate the *law of rational intercepts*, which was earlier demonstrated by René Haüy. Crystal faces, as rational planes, also intersect crystallographic axes. Relative to an origin chosen at the geometric center of the crystal, the ratio of the intersections, or *axial ratios* as they are commonly called, can be determined. These ratios give the *relative* lengths of the unit translations.

The disadvantage of the Weiss notation is that it describes only one of a family of parallel planes. Specifically, it describes the *n*th plane out from the origin where *n* is the product *u* · *v* · *w*. In the example of Figure 3.2, five such planes are between the intercept plane (six planes total) and the origin. The unique plane of the family is the first one out from the origin. Its Weiss indices are (*a/h b/k c/l*) where the integers *h*, *k*, and *l* are the *fractional intercepts* of the plane with the *a*, *b*, and *c* axes, respectively.

Instead of the Weiss fractional indices, the current, more convenient notation for rational planes is to use the integers *h*, *k*, and *l* by themselves. The plane labeled "1" in Figure 3.2 can then be specified uniquely by (*hkl*) as (230). This symbol is read to mean that the plane (230) cuts the *a* axis in two equal parts, cuts the *b* axis in three equal parts, and is parallel to the *c* axis. A three-dimensional lattice and the indices of selected rational planes are shown in Figure 3.3. The set of symmetrically equivalent planes is denoted by ·{*hkl*}.

The direct use of *h*, *k*, and *l* integers to describe rational planes was first suggested in 1824 by Professor William Whewell at Cambridge University. The indices are usually re-

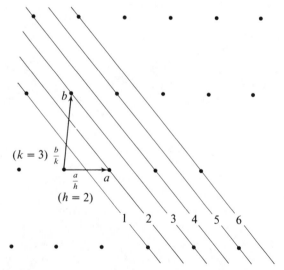

FIGURE 3.2 Two-dimensional lattice illustrating edge view of family of parallel lattice planes that exists between origin and plane with intercept notation (32∞). Intercept notation of first plane out from origin is (*a/n b/k* ∞), and Miller notation is (*hkl*) or (230).

ferred to as *Miller indices*, after William H. Miller who succeeded Whewell at Cambridge and popularized the indices in his crystallography textbook published in 1839. Miller indices are now used universally and will be used exclusively hereafter in this text.

INDICES OF CRYSTAL FACES

The proper assignment of Miller indices to crystal faces is a routine matter if the lattice parameters of a mineral are known or can be determined by x-ray diffraction. Indices can be assigned for crystals with unknown lattices by determining the *relative* intercepts of a crystal face with the crystallographic axes. The first step in this type of indexing is to locate the axes. For crystals with high symmetry, axes are usually chosen to coincide with the highest ranking rotation axis or in directions perpendicular to mirror planes. This general procedure is sufficient for locating axes in all isometric, hexagonal, tetragonal, and orthorhombic minerals. For monoclinic crystals, one crystallographic axis is chosen to coincide with the single twofold rotation axis or with the line perpendicular to the single mirror plane. The other two axes are chosen in the mirror plane and parallel to two prominent faces. In the triclinic system, axes are chosen parallel to any three prominent intersecting faces assumed proportional to the magnitudes of the unit translations.

The second step in the indexing process is to define unit lengths on each of the three crystallographic axes. If a face is present that intersects all three axes (Figure 3.3), those intercepts are chosen to represent unit lengths. The unit lengths are then presumed *proportional* to the magnitudes of the unit translations of the lattice. If more than one such face is developed, the most prominent is chosen. Its Miller indices are then *defined* as (111). These are the indices of the first lattice plane out from the origin and parallel to the actual face to which the indices are assigned. If a (111) face is not developed, two of the more prominent faces that cut different pairs of axes are chosen instead. These will have indices such as (011), (101), and (110) for faces that are parallel to the *a*, *b*, and *c* axes, respectively. Faces that cut only one axis are considered to cut it at unity. These faces will have indices of (100), (010), and (001), depending on whether the single axis cut is *a*, *b*, or *c*, respectively.

In the preceding discussion, faces are presumed to intercept the positive ends of the crystallographic axes relative to the center of the crystal. Faces developed on the "back side" or bottom of the crystal will have negative indices denoting their intersection with the negative ends of the *a*, *b*, and *c* axes. A bar over the appropriate integer denotes a negative intercept.

The third and final step is to index the other faces of the crystal relative to the length of the intercepts of the unit face or faces. Because this

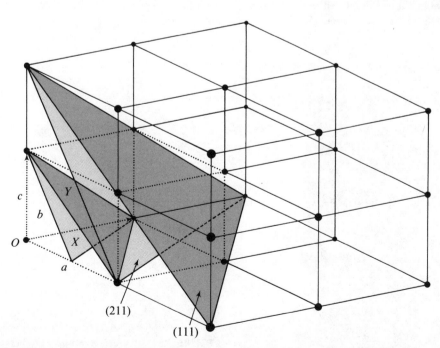

FIGURE 3.3 Three-dimensional lattice with lattice plane (111) shaded, and (211) unshaded. Both planes are described in terms of first planes out from origin, denoted by *Y* (shaded) and *X* (unshaded).

face or these faces define the relative lengths of the unit translations, that is, the axial ratios, all other faces indexed on this basis must also be parallel to lattice planes. If another face cuts an axis at a distance greater than the unit face, it is indexed the same as an imaginary but parallel plane located closer to the center of the crystal (Figure 3.3). The parallel shift does not affect the relative length of the intercepts.

All faces on any crystal can be indexed with the three-digit notation (hkl). Occasionally we find in the literature and in textbooks a four-digit notation ($hkil$) for hexagonal crystals. This resulted from an historical treatment in which three axes, a_1, a_2, and a_3, were used in the plane normal to the c axis. Only two of these axes are necessary. The relationship $a_3 = -(a_1 + a_2)$ defines the third axis. Whenever an hexagonal face is encountered with a four-digit notation, we can convert the

notation to normal usage by the relationship $i = -(h + k)$. For example, in Figure 3.4a the c-axis projection of the prismatic faces of quartz (SiO_2) are labeled in terms of four indices, and in Figure 3.4b in terms of the preferred three axes. Occasionally, a dot in place of i will denote omission of the i index, but even that is superfluous.

STEREOGRAPHIC PROJECTION OF CRYSTAL FACES

Various symmetry elements such as mirror planes, rotation axes, and centers of symmetry are often revealed by the arrangement of faces on crystals. The three-dimensional relationships between symmetry operations and crystal faces are best represented in stereographic projection. The basic concept is the same as that presented in Chapter 2 with the additional application to crystal faces. For this purpose, we plot the pole to the face of interest rather than plot the face itself. The pole is the line perpendicular to the face (also called the face normal) and is designated hereafter as the (hkl) pole. Figure 3.5 shows the poles and their projection as points in the equatorial plane. As before, poles to faces in the lower hemisphere project through the north pole, and poles to faces in the upper hemisphere (top half of crystal) project through the south pole. The resulting two-dimensional projections of the vertical and "top" faces of the faceted cube (Figure 3.5) are shown in Figures 3.6a and 3.6b. In the latter, all faces are shown, but only the "bottom" faces (open circles) are labeled.

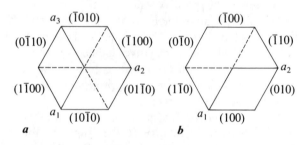

FIGURE 3.4 c axis projection of quartz (SiO_2) with prismatic faces parallel to c axis labeled in terms of (a) four-digit index and (b) preferred three-digit index.

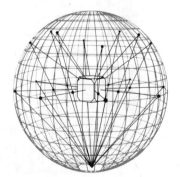
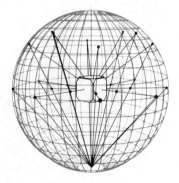

FIGURE 3.5   Poles (face normals) to crystal faces with appropriate Miller indices projected onto equatorial plane using south pole as point of projection.

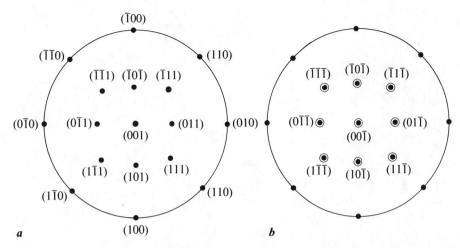

a

b

FIGURE 3.6 **Stereographic projections of crystal faces of faceted cube from Figure 3.5. (a) Vertical and upper** hemisphere face normals shown and labeled. (b) All faces shown with only lower hemisphere faces labeled.

MEASUREMENT OF ANGLES

The true angle between two intersecting planes is always measured in a third plane that is perpendicular to the first two. If this condition were not specified, any of a range of angles from zero to the true value could be measured. The true angle is the maximum value in that range. Measurement of the true angle is analogous to measuring the true dip (angle between the horizontal plane and bedding plane) in field geology, as opposed to the apparent dip. The angle between two crystal faces is given exactly by the angle between their poles and can be measured easily with a contact goniometer (Figure 3.7). The device actually measures the external or outside angle γ between the faces as shown, but the instrument usually has a scale on which to read the complementary angle α as well. External angles are always greater than 180 degrees such that $\gamma = 180 + \alpha$. The angle α is of special interest because it is the angle between the poles of the faces.

The reflecting goniometer is more accurate than the contact goniometer and measures the angle α directly from the rotation between the reflecting positions of faces. This measurement requires that the axis of rotation be parallel with the faces, but this condition is not always the case. A two-circle reflecting goniometer provides an additional axis for situations in which the axis of rotation is not parallel with the faces. Combined with a Wulff net, poles of faces and interfacial angles can be easily and accurately recorded. The horizontal axis of the goniometer is the north-south axis of the sphere of projection (Figure 3.8) located at the center of the Wulff net. Measurements around that axis correspond to angles be-

tween lines of longitude on a globe, and are in terms of the angle ϕ measured from the $\phi = 0$ point located on the right side of the Wulff net. The vertical axis of the goniometer measures the angle ρ, which is plotted on the Wulff net from the center of the net to the equatorial plane. This angle is equivalent to the angle between lines of latitude on a globe.

In practice, a transparent overlay is placed on the Wulff net and fastened by a pin or thumbtack at the center. The overlay can then be rotated relative to the net around the vertical axis. The (001) pole is usually plotted at the center of the net, and in orthogonal crystals the (010) pole is

FIGURE 3.7 **Contact goniometer used to measure interfacial angles on crystals. Angle of 60° measured on internal scale is angle between poles of respective faces.**

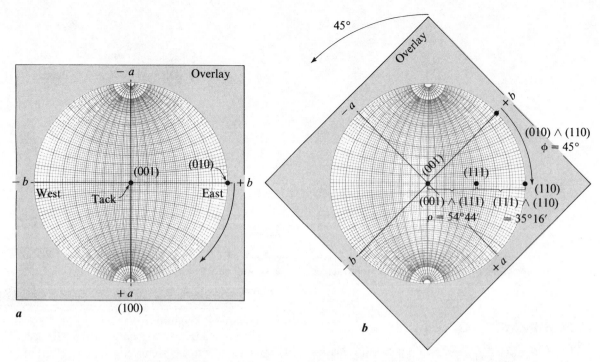

FIGURE 3.8   **(a) Stereographic Wulff net used to project poles to crystal faces. Interfacial angle (001) ∧ (010) = 90°00′. Counterclockwise rotation of transparent overlay to (b) position enables interfacial angles (001) ∧ (111) and (111) ∧ (110) to be plotted along vertical great circle of the net.**

plotted at the east point ($\phi = 0$) and the (100) pole clockwise from that point at the bottom of the net (Figure 3.8a). For example, the (110) and (111) poles of the crystal in Figure 3.5 are at $\phi = 45$ degrees and $\rho = 90$ degrees, and $\phi = 45$ degrees and $\rho = 54$ degrees 44 minutes, respectively. Note that all latitude ρ measurements must be made along great circles. In order to plot the (110) pole of the crystal in Figure 3.5, the transparent overlay is rotated 45 degrees counterclockwise, and the pole is marked directly over the east point ($\phi = 0$) of the net (Figure 3.8b). Since the angle ρ of the (111) pole is also 45 degrees, the overlay is retained in this position, and the angle $\rho = 54$ degrees 44 minutes is plotted from the center of the net toward the equatorial plane along the east-west vertical great circle.

CRYSTAL CLASSES

The purpose of representing crystal faces in stereographic projection is to demonstrate the symmetry relationships between them. This leads directly to the determination of point group symmetries, which are the symmetries of the crystal classes. Determination of crystal class is the first step in a series of studies to analyze the structure of a mineral, and it serves as an excellent criterion in the process of mineral identification.

Once the poles of faces are plotted, the symmetry operations that relate them become apparent. For example, the poles of faces of a zircon ($ZrSiO_4$) crystal (Figure 3.9a) are plotted in Figure 3.9b. The crystal clearly has a strong axial symmetry that must be a fourfold axis of rotation. The faces (211), ($\bar{1}$21), ($\bar{2}\bar{1}$1), and (1$\bar{2}$1) are symmetrically equivalent by this operation (see Table A3.2, Appendix 3). The faces (211) and (2$\bar{1}\bar{1}$), and (2$\bar{1}$1) and (21$\bar{1}$) are related by a horizontal twofold axis that is parallel to a_1 and perpendicular to the (010) pole. The fourfold axis repeats the twofold axis to a position perpendicular to the (100) pole. A different twofold axis, which is in the [110] direction and perpendicular to the (1$\bar{1}$0) pole, relates (211) to (12$\bar{1}$), (121) to (21$\bar{1}$), and so on. It, too, is operated on by the fourfold axis. The two twofold axes are symmetrically nonequivalent, because no symmetry operation will bring one into registry with the other. The symmetry of the faces also requires mirror planes normal to each of the rotation axes as shown. Figure 3.9c summarizes the observed point group symmetry of zircon, and by comparison with Figure 2.21, we see that fourfold axes are present only in the isometric and tetragonal systems. The isometric system is eliminated as a possibility for zircon, because there must also be four threefold axes, which are obviously absent. Of the seven tetragonal point

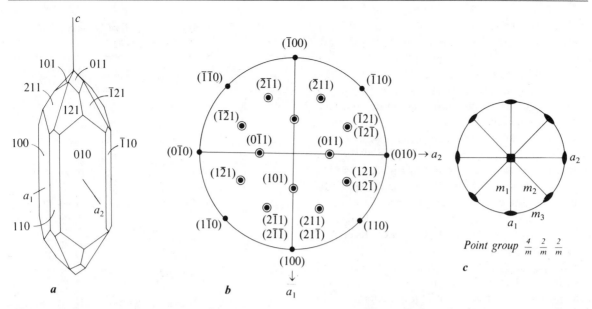

FIGURE 3.9 (a) Zircon crystal (ZrSiO₄). (b) Stereographic projection of zircon crystal faces. (c) Symmetry deduced from projection confirms 4/m 2/m 2/m point group. Interfacial angles are (010) ∧ (110) = 45°00′, (100) ∧ (101) = 47°50′, and (110) ∧ (211) = 31°43′.

groups, only one (4/m 2/m 2/m) satisfies the observed symmetry of zircon.

As another example of crystal class identification, let us consider low quartz (SiO₂). Figure 3.10a shows a typical arrangement of faces and (Figure 3.10b) their stereographic projection. At first glance, a sixfold axis would seem to be present because of the repetition of the six prismatic faces (100), (010), (1̄10), (1̄00), (01̄0), and (11̄0). Closer inspection shows that poles of faces such as

(111) and (151̄) are repeated only three times by a vertical threefold axis and therefore preclude an axis of higher rank. A horizontal twofold axis, which coincides with a_1, relates (151̄) to (5̄61̄) and (511) to (16̄1̄). That axis is repeated by the vertical threefold axis to positions 120 degrees apart. Those positions coincide with a_2 and with the [1̄1̄0] direction (Figure 3.10c). Because the twofold axis is horizontal, it also coincides with the [110] direction exactly 60 degrees from a_1. Thus, the

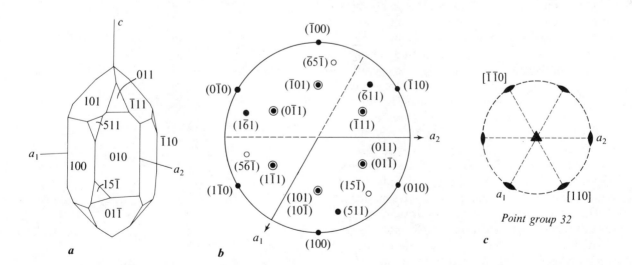

FIGURE 3.10 (a) Low quartz crystal (SiO₂). (b) Stereographic projection of low quartz crystal faces. (c) Symmetry elements deduced from projection confirm 32 point group. Interfacial angles are (100) ∧ (101) = 38°13′, (100) ∧ (011) = 60°00′, (100) ∧ (011) = 66°52′, (100) ∧ (011) = 12°01′, and (101) ∧ (011) = 46°16′.

twofold axis relates (511) to (15$\bar{1}$) and (010) to (100). In the latter case, note that these prismatic faces are *not* related by a vertical sixfold axis as it might first have appeared.

This discussion illustrates the convenience of being able to use a symmetry operation to generate the Miller indices of all symmetrically equivalent faces. The operation is done with stereographic projection by plotting the pole of the face of interest as illustrated earlier in Figure 2.24. The results are summarized in Table A3.2, Appendix 3, in which the equivalent indices are listed.

Once the faces of an unknown crystal are plotted and the symmetry operations are defined, a direct comparison can be made with the point group diagrams in Figure 2.21. An inspection of the structural systematics in Part II shows that the majority of the most common minerals (over 80%) are centric, and almost all are limited to nine point groups. By far the most common symmetries are $2/m$, $2/m\,2/m\,2/m$, $4/m\,\bar{3}\,2/m$, $6/m\,2/m\,2/m$, $\bar{3}\,2/m$, and $4/m\,2/m\,2/m$.

APPARENT SYMMETRY

The true point group symmetry of a crystal is not always displayed by the symmetry of its faces. In fact, the majority of crystals will display higher than true symmetry. The major reason for the higher *apparent symmetry* is the coincidence of crystal faces with symmetry operations.

We have already learned that atoms may be in general or special positions. When an atom is in a special position, it is not repeated by the symmetry operation in that position. Similarly, if a crystal face coincides with a mirror plane, or if a face pole coincides with a rotation axis in a point group, the face is in a *special position*. Consequently, the face is not repeated by the operation of the symmetry element in that position. A crystal face (or pole) that does not coincide with the above symmetry elements is in a *general position* and will be repeated by the operation of those elements. Only those crystals that have faces in general positions can exhibit all of the symmetry operations of the point group, and only those crystals can reveal the true point group symmetry.

This statement has a few important exceptions. Crystal faces in general positions do not always reveal the true point group, and in some cases, cannot reveal the true symmetry. In the monaxial point groups 3, 4, and 6, the symmetry exhibited by faces in the general position will always be higher than the true symmetry, because additional twofold axes or mirrors or both always appear (Table 3.1). The reason for this is shown in Figure

TABLE 3.1. *Apparent and True Monaxial Point Group Symmetry*

Monaxial Point Groups	Apparent Symmetries	Lattice Symmetries
3	$3m$	$\bar{3}\dfrac{2}{m}$
$\bar{3}$	$\bar{3}\dfrac{2}{m}$	$\bar{3}\dfrac{2}{m}$
4	$4mm$	$\dfrac{4}{m}\dfrac{2}{m}\dfrac{2}{m}$
$\bar{4}$	$\bar{4}\dfrac{2}{m}$	$\dfrac{4}{m}\dfrac{2}{m}\dfrac{2}{m}$
$\dfrac{4}{m}$	$\dfrac{4}{m}\dfrac{2}{m}\dfrac{2}{m}$	$\dfrac{4}{m}\dfrac{2}{m}\dfrac{2}{m}$
6	$6mm$	$\dfrac{6}{m}\dfrac{2}{m}\dfrac{2}{m}$
$\bar{6}$	$\bar{6}\dfrac{2}{m}$	$\dfrac{6}{m}\dfrac{2}{m}\dfrac{2}{m}$
$\dfrac{6}{m}$	$\dfrac{6}{m}\dfrac{2}{m}\dfrac{2}{m}$	$\dfrac{6}{m}\dfrac{2}{m}\dfrac{2}{m}$

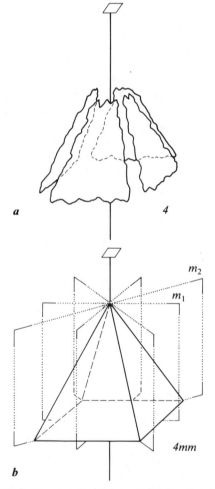

FIGURE 3.11 **(a) Crystal faces of general form in class 4 must be related by the fourfold axis. (b) Face intersections in (a) must be symmetrically equivalent, so additional mirror planes are introduced, 4*mm*.**

3.11. Because crystal faces are planes, they must intersect along a straight edge. Both faces and edges are repeated according to the rotational symmetry of the monaxial point group in question. When the rotation axes are threefold, fourfold, or sixfold, opposite faces and edges will always be related by an apparent mirror plane or by a twofold axis or by both, and thus the symmetry will always appear to be greater than it actually is. The argument here is similar to the one used in Chapter 2 to demonstrate why isogonal point groups have a higher lattice symmetry. In that case, the geometry of the unit cell parallelepiped was responsible for the increased symmetry of the lattice.

If a crystal has faces in special positions only, the apparent symmetry is always one of the more symmetrical point groups isogonal with the true symmetry, except when the true symmetry is already the highest in the isogonal group. An example is the sperrylite ($PtAs_2$) crystal in Figure 3.12a. The stereographic projection of the poles of faces and the interfacial angles of this crystal are shown in Figure 3.12b. The symmetry of the crystal *appears* to be $4/m\bar{3}2/m$ as illustrated by the stereographic projection of the symmetry elements in Figure 3.12c. The true point group symmetry, $2/m\bar{3}$, is shown in Figure 3.12d. Comparison of these three figures reveals that all of the faces (or poles) are in special positions. The (100) pole is coincident with the fourfold axis, the (111) pole is coincident with the threefold axis, and the (110) pole is coincident with the twofold axis. In Figure 3.12d, the (100) pole is coincident with a twofold axis, and the (111) pole is coincident with a threefold rotoinversion axis. Consequently, the symmetry of faces in special positions does not enable us to distinguish between these two point groups for sperrylite.

Sperrylite does occasionally develop faces (Figure 3.13a) that occupy general positions. The faces {321} are all symmetrically equivalent (total of 24 faces) and are plotted in Figure 3.13b. The distribution of the faces clearly eliminates mirror planes, and hence the symmetry of point group $4/m\bar{3}2/m$. We must conclude, correctly, that the point group of sperrylite is $2/m\bar{3}$.

OTHER CRITERIA FOR CLASS DETERMINATION

Apparent symmetry based on the arrangement of crystal faces is so common that other means are frequently required to assign the true point group. Often we can obtain information that helps determine the true symmetry from the details of crystal

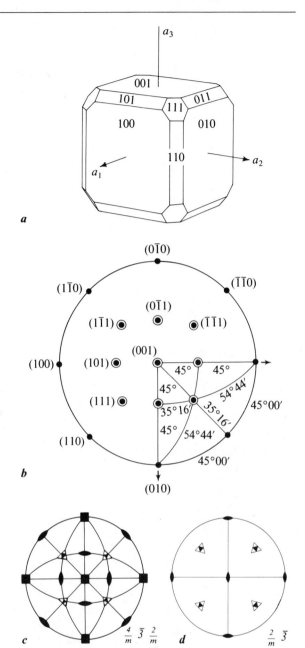

FIGURE 3.12 (a) Sperrylite ($PtAs_2$) crystal. (b) Interfacial angles and face normals of sperrylite in stereographic projection. (c) Stereographic projection of $4/m\bar{3}2/m$ apparent symmetry of sperrylite. (d) Stereographic projection of true point group symmetry $2/m\bar{3}$.

faces. For example, pyrite (FeS_2) commonly occurs as cubes, having $4/m\bar{3}2/m$ apparent symmetry. The faces are usually striated (Figure 3.14) in a manner dependent on a crystallographically controlled growth history. The direction of the striations is rotated 90 degrees on alternating faces, which serves to identify the presumed fourfold axis as a twofold instead. This information automatically excludes the point groups $4/m\bar{3}2/m$, 432, and $\bar{4}3m$, leaving only the acentric 23 point group and the centric $2/m\bar{3}$ point

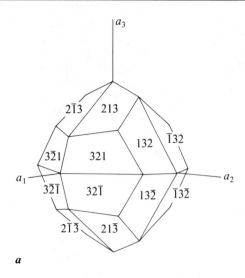

a

b

FIGURE 3.13   (a) Sperrylite crystal with general form
{321}. (b) Stereographic projection of face normals of
sperrylite crystal. Only point group $2/m\bar{3}$ is consistent
with this distribution of faces.

group as possibilities for pyrite. The latter is the
correct one.

The magnitude of many other crystal properties is dependent on structural anisotropy and
hence on the positioning of symmetry elements.
The development of etch pits on crystal faces is
one example. By application of an appropriate
solvent,[1] dissolution will proceed from isolated

---

[1] Depending on crystal solubility, etchants such as water, acetone, or dilute HCl, HNO$_3$, or HF (used on silicates) may be
applied to the crystal surface directly with a cloth, or the
crystal may simply be exposed to the acid vapor.

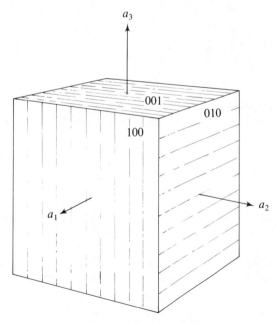

FIGURE 3.14   Cubic crystals of pyrite (FeS$_2$) give
apparent $4/m\bar{3}2/m$ symmetry. Sense of striations
shows that all faces are not symmetrically equivalent.

spots on the crystal face. The rate of dissolution is
controlled by the crystal structure and its symmetry. The shape of the etch pits on the faces and
their relationship with the pits on other faces may
reflect the true point group of the crystal. On
faces perpendicular to twofold axes, the pits will
be parallelograms or rectangles if mirror planes
exist parallel to the twofold axis. In the case of
threefold axes, the pits will have the shape of
equilateral triangles. Square shapes result from a
fourfold axis, and hexagons result from a sixfold
axis. The shape of the pits expresses the fact
that dissolution proceeds faster along one direction in the crystal than the other. Rapid dissolution implies that either the chemical bonds
between atoms in that direction are weaker or
that fewer atoms occupy a given unit area in that
direction. From this viewpoint, symmetry of the
etch pits expresses an important relationship between atoms, namely, that they are arranged in a
symmetric manner.

An example of how the symmetry of etch pits
can be used to distinguish between a crystal's
apparent and true point group is calcite (CaCO$_3$).
This mineral is found occasionally in the form
of hexagonal prisms with an apparent
$6/m2/m2/m$ symmetry (e.g., Figure 1.3). If calcite were truly hexagonal, however, etch pits that
form on the prism faces, either in nature or by the
application of dilute HCl acid, would all point in
the same direction, indicating that the faces were
symmetrically equivalent. In fact, the pits point in
opposite directions on alternating faces. This observation helps us eliminate the apparent sixfold
axis as well as the perpendicular mirror plane. The
true point group of calcite is $\bar{3}2/m$.

Numerous other properties of minerals express point group symmetry, and the detailed study of any symmetry-dependent physical property can aid in the determination of the true point group. Variations in hardness from face to face, or variations in properties such as electrical or thermal conductivity can be useful. Tests for piezoelectricity or pyroelectricity in some cases will reveal the presence or absence of a center of symmetry. The optical properties of minerals in transmitted or reflected light are dependent on symmetry and are described in Chapter 11. The distinction between centric and acentric is often critical to space group determination, as in the example of pyrite given earlier. Studies of the x-ray diffraction pattern of a mineral of unknown space group frequently lead to a situation in which two alternative space groups, one centric and the other acentric, are possible. A separate and independent test is required to make the distinction.

CRYSTAL FORMS

The symmetrically equivalent faces of a crystal define a geometric shape, which is referred to as a *crystal form*. Some forms enclose a crystal on all sides and are called *closed forms*. An example of a closed form is the general form {321} of sperrylite (Figure 3.13a). Other forms may have some directions unterminated, and are called *open forms*. Some crystals consist of a single form, such as the sperrylite crystal, while others consist of several forms. The sperrylite crystal in Figure 3.12a has three forms. The faces symmetrically equivalent to (100) define a cube, a *special form*, because each face is perpendicular to a fourfold axis. The faces symmetrically equivalent to (111) define an octahedron, an eight-sided closed form. The octahedron is also a special form, because each face is perpendicular to a threefold axis. Lastly, the faces symmetrically equivalent to (110) define a dodecahedron, a twelve-sided special form that is also closed. Because of growth mechanisms, growth rates, and other factors, all three forms appear in this particular sperrylite crystal. In principle, however, the shape of the crystal could be that of any of the closed forms, or any combination thereof.

During the historical development of crystallography, especially in the 19th century, a great deal of attention was paid to the study of crystal forms. All crystal forms, almost 150, were given Greek names. A student of that time was required to know each form by name and to identify the forms and their combinations on sight.

During the last half century, the detailed study of crystal forms has diminished, and many of the classical names are no longer in frequent use. Some of the crystal form names do, however, still appear in the literature and are often used to describe minerals. Some form names express the relationship between cleavage faces and are used in that context. For example, fluorite (CaF_2) commonly exhibits *octahedral* cleavage, and pyrite (FeS_2) exhibits *cubic* cleavage. Both names correspond to crystal form names discussed below. Some closed forms correspond in shape to the common coordination polyhedra found in crystal structures and are discussed in Chapter 5.

Mineralogy students should know the common forms, all of which are illustrated above the tables of physical properties in Part II of this text. The *cube* form already mentioned develops in cuprite (Cu_2O), diamond (C), fluorite (CaF_2), galena (PbS), halite (NaCl), and pyrite (FeS_2). The *octahedron* is a common form of diamond, chromite (Cr_2O_3), cuprite, fluorite, galena, magnetite (Fe_3O_4), and native gold (Au). The *dodecahedron*, also called a rhombic dodecahedron because of the rhombus shape of each of its 12 faces, is found in diamond, boracite ($Mg_3ClB_7O_{13}$), garnet ($Fe_3Al_2Si_3O_{12}$), magnetite (Fe_3O_4), sodalite ($Na_4ClSi_3Al_3O_{12}$), and tetrahedrite ($(Cu, Fe)_{12}Sb_4S_{13}$). The *tetrahedron* form consists of four (111) faces and displays a $\bar{4}3m$ point group symmetry. The minerals tetrahedrite (named for the form), boracite, and sphalerite (ZnS) frequently have this form.

A common open form is the *pinacoid*, consisting of two parallel faces related by either a center of inversion, a mirror, or a twofold axis of rotation. A *sphenoid* consists of two nonparallel faces related by a twofold axis. Two nonparallel faces related, instead, by a mirror plane are referred to as a *dome*. A *prism* consists of three or more faces, all parallel to each other and to a common axis. *Prismatic forms* are particularly common in hexagonal and tetragonal minerals such as beryl ($Al_2Si_6Be_3O_{18}$) and zircon ($ZrSiO_4$). A *pyramid* refers to three or more nonparallel faces that intersect at a single point. *Dipyramids* and *disphenoids* refer to pyramid forms joined symmetrically together. A *rhombohedron* has six rhombus-shaped faces. A *scalenohedron* consists of faces having the shape of two scalene triangles—that is, three unequal sides with unequal angles between them—joined across a side. A *pedion* is a form that has only a single face with no symmetrically equivalent face.

Rather than use the shapes and Greek names for identification, modern mineralogists and especially crystallographers have tended to identify and describe crystal forms by their symmetry. One such technique uses the indices of one face of a crystal form. Since each form has a distinct symmetry, knowledge of the point group and the indices of one face of the form is sufficient for a complete description. The indices of a general form can have any h, k, and l values except those that represent special forms. For example, the symbols $\{hkl\}$ $4/m\bar{3}2/m$ denote the general form in the $4/m\bar{3}2/m$ point group. The symbols $\{111\}$ $4/m\bar{3}2/m$ and $\{111\}$ 432 denote the special form of octahedra in those point groups, and $\{111\}$ 23 and $\{111\}$ $\bar{4}3m$ denote tetrahedra. The only ambiguities in this notation are found in the isometric system in which the indices (hkl) can define two different forms in each point group. The forms can be distinguished by stating whether $h > l$ or $h < l$. The indices of all of the crystal forms and their equivalents are given in the last two columns of the crystal form tables in Appendix 4.

The approach proposed here for the identification and definition of crystal forms is based on the pole symmetry of the form faces. The *pole symmetry* is the symmetry of the face normal in the point group, and is analogous to *point symmetry* in space groups. The pole symmetry of the general form of any crystal in an acentric or centric point group is 1 or $\bar{1}$, respectively. The pole symmetries of special forms state the relationship between poles and coincident symmetry operations. A pole symmetry of 2 means that the pole (face normal) occupies the same position as a twofold axis. Multiple entries have the usual meaning of Hermann-Mauguin symbols. That is, a pole symmetry of $2/m$ denotes that the face normal is coincident with a twofold axis and is perpendicular to a mirror plane. The pole symmetry of $\bar{3}m/2$ means that the pole is coincident with a $\bar{3}$ axis and lies in a mirror plane as well as being perpendicular to a twofold axis.

The coincidence of a pole with symmetry operations decreases the number of equivalent faces from that of the general form. In the case of a single coincident symmetry element (e.g., pole symmetry 2), the number of equivalent faces is equal to the equipoint of the general form divided by the rank of the element. If two or more coincident symmetry elements exist (e.g., $\bar{3}2/m$), the number of faces is equal to the equipoint of the general form divided by the product of the ranks of the first two elements. For example, in point group 432 the pole symmetry of $3/2$ indicates the form of an octahedron. The equipoint number of the general form is 24 (Figure 2.21), and hence the general form will have 24 faces. In the special form, the only coincident symmetry operation is

the $\bar{3}$ axis with a rank of 3. The number of faces is therefore $24/3 = 8$, the number of faces in an octahedron. Another example is the pole symmetry of $\bar{3}m/2$ in point group $4/m\bar{3}2/m$, which is also the form of an octahedron. The number of faces in the general form is 48, and two coincident symmetry elements are in the pole symmetry, namely, the $\bar{3}$ (rank = 3) and the mirror plane (rank = 2). The product of ranks is 6, and the number of faces in the special form is $48/6 = 8$. If more than two coincident operations exist, for example the cube form (pole symmetry $4mm/\bar{4}2m$) in point group $4/m\bar{3}2/m$, the third operation m is a consequence of the first two, and only the first two operations need be considered in the determination of the equipoint. This argument is analogous to the one presented in Chapter 2 in which we demonstrate that two rotation axes necessitate the existence of a third rotation axis. The third axis does not modify the number of equivalent positions created by the first two.

In some notation symbols of pole symmetry, a dot is placed between two rotation axes (e.g., $3 \cdot 2$) to indicate the position of the pole somewhere between the axes. The dot symbol is also used between two mirror planes (e.g., $m \cdot m$) to indicate the pole position in the bisector plane between the mirrors. In both cases, the positions are not special because the poles are not coincident with either a mirror or a rotation axis. The poles do coincide, however, with symmetry operations present in the *lattice symmetry* of the point group and thus are said to be in *semispecial positions*. The coincidence of a pole with any of the symmetry elements of the lattice generates a set of faces that defines a *semispecial form*. Semispecial forms have no effect on the equipoint number, but they do modify the geometry of the form because of their coincidence with lattice symmetry operations.

The full symbol for pole symmetries can be rather complex, especially in the high-symmetry point groups. Much of the information is unnecessary. If we use only the symmetry elements essential to the definition of pole symmetry, a short symbol results. The complete symbol, the short symbol, and the plane point symmetries of faces for all possible forms in each system are given in Appendix 4. The apparent symmetries of all special forms and the number of equivalent faces for each is also tabulated.

Let us consider the crystal class $4mm$ as an example of the use of Appendix 4. Figure 3.15 gives the stereographic projection of this class from Appendix 4 and a portion of the crystal form table (Table A4.4) with the five possible

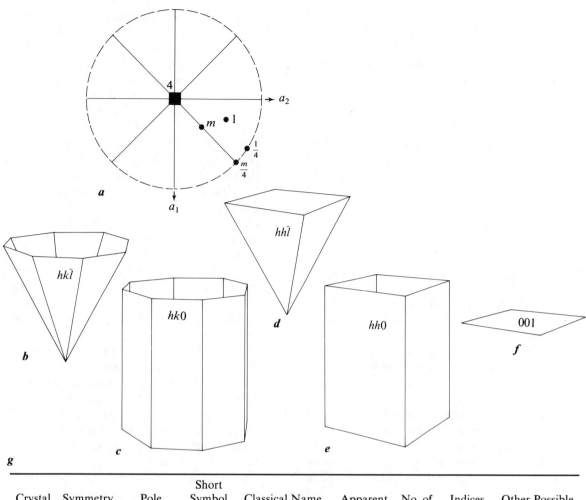

FIGURE 3.15 **(a) Stereographic projection of 4*mm* point group showing representative face normals of the general position, 1, semispecial position, 1/4, special positions, *m*, *m*/4, and 4. The resulting three-dimen-**sional forms are (b) ditetragonal pyramid, 1, (c) ditetragonal prism, 1/4, (d) tetragonal pyramid, *m*, (e) tetragonal prism, *m*/4, and (f) pedion, 4. Information is summarized in (g), a section of the crystal form tables in Appendix 4.

Crystal Class	Symmetry of Faces	Pole Symmetry	Short Symbol of Form	Classical Name of Crystal Form	Apparent Symmetry	No. of Faces	Indices of a Face	Other Possible Indices
4*mm*	1	1	1	Ditetragonal pyramid	4*mm*	8	*hkl*	*hk*$\bar{l}$
	1	$\frac{1}{4}$	$\frac{1}{4}$	Ditetragonal prism	$\frac{4}{m}\frac{2}{m}\frac{2}{m}$	8	*hk*0	
	m	*m*	*m*	Tetragonal pyramid	4*mm*	4	*hhl*	*hh*$\bar{l}$ *h0l* *h0*$\bar{l}$
	m	$\frac{m}{4m}$	$\frac{m}{4}$	Tetragonal prism	$\frac{4}{m}\frac{2}{m}\frac{2}{m}$	4	110	100
	4*mm*	4*mm*	4	Pedion		1	001	00$\bar{1}$

forms, also from Appendix 4. The symbol for the general form is simply 1, denoting coincidence of the pole with the trivial onefold axis of rotation. The symbol *m* indicates that the pole occupies a mirror plane. The symbol *n* (*n* = 4 in this example) indicates that the pole coincides with a rotation axis of rank *n*. Both *m* and *n* represent special forms. In this example, *m* denotes the tetragonal

pyramid {*hhl*} and *n* the pedion {001}. The symbol *m*/4 stands for a special form in which a face normal occupies a mirror plane and is perpendicular to the fourfold axis that occupies the same plane. The short symbol *m*/4 is sufficient to describe the relationship since the position of the fourfold axis in the mirror·plane is implied. The form of the ditetragonal prism is denoted by 1/4.

This symbol means that the pole of each face of the form is perpendicular to the fourfold axis, but unlike the $m/4$ form, they do not occupy mirror planes. The form $1/4$ is semispecial. In the class $4/m2/m2/m$ of this system, which gives the tetragonal lattice symmetry, poles of the form $1/4$ would occupy a horizontal mirror perpendicular to the fourfold axis.

The advantage of this approach to the identification and study of crystal forms is that no criteria other than symmetry are needed. The approach is analogous to the description of atomic positions in space groups, and as such emphasizes the fundamental relationships between crystal form and crystal structure.

VARIATIONS OF IDENTICAL FORMS

In certain classes, the form of a crystal can retain its geometric shape but have variations of handedness or orientation. Such variations are distinguished in three categories: (1) enantiomorphic, or left-handed and right-handed forms, (2) positive and negative forms, and (3) first-order and second-order forms.

1. ***Enantiomorphic forms:*** Crystal forms are said to be *enantiomorphs* when, within the same class, one form is related to the other by an operation of the second sort, that is, either a mirror plane or a center of inversion. Such forms are related as the left hand is to the right hand; they cannot be superposed by the operation of rotation. The terms "right-handed" and "left-handed" are used, therefore, to make this distinction (Figure II.2). Single crystals, for example, quartz, apparently develop either one or the other form during growth with equal frequency. Both right-handed and left-handed forms can never develop in the same single crystal.

   Enantiomorphs are related to each other within the crystal structure by the sense of rotation of screw axes. Screw axes 3_1 and 3_2 are said to be right-handed and left-handed, respectively, because the senses of screw motion are opposite (e.g., Figure 2.10). The space group is $P3_121$ for right-handed quartz and $P3_221$ for left-handed quartz. The geometric forms of the two are identical in all respects except that they are mirror equivalents. This distinction can be made only if a general form is developed. Hexagonal prisms of quartz do not reveal the handedness in the structure, because the poles of faces occupy special positions on the twofold axes. The general form $\{123\}$ of quartz (see point group

32 in Table A4.5, Appendix 4) will reveal the right-handed and left-handed nature as shown in Figure II.2.

The possibility of enantiomorphic pairs is restricted to crystals with space groups containing 3_1, 3_2, 4_1, 4_3, 6_1, 6_2, 6_4, and 6_5 screw axes. Screw axes 2_1, 4_2, and 6_3 are excluded because the resulting faces will always be rotation equivalent. That is, the enantiomorphic relationship can be distinguished in the structure but not in the crystal form. The corresponding axial point groups that lack a center of symmetry or a mirror plane or both are 3, 32, 4, 422, 6, 622, 23, and 432. Of these, the monaxial point groups 3, 4, and 6 do not permit enantiomorphs. The reason for this can be seen from the crystal form tables in Appendix 4. The general forms for these classes are a pedion (1), a sphenoid (2), a trigonal dipyramid (3), a tetragonal pyramid (4), and a hexagonal dipyramid (6). Each of these is already symmetric with respect to mirror planes parallel to the rotation axis, and thus no handedness can be expressed. Handedness can be expressed, however, for the remaining five point groups (432, 23, 622, 32, and 422), all of which are multiaxial.

2. ***Positive and negative forms:*** Crystal forms are said to be *positive* or *negative* if they can be related to each other by the operation of rotation. For example, the positive and negative crystals in Figure 3.16 can be made to coincide by rotation through 180 degrees,

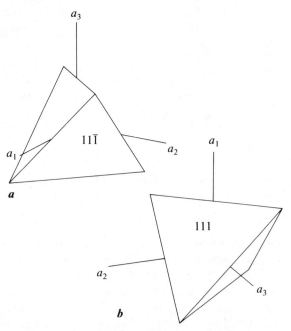

FIGURE 3.16   (a) Positive form of tetrahedron. (b) Negative form of tetrahedron.

whereas they cannot be made to coincide by the operation of a mirror plane. The faces of the tetrahedra are structurally different, and their Miller indices {111} and {11$\bar{1}$} are different. Their structures have, however, the same orientation. This is the opposite of enantiomorphs for which the orientation of the structures is different, but the form faces occupy identical structural positions. Positive and negative crystals may be expressed as special forms or general forms, and are permissible in all point groups containing only rotational symmetry. Point groups that are isogonal with space groups having 2_1, 4_2, and 6_3 screw axes can have positive and negative forms, but they cannot have enantiomorphic pairs.

3. ***First-order and second-order forms:*** The third category is referred to as a "variation of the order of forms." This distinction is based on different orientations of the form without any necessary symmetry relationship between them. The {110} prisms in tetragonal and hexagonal crystals are said to be *first order*, and prismatic faces with {100} indices are said to be *second order*. In orthorhombic crystals, prisms with {0kl}, {h0l}, and {hk0} indices are usually called first, second, and third order, respectively. Fourth-order prisms in monoclinic crystals have {hkl} indices. This usage is a carryover from the classical studies on crystal morphology and is diminishing in importance.

Two additional terms that have some importance in crystal morphology studies are *zone* and *zone axis*. All crystal faces that are parallel to a common axis, or all faces that intersect in parallel edges, are considered to be in the same zone. The common axis is the zone axis, and it is described by the [*uvw*] indices of the zone. For example, the prismatic faces such as (100), (110), and (010) of a tetragonal crystal have the zone axis [001]. As in this example, the faces of a zone need not be symmetrically equivalent.

TWINNING

Two or more identical crystals can be intergrown in such a way that individuals on either side of a common interface are symmetrically related by operations not present with the same orientation in the point group of the individuals. Such intergrowths are referred to as *twins* or *twinned crystals*. Twins are distinctly different from random mechanical intergrowths that are symmetrically unrelated. The latter are common in mineralogy and are sometimes difficult to distinguish from twins if the characteristic symmetry of the twin has not been identified.

Individuals of a twinned crystal must have different crystallographic orientations with respect to each other. The interface between twins, however, is crystallographically equivalent in both individuals. This interface is called the *composition plane*. Rutile (TiO$_2$) commonly displays twinning on the (101) plane. Note in Figure 3.17 how the

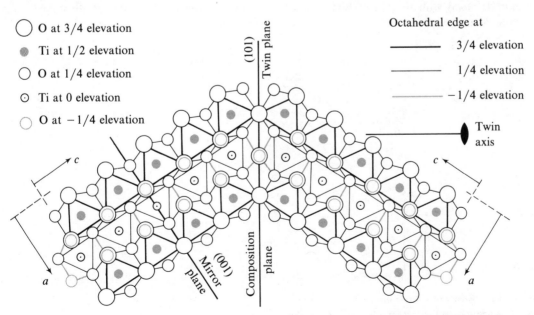

O at 3/4 elevation
Ti at 1/2 elevation
O at 1/4 elevation
Ti at 0 elevation
O at −1/4 elevation

Octahedral edge at
——— 3/4 elevation
——— 1/4 elevation
- - - −1/4 elevation

Twin axis

(101) Twin plane

(001) Mirror plane

Composition plane

FIGURE 3.17   (a) Contact twin (101) in rutile (TiO$_2$) and its structural relationships as viewed in the (010) plane. Six-coordinated Ti shown as octahedra.

crystallographic axes differ in their orientation, and how the Ti and oxygen atoms register in the composition plane. In this example, the composition plane is a mirror plane since the individuals are mirror images. When this is the case, the composition plane is also a *twin plane*, and the two individuals are related by a twofold axis, or *twin axis*. Either the twin plane or the twin axis will operate on one individual to bring it into an equivalent orientation with the other. Neither the twin plane (mirror) nor the twin axis can coincide with identical symmetry operations in the crystal structures. Otherwise, the two individuals would have the same crystallographic orientation and could not be distinguished from a homogeneous single crystal. Twin planes and twin axes are "extra" elements that usually impart to the twin a higher symmetry than the symmetry of the individuals.

The composition plane in a twinned crystal is usually rational, and the relationship between individuals in the twinned crystal can be expressed in terms of point group symmetry. In some twins, for example the Dauphiné twin in quartz (Figure II.6c), the contact between the individuals is not a plane but an irregular surface. In other twins such as in aragonite ($CaCO_3$) (Figure 3.18), the composition plane appears macroscopically to be a twin plane, but in the details of the structure it may actually be a glide plane. The translational component of the glide is not visible, and the composition plane looks like a mirror plane. Fortunately, the detailed structure of twin boundaries can now be better understood with the aid of high energy electron microscopy, and when more data become

available, the importance of symmetry should become even more apparent.

A common classification of twins is based on the symmetry element or elements that operate to relate one member of a twinned crystal to another (Table 3.2). In the example of rutile, the twin axis is perpendicular to the twin plane. This is generally the case in crystals that possess relatively high symmetry and a center of inversion. The reason for this relationship is that in the presence of an inversion center, a twin operation corresponding to a mirror plane will generate a twin axis, and similarly, a twin operation corresponding to a twin axis will generate a mirror plane. In rutile, the composition plane is a rational lattice plane, and as such, it can be indexed as (101). The twin axis, however, is not rational and therefore cannot be indexed. We learned in Chapter 2 that only in isometric crystals can a direction perpendicular to a plane have the same numerical indices as the plane itself. The same is true of the relationship between a composition plane and a twin axis. In most cases in which either the plane or the axis (but not both) is rational, the rational element is used to define the twin. This definition of the twin by the rational element is referred to as the *twin law*.

As an example of how a twin law works, consider plagioclase feldspar ($CaSi_2Al_2O_8$), the most abundant silicate mineral in the earth's crust. The two most common twin laws are called the *albite* and *pericline laws* (see Figures II.13c and II.13d, respectively). The albite twin has a rational twin plane (010), which is also the composition plane, but the twin axis is irrational (Figure

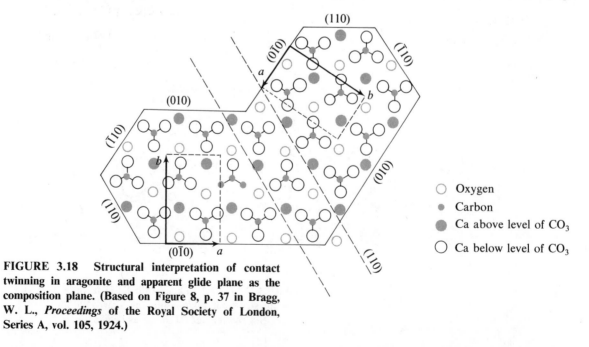

FIGURE 3.18 Structural interpretation of contact twinning in aragonite and apparent glide plane as the composition plane. (Based on Figure 8, p. 37 in Bragg, W. L., *Proceedings* of the Royal Society of London, Series A, vol. 105, 1924.)

○ Oxygen
• Carbon
⬤ Ca above level of CO_3
○ Ca below level of CO_3

TABLE 3.2. *Twin Laws and Twin Elements*

Twin Law	Twin Plane	Twin Axis	Description	Examples
Spinel or octahedral	(111)	[111]	Contact Penetrating Polysynthetic	Spinels, diamond, galena, sphalerite, fluorite, sodalite, tetrahedrite, bornite
Iron cross or dodecahedral	(110)	[110]	Penetrating	Pyrite
Tetrahedrite		[111]	Penetrating	Tetrahedrite
Tetragonal pyramidal	(101) (301)		Contact Cyclic Polysynthetic	Rutile, hausmannite, cassiterite, zircon, chalcopyrite
Tetragonal prismatic	(100) (110)		Contact Penetrating Cyclic	Scheelite, chalcopyrite
Tetragonal sphenoidal	(111)		Contact	Chalcopyrite
Hexagonal pyramidal	(101)		Contact Cyclic	Pyrrhotite
Hexagonal basal	(001)		Penetrating	Hematite, calcite, dolomite
Rhombohedral axial		[001]	Contact Penetrating	Calcite, chabazite
Rhombohedral	(110) (012) (021) (101) (104)		Contact Penetrating	Calcite, hematite, corundum, quartz, chabazite, pyrargyrite
Japanese	(112)		Contact	Quartz
Brazil	(110)		Penetrating	Quartz
Dauphiné		[001]	Irregular Penetrating	Quartz
Orthorhombic prismatic	(110) (130) (230) (012) (014) (031) (032) (101)		Contact Cyclic Polysynthetic	Aragonite, enargite, enstatite, staurolite, marcasite, chalcocite, chondrite, chrysoberyl, columbite, anhydrite, cordierite
Orthorhombic pyramidal	(112) (232)		Contact Penetrating	Chalcocite, staurolite
Orthorhombic basal	(100) (001)		Contact Penetrating	Humites, struvite, calamine, titanite (sphene)
Monoclinic domeal	(101) (011) (011)		Contact Cyclic	Pyroxenes, phillipsite, harmotome
Monoclinic basal	(100) (001)		Contact Penetrating Polysynthetic	Stilbite, gypsum, phillipsite, hornblende, pyroxenes, malachite
Monoclinic pyramidal	($\bar{1}$22)		Contact Cyclic	Pyroxenes
Monoclinic axial		[010]	Contact	Harmotome
Carlsbad	(100) or composition plane (010)	[001]	Contact Penetrating	Orthoclase (pyroxenes)
Baveno	(021)		Contact Polysynthetic	Feldspars
Manebach	(001)		Contact Polysynthetic	Feldspars (pyroxenes)
Mica	~(100) composition plane (001)		Polysynthetic	Muscovite
Albite	(010)		Contact Polysynthetic	Plagioclase
Pericline		[010]	Contact Polysynthetic	Plagioclase
Triclinic basal	(001)		Polysynthetic	Rhodonite

3.19a). The pericline twin has a rational twin axis [010], but an irrational twin plane. The axis in this case is not perpendicular to the composition plane, but parallel to it (Figure 3.19b). The composition plane and the twin plane are not the same. Both twin laws may operate in the same crystal, and in the case of microcline (KSi_3AlO_8), the "grid" or cross-hatched pattern observed with the petrographic microscope (Chapter 11) is diagnostic (Figure 3.20a). Similarly, the presence of twinning according to the albite twin law in plagioclase (Figure 3.20b) provides for immediate identification.

A descriptive classification of twins, based on the pattern of the individual crystals, can be used as a complement to the twin laws. The most common type is the *contact twin* in which the two individuals join along a single composition plane. Calcite ($CaCO_3$), gypsum ($CaSO_4 \cdot H_2O$), and rutile (TiO_2) are examples. Less frequently, individuals of a twinned crystal are intergrown in such a way that they have more than one composition plane or surface. These are called *penetration twins*; pyrite (FeS_2) and staurolite ($Al_9HSi_4Fe_2O_{24}$) are examples. *Cyclic twins* result when several composition planes of individuals are repeated at a constant angular interval. Aragonite ($CaCO_3$) and rutile are examples. *Polysynthetic twins* are commonly developed in plagioclase (e.g., $CaSi_2Al_2O_8$) and consist of alternating parallel crystals referred to as *twin lamellae*. Each lamella is related to its adjoining neighbor by the appropriate twin law.

As mentioned earlier, the effect of twinning may increase the apparent symmetry in crystals for which the composition plane is not obvious. In cyclically twinned aragonite, for example, the apparent symmetry is $6/m2/m2/m$, but the true point group symmetry is $2/m2/m2/m$. Another example is the Brazil twin of quartz (Figure II.6d) in which the apparent symmetry is $\bar{3}2/m$ compared to the real symmetry of 32.

Most twinned crystals form by one of three mechanisms: growth twinning, transformation twinning, or gliding or deformation twinning.

In the first category, *growth twinning* usually originates at the beginning of crystallization and persists throughout the growth process. The development of growth twins in a certain mineral may be accidental or may be a function of temperature-pressure conditions.

In the second category, *transformation twins* develop as a consequence of structural reorganization due to changes in temperature-pressure conditions. For example, high quartz (SiO_2) is transformed to the low quartz structure on cooling below 573 °C. In the process, the Si—O bond lengths and bond angles of high quartz (space group $P6_222$) change to become consistent with the $P3_221$ space group of low quartz. In so doing, two orientations of the low quartz structure, related to each other by a twofold axis around [001], can result. The twin axis is [001] and is referred to as the *Dauphiné* twin law.

In the third category, *gliding* or *deformation twins* develop when directional stresses deform crystals. Such deformations take place along certain crystal structural planes where one portion of the crystal can glide over the other. This mechanical twinning is demonstrated easily in calcite. If a blade is placed over the edge or face of a calcite crystal in an orientation coinciding with a cleavage plane and pressure is applied, glide twinning will develop with {102} as the twin plane.

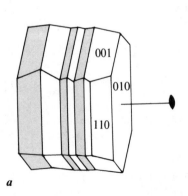

a

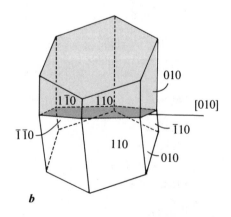

b

FIGURE 3.19 (a) Albite twinning in plagioclase feldspar showing a rational twin plane (010) but an irrational twin axis. (b) Pericline twinning in feldspar showing a rational twin axis [010] but an irrational twin plane.

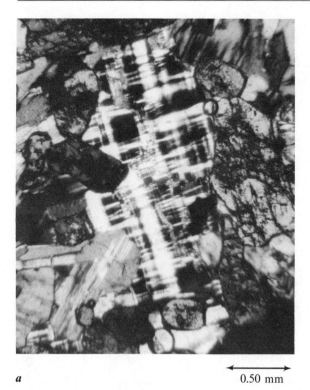

a ⟵ 0.50 mm ⟶   *b* ⟵ 0.20 mm ⟶

FIGURE 3.20 **(a) Characteristic "grid" twinning of microcline due to operation of both albite and pericline laws. (b) Characteristic albite twinning in plagioclase** **feldspar. Photographs taken through a petrographic microscope with crossed polarizers.**

An analogue of glide twinning is *translational glide*. In this case, thin lamellae of the crystal slide along the glide plane under shear stress. The glide plane in calcite is the same as the glide twin plane (012). The lamellae, however, are not related by twinning, because they are all translated in the same direction. Translation glidings can produce striations in calcite crystals. The equivalent glide plane in dolomite is (021), and dolomite displays a pattern of translation glide striation different from the calcite striation pattern.

The scale of twinning ranges from that clearly visible in hand specimens to the submicroscopic. In the latter category, several structural phenomena of unit cell scale (approximately 1–10 nm) can be related to twinning. These phenomena are observed with the aid of x-ray diffraction and electron microscopy. They include various types of polytypism in silicate structures, mosaic structures in sulfides, and microstructures in many different minerals currently being studied.

The mechanics of twinning can also be discussed in terms of energy. As we will learn in Chapter 8, the lowest energy state appropriate to a particular temperature and pressure is always preferred. When a twin forms, it normally does so in the energetically most efficient manner, so the orientation of the individuals about the composition plane will avoid unnecessary distortion of the atomic positions relative to each other. Strained structures generally represent higher energy states and thus are not usually formed. The tendency to minimize the total energy is the reason composition planes are crystallographically equivalent for both twin individuals, and is indirectly the reason the individuals are related by symmetry.

CHAPTER FOUR

SYMMETRY AND ATOMIC BONDING

Throughout the development and application of symmetry concepts, keep in mind that the outward symmetry displayed by a mineral is an expression of more fundamental phenomena. Minerals consist of orderly arrangements of atoms and ions, and many of their physical and chemical properties can be related to the behavior of a relatively small number of outer electrons involved in chemical bonding. A close relationship exists between the symmetrical arrangement of atoms in a structure and the symmetry of chemical bonds that hold the atoms together. We will discover in the next chapter that the basis for predicting crystal structures as symmetrically packed arrays of atoms depends on the intrinsic symmetry possessed by the atoms themselves, yet the atomic symmetry is also to some extent dictated by the geometric arrangement of neighboring atoms.

The combination of chemical bonding and structure explains the physical properties of minerals. Some properties such as compressibility and electrical conductivity have a satisfactory explanation based on bond type and strength, whereas properties such as piezoelectricity and ferromagnetism are more closely related to structure. The evidence of this is the fact that the latter properties are completely destroyed when melting occurs. Melting destroys the structure, but the chemistry remains the same. In this chapter, we will first review the basic architecture of atoms, and

then provide some understanding of how various mineral properties are related to bonding.

THE PERIODIC TABLE OF THE ELEMENTS

The use of symmetry arguments to understand mineral properties better came long before the internal structure of atoms was recognized. Around the middle of the last century, scientists applied a few simple symmetry concepts and set the stage for investigating the subatomic region of matter. A geologist was the first 19th century scientist to recognize translational symmetry in the physical properties of the elements. Alexandre Beguyer de Chancourtois, a professor of geology at the School of Mines in Paris, France, was well trained in mineralogy and had studied the similar behavior of Mn and Fe in pyroxenes, and of Na, K, and Ca in the feldspars of volcanic rocks. This work led him to publish a list in 1862 of all of the known elements of his time in order of their atomic weights. This he did in an unusual manner with a helical graph wound about a cylinder. The cylinder was divided into 16 equal parts (atomic weight of oxygen = 16) around its base and had a scale of atomic weight parallel to its length. As illustrated in Figure 4.1, the known elements of that time defined a helical curve. Those elements with similar properties occupied positions on the same vertical line. Pairs of elements such as Li

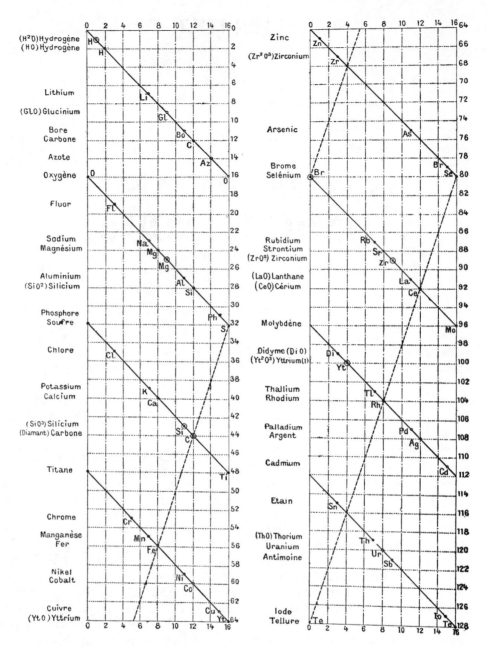

FIGURE 4.1 Portion of the original telluric screw conceived in 1862 by French geologist de Chancourtois. (From de Boisbaudran, Lecoq, and de Lapparent, A.,

1891. Sur une reclamation de priorite en faveur de M. de Chancourtois, relativement aux relations numeriques des poids atomiques. *Comptes Rendus* 112: 77–81.)

(atomic weight = 7) and Na(23), or Mg(24) and Ca(40), differ in weight by the number 16. De Chancourtois called his representation the "telluric screw." Unfortunately, his ideas were not recognized by the French Academy, which apparently criticized his contribution as being more geological than chemical. The basic idea was published in April, 1862, but the graphical representation so important to the periodicity of properties was not published for many years.

Less than a decade later, the Russian chemist Dmitri Mendeleev again recognized the periodicity of physical properties. When he arranged the known elements of his time in order of increasing atomic mass, he noticed a periodic repetition of atoms with similar physical and chemical properties. For example, atoms numbered 3, 11, 19, 37, and 55 (Li, Na, K, Rb, and Cs) were all soft, silvery white metals. These elements were also chemically reactive, forming perfect cubes when combined with chlorine. The resulting compounds were chemically simple (LiCl, NaCl, KCl, RbCl, and CsCl), colorless, and displayed perfect cubic cleavage. Mendeleev also recognized close similarities in the properties of other groups of atoms, such as Be, Mg, Ca, Sr, and Ba. These groups he arranged as columns in his now famous *Periodic Table of the Elements* published in 1869. A revised version published in 1871 contained eight columns or groups of elements and provided the foundation for the current table given on the inside back cover of this book. The rows of Mendeleev's table became known as periods, aptly named because of the periodic repetition of properties and the sense of symmetry inherent in the table.

The symmetry of Mendeleev's arrangement of elements was so compelling that he immediately recognized the absence of some elements. There was a conspicuous hole in the sequence between Ca (atomic mass = 40.08) and Ti (47.90) where he predicted the presence and properties of a yet undiscovered element he called "eka-boron." In 1879, the element was discovered in the mineral euxenite ($CaScTi_4O_8(OH)_4$) and was named scandium (44.96). Another obvious departure from periodicity occurred between the elements zinc (65.37) and arsenic (74.92). Mendeleev predicted two undiscovered elements in this hole. One he called "eka-aluminum" with an atomic mass of 68 and a specific gravity of 5.9. In 1875, it was discovered in trace amounts in the mineral sphalerite (ZnS). The element, named gallium, had an atomic mass of 69.72 and a specific gravity of 5.94 —a remarkably close prediction! The other element Mendeleev called "eka-silicon" with a predicted atomic mass of 72 and a specific gravity of 5.5. The element was discovered in 1886 in argyrodite ($Ag_8S_4 \cdot GeS_2$), a silver ore from Germany. Named germanium, the element had an atomic mass of 72.32 and a specific gravity of 5.47. Again, a close agreement with Mendeleev's predictions.

Although Mendeleev's table demonstrated an ordered arrangement of properties, the fundamental reasons for that periodicity remained unknown. What was it about the internal structure of atoms that made their properties so periodic and predictable? What tools could be used, and what experiments could be devised to bridge the gap between the real, macroscopic world of observations and the subatomic world where the ultimate solutions were thought to lie? The generality of symmetry arguments was proven in the relationship of crystal forms to atomic arrangements. The assumption was that symmetry relationships exist independent of scale. Could it be that the periodicity in Mendeleev's table reflected some periodicity in the internal architecture of the atoms themselves?

THE BOHR MODEL OF THE ATOM

Additional clues came with the recognition that atoms consisted of smaller, more fundamental particles. The discovery that all matter consisted of electrons was made by Sir Joseph John Thomson in 1897, but the specific arrangement of the electrons remained a puzzle. In a series of famous oil droplet experiments worthy of a Nobel Prize in 1923, Robert Andrews Millikan determined the electric charge carried by a single electron. Knowing the ratio of charge to mass as a result of other experiments, Millikan calculated the mass of an electron to be 9.12×10^{-28} g, or about 1/1838 of the lightest known element, hydrogen. He attributed the remainder of the atomic mass of hydrogen to the existence of another subatomic particle, the proton. It has a positive charge of the same magnitude as the negative electron charge, but has a charge-to-mass ratio much smaller than an electron. Calculations show that a proton is 1837 times heavier than an electron, and has very nearly the mass of the hydrogen atom itself. Given these data, Lord Ernest Rutherford in 1911 reasoned that the mass of any atom would be nearly the same as the number of protons in its nucleus. This proved not to be the case. The sum of positive nuclear charges accounted for only about one half of the atomic mass, so clearly there had to be additional, as yet unidentified contributions. Rutherford therefore postulated the existence of neutrons, particles slightly heavier than protons but electrically neutral, that reside with protons in the atom's nucleus.

Finally, a physical model of a simple atom was beginning to emerge. Further experiments by Rutherford in 1911 demonstrated that the diameter of an electron was approximately 2.5×10^{-6} nm[1] and that atomic nuclei consisting of protons and neutrons had effective diameters of 10^{-6} to 10^{-7} nm, yet atoms as a whole were known to be approximately 10,000 times larger, on the order of a tenth of one nanometer in diameter. What a strange model this must have seemed! Hydrogen, the simplest of all atoms, had 1837/1838 of its entire mass concentrated in a tiny nucleus with the remainder represented by a single electron nearly a million nuclear diameters away. In our macroscopic world, the nucleus may be likened to

---

[1] 1 nm (nanometer) = 10^{-9} meters = 10^{-3} microns = 10 Ångstrom.

a golf ball with an approximately equal-sized electron situated 500 m away.

The model was fine for hydrogen, but it could not account adequately for the behavior of the heavier, more complex elements. During Mendeleev's time, scientists knew that every element emitted a characteristic line spectrum when heated to a high temperature. The emitted light consisted of discrete wavelengths, each expressed as a bright line when refracted through a glass prism (Figure 4.2). Each element could be identified by its spectrum, and in fact several new elements, including gallium (predicted by Mendeleev), were first detected by their distinctive line spectra. The mechanism by which atoms performed such mysterious feats was not, however, fully understood.

In 1913, in a quest to understand the line spectrum of hydrogen, Niels Bohr made the fundamental breakthrough in understanding electron arrangement. He postulated that electrons were restricted to specific energy levels arranged at varying distances from the nucleus. When an electron absorbed energy, it jumped out to a higher energy level. When it emitted energy, an electron dropped back to a lower level. Max Planck in 1900 and Albert Einstein in 1905 had already established the fundamental relationship between energy E and frequency ν given by

$$E = h\nu \qquad (4.1)$$

where h is a constant of proportionality referred to as Planck's constant. Since the observed frequencies in line spectra were discrete, Bohr concluded that the energy level of an electron in a hydrogen atom is discrete, or quantized. He then developed the concept of the principal quantum number, having values of 1, 2, 3, and up to infinity, corresponding to possible energy levels in hydrogen (Figure 4.3). Bohr was able to calculate from kinetic energy considerations the radius of the normal or ground state of a single electron. The value of 0.053 nm agreed well with the known dimensions of hydrogen.

THE WAVE MODEL OF THE ATOM

Bohr's model was a monumental achievement for his time. Evidence soon accumulated, however, that could not be adequately explained by a simple model. Lines in the spectra of heavier elements could not be interpreted. Scientists also found that electrons behaved in some instances more like electromagnetic waves than discrete particles. In 1924, the French physicist Louis de Broglie suggested that all particles of matter exhibit properties normally associated with waves. This concept represented another monumental breakthrough, for de Broglie was challenging the classical distinction between waves and particles in much the same manner as Einstein in 1905 had challenged the distinction between mass and energy. Using Einstein's relation

$$E = mc^2 \qquad (4.2)$$

where E is energy, m is mass, and c is the velocity of light in a vacuum, and using Planck's relationship between energy and frequency, de Broglie described the wavelength of an electron by

$$\lambda = \frac{h}{mv} \qquad (4.3)$$

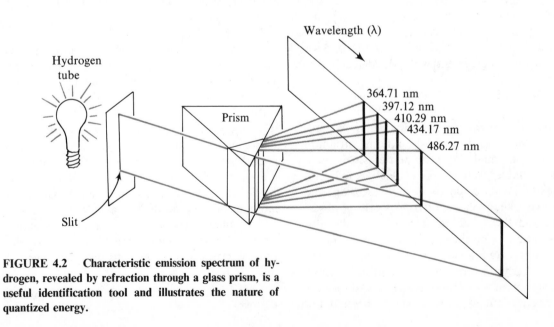

FIGURE 4.2 **Characteristic emission spectrum of hydrogen, revealed by refraction through a glass prism, is a useful identification tool and illustrates the nature of quantized energy.**

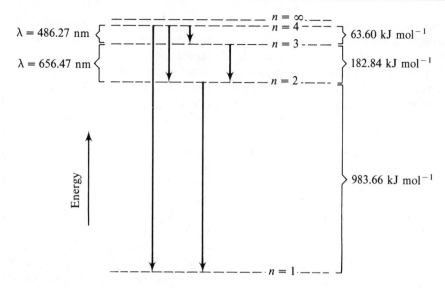

FIGURE 4.3 Possible energy levels of a hydrogen electron. Parameter *n* is the principal quantum number. Compare wavelengths. λ with Figure 4.2.

where λ is the wavelength, *h* is Planck's constant, *m* is the mass of the electron, and *v* is the velocity of the electron.

The wave-particle duality provided fresh insight into the problem of accurately describing the behavior of electrons. Physicists developed the field of quantum mechanics in which electrons are described by wave equations. In 1926, Erwin Schrödinger first proposed a general wave equation for the hydrogen atom. The equation forgoes the classical description used by Bohr and based on Newtonian mechanics. Instead, the Schrödinger equation relates the probability of finding an electron at any given time *t* and place *x* to the mass and potential energy of the particle at that time and place. This approach to electron behavior was more satisfying than a Newtonian approach because the Schrödinger equation treated the position of an electron as a statistical concept, and hence avoided the inherent impossibility of simultaneously measuring both position and velocity of a small particle. The general form of the Schrödinger wave equation for an electron is:

$$\nabla^2 \psi + \frac{8\pi^2 m}{h^2} (E - V) = 0 \qquad (4.4)$$

where ψ is the wave function, *m* is the mass of the electron, *E* is the total energy of the system, *V* is the potential energy of the electron at some specified point, and *h* is Planck's constant. The symbol ∇^2 is for a differential operator;[2] the precise form is unnecessary for our purposes here. Students

should not be dismayed by the apparent complexity of the Schrödinger equation. It simply explains a set of observations in a region where classical Newtonian concepts are not entirely valid. The equation cannot be derived from more basic concepts. It must be considered a law, which so far has led to a correct prediction of experimental findings.

An intuitive picture of the wave equation is provided by considering briefly a simple, one-dimensional sine wave with amplitude ψ at any distance *r* given by

$$\psi = A \sin \frac{2\pi r}{\lambda} \qquad (4.5)$$

where λ is the wavelength of a standing wave, and *A* is its maximum amplitude (Figure 4.4). James

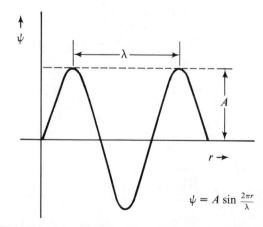

$$\psi = A \sin \frac{2\pi r}{\lambda}$$

FIGURE 4.4 Standing sinusoidal wave. Amplitude ψ at any value of distance *r* may be calculated if maximum amplitude *A* and wavelength λ are known.

[2] $\nabla^2 \psi = \partial^2 \psi / \partial x^2 + \partial^2 \psi / \partial y^2 + \partial^2 \psi / \partial z^2$ for the three-dimensional form.

Clerk Maxwell, in his development of the theory of electromagnetic radiation, had already suggested that the intensity of a wave was given by the square of its amplitude. Schrödinger reasoned by analogy that ψ^2 for an electron was a measure of its density, or probability of being found at any particular point around the nucleus. Solutions of the wave equation in three dimensions for specified energy levels yield wave functions that, when squared, represent a statistical distribution of electron position within a given orbital. The specific calculations are beyond our scope here, but the results show conclusively that the form and symmetry of electron orbitals is far from spherical as Bohr had originally supposed.

THE SPATIAL DISTRIBUTION OF ELECTRONS

With the powerful Schrödinger equation, scientists then had the tool to specify the probability of finding an electron in some small unit of volume at various places around the nucleus. In the case of the hydrogen atom, its single electron does not travel in a fixed orbit around the nucleus. Because of its wave properties, the electron can be visualized as moving in all directions but occupying certain regions of space around the nucleus more frequently than others. Figure 4.5 shows that the greatest probability of finding an electron in a spherical shell of radius r and thickness Δr is when $r = 0.053$ nm, exactly the radius calculated by Bohr in his classic model. Note that the electron distribution does not drop off abruptly from the classic Bohr radius. The distribution falls off

gradually and only approaches zero as r approaches infinity. The consequences of the wave model for the concept of atomic and ionic radii differ considerably from the consequences of the fixed-orbit Bohr model. Instead of a fixed radius, the wave model of Schrödinger gives a statistical radius within which we have a high probability of finding the electron. We can anticipate on this basis that the effective radius of any atom or ion in a crystal structure will be influenced by factors such as the number and kind of its closest neighbors and the polarization of its neighbors.

In Bohr's classic model, the energy of various electron orbits depended only on the principal quantum number n. In the wave model, the full three-dimensionality of specifying the position of an electron is taken into account. Most forms of the wave equation use spherical coordinates consisting of a radial coordinate r, and two angular coordinates θ and ϕ, which specify the position of r relative to the origin. One of the important results when the three-dimensional Schrödinger equation $\psi(r,\theta,\phi)$ is solved for hydrogen is that there are three quantum numbers, one for each dimension of the problem. The *principal quantum number* n is associated with the distance r from the nucleus. It has values of 1, 2, 3, ... The corresponding letter designations are K, L, M, . . . , in the customary chemical notation. The *azimuthal quantum number* l with values of 0, 1, 2, . . . , $n - 1$, is associated with the θ angular coordinate and specifies the position of an electron within one or more subshells located within a specified principal shell. Each subshell is uniquely designated by one of the letter symbols s, p, d, f, and so forth, as shown in Table 4.1. The *magnetic quantum number* m is associated with the remaining angular coordinate ϕ, and serves to specify one or more orbitals within each subshell.

The three quantum numbers n, l, and m follow directly from the solution of the wave equation. The existence of yet another quantum number, the *spin quantum number* s, must be invoked, because electrons by themselves behave as tiny magnetic dipoles. In 1924, Wolfgang Pauli recognized that magnetic moment, like electrical charge, was a fundamental property of all electrons. Electromagnetic theory predicts that such dipoles are generated by electric charges that circulate around an axis. The behavior of the electron, then, is as if it were spinning on an axis. Because the resulting magnetic dipole can only be aligned parallel or antiparallel to the external field, the spin quantum number can only have values of $+1/2$ and $-1/2$.

The size and shape of higher energy orbitals

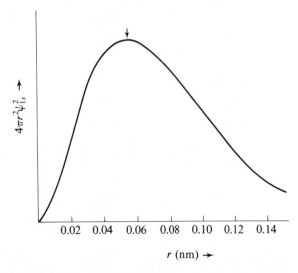

FIGURE 4.5 Radial distribution function for the 1s electron of hydrogen.

TABLE 4.1. *Mutual Relationships Between n, l, and m Quantum Numbers and the Number of Subshell Orbitals*

Principal Quantum Number n	Azimuthal Quantum Number l	Subshell Designation	Magnetic Quantum Number m	Number of Subshell Orbitals
1	0	$1s$	0	1
2	0	$2s$	0	1
	1	$2p$	$-1, 0, +1$	3
3	0	$3s$	0	1
	1	$3p$	$-1, 0, +1$	3
	2	$3d$	$-2, -1, 0, +1, +2$	5
4	0	$4s$	0	1
	1	$4p$	$-1, 0, +1$	3
	2	$4d$	$-2, -1, 0, +1, +2$	5
	3	$4f$	$-3, -2, -1, 0, +1, +2, +3$	7

occupied by the single hydrogen electron follow directly from the solution of the wave equation. Planck's relationship (Equation 4.1) indicates that higher energy levels will be associated with higher frequencies of vibration. Since the various energy levels are quantized, the unique frequencies are related to integer values of nodes in a standing wave. Figure 4.6 illustrates the sequential development of nodes for increasing principal quantum number n. Solutions of the wave equation for higher energy levels will involve nodes that will largely determine the size and shape of orbitals. For example, the regions of maximum electron density for the $3s$ and $4s$ orbitals are spherical surfaces, because they depend only on r, the distance from the nucleus ($l = m = 0$). Figures 4.7a and 4.7b are computer-plotted solutions of the three-dimensional probability function for the $3s$ and $4s$ orbitals, respectively, of hydrogen.

The various angular components (i.e., $l \neq 0$) impart a nonspherical shape to electron orbitals. Examples of $2p(m = 0)$ and $3p(m = 0)$ orbitals are shown as stereo pairs in Figures 4.7c and 4.7d. For $m = +1$ or -1, the electron distributions will be symmetric around the x and y axes of these figures instead of around the z axis.

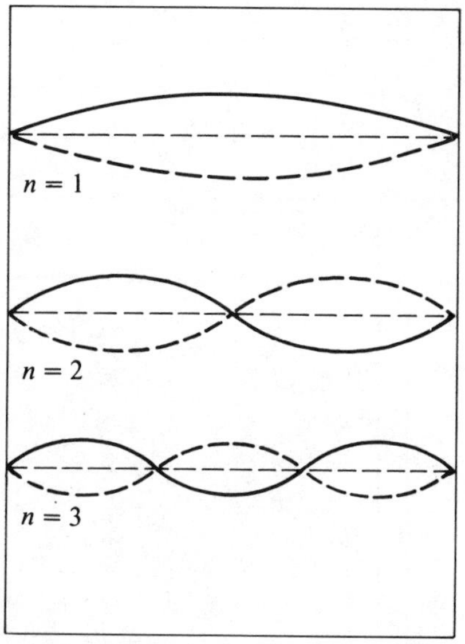

THE PERIODICITY OF ATOMIC PROPERTIES

Pauli's recognition in 1924 that the spin quantum number s must have a value of $+1/2$ or $-1/2$ provided the final piece of evidence necessary to account for the periodicity of physical properties Mendeleev had detected 55 years earlier. Pauli concluded from his study of line spectra that no two electrons of a single atom could have identical sets of quantum numbers. Stated another way, no two electrons can occupy the same spin orbital. This means that the number of electrons in an orbital is limited to two, and these must have spins in opposite directions. If two electrons have magnetic moments or spins in the same direction, they must, therefore, occupy different energy levels. This relationship is known as the *Pauli exclusion principle*.

The energies of the various orbitals are shown in Figure 4.8. Note that the energy differences between higher level orbitals are small compared with the energy differences between lower level orbitals, and note also the overlap among the subshells of the third and higher shells. The order

FIGURE 4.6 **Relationship between principal quantum number n and number of nodes in a standing wave. In solutions of the wave equation, nodes represent regions of no electron density (Figure 4.7).**

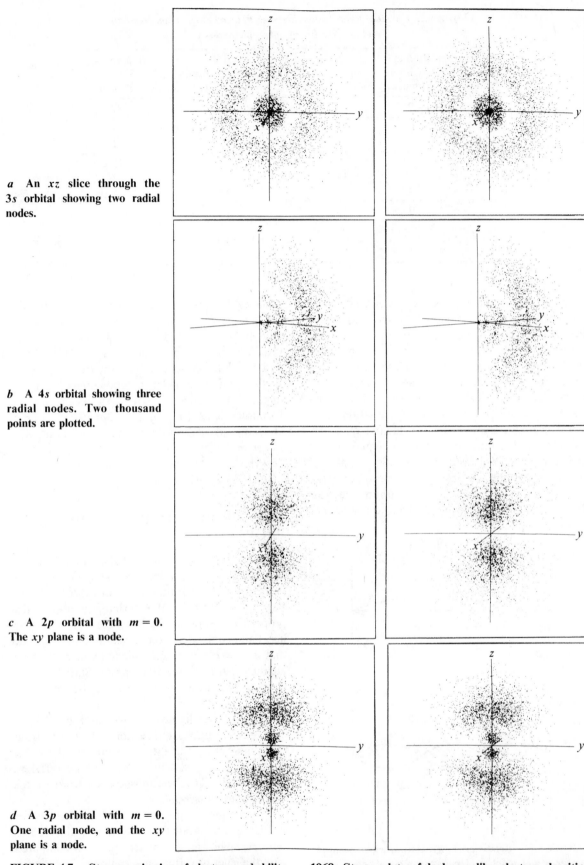

a An *xz* slice through the 3*s* orbital showing two radial nodes.

b A 4*s* orbital showing three radial nodes. Two thousand points are plotted.

c A 2*p* orbital with *m* = 0. The *xy* plane is a node.

d A 3*p* orbital with *m* = 0. One radial node, and the *xy* plane is a node.

FIGURE 4.7 Stereoscopic view of electron probability function for various orbitals. Approximately 3000 electron positions are calculated from the hydrogen wave equation and plotted by computer. (From Cromer, D. T., 1968. Stereo plots of hydrogen-like electron densities. *Journal of Chemical Education* 45:627–628. Reproduced by kind permission of the *Journal of Chemical Education*, copyright © 1968.)

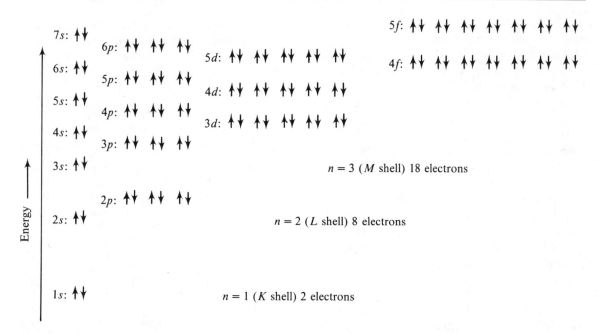

FIGURE 4.8   Relative energies of electrons in various orbitals. Each arrow denotes a single electron and its relative sense of spin. All electron configurations must be built up in order of increasing energy levels.

of subshells is determined by their energies, so the logical buildup of electronic configurations beginning with hydrogen must be accomplished in this order. As the atomic number increases from one element to the next, electrons are simply added to the available orbits having the lowest energy level. Pauli's principle insures that no more than two electrons will occupy each orbital. An additional principle, sometimes referred to as *Hund's rule*, requires that one electron be placed in each of the orbitals of equal energy level before the second electron can be placed in those orbitals. For example, neutral nitrogen (atomic number = $Z = 7$) has seven protons and seven electrons. Figure 4.8 shows that two of the electrons will occupy the K shell ($n = 1$), and five electrons will occupy the eight available levels in the L shell ($n = 2$). Two of the L electrons with opposite spins occupy the $2s$ subshell, and the remaining three L electrons with the same spin occupy half of the three orbitals of the $2p$ subshell. The electron configuration of oxygen ($Z = 8$) is obtained by placing an additional electron in one of the half-filled orbitals of the $2p$ subshell. When the $2p$ subshell orbitals are filled, the L shell is completely filled with eight electrons, and we have the noble gas neon.

In this manner, the entire periodic table is built up. By arranging atoms with similar outer electron configurations in columns (see inside back cover), the explanation for similar properties is finally clear. For example, the soft, silvery white metals Li, Na, K, Rb, and Cs have an outermost single electron in $2s$, $3s$, $4s$, $5s$, and $6s$ orbitals, respec-

tively. Hydrogen ($Z = 1$) and francium ($Z = 87$) have their outer electron in the $1s$ and $7s$ orbitals, respectively, and are therefore in the same column. Note that each of these orbitals has no angular dependence ($l = 0$), and consequently the orbital shapes are spherical. Most of the metallic properties of the alkali metals, such as high electrical and thermal conductivity, metallic luster, softness, malleability, and ductility, can be attributed to this electron configuration. The alkali earth metals, Be, Mg, Ca, Sr, Ba, and Ra, occupy the second column of the periodic table and differ from elements in the first column by having one additional electron in the appropriate orbital. This seemingly small difference is enough to give the alkali earth metals greater hardness and less reactivity.

FORMATION OF IONS

Up to this point, we have been mainly concerned with atoms for which the number of protons and electrons is equal. Such an atom is electrically neutral in contrast to an *ion*, which may have either a net positive charge or a net negative charge depending on whether it has lost or gained electrons. In either case, the formation of an ion from an atom involves an energy change. The energy required to remove an electron from an atom to create a positively charged *cation* is called the *ionization energy*. The ionization energy provides a quantitative measure of the energy with which an isolated gaseous atom binds its electrons. Ionization energies are also an indication of the relative stability of ions. For example, the first numerical column in Table 4.2 gives the energy required to remove the most weakly held electron in the atom. The second column gives the energy required to remove the second most weakly held electron. These values are commonly referred to as the *first and second ionization potentials*, respectively. If we compare these values for the alkali metals (e.g., Li, Na, K, Rb), we see that approximately an order of magnitude ($\times 10$) more energy is required to remove the first inner electron than is required to remove the outer electron. Recall that each of these atoms has a filled outer electron shell plus one additional electron in the next higher orbital. Sodium, for example, has the electron configuration $1s^2 2s^2 2p^6 3s^1$. Potassium has the configuration $1s^2 2s^2 2p^6 3s^2 3p^6 4s^1$. In each case, the outer electron is readily lost, and the monovalent cations Li^{1+}, Na^{1+}, K^{1+}, and Rb^{1+} are formed. The great amount of additional energy required to liberate a second electron indicates the relative stability of these cations and the stability of the noble gas configurations in general. We can see why, for example, a substance with $NaCl_2$ chemistry does not exist. A single Na^{2+} cation would be required to balance the two negative charges associated with the two Cl^{1-} anions.

A similar comparison of the first three ionization potentials of atoms in the second column of the periodic table (e.g., Be, Mg, Ca) shows that relatively little energy is required to remove the first two electrons compared with the energy required to remove a third one. Again, the filled-shell configurations of the noble gases are energetically preferred. The divalent cations Be^{2+}, Mg^{2+}, and Ca^{2+} result. Similarly, the trivalent (e.g., Al^{3+}) and quadravalent (e.g., Si^{4+}) cations may be explained.

Many of the most common elements found in minerals have more than one valence or oxidation state. Divalent iron (Fe^{2+}) is referred to as *ferrous iron* and is common in many silicate and oxide minerals such as fayalite (Fe_2SiO_4) and ilmenite ($FeTiO_3$). Trivalent iron (Fe^{3+}) is referred to as *ferric iron* and is common in more oxidized environments in which minerals such as hematite (Fe_2O_3) are formed. Figure 4.9 illustrates the various oxidation states of the elements and some of the more common oxidation states found in minerals. The elements are listed in order of progressive filling of subshells. The pattern of repetition is striking. The orderly increase in stable valence state from Li^{1+} ($Z = 3$) to N^{5+} (7) reflects the loss of electrons from the $2p$ subshell required to

TABLE 4.2. *First (E_1), Second (E_2), and Third (E_3) Ionization Energies of Elements Through Atomic Number 19*

Atomic Number	Element	E_1	E_2	E_3
2	He	2371.91	5246.74	— —
3	Li	520.07	7296.90	11811.43
4	Be	899.14	1756.86	14844.83
5	B	800.40	2426.72	3658.91
6	C	1086.17	2352.24	4619.14
7	N	1402.06	2856.84	4577.30
8	O	1313.78	3391.55	5301.13
9	F	1681.13	3375.23	6045.88
10	Ne	2080.28	3963.08	6276.00
11	Na	495.80	4564.74	6911.97
12	Mg	737.64	1450.17	7732.03
13	Al	577.39	1816.27	2744.29
14	Si	786.17	1576.53	3228.79
15	P	1062.74	1896.19	2909.97
16	S	999.56	2259.36	3376.49
17	Cl	1255.20	2296.60	3850.77
18	Ar	1520.47	2665.21	3946.77
19	K	418.82	3096.38	4602.40

NOTE: All data are in kilojoules per mole ($kJ\ mol^{-1}$).

achieve the filled-shell configuration of He. Unlike the elements of lower atomic number, nitrogen can either lose electrons to attain the He configuration, or gain them to attain the Ne(10) configuration. Consequently, nitrogen may occur either as an anion or as a cation. The orderly increase in valence from N^{3-} to P^{5+} (15) expresses the progressive loss of $3s$ and $3p$ electrons. The elements P(15), As(33), Sb(51), and Po(84) are situated between noble gases, and because these elements may form either anions or cations, they generate the steplike symmetry apparent in Figure 4.9.

The symmetry breaks down with the transition elements, beginning with Sc(21) and extending through Zn(30). This is because the electrons are added progressively to an unfilled $3d$ subshell rather than to the outer $4s$ subshell (Figure 4.8). When electrons are given up by an atom, they are in general lost first from the subshell having the highest principal quantum number. Consequently, the two lower energy $4s$ electrons are lost before the higher energy $3d$ electrons, and with the exception of Sc^{3+}, each of the transition elements

FIGURE 4.9 Valence states of the elements listed in order of progressive filling of $1s$, $2s$, $2p$, $3s$, $3p$, $4p$, and $5p$ orbitals. Filled squares denote the most common valence state of ions in minerals. Shaded squares denote less common but known valence states.

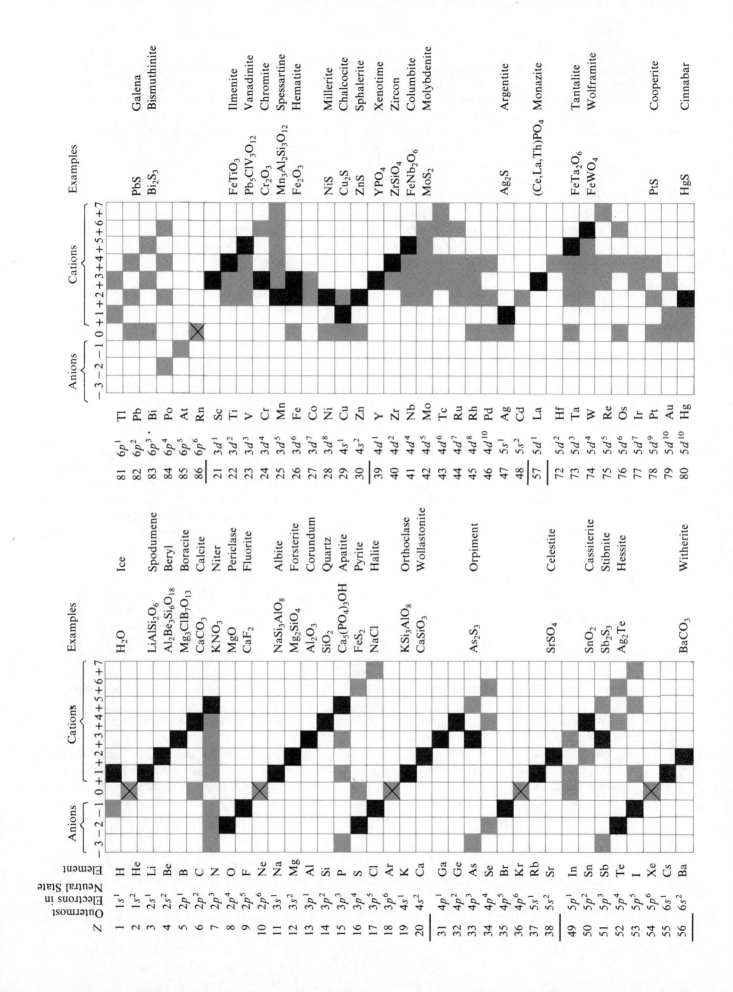

may be divalent. This unusual behavior is of great importance in the explanation of transition metal properties, particularly color in minerals.

We emphasize that the chemical behavior of the elements, by themselves or as they are found in minerals, is largely determined by the outermost electron configuration. The ions Na^{1+}, K^{1+}, and Rb^{1+} occupy the same column in Figure 4.9, and hence have remarkably similar properties. All are characterized by having a single electron in the outer s subshell. The ability to predict those ions that will have similar properties and behavior is important when we consider the effects of ionic substitution and solid solution in minerals.

The energy given up by a neutral atom when it gains an electron and becomes a negatively charged *anion* is called *electron affinity*. This is the energy required to remove an electron from a stable anion. Let us consider the halogens fluorine and chlorine, for example. Their electron configurations are $1s^2 2s^2 2p^5$ and $1s^2 2s^2 2p^6 3s^2 3p^5$, respectively. Each atom lacks a single electron to complete its $2p^6$ or $3p^6$ shell. The attraction of F and Cl for this one extra electron is quite large, which means that the attractive forces that would bind the needed electron are strong, and hence the energy required to remove it is large. Electron affinity values are large for the halogens F, Cl, Br, and I, which form the monovalent anions F^{1-}, Cl^{1-}, Br^{1-}, and I^{1-}. The values are somewhat less for oxygen and sulfur, which tend to form the divalent anions O^{2-} and S^{2-}, and the values are very low or negative for most of the remaining atoms, which tend to form stable cations.

THE SHAPES OF ATOMS AND IONS

We have thus far considered all atoms as essentially spherical in shape. Idealized atomic packing models on which most mineral structures can be developed (Chapter 5) are based on spherical atomic shapes. From a macroscopic point of view, this is not a bad approximation. Measured and calculated densities of close-packed structures are similar.

Our understanding of electron orbitals suggests that atomic shapes are more complicated than the spherical models. In general, the shape of an atom can be considered as the sum of the shapes of the occupied orbitals. The resulting electron densities at any point are given by the sum of the squares of the appropriate hydrogen wave functions, ψ^2, for each occupied orbital. In the case of the alkali metals (e.g., Li, Na, K) discussed earlier, the atoms must have spherical shapes, because the orbitals ($2s$, $3s$, $4s$, and so forth) depend only on

the radial distances from the nucleus. What about atoms with occupied p orbitals? As Figure 4.7 illustrates, electrons in the p orbital do not have spherical symmetry when considered by themselves. For example, elemental boron ($Z = 5$) has two electrons in the spherical $1s$ subshell, two in the spherical $2s$ subshell, and one in one of the dumbbell-shaped $2p$ orbitals. This configuration gives a slightly linear form to the atom. Similarly, elemental carbon ($Z = 6$) has the shape of a somewhat flattened sphere, because only two of the three $2p$ orbitals are occupied. The two dumbbell-shaped orbitals define a plane representing the longest dimension of the atom. In the case of elemental nitrogen ($Z = 7$), the $2p$ subshell is half filled with electrons having the same spin. The geometry of the three occupied p orbitals is such that the sum of the electron probability functions has spherical symmetry.[3] This is because all three orbitals, when occupied, contribute to some extent to the electron probability in the regions between the axes.

In general, atoms with filled and half-filled subshells of electrons will have spherical symmetry. Elemental manganese ($Z = 25$) has filled $1s$, $2s$, $2p$, $3s$, $3p$, and $4s$ subshells, and a half-filled $3d$ subshell. Elemental calcium ($Z = 20$) has five fewer electrons, so the $3d$ subshell is empty. In both cases, the atoms have spherical shapes.

The tendency for stable cations and anions to achieve completed-shell configurations has important implications for their shape. For example, in the stable cation of carbon, the loss of the four electrons necessary to stablize C^{4+} relative to elemental carbon leaves only the $1s$ subshell occupied. This means that the cation will have spherical symmetry, whereas the neutral atom does not. Similarly, the stable cation of boron is B^{3+}, having lost its single $2p$ electron and both $2s$ electrons to achieve the same $1s$ subshell arrangement as C^{4+}. Again, the symmetry of the ion is spherical. Oxygen ($Z = 8$) is of particular interest because it is the most abundant element in the earth's crust. It has a filled $1s$ and $2s$ subshell, but an unfilled $2p$ subshell. The stable anion is O^{2-}, the structural framework of all silicate minerals. By gaining two additional electrons, a filled $2p$ subshell is achieved, and thereby spherical symmetry. A last example of equal importance is silicon ($Z = 14$). Its stable cation is Si^{4+}, achieved by losing two electrons in the $3p$ subshell and two electrons in the $3s$ subshell. A filled $2p$ subshell of six electrons results, giving Si^{4+} a spherical shape.

[3] A proof of this may be found in Johnson, R. C., and Rettew, R. R., 1965. Shapes of atoms. *Journal of Chemical Education* 42:146.

THE SIZE OF ATOMS AND IONS

The size of an atom or ion, usually expressed in terms of its radius, is one of the most important concepts in mineralogy. In the field of crystal chemistry, one factor for determining the likelihood of mutual substitution of one ion for another in a mineral is similarity of the ionic radii. If the radii of two ions are grossly different, chemical substitution of one for the other would distort the structure and is therefore resisted. Accurate values of ionic radii are also important for determining cation/anion radius ratios, which we discuss in Chapter 5. These ratios are useful for the systematic derivation of crystal structures.

Defining the radius of an atom or ion poses some difficult problems. The ψ^2 function plotted in Figure 4.5 illustrates that there is always some probability, however small, of finding an electron even at a very great distance from the nucleus. How then do we decide where the radius is? The concept of a fixed radius of an isolated ion or atom is not useful. We do know, however, that atoms and ions in association with each other in minerals have *effective radii*. All minerals have measurable interatomic distances that can be routinely and accurately determined by x-ray diffraction methods. If we imagine a hypothetical structure consisting of hard, spherical balls of equal size in a close-packed arrangement, knowledge of the interatomic distance allows us to determine the effective radius of each ball (Figure 4.10). This determination is in fact a close approximation of the structures of most elements in the periodic table.

The procedure is not so straightforward for most minerals of geologic interest. Interionic distances can still be accurately determined, but in the common case in which the measured distance is the bond length between two *different* ions, the radius of one ion must be known before we can determine the radius of the other. The radius of an ion also depends on the distribution of its outermost electrons (Figure 4.11). For example, the sulfur in barite ($BaSO_4$) has a valence of $+6$ and a radius of 0.020 nm. Contrast this with the sulfur in galena (PbS), which has a valence of -2 and a radius of 0.172 nm. We see that as electrons are stripped off to produce less negative anions and, eventually, cations, the effective radius decreases.

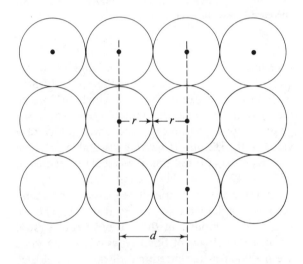

FIGURE 4.10   Closely packed arrangement of rigid, equal-sized spheres is typical of many elements. The interatomic distance *d* equals an atomic diameter 2*r*.

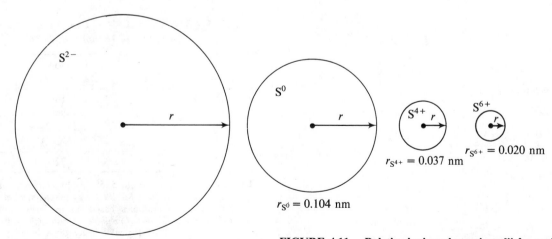

$r_{S^{2-}} = 0.172$ nm

$r_{S^0} = 0.104$ nm

$r_{S^{4+}} = 0.037$ nm

$r_{S^{6+}} = 0.020$ nm

FIGURE 4.11   Relative ionic and atomic radii for various valence states of sulfur. Size varies in proportion to number of outer electrons, being greatest for S^{2-} and least for S^{6+}.

The problem of determining effective radius is complicated further by the fact that ionic radius depends on *coordination number*, the number of closest neighbors to which the ion is bonded. Trivalent Al^{3+} in tetrahedral coordination (four closest neighbors) with oxygen has an ionic radius of 0.047 nm, but its radius in octahedral coordination (six closest neighbors) is 0.061 nm. Similarly, Si^{4+} in tetrahedral coordination with oxygen has a radius of 0.034 nm, whereas in octahedral coordination the radius is 0.048 nm. Figure 4.12 illustrates the variation in radius for some common cations. The larger the coordination number, the larger the radius.

The same relationship is true, but to a lesser extent, for anions. Figure 4.13 shows a slight increase in the ionic radii of S^{2-}, O^{2-}, and F^{1-} with increasing coordination. In other words, the size of the particular ion is not a fixed and predetermined number. It depends on where the ion is in the structure, and on the nature and number of nearest neighbors. With these conditions in mind, we will define *ionic radius* as follows: the radius of a sphere effectively occupied by the ion in a particular structural environment. The definition has some subtleties that warrant further discussion. By defining the radius as that of a sphere, we

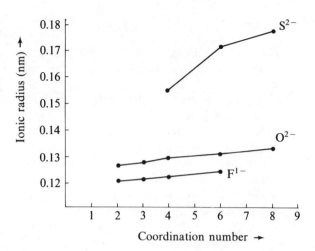

FIGURE 4.13   Variation of effective ionic radius of common anions with coordination number. Rate of radius increase is less than for cations.

take into account the fact that *average* bond distances are used. An example is the radius of Si^{4+} in tremolite, an amphibole in which Si^{4+} is in tetrahedral coordination with oxygen. Table 4.3 gives the four interionic distances with an average value of 0.1632 nm. If we now assume an oxygen (O^{2-}) radius of 0.132 nm, the calculated radius of Si^{4+} is 0.312 nm. This result depends on the assumption that the sum of the cation and anion radii is equal to the interionic distance. The assumption is implicit in the use of effective radii, because any effects of bond types other than ionic are included in the interionic distance and hence in the radius determination.

Table 4.4 lists the effective ionic radii of the most common ions in rock-forming minerals. A

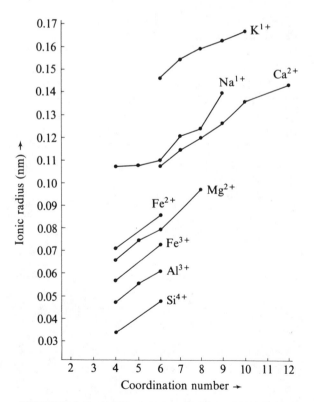

FIGURE 4.12   Variation of effective ionic radius of common cations with coordination number. Effective radius of an ion increases as its number of nearest neighbors increases.

TABLE 4.3.   *Selected Si—O Bond Lengths for Si Tetrahedra (T2) in Tremolite at 24 °C*

Ion Pair	Interionic Distance (nm)
T—O2	0.1616(2)
T—O4	0.1586(2)
T—O5	0.1653(2)
T—O6	0.1672(2)
Mean	0.1632

SOURCE: Data from Table 5 of Papike, James J., Ross, Malcolm, and Clark, Joan R., 1969. Crystal-chemical characterization of clinoamphiboles based on five new structure refinements. *Pyroxenes and Amphiboles: Crystal Chemistry and Phrase Petrology*. Mineralogical Society of America Special Paper 2:121. (Data converted to SI units.)

NOTE: The symbol T refers to the centrally coordinated Si; symbols O2, O4, O5, and O6 refer to the four oxygens at apices of the tetrahedron. Figure in parentheses in interionic distance column is standard deviation of measurement.

TABLE 4.4.   *Selected Ionic Radii of the Most Abundant Ions in Rock-Forming Minerals*

Ion	CN*	Radius (nm)	Example
O^{2-}	2	0.127	Quartz (SiO_2)
	3	0.128	Albite $(NaSi_3AlO_8)$
	4	0.130	Forsterite (Mg_2SiO_4)
	6	0.132	Periclase (MgO)
F^{1-}	4	0.123	Fluorite (CaF_2)
	6	0.125	Villiaumite (NaF)
Na^{1+}	6	0.110	Albite $(NaSi_3AlO_8)$
	9	0.140	Nepheline $(NaSiAlO_4)$
Mg^{2+}	4	0.066	
	6	0.080	Forsterite (Mg_2SiO_4)
Al^{3+}	4	0.047	Anorthite $(CaSi_2Al_2O_8)$
	5	0.056	Andalusite $(AlAlOSiO_4)$
	6	0.061	Kyanite (Al_2OSiO_4)
Si^{4+}	4	0.034	Quartz (SiO_2)
	6	0.048	Stishovite (SiO_2)
S^{2-}	4	0.156	Sphalerite (ZnS)
	6	0.172	Pyrite (FeS_2)
K^{1+}	9	0.163	Microcline (KSi_3AlO_8)
	12	0.168	Muscovite $(KAl_2(OH)_2Si_3AlO_{10})$
Ca^{2+}	8	0.120	Fluorite (CaF_2)
	12	0.143	Perovskite $(CaTiO_3)$
Ti^{4+}	6	0.069	Rutile (TiO_2)
Mn^{2+}	4	0.061	Hausmannite $(MnMn_2O_4)$
	6(HS**)	0.091	Pyrochroite $(Mn(OH)_2)$
Mn^{3+}	6(LS†)	0.066	
	6(HS)	0.073	Groutite $(MnO(OH))$
Fe^{2+}	4(HS)	0.071	Staurolite $(Al_9H\ Si_4Fe_2O_{24})$
	6(LS)	0.069	
	6(HS)	0.086	Fayalite (Fe_2SiO_4)
Fe^{3+}	4(HS)	0.057	Cronstedtite $(Fe_3(OH)_4SiFeO_5)$
	6(LS)	0.063	
	6(HS)	0.073	Hematite (Fe_2O_3)

SOURCE: Data from Whittaker, E. J. W., and Muntus, R., 1970. Ionic radii for use in geochemistry. *Geochimica Cosmochimica Acta* 34:945–956.
NOTE: All radii are in namometers (1 nm = 10 Å).
*CN = coordination number.
**HS = high-spin state.
†LS = low-spin state.

complete list is found in Appendix 5. This is the most current and refined set of data available, based on over 1000 accurate determinations of interionic distances in minerals and other crystalline inorganic compounds. All of the values depend on the choice of 0.132 nm as the oxygen (O^{2-}) radius in octahedral coordination. Why this value and not some other? The value was chosen by Whittaker and Muntus[4] as a compromise between the values 0.126 nm and 0.140 nm. The former value is the basis for the "crystal radii" of ions and is determined from electron density dis-

tributions and other properties of the alkali halides. Because this anion radius is relatively small, the calculated cation values are relatively large. The larger oxygen radius of 0.140 nm is the basis for the "ionic radii" tabulated by Shannon and Prewitt[5] and reprinted in column one, Appendix 5. This value leads to relatively small cation radii.

Both of these sets of values have disadvantages. Neither set is totally consistent with the radius ratio principle discussed in Chapter 5. This principle relates the ratio of the cation and anion radii to the coordination number of the cation. The two

[4]Whittaker, E. J. W., and Muntus, R., 1970. Ionic radii for use in geochemistry. *Geochimica Cosmochimica Acta* 34:945–956.

[5]Shannon, R. D., and Prewitt, C. J.; Effective ionic radii in oxides and fluorides. *Acta Crystallogr.* 25B(1969):925–946, 26B(1970):1046–1048.

sets of radii give large differences in radius ratio that contradict actual observation for some cation-oxygen combinations. An intermediate value (0.132 nm) of the oxygen radius therefore seems more appropriate and is adopted in Table 4.4 and in the second column of Appendix 5.

Table 4.4 also reveals how the radii of some transition metal cations vary with electron spin state, a topic we take up later in this chapter. Ions with a high-spin (HS) electron configuration have larger radii than those with a low-spin (LS) configuration. The correct choice for a particular crystal-chemical problem depends on the crystal field energy as determined by the nature and number of coordinating anions.

Note that each set of radii gives the same cation-anion distances. This is because both sets are based on the same empirical data. The absolute radii, however, do differ as a result of the choice of radius for six-coordinated oxygen. For purposes of determining the extent of chemical substitution on the basis of similar size, relative values seem to work reasonably well. Knowledge of the absolute values is useful in understanding the systematics of mineral structures. If the cations are small compared with a predetermined oxygen radius, the anions will pack together in a close array (touching each other) with the cations filling the voids between. If the cations are large compared with the radius choice for O^{2-}, the anions will not be in contact, and the structure systematics take on an additional complexity. The uncertainty of cation size has led us to consider the symmetry of atomic and ionic packing as the basis for mineral structures, rather than the knowledge of absolute radii.

POLARIZATION

The definition of effective ionic radius relies on the assumption that ions are spherical. If ions were considered in isolation without the influence of neighbors to which they are bonded, their spherical shape would be more probable. In real minerals, however, the electron distribution of an anion may be distorted by a nearby cation so that neither is spherical. This is the phenomenon of *electron polarization*, and it is an important contributing factor in our uncertainty about the absolute radii of ions.

By defining an ionic radius in terms of the effectively occupied sphere, the radius of the sphere is always measured in the bonding direction. That is, the interionic distance is the bonding direction. Figure 4.14a illustrates that if there is polarization of either ion or both, the sum of the

two radii will not equal the interionic distance. This results (Figure 4.14b) in an underestimation of the cation radius if the anion is presumed to have a fixed radius. If the ions are not in close contact but if one still has an influence on the other, the cation radius can be overestimated (Figure 4.14c).

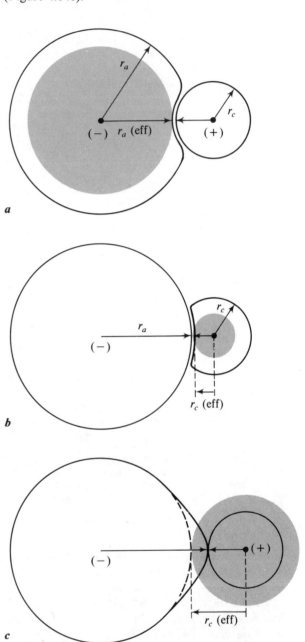

FIGURE 4.14   (a) Polarization of anion charge distribution by a small, highly charged cation. Shaded circle delineates effective anion radius r_a(eff) when cation radius r_c is fixed. (b) Shaded circle delineates effective cation radius when anion radius is fixed. This is the common procedure for determining cation radius, but if polarization exists, the cation radius is underestimated. (c) Situation in which the effective cation radius may be overestimated.

CHEMICAL BONDING BETWEEN ATOMS AND IONS

The previous considerations are essential to an understanding of the important types of chemical bonds in minerals. Elements that have a tendency to acquire electrons rather than lose them are said to be *electronegative*. These elements have large electron affinities and large first ionization potentials (Table 4.2). In contrast, elements that tend to readily lose electrons are said to be *electropositive*. A scale of electronegativities was formulated by R. S. Mulliken in 1934 based on the measured first ionization energies and electron affinities of the elements. He suggested that the average of these two energy terms would be a relative measure of the ability of each element to attract and hold electrons.

Linus Pauling offered an alternative scheme based on the observation that the chemical bonds between atoms with a large difference in their electronegativities are stronger then the bonds between atoms with a lesser electronegativity difference. Pauling actually used single-bond energies in his formulation, but his relative values of electronegativity differed little from those proposed by Mulliken. The electronegativities for most of the elements are given in the periodic table on the inside back cover. Note that values increase from left to right along any row, and from bottom to top in any column of the periodic table.

The differences in bond strengths recognized by Pauling are fundamentally related to bond type. Chemical bonds between strongly electronegative and strongly electropositive ions are said to be principally *ionic*. Halite (NaCl) and sylvite (KCl) are examples. In these ionic crystals, both the anion and the cation have completed outer electron orbitals, and all electrons are paired. When, therefore, the anion and the cation are in contact in a mineral, their electron distributions will not overlap to any extent. The force of attraction that holds the ions together and imparts many characteristic properties to ionic crystals is essentially electrostatic in nature. An important consequence of this is the tendency for each ion to be surrounded symmetrically by as many oppositely charged ions as possible.

Chemical bonds between elements having little or no difference in their electronegativities are said to be principally *covalent* in nature. Diamond (C), silica metal (Si), and moissanite (SiC) are examples. Atoms that tend to form covalent bonds do not have the form of ions with filled electron shells. Instead, each atom in the pair satisfies its need for additional electrons by sharing with the other. Atoms having singly occupied outer orbitals are ideal for covalent bonding, because each atom can accept the single electron of the other, providing the electrons have opposite spins. The resulting electron distributions exhibit as much overlap as possible, and the resulting covalent bond is highly directional and commonly possesses low symmetry. As we will see in a later section, the extent to which outer electron distributions overlap provides an important distinction between ionic and covalent bonds and is reflected in the interatomic distances of minerals.

Keep in mind that purely ionic or covalent bonds are rarely found in minerals. They represent idealized types that possess certain predictable properties with which we can compare real mineral properties. Most minerals exhibit a range of properties that must be interpreted in terms of mixed bond types.

Using differences in electronegativity to predict the relative percent of ionic or covalent character of a particular bond must be done with caution. The differences provide a qualitative guideline for predicting what bonding behavior is likely to be, but exceptions to the rule are notable.

THE IONIC MODEL

In an ideal ionic crystal, electrostatic forces operate to hold the ions together in the structure. To better understand these forces and hence the nature of ionic bonds, consider an isolated Na^{1+} cation separated by a very great distance from a Cl^{1-} anion. In 1787, the famous French physicist Charles Coulomb, known for his contributions to the theory of electricity, formally recognized that opposite charges will be attracted to each other by a force

$$F_A = \frac{k(q^+)(q^-)}{d^2} = \frac{ke^2}{d^2} \qquad (4.6)$$

where q^+ is the magnitude of the positive charge on the cation, q^- is the magnitude of the negative charge on the anion, and d is their interionic separation measured from their centers. The constant k depends on the surrounding medium (e.g., air, vacuum) and the units we choose for measurement. This force F_A is purely electrostatic in nature and will exist between all oppositely charged ions. The attractive force between Na^{1+} and Cl^{1-} is plotted against their interionic separation as the upper curve in Figure 4.15.

Equation 4.6 shows that as the ions are drawn together, the attractive force between them increases. This tendency is counteracted by a repulsive force that operates when d becomes small

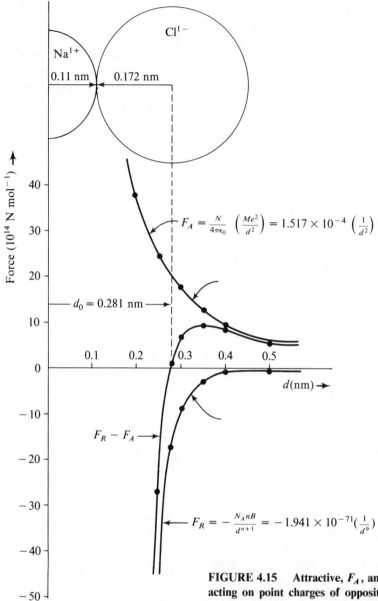

FIGURE 4.15   Attractive, F_A, and repulsive, F_R, forces acting on point charges of opposite sign separated by a distance d (nm). The equilibrium distance d_0 is attained when net force is zero. Example is NaCl.

and the outer electrons of each ion begin to overlap the same space. The electron configurations of Na^{1+} and Cl^{1-} are $1s^2 2s^2 2p^6$ and $1s^2 2s^2 2p^6 3s^2 3p^6$, respectively. Each ion has filled outer orbitals that cannot accommodate additional electrons. Physical overlap of the electron distributions can only be accomplished if some of the electrons are excited to higher orbitals. Because this requires energy, there is resistance. The repulsive force between the closed-shell electrons of each ion is a complicated expression that can be simplified to the form

$$F_R = -\frac{nB}{d^{n+1}} \qquad (4.7)$$

where B is a positive constant depending on the details of the ions and the structure, d is again the interionic separation, and the exponent n is a measure of the number of filled electron shells. Values of n were tabulated by Linus Pauling (1960) and by W. E. Dasent (1970) and are given in Table 4.5. The exponential increase of the repulsive force as the ions impinge is shown by the lower curve in Figure 4.15.

Because interatomic distances are real and reproducible numbers in crystalline NaCl, a balance between the Coulomb attractive forces and the Born repulsive forces is apparently achieved. Attractive forces predominate above the abscissa (Force = 0) in Figure 4.15 where d is greater than

TABLE 4.5. *Values of the Born Exponent n Determined Experimentally From the Compressibilities of Ionic Crystals*

Outer Electron Configuration	n_1*	n_2**	Examples
He $(1s)$	4	5	Li^{1+}, Be^{2+}
Ne $(\ldots 2p^6)$	6	7	F^{1-}, Na^{1+}, Mg^{2+}, Al^{3+}
A $(\ldots 3p^6)$	8	9	Cl^{1-}, K^{1+}, Ca^{2+}
Kr $(\ldots 4p^6)$	9	10	Br^{1-}, Rb^{1+}, Sr^{2+}
Xe $(\ldots 5p^6)$	11	12	I^{1-}, Cs^{1+}, Ba^{2+}

NOTE: These values are preferred in this text, because calculated cohesive energies are closer to measured values.
*Values n_1 from Pauling, L., 1960. *The nature of the chemical bond.* Ithaca, N.Y.: Cornell Univ. Press, p. 509.
**Values n_2 from Dasent, W. E., 1970. *Inorganic energetics.* Harmondsworth, Middlesex: Penguin Books, Ltd., p. 74.

d_0, thus the ions are drawn together. Repulsive forces predominate below the abscissa where d is less than d_0, so the ions are pushed apart. The equilibrium interatomic distance is at d_0 where the net force is zero.

These relationships may also be visualized from an energetic point of view. Any force F that operates through a distance d gives energy. The attractive energy between any two ions of opposite charge is thus given by

$$E_A = \frac{ke^2}{d} \qquad (4.8)$$

where the symbols are the same as in Equation 4.6. The repulsive energy term has a form similar to Equation 4.7. Both terms are represented in Figure 4.16, and their sum gives an energy minimum at d_0. All spontaneous processes proceed in a direction that minimizes the total energy, and the equilibrium state is achieved when that energy is minimized. An input of energy, such as might be accomplished by heating the crystal until it melts or vaporizes, is required to pull the ions apart. An input of energy is also required to push the ions closer together. This could be accomplished experimentally by compressing the crystal under high pressure.

ENERGY OF IONIC CRYSTALS

Why do certain combinations of atoms or ions have a symmetrical and orderly arrangement as crystalline solids rather than existing as separate ions in a liquid or gas? We find the answer to this question in the concept that all spontaneous processes in nature will always proceed in a direction that will decrease the total energy of the system. Given two possible states of a substance under the

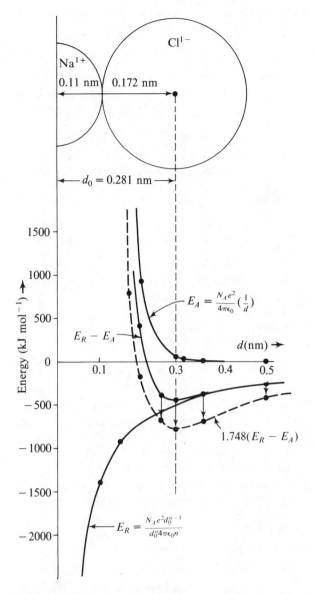

FIGURE 4.16 Coulomb attractive energy E_A and Born repulsive energy E_B acting on point charges of opposite sign separated by a distance d (nm). Equilibrium distance d_0 is attained when net energy is minimized. Additional energy decrease due to Madelung constant (1.748) multiplier shown by arrow. Example is NaCl.

same conditions, the state with the lower energy will be energetically preferred.

In the calculations that follow, the energy terms that contribute to the formation of one mole[6] of crystalline halite (NaCl) are presented in three steps. In the first step, imagine a situation in which Na and Cl exist as free, unattached atoms. Ionization energies from Table 4.2 show that in order to ionize one mole of Na to Na^{1+}, 495.8 kJ

[6] One mole of Na atoms is Avogadro's number (6.023×10^{23}) of atoms. One mole of NaCl is 6.023×10^{23} units of NaCl.

mol^{-1} of energy is required. This energy (E_i) is therefore absorbed by the reaction

$$Na + E_i = Na^{1+} + e^- \qquad (4.9)$$

If the Cl atoms each gain an electron to form Cl^{1-} anions, electron affinity data show that 348 kJ mol^{-1} of energy (E_a) is given up by the reaction

$$Cl + e^- = Cl^{1-} + E_a \qquad (4.10)$$

The difference, 147.8 kJ mol^{-1}, is the net energy that must be added to ionize the original atoms. This is apparent from summing Equations 4.9 and 4.10 to give the reaction

$$Na + Cl + (E_i - E_a) = Na^{1+} + Cl^{1-} \qquad (4.11)$$

in which $E_i > E_a$.

The second step is to form one mole of NaCl ion pairs from the isolated ions above. At this stage, each Na^{1+} interacts with only one Cl^{1-}, and no interaction occurs between ion pairs. Considering for the moment a single Na^{1+} and Cl^{1-}, the Coulomb attractive energy between the ions is

$$E_A = \frac{1}{4\pi\epsilon_0} \left(\frac{e^2}{d} \right) \qquad (4.12)$$

where $1/4\pi\epsilon_0$ is the constant k given in Equation 4.8, and ϵ_0 is the permittivity of a vacuum, having the value 8.854×10^{-12} F m^{-1}. The measured interionic distance d_0 for NaCl is 0.2814 nm or 2.814×10^{-10} m. The unit of electron charge e is 1.602×10^{-19} C. These values yield 8.198×10^{-19} J per ion pair. For one mole of ion pairs, we multiply by N_A (Avogadro's number) to get 4.94×10^5 J mol^{-1} or 494 kJ mol^{-1}. This is the Coulomb energy (Figure 4.16) liberated when the reaction

$$Na^{1+} + Cl^{1-} = NaCl \text{ (solid)} + E \text{ (ion pair)}$$

$$(4.13)$$

proceeds to completion from left to right.

In this second step, a repulsive energy will be absorbed by the ion pair and will tend to offset the attractive energy. For one mole,

$$E_R = \frac{N_A B}{d_0^n} = \frac{N_A}{d_0^n} \left(\frac{e^2 d_0^{n-1}}{4\pi\epsilon_0 n} \right) \qquad (4.14)$$

The expression for B is derived by differentiating Equation 4.12 with respect to d and setting the expression equal to zero at the equilibrium d_0 where the attractive and repulsive forces must be equal. The Born exponent n for NaCl has the value 8 as an average of Na^{1+} (7) and Cl^{1-} (9) from Table 4.5. Inserting the appropriate values,

$$E_R = \frac{(6.023 \times 10^{23})(2.566 \times 10^{-38})}{(2.814 \times 10^{-10})(100.48)(8.854 \times 10^{-12})}$$

$$= 61.73 \text{ kJ mol}^{-1} \qquad (4.15)$$

This value is about 12% of the Coulomb attractive energy (494 kJ mol^{-1}) calculated earlier. Taking this value into account, the *net energy release* upon forming one mole of NaCl ion pairs from the constituent ions (Equation 4.13) is 432.3 kJ mol^{-1}. The ion pairs are thus more stable than the ions by themselves, and are in fact more stable than the original atoms in step one.

The third and final step provides an energy contribution that stabilizes the crystal even further. In every ionic crystal, each ion is acted upon by all other ions present. Figure 4.17 illustrates a hypothetical one-dimensional NaCl crystal. If we focus our attention on a centrally located Na^{1+}, its two closest Cl^{1-} neighbors at a distance d contribute an attractive energy given by the first term in parentheses in Equation 4.16. The next closest neighbors are Na^{1+} cations at a distance $2d$. They contribute a repulsive Coulomb energy given by the second term in the parentheses, and so on as follows:

$$E = \frac{1}{4\pi\epsilon_0} \left(-\frac{2e^2}{d} + \frac{2e^2}{2d} - \frac{2e^2}{3d} + \frac{2e^2}{4d} + \cdots \right)$$

$$(4.16)$$

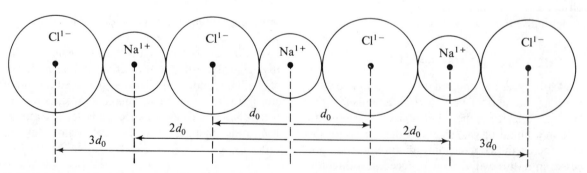

FIGURE 4.17   Hypothetical one-dimensional crystal of NaCl illustrating Madelung contribution by an ordered array of ions. This array is more stable than an equivalent number of isolated ion pairs.

In this expression, the attractive terms are given a negative sign to indicate the liberation of energy, and the repulsive terms are positive to denote absorption of energy. This expression can be rewritten as:

$$E = \frac{-2e^2}{4\pi\epsilon_0 d}\left(1 - \frac{1}{2} + \frac{1}{3} - \frac{1}{4} + \frac{1}{5} - \frac{1}{6} + \cdots\right)$$

If we sum the first ten terms in parentheses, we get

$$E = -\frac{2e^2}{4\pi\epsilon_0 d}(0.668) = -\frac{1.336e^2}{4\pi\epsilon_0 d} \qquad (4.17)$$

which clearly shows that the energy of bonded ions in the one-dimensional crystal is less than that of the same ions considered as ion pairs. The crystal must, therefore, be a more stable form.

The same approach may be extended to two and three dimensions with similar results. In halite, each Na^{1+} is surrounded by six Cl^{1-} at a distance d_0. Such an arrangement is known as octahedral coordination, because each Cl^{1-} occupies one of the six apices of a regular octahedron (eight faces). Twelve additional Na^{1+} cations are at a distance $2d_0$, eight Cl^{1-} anions at $3d_0$, and so forth. These terms sum over the entire crystal to give:

$$E_M = -6\frac{e^2}{4\pi\epsilon_0 d_0} + 12\frac{e^2}{4\pi\epsilon_0\sqrt{2}\,d_0}$$

$$-8\frac{e^2}{4\pi\epsilon_0\sqrt{3}\,d_0} + \cdots$$

$$= -\frac{e^2}{4\pi\epsilon_0 d_0}(1.74756) \qquad (4.18)$$

The number in brackets is referred to as the Madelung constant, named after the scientist Erwin Madelung who first did the calculation in 1918. Its value depends on the particular geometric arrangement of ions in the crystal. For the example of NaCl, the value of 1.74756 is interpreted to mean that the ionic solid has approximately 75% lower energy than the equivalent number of ion pairs. The Madelung constant is thus essentially a correction factor, and when applied to the numerical value obtained earlier (Equation 4.15 subtracted from Equation 4.12), yields a total crystal energy of 755.6 kJ mol^{-1}. Table 4.6 lists Madelung constants for several structure types.

The whole point of these calculations is to demonstrate that a collection of ions bonded together in an orderly array possesses a lower energy than a random array of isolated atoms. The energy released when the isolated ions condense to form a crystalline solid is referred to as the *crystal energy*[7] or *cohesive energy*. The term "structural energy" would be equally appropriate and perhaps more descriptive.

EFFECTS OF SYMMETRY

An important point to recognize in the preceding discussion is that the energy decrease imparted to bonded ion pairs in a three-dimensional structure is symmetry dependent. The Madelung constant is a pure number that depends on the geo-

---

[7]The term "lattice energy" is also commonly used. We avoid it here because the energy contributions that stabilize a crystal are more closely related to the structure than to translational symmetry (the lattice).

TABLE 4.6. *Madelung Constants (M) for Some Common Structure Types*

Structure Type	M	Mineral Examples
AB octahedral network	4.816	Rutile (TiO_2), cassiterite (SnO_2), ilmenite ($FeTiO_3$), corundum (Al_2O_3)
ABC octahedral network	1.747565	Halite (NaCl), periclase (MgO), galena (PbS)
AB tetrahedral network	1.641322	Wurtzite (ZnS), zincite (ZnO), greenockite (CdS)
ABC tetrahedral network	1.638055	Sphalerite (ZnS), chalcopyrite ($CuFeS_2$), bornite (Cu_5FeS_4)
SS cubic network	5.038785	Fluorite (CaF_2), uraninite (UO_2), thorianite (ThO_2)
CN = 2 network	4.442475	Cuprite (Cu_2O)
$\sqrt{1/3}$ *ABC*-type	1.762675	CsCl, CsBr, CsI, NH_4Cl
High quartz type	4.4394	High quartz (SiO_2)
CdI_2-type	4.71	CdI_2

NOTE: Nomenclature for structure types discussed in Chapter 5 and tabulated for common minerals in Part II.

metric arrangement of ions about some arbitrarily chosen point. The symmetry of space around any ion in the structure will be different for different structures. In one instance there may be six closest neighbors, and in another instance there may be eight.

The numerical evaluation of Madelung constants is tedious, and values remain unknown for most minerals because of lack of convergence in the successive terms of Equation 4.17. This is ample justification for using instead the systematic *structural* classification of minerals presented in the next chapter. Keep in mind that minerals as diverse in properties as halite (NaCl), galena (PbS), and periclase (MgO) all have the same *ABC* octahedral network structure, and hence all must have the same Madelung constant. Their total crystal energies will be different because of variations in charge and interionic distance, but the decrease in energy imparted by the symmetry of ion arrangement will be the same.

From an energetic viewpoint, we can readily understand why minerals are solid aggregates of ions rather than something else. A moment's thought reveals, however, that a random, disordered array of ions would lead to a numerical Madelung constant by the same procedure followed earlier. Why should crystals have such an orderly and symmetrical arrangement of ions? Part of the reason is that each ion in a structure will adjust its position relative to other ions until the following conditions are fulfilled:

1. Anions and cations are as close together as possible to maximize the Coulomb attractive forces.
2. Ions of the same charge are as far apart as possible to minimize the Coulomb repulsive forces.
3. Electrons of outer orbitals (valence electrons) are as far apart as possible to minimize Born repulsive forces.

All three factors operate to minimize the total energy of the crystal. In the process, interionic bond angles tend to be 180 degrees rather than some lesser angle. As Figure 4.18 illustrates, the 180-degree configuration of a hypothetical double ion pair has lower energy than the bent configuration. The Coulomb attractive forces are the same in each case, but the Coulomb repulsive forces between the two Cl^{1-} are greater for the bent configuration. Therefore, the bent configuration has a higher energy and is relatively unstable.

Given this background, we should not be surprised that many simple ionic minerals crystallize with cubic structures. That ions do have orderly and highly symmetric arrangements and that

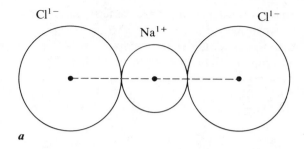

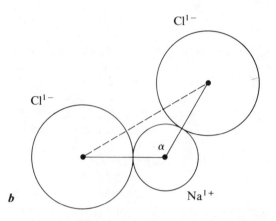

FIGURE 4.18   Hypothetical double ion pair consisting of Na^{1+} and two Cl^{1-} ions. The 180° configuration of (a) has a lower potential energy than the bent configuration of (b) because of the Cl^{1-}-Cl^{1-} repulsion.

cleavage is a planar feature follow directly from the more fundamental fact that the lowest energy configuration is preferred.

PROPERTIES OF IONIC CRYSTALS

One important consequence of ionic bonding and of the attractive and repulsive nature of the electrostatic forces involved is that every cation will attract as many anions around it as possible. This results in highly symmetrical coordination, which is central to the arrangement of ions in the structure as a whole. Ions so arranged resist being forced from their positions of minimal energy by external stresses. Ionic crystals tend, therefore, to be brittle. They commonly break with ease, and usually break along well-developed cleavage planes. This behavior is in sharp contrast to the ductile behavior of many metals, even though the symmetrical arrangements of atoms may be the same. Any attempt to slide one part of an ionic crystal by another part places like charges next to each other, and repulsion occurs.

Another distinctive property of ionic crystals is their poor electrical conductivity. This is because the bonding electrons are localized about each nucleus and are not free to move through the crystal. When the crystal is melted, the ions are freed from each other, and the resulting liquid will then conduct electrons.

Other properties are more closely related to bond strength rather than to bond type. A strong ionic bond is one in which both the attractive and repulsive forces between the ions are large. Put another way, a strong bond has a deep potential energy well on the graph in Figure 4.16 and is therefore difficult to remove. Equation 4.12 tells us that attractive energy increases as the square of the ion charge or valence increases, so highly charged cations and anions will generally form strong ionic bonds. A good example is the comparison of halite (NaCl) with a Mohs hardness of 2.5 and periclase (MgO) with a hardness of 6. They both have the same structure with six closest neighbors. The divalent character of Mg^{2+} and O^{2-} results in a much stronger bond. Attractive energy also varies as the inverse of the interionic distance, so small, highly charged ions form even stronger bonds.

In contrast to the bond in a hypothetical isolated ion pair, the strength of a single bond in a real mineral structure depends on the number of closest neighbors around any particular ion. For example, if a central cation has four anions bonded to it, one fourth of the electron charge associated with the free cation must be distributed equally among the anions. If the same cation has six closest anions, however, the interionic bond will be weaker, because the cation charge is distributed over six bonds rather than four. For this reason, ionic bond strength can be approximated by dividing the cation valence by the number of close, equally distant anions to which the cation is bonded.

A general relationship exists between the strength of the chemical bonds in a mineral and the mineral's hardness. A uniform distribution of strong bonds produces a hard mineral. The properties of solubility and melting point are also generally related to bond strength. Minerals with strong ionic bonds tend not to dissolve easily, and in many instances they have high melting points.

THE COVALENT MODEL

A fundamental part of the previous model of an ideal ionic bond is the ease with which the participating atoms become ions. This ease of ionization is enhanced by large ionization energies and electron affinities, or put another way, by large differences in electronegativity. Elements that have little or no difference in electronegativity, and hence have no tendency to permanently transfer an electron from one to the other, behave in a different manner. Take any of the diatomic gas molecules, H_2, N_2, O_2, F_2, or Cl_2, as examples. Both atoms of a bonded pair could form stable anions (Cl^{2-}, F^{1-}, O^{2-}) or cations (H^{1+}, N^{3+}), but because both will have the same electron charge, they will not be attracted by electrostatic forces. Instead, each participating atom satisfies its need for a noble gas configuration by *sharing* the available electrons with its partner. Fluorine has an electron configuration of $1s^2 2s^2 2p^5$ (Figure 4.19) and needs one additional electron in its $2p_z$ orbital to achieve the $2p^6$ configuration of Ne. A neighboring F to which a chemical bond could form will have the same need, so by sharing the needed electron both can achieve the $2p^6$ configuration half of the time. Because the two outer electrons share the same orbital, they must have opposite spins as required by the Pauli exclusion principle.

The mechanical wave interpretation of electron "sharing" is an increase in electron density in the internuclear region. This means that the probabil-

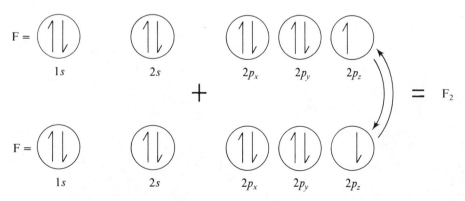

FIGURE 4.19 **Electron configurations for F atoms that combine to form a stable F_2 molecule. A $2p^6$ Ne configuration is achieved by sharing a single electron in** the $2p_z$ orbital. Note opposite spin of $2p_z$ electrons required for covalent bonding.

ity of finding an electron between two nuclei is greater in the case of a covalent bond than in the case of an ionic bond. The physical reality of this important difference is illustrated in Figure 4.20. An electron density map of the (100) plane of FeO shows virtually no electron overlap. The electron density map of the (100) plane of pyrite (FeS_2), however, reveals a substantial increase in electron density in the internuclear region. These "maps" are constructed from x-ray diffraction data obtained from small, single crystals of these minerals. The electron density contours are in units of electrons/nm^3.

The electronegativities in the periodic table on the inside of the back cover show that we should expect Fe—O bonds to be largely ionic, whereas Fe—S bonds should be largely covalent. The difference between the electronegativity of S (2.5) and Fe (1.8) is 0.7, which is much smaller than the electronegativity difference between oxygen (3.5) and Fe (1.8). This suggests that if Fe and S were covalently bonded, their observed interionic distance would be less than the calculated value. We

learned earlier that the radii of Fe^{2+} ($r = 0.086$ nm) and S^{2-} ($r = 0.172$ nm), as inferred separately from ionic compounds, results in an interionic distance of 0.258 nm. The measured interionic distance is 0.226 nm. Clearly the Fe and S atoms in pyrite are closer together than simple ions bonded together by electrostatic forces.

The wave function that describes how a shared electron is distributed in a covalent bond is called a *molecular* or *hybrid orbital*. Like atomic orbitals, only two electrons can occupy a hybrid orbital, and then only if the electrons have opposite spins. Hybrid orbitals are classified by the atomic orbitals into which they separate when the atoms are separated. Consequently, we can picture a hybrid orbital as being built up of two overlapping atomic orbitals. The physical significance of a hybrid orbital is the increased electron density between the nuclei due to the "overlap." For example, recall earlier that the wave function for p orbitals (e.g., Figure 4.7c) has maximum density along mutually perpendicular coordinate axes. This imparts a distinctive "dumbbell" shape to the

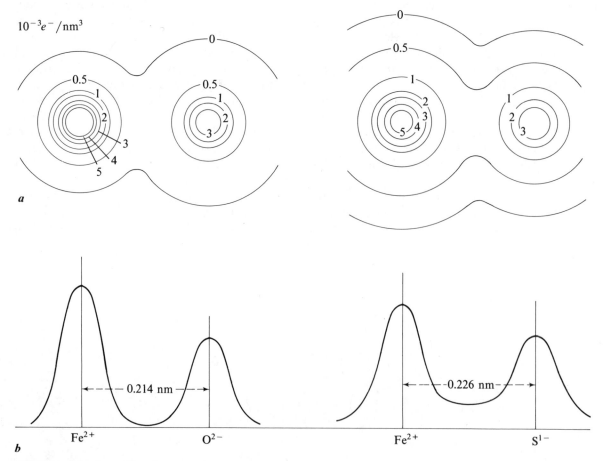

FIGURE 4.20   Electron density maps of (a) (100) plane of FeO and (b) (100) plane of pyrite (FeS_2). Electron density contours are in units of $10^{-3}\ e^-/nm^3$.

orbital in each direction. In fluorine, only the ends of the dumbbells overlap, and only in one direction, z. Two atoms bonded together in this manner may be rotated with respect to each other, and the internuclear electron distribution will remain unchanged. Covalent bonds of this sort that possess *axial symmetry* are called σ (*sigma*) *bonds*. The fixed position of the resulting molecular orbital is responsible for the strong directional nature of covalent bonds. Actual molecules can and do form as a result. In contrast, the concept of a molecule has no basis in ionic minerals because of the difference in bond type.

The minerals diamond and graphite are examples of how we apply these concepts. Elemental carbon in its ground state has a $1s^2 2s^2 2p^2$ electron configuration (Figure 4.21a). In an excited state, the carbon atoms in diamond have four sp^3 hybrid orbitals of equal energy that are formed by combining one s orbital (from $2s^2$) and the three p orbitals (Figure 4.21b). These hybrid orbitals, being of equal energy, position themselves about the carbon nucleus such that their respective electrons

are as far away from each other as possible. This gives the tetrahedral arrangement shown in Figure 4.21c. In the case of diamond, each sp^3 orbital overlaps with an sp^3 orbital of an adjacent carbon to form a three-dimensional network that makes up the structure. Because of the high symmetry of the sp^3 orbitals and the regular tetrahedral coordination that results, diamond possesses remarkable properties that include great hardness.

In graphite, only two of the three p orbitals of carbon in effect hybridize with the $2s$ electrons. This arrangement gives three sp^2 hybrid orbitals instead of the four found in diamond. These three form strong sp^2-sp^2 bonds in a two-dimensional sheet. The remaining orbital ($2p_z^1$) contains a "lone-pair" electron that "sticks out" above and below the sheet much like a normal $2p_z$ orbital. This electron, however, will overlap sideways with $2p_z^1$ orbitals of adjacent carbon atoms within the strongly bonded sheet. These electrons form π (pi) bonds because of their sideways overlap. An important property of π bonds is related to their symmetry. Any attempt to rotate one of the car-

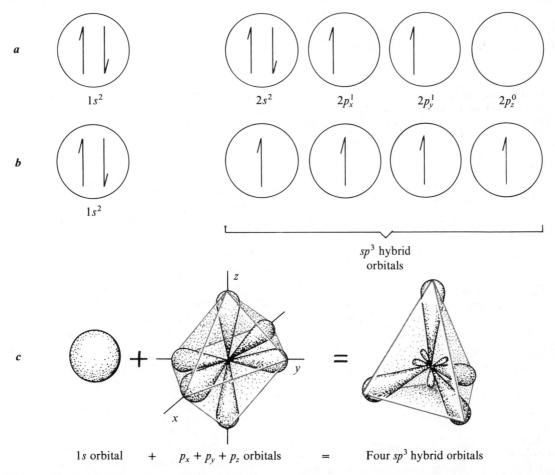

FIGURE 4.21 Schematic electron configurations of the ground state (a) and excited state (b) of carbon in diamond. (c) Three-dimensional structure of sp^3 bonds in diamond.

bon atoms relative to another along the π bond axis in the plane of the carbon sheet will decrease the sideways overlap of the $2p_z$ orbitals, and hence will be resisted.

Another important property of π bonds is that they effectively decrease the interatomic distance between carbon atoms in the sheet. The additional attractive forces over and above those provided by σ bonds (sp^2-sp^2) within the sheet draw the carbon atoms closer together. Because of their shorter interionic distance, the C—C bonds within the sheets of graphite are actually stronger than the C—C bonds in diamond. The softness of graphite is imparted by the very weak bonds between the sheets.

Hybridization can also involve d orbitals and all combinations of s, p, and d orbitals. A summary of the symmetrical types of hybrid orbitals is given in Table 4.7. Each type is associated with a distinctive coordination polyhedron with a symmetry and shape reflecting the most energetically favorable bond angles. These polyhedra and the various ways in which they may be arranged in a three-dimensional structure are the subject of the next chapter. We will discover that several bond types form polyhedra that cannot be accommodated in simple symmetrically packed structures. Other polyhedra, such as the dsp^2 and d^4sp orbitals, are unique to covalent bonding. These polyhedra are not found in ionic crystals.

PROPERTIES OF COVALENT CRYSTALS

One consequence of the purely covalent model is that the chemical bonds are highly directional. This may or may not influence the packing arrangement of atoms in the structure. If the orbital arrangement is highly symmetrical, as with tetrahedral (e.g., diamond) or octahedral symmetry, then the structure of such covalent minerals may be understood on the basis of symmetrically packed geometry. In this case, the symmetry of bonding in minerals with covalent bonding may be the same as in minerals with ionic bonding. If, however, the orbital arrangement requires plane tetragonal or pentagonal symmetry (Table 4.7), then simple close-packed structures do not follow.

Covalent minerals exhibit a broad range of physical properties, and attempts to relate those properties to bonding characteristics independent of the mineral structure are somewhat artificial. For example, diamond is the hardest known substance, and graphite is one of the softest. Both have exactly the same chemistry, and both are held together by covalent bonds. Graphite has good cleavage on (001), expressing the fact that π bonds are relatively weak, yet graphite has a high melting point (> 3000 K), which reflects the energy required to break the strong sp^2-sp^2 hybrid bonds within each sheet. Graphite is black and submetallic in appearance, and is a fair conductor of electricity. Diamond, on the other hand, is transparent and nonmetallic, and a poor electrical conductor. These differences are attributed to the presence of the delocalized π electrons, absent in diamond, that are between (001) sheets of graphite.

As we emphasized earlier, most minerals have bond characteristics that seem to be a mixture of both ideal covalent and ideal ionic models. Therefore, properties such as hardness, solubility, and melting point are better understood when considered in terms of average bond strength rather than bond type. The strength of the ideal covalent bond is essentially determined by the *extent of overlap* of atomic orbitals. A large overlap means a stronger bond and implies a shorter interatomic distance, which is the parameter that also affects the strength of an ideal ionic bond. All other factors being equal, short bonds tend to be strong bonds quite independent of bond type.

We can combine the effects of bond strength and bond structure by considering how the hard-

TABLE 4.7. *Summary of Symmetrical Hybrid Orbitals*

Type of Hybridization	Arrangement of the Hybrid Orbitals	Coordination Number
sp	Linear	2
dp	Linear	2
sp^2	Plane trigonal	3
dp^2	Plane trigonal	3
ds^2	Plane trigonal	3
d^2p	*Pyramid, trigonal	4
sp^3	Tetrahedral	4
d^3s	Tetrahedral	4
dsp^2	**Plane tetragonal	4
dsp^3	*Dipyramid, trigonal	5
d^3sp	*Dipyramid, trigonal	5
d^4s	*Pyramid, tetragonal	5
d^2sp^3	Octahedral	6
d^4sp	**Trigonal prism	6
d^3sp^3	*Dipyramid, pentagonal	7

*Arrangements are not possible in simple close-packed structures.

**Arrangements are unique to covalent bond types.

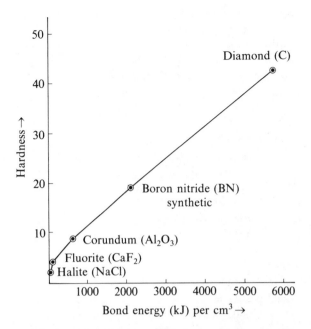

FIGURE 4.22   Hardness of some common minerals and synthetic abrasives versus bond strength per unit volume. Data converted to SI units. (After Bundy, F. P., 1974. Superhard materials. *Scientific American* 231:64.)

ness of minerals varies with bond strength per unit volume. The result is a measure of the density of the bonds that must be broken when a hardness test is applied. Figure 4.22 shows that a high bond strength density correlates with high hardness. The hardest minerals, such as diamond, also have a highly symmetric arrangement of bonds in their structures. Contrast the bond arrangement in diamond with graphite in which the arrangement of the strong sp^2-sp^2 hybrid bonds is confined to carbon sheets.

THE METALLIC BOND MODEL

Many minerals have physical properties that cannot be explained in terms of ideal ionic or covalent bond types. For example, the malleability of native gold or silver and the ability of metals such as native copper to conduct electricity are properties that are inconsistent with the previous model of electrons being tightly held or localized to atomic nuclei. The concept of an *ideal metallic bond* requires that the bonding electrons be highly delocalized and free to move from one atom to the next within the structure. Around the turn of the century, pioneering research by the German physicist Paul Karl Drude resulted in the "free electron" theory of metals that pictures positively charged atom cores and their inner electrons held together by an interatomic "glue" of loosely bonded outer electrons. The minerals we refer to as "metals" consist of single elements that have the following important properties: (1) a small number of outer electrons resulting in a closed-shell configuration not readily attainable by the sharing of electrons in normal covalent bonding, and (2) low first ionization energies (Table 4.2), so the element will readily give up an electron to a highly mobile electron "glue" that surrounds the atom cores.

Figure 4.23 illustrates how the elements may be classified as metals, semimetals, or nonmetals based on these properties. As discussed earlier, when metallic elements form ions, they are nearly always cations. Nonmetals, in contrast, generally require only a small number of electrons to achieve a noble gas configuration, so nonmetals typically form covalently bonded molecules (e.g.,

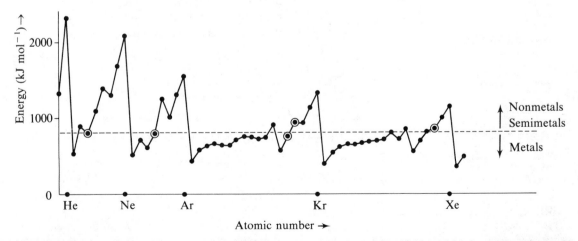

FIGURE 4.23   First ionization energy versus atomic number separates the metals from the nonmetals and semimetals (circled) at approximately 800 kJ mol^{-1}.   **Energies less than this favor the formation of metallic bonds and promote electrical and thermal conductivity.**

Cl_2, and F_2). As a consequence, the electrons of nonmetals become localized and are not available for conduction of electricity or heat. The non-metal elements are thus referred to as electrical and thermal *insulators*. Some typical examples are P, S, and Cl. The semimetals are intermediate in their behavior, having approximately half-filled outer electron shells. Their *semiconducting* properties make them extremely useful for various electronic devices such as transisters and diodes. The elements Si and Ge are good examples of effective semiconductors.

We gain a deeper understanding of the nature of the metallic bond by considering the approach of identical metal atoms in their ground states and the interaction of their electron orbitals. At infinite separation, their wave functions and, therefore, the energies of their electrons are identical. If we take Na as an example, the electron configuration is $1s^2 2s^2 2p^6 3s^1$ with a single, unpaired electron in the outer $3s$ orbital. As the atoms approach closely, mutual perturbation of orbitals occurs such that new molecular orbitals form with slightly different energies. This behavior is a direct consequence of the Pauli exclusion principle, which guarantees that only two electrons can occupy precisely the same quantum state. In general, if we start with N_A Na atoms and bring them together, N_A molecular orbitals will result for the aggregate. Figure 4.24 depicts this process and the formation of *energy bands* of various widths depending on how close the atoms are to each other. The Na atoms have filled $1s$, $2s$, and $2p$ orbitals,

so the resulting energy bands are also filled. The $3s$ band is only half filled, and the $3p$ and higher bands are empty. Just as the $3s$ orbital is considered the *valence orbital*, so the $3s$ band is considered the *valence band*. In the case of Na, the $3s$ band is also a *conduction band*, because more energy levels are in the band than there are electrons. If we take a cubic centimeter volume of Na metal, it contains on the order of 10^{22} atoms. The number of unoccupied energy levels in the $3s$ band is so large that all energies in the band can be regarded as continuously accessible to electrons. When a voltage is applied across Na metal, the electrons conduct readily in the $3s$ band. Any electron in this band is the property of the crystal as a whole and serves to bind many nuclei together.

Figure 4.25 illustrates the potential energy "well" that essentially keeps the electrons attached to the metal. Within the confines of the crystal, each atomic nucleus generates a deep "subwell" because of the Coulomb attractive forces between the nuclei and the electrons. Electrons in the lower energy bands are localized and do not contribute to electrical conduction.

Armed with these principles, we might expect the metal Mg ($1s^2 2s^2 2p^6 3s^2$) to behave differently inasmuch as the $3s$ band is totally occupied. In the case of Mg metal, however, the conduction band $3p$ actually overlaps the valence band $3s$ so conduction can occur. In electrical insulators, a large energy gap exists between the valence and conduction bands, so even under an applied voltage, electrons cannot populate the conduction band. In the example of semiconductors, or semimetals such as Si and Ge, the energy gap is small and can be overcome by heating. The same effect can be achieved by subjecting such materials to pressure. The width of the various bands, and hence the magnitude of the energy gaps between them, depends on how close the atoms are to each other. Since the effect of pressure is to force the atoms closer together, we can expect the conducting properties of metals to be enhanced at greater depths in the earth.

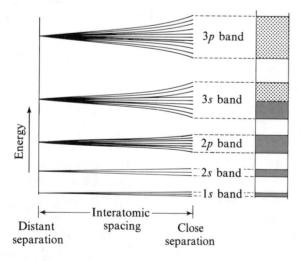

FIGURE 4.24 Successive development of energy bands in Na metal as Na atoms approach each other from a distant separation. N sublevels with two electrons each will generally develop from a close-packed array of N atoms.

PROPERTIES OF METAL CRYSTALS

The explanation of the high electrical and thermal conductivity of metal crystals by band theory is a direct consequence of the metallic bond. The electrical conductivity of metals such as Ag and Cu is 15 to 20 orders of magnitude greater than that of nonmetal insulators. Electrical conductivity probably exhibits the largest range in values of any mineral property.

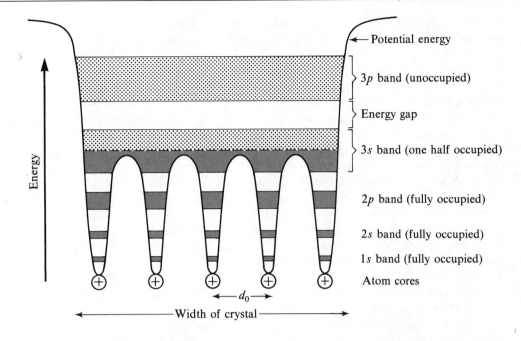

FIGURE 4.25   Potential energy well for Na metal bounded by crystal surfaces. Internal subwells indicate positions of positively charged atom cores and localized electrons in the $1s$, $2s$, and $2p$ bands. The $3s$ band is the conduction band when voltage is applied across the metal.

Not so obvious as electrical conductivity are the exceptional chemical properties that are derived from the nonspecific nature of the metallic bond. Because the conduction electrons are a property of the crystal as a whole, metals are essentially free of the valence requirements that control chemical substitutions in ionic minerals. This is why so many different metal alloys can be made, and why metals can be welded together with a torch.

The distinctive luster of metals and their opaqueness to visible light are also properties related to bonding. Free or nearly free electrons have the property of absorbing and then radiating back most of the light energy that impinges on them. The essentially continuous range of energy levels in the conduction and higher level bands makes this possible. In Chapter 11 we will use as a routine identification criterion the opaqueness of metals and of certain metallic oxides and sulfides to visible light.

Based on the metallic bond model, we would predict that the attractive forces, and hence the cohesive energy of metals, are low. This is in fact true. Only the mobile electron "glue" keeps the positively charged atom cores from flying apart due to electrostatic repulsion. The properties that reflect these relatively weak binding forces are the softness of most metals, their low melting points, and their high malleability and ductility.

The metallic bond has important consequences on the origin of metal structures. Because the bonds are nondirectional, and because in pure metals the atoms are all the same size, the structures of metals can be derived from the purely geometric approach of symmetrically packing spheres (Chapter 5). In this context, we then proceed to discuss physical properties of minerals that seem to be more closely related to their structures. Keep in mind, however, that the nature and symmetry of the mineral structure itself may be due to properties of the chemical bond that holds the atoms or ions together.

THE MOLECULAR BOND MODEL

The fourth and final bond type we need to consider deals with those atoms that have no tendency to lose, gain, or even share valence electrons with other atoms in the structure. The ideal examples, although of little mineralogic importance, are the crystalline forms of the inert gases and the crystalline forms of hydrogen, nitrogen, and oxygen, all of which exist at very low temperatures only. In view of their stable electron configurations, what is it that holds the atoms in these crystals together? The bonding forces cannot be strong, because the stable forms of these elements at room temperature are gases.

The answer lies in the asymmetric charge distributions of the atoms. When the electron motions are averaged over a period of time, these atoms are spherically symmetric. At any given instant, however, each atom will have a small but real electric dipole moment as a result of the higher electron density on one side of the nucleus than on the other. The instantaneous dipole moment produces an electric field, which acts on the electron distribution of neighboring atoms in such a manner that they, too, become instantaneous dipoles. The resulting interaction of dipoles is known variously as *van der Waals attraction* (after Dutch physicist Johannes van der Waals), as the *London attraction* (after Fritz London, who derived the quantum mechanical theory for this attraction in 1930), or simply as *fluctuating dipole interaction*. The net force is attractive, and varies with the inverse sixth power of the interatomic distance.

Van der Waals attractive forces are extremely important to the understanding of *molecular crystals*. We will consider several such crystals of mineralogic interest in Chapter 5. They consist of distinct molecules or discrete groups of atoms bonded to each other by these weak intermolecular forces. Bonding *within* the molecule can be very strong, as in the carbon sheets of graphite, but physical properties like cleavage are dictated by the weak bonds that hold the sheets together. Van der Waals bonding makes a small contribution to the total bonding in all minerals, but it commonly is ignored when calculating crystal energies based on other bond types.

CRYSTAL FIELD THEORY

At this point, we might suppose that most of the important properties of minerals could be qualitatively explained by combinations of the various ideal bond types. As with all models, however, revisions and improvements must be made when a new, unexplained property is discovered. Crystal field theory was developed in the late 1920s to account for the unique properties of color and magnetic behavior typical of minerals with a chemistry that includes transition metal ions. Elements of the first transition series include Sc ($Z = 21$) through Zn ($Z = 30$). The most important of these from a geological perspective is Fe, a major constituent of the earth's core and mantle. The transition elements together comprise approximately 40% of the earth's total mass.

All of the first group of transition elements are characterized by various numbers of $3d$ electrons. The explanation of how $3d$ orbitals of a transition metal cation interact with the electron orbitals of immediately surrounding anions is the principal goal of crystal field theory. These coordinating anions are commonly referred to as *ligands*. Their net negative charges generate an electric field, called the *crystal field*, or ligand field, which has a symmetry and shape dependent on the number of anions, their charge, and their distance from the cation. The bonding electrons of the central cation are very sensitive to these close anion neighbors, and will adjust their energy levels in response to the neighboring anions. The crystal field concept is essential to the understanding of how crystal structure controls the property of color discussed later in this chapter.

The basic assumption of crystal field theory is that the attractive and repulsive forces that hold ions together are electrostatic in nature. We will, therefore, ignore for the moment the possible effects of covalent bonding, and imagine a transition metal ion in free space. Five $3d$ orbitals are available for electrons, and in the absence of neighboring ions, all have the same energy. Each orbital has a distinctive symmetry (Figure 4.26) dependent only on the possible quantum number combinations. Two of these orbitals, $3d_{z^2}$ and $3d_{x^2-y^2}$, have their maximum electron density along the x, y, and z coordinate axes. The remaining three orbitals, $3d_{xy}$, $3d_{xz}$, and $3d_{yz}$, have their greatest electron density directed between the coordinate axes.

Now consider the simultaneous convergence of six anions along each of the coordinate axes. Electrostatic repulsion between the anion orbitals and the centrally located cation orbitals raises the energy levels of the $3d_{x^2-y^2}$ and $3d_{z^2}$ orbitals directed along those axes, whereas the orbitals that are most separated from the anions experience less repulsion and have lower energy. This phenomenon gives rise to the term *crystal field splitting*, which means exactly that. The electron distribution about the anions (i.e., the crystal field) literally "splits" the $3d$ energy levels of the central cation. The split is not equal, however, as three of the five orbitals are in the more stable or lower energy state. The lower energy level (Figure 4.27) is commonly referred to as the t_{2g} energy state, a term borrowed from group theory. The $d_{x^2+y^2}$ and the d_{z^2} orbitals at the higher energy level are referred to as the e_g group orbitals. Because there are five orbitals, the energy of the e_g group orbitals is 3/5 above the mean energy, whereas the t_{2g} group orbitals are 2/5 below the mean. The magnitude of the crystal field splitting in units of energy (kJ/mole) is designated as Δ_o for an octahedrally coordinated cation. The effect of the energy split is to stabilize the cation energy by 2/5 (Δ_o) for each electron in a t_{2g} orbital and to destabilize the cation by $3/5(\Delta_o)$ for each electron

FIGURE 4.26   Stereographic projections of selected $3p$ orbitals and their idealized shapes. Each projection based on 3000 computer calculations of electron position from the hydrogen wave equations. (From Cromer, D. T., 1968. Stereo plots of hydrogen-like electron densities. *Journal of Chemical Education* 45:628. Reproduced by kind permission of the journal.)

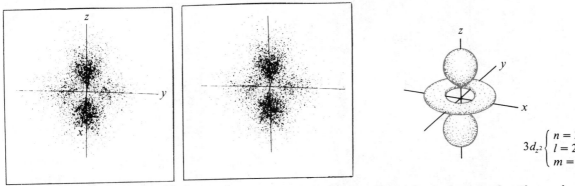

$3d_{z^2}\begin{cases} n = 3 \\ l = 2 \\ m = 0 \end{cases}$

a $3d_{z^2}$ orbital. The main electron density is along the z axis. A doughnut-shaped density is centered on the xy plane.

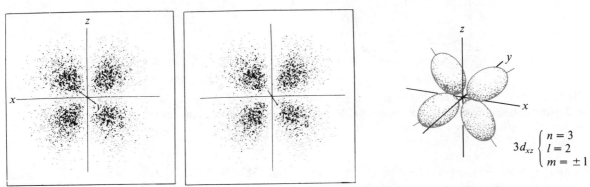

$3d_{xz}\begin{cases} n = 3 \\ l = 2 \\ m = \pm 1 \end{cases}$

b $3d_{xz}$ orbital. Two lobes of high electron density centered on xz plane, and xy and yz planes are nodal planes.

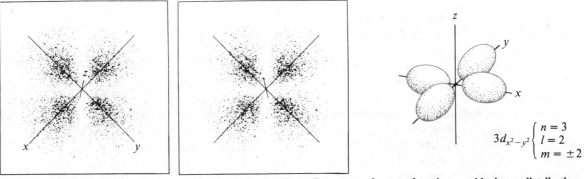

$3d_{x^2-y^2}\begin{cases} n = 3 \\ l = 2 \\ m = \pm 2 \end{cases}$

c $3d_{x^2-y^2}$ orbital. Maximum electron density is in xy plane. For $m = -2$, wave function would give a distribution rotated 45° about z axis.

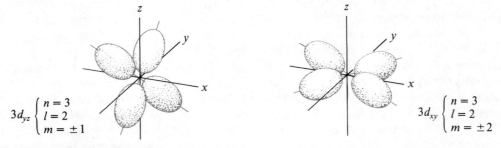

$3d_{yz}\begin{cases} n = 3 \\ l = 2 \\ m = \pm 1 \end{cases}$

$3d_{xy}\begin{cases} n = 3 \\ l = 2 \\ m = \pm 2 \end{cases}$

d   Idealized $3d_{yz}$ orbital. Maximum electron density has the form of four lobes in yz plane.

e   Idealized $3d_{xy}$ orbital. Maximum electron density has the form of four lobes in xy plane.

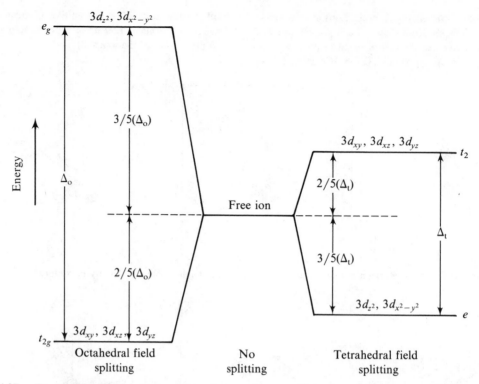

FIGURE 4.27 **Crystal field splitting energy for octahedral and tetrahedral coordination environments. Five $3d$ orbitals are separated into a high energy group e_g and a** **low energy group t_{2g} in response to local variations in electron charge distribution.**

in an e_g orbital. Any net stabilization of the cation will depend on the exact number of $3d$ electrons and how they are distributed in the two energy levels.

We can make a similar analysis of transition metal ions in tetrahedral (fourfold) coordination. In this case, the surrounding anions approach along the [111], [$\bar{1}\bar{1}1$], [$\bar{1}1\bar{1}$], and [$1\bar{1}\bar{1}$] directions.[8] This gives rise to strong electron repulsion of cation orbitals that have maximum electron densities directed between the coordinate axes. Consequently, the energy levels of the $3d_{xy}$, $3d_{xz}$, and $3d_{yz}$ orbitals are raised, and those of the $3d_{x^2-y^2}$ and $3d_{z^2}$ orbitals are lowered. The splitting order is exactly the opposite of that found in octahedral coordination. The two groups of orbitals are designated t_2 (higher energy) and e (lower energy). The difference between them ($t_2 - e$) is called the tetrahedral crystal field splitting parameter Δ_t (Figure 4.27). In general, Δ_t will be smaller than the octahedral parameter Δ_o for each transition metal ion, the net effect being that most transition metal cations show a slight preference for octahedral coordination, because they can achieve a lower energy state in that environment. The exceptions

are Sc^{3+} and Zn^{2+} at the beginning and end of the series, respectively, and Mn^{2+} and Fe^{3+} in the middle of the series. These ions exhibit no site preference.

SPIN STATES OF TRANSITION METAL IONS

Let us now treat each transition metal ion in sequence, assuming a simple sixfold (octahedral) coordination. Trivalent titanium (Ti^{3+}) has an electron configuration of $1s^2 2s^2 sp^6 3s^2 3p^6 3d^1$. Its single $3d$ electron must occupy a t_{2g} orbital. The crystal field stabilization energy (CFSE) is $2/5$ (Δ_o). Both Ti^{2+} and V^{3+} have two $3d$ electrons (Table 4.8), which will impart $2(2/5(\Delta_o)) = 4/5(\Delta_o)$ CFSE to those cations. Divalent vanadium (V^{2+}) and Cr^{3+} each have an additional $3d$ electron that will fill another t_{2g} orbital with its spin parallel to the others, according to Hund's rule. Divalent chromium (Cr^{2+}) and Mn^{3+} require some additional considerations. Both have four $3d$ electrons that must be distributed over the five orbitals shown in Figure 4.28a. The first three electrons will go into the t_{2g} level without pairing, because that level has the lowest energy and paired electrons in the same orbital tend to repel each other. Where do we place the fourth elec-

---

[8] These are the alternate vertices of a cube with $x = $ [100], and so forth.

TABLE 4.8. *Transition Metal Ions and Their Low and High Spin Electronic Configurations*

Transition Metal Ion	Number of $3d$ Electrons	Low Spin State					Number of Unpaired Electrons	High Spin State					Number of Unpaired Electrons
		d_{xy}	d_{yz}	d_{xz}	d_{z^2}	$d_{x^2-y^2}$		d_{xy}	d_{yz}	d_{xz}	d_{z^2}	$d_{x^2-y^2}$	
Ti^{3+}, V^{4+}	1	↑	—	—	—	—	1	↑	—	—	—	—	1
Ti^{2+}, V^{3+}	2	↑	↑	—	—	—	2	↑	↑	—	—	—	2
V^{2+}, Cr^{3+}	3	↑	↑	↑	—	—	3	↑	↑	↑	—	—	3
Cr^{2+}, Mn^{3+}	4	↑↓	↑	↑	—	—	2	↑	↑	↑	↑	—	4
Mn^{2+}, Fe^{3+}	5	↑↓	↑↓	↑	—	—	1	↑	↑	↑	↑	↑	5
Fe^{2+}, Co^{3+}	6	↑↓	↑↓	↑↓	—	—	0	↑↓	↑	↑	↑	↑	4
Co^{2+}, Ni^{3+}	7	↑↓	↑↓	↑↓	↑	—	1	↑↓	↑↓	↑	↑	↑	3
Ni^{2+}, Pt^{2+}	8	↑↓	↑↓	↑↓	↑	↑	2	↑↓	↑↓	↑↓	↑	↑	2
Cu^{2+}, Ag^{2+}	9	↑↓	↑↓	↑↓	↑↓	↑	1	↑↓	↑↓	↑↓	↑↓	↑	1
Cu^{1+}, Zn^{2+}	10	↑↓	↑↓	↑↓	↑↓	↑↓	0	↑↓	↑↓	↑↓	↑↓	↑↓	0

NOTE: A difference between high-spin and low-spin radii exists only for ions having four to seven $3d$ electrons.

tron? In the absence of crystal field splitting, it would go as an unpaired electron in a fourth orbital. If the crystal field is split, however, the electron can jump to a higher energy level e_g in which spins may remain parallel, or it can pair with one of the lone electrons in the t_{2g} level. How

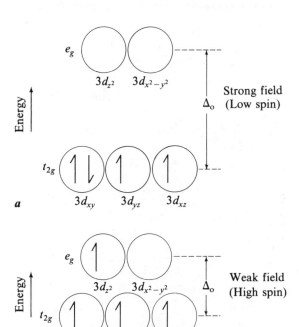

a

b

FIGURE 4.28 (a) Strong field (low-spin) electron configuration when the crystal field splitting Δ_o is large compared to the repulsive energy of two paired electrons. A small effective ionic radius results. (b) Weak field (high-spin) electron configuration when the crystal field splitting is small compared to paired electron repulsion. A large effective radius results.

the electron will act depends on the relative magnitudes of the repulsion energy for paired electrons and the crystal field energy. If Δ_o (or Δ_t as the case may be) is large, as it would be if the degree of repulsion between the crystal field and the $3d$ electrons is large, pairing the electron in the t_{2g} level (Figure 4.28a) will be energetically favorable. If Δ_o is small, however, less energy may be required to push the electron up to the e_g level than to pair it off at the t_{2g} level, as shown in Figure 4.28b.

If energy considerations favor placing electrons in the e_g level before six electrons are paired in the t_{2g} level, we refer to a *strong field configuration*. If Δ_o is small and electrons fill orbitals without pairing in the normal way, we refer to a *weak field configuration*. These are also called "low spin" and "high spin" states, respectively.

The possibility of either a high or low spin state exists only for those ions in the central part of the transition series, namely, those with d^4, d^5, d^6, and d^7 outer orbital electrons. In Table 4.8, these are Cr^{2+}, Mn^{2+}, Mn^{3+}, Fe^{2+}, Fe^{3+}, Co^{2+}, Co^{3+}, and Ni^{3+}. Whether any one of these cations is in the high spin (larger radius) or low spin state makes a significant difference in the effective ionic radius. The choice is determined by which state represents the lowest energy configuration under the temperature and pressure conditions of interest. Normally the crystal field splitting energy is less than the repulsion energy of paired electrons, and the ions prefer the high spin state. Important exceptions can occur. For example, one of the most important ionic substitutions in crystal chemistry is the replacement of Fe^{2+} by Mg^{2+} and vice versa. At low pressure in the upper crust of the earth, Fe^{2+} is normally in the high spin state in which its effective radius (0.077 nm) is close to that of Mg^{2+} (0.072 nm). At high pressure in the earth's mantle, however, Δ_o can be large due to compression of the crystal field, and Fe^{2+} could prefer the smaller effective radius (0.061 nm) associated with the low spin state. The possible difference in radius between Fe^{2+} and Mg^{2+} under these conditions can drastically limit the extent of solid solubility (Chapter 8) in olivines, spinels, garnets, and other minerals that normally exhibit complete Fe^{2+}-Mg^{2+} substitution.

COLOR IN MINERALS

All sensations of light intensity and color that reach the human eye represent but a small part of the total electromagnetic spectrum (Figure 4.29). Most people can "see" color only over a range of wavelengths from about 390 nm (violet) to about 770 nm (red). All wavelengths below this range are invisible and have higher energies that can penetrate human tissue and do irreparable damage. Wavelengths above this range are also invisible but have lower energies characteristic of infrared, television, and radio bands.

Light of a single wavelength is referred to as *monochromatic*. Each wavelength is associated with a characteristic frequency ν, which is a measure of the number of waves per second that pass a fixed point along the propagation direction. The product of a wave's frequency and its wavelength is equal to the wave's velocity. This constant, known as the speed of light, has the following value when measured in a perfect vacuum:

$$\nu\lambda = c = 2.997924 \times 10^8 \text{ ms}^{-1} \approx 3 \times 10^8 \text{ ms}^{-1}$$

$$(4.19)$$

All electromagnetic radiation travels with this velocity in a vacuum, or approximately 7.5 times around the earth per second. Given the value of either the wavelength or the frequency, the other may be calculated. For example, monochromatic violet light ($\lambda = 390$ nm) in a vacuum has a frequency of about 7.6×10^{14} cycles per second, or 760 trillion vibrations per second.

There are two general causes of color in minerals. One is a physical-optical effect having to do with the physical dispersion and refraction of light, which we will discuss briefly at the conclusion of this section. The other cause is far more common, and involves the selective absorption of certain components from the visible spectrum and

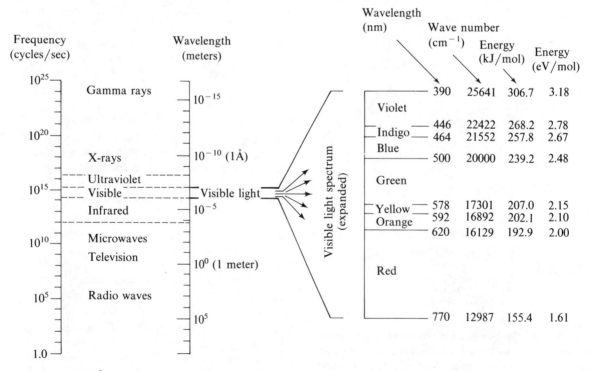

NOTE: 10,000 Å = 1000 nm = 28.59 kcal/mol = 119.621 kJ/mol = 1.24 eV/mol

FIGURE 4.29   The electromagnetic spectrum with visible region expanded to show relationships between color, wavelength, wave number, and energy.

the transmission of the remaining spectral components, which our eyes detect.

Light that contains the full range of wavelengths in the visible region (Figure 4.29) is called "white light," because that is how the eye perceives it. If a mineral passes all wavelengths of white light simultaneously, the mineral appears white or colorless. The clear varieties of quartz and calcite are familiar examples. On the other hand, if all visible wavelengths are absorbed, no transmission of light can occur, and the mineral appears black. *Selective absorption* comes about whenever the energy required to excite an electron from one energy state to another within a mineral corresponds to one or more of the visible wavelengths. When that correspondence occurs, the component is removed by absorption, and the remainder of the spectrum is transmitted. For example, the typical red color of almandine garnet (Figure 4.30) is due to selective absorption of the violet through orange region of the visible spectrum, and the typical green color of olivine is caused by absorption at both ends of the visible

spectrum, allowing only the middle region (green to yellow) to be transmitted.

The excitation of an electron from one state to another is called an *electronic transition*. Many different types of electronic transitions occur, but the ones of importance to the production of color are those that require energies within the visible spectrum. The most important electronic transitions in this regard are among the first row transition elements, namely, Ti, V, Cr, Mn, Fe, Co, Ni, and Cu. The second and third row transition elements and the rare earth elements also exhibit electronic transitions within the visible spectrum, but because of their relative low abundance, they are less important.

Before proceeding further, some important relationships that center around Equation 4.1 should be developed. Combining Equations 4.1 and 4.19, we have

$$E = h\nu = \frac{hc}{\lambda} = hc\bar{\nu} \qquad (4.20)$$

where E is energy, h is Planck's constant, and $\bar{\nu}$ is the reciprocal of the wavelength commonly re-

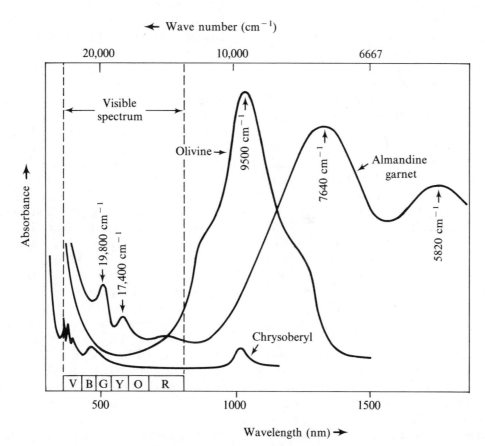

FIGURE 4.30   Absorption spectra of Fe^{2+}-bearing olivine, almandine garnet, and Fe^{3+}-bearing chrysoberyl. Absorption peaks labeled according to wave number. Relative transmission occurs in regions of low absor- bance. (After Loeffler, B. M., and Burns, R. G., 1976. Shedding light on the color of gems and minerals. *American Scientist* 64:636–647.)

ferred to as the wave number. As the wavelength becomes shorter, both the energy and the wave number become larger. Given the wavelength, the wave number, or the frequency, we can calculate the energy. In the case of electronic transitions, we are interested in the energy difference between two electron levels and in either the wavelength, wave number, or frequency of light consequently absorbed.

Among the transition elements, we observe two important types of electronic transitions: (1) the *crystal field transitions* that take place between split 3d orbitals of single ions, and (2) the *charge transfer transitions* that take place between neighboring ions in the same mineral. In both cases, the transitions are induced by white light. These two transition types account for most color in minerals.

Returning to Figure 4.27 will help illustrate some of the factors of crystal field transition that influence the development of color. The specific values of the various crystal field splitting energies, which determine the exact region of the visible spectrum to be absorbed, depend on (1) the particular transition metal ion and its oxidation state, (2) the anion coordination and the cation-anion bond type, and (3) distortions of the coordination polyhedra from ideal tetrahedra, octahedra, cubes, and so forth. These three factors are discussed below. Seemingly small differences in any one of them can dictate whether a mineral is red, yellow, or blue, or whether it will have any color at all.

1. Only those transition metal ions that have partially filled 3d orbitals can have crystal field transitions. Both Zn^{2+} and Cu^{1+} have a full complement of ten 3d electrons, so they have no vacant orbitals to accommodate excited electrons (see Table 4.8). Similarly, Ti^{4+} and Sc^{3+}, which are not included in Table 4.8, have no 3d electrons at all and therefore cannot selectively absorb energy by this mechanism.

   Among the remaining ions, only those transitions that preserve the number of unpaired electrons are generally observed. Transitions that change the number of unpaired electrons are referred to as "spin-forbidden," because they cause a change in the total spin of the cation and, consequently, a theoretically unallowed change in magnetic and related properties. According to quantum mechanics, such transitions ought not to occur, but they are observed due to coupling effects between orbitals. Although these "forbidden" transitions are generally weaker and at higher

energies than the usual "spin-allowed" transitions, they can be the major cause of color in some minerals. For example, the typical red color of almandine garnet ($Fe_3Al_2Si_3O_{12}$) (Figure 4.30) is not due to the major spin-allowed transition in ferrous iron (7640 cm^{-1}), because that absorption is well beyond the visible range. The less intense, spin-forbidden transitions, however, absorb in the shorter wavelength region of the visible spectrum and allow the longer red wavelength to pass.

A good example of the effect of the oxidation state on mineral color is the contrasting behavior of ferric (Fe^{3+}) and ferrous (Fe^{2+}) iron. The typical light yellow color in the mineral chrysoberyl ($Fe^{3+}Al_2BeO_4$) can only be due to ferric iron, yet a look at Table 4.8 shows that Fe^{3+} in the high-spin state has five unpaired electrons. Any transition must involve pairing and hence a change in the number of unpaired electrons. Figure 4.30 shows how the resulting "spin-forbidden" transitions absorb the violet and blue components of the visible range.

2. In garnet and olivine (Fe_2SiO_4), we have examples of the effect of anion coordination on color. Each Fe^{2+} cation in garnet is surrounded by eight oxygen anions in an arrangement that defines the apices of a distorted cube. In olivine, the six oxygens around each Fe^{2+} define an octahedron in which the cation-anion distance (0.212 nm) is relatively short compared with the same pair in garnet (0.222 nm). This means that the d electrons about Fe^{2+} in garnet experience a weaker crystal field than in olivine, which results in a diminished crystal field splitting in olivine. These theoretical predictions are borne out by observation. The major absorption peak of Fe^{2+} in olivine is at higher energy (higher wave number, see Figure 4.30) compared with the peak position in garnet.

Examples of the effect of bond type on color are ruby and emerald. Ruby is Cr^{3+}-bearing corundum (Al_2O_3); emerald is Cr^{3+}-bearing beryl ($Be_3Al_2Si_6O_{18}$). The transition metal cation of interest is trivalent chromium (Cr^{3+}), which substitutes for trivalent aluminum (Al^{3+}) in similar octahedral sites in both minerals. The cation, its oxidation state, and its anion coordination are essentially the same in both minerals, yet ruby is red and emerald is green. Why? The reason most probably lies in the difference in bond type in the two minerals. Corundum is an oxide mineral with predominantly ionic bonds between the cations Al^{3+} and Cr^{3+} and oxygen, whereas beryl is a silicate mineral in which the bonding appears to have a more covalent character, the anion electrons being somewhat delocalized over the structural framework. The consequence for the production of color is that the crystal field splitting is greater in corundum because of the greater electron density around the oxygens. The major absorption due to transitions in Cr^{3+} is thus at a higher energy ($18,000$ cm^{-1}) in ruby, absorbing green and transmitting red. In emerald, the relevant transition is at lower energy ($16,100$ cm^{-1}), absorbing red and transmitting green.

3. The principles of crystal field splitting discussed earlier in this chapter were illustrated by perfectly symmetric crystal fields produced by perfect coordination polyhedra. Many common minerals have distorted crystal fields, because the bond lengths between the transition metal cations and their surrounding oxygens are not identical. The coordination polyhedra defined by those bond lengths are therefore distorted and have lower symmetry. Instead of two different energy levels, there may be as many as five levels, each giving rise to an electronic transition. For example, Fe^{2+} may occupy two different octahedral sites, one distorted and the other not, within the same mineral, and will thus exhibit a more complicated absorption spectrum. In the olivine spectrum of Figure 4.30, the major absorption at 9500 cm^{-1} is due to Fe^{2+} in a distorted octahedral site, whereas the shoulders on the peak at approximately 8300 cm^{-1} and $11,700$ cm^{-1} are due to transitions in a different, more symmetric octahedral site.

In many minerals, *charge transfer transitions* take place when valence electrons transfer back and forth between adjacent ions. The electrons become delocalized in the sense that the valences of the single ions may be "smeared out" and no longer characteristic of the single-valence states. In such cases, crystal field theory cannot describe the observed electronic transitions. Instead, we can think of the electrons as occupying molecular orbitals, and describe their behavior by molecular orbital theory.

Several important charge transfer transitions (also called molecular orbital transitions) have energies within the visible range and therefore cause selective absorption. One of the most important examples is the charge transfer between Fe^{2+} and Fe^{3+} in the micas, amphiboles, pyroxenes, and tourmalines. A characteristic of the absorption spectrum measurements is that the intensity of absorption is greater when the mineral is in a particular orientation relative to a polarized light beam and weaker when the mineral is in other orientations. This phenomenon gives rise to the important property of *pleochroism*, discussed in Chapter 11. The directional nature of the absorption is an important property that distinguishes charge transfer transitions from crystal field transitions. In the former, the neighboring ions involved in the electron transfer may be arranged in a sheet as in micas or in chains within the structure as a whole.

The $Fe^{2+} \rightarrow Fe^{3+}$ charge transfer may also cause complete opacity to visible light. The classic example is magnetite (Fe_3O_4 or $Fe^{2+}Fe_2^{3+}O_4$) in which the valence electrons of both cations are delocalized to such an extent that a broad range of charge transfer energies spans the entire visible range of the spectrum.

The charge transfer of $Fe^{2+} \rightarrow Ti^{4+}$ may be important for selective absorption in certain pyroxenes, in kyanite, in certain micas and tourmalines, and in sapphires, causing their blue color. The uncertainty results because many of these minerals also have Fe^{3+} and perhaps other transition metals, which makes the distinctions between the various absorption mechanisms difficult. For the examples listed in Table 4.9, more than one mechanism may be operative.

Other important electronic transitions within minerals cause color but do not involve transition metals. During growth, fluorite (CaF_2) develops vacancies where F^{1-} normally would be. To preserve electrical neutrality, a single electron occupies the site and undergoes various transitions that absorb all but the violet end of the visible spectrum. Such anion vacancies, when occupied by an electron, are referred to as *electron color centers*.

In contrast to electron color centers, *hole color centers* arise when an electron is missing from a location normally occupied by an electron pair. For example, the color of smoky quartz is caused by such a center. Most quartz has small amounts of Al^{3+} substituted for Si^{4+}, and the charge balance is maintained by an appropriate amount of Na^{1+} or H^{1+}. When irradiated by x-rays or gamma rays, one electron of the pair associated with the Al—O bond is displaced, allowing the remaining electron to undergo transitions that cause the smoky color. The displaced electron becomes "trapped" elsewhere in the structure, but in fact can be returned to the hole center by heating to around $400\,°C$. During this operation, the smoky color disappears. Artificial radiation can cause a variety of colors in diamond (green, yellow, brown), topaz, and other minerals.

The same mechanism operates to produce the purple color of amethyst, but involves Fe instead of Al. In this case, heating causes a change in color either to yellow or green. Upon irradiation, the normal color of amethyst returns.

In each of these examples involving color cen-

TABLE 4.9. *Common Examples of the Three Principal Mechanisms by Which Color Is Generated in Minerals*

		Crystal Field Transitions		
ABSORBING ION	MINERAL NAME	FORMULA	COLOR	VARIETY NAME
V^{3+}	Grossular	$Ca_3Al_2(SiO_4)_3$	Green	Tsavorite
V^{3+}	Zoisite	$CaAl_3HO_2SiO_4Si_2O_7$	Blue	Tanzanite
Cr^{3+}	Beryl	$Al_2Si_6Be_3O_{18}$	Green	Emerald
Cr^{3+}	Uvarovite	$Ca_3Cr_2Si_3O_{12}$	Green	
Cr^{3+}	Chrysoberyl	$BeAl_2O_4$	Green/red	Alexandrite
Cr^{3+}	Corundum	Al_2O_3	Red	Ruby
Mn^{3+}	Spessartine	$Mn_3Al_2(SiO_4)_3$	Pink, yellow	
Mn^{3+}	Beryl	$Al_2Si_6Be_3O_{18}$	Pink	Morganite
Fe^{3+}	Andradite	$Ca_3Fe_2(SiO_4)_3$	Green	Demantoid
Fe^{3+}	Chrysoberyl	$BeAl_2O_4$	Yellow	
Fe^{2+}	Olivine	$(Fe, Mg)_2SiO_4$	Green	Peridot
Fe^{2+}	Almandine	$Fe_3Al_2(SiO_4)_3$	Red	
Co^{2+}	Spinel	$MgAl_2O_4$	Blue	
Ni^{2+}	Bunsenite	NiO	Green	
Cu^{2+}	Dioptase	$Cu_6Si_6O_{18} \cdot 6H_2O$	Green	
Cu^{2+}	Turquoise	$CuAl_6(PO_4)_4(OH)_8 \cdot 4H_2O$	Blue	

		Charge Transfer Transitions		
ION PAIR	MINERAL NAME	FORMULA	COLOR	VARIETY NAME
$Fe^{2+} \rightarrow Fe^{3+}$	Vivianite	$Fe_3(PO_4)_2 \cdot 8H_2O$	Blue green	
$Fe^{2+} \rightarrow Fe^{3+}$	Biotite		Brown green	
$Fe^{2+} \rightarrow Fe^{3+}$	Beryl	$Al_2Si_6Be_3O_{18}$	Blue yellow	Aquamarine
$Fe^{2+} \rightarrow Fe^{3+}$	Cordierite	$(Fe, Mg)_2Al_4Si_5O_{18}$	Blue	Iolite
$Fe^{2+} \rightarrow Ti^{4+}$	Corundum	Al_2O_3	Blue	Sapphire
$Fe^{2+} \rightarrow Ti^{4+}$	Kyanite	Al_2OSiO_4	Blue	
$O^{2-} \rightarrow Cr^{6+}$	Crocoite	$PbCrO_4$	Orange	
$O^{2-} \rightarrow V^{5+}$	Vanadinite	$Pb_5(VO_4)_3Cl$	Orange	
$O^{2-} \rightarrow Fe^{3+}$	Beryl	$Al_2Si_6Be_3O_{18}$	Yellow	Heliodore

		Color Center Transitions		
CENTER	MINERAL NAME	FORMULA	COLOR	VARIETY NAME
Electron	Fluorite	CaF_2	Violet	
Hole-Al^{3+}	Quartz	SiO_2	Brown, black	Smoky quartz
Hole-Fe^{3+}	Quartz	SiO_2	Violet	Amethyst

ters, energy in the form of heating is required to return the mineral to its original state. In some irradiated beryl and topaz, the visible spectrum has sufficient energy to cause the colors to fade over periods of days. A related phenomenon is *fluorescence*, whereby a mineral will emit visible light in response to being irradiated at a higher energy level, as with an ultraviolet lamp. Examples of fluorescent minerals include scheelite, fluorite, and willemite. If the visible emission lasts for hours or days, the mineral or material is said to be *phosphorescent*.

The final category of color-producing mechanisms does not involve electrons at all. Physical scattering of light, discussed in more detail in Chapters 10 and 11, usually occurs because of very fine particles in the mineral. The properties of opalescence and pearly luster are caused by scattering of this sort. Chatoyancy as observed in cat's-eyes and tiger's-eyes and asterism in star corundum, garnet, and quartz are also caused by physical scattering. The quality of fire observed in opal, diamond, zircon, and rutile is not caused by random scattering but rather by diffraction and subsequent interference of light waves.

ADDITIONAL READINGS

Kittel, C., 1966. *Introduction to solid state physics*. New York: John Wiley and Sons.

Nassau, K., 1978. The origin of color in minerals. *American Mineralogist* 63:219–229.

Tosi, M. P., 1965. Cohesion of ionic solids in the Born model. In *Solid State Physics*. Seitz, F., and Turnbull, D., eds. New York: Academic Press.

CHAPTER FIVE

CRYSTAL STRUCTURES

The concepts of symmetry presented in Chapters 2 and 3 and the principles of chemical bonding in Chapter 4 are joined in the study of crystal structures. An understanding of why and how atoms or ions combine in highly symmetric, ordered arrangements is fundamental to all physical and chemical properties of minerals. The most obvious properties such as cleavage, crystal form, point group symmetry, and even the chemical formula depend on how atoms are arranged in three dimensions. The relative thermal and pressure stabilities of minerals, and hence their value in reconstructing geologic history, are highly dependent on crystal structure. In the case of polymorphism, the same chemical constituent (e.g., SiO_2) can have two or more structures (e.g., quartz, tridymite, cristobalite, coesite, and stishovite) depending on the conditions of formation. We must therefore have a thorough understanding of the basic principles that underlie crystal structures, for without them, our appreciation of why and how minerals behave in the way they do is severely limited.

The first step toward this objective is to understand the known structures of minerals and the systematics of the relationships among them. The principal sources of information are crystal structure determinations (Chapter 10), which are now available for all of the common minerals. Because there are thousands of different crystal structures, even the most efficient description of each is impossible within a single volume. The alternative is a classification scheme based on fundamental and natural properties of crystal structures. That is, the criteria used for the classification must depict fundamental concepts related to the bonding, the energy, and the geometry of crystal structures. At the same time, a successful classification of crystal structures must be simple, include all structures, and offer a quick and easy visualization of important features.

In an attempt to provide such a classification system, we discuss the crystal structures of minerals in terms of the types and symmetries of their atomic bonds. Three end-members will be identified, between which an almost continuous range of properties exists.

1. Polyhedral-Frame Structures Most of the important rock-forming minerals, notably the silicates, are in this category. All of the structures are a direct consequence of predominantly ionic bonds between the constituent ions. As a result of the bonding, anions tend to group around cations in a highly symmetric manner to define coordination polyhedra. Common examples are the silica tetrahedron ($(SiO_4)^{4-}$) and divalent cation octahedra (e.g., $(MgO_6)^{10-}$). By sharing apical oxygens, these polyhedra link together in various ways to define a structural frame that possesses at least half of the total bonding energy of the mineral. The resulting frame is relatively strong and has an important influence on most physical and chemical properties.

2. Symmetrically Packed Structures In these structures, either the bonds between atoms are *nondirectional*, or the bond directions are *highly symmetrical*. The metallic bond meets this requirement as do many examples of ionic and covalent bonds. Atoms that are constrained by this type of bonding form highly symmetrical structures in which the atoms are packed together in various symmetrical ways.

The symmetrically packed structures are divided into monatomic and multiatomic groups. Most native metals are monatomic. Their atoms are all in symmetrical sites and their structures can be analyzed in terms of symmetrical sheets and their symmetrical stackings. If the atoms are in contact in and between the sheets, a highly efficient packing called *closest packing* results. If atoms lose contact within the sheets but retain contact between the sheets, the atoms are called *close-packed*.

In the multiatomic group we find both cations and anions. Many oxides, sulfides, hydroxides, halides, and even most of the important silicates considered as polyhedral-frame structures are in this category. The anions are in symmetrically packed sites, and the cations occupy the voids between them. The anions are not in contact within or between the sheets. The *symmetry of the anion packing* is the basic characteristic.

3. Molecular Structures These structures are composed of atoms characterized by strongly directional and low symmetry bonds. These *asymmetric* bonds form strong clusters, chains, and layers of atoms that behave as discrete units. These units are connected by much weaker residual bonds to form three-dimensional networks.

Figure 5.1 is a simplified illustration of the three major categories of structure type. Molecular structures are relatively unimportant in mineralogy and will not be discussed in great detail. As the majority of mineral structures are symmetrically packed or polyhedral-frame or both, we will describe thoroughly the principles that underlie these structures, and we will then give a relatively concise and systematic development for both structure types. Distortions of symmetrically packed structures due to *bond asymmetry* are found in certain simple and complex sulfides and in various oxygen compounds. These "molecular" distortions due to bond symmetry are shown in Figure 5.1 as transitional from symmetrically packed and polyhedral-frame structures to molecular structures.

In addition, Figure 5.1 shows how many of the common rock-forming minerals are related to the ideal structure types. The basis for this classification will not become apparent until later in the chapter. Our purpose now is to illustrate some structural similarities between certain groups of minerals, for example, the clays, micas, pyroxenes, and amphiboles. Each group can and should be discussed in its own right because of its importance in petrology, geochemistry, geophysics, and related disciplines. This we do in the appropriate sections of Part II of the text. These individual classifications are important, but they lack the generality and breadth of application that we consider in this chapter.

The relative position of the various polymorphs of silica in Figure 5.1 illustrates another general advantage of a comprehensive structural classification. The relative significance of symmetrical packing in these structures is expressed by the symmetrical packing index, SPI, explained later in this chapter. The index is related to an adjusted packing efficiency and the density of the symmetrically packed anions of these structures. It decreases toward the polyhedral-frame corner of the triangular diagram.

We will learn in Chapter 8 that minerals respond to increased pressure in the earth by reducing their volume. For any particular mineral, that reduction has a limit beyond which a denser atomic packing arrangement is necessary. The sequence stishovite-coesite-quartz-tridymite, which is based *solely* on structural considerations, is thus a sequence consistent with a response to decreasing pressure. Similar generalities can also be made with respect to temperature, entropy, and symmetry. The relationships between the structure of a mineral and its stability in various pressure and temperature environments are far more complicated than can be represented simply along a line in a diagram. Before we can begin to understand these relationships, the principles that underlie the systematic description of crystal structures must first be understood.

POLYHEDRAL-FRAME STRUCTURES

We learned in the previous chapter that as two oppositely charged ions are brought together, their equilibrium interionic distance will be determined by the balance of electrostatic attractive forces between their outer electron charges and repulsive forces between their nuclei. The ions do not ap-

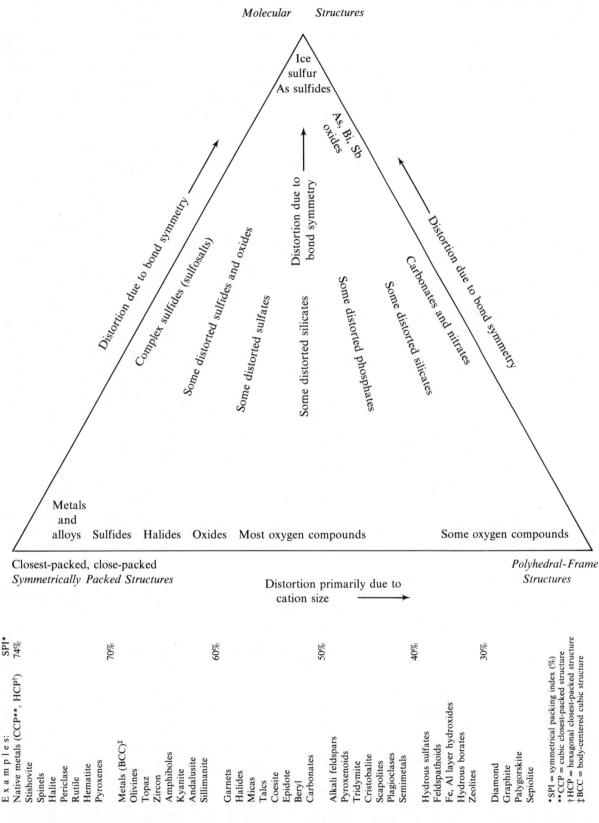

FIGURE 5.1 Summary of the crystal structures of minerals in terms of three idealized structure types. Most of the common rock-forming minerals lie along the base **between symmetrically packed and polyhedral-frame structures. Examples tabulated along base arranged in order of decreasing packing efficiency from left to right.**

proach closer than this distance. In three dimensions, ions with relative positions that follow the principles of ionic bonding tend to form highly symmetric polyhedra that have the same interionic distances. Tetrahedra and octahedra are the most common structure types, but triangles, cubes, and other types are also important. These coordination polyhedra link together in various ways to form the polyhedral-frame structures. They include all of the rock-forming silicates as well as many borates, sulfates, phosphates, tungstates, oxides, and hydroxides.

Because the bonding is primarily ionic, we have definite principles that we can follow to understand the polyhedral-frame structures. These principles were formulated in the late 1920s by Linus Pauling and W. L. Bragg who used them to determine many of the first crystal structures. They are now known as the five *Pauling's rules.*

> *Rule 1. Interatomic Distances.* A coordination polyhedron of anions is formed about each cation, the cation-anion distance being determined by the radius sum and the coordination number of the cation by the radius ratio.

When the bonding is dominantly ionic, each cation in a structure will tend to attract or *coordinate* as many anions as will fit around it. The number of anions that will fit and still retain equal distances between themselves and between the central cation is called the *coordination number*, abbreviated CN. Tetrahedra have CN = 4, octahedra have CN = 6, and so on.

If we focus our attention on a particular cation, its coordination number is determined solely by the equilibrium interionic distance to the coordinated anions. If the distance is short, fewer anions can coordinate than if the distance is large. Ionic bonds that are relatively short are also relatively strong, because the cation generally has a high charge and hence a small radius. This is the reason why coordination polyhedra with small values of CN (e.g., triangles, tetrahedra) play such an important role in determining most mineral properties. The bonds are not easily broken.

Carbon (C^{4+}, $r = 0.008$ nm) and silicon (Si^{4+}, $r = 0.034$ nm) when coordinated with oxygen (O^{2-}, $r = 0.132$ nm) provide good examples of how ionic radii dictate the type of coordination polyhedron. In the one case, the carbonate minerals such as calcite ($CaCO_3$) form. In the other case, the silicates such as forsterite (Mg_2SiO_4) result. Figure 5.2a illustrates the example of the smaller C^{4+}. In order for the C—O bonds to be of equal length and the O—O bonds to be of equal length, the CN must be 3. This condition is equivalent to having all of the ions in contact with each other. The ratio of the cation radius to the anion radius (R_c/R_a), known as the *radius ratio*, has a value of 0.155 in this case. This is the smallest possible value for threefold coordination, because the anions are already in contact. The cation cannot be smaller, or it would lose contact with the anions and would thus be in violation of the ionic bonding requirements. The cation can be larger, but that would require that the anions be pushed apart, losing contact with each other. We now understand why the range of radius ratio values consistent with a particular CN extends from the minimum value given in Table 5.1 to infinity.

In practice, the radius ratio value does not extend much beyond the minimum value for a particular CN. As the cation radius increases, the anions separate far enough from each other to allow another anion to fit in. When the radius ratio increases to 0.225, four anions will fit around a central cation to form a perfect tetrahedron

TABLE 5.1. *Coordination Number, Coordination Polyhedra, and Ranges of Permissible R_{cation}/R_{anion}*

| CN | Polyhedron | | R_c/R_a | Examples |
	IONIC	COVALENT		
2	Line		0–∞	Cuprite (oxide)
3	Triangle		0.155–∞	Carbonates, nitrates
4	Tetrahedron		0.225–∞	Silicates
4		Square	0.414–∞	Cooperite (sulfide)
5		Tetrahedral pyramid	0.414–∞	Millerite (sulfide)
6	Octahedron		0.414–∞	Halite (halide)
6		Trigonal prism	0.529–∞	Molybdenite (sulfide)
8		Square antiprism	0.625–∞	Scheelite (tungstate)
8	Cube		0.732–∞	Fluorite (halide)
12		Cuboctahedron	1.00–∞	Metals, alloys
12		Disheptahedron	1.00–∞	

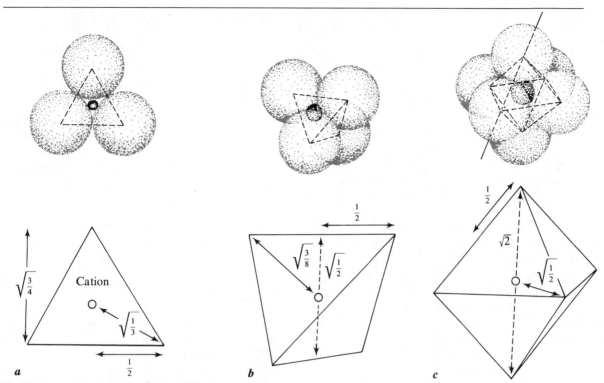

FIGURE 5.2 **Geometry of (a) trigonal, (b) tetrahedral, and (c) octahedral coordination polyhedra. Fractions based on an anion-anion distance of unity.**

(Figure 5.2b). The four anions are in mutual contact, and all are in contact with the cation. This is exactly the situation with Si^{4+} and its coordination to oxygen to form the silica tetrahedron. The radius ratio value in this case is 0.257, clearly too large for CN = 3 but too low for sixfold coordination (Figure 5.2c). Ionically bonded cations always have a tendency to form the highest coordination possible.

The type of coordination polyhedron that forms for a particular cation and group of anions is not dependent solely on the radius ratio. The type and symmetry of bonding are important. Minerals such as the silicates, in which the bonding is principally ionic, tend to form tetrahedra, octahedra, and cubes almost exclusively. Minerals such as the sulfides, in which the bonds are dominantly covalent and highly directional, commonly form square coordination (e.g., cooperite, Table 5.1) or other less symmetrical polyhedra. In these minerals, the central cation coordinates to fewer anions than would ideally fit because of the directional nature of the covalent orbitals.

Rule 2. Electrostatic Valency Principle. In a stable coordination structure, the total strength of the valency bonds that reach an anion from all neighboring cations is equal to the charge of the anion.

This rule is a direct consequence of ionic bonding. Because the bonds are electrostatic, the total bonding capacity of a cation is proportional to its charge. A measure of the electrostatic strength of a single bond is simply the total cation charge Z divided by the total number of anions to which the cation must bond, namely, the coordination number. In the structure of rutile (TiO_2), for example, Ti^{4+} is in octahedral coordination with oxygen. Each Ti—O bond has a strength of $Z/CN = 4/6 = 2/3$ (Figure 5.3). Each oxygen in the structure has three neighboring Ti^{4+} cations such that their collective bond strength ($3 \times 2/3$) equals the oxygen charge of -2.

Another example of Pauling's second rule is talc ($Mg_3(OH)_2Si_4O_{10}$). Cations of Mg^{2+} are in sixfold coordination (Figure 5.4a), and Si^{4+} has the usual fourfold coordination with oxygen (Figure 5.4b). The structure has two symmetrically

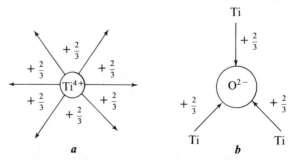

FIGURE 5.3 **Distribution of (a) the valence of Ti^{4+} toward the six-coordinated oxygens and (b) the total bonds reaching each oxygen in rutile (TiO_2).**

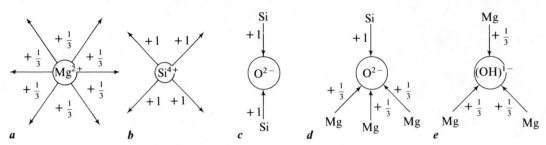

FIGURE 5.4   Distribution of (a) octahedrally coordinated Mg^{2+} and (b) tetrahedrally coordinated Si^{4+} valences. Total bonds reaching the two oxygens (c and d) and the hydroxyl (e) balance the negative anion charge in talc $(Mg_3(OH)_2Si_4O_{10})$.

nonequivalent oxygens, one bonded to two Si^{4+} (Figure 5.4c), and the other to one Si^{4+} and three Mg^{2+} (Figure 5.4d). The $(OH)^{1-}$ is bonded to three Mg^{2+} cations (Figure 5.4e). In each case illustrated in Figures 5.4c, 5.4d, and 5.4e, the anion charge is neutralized by the sum of the bond strengths from neighboring cations in conformity with the second rule.

One consequence of these relationships is that anions tend to be locally charge-balanced. That is, no excess negative charge is associated with an anion once it is bonded to its closest cations. The mineral as a whole must, therefore, be charge-balanced, which is true of all minerals.

Numerous exceptions to the local charge-balance concept exist. One is andalusite $(AlAlOSiO_4)$ in which three of the four symmetrically different oxygens have net charges locally (Figure 5.5). The reason for the exception is probably the unusual fivefold coordination of one of the Al^{3+} cations and the consequent bond strength of 3/5 (Figure 5.5a) instead of the usual

value of 1/2 for six-coordinated Al^{3+} (Figure 5.5b). The Si^{4+} has the usual fourfold coordination (Figure 5.5c) with the oxygens shown in Figures 5.5d–5.5g. The first of these oxygens is also bonded to the Al^{3+} shown in Figure 5.5a, and consequently, this oxygen has a net $+1/5$ charge. The two oxygens in Figures 5.5e and 5.5f are symmetrically equivalent, and each has a net $+1/10$ charge. The oxygen in Figure 5.5h has no charge imbalance, whereas the oxygen in Figure 5.5g has an excess charge of $-2/5$. Overall charge balance in the structure is achieved because the sum of the excess charges on the oxygens is zero.

According to Pauling's second rule, the number and kinds of coordination polyhedra that can meet together at a point are severely limited. For example, no more than two Si^{4+} tetrahedra can share a common oxygen, even though radius ratio considerations alone would permit three, four, or more Si^{4+} to coordinate to a central oxygen. Each Si—O bond contributes an electrostatic strength

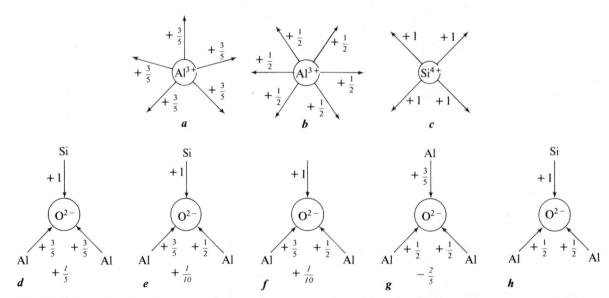

FIGURE 5.5   Violation of the electrostatic valency rule in structure of andalusite (Al_2SiO_5). (a–c) Distribution of cation valence. (d–g) Sum of bonds reaching four oxygens does not balance negative oxygen charge. (h) Fourth oxygen is balanced by two Al and one Si bond.

of $4/4 = 1$, so two Si—O bonds will just satisfy the -2 charge of the oxygen. Similarly, exactly three divalent cation octahedra will share a common oxygen with a Si^{4+} tetrahedron. If there are fewer than three, a net negative charge remains on the oxygen. If more than three, a net positive charge remains on the oxygen. The same is true for two trivalent cation octahedra and a Si^{4+} tetrahedron.

These restrictions on possible combinations of polyhedra are important in explaining the uniqueness of both the structures and chemistries of ionic crystals. As examples of possible combinations, one Si^{4+} tetrahedron and three Mg^{2+} octahedra share a common oxygen in the minerals forsterite (Mg_2SiO_4) and phlogopite ($KMg_3(OH)_2Si_3AlO_{10}$). One Si^{4+} tetrahedron and two Al^{3+} octahedra share a common oxygen in the minerals kyanite (Al_2OSiO_4), muscovite ($KAl_2(OH)_2Si_3AlO_{10}$), and topaz ($Al_2F_2SiO_4$).

Nearly all silicate minerals tend to conform to Pauling's first two rules despite the fact that the Si—O bond appears to be about 50% covalent based on electronegativity arguments.[1] The mea-

---

[1] For an interesting discussion of this problem, see Pauling, L., 1980. The nature of silicon-oxygen bonds. *American Mineralogist* 65:321.

sured Si—O bond lengths in actual structural refinements are also consistent with ionic radii for Si^{4+} and O^{2-} ions, so the extent to which the bond is covalent is debatable. The important fact as far as the silicate structures are concerned is that the Si—O bonds *behave* as if they are ionic.

Rule 3. Sharing of Polyhedral Elements, I. The existence of edges, and particularly of faces, common to two coordination polyhedra decreases the stability of ionic structures.

This rule is again a direct outgrowth of the electrostatic forces that bind ionic crystals together. The most stable configuration occurs when two polyhedra share only a corner, because then the two central cations, which tend to repel each other, are as far apart as possible. The first column of Table 5.2 lists these maximum distances for various polyhedra when the O—O distance is arbitrarily taken as unity. In each case, there is also a minimum distance between cations with respective polyhedra sharing one common corner. This distance is determined by rotating the polyhedra toward each other until their oxygens are in contact. The cations can come no closer unless either edges or faces of the polyhedra are shared.

In the case of edge-sharing (Table 5.2), the

TABLE 5.2. *Maximum and Minimum Cation-Cation Distances for Various Coordination Polyhedra*

Polyhedron	Corner-Sharing MAXIMUM	Corner-Sharing MINIMUM	Edge-Sharing MAXIMUM	Edge-Sharing MINIMUM	Face-Sharing
Triangle	$\sqrt{4/3} = 1.15$	$1/3 = 0.33$	$\sqrt{1/3} = 0.58$	$1/6 = 0.17$	—
Tetrahedron	$\sqrt{3/2} = 1.22$	1.00	$\sqrt{1/2} = 0.71$	$2/3 = 0.67$	$\sqrt{1/3} = 0.58$
Octahedron	$\sqrt{2} = 1.41$	$1 + \dfrac{2\sqrt{3}-2}{6-\sqrt{3}} = 1.34$	1.00	—	$\sqrt{2/3} = 0.82$
Cube	$\sqrt{3} = 1.73$	$\sqrt{3/2} = 1.22$	$\sqrt{2} = 1.41$	$\sqrt{1/4} + \sqrt{3/4} = 1.37$	1.00

shorter cation-cation distance decreases the stability of the structure because of the additional energy required to keep the cations in such close proximity. For the same reason, the face-sharing geometry is the least stable.

> **Rule 4. Sharing of Polyhedral Elements, II.** In a crystal containing different cations, those with large valence and small coordination tend not to share polyhedral elements with each other.

This statement is a corollary of rule 3, emphasizing the fact that highly charged cations will tend to be distributed as far apart as possible in a crystal structure. The effect of rule 4 will be stronger for cations having low coordination than for those with higher coordination, even though the cation charge may be the same. For example, Al^{3+}, because of its ionic radius relative to oxygen, may have either fourfold or sixfold coordination. Table 5.2 shows that the centers of corner-shared octahedra are farther apart than those of corner-shared tetrahedra. The centers of corner-shared cubes are farther apart still. The same rule applies for edge-sharing and face-sharing polyhedra: the higher the CN, the farther apart are the centrally coordinated cations and hence the edge-sharing polyhedra will be more common than face-shared polyhedra.

Pauling's Rule 4 is well illustrated by the silicate minerals. Among the hundreds of stable silicates, we do not find one example of either edge-sharing or face-sharing of Si tetrahedra. We do, however, find numerous examples of divalent cation octahedra that share edges, for example, the Ti octahedra in rutile (TiO_2, Figure II.93). We also find examples of face-sharing octahedra such as the Fe^{3+} octahedra in hematite (Fe_2O_3, Figure II.91). In every case, the tendency for sharing polyhedral elements increases as the coordination number increases and as the cation charge decreases.

> **Rule 5. Principle of Parsimony.** The number of essentially different kinds of constituents in a crystal structure tends to be small.

This fifth and final rule is a consequence of all the previous rules. The number of different ions and of different symmetrical coordination polyhedra that can form in a crystal is limited. Usually we find no more than two or three different coordination polyhedra in a crystal structure. The number of crystallographically different sites is, therefore, small, and in small integer ratios to each other. This is the fundamental reason why the various cations and anions in chemical formulas are generally in small integer ratios to each

other. Their relative abundance is controlled by the available sites in the structure.

CLASSIFICATION OF POLYHEDRAL-FRAME STRUCTURES

On the basis of Pauling's rules we can now understand why minerals with predominantly ionic bonds tend to form polyhedral-frame structures. Cation repulsion and the tendency of coordination polyhedra not to share edges and faces force the polyhedra away from each other to form an open frame consisting of various polyhedral linkages. These linkages, called the *primary frame*, are the backbone of the structure. They must contain at least one half of the anion bonding strength. The frame is therefore relatively strong, resisting deformation and controlling physical properties such as cleavage and crystal form.

The simplest polyhedral-frame structure is exemplified by quartz (SiO_2), in which every silicon is four-coordinated to oxygen and every oxygen is two-coordinated to silicon. A close examination of Figure II.2 shows a three-dimensional network of tetrahedra, each linked to another by the sharing of a common apical oxygen. In this case, the frame contains all of the anion bonding strength, and has no net charge. Primary frames become less extensive and develop a net negative charge as the number of shared tetrahedral oxygens decreases, or as the central cations are replaced by ones of lower valence. In the silicate minerals, several very important primary frames result. These include the network silicates (e.g., quartz and feldspar), the layer silicates (e.g., micas and clays), the chain silicates (e.g., pyroxenes and amphiboles), the ring silicates (e.g., tourmaline), and the isolated group silicates (e.g., epidote and olivine).

These basic types of tetrahedral-frame structures are summarized in Tables 5.3, 5.4, 5.5, and 5.6. In each case, the primary frame is identified as having one half or more of the oxygen valence bonded to silica. The remaining oxygen shows up as a net negative charge on the frame itself, which is then available for bonding to other cations. For example, the primary frame of the pyroxenes is a single infinite chain that has an $(SiO_3)^{2-}$ repeat unit. Of the six valence electrons, four are required to balance the quadrivalent silicon, leaving a net -2 charge per Si on the frame. The equivalent of one divalent cation, most commonly Fe^{2+}, Mg^{2+}, and Ca^{2+}, outside the primary frame can then contribute the $+2$ charge per Si required for electrical neutrality. Because these cations are usually in octahedral coordination (CN = 6), six octa-

TABLE 5.3. *Classification of Zero-Dimensional Tetrahedral Frames*

Groups	Sharing Coefficient	RU	Charge of Frame	Examples	Figure 5.6*
Single tetrahedra	1.000	1	$(SiO_4)^{4-}$	Garnets	**a**
Double tetrahedra	1.250	2	$(Si_2O_7)^{6-}$	Ilvaite	
	1.250	2	$(Si_2O_7)^{6-}$	Tilleyite	**b**
	1.250	2	$(Si_2O_7)^{6-}$	Thortveitite	
Triple tetrahedra	1.333	3	$(Si_3O_{10})^{8-}$	Kinoite	**c**
Mixed **a** + **b**	1.167	3	$(Si_3O_{11})^{10-}$	Epidote	**d**
Mixed single and quintuple tetrahedra	1.400	6	$(Si_5AlO_{20})^{17-}$	Zunyite	**e**
Single ring	1.500	3	$(Si_3O_9)^{6-}$	Benitoite	**f**
	1.500	4	$(Si_4O_{12})^{8-}$	Boatite	**g**
	1.500	6	$(Si_6O_{18})^{12-}$	Tourmaline	**h**
	1.500	6	$(Si_6O_{18})^{12-}$	Dioptase	
	1.500	8	$(Si_8O_{24})^{16-}$	Muirite	**i**
Double ring	1.750	4	$(Si_8O_{20})^{8-}$	Ekanite	**j**
Mixed rings	1.500	9, 3	$(Si_{12}O_{36})^{24-}$	Eudyalite	
Complex	1.500	10	$(Si_8B_2O_{24})^{10-}$	Axinite	**k**

*Examples are illustrated in Figure 5.6.

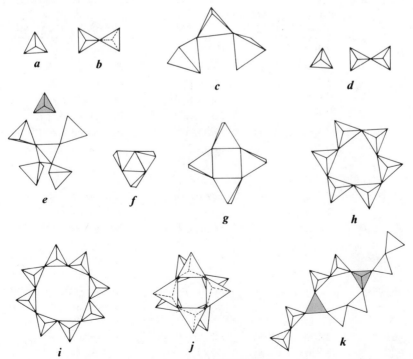

FIGURE 5.6 Linkage patterns of tetrahedra within zero-dimensional frames. Mineral examples tabulated in Table 5.3.

hedral bonds ($6 \times 1/3$) are required to neutralize the frame.

In some polyhedral-frame structures the space available for additional cations outside the frame may be restricted by the linkage pattern of tetrahedra. In feldspars the coordination of K^{1+}, Na^{1+}, and Ca^{2+} is irregular for this reason. If there are no such restrictions, the additional cations will form regular polyhedra and may link together to form a *secondary frame*. In pyroxenes, for example, the primary frame is a single chain of tetrahedra, whereas the secondary frame consists of a double chain of regular and distorted octahedra (Figure II.33).

The chemical formulas of minerals as written in this book convey not only the elemental composition in the appropriate proportions but also the positions of each atom in the structure. For example, the formula of prehnite is $CaAl(OH)_2Si_3AlO_{10}$. The *primary frame* consists of Si and all elements to the right of Si in the formula. This particular frame (Figure II.28) is characteristic of the layer silicates (p. 331) and consists of Si^{4+} and Al^{3+} bonded to four closest oxygen ions within a two-dimensional layer structure (Table 5.5). The primary frame can be written as $(Si_3AlO_{10})^{5-}$ to denote the excess negative charge that must be neutralized by bonding with other cations in the structure.

The *secondary frame* of the prehnite structure consists of the unit $Al(OH)_2$ in which two of the six octahedrally coordinated anions of Al^{3+} are hydroxyl. The remaining four anions are oxygens that are shared with the primary frame. Depending on the structure, the secondary frame may consist of isolated polyhedra linked to the primary frame, or coherent chains or layers of polyhedra.

The final part of the chemical formula is the *ternary frame*, which almost always consists of those cations with irregular coordination that occupy the space available between the primary and secondary frames. In prehnite, Ca^{2+} is the only cation forming the ternary frame, and is placed first in the formula to convey that fact. In all polyhedral-frame structures, the primary frame is always present, but either the secondary or ternary frame may be absent.

The distinctive features of polyhedral-frame structures were recognized within a decade after the first mineral structures were determined. Because the silicates are the most important of the polyhedral-frame structures, they formed the basis for the earliest classifications. Both F. Machatschki (1928) and W. H. Bragg (1930) re-

TABLE 5.4.   *Classification of One-Dimensional Tetrahedral Frames*

Groups	Sharing Coefficient	RU (loops)	Charge of Frame	Examples	Figure 5.7*
Single chain	1.500	1	$(SiO_3)^{2-}$	"Iscorite"	**a**
	1.500	2	$(Si_2O_6)^{4-}$	Pyroxenes	**b**
	1.500	3	$(Si_3O_9)^{6-}$	Pectolite	
	1.500	3	$(Si_3O_9)^{6-}$	Wollastonite	**c**
	1.500	4	$(Si_4O_{12})^{8-}$	Pentagonite	**d**
	1.500	5	$(Si_5O_{15})^{10-}$	Rhodonite	**e**
	1.500	6	$(Si_6O_{18})^{12-}$	Stokesite	**f**
	1.500	7	$(Si_7O_{21})^{14-}$	Pyroxmangite	**g**
Double chain	1.750	1(4)	$(SiAlO_5)^{2-}$	Sillimanite	**h**
	1.625	2(6)	$(Si_4O_{10})^{4-}$	Amphiboles	**i**
	1.600	5(6–8)	$(Si_{10}O_{28})^{16-}$	Inesite	**j**
Triple chain	1.667	2(6)	$(Si_6O_{17})^{10-}$	Jimthomsonite	**k**
Winged chain	1.500	2 + 2	$(Si_4O_{12})^{8-}$	Astrophyllite	**l**
	1.500	4 + 2	$(SiAl_3O_{12})^{11-}$	Sapphirine	**m**
Complex	1.583	(2–6)	$[Si_{12}O_{31}(OH)]^{17-}$	Howieite	**n**
	1.750	(4–6)	$(Si_8O_{20})^{8-}$	Narsarsukite	**o**
	1.625	5(2–5–5)	$[Si_2B_2O_{10}(OH)]^{7-}$	Hellandite	**p**

*Examples are illustrated in Figure 5.7.

ferred to single chain, double chain, layer, and network silicates. This classification is still in use, although it has been greatly extended. Several Latin and Greek names are also used frequently to describe the same features. Single tetrahedral structures are called orthosilicates or nesosilicates, and double tetrahedral structures are called diorthosilicates or sorosilicates. Structures with rings of tetrahedra are called cyclosilicates. Those with single or double chains of tetrahedra are called inosilicates, and those with layers of tetrahedra are referred to as phyllosilicates.

A more extensive and general classification of polyhedral structures (Tables 5.3, 5.4, 5.5, and 5.6) is based on the following five features:

1. Coordination polyhedron of the frame
2. Dimensional extent of the frame
3. Polyhedral groups within the frame
4. Number of shared polyhedral corners
5. Number of polyhedra in an asymmetric unit or polyhedral loop

1. The kind of *coordination polyhedron* that makes up a structural frame is the most fun-

damental feature. The frame will consist of only one kind of polyhedra, most commonly either tetrahedra or octahedra. All cations that occupy the frame are considered part of it. The best examples are the tetrahedral frames of the silicate minerals.

2. The next most important feature of the polyhedral frame is its *dimensional extent*. A single polyhedron (e.g., as in olivine, Mg_2SiO_4) or a small group of polyhedra (e.g., as in epidote, $Ca_2Al_2FeHO_2SiO_4Si_2O_7$) are considered to be *zero-dimensional* (Table 5.3), because the geometric pattern of polyhedral linkage is terminated in all directions. A single or double chain of tetrahedra, such as in the pyroxenes and amphiboles, is considered *one-dimensional* (Table 5.4), because the linkage pattern extends indefinitely in a single dimension. A layered pattern of tetrahedra, such as in the micas and clays, is *two-dimensional* (Table 5.5). A network of tetrahedra extended indefinitely in all directions is *three-dimensional* (Table 5.6).

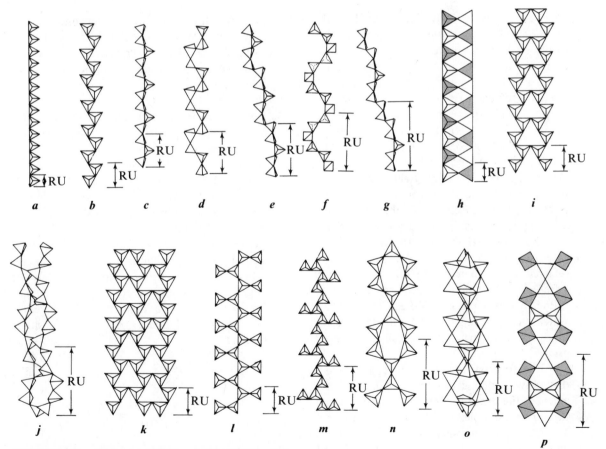

FIGURE 5.7 **Linkage patterns of tetrahedra within one-dimensional frames. Mineral examples tabulated in Table 5.4.**

TABLE 5.5.　*Classification of Two-Dimensional Tetrahedral Frames*

Groups	Sharing Coefficient	Periodicity	Charge of Frame	Examples	Figure 5.8*
Single layer	1.625	12	$[Si_5O_{14}F_2(OH)]^{13-}$	Zeophyllite	**a**
	1.750	4-6-8	$(Si_2O_5)^{2-}$	Dalyite	**b**
	1.750	4-8	$(Si_2O_5)^{2-}$	Apophyllite	**c**
	1.750	4-8	$(Si_2O_5)^{2-}$	Cavansite	
	1.750	4-8	$(Si_2O_5)^{2-}$	Gillespite	**d**
	1.750	4-8	$(SiBO_5)^{1-}$	Datolite	
	1.750	5-8	$(SiO_2OH)^{1-}$	Okenite	**e**
	1.750	6	$(Si_2O_5)^{2-}$	Clays	**f**
	1.750	6	$(Si_3AlO_{10})^{5-}$	Micas	
	1.750	6	$(Si_3AlO_{10})^{5-}$	Prehnite	
	1.750	6-6-6	$(Si_2O_5)^{2-}$	Palygorskite	**g**
	1.833	5	$(SiAl_2O_7)^{8-}$	Gehlenite	**h**
	2.75	3	$(SiAl_2Mn_4O_{14})^{10-}$	Catoptrite	**i**
Double layer	1.875	4-6-8	$(Si_4O_9)^{2-}$	Carletonite	**j**
	2.000	4-6	$(Si_3AlO_8)^{1-}$	Cymrite	
Complex layer	1.833	3-6-6	$(Si_{17}AlO_{42})^{13-}$	Zussmanite	**k**

*Examples are illustrated in Figure 5.8.

3. The *polyhedral groups* within each *n*-dimensional linkage pattern are usually distinctive. For example, among the zero-dimensional frames are many different groups consisting of one or more polyhedra in different arrangements. The *group* may consist of a *pair* of tetrahedra (Figure 5.6b) as in ilvaite (Table 5.3), a *ring* of tetrahedra as in tourmaline (Table 5.3), or more complicated combinations. A ring can be single or double, or it may have additional attached polyhedra, in which case the term *winged* is used. We refer also to single chains, double chains, triple chains, and other multiple chains as distinctive groups. If the pattern is too complicated to describe in these terms, it is simply referred to as *complex*. In other cases, two different groups may be present in a single frame. An example is epidote (Figure II.62), which has both single and double tetrahedral groups. Such structures consist of *mixed groups*.

Although a particular structure may have a distinctive group geometry, the dimensional extent is the overriding criterion for classification. For example, beryl (Figure II.17 and Table 5.6) has prominent six-membered rings within its structure and is commonly referred to as a ring silicate. The rings are connected to each other, however, by Be tetrahedra that give the frame a three-dimensional extent. Cordierite ($Mg_2Si_5Al_4O_{18}$) is another mineral usually referred to as a ring silicate. Both are more correctly considered as network structures based on the three-dimensional extent of the frame.

4. Polyhedral frame structures, especially the silicates, may be further characterized based on the *numbers of shared polyhedra*. The range of structures and the number of tetrahedra shared in each frame are apparent from Tables 5.3 through 5.6. Isolated tetrahedral structures such as forsterite (Mg_2SiO_4) with no shared tetrahedra are at one end of the range, and the three-dimensional network structures such as quartz (SiO_2) are at the other end. In order to describe adequately the degree of sharing between these ends, a *sharing coefficient S* is defined as the average number of polyhedra that share a single corner:

$$S = \frac{\sum im_i}{C(CN)} \qquad (5.1)$$

$$= \frac{m_1 + 2m_2 + 3m_3 + \cdots + im_i}{C(CN)}$$

where m_1 is the number of tetrahedral corners in the asymmetric unit of the frame that share with no other tetrahedra, m_2 is the number of

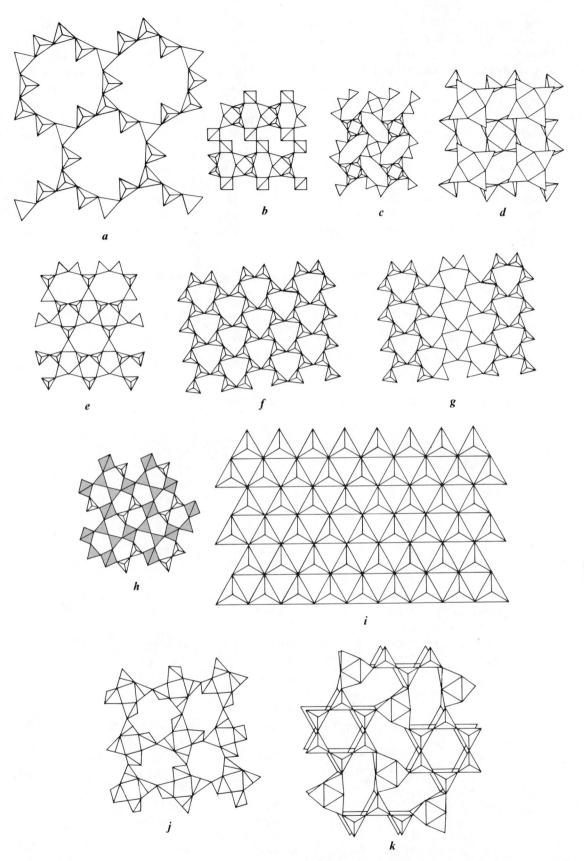

FIGURE 5.8 Linkage patterns of tetrahedra within two-dimensional frames. Mineral examples tabulated in Table 5.5.

tetrahedral corners that share with one other tetrahedron, m_3 is the number of corners that share with two other tetrahedra, and so forth. The symbol C is the number of cations in the asymmetric unit and CN is their coordination number. The denominator in Equation 5.1 for tetrahedral frames is equal to the total number of tetrahedral corners in the asymmetric unit.

We learned earlier in this chapter that the corner oxygen of a Si^{4+} tetrahedron may be shared with at most one other Si^{4+} tetrahedron in the common rock-forming silicates. The value of m_3 and higher terms in Equation 5.1 will therefore be zero when only Si^{4+} tetrahedra are considered. A few rare minerals have lower valent cations (e.g., Al^{3+}, Mn^{2+}, Fe^{2+}, Zn^{2+}, Be^{2+}) in tetrahedral coordination and more than two of these may share a common apical oxygen. A specific example is phenacite ($SiZn_2O_4$) in which two Zn^{2+} tetrahedra and one Si^{4+} tetrahedron all share a common corner (oxygen).

As an example of how to calculate the sharing coefficient, let us use the amphibole edenite ($NaCa_2Mg_5(OH, F)_2$- Si_7AlO_{22}). The tetrahedral frame is illustrated in Figure 5.7i, and we see that the asymmetric unit has two

TABLE 5.6. *Classification of Three-Dimensional Tetrahedral Frames*

Groups	Sharing Coefficient	Periodicity	Charge of Frame	Examples	Figure (Part II)
Single network	1.875	4-6-8	$[Si_3AlO_8(OH)]^{2-}$	Ussingite	
	1.938	6-6-6	$(Si_6(Fe,Mg)_2O_{15})^{2-}$	Pellyite	
	2.000	4-5-6-8	$(Si_3AlO_8)^{1-}$	Marialite	II.16
	2.000	4-5-6-8	$(SiAlO_4)^{1-}$	Meionite	
	2.000	4-6-8-8	$(SiO_2)^0$	Quartz	II.2
	2.000	4-6-8-9	$(SiO_2)^0$	Coesite	II.5
	2.000	6	$(SiO_2)^0$	Cristobalite	II.4
	2.000	6-6	$(SiO_2)^0$	Tridymite	II.3
	2.000	4-6-8-10	$(Si_3AlO_8)^{1-}$	Orthoclase	II.9
	2.000	4-6-8-10	$(SiAlO_4)^{1-}$	Anorthite	II.11
	2.000	4-6-8-10	$(Si_2AlO_6)^{1-}$	Analcime	II.14
	2.000	4-6-9	$(Si_2BeO_6)^{2-}$	Beryl	II.17
	2.000	4-6-6-9	$(Si_2AlO_6)^{1-}$	Osumilite	
	2.000	6-6	$(SiAlO_4)^{1-}$	Nepheline	II.15
	2.000	6-8-8	$(SiAlO_4)^{1-}$	Eucryptite	
	2.000	4-6-12	$(SiAlO_4)^{1-}$	Sodalite	II.18
	2.000	4-6-8-12	$(Si_2AlO_6)^{1-}$	Chabazite	
	2.000	4-6-6-10	$(Si_2AlO_6)^{1-}$	Laumontite	
	2.000	4-8-9	$(Si_3Al_2O_{10})^{2-}$	Natrolite	
	2.000	4-8-8-12	$(Si_{11}Al_5O_{32})^{5-}$	Phillipsite	
	2.000	4-4-5-6-8	$(Si_{13}Al_5O_{36})^{5-}$	Stilbite	II.20
	2.000	4-5-6-8	$(Si_7Al_2O_{18})^{2-}$	Heulandite	
	2.375	3-4-6	$(Si_2Be_2O_7)^{4-}$	Barylite	
	2.750	3-4-6	$[Si_2Be_4O_7(OH)_2]^0$	Bertrandite	
	2.750	3-4-6-8	$[Si_2Zn_4O_7(OH)_2]^0$	Hemimorphite	
	3.000	3-4-5	$(SiBe_2O_4)^0$	Phenacite	
	3.000	3-4-5	$(SiZn_2O_4)^0$	Willemite	5.6
Double network	1.625	14-20	$(Si_4O_{11})^{6-}$	Neptunite	

tetrahedra. Three corners share with no other tetrahedra ($m_1 = 3$), and five corners share with one other ($m_2 = 5$). There are two cations, and thus

$$S = \frac{3 + 2(5)}{2(4)} = \frac{13}{8} = 1.625$$

We can calculate the sharing coefficient directly from the chemical formulas of all polyhedral-frame structures except those in which the difference between the smallest and the largest number of tetrahedra sharing a corner is larger than one. If the number of cations in tetrahedral coordination is C, and the number of anions is A, then

$$S = 2n + 1 - \frac{A}{4C}(n^2 + n)$$

where n is the integer part of the ratio $4C/A$. Our previous example of edenite has 8 cations and 22 oxygens such that

$$\frac{4C}{A} = \frac{32}{22} = 1.455$$

The integer of that ratio is 1. Hence,

$$S = 2 + 1 - \frac{22}{32}(1 + 1) = 3 - \frac{22}{16} = 1.625$$

The sharing coefficient is more than a criterion for classification. It is also a qualitative guide to the relative chemical and mechanical stabilities of silicate minerals. The key again is the integrity of the frame. A low sharing coefficient means that less of the total bonding energy (still at least one half) is associated with strong Si—O bonds and more of the bonding energy is associated with weaker octahedral or other cation-oxygen bonds. Silicate minerals with low sharing coefficients are among the first to break when exposed to changes in chemical and mechanical environments. A low sharing coefficient is one reason why the olivines ((Mg,Fe)SiO_4), for example, break down so easily in the weathering environment, and why quartz (SiO_2) with a high sharing coefficient is, in contrast, so resistant. This behavior is a direct consequence of the relative numbers of shared tetrahedra within the primary frame, a feature commonly referred to as *polymerization*, and is analogous to similar linkages involving carbon in organic chemistry. Silicates with a large sharing coefficient are highly polymerized; minerals with lower values are less polymerized.

In listing the net charge on the frames in Tables 5.3–5.6, we presume that the tetrahedra are occupied only by Si^{4+}. Most silicates permit Al^{3+} (and less frequently other cations) to substitute for Si^{4+} in tetrahedral coordination. This substitution will increase by -1 the frame charge for each introduced Al^{3+}. Additional cations in the secondary frame are required to balance the added negative charge. In the alkali feldspars (KSi_3AlO_8, $NaSi_3AlO_8$), for example, the occupancy of every fourth tetrahedron by Al^{3+} is balanced by a monovalent alkali cation in the secondary frame. In the calcic feldspars ($CaSi_2Al_2O_8$), every other tetrahedron is occupied by Al^{3+}, and a divalent cation Ca^{2+} provides for electrical neutrality.

5. The *number of polyhedra* in translational units of one-dimensionally extended frames provides an additional criterion for distinguishing frames that otherwise would appear similar. In the one-dimensional frames, the number of tetrahedra in the translationally asymmetric unit, referred to as the *repeat unit* (RU), ranges within the single chain structures from 1 in synthetic "iscorite" ($Fe_6O_4SiFeO_6$) to at least 12 in alamosite ($Pb_3Si_3O_4$) (see Figures 5.7a to 5.7g). The most common of these single chain structures are the pyroxenes with two tetrahedra per repeat unit. The polyhedral group in this case is a *single chain*. In the amphiboles, the polyhedral group is a *double chain* with the number of tetrahedra per repeat unit still two. The asymmetric unit therefore has four tetrahedra, because the chain is doubled. The repeat units in the double chain silicates range from one tetrahedron as in sillimanite ($AlSiAlO_5$) to at least five tetrahedra as in inesite ((Mn,Ca)$_9(OH)_2$ $Si_{10}O_{28} \cdot 5H_2O$) (Figure 5.7j) and possibly higher, although these silicates are as yet unknown.

In two-dimensional and three-dimensional frames, the number of tetrahedra per repeat unit is less conspicuous than in chains. In these structures, the number of tetrahedra in a *loop* provides a more visual and meaningful criterion of polyhedral structure classification. A loop of tetrahedra differs from a ring in that the ring is an isolated unit, whereas the loop is part of an extended frame. Although this criterion is designed for classification of two-dimensional and three-dimensional frames, it is also useful in the description of multiple and complex chains. For example, six-membered loops characterize the amphiboles (Figure 5.7i), and four-membered loops characterize the frame of sillimanite (Figure 5.7h).

Among the layer silicates, six-membered loops such as those found in the structures of clays, micas, and chlorites are the most common. Three-membered, four-membered, and five-membered loops are also known but are less common (Figure 5.8).

In the three-dimensional network structures, loops are characteristic, and the num-

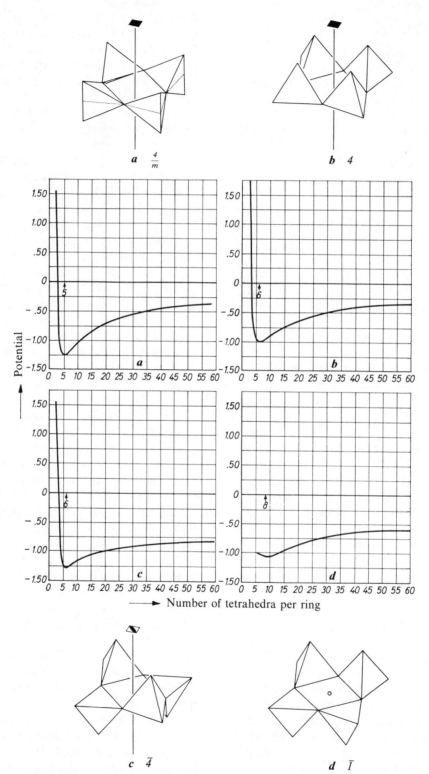

FIGURE 5.9 **The four different symmetries of tetrahedral rings used in calculation of the electrostatic energies. (After Zoltai, T., and Buerger, M. J., 1960. The relative** energies of rings of tetrahedra. *Zeitschrift für Kristallographie* **114:7.)**

ber of tetrahedra comprising them can be used to distinguish various primary frames. For example, the polymorphs of silica (SiO_2), with the exception of stishovite, have the same group geometry, the same frame dimensionality, the same sharing coefficients, and the same chemistry (see Table 5.6). However, the number of tetrahedra per repeat unit combined with the number of tetrahedra in each of several loops differs. Cristobalite and tridymite, for example, have four tetrahedra per repeat unit, but in cristobalite the tetrahedra are distributed in a single six-membered loop, and in tridymite they are distributed in two different six-membered loops.

Although the range of numbers of tetrahedra per loop in the different silicates is great, most loops have either four or six members. Calculations of the relative ionic energies of rings (or loops) with different numbers of tetrahedra related by various symmetry operations help explain why the members of tetrahedra per loop are small. Figure 5.9 illustrates four ring symmetries and the variations of energy with number of tetrahedra per ring for each type. The minimum energy state is achieved for $4/m$, 4, and $\bar{4}$ symmetries by five-membered, six-membered, and six-membered loops, respectively (Figures 5.9a–5.9c). The $\bar{1}$ symmetry (Figure 5.9d) yields eight tetrahedra per loop as the minimum energy state.

SYMMETRICALLY PACKED STRUCTURES

All crystal structures that can be constructed by the symmetrical stacking of symmetrical sheets of atoms or ions are called *symmetrically packed*. Of the two categories of such structures, the *monatomic* structures include minerals that have atoms joined by metallic or covalent bonds. These atoms may be arranged in either the close-packed or closest-packed arrays.

The notion of an array of atoms, all of which are in mutual contact, goes back to the early 17th century. Johannes Kepler recognized the two unique arrangements of spheres in a plane, given the constraint that the spheres be "packed" together as closely as possible (Figure 5.10a). He also undoubtedly envisioned the two-dimensional structure of snowflakes as a progressive buildup of spheres arranged in this manner (Figure 5.10b).

In the multiatomic category, the symmetry of the close-packed and closest-packed structures is retained in the arrangement of the anions, but the

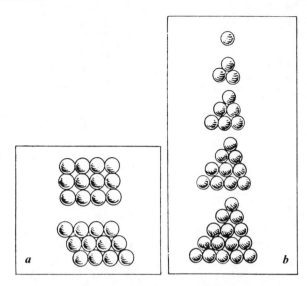

FIGURE 5.10 Johannes Kepler's notion of (a) how spheres could be arranged in a plane, and (b) how spheres might be arranged to explain the form of snowflakes. From Kepler, J., 1611. Two planar arrangements of equal spheres found in close-packing. (Figure in: A new year's gift, or on a hexagonal snowflake. Schneer, C. J., ed. Copyright © 1977 by Dowden, Hutchinson & Ross, Inc., Stroudsburg, Pa. Reprinted by kind permission of the publisher.)

aspect of "closeness" is no longer applicable because no bonding occurs between the anions. We introduce the term *symmetrically packed* to eliminate the misleading implication of the close-packed label. The coherence of the structures is provided by bonds between the symmetrically packed anions and the cations located in the voids between them. Many of the minerals already discussed as polyhedral-frame structures are included in the multiatomic group even though their bonding is primarily ionic. The dual classification exists because the common ionic coordination polyhedra are identical with the common coordination pattern of the voids.

The first person to view structure systematics seriously as symmetrically packed arrays was William Barlow. In 1883 he predicted several "close-packed" structures based on the stacking of various types of sheets (Figure 5.11). He consistently referred to these structures as "symmetrical arrangements" rather than as close-packed. His "plans" *a* and *b* illustrate the reason for his choice of terms. The atoms have a highly symmetrical arrangement but are by no means packed closely together. From his plans *c* and *d*, which are identical with Kepler's two arrangements, Barlow derived the simple cubic (SC) structure, the hexagonal closest-packed (HCP) structure, and the cubic

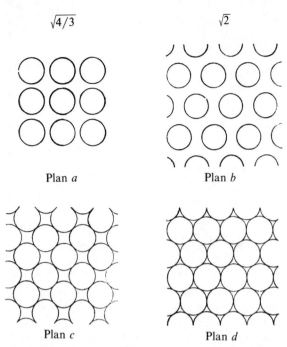

$\sqrt{4/3}$ $\sqrt{2}$

Plan *a*   Plan *b*

Plan *c*   Plan *d*

FIGURE 5.11   Symmetrical arrangements of spheres in a plane envisioned by William Barlow in 1883. Plans *a* and *c* have *p4mm* plane group symmetry; plans *b* and *d* have *p6mm* symmetry. (From Barlow, W., 1883. Probable nature of the internal nature of crystals. *Nature* 29:186–187.)

closest-packed (CCP) structure. All are widely recognized today. He emphasized that the closeness of the atoms is less important than the symmetry of the sheets. Both his plans *b* and *d* have the same *p6mm* plane group symmetry, even though the "closeness" of the atoms is obviously different. Plans *a* and *c* have the same plane group symmetry, namely, *p4mm*, yet the interatomic distances are very different. Barlow discovered that by separating the atoms in a sheet while retaining the plane group symmetry, he could derive new structures. By increasing the distance between the atoms in his plan *c* so that the sides of the squares bear to the diameter of the spheres the ratio $2:\sqrt{3}$, Barlow created a new structure, which today we call a body-centered cubic (BCC) structure. On the other hand, if we separate the atoms in plan *d* so that the sides of the equilateral triangles bear to the diameter of the spheres the ratio $\sqrt{2}:1$, then the stacking will create a simple-cubic structure, which may also be constructed by the stacking of the plan *c* sheet.

SYMMETRICAL SHEETS

The two most common and important types of symmetrical sheets are called *hexagonal* and *tetragonal*. They have *p6mm* and *p4mm* plane group

symmetry, respectively, and will hereafter be oriented as in Figure 5.12. Other types of symmetrical sheets have different unit meshes or lower symmetry or both. These are important in some crystal structures, but for now we will consider only the hexagonal and tetragonal sheets in detail.

If the atoms of a symmetrical sheet are all in mutual contact, the sheet is considered *dense*, or close-packed. If the atoms are not in contact but are symmetrically arranged, the sheet is *open*, or open-packed. Barlow's plan *a* and *b* arrangements are open, with the separations between atoms being equal to $\sqrt{4/3}$ and $\sqrt{2}$ times the atom diameter, respectively. These values, *n*, as well as others such as $\sqrt{3/2}$ and $\sqrt{8/3}$, will yield unique and symmetrical structures when stacked in three dimensions. The parameter *n* is identical to interatomic distance relative to a value of unity ($n = 1$) where atoms are in contact. With few exceptions, all symmetrically packed mineral structures consist of either hexagonal or tetragonal sheets.

In dense hexagonal and tetragonal sheets (Figure 5.12), the length of unit translations a_1 and a_2 in each mesh is equal to the sum of atomic radii ($n = 1$). The parameter *n* defines the openness of the sheets, and is given simply by

$$n = \frac{a}{D}$$

where *a* is the unit translation of the sheet and *D* is the diameter of the atom. As *n* increases in the open sheets, the two-dimensional packing efficiency decreases rapidly. The packing efficiency (PE) of a sheet can be expressed as the percentage of the area occupied by atoms (represented by circles in a two-dimensional projection). For hexagonal sheets,

$$\text{PE} = \frac{100(1/4\pi D^2)}{a^2\sqrt{3}/4} = \frac{100(\pi D^2)}{2a^2\sqrt{3}} = \frac{100\pi}{2\sqrt{3}\,n^2}$$

$$(5.2)$$

and for tetragonal sheets,

$$\text{PE} = \frac{100(1/4\pi D^2)}{a^2} = \frac{100(\pi D^2)}{4a^2} = \frac{100\pi}{4n^2}$$

$$(5.3)$$

Table 5.7 gives the packing efficiency for hexagonal and tetragonal sheets for selected values of *n*. The values given are of some common symmetrically packed structures. The atoms in the dense hexagonal sheet are arranged in the most efficient manner possible, occupying 90.69% of the area. The most efficient tetragonal arrangement has only a 78.54% efficiency.

The *p6mm* plane group has one sixfold axis,

TABLE 5.7. *Packing Efficiency of the Hexagonal and Tetragonal Sheets in Two-Dimensional Projection for Selected Values of n*

	Packing Efficiency	
$n =$	HEXAGONAL SHEET	TETRAGONAL SHEET
1 (dense)	90.69%	78.54%
$\sqrt{4/3}$	68.02	58.90
$\sqrt{3/2}$	60.46	52.36
$\sqrt{2}$	45.34	39.27
$\sqrt{8/3}$	34.01	29.45

two threefold axes, and three twofold axes per unit mesh (Figure 5.12a). The threefold axes are symmetrically equivalent, as are the twofold axes, but now we give them separate labels as shown in Table 5.8. Each axis operates on the others such that twofold axes will occupy points halfway between sixfold axes located at each corner. In the *p4mm* tetragonal sheet (Figure 5.12b), two symmetrically distinct fourfold axes and two equivalent twofold axes occupy each unit mesh.

In each of these sheets, every atom has the same symmetrical environment. Comparison of Figures 5.12a and 5.13a shows that each atom in the hexagonal sheet occupies the site of a sixfold axis. The symmetry of two-dimensional space around each atom must be identical, because their

TABLE 5.8. *Labels, Ranks, and x, y Coordinates of Rotation Axes in Hexagonal and Tetragonal Sheets*

Close-Packed Sheet	Symbol	Rank of Rotation Axis	Coordinates	
			x	y
Hexagonal	*A*	sixfold	0	0
	B	threefold	2/3	1/3
	C	threefold	1/3	2/3
	D	twofold	1/2	1/2
	E	twofold	1/2	0
	F	twofold	0	1/2
Tetragonal	*S*	fourfold	0	0
	D	fourfold	1/2	1/2
	E	twofold	1/2	0
	F	twofold	0	1/2

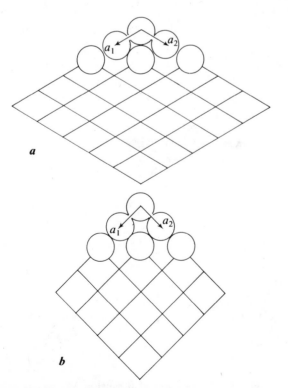

FIGURE 5.12 Unit meshes of (a) dense hexagonal sheet, and (b) dense tetragonal sheet.

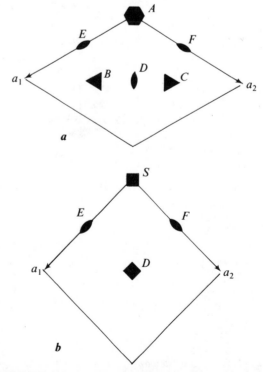

FIGURE 5.13 Notation of rotation axes present in (a) hexagonal sheet, and (b) tetragonal sheet.

locations are equivalent by translation. The voids between atoms are on the threefold axes. In the tetragonal sheet (compare Figures 5.12b and 5.13b), all atoms are on the same fourfold axes, and the voids are on the other fourfold axes. Each atom has four closest neighbors in the tetragonal sheet, whereas in the hexagonal sheet each has six closest neighbors.

SYMMETRICAL STACKING OF DENSE SHEETS

The structures that result from the stacking of dense sheets are restricted to the elements and thus are relatively unimportant in mineralogy. What is important, however, are the principles from which we derive these structures. Many com-

mon rock-forming minerals have anion packings symmetrically identical to element structures, the main difference being the separation of the anions in the minerals and the manner in which the voids are filled. Because the voids are unoccupied in the elements, they provide relatively simple models that will aid in our understanding of more complex structures. Examples of such complex structures include most of the sulfides, oxides, and hydroxides, as well as the common rock-forming silicates such as olivines, pyroxenes, amphiboles, and micas.

To achieve a symmetrical environment for atoms in three dimensions, the symmetrical sheets must be stacked one upon the other in a symmetrical manner. In order to yield a highly symmetrical relationship between each atom and its neigh-

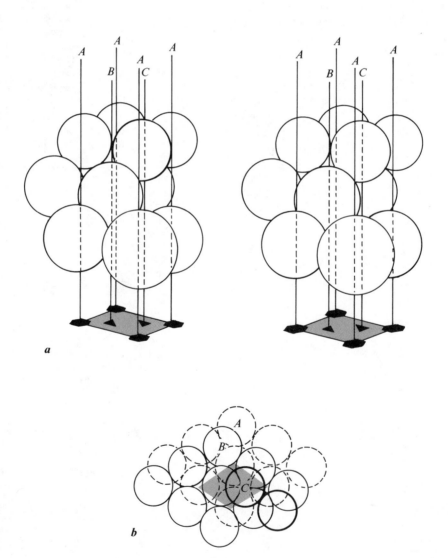

FIGURE 5.14 **(a) Stereoscopic view of** *ABC* **stacking of dense hexagonal sheets. Atoms in** *B* **and** *C* **sheets overlie the** *B* **and** *C* **rotation axes of reference sheet** *A*.

(b) Plan view of how atoms of *B* **sheet occupy the voids (***B* **site) of reference sheet** *A*. **Shaded region is one unit mesh.**

bors, the atoms of one sheet must be located directly over the rotation axes of the *first sheet*, called hereafter the *reference sheet*. Taking a sequence of hexagonal sheets as an example, the reference sheet is always labeled *A* to denote the atomic positions directly over sixfold axes of rotation. An identical second sheet can occupy any of several positions, depending on which rotation axis of the reference sheet the atoms of the second sheet will directly overlie. If the atoms of the second sheet overlie one of the threefold axes, the atoms fit into the voids of the reference sheet. Successive repetition of these two sheets creates a dense arrangement called the *hexagonal closest-packed* (HCP) structure[2] (Figure 5.14a). We will call this arrangement simply the *AB* stacking se-

quence in which the symbol *B* denotes that atoms of the second sheet overlie the *B* position (Figure 5.13a) of the reference sheet. This stacking is also illustrated stereoscopically in the first two sheets of Figure 5.14a. Note that the sixfold axis in the plane group symmetry of the reference sheet is degraded by the addition of the second sheet to a threefold axis (Figure 5.14a). The symmetries of the *A*, *B*, and *C* positions thus correspond in three dimensions to threefold axes of rotation. If the atoms of the third sheet are not over the *A* position of the reference sheet but instead occupy the threefold axis labeled *C*, an *ABC* stacking sequence is generated (Figure 5.14b). An example of an *AB* stacking sequence is the structure of elemental magnesium (Figure 5.15). Native copper (Figure 5.16), gold, and silver are minerals that have the *ABC* stacking sequence. This sequence is also called the *cubic closest-packed* (CCP) structure.

[2] In figure 5.13 and most subsequent structures, the radii of atoms are reduced about their geometric centers for the sake of clarity.

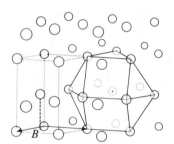

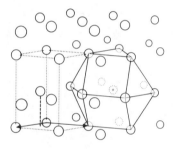

FIGURE 5.15   Crystal structure of Mg. *AB* stacking is hexagonal closest-packed (HCP) with CN = 12 and pack- ing efficiency of 74.05%. Coordination polyhedron in solid lines, unit cell in dotted lines.

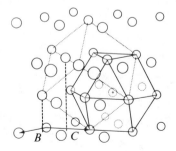

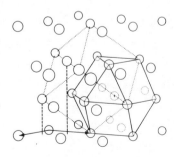

FIGURE 5.16   Crystal structure of Cu. *ABC* stacking is cubic closest-packed (CCP) with CN = 12 and packing efficiency of 74.05%. Coordination polyhedron in solid lines, unit cell in dotted lines. Compare with Figure 5.14a.

Hexagonal Sheets

	n = 1.0	$\sqrt{4/3}$ 1.1	$\sqrt{3/2}$ 1.2	$\sqrt{12/7}$ 1.3	$\sqrt{2}$ 1.4	$\sqrt{9/4}$ 1.5	$\sqrt{8/3}$ 1.6	$\sqrt{3}$ 1.7	1.8	$\sqrt{15/4}$ 1.9	$\sqrt{4}$ 2.0
ABC n =	1 ———				$\sqrt{2}$ ———	$\sqrt{8/3}$ ———					
CN =	12	6(+)		6	6(+)	8(+6 at $\sqrt{4/3}$)					
Coordination polyhedron:	Cuboctahedron	Tall trigonal antiprism		Octahedron	Stubby trigonal prism	Cube					
Identical with:	*SD*					$\sqrt{4/3}$ *SD*					
Example:	CCP 74.05%	RCP		SC 52.36%	RCP	BCC 68.02%					
AB n =	1 ———		$\sqrt{12/7}$ ———			$\sqrt{9/4}$ ———		*$\sqrt{3}$ *AB* ———		$\sqrt{15/4}$	
CN =	12	6(+)	6	6(+)	8			5			8
Coordination polyhedron:	Disheptahedron	Tall trigonal prism	Equidimensional trig. prism	Stubby trigonal prism	Trigonal prism + dipyramid			Trigonal dipyramid			Tetragonal prism
Example:	HCP 74.05% 1.03 – Be		64.70%	52.36%	53.74%			Graphite 40.31%			30.23%
Other stackings:	*ABAC*: Double HCP: 4H: La; *ABABCBC4C*: 9R: Sm; possible sequences										
ADEF n =	1 ———				$\sqrt{2}$ ———					$\sqrt{15/4}$ ———	*$\sqrt{4}$ *ADE*
CN =	12	4(+)		4	4(+)					6	8
Coordination polyhedron:	Rhombic prism + disphenoid	Tall rhombic disphenoid		Tetragonal disphenoid	Stubby rhombic disphenoid					Rhombic pyramids (2)	Tetragonal prism
Example:	γ-Pu 69.81%			42.75%						64.49%	30.23%
AD n =	1 ———				$\sqrt{2}$ ———					$\sqrt{15/4}$ ———	*$\sqrt{4}$ *AD*
CN =	10	4(+)		4	4(+)		4			6	4
Coordination polyhedron:	Tetragonal prism + dipyramid	Tall rectangle		Square	Stubby rectangle		Rectangle			Elongated hexagon	Square
Identical with:	$\sqrt{3/2}$ *SD* Pa 69.81%										
Example:				42.75%			40.3%			64.49%	30.23%
AABB n =	1 ———						$\sqrt{8/3}$ ———	$\sqrt{3}$			
CN =	10		4(+)			4	4	4			
Coordination polyhedron:	Ditrigonal pyramids (2) & pedion		Tall trigonal pyramid			Stubby trigonal pyramid	Tetra-hedron	Trigonal pyramid (center in base)			
Examples:	33.28%						Hex. Si				
Other stackings:	$\sqrt{8/3}$ *AABBCC*: Diamond, identical with $\sqrt{8/3}$ *SEDF*										
Example:							34.01%	40.31%			
A n =	1 ———			2(+) Line ———							no limit
CN =	8										
Coordination polyhedron:	Hexagonal dipyramid			Line							
Identical with:	*SE*										
Example:	SH 60.46%										

EXAMPLES for RCP:

Tall RCP: $n = \sqrt{4/3}$ Hg, 1.40 – Sb; Stubby RCP: $n = 1.50$ As

EXAMPLES: $\sqrt{3}$ *ABAC*···· graphite *H*; $\sqrt{3}$ *ABACBC*···· graphite *R*

Tetragonal Sheets

EXAMPLES for BCT:
Tall BCT: n = 1.03, γ-Mn, Stubby BCT: n = 1.27 HgII

SD n =	1		√4/3		√3/2
CN =	12	8(+)	8(+6 at √4/3)	8(+)	10
Coordination polyhedron:	Cuboctahedron	Tall tetragonal prism	Cube	Stubby tetragonal prism	Tetragonal prism + dipyramid
Identical with:	ABC		√8/3 ABC		AD
Example:	[CCP] 74.05%	[BCT]	[BCC] 68.02%	[BCT]	Pa 69.81%

SEDF n =	1		√8/3		√15/4	*√4 SED
CN =	8	4(+)	4	4(+)	4(+)	4 and 8
Coordination polyhedron:	Tetragonal disphenoid + scalenohedron	Tall tetragonal disphenoid	Tetrahedron	Stubby tetragonal disphenoid	Tetragonal disphenoid	Square and octahedron
Identical with:	—		√8/3 AABBCC			
Example:	66.57%		Diamond 34.01%		55.81% β Sn	26.18% Ge III (Dist.)

SE n =	1		√2		√8/3	√15/4	*√4 SE
CN =	8	4(+)	4	4+	4	6	4
Coordination polyhedron:	Hexagonal dipyramid	Tall rectangle	Square	Stubby rectangle	Rectangle	Elongated hexagon	Square
Identical with:	A						
Example:	[SH] 60.46%		37.02%	—	34.81%	55.81%	26.18%

SSDD n =	1		√2
CN =	9	5(+)	5
Coordination polyhedron:	Tetragonal pyramids (2) + pedion	Tetragonal pyramid	Tetragonal pyramid (center in base)
Example:	56.12%		30.66%

SDEF n =	1		√12/7		√2
CN =	10	6(+)	6	6(+)	8
Coordination polyhedron:	Orthorhombic domes, pyramids, pedion	Thick triangular prism	Trigonal prism	Thin triangular prism	Tetragonal disphenoid + prism
Example:	66.57% 1.09-γ-Ga	1.33-β-Ga	53.87%		52.36%

S n =	1		no limit
CN =	8	2(+)	
Coordination polyhedron:	Octahedron	Line	
Identical with:	√2 ABC		
Example:	[SC] 52.36%		—— no limit

NOTE: CCP = Cubic closest-packed, BCC = Body-centered cubic, BCT = Body-centered tetragonal, SC = Simple cubic, HCP = Hexagonal closest-packed, RCP = Rhombohedral close-packed, SH = Simple hexagonal, % = Packing efficiency, percent.

*In these open close-packed sheets, the distance between A (or S) atoms in the reference sheet is such that the atoms of the second (and third) sheet fit within the first and are at unit distance from the atoms of the latter. Consequently, the first and second (and third) sheets constitute one single sheet.

If the dense hexagonal sheets were to be arranged in the sequence ACB, the arrangement would not differ from the ABC sequence. Similarly, the sequence AD is the same as AE or AF. Only eight simple stacking sequences of dense hexagonal sheets yield highly symmetric and identical environments for every atom. These sequences are: A, AB, ABC, AD, ADE, $ADEF$, $AABB$, and $AABBCC$.

That elements such as copper, gold, and silver have close-packed structures has long been recognized. Similar stacking structures exist for many of the elements, some of which are not generally recognized. For example, γ-plutonium (Pu) has an idealized $ADEF$ stacking, which is found also in the minerals cinnabar (HgS, Figure II.78) and hessite (Ag_2Te).

The stacking of dense tetragonal sheets proceeds in the same way. The reference sheet is always labeled S to denote that all atoms in that sheet are on the same fourfold axis. Note that the first letter of the stacking sequence (either A or S) indicates whether the structure consists of hexagonal or tetragonal sheets. If the atoms of the second sheet of a tetragonal stacking sequence are over the other fourfold axes, and if that two-sheet sequence is repeated as the asymmetric unit, the sequence is simply SD. If the atoms of the second sheet occupy instead the twofold axes of the reference sheet, the stacking sequence is SE. The sequence SF is the same as SE, because they can be made equivalent by fourfold rotation and relabeling of the a_1 and a_2 axes. Only six sequences of dense tetragonal sheets yield equal and symmetrical coordination for all atoms. These are: S, SD, $SDEF$, SE, $SEDF$, and $SSDD$.

The symmetrical environment around each atom, whether in a stacking sequence involving dense hexagonal sheets or dense tetragonal sheets, can be characterized by the number and the arrangement of closest neighbors. For ABC-type structures (Figure 5.16), each atom of the reference sheet A has six closest neighbors within the sheet. Three atoms in the overlying B sheet are also closest neighbors, as well as three atoms in the directly underlying C sheet. The coordination number is therefore 12, and the coordination polyhedron is a cuboctahedron. The other types of polyhedra for various stacking sequences of hexagonal and tetragonal dense sheets are listed in the first column of Table 5.9. Included in the table are only those packing sequences that yield the same type of coordination polyhedron for *all atoms* in a given structure. This should not be interpreted to exclude other stacking sequences. Indeed, other sequences are possible. In those

structures, however, atoms in different types of sheets will have different coordinations. We will consider such structures later as variations of the 15 basic stacking types. Some of these variations will be discussed in this chapter under the heading of mixed stacking sequences.

SYMMETRICAL STACKING OF OPEN SHEETS

Identical open sheets may also be stacked in three dimensions to yield symmetrical coordination of atoms. The distance between atom centers, n, may have any value up to a limit determined by the following: (1) When the void size of the reference sheet equals the size of directly overlying atoms in an adjacent sheet, the second sheet "collapses" into the first, and (2) if n reaches such a value that the atoms in the two adjacent sheets above and below the reference sheet come in contact across the voids of the reference sheet, then any further increase in n is not permissible.

The limiting values of n for the stacking of hexagonal and tetragonal sheets depend on the symmetry of the sheet and the type of stacking. The limitation for hexagonal stacking is $n = \sqrt{3}$ for AB, ABC, $AABB$, and $AABBCC$ type structures. The minimum dimension of voids in these structures is $\sqrt{3} \times \sqrt{1/3} = 1$, the sum of the radii of two atoms in contact. Atoms of a B sheet can thus fall into the voids of an A or C sheet whenever n is equal to or exceeds $\sqrt{3}$. For AD, ADE, and $ADEF$ stackings of tetragonal sheets, the maximum value beyond which collapse to a single sheet occurs is $n = \sqrt{4} = 2$. For SD, $SDEF$, and $SSDD$ stackings of tetragonal sheets, the limiting value of n is $\sqrt{2}$, and for SE and $SEDF$ stackings, it is at $n = \sqrt{4} = 2$.

The second limitation may occur at a lower value of n than that established for the first limitation. In AB stacking, for example, atoms of every other equivalent sheet (A and A, or B and B in $ABABAB \ldots$) will come in contact when the distance between the sheets becomes unity (i.e., $2\sqrt{1 - n^2/3}$). When this happens, $n = \sqrt{9/4}$, and no further increase in n is possible. The same type of limit is reached at $n = \sqrt{8/3}$ for ABC, at $n = \sqrt{3/2}$ for SD, at $n = \sqrt{3}$ for AD and SE, at $n = \sqrt{33/9}$ for ADE, and at $n = \sqrt{15/4}$ for $ADEF$ and $SEDF$ stacking types. Because these maximum values are less than those set by the first limitation, none of these sheets can be sufficiently open to enable another sheet to collapse within it. The only stacking types in which the first limitation imposes the maximum value of n are $AABB$,

AABBCC, *SDEF*, and *SSDD*. These restrictions are summarized in Table 5.9.

Although the stacking of hexagonal and tetragonal sheets will yield uniform and symmetrical coordination polyhedra for all permissible values of n, only a few special values of n will yield coordination polyhedra of high symmetry. For example, Table 5.9 shows that perfect octahedral coordination (CN = 6) is present in the $\sqrt{2}\,ABC$ stacking type, which is equivalent to *SS* stacking. Perfect cubic coordination (CN = 8) is present in the $\sqrt{8/3}\,ABC$ (or $\sqrt{4/3}\,SD$) stacking type. Tetrahedral coordination (CN = 4) is present in the $\sqrt{8/3}\,AABBCC$ (or $\sqrt{8/3}\,SEDF$) stacking type. These and other common coordination polyhedra are possible only in the stacking of open sheets. The appearance of fractional square root values of n is not accidental. These values represent simple but fundamental components of symmetrical coordination polyhedra. Specifically, they give the edge lengths of polyhedra in which the center to corner distances are defined as unity.

Instead of deriving the special values of n by considering all of the major types of coordination polyhedra, the same result can be achieved by considering only one polyhedron, namely, the cube. The cube is unique as a coordination polyhedron, because it represents the symmetry of both the hexagonal and tetragonal sheets. The (001) section of a cube has atoms arranged in a tetragonal sheet with $p4mm$ symmetry, whereas the (111) section has atoms arranged in a hexagonal sheet. By analyzing the geometry of these sections and the relationships between them, we not only can derive the special values of n, but we can also demonstrate the equivalence of certain tetragonal and hexagonal stacking sequences.

The lengths of selected components of a cube with unit edge length are illustrated in Figure 5.17. The (001) face of this cube represents the unit mesh of a dense tetragonal sheet and the (111) plane the unit mesh of an open hexagonal sheet with unit translation equal to $\sqrt{2}$. Let us consider the corners of the cube to be atoms in symmetrical packing, which can be interpreted as an *SS* stacking of dense tetragonal sheets, or as an *ABC* stacking of an open hexagonal sheet with $n = \sqrt{2}$. For the sake of easy visualization of these two types of stackings, the corners of the cube are labeled A, B, and C in Figure 5.17. Figure 5.18 is a simple demonstration that the S and the $\sqrt{2}$ *ABC* structure are identical. This structure is called *simple cubic* (SC) and is found in the element α-polonium. Cooperite (PtS) and parkerite ($Ni_3Bi_2S_2$) are mineral examples (Figure 5.18). The tetragonal sheet is in the horizontal plane of this diagram. The corresponding hexagonal sheet is illustrated in the (111) plane of the isometric unit cell. Fluorine (F^{1-}) anions in the mineral fluorite (CaF_2) are stacked in this manner, with Ca^{2+} occupying one half of the voids (Figure II.90).

The other geometric components of the cube shown in Figure 5.15 correspond to interatomic distances in various stackings of the hexagonal and tetragonal sheets. These values are relative to the edge length of the cube. That is, if the edge of the cube is unity, the atom-to-atom distance between the S and D tetragonal sheets is $\sqrt{3/4}$ and between S, E, or F sheets is $\sqrt{1/2}$. Between A and B or C hexagonal sheets, the corresponding distance is unity, and between A and D or E or F sheets, it is $\sqrt{5/4}$. Atoms between adjacent sheets are in contact, and the sum of their radii is taken as unity. The lengths of these components must

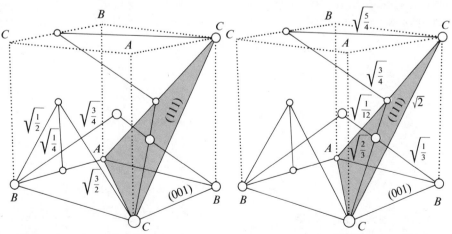

FIGURE 5.17   Geometric elements of a cube that are related to the stacking sequences of hexagonal and tetragonal open sheets. Dimensions related to hexagonal sheet are given on right half of diagram; dimensions related to tetragonal sheet are on left. Cube edges are unity.

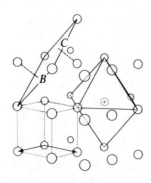

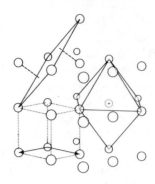

FIGURE 5.18 Crystal structure of α-polonium, stacking as either $\sqrt{2}$ *ABC* or *S*. Simple cubic (SC) structure has CN = 6 and packing efficiency of 52.36%. Coordination polyhedron in solid lines, unit cell in dotted lines.

consequently be equal to unity, and the corresponding n for the hexagonal and tetragonal sheets is calculated on that basis.

If the center to corner distance (shown in Figure 5.17 as $\sqrt{3}/4$) is taken as unity for an *SD* stacking, n of that tetragonal sheet will be $1/\sqrt{3}/4 = \sqrt{4/3}$. The corresponding n for the hexagonal sheet is then $\sqrt{2}\sqrt{4/3} = \sqrt{8/3}$. That is, the $\sqrt{4/3}$ *SD* and $\sqrt{8/3}$ *ABC* structures are identical and are called *body-centered cubic* (BCC). An example among the elements is the structure of tungsten (W) illustrated in Figure 5.19. This type of stacking is found also in several mineral structures such as native iron, petzite (Ag_3AuTe_2), argentite (Ag_2S), and acanthite (Ag_2S).

The equivalence of the dense *SD* and *ABC* structures can be demonstrated by the analysis of the cube's geometry. If we increase the size of the cube such that the (001) face will have the area of two dense tetragonal unit meshes, all faces of the cube will have an atom at the center (Figure 5.20). The cube now represents two *SD* stacking sequences. The arrangement is the *cubic closest-packed* (CCP) structure or *face-centered cubic* (FCC) structure (Figure 5.20a). The edge of this cube is $\sqrt{2}$, and its face diagonal is equal to 2. The (111) plane of this face-centered cube corresponds to two unit meshes of the dense hexagonal sheet, and the structure can be described as an *ABC* stacking, as demonstrated in Figure 5.20b. In summary, the CCP (or FCC) structure can be described by either *SD* or *ABC* stacking.

Another interesting equivalence of structures derivable with both the hexagonal and tetragonal sheets is that of diamond (C). If in the cube in Figure 5.17 the corner to center distance is taken to equal unity, a cube with $\sqrt{4/3}$ edge length results. In this $\sqrt{4/3}$ *SD* structure, the coordination is body-centered cubic. If we omit alternating atomic sites, the larger tetragonal unit mesh will have as its edge the diagonal of the cube, $\sqrt{2}\sqrt{4/3} = \sqrt{8/3}$. The *SEDF* stacking of that sheet gives the diamond structure. The same structure can be produced by the *AABBCC* stacking of the hexagonal sheet, which is in the (111) plane of the $\sqrt{4/3}$ cube or in the (101) plane of the $\sqrt{8/3}$ cube. Figure 5.21 illustrates the diamond struc-

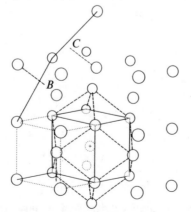

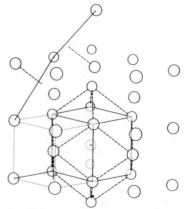

FIGURE 5.19 Crystal structure of tungsten, stacking as either $\sqrt{8/3}$ *ABC* or $\sqrt{4/3}$ *SD*. Body-centered cubic (BCC) structure has CN = 4 + 4 with packing efficiency of 68.02%.

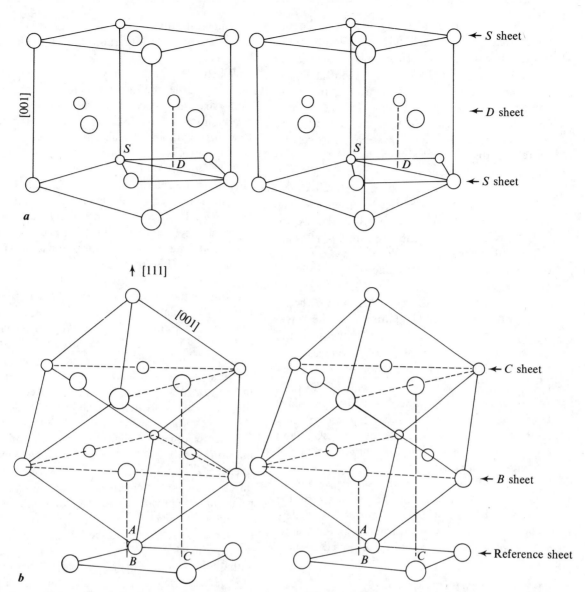

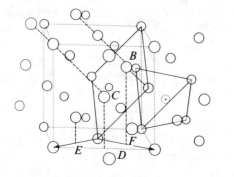

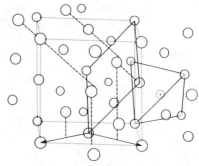

FIGURE 5.20 (a) *SD* stacking in [001] direction of CCP structure. *S* sheet is at base (001), *D* sheet contains face-centered atoms, and *S* sheet is at top. Hexagonal sheets are stacked in [111] direction of body diagonal. (b) *ABC* stacking in [111] vertical direction. Equivalent *SD* stacking in [001] direction.

FIGURE 5.21 Crystal structure of diamond. Stacking sequence is either $\sqrt{8/3}$ *AABBCC* or $\sqrt{8/3}$ *SEDF*. Structure has CN = 4 and packing efficiency of 34.01%. Coordination polyhedron in solid lines is a tetrahedron.

ture with the horizontal tetragonal sheet. The location of the hexagonal unit mesh is also indicated. The $\sqrt{8/3}$ *AABB* structure is the hexagonal equivalent of diamond and is called lonsdaleite.

As noted earlier, open sheets are not limited to these special values of *n*. In fact, Table 5.9 shows that many structures have intermediate *n* values. Their coordination polyhedra are less regular and more variable. For example, hexagonal sheets with *n* between 1 and $\sqrt{2}$ and between $\sqrt{2}$ and $\sqrt{8/3}$ yield, in *ABC* stacking, the well-known *rhombohedral close-packed* (RCP) structures. The coordination polyhedra in those structures for which *n* is less than $\sqrt{2}$ (e.g., Hg with $n = 1.15$) are tall rhombohedral antiprisms, that is, an octahedron that is flattened along one of its [111] axes. In those structures for which *n* exceeds $\sqrt{2}$ (e.g., Ti II with $n = 1.63$), the coordination polyhedra are short rhombohedral antiprisms. In the stacking of open tetragonal sheets, *n* between 1 and $\sqrt{4/3}$ (e.g., $\gamma = $ Mn with $n = 1.03$) produces coordination polyhedra with the shape of tall tetragonal prisms. For *n* between $\sqrt{4/3}$ and $\sqrt{3/2}$ (e.g., Hg II with $n = 1.26$), the polyhedra are short tetragonal prisms. These structures are frequently referred to as *body-centered tetragonal* (BCT) and are useful for describing the anion packing of numerous minerals.

Table 5.10 summarizes the simple symmetrically packed structures of the elements in the periodic table. Most of the entries are widely recognized as close-packed structures. Some, however, have not heretofore been interpreted as such.

MIXED AND EXTENDED STACKING SEQUENCES

In all of the symmetrically packed structures considered so far, a necessary condition is that the coordination of all atoms be the same throughout the structure. This is not the case when *AB* and *ABC* stackings are mixed in a single crystal structure. The most common example is the so-called double HCP structure in which *AB* and *ABC* sequences are mixed to produce the *ABAC* sequence. The stacking sequence of oxygen anions in the mineral topaz ($Al_2F_2SiO_4$) is an example (Figure II.59). Atoms in the *A* sheet have the coordination of the *ABC* sequence, whereas atoms in the *B* and *C* sheets have the coordination of the *AB* sequence. The asymmetric unit in such mixed stacking sequences can be extensive, as in the element samarium (Sm), which has an *ABABCBCAC* stacking. This stacking sequence also describes the arrangement of sulfur anions in

the mineral smythite (Fe_3S_4). Five of the sheets have *AB*-type coordination and four have the *ABC* type.

When the number of sheets in a stacking sequence is large, the same sequence, and hence the same structure, may be represented by different combinations of symbols. The double HCP structure, for example, can also be written as *ABCB* rather than *ABAC*. The symbols are different only because a different sheet of the sequence was chosen as the reference. That is, seemingly different stacking sequences may be *translationally equivalent*. If sheet *C* of the *ABAC* sequence is taken as the reference sheet, it becomes *A* by definition, and the other sheets must be labeled relative to it. Figure 5.22a illustrates how the *A* sheet becomes *B*, and the *B* sheet becomes *C*. The stacking sequence then becomes *BCBA*. In addition to translational equivalence, stacking sequences may also be *rotationally equivalent*. As shown in Figure 5.22b, the structure does not change if the choice of the unit mesh in the reference sheet is rotated 60 degrees around the three threefold axes *A*, *B*, and *C*. Accordingly, there are six equivalent versions (Table 5.11) of stacking symbols for the double HCP structure.

The structure of the element samarium has nine differently oriented sheets in its stacking sequence. That sequence can be described in six different ways, which are all translationally and rotationally equivalent. In order to eliminate this ambiguity, a *preferred stacking sequence* should be defined and used. The unique stacking sequence is the one in which the letter symbols are given in alphabetical order. For the double HCP structure, the preferred sequence is *ABAC*, and for the samarium structure it is *ABABCBCAC*.

The stacking sequences of some mixed-sheet structures can be extensive and may contain several hundred sheets per asymmetric unit. In some instances, the stacking may be completely *random*, in which case no stacking sequence can be determined and the structure is said to be *disordered*. There are transitional cases between *ordered and disordered* stackings, and partially ordered stackings can occur.

Two minerals that have identical composition but have different crystal structures are called *polymorphs*. A subdivision of polymorphism, in which the difference in structures is due solely to the stacking of essentially identical sheets, is *polytypism*. That is, *AB* and *ABC* structures of identical composition are *polytypes*, such as the hexagonal (*AB*) and isometric (*ABC*) structures of nickel. In this example, the number of sheets per stacking sequence differs. As the number of sheets

TABLE 5.10. Simple Symmetrically Packed Structures of the Elements

← Primarily molecular structures →

NOTE: metalloids dotted
Greek letters: alternate or temperature dependent phases
Roman numerals: pressure dependent phases

Distorted structures (distortions less than 2% are neglected)

SD (or ABC) CCP	In	Diamond	●ADEF	γ-Pu	Hexagonal Si
nSD	BCT	β-Sn	√8/3 AABB	RCP	Graphite 2H
√3/2 SD Pa	GeIII	√3 ABAC	Hg	Graphite 3R
√4/3 SD BCC	γ-Ga	√3 ABACBC	
S SC		nABC	HCP
nSSDD		√3 ABC	HCP
nSEDF		ABC variable stacking	La
√8/3 SEDF (or √8/3 AABBCC)			Sm
√15/4 SEDF			
√4 SEDF			
nSDEF			

Legend structures: AB, nAB, ABAC, ABABCBCAC

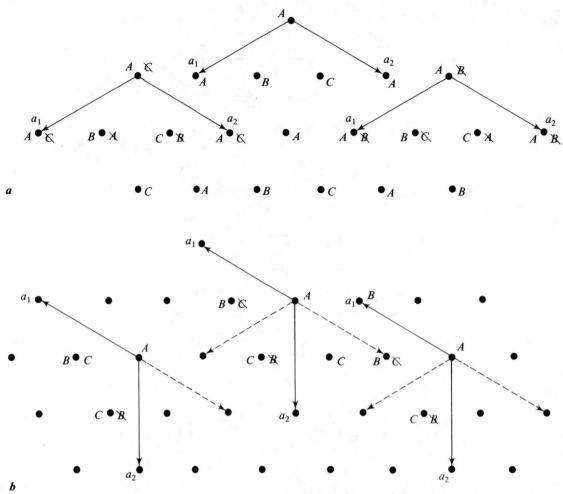

a

b

FIGURE 5.22 (a) Translational equivalence of reference sheets in *A*, *B*, and *C*, and (b) rotational equivalence.

increases in the asymmetric unit of a structure type, the possibility for additional polytypes increases. For example, the samarium stacking of *ABABCBCAC* is not one of the six equivalent versions in Table 5.11, and therefore must be a polytype. We can derive the number of permissible polytypes for the same number of sheets per stacking sequence (Table 5.12). Clearly the num-ber of polytypes of hexagonal sheets is a large number, and if the number of sheets per sequence is not limited, the number of polytypes is unlimited. Within the limits of this table (10 sheets) are 39 possible polytypes.

The study of stacking sequences and polytypes of hexagonal sheets is of little practical importance to mineralogy. This type of study can, how-

TABLE 5.11. *Translationally and Rotationally Equivalent Stacking Sequences of Double Hexagonal and Samarium Stackings*

Operation for Equivalence	Transformation	Equivalent Stacking DOUBLE HEXAGONAL				Sequence Symbol SAMARIUM								
Preferred Symbol	*ABC*		*A*	*B*	*A*	*C*		*A*	*B*	*A*	*B*	*C*	*B*	*C*	*A*	*C*
Translation to *B*	*CAB*		*A*	*C*	*B*	*C*		*A*	*C*	*A*	*B*	*A*	*B*	*C*	*B*	*C*
Translation to *C*	*BCA*	*A*	*B*	*C*	*B*		*A*	*B*	*C*	*B*	*C*	*A*	*C*	*A*	*B*	
Rotation around *A*	*ACB*		*A*	*C*	*A*	*B*		*A*	*C*	*A*	*C*	*B*	*C*	*B*	*A*	*B*
Rotation around *B*	*CBA*	*A*	*C*	*B*	*C*		*A*	*C*	*B*	*C*	*B*	*A*	*B*	*A*	*C*	
Rotation around *C*	*BAC*		*A*	*B*	*C*	*B*		*A*	*B*	*A*	*C*	*A*	*C*	*B*	*C*	*B*

TABLE 5.12. *Symmetrically Unique Stacking Sequences Containing Hexagonal Sheets*

No. of Sheets in a Sequence	Permissible and Nonequivalent Sequences			
2	*AB*			
3	*ABC*			
4	*ABAC*			
5	*ABABC*			
6	*ABABAC*	*ABACBC*		
7	*ABABABC*	*ABABCAC*	*ABACABC*	
8	*ABABABAC*	*ABABACAC*	*ABABACBC*	*ABABCABC*
	ABABCACB	*ABACABCB*	*ABACBABC*	
9	*ABABABABC*	*ABABABACB*	*ABABACABC*	*ABABCABAC*
	ABABCACBC	*ABABCBCAC*	*ABACBACBC*	
10	*ABABABABAC*	*ABABABACAC*	*ABABABACBC*	*ABABABCABC*
	ABABABCBAC	*ABABACABAC*	*ABABACACBC*	*ABABACBABC*
	ABABACBCBC	*ABABCABCAC*	*ABABCABCBC*	*ABABCACBAC*
	ABACABABCB	*ABACABACBC*	*ABACABCABC*	*ABACBACABC*

ever, be indirectly useful in understanding poly-types that exist in many mineral structures in which the anions are in hexagonal sheets.

STUFFED DERIVATIVES OF THE CLOSE-PACKED STRUCTURES

In this category of crystal structures, the anions are generally not in contact with each other, and the voids between them are occupied by cations. Such structures are called *stuffed derivatives*, because they are derived from the close-packed structures by "stuffing" the voids with cations. The only important similarity to the close-packed structures is retention of the symmetry of the packing pattern. The aspect of "closeness" of packing does not exist in the stuffed derivatives, but because the packing pattern is still highly symmetrical, they are considered as symmetrically packed structures.

Like the close-packed structures, the structures of the stuffed derivatives are based on the stacking of hexagonal and tetragonal (and other lower symmetry) sheets of anions. The region of the structure between the central planes of adjacent sheets is referred to as a *layer*, and we will use the term layer exclusively from now on to distinguish that specified region from *sheets* of anions. Voids within the layers as well as within the sheets are called *primary voids* and *secondary voids*. The primary voids are regular coordination polyhedra with the anions of adjacent sheets at their apices. The secondary voids are either faces or edges of the primary voids. Most of these secondary voids can be occupied by cations.

Stuffed Derivatives of Hexagonal Anion Sheets

The stuffed derivatives of the symmetrically

packed structures consist of stacking sequences of anion sheets. The symmetry of the hexagonal sheets was summarized earlier in Figure 5.13a. In Figure 5.23, the *AB layer* is that region of the structure between the *A* and *B* anion sheets. It contains three distinct primary voids. Two are similar but differently oriented tetrahedra, and the other is an octahedron. A *layer unit cell* is defined by the hexagonal unit mesh of the anion sheet, a_1 and a_2, in two dimensions, and by the shortest translation between two sheets, *c*, in the third dimension. By this definition, two tetrahedral and one octahedral void are in every layer unit cell.

In the *AB* layer, the two tetrahedral voids have their centers on the *A* and *B* rotation axes of the reference sheet *A* (Figure 5.23). The octahedron, positioned with two triangular faces in the *A* and *B* sheets, has its center directly over the *C* rotation axis of the reference sheet *A*. The centers of the primary voids therefore occupy the threefold rotation axes used in the stacking of hexagonal sheets.

The size of the voids and hence the kinds of cations that can occupy them depends on the size of the anion in the sheets. If we take the diameter of the anions as unity and assume for the moment that they are all in mutual contact as in closest-packed structures, the minimum radius ratio values in Table 5.1 apply. Cations with a radius of 0.414 nm would fit perfectly in the primary octahedral voids, and cations with a radius of 0.225 nm would fit perfectly in the primary tetrahedral voids. The secondary voids located in the triangular faces of the primary voids have threefold coordination and could accommodate cations with a radius of 0.155 nm. This example has no secondary void at the edges of the primary voids because the anions are assumed to be in contact.

In all stuffed derivative structures, larger cations can be accommodated because the anions are

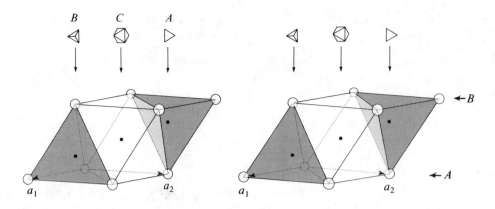

FIGURE 5.23   Stereoscopic view of an *AB* layer unit cell and position of the primary voids: two tetrahedra (shaded) and an octahedron.

not required to be in mutual contact. The size of both the primary and secondary voids can be increased by separating the anions while retaining the symmetry of the packing.

The relative orientations of the octahedral and two tetrahedral voids, as well as the position of their centers in the layer unit cell in Figure 5.23 and Table 5.13, depend on the relative positions of the anion sheets. The orientation and positioning are different in *BC* and *CA*, or in *BA*, *CB*, and *AC* layers. These relationships are summarized in Table 5.10. In each case, the orientations of the polyhedra in the table are in accord with the unit mesh of Figure 5.13. As an example, the first row of Table 5.13 corresponds to the *AB*

TABLE 5.13.   *Relative Orientation of Primary Voids in Layers of Various Stacking of Hexagonal Sheet**

Between or in Close-Packed Sheets	Orientation of Voids			Symmetrical Packing Symbol**
	A	*B*	*C*	
AB	▷	◁	⬡	$A(---)B \ldots$
BC	⬡	▷	◁	$B(---)C \ldots$
CA	◁	⬡	▷	$C(---)A \ldots$
BA	▷	◁	⬡	$B(---)A \ldots$
CB	⬡	▷	◁	$C(---)B \ldots$
AC	◁	⬡	▷	$A(---)C \ldots$

*Anions restricted to locations of threefold axes.
**Corresponding symmetrical packing symbol.

layer in Figure 5.23. The tetrahedron occupying the *A* position in the layer unit cell points down (i.e., a triangular face of the tetrahedron is part of the *B* anion sheet, and the remaining apex is part of the reference sheet *A*), the tetrahedron occupying the *B* position points up, and the octahedron in the *C* position has its upper triangular face pointed at the *B* tetrahedron. In the *CA* layer (third row, Table 5.13), the octahedron is in the *B* position, and its triangular face points toward the "up-pointing" tetrahedron in the *A* position.

All of the symmetrically packed, stuffed derivative structures can be described by a *symmetrical packing symbol* that gives the stacking sequence of anion sheets and the fractional cation occupancy of the voids between them (Table 5.13). If all of the tetrahedral voids are fully occupied by cations, a continuous layer of up-pointing and down-pointing tetrahedra is obtained (Figures 5.24a, 5.24b, and 5.24c) for *AB*, *BC*, and *CA* layers, respectively. For the *AB* stacking sequence, the symmetrically packed structural symbol $A(11-)B(11-)$ denotes this fact. The number one in the first two places between the first parentheses means full occupancy of both tetrahedra, and a dash in the third place means zero occupancy of the octahedron. The sequence of places between the parentheses is always as indicated in Table 5.13, namely, the positions of voids over the *A*, *B*, and *C* rotation axes of the reference sheet, respectively. The second set of parentheses gives the void filling between the *B* and *A* sheets of the *ABAB* ... sequence. If the stacking sequence were *ABC*, and again only the tetrahedra were occupied, the symmetrical packing symbol would be $A(11-)B(-11)C(1-1)$.

Full occupancy of the octahedral voids results in the *octahedral layers* of Figures 5.24d, 5.24e,

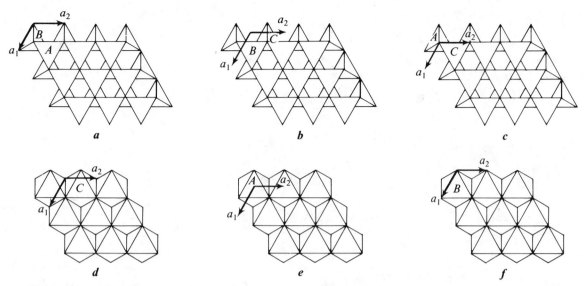

FIGURE 5.24 Tetrahedral layer in (a) an AB, (b) a BC, and (c) a CA layer. Octahedral layer in (d) an AB, (e) a BC, and (f) a CA layer. Symbols a_1 and a_2 are unit translations.

and 5.24f for AB, BC, and CA layers, respectively. Full octahedral occupancy in the ABC stacking sequence is described by the packing symbol $A(--1)B(1--)C(-1-)$. This in fact is the structure of halite (NaCl) shown in Figure II.88. The hexagonal sheets of Cl^{1-} anions are arranged in the ABC stacking sequence, and the Na^{1+} cations occupy only the octahedra. The Cl^{1-} anion sheets are in the (111) plane of the lattice. The isometric unit translations a_1 and a_2 are in the horizontal plane. The remaining unit translation a_3 is vertical and equal to 0.56404 nm in the real structure. On that basis, the Cl—Cl distance, equivalent to the length of the octahedral edge in Figure 5.24, is $0.56404(\sqrt{1/2})$ or 0.399 nm. If the Cl^{1-} anions were actually in contact, the octahedral edge would be exactly twice the Cl^{1-} radius, or 2×0.172 nm = 0.344 nm. We conclude, therefore, that the Cl^{1-} anions are not in contact. Moreover, if they were in contact, then the octahedral void would accommodate a cation with a radius of only 0.344 nm $\times (\sqrt{1/2} - 1/2) =$ 0.071 nm, which is smaller than the 0.110 nm radius of Na^{1+}. Given the actual Cl—Cl bond length of 0.399 nm, the octahedral void accommodates a cation of radius 0.399 nm $\times \sqrt{1/2} - 0.172$ nm = 0.110 nm, exactly the known radius of Na^{1+}.

We can visualize why structures such as halite (NaCl) are referred to as stuffed derivatives of the close-packed structures. They are derived from the close-packed structures by stuffing a cation into voids (octahedral in the case of halite) that are not large enough to accommodate cations when the anions are in contact. The anions must therefore lose contact with each other, but in so doing they retain their highly symmetrical arrangement in sheets.

Other examples of structures with completely filled octahedral layers are niccolite (NiAs) and pyrrhotite (FeS). Their symmetrically packed structure symbol is $A(--1)B(--1)$. The pyrrhotite structure is illustrated in Figure II.74. The As^{1-} anions occupy the A and B hexagonal sheets, and the Ni^{1+} cations occupy the AB and BA layers.

Examples of structures with tetrahedral layers (octahedral voids are empty) are sphalerite (ZnS, Figure II.68) and its polymorph wurtzite (ZnS, Figure II.71). In both structures, one of the two different types of tetrahedral voids is fully occupied by Zn^{2+}. The structures differ, however, with sphalerite having an ABC stacking sequence and wurtzite having an AB sequence. The full stacking symbol for sphalerite is $A(-1-)B(--1)C(1--)$, and for wurtzite it is $A(-1-)B(1--)$.

Many symmetrically packed structures have the tetrahedral and octahedral voids only partially occupied. In such cases a fraction is used to express the *average occupancy* of the voids. Olivine $((Fe,Mg)_2SiO_4)$, for example, has the packing symbol $A(\frac{1}{8}\frac{1}{8}\frac{1}{2})B(\frac{1}{8}\frac{1}{8}\frac{1}{2})$ indicating an AB stacking sequence of oxygen sheets (Figure II.51). Only a fraction, one eighth, of each of the two types of tetrahedra in both the AB and BA layers is occupied by Si^{4+}, and only one half of the octahedral voids are occupied by Fe^{2+} or Mg^{2+} or both, as shown in Figure 5.25a. Note that olivine was described earlier as a polyhedral-frame structure comprised of isolated Si^{4+} tetrahedra joined together by Mg^{2+} or Fe^{2+} octahedra or both. That

FIGURE 5.25 Layers in the three polymorphs of M$_2$SiO$_4$. (a) Mixed octahedral-tetrahedral layer in the (100) plane of olivine. (b) *AB* octahedral and (c) *BC* octahedral-tetrahedral layers in the {111} planes of the spinel structure. (d) Mixed octahedral-tetrahedral layer in the {021} planes of ringwoodite, and (e) alternating octahedral and octahedral-tetrahedral layers in the {101} planes of ringwoodite structure.

description has the advantage of being easily visualized in three dimensions, but it lacks the structural details that would facilitate rigorous comparisons with other structures. The case in point are the three known polymorphs of (Mg,Fe)$_2$SiO$_4$. Olivine is stable at pressures up to about 10 GPa (100 kbar) above which it converts to ringwoodite (previously known as "modified spinel"). That mineral is stable up to about 15 GPa (150 kbar) above which the spinel form of (Mg,Fe)$_2$SiO$_4$ is stable (Figure II.97). This pressure corresponds to depths in the earth on the order of 450 km. The symmetrical packing symbols for all three polymorphs are compared in Table 5.14.

Table 5.14 also illustrates both the equivalence and nonequivalence of different orientations of the hexagonal sheets within a structure. In an *ABC* closest-packed structure, a few orientations of the hexagonal sheets, corresponding to the few lattice planes, are parallel with the symmetrically

TABLE 5.14. *Summary of Symmetrical Packing Symbols of Olivine, Ringwoodite (Modified Spinel), and Spinel*

Mineral	Symmetrical Packing Symbol
Olivine	
in (100)	$A(\frac{1}{8}\frac{1}{8}\frac{1}{2})B(\frac{1}{8}\frac{1}{8}\frac{1}{2})$
Ringwoodite	
in {021}	$A(\frac{1}{8}\frac{1}{8}\frac{1}{2})B(\frac{1}{2}\frac{1}{8}\frac{1}{8})C(\frac{1}{8}\frac{1}{2}\frac{1}{8})$
in {101}	$A(--\frac{3}{4})B(\frac{1}{4}\frac{1}{4}\frac{1}{4})C(-\frac{3}{4}-)A(\frac{1}{4}\frac{1}{4}\frac{1}{4})B(\frac{3}{4}--)C(\frac{1}{4}\frac{1}{4}\frac{1}{4})$
Spinel	
in {111}	$A(--\frac{3}{4})B(\frac{1}{4}\frac{1}{4}\frac{1}{4})C(-\frac{3}{4}-)A(\frac{1}{4}\frac{1}{4}\frac{1}{4})B(\frac{3}{4}--)C(\frac{1}{4}\frac{1}{4}\frac{1}{4})$

equivalent {111} plane. These four orientations of the sheets and the stacking patterns are still identical in the isometric structure of spinel. The occupancy pattern of the layers is shown in Figures 5.25b and 5.25c. The symmetry of ringwoodite, on the other hand, is orthorhombic. The (021) and the (02$\bar{1}$) lattice planes are identical. The (101) and (10$\bar{1}$) orientations of the hexagonal sheets and layers are also. The two {021} orientations and the two {101} orientations are not equivalent to each other, however. The packing symbols in ringwoodite are consequently different in the two orientations of the hexagonal sheets. The packing symbol in the {021} plane resembles the void-filling pattern of olivine (Figure 5.25d), whereas the packing symbol in the {101} plane is similar to that of spinel (Figures 5.25e and 5.25f).

The most common stackings of the hexagonal sheet are the AB and the ABC types, and other combinations of A, B, and C. One such complex stacking is the $ABCB$ sequence found in the mineral topaz ($Al_2(F,OH)_2SiO_4$), which has the structure symbol $A(\frac{1}{6}-\frac{1}{3})B(\frac{1}{3}-\frac{1}{6})C(\frac{1}{3}-\frac{1}{6})B(\frac{1}{6}-\frac{1}{3})$. Another stacking of hexagonal anion sheets using only threefold axes is the AA type. This is found in the structure of molybdenite (MoS_2). In this stacking the voids are trigonal prisms and are filled by Mo.

Hexagonal sheets may also be stacked on the twofold rotation axes D, E, or F. The structure symbols of cinnabar (HgS), for example, are $A(--1)D(--1)E(--1)$. The Hg^{1+} occupy secondary voids in this structure (Figure II.78). The coordination polyhedra resemble octahedra but have the lower $2/m$ symmetry of antiprisms. In each of these voids, two bonds are shorter ($\sqrt{3}/8$ = 0.612) and four are longer ($\sqrt{5}/8$ = 0.791), as can be derived from Figure 5.26.

Stuffed Derivatives of Tetragonal Anion Sheets

The stacking of tetragonal sheets over the fourfold axes A and D creates voids identical to those formed by the ABC stacking of hexagonal anion sheets. The ideal anion packings of the ABC and SD symmetrically packed structures are thus identical, being exactly analogous to the corresponding two orientations of the cubic close-packed structures shown in Figure 5.18.

Figure 5.26 illustrates the SD layer and the relative orientations of the tetrahedral and octahedral voids associated with the layer. The center of the octahedral void is in the sheet, however, rather than in the layer between stacked hexagonal sheets. This fact is expressed in the structure symbol by placing the digit or fraction denoting occupancy of the octahedral void in front of the parenthesis. The fractional occupancy of tetrahedral voids is still inside the parentheses, because the tetrahedra have their centers between the sheets (Table 5.15).

A few unique structures can be described only by the stacking of open ($n = \sqrt{4/3}$) tetragonal sheets. Several of the sulfides and arsenates, in which the cation to anion ratio is in excess of $2:1$, are in this category (e.g., domeykite, Cu_3As). Their structures can be described by the $\sqrt{4/3}$ SD stacking. In the normal SD stacking, the layer unit cell has only two tetrahedra per anion, but the $\sqrt{4/3}$ SD stacking has six pseudotetrahedra (tetragonal bisphenoids) per anion.

STRUCTURE SYMBOLS AND CHEMICAL COMPOSITION

A direct relationship exists between the struc-

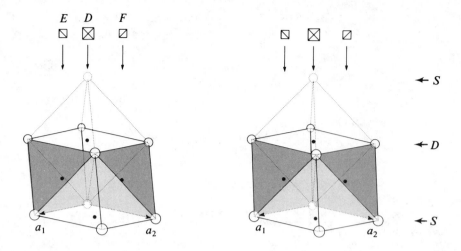

FIGURE 5.26 **Stereoscopic drawing of the positions of primary voids in an *SD* layer.**

TABLE 5.15. *Octahedral and Tetrahedral Voids in SD Stacking of Tetragonal Anion Sheet*

Between or in Close-Packed Sheets	Orientation of Voids				Symmetrical Packing Symbol
	S	*D*	*E*	*F*	
DSD	⬦		◻	◻	$D \ldots S{-}({-}{-})D$
SDS	⬦		◻	◻	$S \ldots D{-}({-}{-})S$

NOTE: The center of the octahedron is within the anion sheet and not between the sheets, as in the stacking of the hexagonal sheets. This occupancy fraction is given in front of the parenthesis.

ture symbol of a mineral and its chemical composition. Every layer unit cell has one anion and a definite number of voids, which may be wholly or partially occupied by cations. Most mineral structures consist of simple hexagonal sheets between which are exactly two tetrahedral and one octahedral void. If we know the fractional occupancy of the voids as given in the structural symbol, we know the cation/anion ratios in the chemical formula.

For example, the symmetrical packing symbol of olivine (($Mg,Fe)_2SiO_4$) (Figure II.51) is $A(\frac{1}{8}\frac{1}{8}\frac{1}{2})B(\frac{1}{8}\frac{1}{8}\frac{1}{2})$. We know that Si will be in tetrahedral coordination and that the divalent cations will be in octahedral coordination. Table 5.13, row one, reminds us that the first two entries in the *AB* layer are tetrahedra and the third entry is an octahedron. Row four of Table 5.13 indicates that the first two entries in the *BA* layer are also tetrahedra, and the third entry is an octahedron. Therefore, in both layers, one eighth of each of the two tetrahedral voids and one half of the

octahedral voids are occupied. The chemistry of one layer unit cell is then

$$(Mg,Fe)_{1/2}Si_{1/4}O \quad \text{or} \quad (Mg,Fe)_2SiO_4$$

Another example of the same chemistry in a different structure is the spinel polymorph of olivine. Its structural formula is $A(--\frac{3}{4})B(\frac{1}{4}\frac{1}{4}\frac{1}{4})$ $C(-\frac{3}{4}-)A(\frac{1}{4}\frac{1}{4}\frac{1}{4})B(\frac{3}{4}--)C(\frac{1}{4}\frac{1}{4}\frac{1}{4})$. In this case, the fractional occupancies within the second *ABC* sequence are different from those in the first, and we must consider two layer unit cells in order to represent the full chemistry. Summing the tetrahedral and octahedral sites we get

$$Mg_{4/4}Si_{1/2}O_2 \quad \text{or} \quad Mg_2SiO_4$$

DISTORTION OF SYMMETRICALLY PACKED STRUCTURES

The ideal packing symmetry of anions may be modified by distortion. In a simple structure such as halite, the size of the voids was increased by the uniform expansion of the anion (Cl^{1-}) structure in order to accommodate the larger cation (Na^{1+}). In more complicated structures in which only selected voids are occupied, the immediately surrounding anions will also be displaced, but the others will not. This causes a distortion and reduction of the normal high symmetry of the *ABC* or *SD* anion structure.

An example of such structural distortion is magnetite (Fe_3O_4), having a spinel structure with *ABC* or *SD* anion stacking of oxygen. The point group symmetry is $4/m\bar{3}2/m$. At low temperature, the single Fe^{2+} and the two Fe^{3+} are ordered in such a fashion that the oxygens are displaced to give the structure an orthorhombic $2/m2/m2/m$ symmetry. In order to understand how this happens, we must consider the structure of spinel minerals in detail.

The generalized chemical formula of spinel minerals is M_2TO_4, which has twice as many occupied octahedral voids *M* as tetrahedral voids *T*. This follows from the symmetrical packing symbol

$$A\left(--\tfrac{3}{4}\right)B\left(\tfrac{1}{4}\tfrac{1}{4}\tfrac{1}{4}\right)C\left(-\tfrac{3}{4}-\right)A\left(\tfrac{1}{4}\tfrac{1}{4}\tfrac{1}{4}\right)B\left(\tfrac{3}{4}--\right)C\left(\tfrac{1}{4}\tfrac{1}{4}\tfrac{1}{4}\right)$$

which shows that the structure consists of *AB*, *BC*, and *CA* alternating layers. Reference again to Table 5.13 and Figure 5.23 shows which of the voids are occupied in each layer. Each void corresponds to a potential cation site and a variety of cations (e.g., Fe, Al, Mg, Mn, Ti, Cr, Co, Cu, Zn, Si, Ge) may occupy these sites depending on the bulk chemistry of the rock. All spinels contain two different cations, either of two different elements or of two different valences (e.g., Fe^{2+} and Fe^{3+}).

The two different cations are either in a $2:1$ ratio or close to it. If the more abundant cation is in the *M* site and the less abundant in the *T* site, the structure is referred to as *normal*. If half of one cation occupies the *T* sites and half shares the *M* sites with the other cation, the structure is referred to as *inverse*.

If the oxygen of spinel structures is in a close-packed *ABC* structure, and if the radius of O^{2-} is 0.128 nm, then the radii of the cations in the *M* and *T* sites are 0.053 and 0.029 nm, respectively. As the distance between the oxygen anions increases uniformly, the voids expand and larger cations can be accommodated. The radius ratio of the two cations ($0.053/0.029 = 1.84$) must, however, remain the same to retain the open hexagonal sheets of oxygens and the octahedral and tetrahedral symmetries of the voids. No spinel structures containing cations with that radius ratio are known. The relative size of the two voids must be modified by a localized displacement of the oxygen common to the two polyhedra. If that shift is limited to the direction of the *T*—O bond, which is in the mirror plane containing the three-fold axes, the symmetry of the structure and the symmetry of the tetrahedral void remain unaltered. The symmetry of the octahedral void, however, is changed to that of a rhombohedral antiprism.

The solid lines in Figure 5.27 outline a pair of tetrahedral voids and an octahedral void in an expanded *ABC* structure of oxygen. The radius ratio of the *M* and *T* voids is 1.84. If the oxygen shared between two different polyhedra is dis-

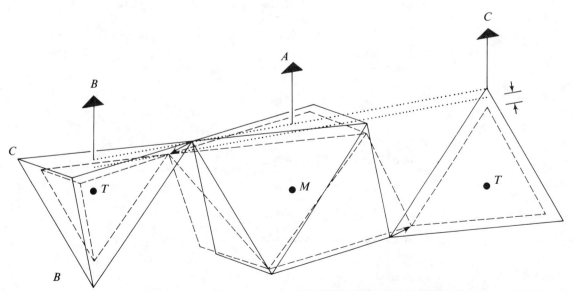

FIGURE 5.27 Two tetrahedral and an octahedral void in an expanded *ABC* structure of spinel.

placed toward *T*, the tetrahedral void retains its symmetry but shrinks in size. The octahedral void, on the other hand, changes shape to a stubby antiprism and becomes smaller. The new shapes and sizes of the voids are shown with dashed lines. Note that the radius ratio of the *M* and *T* voids is lower than 1.84 in the new model and that the oxygen sheet is no longer a plane. The up-pointing and down-pointing apices of the tetrahedra are, respectively, below and above the plane defined by the corners of the *M* polyhedra. Figure 5.27 is close to the deformed pattern of oxygens found in the mineral spinel (Al_2MgO_4). In this mineral, the radius ratio of Al/Mg is 0.92, and the *M*—O and *T*—O distances are 0.191 and 0.196 nm, respectively.

The combination of the uniform expansion of the oxygen structure and the nonuniform displacement of the oxygens can create a large variety of void sizes and radius ratios. The pattern of the oxygen expansion and displacement can be analyzed mathematically and equations derived for the accommodation of almost any combination of cation pairs. The relevant equations are given in Table 5.16. In addition to the derivation of the spinel structure for different compositions, the crystallographic and crystal chemical data for these minerals can be calculated for normal and inverse spinel structures and for various percentages of invertness. The basic structural data of the normal and inverse structures of Fe-silicate and Mg-silicate spinels are tabulated in Table 5.17. Not only does the distortion of the *M* voids change with the changing percentage of invertness, but the density of the structure changes also.

The notable increase in the density of the M_2SiO_4 polymorphs in the olivine, ringwoodite, and spinel structures ($D = 4.71$, 5.04, and 5.17, respectively, for Ca_2SiO_4) can only be explained by the differences in the permissible distortion patterns of these three structures. In olivine, all of the layers contain both octahedra and tetrahedra, but in the spinel structure, one of the layers contains only octahedra. The contraction or expansion of the octahedra in the olivine structure is restricted by the intermixing of the rigid Si^{4+}

TABLE 5.16. *Equations for the Derivations of the Spinel Structures Containing Different Cations*

T—O and *M*—O distances from known *a* and *x* values:

$$T—O = a\sqrt{3}\left(x - \frac{1}{8}\right)$$

$$M—O = a\sqrt{3}\sqrt{(x - 1/3)^2 + 1/72}$$

a and *x* from known *T*—O and *M*—O distances:

$$a = 40(T—O)\sqrt{\frac{1}{363}} + 8\sqrt{\frac{1}{11}(M—O)^2 - \frac{8}{363}(T—O)^2}$$

$$x = \frac{T—O}{a\sqrt{3}} + \frac{1}{8}$$

O–O distances and O—*M*—O angles:

$$(O—O)_T = (T—O)\sqrt{\frac{8}{3}}$$

$$(O—O)_{Ms} = a\sqrt{\frac{1}{2}} - (O—O)_T$$

$$(O—O)_{Mu} = a2\sqrt{(x - 1/4)^2 + 1/32}$$

$$(O—M—O)_s = 2\sin^{-1}\frac{(O—O)_{Ms}}{2(M—O)}$$

$$(O—M—O)_u = 2\sin^{-1}\frac{(O—O)_{Mu}}{2(M—O)}$$

NOTE:
x = fractional coordinate of anion
T = tetrahedral
M = octahedral
s = shared edge (or opposite to)
u = unshared edge (or opposite to)

TABLE 5.17. *Basic Crystallographic Data of Fe$_2$SiO$_4$ and Mg$_2$SiO$_4$ Spinels in Normal, Inverse, and Intermediate Structures*

% inverse	T—O	M—O	(O—O)$_T$	(O—O)$_{Ms}$	(O—O)$_{Mu}$	O—M—O(°)	a Unit Translation	x_{oxygen}	Density (mg/m^3)
Mg$_2$SiO$_4$									
0	0.1640	0.2100	0.2678	0.3063	0.2874	86.3–93.6	0.8119	0.242	3.49
20	0.1686	0.2068	0.2753	0.2980	0.2868	87.8–92.2	0.8108	0.245	3.50
40	0.1732	0.2036	0.2828	0.2896	0.2862	89.3–90.7	0.8096	0.248	3.52
48.4*	0.1751	0.2023	0.2859	0.2861	0.2860	90–90	0.8091	0.250	3.53
60	0.1778	0.2004	0.2903	0.2811	0.2857	90.0–89.1	0.8081	0.252	3.54
80	0.1824	0.1972	0.2979	0.2724	0.2852	92.6–87.4	0.8064	0.256	3.56
100	0.1870	0.1940	0.3054	0.2635	0.2848	93.5–85.5	0.8045	0.259	3.59
Fe$_2$SiO$_4$									
0	0.1640	0.2150	0.2678	0.3155	0.2921	85.6–94.4	0.8249	0.240	4.82
20	0.1714	0.2113	0.2799	0.3049	0.2926	87.6–92.4	0.8271	0.245	4.79
40	0.1788	0.2076	0.2920	0.2941	0.2931	89.8–90.2	0.8289	0.250	4.75
41.8*	0.1795	0.2073	0.2931	0.2932	0.2932	90–90	0.8292	0.250	4.75
60	0.1862	0.2039	0.3041	0.2829	0.2936	92.1–87.9	0.8302	0.254	4.73
80	0.1936	0.2002	0.3161	0.2715	0.2943	94.6–85.4	0.8311	0.260	4.73
100	0.2010	0.1965	0.3282	0.2597	0.2950	97.3–82.7	0.8315	0.265	4.71

NOTE: s = shared edges, u = unshared octahedral edges, x = variable coordinate of the oxygen ion.
*Intermediate structures on the average undistorted.

tetrahedra. In the spinel structure's octahedral layer, the octahedra can contract or expand without interference from tetrahedra.

SYMMETRICAL PACKING INDEX

The fundamental difference between close-packed and symmetrically packed structures is that in the former the atoms are in contact, but in the latter the anions are separated by some distance. This expansion of the distance between anions is due to the occupancy of the voids by larger than ideal-sized cations. The packing efficiency, given by the percentage of space occupied by the atoms, is a function of the structure type in the close-packed structures. The highest known packing efficiency value is 74.05% for the AB and ABC structures. The packing efficiency of the AD and $\sqrt{4/3}\,SD$ structures is 69.81% and 68.02%, respectively. Due to the expansion of the distance between anions, the packing efficiency of symmetrically packed structures is much lower than that of the structurally related close-packed structures. The equivalent close-packed packing efficiency value may be recovered from the symmetrically packed structure by disregarding the space increase due to the expansion of the anion structure. In undistorted ideal structures, this procedure yields the same percentage as the equivalent close-packed structure. The expansion of the anion structure is uniform, however, in only a few cases in which one type of void is fully occupied

by the same cation. In all other symmetrically packed structures, the expansion of the anion structure is not uniform, because the anion sheets will be distorted to accommodate the vacancy of some voids and the dimensions of cations in other voids. The calculation of the corresponding close-packed packing efficiency from symmetrically packed structures (by disregarding the space increase due to expansion) will therefore not equal the packing efficiency of the equivalent close-packed structure. The procedure will, however, give us a numerical value that expresses the general distortion of the ideal symmetrical packing. For this reason, the calculated packing efficiency serves as an index of the degree of symmetrical packing. The use of the symmetrical packing index (SPI) can be extended to include polyhedral-frame structures as well. The SPI of several major minerals is given in Figure 5.1 for comparison.

The value of the SPI can be calculated from the volume V of the unit cell, the number of anions A, and the *average* radii R_a of the anions and cations in tetrahedral (T, R_t) and octahedral (M, R_m) coordination. The ionic radii, adjusted for proper coordination number, are taken from Appendix 5.

The volume of the layer unit cell of an ideal close-packed model, with hexagonal sheets in A, B, or C positions, is:

$$V = \sqrt{\tfrac{1}{2}}\, a^3 \quad \text{or} \quad \sqrt{\tfrac{1}{2}}\, 8R^3 \qquad (5.4)$$

The volume of the layer unit cell of a symmetrically packed structure can be calculated by deriving the average value of *a* (magnitude of the unit translation of the layer unit cell) from the average ionic radii:

$$a_{\text{tetrahedral}} = (R_t + R_a)\sqrt{\tfrac{8}{3}} \qquad (5.5)$$

$$a_{\text{octahedral}} = (R_m + R_a)\sqrt{2} \qquad (5.6)$$

If the tetrahedral and octahedral values are different, their weighted average should be used to calculate the volume of the layer unit cell. The difference between the volumes of the two layer unit cells (symmetrically packed and close-packed) is the volume increase due to the expansion of the anion structure. This value is multiplied by the number of anions *A* in the unit cell and then deducted from the volume of the unit cell. The resulting volume divided into the volume of an average anion yields the value of the SPI. A summary of the calculation is:

$$\text{SPI} = \frac{\tfrac{4}{3}\pi R_a^3}{\dfrac{V}{ZA} - \sqrt{\tfrac{1}{2}}\left\{W^3 - 8R_a^3\right\}} \qquad (5.7)$$

in which

$$W = \frac{T\sqrt{\tfrac{8}{3}}\,(R_t + R_a) + M\sqrt{2}\,(R_m + R_a)}{T + M}$$

where *A* is the number of anions, *T* is the number of tetrahedral cations times 4, *M* is the number of filled octahedra times 6, *Z* is the number of chemical formula per unit cell, *V* is the volume of the unit cell, R_a is the anion radius, and R_t and R_m are the ionic radii of the two cations.

Although Equation 5.7 is limited to close-packed structures containing hexagonal sheets in *A*, *B*, or *C* positions, Equation 5.7 has broader application to other structure types, as only the ratio in volumes is considered. The use of the equation, however, is restricted to structures containing tetrahedrally or octahedrally coordinated cations or both. The calculation of the SPI is more accurate if actually observed cation-anion distances are used.

The values of the SPI are given for all ionic or dominantly ionic minerals in Part II. These calculations are based on the appropriate ionic radii obtained from Appendix 5 and are therefore approximate. In the case of the sulfides, the calculation of the SPI is not feasible because the atomic

radii in these minerals are variable due to their strong covalent and partial metallic bonding.

The value of the SPI is influenced by other factors besides the polyhedral distortion of the structures. One such factor is the compression of anions in high-pressure minerals. For example, the SPI of stishovite (high-pressure SiO_2) is 83%. That value is higher than the maximum observed value of closest-packed structures (74.05%). Since the ionic radii used for the stishovite calculation came from standard values (Table 4.4), the high SPI indicates that the radii of the ions in this structure must be smaller at high pressure. Similarly, the pressure-sensitive polymorphs of aluminum silicates (andalusite,[3] sillimanite, and kyanite) have increasingly higher SPI values (58%, 62%, and 71%, respectively) corresponding to their increasingly higher crystallization pressures.

MOLECULAR STRUCTURES

Molecular structures are characterized by strong chemical bonds between atoms that form molecular units. These bonds have relatively low symmetry and are almost always covalent. The molecular units are bonded together by much weaker residual forces that give rise to low melting point, lack of hardness, and the other characteristics discussed in Chapter 4. The molecular units may be equidimensional in shape (e.g., realgar, Figure II.87) or may consist of chains or layers of various extent.

The mineral kingdom has only a few truly molecular crystals. The minerals ice (H_2O), native sulfur (S), realgar (AsS), and orpiment (As_2S_3) are among them. A larger number of minerals are partially molecular in nature. In many ways they are more similar to the idealized symmetrically packed and open-frame structures, but they are distorted by strong directional bonds. Examples are pyrite (FeS_2), marcasite (FeS_2), the complex sulfides (also referred to as the sulfosalts), and many simple sulfide minerals that contain As, Sn, Bi, Se, and Te. Recall from Chapter 4 that these elements are nonmetals (Figures 4.9 and 4.23) and require only a small number of electrons to attain a noble gas configuration. They tend to form covalent bonds.

[3] The five-coordinated Al was assumed to be an average of a tetrahedron and an octahedron in this calculation.

CHAPTER SIX

MINERAL PHYSICS AND SYMMETRY

The subject of mineralogy deals with the origin of mineral properties as well as mineral description and identification. Chapter 4 dealt with properties perhaps more closely related to chemistry than to physics. In this chapter, we deal with properties that are closer to physics in the sense that they depend more on the crystal structure of minerals than on chemical bonding. Earlier, in considering chemical properties, we were restricted to a submicroscopic world, but the introduction of crystal structure systematics in the previous chapter opened up an entirely new point of view on the macroscopic scale. Most physical properties have their ultimate explanation on an atomic scale, for the specific arrangements of atoms and the symmetry relationships between them set mineral properties apart from those of chemically equivalent liquids and vapors. We know from previous chapters, however, that the symmetry arrangements on the atomic scale are expressed on the macroscopic scale where we make most of our physical observations. Explanations of mineral behavior can consequently be offered from two vastly different points of view. Both are important for a full understanding of mineral properties.

THE MACROSCOPIC POINT OF VIEW

Before we proceed on the macroscopic scale, let us define a *physical property* as any observable or measurable response by a mineral to some external cause. If we treat color as a property, our eye is measuring a response by the mineral to incident, visible light rays. If we treat a state of physical deformation as a property of a mineral, it can be a response by the mineral to an external stress field. Physical properties thus become "effects" in the sense of a "cause and effect" relationship. Four cause and effect relationships are illustrated at the apices of the tetrahedron in Figure 6.1. The common mechanical properties of minerals involve stress as a cause and strain as an effect. When uniform pressure is applied to a mineral, it responds by reducing its volume. The effect of compression may be uniform in all directions, or anisotropic, depending on the crystal symmetry. In any case, the reduction in volume is a response on the part of the mineral to an externally imposed stress. All of the various elastic properties of minerals such as bulk modulus, Young's modulus, and shear modulus are included in this category.

The common thermal properties of minerals include heat capacity, thermal conductivity, and a host of other thermodynamic properties. When a mineral is heated, the response is recorded as a rise in entropy, a concept considered in greater depth in Chapter 8. Similar relationships exist between applied magnetic or electrical fields and the resulting magnetic or electrical properties. An investigation of any of these properties on the

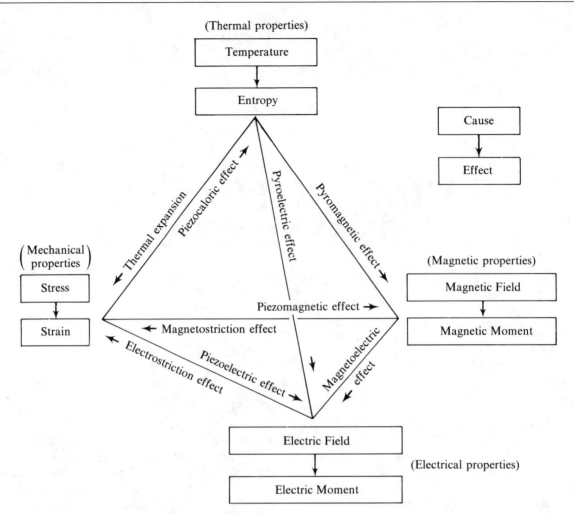

FIGURE 6.1   Tetrahedron of physical properties illustrating cause and effect relationships. Coupled effects involve a cause at one apex and an effect at another in direction of arrow.

macroscopic scale deals with the measurable effect itself and any symmetry restrictions imposed upon the property by the mineral.

The property tetrahedron of Figure 6.1 also illustrates several important coupled effects that involve a cause at one apex and one or more effects at other apices. The piezoelectric effect, for example, involves the production of an electric field in certain minerals when subjected to mechanical stress. Similarly, when a mineral is heated, the mineral can respond by changing its entropy and its volume. This gives rise to the property of thermal expansion. Heating of some minerals may also produce an electric field, and hence the property of pyroelectricity so important to the design of solar cells.

SYMMETRY CONSTRAINTS

Our observation of the effects in each of these examples provides some useful information regarding the symmetry of the responding mineral.

Consider, for example, the property of thermal expansion. The observable effect when we heat a mineral is that the mineral expands and becomes less dense. Suppose we take a garnet $(4/m\bar{3}2/m)$ and from it grind a perfect sphere. We then measure its diameter at room temperature. Upon heating the garnet sphere uniformly to a higher temperature and making the appropriate measurement, we find that the spherical symmetry is unchanged but that the diameter has increased. That is, the garnet has expanded equally in all directions. If, however, we perform the same experiment with a perfect sphere of zircon $(4/m2/m2/m)$, we find on heating that the sphere becomes an ellipsoid of revolution with two unequal dimensions. Our conclusion is that the zircon crystal has expanded more in one direction than in another. Why? The answer from a macroscopic point of view depends on the point group symmetry of the crystal involved and on any intrinsic symmetry possessed by the cause. This relationship is illustrated in Figure 6.2a for a

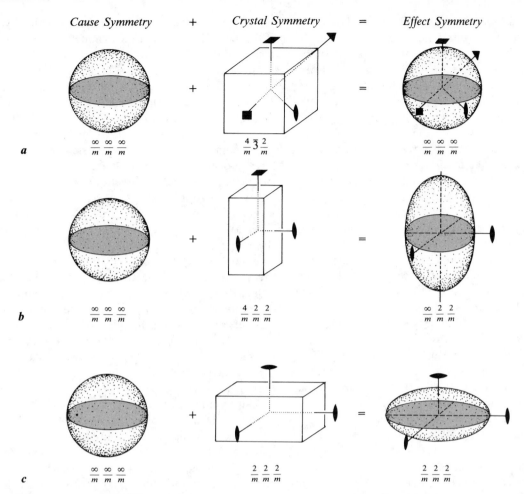

Cause Symmetry   +   *Crystal Symmetry*   =   *Effect Symmetry*

FIGURE 6.2 Relationships between cause symmetry (first column), crystal symmetry (second column), and effect symmetry (third column). Mineral properties can have symmetry no less than that of (a) a sphere if the mineral is isometric, (b) an ellipsoid of revolution if the mineral is tetragonal or hexagonal, and (c) a triaxial ellipsoid if the mineral is orthorhombic, monoclinic, or triclinic.

cubic mineral and in Figure 6.2b for a tetragonal mineral. In both cases, the "cause" has the infinite symmetry of a sphere, because in the absence of any regular atomic structure such as would be found in a gas or a liquid, atomic vibrations in response to rising temperature expand the substance by the same amount in all directions. Note that the cause symmetry is the same in both examples, but the crystal symmetries are different, and hence the observable effect symmetries are different.

The general principle that relates the symmetry of an observable effect to the symmetry of the cause and the symmetry of the medium (a mineral in this case) is sometimes referred to as *Neumann's principle*.[1] Pierre Curie, although noted more in

---

[1] Franz Ernst Neumann (1798–1895), a professor of physics and mineralogy in Königsberg, Germany, is commonly cited as the originator of this principle. A careful reading of Neumann's collected works has, however, failed to reveal the original formulation.

the history of science for his research on radioactivity, was apparently the first person to formulate the general relationship between cause and effect symmetry. Hereafter we will refer to this general relationship simply as a *symmetry argument*, because the formulation is really an outgrowth of the definition of symmetry. The symmetry argument states that *the minimum symmetry of an effect is equal to the combined symmetry that exists in both the cause and the medium*. Stated another way, the effect can have *more* symmetry than the collection of symmetry elements present in both the cause and the crystal, but it cannot have less. Some properties, such as those of the optical indicatrix discussed in Chapter 11, commonly have higher symmetry than the cause or the mineral. This is also true of the thermal expansion of garnet (e.g., $Fe_3Al_2Si_3O_{12}$) and zircon ($ZrSiO_4$) considered here. In the case of garnet, the four threefold axes are common to both the cubic crystal and the sphere (cause). The *minimum* symmetry of the effect is therefore that of a cube, although higher symmetries are permissible. In fact, the effect symmetry is that of a sphere. In the case of zircon, the tetragonal symmetry of the crystal is contained within the infinite symmetry of the cause, but the opposite is clearly not true. The only symmetry elements *common* to both the zircon and the cause are the single fourfold axis, the mutually perpendicular twofold axes, and the mirror planes. The minimum effect symmetry we can predict is $4/m\,2/m\,2/m$. As Figure 6.2b illustrates, the effect symmetry is that of an ellipsoid of revolution. Given the same cause symmetry, all tetragonal, trigonal, and hexagonal minerals will produce the same effect symmetry.

In the examples so far considered, the minimum effect symmetry can be predicted if the cause symmetries and the crystal symmetries are known. This is extremely useful to researchers in materials science who are engaged in the design or choice of crystalline substances that will produce a desired effect. For example, we will learn in a later section that the piezoelectric effect and other polar properties are possible only for noncentric crystals. In the design of piezoelectric devices, researchers can therefore automatically eliminate as material choices all minerals that possess a center of symmetry.

A mineralogist concerned with the identification of minerals applies symmetry arguments in a slightly different way. Usually the cause symmetry and the property symmetry are known, and we wish to place constraints on the possible symmetries possessed by an unknown mineral. In this context, the symmetry argument states that the *maximum* symmetry a crystal can possess is the combination of symmetry elements common to both the cause and the effect. Returning to the thermal expansion of zircon, the symmetry in common to both the cause and effect is $\infty/m\,2/m\,2/m$. As this is the maximum symmetry the mineral can have, the mineral cannot possibly be isometric.

Because the symmetry argument places only a maximum on possible crystal symmetries, a mineral with an effect symmetry of a sphere can in principle have any point group symmetry; the argument sets no restriction on the minimum symmetry. This condition is exactly analogous to the restrictions imposed on crystal symmetries by lattice geometry as discussed in Chapter 2. For example, a crystal that has lattice parameters consistent with an hexagonal coordinate system can have lower symmetry. Most minerals with spherical effect symmetries are in fact isometric because all isometric crystals *must* have spherical

effect symmetries. Similarly, minerals with physical properties that yield the symmetry of an ellipsoid of revolution cannot be isometric, and are in most cases either tetragonal or hexagonal (or trigonal). They can, however, be either orthorhombic, monoclinic, or triclinic.

Figure 6.2c illustrates the third type of effect symmetry commonly observed in minerals. All orthorhombic, monoclinic, and triclinic minerals have properties that vary in magnitude in three mutually perpendicular directions. The effect symmetry is described as a triaxial ellipsoid. The thermal expansion of forsterite (Mg_2SiO_4) provides a good example. When a perfect sphere of the mineral is heated, the extension of the structure along each of its crystallographic axes is not the same. The resulting triaxial ellipsoid may be oblate (flattened) or prolate (vertically extended), depending on the mineral. In any case, the observable effect has orthorhombic symmetry. If we observed an unknown mineral to have orthorhombic thermal expansion, we could restrict its identity to having an orthorhombic, monoclinic, or triclinic lattice.

The symmetry argument we present here is very general in its application. Recall from Chapter 3 that the apparent symmetry of a crystal based on its crystal form can be greater than the real point group symmetry, but never less. The minimum symmetry of crystal forms is thus the true symmetry of the crystal, but higher symmetries are possible.

Not surprisingly, physical properties other than external morphology of crystal faces have symmetry. Recall that the concept of translational symmetry insures that the point group symmetry inherent to the mineral will be expressed everywhere in that mineral. Symmetry related properties such as thermal expansion, density, electrical conductivity, and many others will have the same value for a small chip taken from the corner of a crystal as for the center of the crystal or for the crystal as a whole. We will see in the next section that both physical and chemical properties viewed on an atomic scale are subject to symmetry constraints, mainly because most chemical bonds are directed from one atom to the next and are therefore dependent on the positions of the atoms.

THE ATOMISTIC POINT OF VIEW

One of the principal objectives of modern research in mineralogy is to relate the physical and chemical properties of minerals to the actual arrangement of the atoms in their crystal structures. Crystallography has traditionally played a central role in achieving this objective. By the use of x-ray diffraction techniques, the three-dimensional distribution of electrons, of atoms, and of ions can be determined for the asymmetric part of the unit cell. This procedure is relatively straightforward with the use of automated diffractometers and modern digital computers. Application of space group symmetry extends the asymmetric unit to the complete unit cell, and translational symmetry then extends the structure over the entire specimen.

Given the electron distribution, we can locate atom centers and calculate interatomic distances. The shapes of coordination polyhedra, and the manner in which they are linked, are then determined. This basic information is essential to the understanding of structural anisotropy and of why the magnitude of various physical properties varies with direction in most crystals.

Most of the classic research in structure determination has been conducted with crystals at room temperature and pressure. This is entirely appropriate for the purpose of relating structure to physical properties under those conditions, but earth scientists who are concerned about the interior of the earth are interested in structure-property relations at much higher temperatures and pressures. For example, the density and various elastic properties of the earth's interior are inferred directly by geophysicists from the analysis of seismic wave propagation. The correct identity of materials at depth is greatly facilitated by a knowledge of how these properties may vary with increased pressure and temperature.

For these reasons, a recent and important trend has developed in mineralogical research to determine mineral structures at elevated temperatures and pressures. Specially designed heaters and pressure cells now permit the study of crystals at temperatures over 1200 K and at pressures over 150 GPa (1.5×10^6 bars). Results show how interatomic distances and angles change with temperature and pressure, and thus provide valuable insight into the actual atomic mechanisms by which certain properties change.

THE DIMENSIONALITY OF PROPERTIES

As minerals are three-dimensional objects, a complete description of any particular property requires knowledge of how that property varies with direction. If the magnitude of the property does not change with crystallographic direction, the effect symmetry is that of a sphere (Figure 6.2a), and the mineral is said to be *isotropic* with respect to that property. If a property does vary with direction, the mineral is said to be *anisotropic* with respect to that property. A single mineral may be isotropic with respect to some properties and anisotropic with respect to others.

Properties such as density and temperature have no direction in a crystal. A complete description requires only a number and an appropriate choice of units. Such *scalar properties* are mathematically referred to as *zero-rank tensors*. A *tensor* in the present context is a quantity that specifies how physical properties vary in one or more directions. Its value in more advanced treatments of mineralogy and crystal physics is the ease with which more than three dimensions can be treated.

Many other properties exist for which both the magnitude and direction must be specified to give a complete description. The resulting vector property is referred to as a *first-rank tensor*. In the three-dimensional coordinate system that every mineral has, the projection of this vector on each of the three axes generates three coefficients. In general, we need to know all of these coefficients in order to relate the property to the crystal structure.

The number of coefficients required for a complete property description increases as the relationship between the property and the structure becomes more complex. A commonly cited example is electrical conductivity. Ohm's law states that the electric current density **J** is proportional to the electric field **E**. The constant of proportionality is called the electrical conductivity. The current produced by the field will in general be different in three mutually perpendicular directions through a crystal. The component of the total current in each of these directions requires a vector quantity, each of which has three conductivity coefficients σ. In all, 3^2, or nine, coefficients are required to fully describe the variation of electric current as a function of direction.

$$\mathbf{J}_x = \sigma_{xx}\mathbf{E}_x + \sigma_{xy}\mathbf{E}_y + \sigma_{xz}\mathbf{E}_z$$
$$\mathbf{J}_y = \sigma_{yx}\mathbf{E}_x + \sigma_{yy}\mathbf{E}_y + \sigma_{yz}\mathbf{E}_z$$
$$\mathbf{J}_z = \sigma_{zx}\mathbf{E}_x + \sigma_{zy}\mathbf{E}_y + \sigma_{zz}\mathbf{E}_z$$

Properties such as electrical conductivity are referred to as *second-rank tensors*. The rank *n* of any tensor is given by the maximum number of coefficients 3^n necessary to describe the property. Second rank tensors relate two vector quantities (**J** and **E** in the example above), whereas third-rank tensors (27 coefficients) relate a second-rank tensor to a vector. Fourth-rank tensors, required to describe the full generality of elastic properties, relate two second-rank tensors.

Fortunately, the presence of crystal symmetry greatly reduces the complexity of tensor representation. The symmetry operations given by a point group make certain directions through a crystal equivalent. For instance, in the example of electrical conductivity, only six independent coefficients exist, because the effect in one direction is the same as in the opposite direction (e.g., $\sigma_{xy} = \sigma_{yx}$). This is the situation in the absence of crystal symmetry. In the presence of crystal symmetry, a single twofold axis will reduce to four the number of required coefficients. Additional symmetry elements reduce the number still further until, finally, in the isometric system only a single parameter is needed. The algebraic formalism by which this reduction is accomplished is beyond our scope here, but it is a standard approach in more advanced treatments. The important point is that physical properties have symmetry, and the rank of the tensor required to describe the property depends on the mineral symmetry.

One approach to discussing the various structurally related properties is in terms of their rank. The simplest properties are discussed first, and then additional more complex properties are introduced. In mineralogy, however, some properties are far more important and commonly used than others. We will restrict our attention to these.

THERMAL EXPANSION

The two most important properties with respect to mineral stability in the earth's crust and mantle are thermal expansion and compressibility. The final word on stability, alluded to in Chapter 5, is an energetic problem, which we will discuss in greater depth in Chapters 8 and 9. An important contribution to the total energy at any temperature and pressure is the state of atomic vibration and distortion under those conditions. As the temperature is raised, minerals expand, and the ions within the structure adjust to the increased energy

state by increasing their amplitudes of vibration. A standoff exists between this tendency of the ions to separate from each other and the interacting attractive and repulsive forces that tend to maintain the ions in their equilibrium positions. Once the kinetic energy exceeds the attractive forces, either the mineral forms a different structure (polymorph) or melts. The exact mechanism by which this happens is one that involves the position of ions or atoms in the structure and the symmetry constraints placed upon them.

The linear coefficient of thermal expansion α, at constant pressure P, is given by

$$\alpha_P = \frac{\Delta \ell}{\ell \Delta T}$$

where ℓ is the original length, and $\Delta \ell$ is the change in length through a temperature interval ΔT. In practice, it is a simple measurement to make. We need only measure a change in length with a mechanical dilatometer, or as is more commonly done in mineralogy, measure the lattice parameters as a function of temperature. The standard procedure employs the methods of x-ray diffraction discussed in Chapter 10. Given the change in lattice parameters, the change in volume with temperature change is easily calculated. This gives the volume coefficient of thermal expansion $\alpha_P(V)$ as

$$\alpha_P(V) = \frac{\Delta V}{V \Delta T}$$

We discussed the macroscopic aspects of thermal expansion in the introduction to this chapter. Expansion creates a deformation in the unit cell that may be represented by one of the three effect symmetries given in Figure 6.2. These are strain ellipsoids commonly used to describe the state of mechanical deformation. In minerals having orthorhombic or higher symmetry, the principal axes of the ellipsoid are constrained by symmetry to coincide with the crystallographic axes. The strain ellipsoid is therefore totally defined if

TABLE 6.1.   *Unit Cell Dimensions of an Orthopyroxene at Various Temperatures at Room Pressure*

Unit Cell Dimensions	Temperature (°C)					
	20	175	280	500	700	850
a (nm)	1.8337	1.8364	1.8371	1.8429	1.8483	1.8546
b (nm)	0.8971	0.8988	0.9000	0.9028	0.9053	0.9081
c (nm)	0.5232	0.5238	0.5242	0.5260	0.5280	0.5298
V (nm³)	0.8607	0.8649	0.8667	0.8751	0.8835	0.8923

SOURCE: Data from Table 2 of Smyth, J. V., 1973. An orthopyroxene structure up to 850 °C. *American Mineralogist* 58:636–648. (Data converted to SI units.)

the linear coefficients of thermal expansion are known in each of the lattice directions *a*, *b*, and *c* as a function of temperature. The data in Table 6.1 are from studies of an orthopyroxene and are fairly typical. They yield linear thermal expansion coefficients α on the order of 10^{-5} per degree centigrade.

For minerals of lower symmetry, that is, monoclinic and triclinic, the same constraint does not generally hold. The strain upon expansion may still be described by an ellipsoid, but its axes need not coincide with the crystallographic axes. The only exception is the monoclinic $2/m$ point group in which the twofold axis must coincide with one of the ellipsoid axes. The symmetry constraints for an orthorhombic and a monoclinic pyroxene are compared in Figure 6.3. In the case of the orthorhombic pyroxene (Figure 6.3a), thermal expansion is greater along the *c* axis than along the *a* axis, and the symmetry axes of the resulting strain ellipse correspond to the crystallographic axes. In the case of the monoclinic pyroxene (Figure 6.3b), the direction of maximum thermal expansion r_1 does not correspond to any of the crystallographic axes.

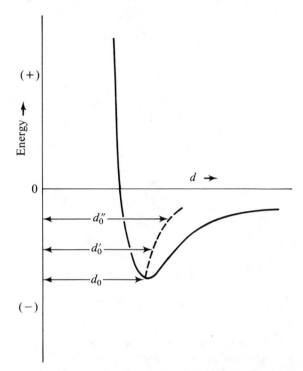

FIGURE 6.4 Relationship between interatomic distance *d* and temperature based on bonding energy. Interatomic distance at absolute zero is at d_0. As temperature rises, mean amplitude of atomic vibration increases, energy increases, and interatomic distances increase from d_0 to d'_0 to d''_0.

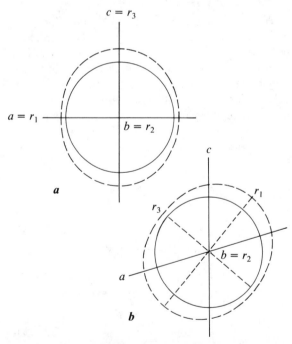

FIGURE 6.3 Symmetry constraints on the orientation of the strain ellipsoid due to thermal expansion in (a) orthorhombic crystal, and (b) monoclinic crystal. r_1, r_2, and r_3 are the components of maximum, intermediate, and minimum strain, respectively. (Example of orthopyroxene and clinopyroxene after Ohashi, Y., and Burnham, C. W., 1973. Clinopyroxene lattice deformations: the roles of chemical substitution and temperature. *American Mineralogist* 58:846.)

Volume increases as temperature increases because the interatomic distances increase as the amplitudes of vibration of each atom or ion increase. The reasons for this are illustrated in Figure 6.4. Recall from Chapter 4 that the electrostatic attractive energy of a cation for an anion varies as the inverse of the distance between them, whereas the Born repulsive forces vary as the inverse of the sixth to the eighth power of the interionic distance, depending on the number of outer electrons. The energy curve in Figure 6.4 is thus strongly asymmetric, being steeper on the nucleus side. This imparts a marked asymmetry to the total energy curve. The bottom of the energy well represents the energy of an ion pair at absolute zero temperature. Under that condition, the ions have no vibrational energy due to thermal energy, and the interionic distance d_0 is well defined. As the temperature increases, each ion begins to oscillate about a mean position determined by a balance between the attractive and repulsive forces. As illustrated by the asymmetric energy curve, the ions tend to vibrate outward against the attractive forces rather than inward against the stronger repulsive force. The interionic distance therefore increases as the temperature is raised, and we observe thermal expansion.

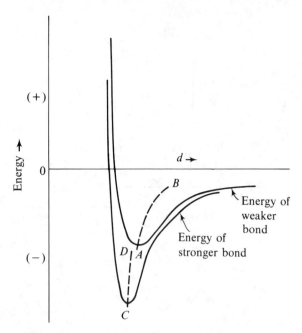

FIGURE 6.5   Strong chemical bonds are characterized by deeper and more symmetric energy wells. Thermal expansion *CD* of stronger bond is less for the same temperature increase than expansion *AB* of weaker bond.

We can further reason from these principles that strong bonds will exhibit less thermal expansion than weak bonds. Figure 6.5 illustrates the general case. Other conditions being equal, a strong bond offers more resistance to deformation and consequently has a deeper and more symmetric energy well. Note that if the well were perfectly symmetric, no expansion would occur. The only constraint on the symmetry of the well is that the repulsive limb have a negative slope and the attractive limb have a positive slope. If the attractive limb were steeper, a bond could conceivably contract upon heating.

For the same reasons, minerals that have highly charged or highly electronegative ions or both will generally have small thermal expansions. Minerals with strong covalent bonds, for example, diamond, have small expansion coefficients. A metal cation-oxygen bond with low coordination should also expand less than the same bond in a high coordination site, because the total electrostatic energy associated with a particular cation is distributed among fewer bonds when the coordination is low. This is the basic relationship given by the *Pauling electrostatic bond strength* equation:

$$\text{bond strength} = \frac{Ze}{\text{CN}}$$

where *Ze* is the cation charge and CN refers to the cation coordination number. We would expect the thermal expansion behavior of tetrahedral

structures to differ from that of purely octahedral structures. Most of the common rock-forming silicates consist of both tetrahedra and octahedra, and we can anticipate the details of their expansion to be more complex.

Figure 6.6 shows the thermal expansion for various structures. The behavior of low quartz may seem surprising at first because of its rapid expansion. Because it consists exclusively of Si^{4+} tetrahedra, we would anticipate minimal expansion of the strong Si—O bonds. In this case, the rapid expansion is due to rotation of tetrahedra about the twofold axis until, at a temperature of 573 °C, the high quartz structure is stabilized. Rotation is accomplished without loss of symmetry ($P3_12$), because the apical oxygens act as pivotal points on which the structure can expand from its somewhat collapsed form (Figure 6.7a). Above 573 °C, high quartz expands very little. This behavior is more in line with the anticipated behavior of Si—O bonds. The Si tetrahedra in high quartz can no longer rotate, because their positions in the structure are fixed by the $P6_422$ symmetry. Two twofold axes are now passing through each tetrahedron and preventing their rotation. Thermal expansion is therefore accomplished by Si—O bond expansion.

A general rule was formulated[2] that the thermal expansion of structures in which polyhedra orientations are fixed by symmetry is always less than the thermal expansion in structures in which polyhedra are rotated away from highly symmetrical positions. Any structure in which the coordination polyhedra share edges or faces are generally

[2] Megaw, Helen D., 1971. Crystal structures and thermal expansion. *Materials Research Bulletin* 6:1007–1018.

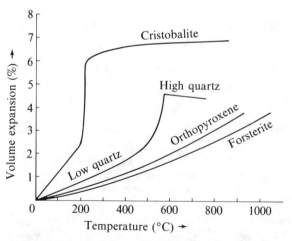

FIGURE 6.6   Volume expansion in percent of room-temperature volume for selected silicate minerals.

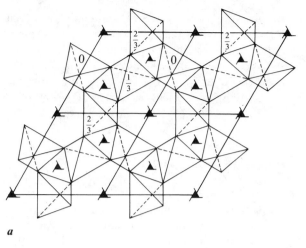

a

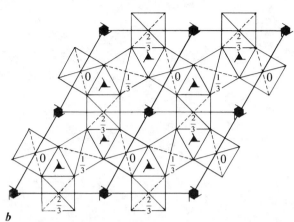

b

FIGURE 6.7   Structures of (a) low quartz and (b) its high-temperature polymorph, high quartz. Rotation of tetrahedra in low quartz is permissible by the $R3_121$ **symmetry. Rotation of tetrahedra is impossible in high quartz because of the higher $P6_422$ symmetry of the space group.**

constrained by symmetry not to expand by rotation (Figure 6.7b).

Among the collapsed structures in which tetrahedra share corners are the minerals low quartz (SiO_2), cristobalite (SiO_2), leucite (KSi_2AlO_6), sodalite ($Na_4ClSi_3Al_3O_{12}$), and perovskite ($CaTiO_3$). The structures of these minerals demonstrate rapid initial expansion, whereas structures in which tetrahedra or octahedra share edges or faces do not. Some of the minerals in the latter category are forsterite (Mg_2SiO_4), corundum (Al_2O_3), periclase (MgO), and fluorite (CaF_2). These are much more representative of the normal thermal expansion of most minerals. Figure 6.6 shows that for the temperature range of 25 to 1000 °C, most minerals expand on the order of 3 to 4% of their cell volumes at 25 °C.

We now turn to the thermal expansion behavior of the common rock-forming silicates, which have structures consisting of both Si tetrahedra and various metal cation octahedra. In this category are all pyroxenes, amphiboles, garnets, olivines, and many others. Modern methods of crystal structure refinement yield the precise shape and dimension of each coordination polyhedron. When the structure is determined at different temperatures, the change in the cation-oxygen bond lengths and the shape of polyhedra can be followed as functions of temperature. This is the most important and productive technique used today for studying the mechanism of thermal expansion.

The results of such studies conducted since 1973 show that many silicates accommodate ther-

mal energy by *differential polyhedral expansion*. This mechanism is based on the experimental observation and theoretical prediction that the weaker cation-oxygen bonds in octahedra will expand more than the stronger Si—O bonds in tetrahedra. In the case of the orthopyroxene considered earlier (Table 6.1), the cation-oxygen bonds in the two octahedra, *M1* and *M2*, expand as expected, and the Si—O bonds appear to shorten as the temperature is raised (Figure 6.8). This shows that weaker bonds expand more than stronger bonds. In the case of the orthopyroxene, the decreased bond strength of octahedra at elevated temperature is balanced by an increase in bond strength in the tetrahedra, and hence the apparently shorter bond length.

In spite of differences in behavior of various coordination polyhedra upon expansion, we can make some generalizations about average thermal expansion in many different structures. Mean thermal expansion coefficients for individual polyhedra are plotted in Figure 6.9 against Ze/CN, the Pauling bond strength notation. Large monovalent cations (e.g., Na^{1+}) with high coordination are weakly bonded to oxygen, whereas small, more highly charged cations (e.g., Al^{3+}) with low coordination are strongly bonded. A straight line fit to these data yields an expression (Figure 6.9) that can be used to predict the thermal expansion behavior of polyhedra in minerals not so well studied.

Differential polyhedral expansion, which we will consider further in Chapter 8, has important consequences for the thermal stability of minerals.

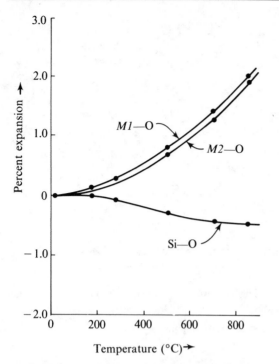

In nearly all of the silicate minerals in which tetrahedral and octahedral layers of cations alternate, an increase in temperature causes a progressive distortion of the structures and a misfit of the layers. The strain that develops so increases the total energy that finally a different structure is energetically favored. The new structure will be relatively unstrained and will have a lower energy.

COMPRESSIBILITY

Compressibility, like thermal expansion, is an elastic property of minerals. When a mineral is subjected to a uniform pressure on all sides, it responds by decreasing its volume. If an isometric crystal is ground to a spherical shape and then subjected to such a pressure,[3] it will retain its spherical shape upon compression, but if a sphere of any tetragonal or hexagonal mineral is compressed, the change in volume is *anisotropic*. That is, the linear compressibility is more in one direction than in the other, and an ellipsoid of

FIGURE 6.8 **Variation of octahedral (*M1*—O and *M2*—O) and tetrahedral (Si—O) bond lengths with temperature for a single orthopyroxene crystal. (Data from Smyth, J. V., 1973. An orthopyroxene structure up to 850 °C. *American Mineralogist* 58:643.)**

[3]The term *hydrostatic* is commonly used in this context to describe the homogeneous pressure exerted by a medium (e.g., water) having no strength.

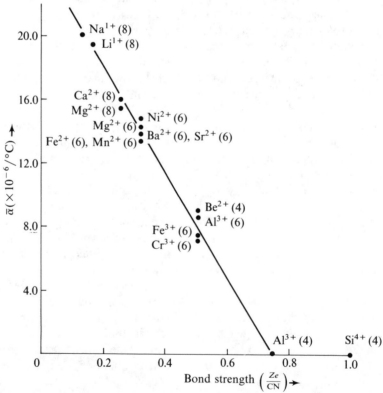

FIGURE 6.9 **Variation of mean thermal expansion coefficients of various cation polyhedra with Pauling bond strength. A straight line fit to data (ignoring Si⁴⁺) yields $\bar{\alpha} = 24.68 - 32.9 \, (Ze/\text{CN}) \times 10^{-6}/°\text{C}$. Numbers** in parentheses are CN. (Data and fit from Hazen, R. M., and Prewitt, C. T., 1977. Effects of temperature and pressure on interatomic distances in oxygen-based minerals. *American Mineralogist* 62:309–315.)

revolution (Figure 6.2b) results. Spheres of all orthorhombic, monoclinic, and triclinic minerals compress to triaxial ellipsoids (Figure 6.2c) with principal axes proportional to the linear compressibilities in those directions. We can consider spheres of anisotropic minerals to have "soft" regions, which are more compressible and which are represented by the shortest dimensions of the resulting ellipsoids. Elastically "stiff" directions are the least compressible and are represented by the maximum axes of the ellipsoids. The term "elastic stiffness coefficients" is commonly used in geophysics and crystal physics to refer to the matrix of coefficients in a fourth-rank tensor required to describe elastic problems in general. The orientations of strain ellipsoids relative to crystallographic axes are the same as those that result from thermal expansion (e.g., Figure 6.3).

The coefficient of linear compressibility β_T is determined at constant temperature (usually 25 °C) and is defined by

$$\beta_T = -\frac{\Delta \ell}{\ell \Delta P}$$

Its values are given as a pure number per unit of pressure. Typical values range from -10^{-13} to -10^{-12} per pascal (-10^{-4} to -10^{-3} GPa). Volume compressibility $\beta_T(V)$ is more closely related to the bulk mineral behavior, and is defined as

$$\beta_T(V) = -\frac{\Delta V}{V \Delta P} = -\frac{1}{B_T}$$

The parameter B_T, called the *bulk modulus*, is widely used in geophysics and crystal physics in place of compressibility. The two terms are simply negative reciprocals of each other. Typical values of B_T range from 20 GPa (200 kbar) for the relatively soft alkali halides such as sylvite (KCl) and halite (NaCl) to 300 to 400 GPa (3000 to 4000 kbar) for the relatively stiff oxides such as corundum (Al_2O_3) and stishovite (SiO_2).

The advent of high-pressure single crystal diffraction studies using the diamond anvil cell has made it possible to measure interatomic distances at elevated pressures. Such research has put us in a much better position to evaluate the actual mechanisms by which compression takes place. Figure 6.10 shows the results of such a study conducted on forsterite (Mg_2SiO_4). Like orthopyroxene, forsterite has two octahedral sites and one tetrahedral site in the orthorhombic structure. Both of the octahedral cation-oxygen bonds are relatively soft compared with the Si—O bonds. The compression of the Mg—O octahedra will contribute more to the overall volume reduction than the Si—O tetrahedra. Thus, *differential polyhedral compression* operates in just the opposite

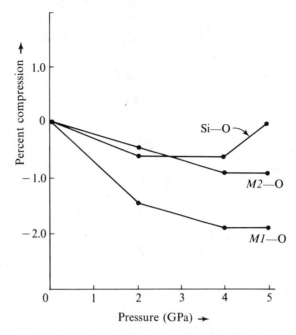

FIGURE 6.10 **Variation of octahedral (*M1*—O and *M2*—O) and tetrahedral (Si—O) bond lengths with pressure for a single crystal of forsterite (Mg_2SiO_4). (Data from Hazen, R. M., 1976. Effects of temperature and pressure on the crystal structure of forsterite. *American Mineralogist* 61:1280–1293.)**

direction from differential polyhedral expansion, which takes place upon heating.

Referring back to Figure 6.4 further illustrates the behavior of individual atoms or ions upon compression. Although the attractive and repulsive forces acting upon ions are exactly balanced at the interatomic distance d, the strong asymmetry of the energy distribution curve shows marked resistance to compression. The Born repulsive forces are several orders of magnitude greater than the Coulomb attractive forces, so as compression proceeds, further compression is even more difficult. Part of this resistance to compression is due also to the fact that, as the Pauli exclusion principle states, more than two electrons cannot occupy the same energy position.

In view of the repulsive forces that resist compression, we might expect some relationship to exist between the bulk modulus, or compressibility, and the density of electrons between two bonded ions. The charge density can be represented by Ze/d^3 where Ze is the charge of the centrally coordinated cation of a polyhedron, and d is the interatomic distance to one of the coordinating oxygens. A plot of mean polyhedral compression $\bar{\beta}$ and charge density shown in Figure 6.11 has striking similarities to thermal expansion behavior shown in Figure 6.9. Highly charged cations with low coordination not only resist ther-

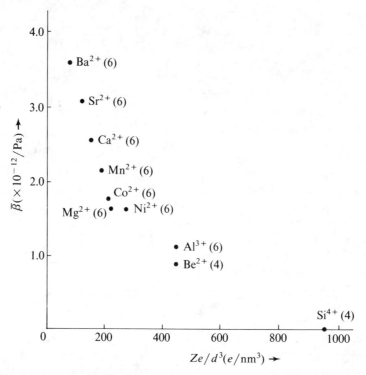

FIGURE 6.11 Variation of mean linear compressibility of various coordination polyhedra with electrostatic charge density. Numbers in parentheses are CN. (Data from compilation by Hazen, R. M., and Prewitt, C. T., 1977. Effects of temperature and pressure on interatomic distances in oxygen-based minerals. *American Mineralogist* 62:309–315.)

mal expansion, but they are also relatively incompressible. In contrast, cations with a low charge and high coordination are both highly expandable and compressible. We can use these generalizations to predict the behavior of other polyhedra for which measurements either have not been made or cannot be made, and to predict the behavior of minerals that consist of such polyhedra.

PIEZOELECTRIC EFFECTS

The property of piezoelectricity refers to the development of a momentary electric current when crystals are squeezed suddenly in certain directions. The strain caused by the squeezing is very small and purely elastic. When the stress is removed, the crystal is restored to its original form without damage. The flow of electricity is caused when an initially uniform charge distribution among cations and anions is distorted such that opposite charges either move directly away from each other or directly toward each other. If both charges move in the same direction, their effects cancel, and no current exists. Current can flow only when opposite charges move in opposite directions and only when the charges are actually in motion.

We can arrange a simple demonstration of piezoelectricity with a low-power neon glow bulb, some electrical leads, a bit of aluminum foil, two thumbtacks, and a crystal of Rochelle salt (potassium sodium tartrate tetrahydrate). The crystal is not a mineral, but it is grown readily from solution and is one of the best piezoelectric materials. When the apparatus is assembled as in Figure 6.12, a flash of light will appear when the

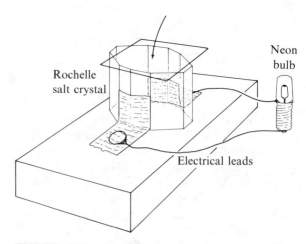

FIGURE 6.12 A sudden pressure (arrow) on a Rochelle salt crystal connected by electrical leads to a neon bulb will produce a flash of light.

crystal is tapped with a hammer. Current flows momentarily in one direction when the crystal responds to the blow, and then flows in the opposite direction as the crystal returns to its normal state. Consecutive taps will produce consecutive flashes of light.

To understand in more detail how this property works, consider the structure of the piezoelectric mineral sphalerite (ZnS). One positively charged Zn^{2+} cation and four negatively charged S^{2-} anions are in tetrahedral coordination as shown in Figure 6.13. Likewise, each S^{2-} anion is tetrahedrally coordinated to four Zn^{2+} cations. Sphalerite is isometric with the $\bar{4}3m$ point group. The Zn^{2+} and S^{2-} ions occupy special positions in the mirror planes. The shortest S—S distances are parallel to the shortest Zn—Zn distances and are in the $\langle 110 \rangle$ direction (Figure 6.13a).

When this structure is compressed in the [110] direction as shown in Figure 6.13b, the S—S and Zn—Zn distances momentarily compress in the [110] direction and expand in the [$\bar{1}$10] direction. The regular isometric ZnS_4 tetrahedron is consequently distorted. The tetrahedral edges along [$\bar{1}$10] become longer, and those along [110] become shorter (Figure 6.13b). The inclined edges change their direction and become slightly longer, and the isometric tetrahedron momentarily takes on an $mm2$ symmetry. Because of this deformation, the centrally coordinated Zn^{2+} and S^{2-} momentarily move toward each other, causing a downward flow of electricity. Note that the flow of electricity is perpendicular to the direction of stress. When the movement of the ions stops, the flow of electricity stops, and when the ions move back to their original positions, the flow of electricity starts again in the reverse direction. The positive and negative poles are thus reversed.

The opposite phenomenon, called the *electrostriction effect*, or the converse piezoelectric effect, is caused by subjecting a suitably oriented crystal to an electric current. In the case of sphalerite, positive and negative electrodes attached to the top and bottom of the crystal (Figure 6.13), respectively, will cause a downward deflection of the Zn^{2+} and an upward deflection of the S^{2-} when the circuit is connected. The crystal changes shape slightly, compressing between the electrodes. When the circuit is opened, the crystal expands in that direction, returning to its original dimension.

The piezoelectric effect and the electrostriction effect are possible only in minerals having noncentric point groups with the exception of the 432 point group, which, due to its high symmetry, can exhibit no piezoelectric effect (Table 6.2). These point groups are sometimes called the 20 piezoelectric crystal classes. In order for an electric current to flow, the charge distribution must be acentric. For that reason, piezoelectric minerals cannot have a center of symmetry, otherwise the current generated in one direction would be cancelled by an opposite but symmetrically equivalent current. A *polar direction* in a crystal is a direction that has no symmetrically equivalent opposite direction. Only those directions are potential *electric axes*, directions in which current is permitted by symmetry to flow. In the 11 centric point groups, *all* directions through the crystal are nonpolar. In the remaining 20 point groups, some directions are polar and others are not. These are summarized in Table 6.2.

Among the common minerals, sphalerite (ZnS), tourmaline $(NaMg_3Al_6(OH)_4B_3O_9Si_6O_{18})$, low quartz (SiO_2), topaz $(Al_2(F, OH)_2SiO_4)$, and low boracite $(Mg_6Cl_2B_6B_8O_{25})$ have readily measurable piezoelectric effects.

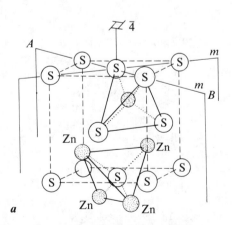

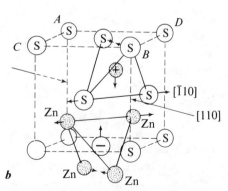

FIGURE 6.13   (a) Structure of sphalerite (ZnS). Dashes outline unit cell and the four ZnS_4 tetrahedra within. Symmetry is $\bar{4}3m$. (b) Effect of compression along [110] *AB* and the relative motion of Zn^{2+} (stipled) and S^{2-} (open) ions when an electric charge is produced.

Crystal System	Point Group PIEZOELECTRIC	PYROELECTRIC	Nonpolar Directions	Unique Polar Axis
Triclinic	1	1	(all)	1
Monoclinic	2	2	$\perp 2$	2
	m	m	$\perp m$	$\|m$
Orthorhombic	$m2m$	$m2m$	$\perp 2$	2
	222	—	$\perp 2$	—
Tetragonal	4	4	$\perp 4$	4
	$\bar{4}$	—	$\perp$ or $\|\bar{4}$	—
	$4mm$	$4mm$	$\perp 4$	4
	422	—	$\perp 4$ or $\perp 2$	—
	$\bar{4}2m$	—	$\perp \bar{4}$ or $\perp 2$	—
Hexagonal	3	3	(all)	3
	$3m$	$3m$	$\perp m$	3
	32	—	$\perp 2$	—
	6	6	$\perp 6$	6
	$\bar{6}$	—	$\perp \bar{6}$	—
	$6mm$	$6mm$	$\perp 6$	6
	622	—	$\perp 6$ or $\perp 2$	—
	$\bar{6}2m$	—	$\perp 2$	—
Isometric	23	—	$\perp 2$	—
	$\bar{4}3m$	—	$\perp$ or $\|\bar{4}$	—

PYROELECTRIC EFFECTS

A crystal is said to be *pyroelectric* if a displacement of positive and negative charge carriers is caused by a change in temperature. The effect is proportional to the magnitude of the temperature change, which renders the crystal useful for infrared heat detectors. Like piezoelectricity, pyroelectricity is highly dependent on crystal symmetry but with the important difference that the effect is restricted to just those point groups with a *unique* polar direction. This means that an electric current can be generated in one direction only, and that direction cannot be related to any other direction by any operation of symmetry elements in the point group.

Only 10 of the 21 piezoelectric point groups possess a unique polar direction (Table 6.2). They are often referred to as the 10 pyroelectric point groups. A crystal that belongs to one of these 10 classes can be both pyroelectric and piezoelectric. Because of the unique polar axis, minerals in these classes may also be piezoelectric under hydrostatic pressure. The other 10 point groups display piezoelectricity under directional pressure only.

We can derive the possible symmetries of pyroelectric and piezoelectric crystals from the symmetry argument, or Neumann principle. The cause is heat or pressure, the medium is the crystal, and the effect is electricity. The symmetry of heat and hydrostatic pressure is $\infty/m\,\infty/m\,\infty/m$, and of directional pressure is $\infty/m\,2/m\,2/m$. The symmetry of electric currents is ∞mm, noncentric. Consequently, those crystals that have symmetries consistent with ∞mm may be pyroelectric or piezoelectric or both under hydrostatic pressure, and those with symmetries ∞mm or $\infty 22$ or a combination of the two may be piezoelectric under directional pressure.

MAGNETIC PROPERTIES

Certain minerals have the property of behaving like a magnet. This behavior is caused by the net interaction of magnetic dipoles on an atomic scale. We know the scale is atomic, because even the smallest fragments of a magnetic substance still possess the property. Thus, for a mineral to be magnetic means that the atoms must be arranged or must interact in such a way that they do not cancel out each other's magnetism.

Not all atoms and ions can possess a magnetic moment. In Chapter 4, we learned that every electron possesses two magnetic quantum numbers, one to describe the orbital behavior around the nucleus and the other to describe the spin behavior of the electron itself. Overall, the situation is analogous to our spinning earth orbiting about the sun, except that in an atom, two electrons may occupy the same orbit if their spins are in opposite directions. Both orbital and spin motions generate magnetic dipoles. In the case of orbital motion, an electric charge accompanies an electron through its orbit much as an electric current flows through a loop. A current loop generates a tiny magnetic field exactly as if a bar magnet were placed perpendicular to the plane of the loop. The resulting *magnetic dipole* has positive and negative poles like a bar magnet. We refer to the intensity of the magnetic field as the *magnetic moment* of the atom. The moment produced by a single orbiting electron is defined as one Bohr magneton (μ_B) $= 0.927 \times 10^{-23}$ Am^2, which is the product of the current loop area (units of m^2) and the electron charge (units of amperes, A).

In an isolated ion, the net magnetic moment is the sum of both the orbital and spin moments of each electron. If an electron shell has equal numbers of electrons with opposing spins, no net magnetic moment can occur. Every orbital in that case has its full complement of two electrons, orbiting and spinning in opposition to each other. A net magnetic moment can be generated only in atoms or ions that have incomplete electron shells. In minerals, the outermost valence electrons of ions cannot be considered as unshared, because they are involved in chemical bonding. The only unshared electrons of importance to magnetic properties are those in the 3d orbitals of the first transition series (Sc, Ti, . . .) and those in the 3f orbitals of the second transition series (La, Ce, . . .). The 3d transition metals are the most important, because they include some of the most important mineral components, namely, Fe, Mn, Ti, and Cr.

In minerals with 3d transition metals, the orbital moment is much less important to the mineral's overall magnetic properties than the spin moment, because the 3d electrons and their orbits are affected by the crystal field of neighboring ions. In pure metals, 3d electrons also have great mobility, which makes their orbital moments less important than their spin moment. The spin magnetic moment, however, is unaffected by bonding. This consideration is less important in the 4f transition series, because those unpaired electrons are well inside the atom and well shielded by the outer valence electrons.

The magnitude of an atomic moment is directly proportional to the number of unshared 3d electrons. This is a consequence of both Hund's rule and the Pauli exclusion principle (Chapter 4). Hund's rule specifies that electrons will fill energy levels with the same spin until half of the electrons are in place, and then, according to the Pauli principle, the remainder of the electrons will fill the orbits with opposite spins. Table 6.3 illustrates how this happens for the 3d transition atoms and ions. The number of 3d electrons increases from left to right, each electron having an "upward" spin, until the Cr, Mn, Mn^{2+}, and Fe^{3+} electron configurations are achieved, which have five electrons with unopposed spins. The magnetic moment thus increases from 1μ_B to 5μ_B in the first half of the series. The sixth electron must enter with an opposing spin, which reduces the magnetic moment to 4μ_B, and so on as additional electrons fill in. As a result, Fe^{3+} and Mn^{2+} have the strongest dipoles among the transition metal cations, followed by Fe^{2+}, Mn^{3+}, Mn^{4+}, Cr^{3+}, Co^{2+}, and Ni^{2+}, in that order.

A big difference exists between the magnetic behavior of an isolated ion and the magnetic behavior of an aggregate of ions. Cr atoms, for example, are among the strongest dipoles, but Cr metal is essentially nonmagnetic. The individual spin moments of 3d electrons interact in several ways. (1) If a mineral has a random distribution of moments, no net moment exists, and the material is said to be *paramagnetic*. An external magnetic field will produce some alignment of dipoles, but this disappears when the field is removed. (2)

TABLE 6.3. *Magnetic Moments in Bohr Magnetons (μ_B) of the First Transition Metal Elements, Their Number of 3d Electrons, and Their Unpaired Spins*

Elements	K	Ca	Sc	Ti	V	Cr	Mn	Fe	Co	Ni	Cu	Zn
Ions		Sc^{3+} Ti^{4+} V^{5+}	Ti^{3+} V^{4+}	Ti^{2+} V^{3+}	Cr^{3+} Mn^{4+} V^{2+}	Cr^{2+} Mn^{3+}	Mn^{2+} Fe^{3+}	Co^{3+} Fe^{2+}	Co^{2+}	Ni^{2+}	Cu^{2+}	Cu^{1+} Zn^{2+}
Number of 3d electrons	0	0	1	2	3	4	5	6	7	8	9	10
Spin directions	—	——	↑	↑↑	↑↑↑	↑↑↑↑	↑↑↑↑↑	↑↑↑↑↑ ↓	↑↑↑↑↑ ↓↓	↑↑↑↑↑ ↓↓↓	↑↑↑↑↑ ↓↓↓↓	↑↑↑↑↑ ↓↓↓↓↓
Magnetic moment	0	0	1μ_B	2μ_B	3μ_B	4μ_B	5μ_B	4μ_B	3μ_B	2μ_B	1μ_B	0

A mineral in which the atomic moments are all aligned, as in ilmenite ($FeTiO_3$), is *ferromagnetic*. When an external field is imposed, the dipoles interact by a process called exchange coupling that enables the imposed field to remain "locked in" after the magnet is removed. (3) A mineral in which alternate atoms have oppositely directed moments is *antiferromagnetic*. An example is Cr metal, which has no net magnetization. (4) If the dipoles of alternate atoms point oppositely but have different magnitudes, an incomplete cancellation occurs, and a residual moment will remain. Such materials are called *ferromagnetic*. The best known example is lodestone (magnetite, Fe_3O_4) in which the divalent and trivalent iron have different moments because Fe^{3+} has one less $3d$ electron. Pyrrhotite (FeS) is another common ferromagnetic mineral.

Most of the common minerals have paired electrons with opposing spins. They are devoid of transition metal ions, and have no magnetic behavior at all. These minerals are called *diamagnetic*.

In most magnetic minerals, the dipole moments of the individual atoms prefer to be aligned in certain crystallographic directions. Volumes of aligned dipoles define *magnetic domains*, which have walls that can migrate through a crystal in response to external magnetic fields. The study of magnetic domains is an important area of research in the computer industry.

MECHANICAL PROPERTIES AND DEFECTS

Up to this point in our discussion of mineral properties, we have been learning about "perfect" crystals in which atoms are always perfectly positioned and related to each other by perfect translational symmetry. A perfect crystal cannot, in fact, diffract x-rays. C. G. Darwin postulated correctly in 1914 that all crystals must contain some defects. Accordingly, the most perfect crystal we can conceive of is the *ideally imperfect crystal*, which contains an ideally low density of defects. The nonexistence of the perfect crystal is further supported by a number of other observations. A. R. Griffith demonstrated by a series of experiments in 1921 that the reason for the low strength of crystals and glasses is the presence of structural defects, and he cited especially those defects which affect the quality of the crystal's surface structure. Similarly, many other common properties of crystals are not explainable if their structures were perfect. For example, plastic deformation of single crystals is due entirely to the presence of defects, or dislocations, that allow the differential movement of planes of atoms.

The important kinds of defects in mineral structures are classified according to their dimensions. *Point defects* have zero dimension, *line defects* have one dimension, *planar defects* have two dimensions, and *spatial defects* have three dimensions.

No mineral is perfectly pure in the sense that the only atoms present in the structure are those given by its ideal chemical formula. Impurity elements will always be present, perhaps in only one or two parts per million, even if the mineral in question exhibits no measurable solid solution. If a foreign atom occupies a regular site within the crystal structure, a *substitutional defect* results. If it occupies an interstitial region between sites, an *interstitial defect* results. Both of these point defects are more probable at elevated temperature, because the thermal vibrations of the atoms are greater, enabling them to occupy positions in the structure that would not be possible at lower temperature. When the amplitude of vibration of an atom is high enough, the atom can detach itself from a regular site, move elsewhere, and leave behind a vacancy. If the atom moves to an interstitial site, the vacancy is called a *Frenkel defect*. Overall charge balance is preserved, but the balance becomes less evenly distributed. If a cation moves to the surface of a crystal and is removed, electrical neutrality can be maintained by creating an anion vacancy, which is called a *Schottky defect*.

Both Frenkel and Schottky point defects are present in every real crystal. At 1000 K, one out of every 10^4 to 10^5 sites in the structure will be vacant. Many fewer sites will be vacant at lower temperatures. This is why solid state diffusion is so enhanced at elevated temperature. A common mechanism of diffusion is by the migration of vacancies.

The most important line defect is the *dislocation*, discovered independently in the early 1930s by G. I. Taylor, E. Orowan, and M. Polanyi. A dislocation is a line internal to a crystal along which the lattice is disrupted. Figure 6.14a illustrates an *edge* dislocation that extends normal to the page. This line represents the leading edge of an extra plane of atoms in effect inserted between two normal planes. The lattice is everywhere normal except in the immediate vicinity of the dislocation where the bonds are strained and the crystal is elastically deformed. This local deformation enables us to observe dislocations with an electron microscope.

When a crystal is deformed so that its shape is

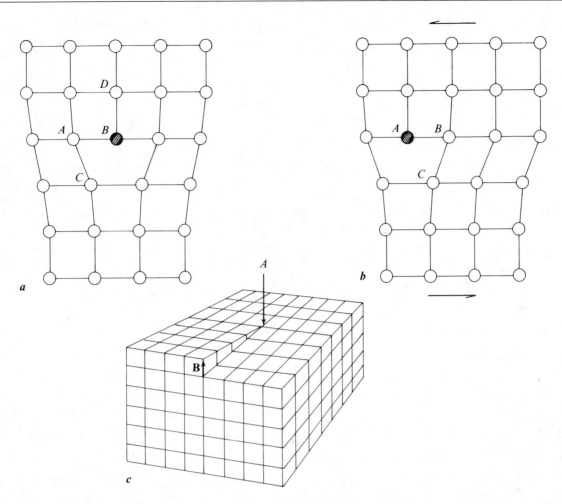

FIGURE 6.14 **(a) An edge dislocation extends perpendicular to the page and emerges just below atom *B*. (b) Edge dislocation migrates one unit translation to the left in response to shear stress. (c) Screw dislocation axis is shown at *A*, and the Burgers vector at B.**

permanently changed, we refer to *plastic flow*. This can occur without brittle rupture of the crystal due to the movement of atoms along dislocations or, put a different way, due to the movement of dislocations through the crystal. Figures 6.14a and 6.14b illustrate how plastic flow occurs. The initial position of the edge dislocation projecting from the page just below atom *B* changes, so the edge emerges just below atom *A* in response to a shear stress. The process involves bond stretching and compression to such an extent that the bonding of atom *B* to atom *C* becomes the energetically preferred state, rather than the atoms remaining unattached. Fewer bonds need be broken by this mechanism, which explains why real crystals are orders of magnitude weaker than theoretically perfect ones.

The displacement in Figure 6.14 is one unit translation in the direction normal to the dislocation, so the motion of the dislocation is effectively in unit steps parallel to itself. The dislocation will continue to migrate until it is stopped by a grain boundary or by some internal obstacle. In the latter case, the dislocation can bypass the obstacle by a process known as *climb*, caused by the migration of vacancies to replace the leading atoms of the dislocation (e.g., atom *B* in Figure 6.14a). When atom *B* diffuses away, the leading edge of the dislocation is effectively identified with atom *D*.

Another important line defect is the *screw dislocation* (Figure 6.14c). It has great importance in the process of nucleation and crystal growth, discussed in the following chapter. Screw dislocations differ from edge dislocations in that the displacement vector, called the *Burgers vector*, **B**, is parallel to the dislocation rather than normal to it. Dislocation combinations are also possible with edge and screw components.

Planar defects include grain boundaries, twin planes, and antiphase boundaries, abbreviated APB. Like twin boundaries, APBs form internally

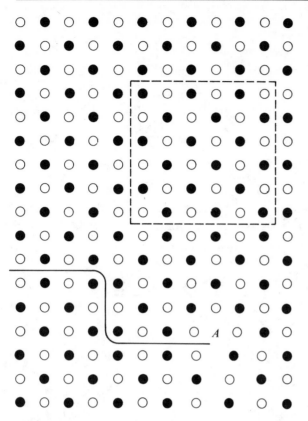

FIGURE 6.15　An imperfect crystal consisting of an antiphase domain enclosed by antiphase boundaries (dashed lines). Solid line is an antiphase boundary terminating at edge dislocation (point *A*).

in crystals. Unlike twin boundaries, APBs are the interfaces between arrays of atoms that are out of register with their neighbors. The APBs represent a type of atomic disorder in which the translational symmetry of a perfect crystal is disrupted. Normally, the chemical compositions on either side of an APB are identical, and they may or may not have the same degree of internal ordering. If the composition on one side is substantially disordered or has a different structure, then it is considered a different mineral.

Antiphase boundaries enclose *antiphase domains* as illustrated in Figure 6.15. If an APB does not close on itself in this manner but instead terminates inside the crystal, it must do so at a dislocation (point *A*).

CLEAVAGE

The most common and easily observed mechanical property of minerals is the development of cleavage. Cleavage along nearly perfect planar surfaces has little to do with crystal imperfections or defects, but depends instead solely on the orderly arrangement of atoms on a larger scale and

on the bonding forces between them. We know that cleavage planes are interfaces across which the chemical bonding is weakest, otherwise the mineral would not separate there. Planes of weakest bonding cleave best. Planes of stronger bonding cleave less easily.

Auguste Bravais pointed out in 1860 that cleavage depends on the balance between cohesive forces that act across the plane and those that act within the plane. Interionic forces between atoms decrease as their interatomic distances increase. As atoms come closer together within a plane, the cohesion or strength within the plane increases. In ionic crystals, more of the electrostatic bond energy associated with a centrally coordinated ion in the plane is tied up with these closer neighbors than with the atoms across the plane. The interplanar cohesion is consequently less, and the

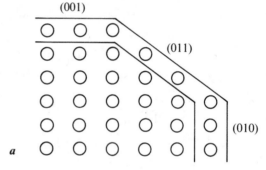

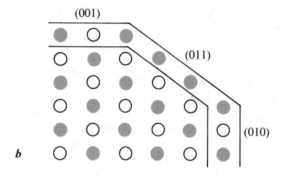

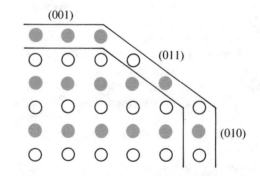

FIGURE 6.16　Three hypothetical structures illustrating the factors that determine ease of cleavage.

plane is a likely site for cleavage. Metals do not behave in this way, because they have no fixed ion charge to be locally satisfied and no anisotropy in bond strength. Thus, metals have no cleavage.

In the simplest crystal structures, we can predict that the most prominent cleavages possessed by a crystal will be parallel to planes having the greatest density of ions. These planes commonly have a high density of lattice points as well and therefore have relatively simple Miller indices. We can generally expect the face (111) to have potentially better cleavage than, for example, the face (432). This reasoning assumes that the bonding across the plane is necessarily weaker, which may not always be true. Simple lattice models must be considered in terms of the specific ions in the structure and the bonds they form.

For example, Figure 6.16a illustrates a hypothetical structure in which all of the atoms are identical, as in a metal. The interplanar spacings are in the order $d_{010} > d_{001} > d_{011}$. The density of atoms within each plane is in the same order, which leads us to predict that cleavage will be easiest on (010), more difficult on (001), and most difficult on (011). In fact this is not the case, because the bond energy per unit area, known as the *cleavage energy*, is the same on each face. Even though the density of atoms in (010) is greater than in the other planes, the decreased interplanar bond strength associated with each atom pair is exactly compensated for by the larger number of interplanar bonds.

Now look at how the specific distribution of different ions affects this model. In Figures 6.16b and 6.16c, the (010) planes are quite different. Ions of opposite charge alternate in both, and in that sense the cohesive strengths within the planes are the same. In the latter (Figure 6.16c), however, cleavage is more likely, because adjacent ions across the interface are of the same charge and therefore tend to repulse each other. In contrast, in Figure 6.16b opposite ions across (010) have opposite charges and therefore attract each other.

CHAPTER SEVEN

CRYSTAL GROWTH AND DEFECTS

Crystal growth is one of the most interesting and practical aspects of mineralogy. We learned in the previous chapter how and why certain minerals exhibit special electrical, magnetic, and conducting properties. Mineralogy owes much of this knowledge to the research efforts that accompany the current era of solid state electronics and communications. With the demand for synthetic quartz oscillator plates in the 1930s, crystal growth technology became a separate field of study. The demand increased for larger and purer crystals for oscillators and laser materials, for semiconductors in solid state detectors (Si and Ge) and light meters (CdS), and for memory chips (various magnetic oxides) in computers. Nature could supply neither the quality nor the control on chemistry to meet these needs, and hence methods were developed to produce the crystals artificially.

The net effect of research in crystal growth has been to greatly enhance our understanding of how crystals grow in nature. Laboratory studies on crystal growth mechanisms, kinetics, and rates of growth are active areas of research today. Because we now have the capability of carefully controlling the chemistry of a growing crystal, the separate effects of chemistry and structure on a physical property can be distinguished. Natural crystals almost always contain small concentrations of impurity elements, which make an ambiguous contribution to a particular property such as color. Crystal growth studies also help establish the range of pressure and temperature over which a mineral is stable. These ranges are important to petrologists whose principal interest is understanding the conditions under which rocks form.

WHY CRYSTALS GROW

Before investigating how crystals grow, we need to have some understanding of why they grow. The detailed reasoning is deferred until the next chapter in which some important energetic concepts are developed. For now we are concerned only with the fact that crystals *should* form from various liquids and gases, providing the crystals represent a lower energy state. Crystals grow over a limited range of pressure and temperature that differs for every mineral. Outside that range of conditions, the mineral is said to be metastable rather than stable, because either a liquid or a gas will have the lower energy and will thus form spontaneously.

For example, Figure 7.1 shows the stability fields for the mineral ice, H_2O water, and H_2O vapor. Chemistry, an important factor in chemically more complex systems, does not change in this example, so the only factors that control whether an ideal ice crystal will grow or not are pressure and temperature. Most people have witnessed the formation of ice crystals when H_2O passes from the stability field of water into the stability field of ice. This is the process of *crystalli-*

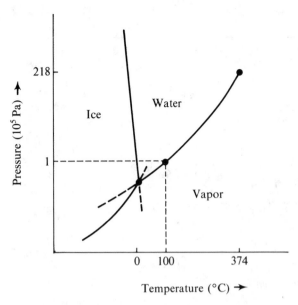

FIGURE 7.1 **Pressure-temperature phase diagram for H₂O showing the stability fields of ice, water, and vapor.**

zation. The opposite operation, namely, transforming from a crystal to a liquid, is *melting*. These two are not simply the reversal of the same process, because crystals, when they melt, always do so at exactly the melting temperature, whereas a liquid may not begin to crystallize until it is several degrees below its crystallization temperature. The reason for this behavior is that crystals must first nucleate in order to grow from a liquid, even though the pressure-temperature conditions may be suitable for growth. This is not a problem when crystals melt.

Crystals are not restricted to growing from a liquid. Some of the finest single crystals in nature form directly from a gas by crystallization. The opposite operation is called *sublimation*. The formation of ice crystals in the form of hoarfrost takes place at temperatures well below the stability field of water.

Most minerals form and grow in chemically more complicated environments than those of the previous examples—environments in which *solutions* of two or more substances are involved. We know that salt (NaCl) crystals will dissolve in normal tap water. When this happens, the crystals disappear and the homogeneous solution that remains is just salty water. If more crystals are added and they dissolve, the solution changes its chemistry, becoming saltier. Without changing the temperature or pressure, a point is reached at which no more crystals can dissolve, and the solution is said to be *saturated*. Prior to this point, the solution was *undersaturated*. Clearly a solution

must be saturated in order for crystals to grow, otherwise they would dissolve.

An undersaturated solution can be made saturated in two simple ways so that crystals can grow. For example, an undersaturated solution of NaNO₃ and water (point *A*, Figure 7.2a) at constant pressure and temperature can be evaporated, causing the solution to become progressively depleted in H₂O until the saturated composition (point *B*) is reached. Continued evaporation changes the chemistry further toward pure NaNO₃ until eventually only solid crystals remain. Providing suitable nuclei are present for growth, the first few crystals that form at *B* are immersed in the saturated solution, and the proportion of crystals to saturated solution gradually increases as pure water is removed by evaporation. This is the region, between point *B* and pure NaCl, in which crystals of NaNO₃ have a suitable *chemical* environment to grow.

The amount of salt that can be dissolved in a given volume of water depends on the temperature and pressure. At constant pressure, the saturated composition becomes saltier as the temperature is increased, and thus more salt can be dissolved in hot water than in cold. The locus of saturated solution chemistries as a function of temperature at atmospheric pressure is called the *solubility curve* (Figure 7.2b). To the left of the curve, the solution is everywhere undersaturated. On the curve and everywhere to the right of the curve the solution is saturated provided nucleation problems do not occur. If nucleation problems do occur and crystals fail to form, the solution is said to be *supersaturated*.

A simple example of how solubility curves are used in crystal growing is illustrated by the solution at *C* (Figure 7.2b), consisting of 120 g of NaNO₃ totally dissolved in H₂O at 100 °C. Without changing the overall chemistry (i.e., no evaporation), the solution is allowed to cool until the saturation temperature at *D* is reached. There the first crystals ideally form, and as temperature is lowered further, the crystals continue to grow.

These principles apply to crystals that form from very high-temperature igneous melts as well as from solutions. In these cases, the solubility curve goes by a different name, the *liquidus*, but it conveys exactly the same information. Another common situation is for two crystalline substances, for example, the pyroxenes diopside and enstatite, to have limited solid solubility toward each other. In these cases, the solubility curve is called a *solvus* (Chapter 8).

The same general principles apply to crystallization from a vapor. At constant pressure, increas-

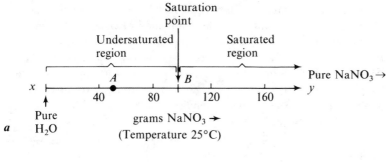

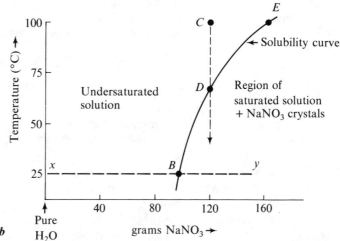

FIGURE 7.2 **(a) Solubility of NaNO₃ in H₂O at atmospheric pressure and 25 °C. Composition at *A* is an undersaturated solution; composition at *B* is saturated with crystalline NaNO₃. All compositions in the saturated region consist of saturated solution at *B* plus** **NaNO₃ crystals in varying proportions from pure solution at *B* to pure crystals off the diagram to the right. (b) Solubility curve *BDE* for NaNO₃ in H₂O liquid. The positive slope indicates that solubility of NaNO₃ in water increases as temperature rises.**

ing the temperature causes more solid to be dissolved in a vapor in a manner analogous to the solid and liquid relationships just considered. Upon cooling, a saturated vapor will crystallize once the problems of nucleation are overcome. The resulting crystals are among the finest nature provides. They often grow on the walls of rock cavities, free from the effects of solid or fluid surroundings. Many of the most beautiful museum specimens are formed in this way.

The second important growth mode is entirely in the solid state and is the way in which metamorphic rocks change their mineralogy in response to changing pressure and temperature. Energetically, the driving mechanism is the lowering of the total energy state appropriate to the rock composition and the conditions of *recrystallization*. We use the latter term in this context, because the rock is already crystallized when these changes take place. Nucleation is less important in recrystallization than it is in crystal growth from a pure fluid or vapor, but *diffusion*, a measure of how mobile the ions are over some distance, becomes more important. In the solid state, ions

have very little mobility due to their physical confinement within crystalline structures, and growth rates are therefore slow. The newly formed crystals are usually adjacent to others in such a way that perfect euhedral shapes are not common.

CRYSTAL GROWTH AND LATTICE GEOMETRY

A distinctive feature of many crystals is their ability to grow to predictable shapes that are bounded by flat planes. We learned in Chapter 3 how external morphology is closely related to internal structure. The earliest systematic work on this relationship was that of René Just Haüy and Auguste Bravais. At a time when the first crystal structures were yet to be determined, they relied on the concept of the lattice to explain form development in crystals. Bravais in 1860 stated in effect that *the most prominent faces of a crystal are those that are parallel to internal planes having the greatest density of lattice points*. This statement became known as the "law of Bravais" and dominated scientific thought on crystal growth for more than a half century. Bravais introduced the term *reticular density* to describe the concentration of lattice points in a rational plane. A high reticular density indicates a large number of lattice points per unit area, or put another way, a small unit mesh. He further recognized that high reticular density in a plane is associated with a large interplanar spacing perpendicular to the plane, and thus, the smaller the spacing, the lower the reticular density. This is simply a consequence of the fact that for any given lattice, the product of the unit mesh area A of a particular plane and the interplanar spacing d_0 is a constant:

$$Ad_0 = \text{Constant} = V$$

where V is the volume of the reduced cell. The two-dimensional analogue is shown in Figure 7.3. Four selected lattice planes illustrate the relationship between lattice spacing and the density of lattice points in the plane of interest. That is, (103) has the least density and the smallest interplanar spacing, and (001) has the greatest density of lattice points and the largest spacing.

Georges Friedel demonstrated in 1907 the empirical basis for Bravais's hypothesis. The reticular density A can be calculated[1] for any face of a crystal, providing we know its Miller indices and lattice symmetry. For a given crystal, these numbers, when tabulated in order of decreasing magnitude, correlate surprisingly well with the de-

creasing importance of faces. By "importance" is meant both the frequency of occurrence and the area occupied by a face relative to other faces. Friedel and those who followed amassed a large number of observations that further documented the general rule that important or prominent faces on a crystal tend to have simple Miller indices (e.g., (001) would be more prominent than (103) in Figure 7.3). This general rule is sometimes referred to as *Haüy's law*.

Like all generalizations, important exceptions appeared as more data were accumulated. Based on Bravais's law, pyrite (FeS_2) was expected to show the faces {100}, {110}, and {111} in that order of decreasing importance. The observed order of importance is in fact {111}, {100}, and {210}. Similarly, in the Bravais sequence of faces in low quartz (SiO_2), the basal pinacoid (001) appears most important. Observation shows that {100}, {101}, {110}, {102}, and {111} are far more important. In view of these flagrant exceptions, Paul Niggli in 1919 and J. D. H. Donnay and D. Harker in 1937 extended the law of Bravais to include the additional lattice planes introduced by the translational components of glide planes and screw axes. These elements are present in the space group symmetry but not in the 14 Bravais space lattices. In the case of a glide plane, the distance between planes that intersect the glide plane is halved. In Figure 7.4, the c glide doubles the number of (001) lattice planes. This will affect the interplanar distance of all other planes except {100}, thus {100} gains relative importance as a prominent face. In the case of

[1] $A_{hkl}^2 = h^2 A_{100}^2 + k^2 A_{010}^2 + l^2 A_{001}^2 + 2(hk A_{100} A_{010} \cos \nu + kl A_{010} A_{001} \cos \lambda + lh A_{001} A_{100} \cos \mu)$ where $\nu = (100) \wedge (010)$, $\lambda = (010) \wedge (001)$, $\mu = (001) \wedge (100)$, and $A_{100} = bc \sin \alpha$, $A_{010} = ca \sin \beta$, $A_{001} = ab \sin \gamma$.

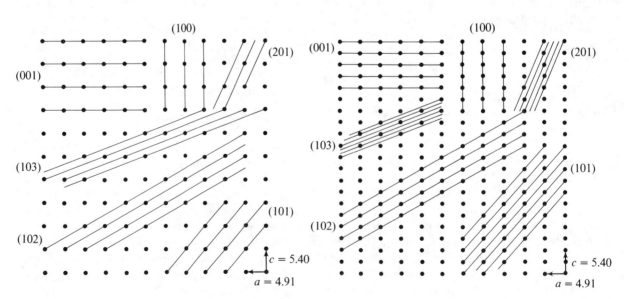

FIGURE 7.3   Various (*h0l*) lattice planes show inverse relationship between lattice spacing d_{h0l} and reticular density.

FIGURE 7.4   Effect of a c glide on various (*h0l*) planes. d_{001} is halved, which doubles the number of (001) lattice planes. {100} is unaffected.

pyrite (FeS_2) with space group *Pa*3, the *a* glide has a translational component of $a/2$, which generates new lattice planes, including those parallel to (010). The relative importance of (010) therefore decreases, because its interplanar spacing is halved to $a/2$. Note, however, that the (100) spacings remain unchanged, so the (100) face becomes more important relative to the other faces, which is in conformity with observation.

In the case of low quartz (SiO_2) with space group $P3_121$ (or $P3_221$), the 3_2 screw axis parallel to the *c* axis generates additional lattice planes between and parallel to (001). Because the interplanar distance is halved, the relative importance of (001) as a face decreases, which is more consistent with observation.

The relevance of lattice geometry to crystal growth can be summarized by stating Bravais's law in a slightly different form. *The rate at which a crystal face grows is, in general, inversely proportional to the interplanar spacing of that face.* The truth of this statement is not at all obvious, particularly since no provision is made for types of atoms, their ionic charge, or bonding properties, yet the empirical relationship between lattice spacing and the relative importance of face development must have some rational basis. Crystal faces become prominent by the process of growth, so any fundamental relationship between prominence and interplanar spacing should also bear on crystal growth.

This reasoning is borne out by the following model. Consider a growing crystal (Figure 7.5) with (100), (010), and (001) faces that have different interplanar spacings. Suppose that growth takes place by addition of primitive unit cells, one by one, to each face. The strength of attachment is proportional to the area of attachment, because the larger the common surface area between crystal face and unit cell, the more bonds that will be formed. Unit cells will also tend to arrange themselves upon attachment so that their lattice planes register with those of the crystal. As Figure 7.3 shows, (100) of the unit cell is aligned with (100) of the crystal and so forth. Because each cell is primitive, its volume is simply the product of the area of attachment and the interplanar spacing. The area of attachment must, therefore, be inversely proportional to the interplanar spacing. The unit cell most likely to attach (and hold) in Figure 7.5 is on the (100) face, because its area of attachment is largest. Stated another way, the unit cell will more frequently attach itself to the crystal face that has the smallest interplanar spacing. Such faces tend to grow faster, because their low reticular density enables a growth layer to be completed in a shorter period of time.

Once the relative growth rates have been established, the relative importance of faces is determined, because rapidly growing faces tend to grow themselves out of existence. Faces that have larger interplanar spacings grow more slowly and eventually dominate the external form. Figure 7.6a illustrates a hypothetical isometric crystal in which all the faces grow at the same rate perpendicular to the *c* axis. After some period of time, the crystal grows to the shape (1) outlined at the end of each growth vector. Successive growth layers (2, 3, and 4) preserve the form of the original crystal. Figure 7.6b illustrates a real isometric crystal with faces that do not all grow outward at the same rate. The faces (010), (100), and (0$\bar{1}$0) grow at the same rate, but faces (110) and (1$\bar{1}$0) grow faster because they have shorter interplanar distances. After some period of time, {110} faces grow to a thickness twice that of {100} faces. The crystal shape (1) has changed slightly because {110} has become less prominent. The successive growth layers (2, 3, and 4) show that {110} is gradually growing out of existence and will disappear entirely when the crystal reaches the form of a square prism consisting of {100} faces only.

Remarkably, the external form of the great majority of minerals can be predicted by Bravais's law. This must mean that the lattice symmetry is an important factor in determining the final shape of a growing crystal. Important exceptions to this generality, however, indicate that other *structural* components (in addition to the translational component, the lattice) as well as other factors operate during growth. We know (e.g., Figure 1.3) that crystals of the same mineral, all of which have the

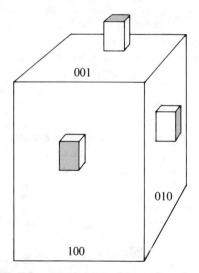

FIGURE 7.5   Attachment of growth units with the shape and dimensions of primitive unit cells.

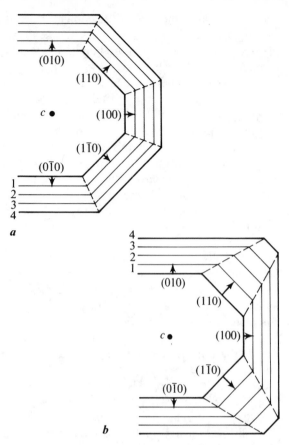

FIGURE 7.6   **(a) Hypothetical isometric crystal with (*hk*0) faces growing at same rate. (b) Real isometric crystal in which {110} faces grow faster than others and eventually disappear.**

same lattice symmetry, can have different forms. We also know that in noncentric crystals $\{hkl\}$ faces may grow at rates different from $\{\bar{h}\bar{k}\bar{l}\}$ faces, even though their reticular densities and interplanar distances are identical.

Another predictable difficulty with the Bravais model is the absence of any distinction between the development of crystal faces and cleavage faces. In general, they are not developed with equal prominence. Recall from Chapter 6 that cleavage faces develop along planes of least chemical bonding and therefore should have little relationship to the surface phenomena associated with crystal growth. Clearly the broader aspects of the structure, namely, the identity and location of individual atoms relative to one another, have to be considered in addition to the lattice.

EQUILIBRIUM FORMS

A complete understanding of the relationship between form and structure is elusive, and debate and discussion on the matter still appear in the professional literature today. One issue is whether

crystals have *equilibrium forms* or *growth forms* or some combination of both related in a way that still is not well understood. The concept of equilibrium forms was especially popular around the turn of the century when the importance of minimum energy states was first recognized, but then as research on specific growth mechanisms and defects emerged, we learned that the world of real crystals is different from that of perfect crystals to which energetic arguments are most applicable. Real crystals with all of their structural imperfections have shapes and properties that may be more closely related to growth forms than to equilibrium forms. Current thinking is that the equilibrium form is more important in the stages of nucleation and initial growth and that growth mechanisms dominate thereafter.

The basic principle underlying the equilibrium form concept is that the crystal at any stage of growth has achieved its lowest energy state. All aspects of the energy have to be considered, including an important contribution from the total surface area. This energy contribution is small, however, compared with the remainder of the crystal energy related to its mass and volume, especially when the crystal is large. J. Willard Gibbs in 1875 was the first to recognize that a substance has achieved its equilibrium form when the contribution of surface area to total energy has been minimized. For simple substances like fluids, the ratio of surface area to mass takes on the lowest possible value at equilibrium, which is why liquids "ball up" as tiny spheres rather than cubes or rectangular blocks. For any substance, a sphere has the minimum surface area per mass. For this reason, metamorphic rocks tend to become coarser grained with progressive recrystallization, because an aggregate of large grains has a smaller ratio of surface area to mass than an aggregate of small grains.

Crystals, because of their ordered arrangement of atoms, are more complicated with respect to surface properties than fluids. Gibbs formulated an expression for a crystal that took into account the difference in surface structure on each face. This was necessary, because the surface contribution to the total energy varies from face to face. Were this not the case, crystals would form as spheres rather than plane-bounded shapes. Gibbs wrote:

$$(G)_{P,T,x} = \gamma_1 A_1 + \gamma_2 A_2 + \gamma_3 A_3 + \cdots$$
$$= \text{minimum} \qquad (7.1)$$

where $(G)_{P,T,x}$ is a measure of the crystal energy at constant pressure, temperature, and composition x. A_1, A_2, . . . , are the surface areas of the various faces, and γ_1, γ_2, . . . , are the specific surface-free energies of each of the faces. The latter terms are equivalent to surface tension and have units of energy per unit area. Essentially, the γ value is a measure of the energy required to remove a layer of atoms from the crystal plane of interest, and since the strength and distribution of chemical bonds per unit area differ from plane to plane, so do the γ values.

Equation 7.1 gained greater physical significance from the work of Pierre Curie in 1885 and later through research by Georg Wulff who in 1901 showed that

$$\frac{\gamma_1}{h_1} = \frac{\gamma_2}{h_2} = \frac{\gamma_3}{h_3} = \cdots = \text{constant} \qquad (7.2)$$

where h_1, . . . , is the radial distance of each face from the center of the crystal. From this relationship, we understand that faces with large values of γ will be effectively characterized by rapid growth, so the corresponding h value will also be large. This must be the case if the ratios in Equation 7.2 are to remain constant.

A graphical representation of the γ/h ratio is called a *Wulff plot* and is shown in Figure 7.7 for forsterite (Mg_2SiO_4). The radial distances h are varied for all faces in a zone until they are propor-

tional to the respective values of γ. The crystal form that corresponds to the lowest surface energy can be found by choosing, at any arbitrary stage of growth, the set of planes enclosing the smallest area in the projection. We are, in effect, choosing only that region of the Wulff plot that can be reached from the origin without crossing other growth planes. For example, in forsterite (Mg_2SiO_4), faces are drawn normal to the γ vectors out to arbitrary distances, and then only their interior parts are considered as the set of bounding planes. Theoretically, the {120} faces do not appear on forsterite, because they are entirely exterior to the other faces on the plot.

NUCLEATION

These relationships between form and structure are sound in principle, providing the crystal surfaces make a significant contribution to the total crystal energy. As crystals increase their volume during growth, surface energy effects become less important, because the surface area to mass ratio becomes progressively smaller. At this stage, several other growth factors such as environment, impurities, and structural defects can become energetically more important and can dominate the growth.

During nucleation, however, the surface energy effects are extremely important. We have a hint of this from the fact that liquids can be *supercooled*, that is, cooled several degrees lower than the crystallization temperature at which the first crystals should form. Aqueous solutions can also be *supersaturated* when crystals ought to be stable. We could almost surmise that the crystals need a little energetic boost before they proceed to form in the way they should.

Our surmise is borne out by the fact that both supercooled liquids and supersaturated solutions can be made to crystallize spontaneously by adding a nucleus on which the crystals can grow. In laboratory crystal growth, the nucleus can be a seed crystal of the same substance, or in some cases even an unwanted substance such as a dust particle. We can make a supersaturated solution in a beaker crystallize by simply tapping the beaker with a hard object. If no preexisting nuclei are present or if none can be added, we can increase the degree of supercooling or supersaturation until enough extra energy is stored up to enable spontaneous crystallization to occur.

The basic reason for this behavior is that crystal nuclei must be *made* to grow, because the nuclei surface areas are large relative to their volume. The surface contribution to the total en-

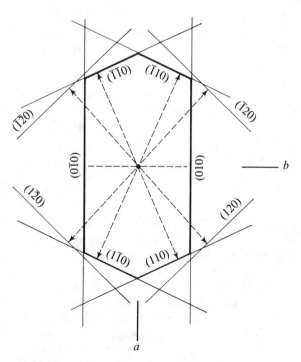

FIGURE 7.7 Wulff plot of stable growth faces of forsterite. (After J. 'T Hart, 1978. The structural morphology of olivine. II. A quantitative derivation. *Canadian Mineralogist* **16:550. Courtesy of the Mineralogical Association of Canada.)**

ergy will always be proportional to $4r^2$, an area, whereas the volume contribution is proportional to $4/3r^3$. The total energy of the crystal can be represented as

$$G_{\text{total}} = \gamma 4r^2 - G_V \tfrac{4}{3}r^3 \qquad (7.3)$$

where γ is the surface energy per unit area, and G_V is the crystal energy per unit volume. We assume that the pressure, temperature, and crystal composition are all constant. Equation 7.3 is formulated in terms of a sphere, but we could use any other shape.

Figure 7.8 illustrates the physical significance of Equation 7.3. For crystals larger than a certain size given by the *critical growth radius* r_c, growth will proceed spontaneously, because lower energy states result. For smaller crystals, however, including nuclei with radii of less than r_c, the first term in Equation 7.3 dominates the total energy change. The nucleation stage of crystal growth must proceed in a direction of increasing crystal energy, which is resisted by nature. The extra energy that must be provided to initiate crystallization is called the *nucleation energy*. Once that is

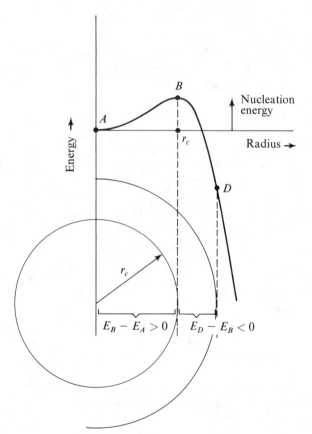

FIGURE 7.8 Variation of crystal energy with radius for very small crystals. Surface area contributions dominate the total energy at radii less than the critical growth radius r_c.

provided and the crystal exceeds the critical growth radius, growth continues spontaneously with a progressive decrease in energy.

DEFECTS AND GROWTH MECHANISMS

The problem of nucleation does not end once a crystal has grown beyond its critical growth radius. Throughout the growth process, each crystal plane, edge, and corner must extend itself on an atomic scale by the addition of atoms, ions, or larger building blocks. The attachment process is most difficult on well-developed planar surfaces, because the probability of a growth unit "sticking" to such surfaces is less than at edges and corners. This is because of the strong chemical bonds that exist between the atoms at the surface, resulting in a consequent decrease in potential bonds to atoms of a new growth layer.

Once a growth layer has started to form on an ideal crystal face, finishing that layer is much easier than starting a new layer. Figure 7.9 illustrates why. A growing layer (Figure 7.9a) generally will have corners and edges where idealized growth blocks can attach two or three of their bonding surfaces. This type of growth takes place when the surface is saturated with respect to the surrounding fluid or vapor solution. When that growth surface is completed, however, additional energy is required to nucleate new growth units (Figure 7.9b). The solution then becomes locally supersaturated (or supercooled) until the necessary energy is provided.

Based on what we know about perfect crystals, the fact that real crystals grow as easily and rapidly as they do is rather surprising. Crystal faces actually grow far faster than theoretical predictions would suggest. This observation led F. C. Frank, an English physicist, to postulate in 1949 that crystal faces do not grow layer by layer as outlined above, but rather that a single layer grows continuously in the form of a spiral. For this to happen, the crystal must possess an internal imperfection in atomic order called a *screw dislocation*. This type of defect (Figure 7.9c) is caused by the displacement of part of the crystal structure one translation unit relative to another in such a way that the displacement terminates along a line (AB) perpendicular to the growing face. The crystal is perfectly normal everywhere except along the dislocation axis, where it is elastically strained and the lattice symmetry disturbed.

This type of crystal defect is called a screw dislocation because a small imaginary person starting at point x (Figure 7.9c) and walking

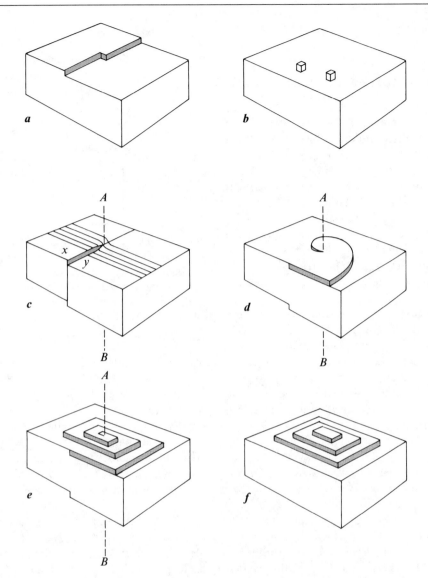

FIGURE 7.9 (a) Growing crystal surface on which partially formed growth layer provides a likely nucleation site. (b) Completed growth layer on which new growth layer must nucleate. (c) Screw dislocation along disloca-tion line *AB*. (d) Spiral growth along leading edge of screw dislocation from (c). (e) Spiral growth layers when growth rates differ from one direction to another. (f) Isolated growth layers due to paired screw dislocations.

clockwise on the growth surface would end up after 360 degrees at point *y*, exactly one transla-tion unit below the starting point. Another 360 degrees on the same surface and in the same direction would terminate two units below the starting point, and so on. If we reverse the opera-tion, crystal growth can be visualized realistically as proceeding up a spiral ramp, always with a leading edge suitable for nucleating new growth units. The rate of growth is the same everywhere along the leading edge, so the outer part of the edge lags behind the central part after some period of time (Figure 7.9d). After continued growth, a continuous spiral ramp, cone-shaped in cross section, is generated around the original dislocation still preserved inside the crystal. In this

manner, the energetic problem of two-dimensional nucleation is overcome, and the crystal face grows faster than would otherwise be expected.

In the event that a crystal face grows at differ-ent rates laterally, as is usually the case, the plan view of the spiral growth is no longer a continuous coil but is instead broken up into straight seg-ments (Figure 7.9e). Other variations can be caused by growth on a leading edge being pinned by dislocations at both ends. Visualize the mirror image of Figure 7.9c attached on the left side of the crystal with the right side climbing clockwise as shown in Figure 7.9d and with the imaginary left side climbing counterclockwise. The growth layers essentially double back on each other to make a self-contained layer. If growth is at different rates laterally, the successive layers can appear as shown in Figure 7.9e.

CRYSTAL GROWTH AND STRUCTURE

After thinking about how real crystals depart from their perfect models, let us now question the reality of the idealized building blocks from which perfect crystals are constructed. The building block analogy is a reasonable model for most of the rock-forming minerals. The silicates in particular have very strong Si—O bonds that show signs of having formed in silicate melts before crystallization temperatures were reached. As temperature lowers, a gradual increase in viscosity (a measure of resistance to flow) signals the creation of Si—O groupings that will ultimately be the building blocks for crystal growth.

These organized arrays of atoms are generally considered small enough to be repeated by translational symmetry yet large enough to retain the stoichiometric composition of the mineral. If the unit cell of the mineral is primitive and if no glide planes or screw axes are present, the building blocks will have the volume of unit cells as depicted in Figure 7.5.

Once formed and in contact with a growing crystal, our building blocks, having some degree of organization already built into them, do not randomly position themselves on a surface. They have the capability of aligning themselves perfectly with the existing crystal such that the lattice and other symmetry elements are in registry. In 1955, P. Hartman and W. G. Perdok introduced the concept of the *periodic bond chain*, abbreviated PBC, which helps explain this growth behavior. Periodic bond chains are chains of building blocks held one to another by chemical bonds that are stronger than those bonds holding one such unit to the crystal surface. In silicates, the strong Si—O bond is the backbone of the PBCs, but other atoms and bonds may be involved as well.

Periodic bond chains are arranged on growing crystal faces in three simple ways (Figure 7.10). If two or more PBCs lie in the same face, the face is called a flat (*F*) face. Each new building block that attaches to this surface is strongly bonded in at least two directions to the existing PBCs. An important consequence is that the growth layer as a whole has a relatively small *attachment energy*, or put another way, the growth layer is not strongly bonded to the underlying layer. The exterior surface of a newly formed growth layer is also a surface of *least bonding*, because much of the bonding energy is tied up in the PBCs *within* the growth layer. *F* faces, as surfaces of least bonding,

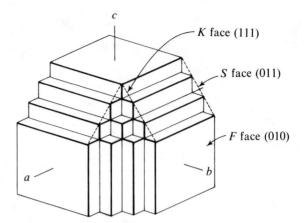

FIGURE 7.10 **Periodic bond chains (PBCs) and the development of *F*, *S*, and *K* faces. PBCs shown diagrammatically in bold lines. (100), (010), and (001) are *F* faces having two parallel PBCs. (101), (011), and (110) are *S* faces having one parallel PBC. (111) is a *K* face with no parallel PBCs.**

attract new building blocks very slowly by either two-dimensional nucleation or screw dislocation. They are always the most prominent faces on crystals.

Stepped (*S*) faces contain a single PBC and grow faster than *F* faces, because the probability of a building block landing and sticking on an *S* face is greater (Figure 7.10). *S* faces grow by one-dimensional nucleation, and their attachment energies are larger than those of *F* faces. *S* faces are also less prominent than *F* faces.

Kinked (*K*) faces (Figure 7.10) are rarely observed on crystals because they grow very rapidly with no nucleation problems. The surface so formed has no PBCs and has a very high attachment energy. Note also that *K* faces are rough compared to *S* and *F* faces. The nucleation of new growth layers is therefore accomplished easily on *K* faces.

We can apply the PBC concept to achieve a practical understanding of the external form of crystal structures. Consider for a moment the flat platy habit of micas and the fact that the $(Si_2O_5)^{2-}$ sheets are in the (001) plane. This is an *F* face, which means that the growth rate normal to the face is slow compared with growth in other directions. Compare also the growth forms of the single-chain pyroxenes with the double-chain amphiboles. The $(SiO_3)^{2-}$ and $(Si_4O_{11})^{6-}$ chains are natural PBCs, so we would expect relatively rapid growth parallel to them and relatively slow growth in the perpendicular direction.

A more sophisticated approach is to calculate by computer methods the attachment energy of every possible plane in a crystal structure and then to predict what the final growth form will be. The energy values obtained are closely related to surface-free energies, γ, discussed earlier. In fact, the [001] zone of forsterite (Mg_2SiO_4) in Figure 7.7 was predicted not from γ values but from attachment energies used in their place. A useful feature of the calculations is that we must make some assumptions about the ionic or covalent nature of the chemical bonds that must be broken to obtain the attachment energies. An assumption can then be tested by comparing the predicted form, as seen in a Wulff plot, with the actual crystals. In this example (Figure 7.7), Si—O bonds were assumed to be entirely ionic. When we assume that the bonds are entirely covalent, the {120} faces have a relatively lower attachment energy and generate stable faces on a Wulff plot, in contrast to the ionic model. In reality, the Si—O bond is approximately 50% ionic and 50% covalent.

We can now better understand the physical differences between an analysis of crystal form and growth based on the lattice and an analysis of form and growth based on the atomic structure. The Bravais concept, by itself and as modified by Donnay and Harker, assumes that if the lattice points are occupied by identical atoms, the bonding energy *per unit area* will be the same for all faces. Even though the interplanar spacing for a particular lattice plane may be large, according to the Bravais concept, the plane's reticular density is correspondingly greater. From a bonding viewpoint, this means that although the bonds are weak because the atoms are far apart, more bonds exist. For closely spaced planes, the bonds are stronger, but we observe fewer of them. The net result is that the total bond strength that must be exceeded to produce cleavage is the same for all lattice planes. The fact that cleavage planes exist at all means that the bonding energy per unit area *cannot* be the same throughout the crystal.

When we take into account the total structure, the well-developed crystal faces, namely, *F* faces, should also be good cleavage planes. The crystal faces are surfaces of least bonding and as such should be more easily cleaved than either *F* or *K* faces. In reality, this is not generally true, because the development of growth faces depends on many factors such as nucleation rates and the presence or absence of defects, whereas the development of cleavage depends only on the internal atomic structure.

GROWTH FORMS

One important factor that controls the growth rate of faces is the nature of the solution or liquid from which the crystal grows. When a crystal solidifies, it has a lower energy than an equal mass of the pure liquid, or solution, from which it formed. Even though the crystal and the solution are at the same temperature, the atoms within the crystalline structure are no longer free to move around and are therefore less energetic than identical atoms in the solution. The difference in energy states is called the *latent heat of crystallization*. This heat is always given up during crystallization and is always absorbed during the melting of crystals. The latent heat associated with crystallization from a vapor is greater than that involving a solution, because the atoms in the vapor state are even more energetic.

When atoms, ions, or other building units are deposited on a crystal surface during growth, the latent heat is transferred to the solution at the crystal interface. Unless this energy is removed or passed on to the larger body of solution, further crystallization is impaired, because the solution atoms are too energetic. A good example of this phenomenon is the freezing of a lake in winter. Crystallization of ice from water occurs rapidly in the cold air. As freezing proceeds downward, the rate of crystallization slows progressively, because the latent heat associated with the additional freezing diffuses at a progressively slower rate through the accumulating ice. In other words, the *rate* of crystallization depends on how rapidly the latent heat can be removed from the growing crystal surface.

The rate of latent heat removal is not the same everywhere on a crystal surface. For example, latent heat is dissipated more rapidly from the corner of a cube-shaped crystal than from its edges or faces, because the volume of solution that is capable of absorbing the latent heat per unit growth volume of crystal is greater for a cube corner than for a cube edge. The cube face dissipates heat least effectively and therefore should grow more slowly than either edges or corners.

The local buildup of undissipated latent heat on *F* faces and the exceptionally rapid growth of edges and corners can promote *dendritic* growth. *Dendrites* are common on natural and synthetic crystals and are recognizable by their fine, treelike protuberances and branching growth forms. The beautiful patterns of dendritic ice on window panes are a familiar example. Most of the common rock-forming minerals will be dendritic if the local crystallization rate is sufficiently high. This usually occurs when the solution becomes greatly supercooled before nucleation and growth commence.

A closely related factor that influences the

growth of dendritic forms is the presence of concentration gradients in the crystallizing solution or liquid. A growing crystal must have a continually replenished source of fresh solution if growth is to proceed at a constant rate. As a crystal face grows from a saturated solution, ions or other building blocks are removed from solution, and the solution can become locally undersaturated unless diffusion from the surroundings restores the necessary growth concentration. Crystal growth experiments have demonstrated that the degree of supersaturation or supercooling is often greater at corners, less on edges, and least on crystal faces (Figure 7.11). The consequence is that lines of constant concentration are closer together at corners, which implies that a stronger concentration gradient exists there than at edges and faces. The dotted lines in Figure 7.11 represent diffusion paths through the solution, drawn perpendicular to the equiconcentration lines (solid). These flow lines are more concentrated at corners and edges (Figure 7.11a) than at faces, and are least concentrated in receded areas (Figure 7.11b).

The consequences of both variable heat dissipation and concentration gradients on dendritic form are schematically represented in Figure 7.12.

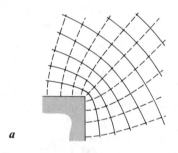

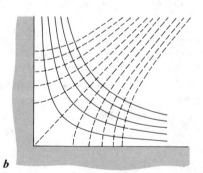

FIGURE 7.11 **Equiconcentration lines (solid) and diffusion flow lines (dashed) in solution growing next to (a) exterior corner and (b) receding corner of crystal. Shaded area denotes crystal. (From McLachlin, D., 1978. Progress in crystal-growth theory.** *Canadian Mineralogist* **16:422. Courtesy of the Mineralogical Association of Canada.)**

FIGURE 7.12 **Progressive changes in shape (a–d) of an ideal cubic crystal with increasing degree of supersaturation. (Redrawn with permission from Elwell, D., and Scheel, H. J., 1975. Crystal growth from high temperature solutions. London: Academic Press Inc., p. 263. Copyright © 1975 Academic Press Inc. [London] Ltd.)**

In some cases, crystallization proceeds so fast that *hopper* crystals, characterized by strong reentrant angles or even hollow centers, form. Halite (NaCl), feldspar (e.g., $CaSi_2Al_2O_8$), and pyroxene (e.g., $CaMgSi_2O_6$) are examples.

COMPOSITIONAL ZONING

A common feature of many minerals (e.g., feldspars, pyroxenes, amphiboles, garnets) is concentric compositional variation from the core of the crystal outward. This feature is in principle possible for any mineral that can have solid solution variations with respect to the various chemical ingredients present in the original solution. The cores of such crystals represent the initial growth, and the rims represent the last growth. Quite frequently the variation in composition is expressed as a change in color as in the zoned hornblende shown in Figure 7.13a. In most cases, the changes are reflected by continuous changes in optical properties such as extinction angle and refractive index as shown by the zoned plagioclase, also in Figure 7.13a.

Compositional zoning occurs for different reasons, all related to the changing chemistry of the crystallizing solution or fluid. In some cases, when diffusion processes in the solution are slower than the crystallization rate, the solution becomes selectively depleted in some elements, and subsequent

a   0.10 mm

b   0.30 mm

FIGURE 7.13   **(a) Compositional zoning of hornblende (left center) and of plagioclase feldspar from Ca-rich core to Na-rich rim (lower right) in tertiary dike from Alaska. (b) Hourglass zoning in titan-augite phenocryst in Hawaiian lava.**

crystallization produces depleted growth layers. This can happen under conditions of essentially constant pressure and temperature. In other cases, the saturation composition of both the liquid and the crystal change systematically with pressure and temperature. Calcic plagioclase, which commonly crystallizes first at relatively high temperature, becomes mantled by progressively more sodic plagioclase as the temperature drops. The core composition is no longer the saturated one with respect to the remaining lower temperature liquid, but it is preserved metastably, because the crystallization rate was faster than the solid-state diffusion rate within the zoned crystal.

In some minerals, especially the calcic clinopyroxenes, epidote, staurolite, and andalusite, chemical zoning may develop in complex *sectors* within a single crystal. The finest examples display a perfect *sector zoning* (Figure 7.13b). The faces developed in such crystals are normal *F* faces, which are predictable from principles previously discussed. Current thought is that either the different growth rates of the faces or their preferential uptake of selected ions during rapid crystallization is responsible for sector zoning. Most crystals in these mineral groups do not exhibit sector zoning, even though most of the pyroxenes, for example, crystallize from quite similar high-temperature igneous environments. This suggests that sector zoning develops under unusual circumstances, perhaps in very stagnant or viscous melts in which exceptional concentration gradients can persist during growth.

EFFECTS OF IMPURITY ATOMS ON GROWTH

In the case of sector zoning, the observable change in color from one sector to another may be due to relatively small changes in concentration of a minor or even trace element. In pyroxenes, Ti^{4+} or Fe^{3+} or both may be responsible for coloration changes, although the ions may be present in relatively small amounts. In experimental crystal growth studies, a few parts per thousand concentration of an impurity element can cause substantial changes in growth form or habit. A logical interpretation is that the presence of impurities has an effect on growth rate, and that the effect must vary from face to face depending on which surface adsorbs the impurities.

The physical processes that determine the effect of impurities on growth rate are not well understood. One clue is suggested by the PBC model discussed earlier. Given a growing crystal on which *F*, *S*, and *K* faces are developed, impurity atoms are more likely to attach themselves to the K faces because of the relatively high attachment energy. As a result, both S and K faces may be converted to F faces (Figure 7.12), because the impurity atoms effectively smooth out the face to the extent that the attachment energy is decreased. This almost always causes a decrease in growth rate of the adsorbing face and enhances its dominance in the final growth form. In some cases, the effect of impurity elements is to produce certain faces that would not normally be present if the crystal were grown from the pure solution or liquid.

Impurities can be of several types. Carbonaceous material and graphite are frequently incorporated in metamorphic minerals that grow entirely in the solid state. The chiastolite cross in metamorphic andalusite is a common example.

EPITAXIAL OVERGROWTHS

Newly formed minerals, whether from igneous, metamorphic, or low-temperature aqueous environments, will frequently nucleate and grow on previously existing crystals. The two minerals need not be the same, but they usually have close structural similarities. If the lattice of the overgrowth has approximately the same shape and dimensions as that of the host and if the two are interfaced along such lattice planes, one mineral is said to be *epitaxial* on the other.

The formation of epitaxial overgrowths involves the same principles that govern normal crystal growth. When a solution becomes supersaturated due to change in temperature, for example, crystals will nucleate and grow best where their chances of "sticking" are greatest. A preexisting structure with a similar lattice is an ideal host, far better energetically than spontaneous nucleation or overgrowing a structurally dissimilar host. Occasionally, the structural selectivity of the overgrowth on the host or substrate is so great that only certain faces are affected. In other cases of structural misfit, only small crystallites will form to reduce the strain that would accompany a laterally continuous overgrowth.

During polymorphic transitions (Chapter 8) in which the only differences between the two minerals are structural, the overgrowth (or sometimes an intergrowth) adopts the lattice geometry of the host. The term *syntaxial* is used to describe such relationships. In more complicated systems involving several reactant and product minerals in a chemical reaction, the newly formed minerals may inherit the structural orientations of the reactant minerals. The term *topotaxial* is used to describe

this phenomenon. In these cases, chemical differences between the minerals exist, and the solid-state diffusion of cations has essentially taken place within an undisturbed structural framework.

ASBESTIFORM GROWTH

All *crystallization habits* are consequences of the conditions under which the crystal was grown. In most cases, the differences in growth conditions are relatively minor, and the differences between physicochemical properties of the mineral in the various habits are similarly minor. Most of the differences in properties are limited to variations in size and shape. In only one crystallization habit are the physical properties of the mineral fundamentally altered so the mineral has a unique set of properties and an appearance that does not resemble the mineral in the other crystallization habits. Minerals crystallized in the *asbestiform habit* constitute a distinct category of natural substances, for their common characteristic properties are unrelated to their different mineralogical identities.

Theoretically, any mineral can crystallize in the asbestiform habit, although some do so more readily than others. There are five minerals that are commonly known to form large, commercial deposits of asbestiform material. These five are chrysotile (an asbestiform variety of serpentine), crocidolite (an asbestiform variety of riebeckite), anthophyllite, actinolite-tremolite, and cummingtonite-grunerite. On the other hand, some lesser known minerals such as palygorskite almost always crystallize in this habit, and a large number of other minerals occasionally crystallize in the habit (e.g., tourmaline, sepiolite, micas, brucite, graphite, halloysite, gypsum), but are not known to constitute commercial asbestos deposits.

In the asbestiform habit, minerals grow rapidly in fine, hairlike crystals that resemble organic fibers in appearance. They are strong and flexible. The tensile strength of these fibers is enhanced by more than an order of magnitude to values above that of any mineral crystallized in other habits. Their surface structures are less flawed and are more resistant to acids than the same minerals in other habits. These are the unique properties of asbestos, which gives the material its unique industrial value.

The crystal structures of these fibers are identical with the structures of the same mineral crystallized in other habits. They differ, however, in secondary structural features. In the asbestiform habit, some of the fibers of the layer silicates (e.g., serpentine, Figure II.29) or other layer structures (e.g., graphite) have their layered structural units curled up in scrolls or tubes. Still others show unusual defects in their basic structures (e.g., amphibole asbestos), and some may display no noticeable structural variation. All, however, are characterized in the asbestiform habit by the lack of extensive surface defects and by fibrous shape.

In good quality asbestos fiber the density of surface defects is low. The quality of the fiber, its strength, and flexibility are inversely proportional to the density of surface defects in nontubular fibers. That is, a complete gradation is possible between elongated crystals and asbestos fibers, including a gradation of the asbestos properties.

Synthetic equivalents of asbestos fibers are known as *whiskers*. Whiskers are grown under controlled conditions and have the appearance and the properties of asbestos. The tensile strength of some exceptionally good whiskers approaches the theoretical maximum strength of the substance. The enhanced strength and flexibility of whiskers are attributed to the lack of surface defects. Whiskers of many inorganic compounds have been grown, including some that have mineral equivalents (e.g., brucite, corundum, perovskite).

OTHER DEFECTS

The role of defects (other than screw dislocations) with respect to physical properties is discussed in Chapter 6. For growth processes, other types of defects are important as nucleation sites. If the strain energy due to the misfit of lattices on epitaxial overgrowths is too large, the overgrowth mineral may not nucleate at all. Normally, nucleation does occur, but on various two-dimensional or three-dimensional defects.

The most effective defect for this purpose is a three-dimensional grain boundary, because the strain due to lattice misfit can be relieved by minute grain boundary adjustments. Linear or two-dimensional defects such as screw dislocations are less effective in this respect, because lattice strains due to misfit may still be retained. Point defects, either Schottky or Frenkel type (Chapter 6), are still less effective. In all cases, some degree of supercooling or oversaturation seems to be required before the growth process can proceed.

ADDITIONAL READINGS

Bunn, C. W., 1964. *Crystals: their role in nature and science*. New York: Academic Press.

Hartman, P., ed. 1973. *Crystal growth, an introduction*. Amsterdam: North-Holland.

Holden, A., and Singer, P., 1971. *Crystals and crystal growing*. London: Heinemann.

Schneer, C. J., ed. 1977. *Crystal form and structure*. New York: Academic Press.

CHAPTER EIGHT

MINERAL CHEMISTRY AND STABILITY

The treatment of the chemical aspects of mineralogy is commonly separated from the topics of crystallography and symmetry. This is convenient for the purposes of organization, but the separation tends to obscure fundamental concepts relating the chemistry of a mineral and its structure. By examining the basic differences between the crystalline and liquid or gaseous states of matter, we gain some insight into the relationship between mineral chemistry and structure. We know empirically that an infinite number of possible chemical combinations can be prepared by mixing chemical reagents from a laboratory. These are essentially mechanical aggregates of single substances, each of which has its own chemistry and structure. We also know that each of these mixtures will melt if heated to sufficiently high temperature. Each liquid so produced has uniformly homogeneous properties. An unlimited and continuous range of chemistries can be produced in this manner. None of these substances are minerals, however, because all lack a single, geometrically ordered arrangement of atoms. We learned in Chapter 5 that only a finite number of geometric arrangements of atoms or ions may exist in minerals. Furthermore, the ratios of atoms or ions in a standard chemical formula are usually small integers due to the multiplicity of equipoints in the structure. This property of structure and symmetry distinguishes all crystalline solids from chemically equivalent liquids and gases, and constrains the possible chemistries minerals may have.

THE CONCEPT OF COMPOSITION SPACE

An ability to visualize one-dimensional, two-dimensional, and three-dimensional patterns is an essential tool in crystallography. An ability to represent the chemistry of minerals in these or higher dimensions is of equal importance. Mineral chemistry is conveniently represented in terms of chemical *components*. If all of the naturally occurring elements were chosen as components, the resulting composition space would have 98 dimensions. Fortunately, the relatively simple chemistry of most minerals requires fewer components. For example, any two points in composition space define a straight line. Figure 8.1 shows such a line with end points arbitrarily identified as the components MgO and SiO_2. Any mechanical mixture of these components can be represented on the line, and conversely, any mixture on the line can be represented in terms of some proportion of MgO and SiO_2. The choice of these components is arbitrary, however, because any two points on the line will suffice to describe any other point on the line. The minimum number of components in this example must be two, regardless of chemical identity. A third component is unnecessary for the definition of the line.

Similarly, only three points in composition space are required to define a plane. Once the chemistry of the points is identified, all other points in the plane may be expressed in terms of these choices. A fourth point in the plane would be redundant as far as the definition of the plane is concerned. A fourth point on either side of the plane will define a volume, which corresponds to a four-component system once each point has been chemically identified. Additional components obviously exceed the dimensionality in which we are accustomed to operating, but fortunately, powerful mathematical techniques, particularly those of linear algebra, enable mineralogists to surmount this difficulty.

Consider the compositional relationships between the minerals periclase (MgO), forsterite (Mg$_2$SiO$_4$), enstatite (MgSiO$_3$), and quartz (SiO$_2$). All four minerals lie on the same straight line in composition space and can be represented by a single *binary composition diagram* (Figure 8.1a). In this example, our choice of components MgO and SiO$_2$ happens to correspond to the chemistry of the minerals periclase and quartz, but this is only a matter of convenience. In order to describe the other minerals along this line in terms of MgO and SiO$_2$, we must adopt a *unit of measurement* and then use the unit consistently in all our calculations. Moles[1] of oxide components are a useful unit choice, because the formula of any mineral

can be readily written in these terms simply by inspection. For example,

$$1\ Mg_2SiO_4 = 2\ MgO + 1\ SiO_2$$

or 1 mole of forsterite is equivalent to 2 moles of MgO plus 1 mole of SiO$_2$. From this expression, the *mole fraction* of SiO$_2$ in forsterite may be calculated as the number of moles of SiO$_2$ in the formula unit divided by the total sum of oxide moles, that is,

$$X_{SiO_2}^{forsterite} = \frac{n_{SiO_2}}{n_{SiO_2} + n_{MgO}} = \frac{1}{2+1} = \frac{1}{3}$$

If we now choose to plot the chemistry of forsterite in terms of the component SiO$_2$, we do so on a scale from zero to unity. MgO contains zero moles of SiO$_2$, so the mole fraction of SiO$_2$ in MgO must be zero. The mole fraction of the component SiO$_2$ in SiO$_2$ is unity since pure SiO$_2$ contains no MgO. The equation then means that the mole fraction of the component SiO$_2$ in forsterite is one third along the line toward SiO$_2$ from MgO (Figure 8.1b). It is important to remember that SiO$_2$ as a chemical component has no mineralogical significance. Clearly no quartz is in forsterite.

We can just as easily plot the chemistry of forsterite in terms of the component MgO:

$$X_{MgO}^{forsterite} = \frac{2}{2+1} = \frac{2}{3}$$

Now the scale on which we plot runs from zero at SiO$_2$ to unity at MgO (Figure 8.1c). Note that the sum of the mole fractions for any mineral is equal to unity, that is,

$$X_{MgO}^{forsterite} + X_{SiO_2}^{forsterite} = 1$$

which is simply to say that the sum of the parts is equal to the whole. The position of any mineral on the line may be established with the use of a single mole fraction. Other choices of units of measurement may be desirable in certain circumstances. Weight percent or atomic percent, rather than mole percent, are commonly used.

Now imagine a laboratory crucible in which we place varying amounts of MgO and SiO$_2$. If only one mole of MgO is present, we can specify that the *bulk composition* of the material in the crucible is at the point MgO on the binary (two-component) diagram shown in Figure 8.1a. If one mole of SiO$_2$ is now added to the crucible in addition to the one mole of MgO already present, the bulk composition changes to the midpoint on the MgO-SiO$_2$ diagram, corresponding to equal molar quantities of each component. When this bulk composition is heated to 1400 °C under controlled laboratory conditions, these two com-

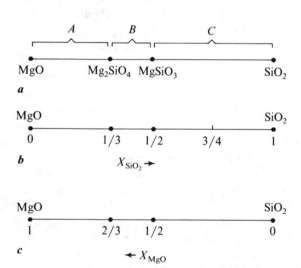

a

b $X_{SiO_2} \rightarrow$

c $\leftarrow X_{MgO}$

FIGURE 8.1   (a) Positions of periclase (MgO), forsterite (Mg$_2$SiO$_4$), enstatite (MgSiO$_3$), and quartz (SiO$_2$) expressed in a single binary composition diagram. **(b)** Mole fraction of SiO$_2$ (X_{SiO_2}) in forsterite and enstatite. **(c)** Mole fraction of MgO (X_{MgO}) in forsterite and enstatite. Regions *A*, *B*, and *C* represent bulk compositions for which no corresponding single homogeneous minerals exist.

[1] See Part II, Appendix 1, on units and constants for the definition of a mole.

ponents combine completely to form enstatite, a pyroxene. This is possible because the bulk composition is appropriate for the formation of enstatite by the chemical reaction $MgO + SiO_2 = MgSiO_3$. Had the original composition, consisting of only one mole of MgO, been heated to 1400 °C under identical conditions, enstatite obviously could not have formed, because the bulk composition would not have been appropriate.

Now place in an empty crucible two moles of MgO and one mole of SiO_2. This bulk composition corresponds to a unique point on the composition diagram at a distance one third from MgO toward SiO_2, or equivalently, at a distance two thirds from SiO_2 toward MgO. Upon heating the crucible and its contents to 1600 °C in the laboratory, we form forsterite by the reaction $2MgO + SiO_2 = Mg_2SiO_4$. Again, all of the reactants are used up, and we are left with pure forsterite.

For each of our previous choices of bulk composition, a corresponding mineral of exactly the same composition existed. Both $MgSiO_3$ and Mg_2SiO_4 have chemistries that, *in combination* with the space group symmetry of enstatite and forsterite, produce the unique physical and optical properties so characteristic of those minerals.

To amplify this concept, consider a bulk composition in our crucible corresponding to three moles of SiO_2 and one mole of MgO. On the binary diagram (Figure 8.1a), this corresponds to a point three fourths toward SiO_2 from MgO and within the region labeled *C*. What happens to the mixture when heated to temperatures at which the appropriate chemical reactions can take place? We discover that the product of our experiment is not one but two minerals: enstatite ($MgSiO_3$) and cristobalite (SiO_2) in equal amounts. The same result would have been achieved with any bulk chemistry within region *C*, with only the proportions of enstatite and cristobalite differing. Why did a *single mineral* with the above chemistry not form? Because for that particular chemistry (i.e., $MgSi_2O_5$), no ordered, symmetrical arrangement of atoms exists that would be more stable than the mixture of enstatite and cristobalite. This phenomenon would not arise if we were dealing with liquids or gases, so the reason must lie in the nature of the crystalline state.

We gain further insight into how the chemistry of minerals is controlled by crystalline structure by plotting the composition of each of the minerals in Figure 8.1 in units that are more closely related to crystal structures. The mineral periclase (MgO) has an *ABC* octahedral network structure consisting entirely of Mg octahedra. The structures of quartz and cristobalite (SiO_2) consist entirely of Si^{4+} tetrahedra. By choosing as a unit of measurement the *number of cations in tetrahedral coordination*, we establish a scale from zero (no tetrahedra, 100% octahedra) to unity (100% tetrahedra, no octahedra). Figure 8.2 plots the four minerals in terms of the fraction of cations in tetrahedral coordination, X_T. Enstatite ($MgSiO_3$) is a chain silicate in which half the cations (Si) are in tetrahedral coordination and the other half (Mg) are in octahedral coordination. Forsterite is an isolated tetrahedral silicate with twice as many octahedra as tetrahedra. Within the compositional limits defined by MgO and SiO_2, no other stable geometric arrangements of Mg^{2+} octahedra and Si^{4+} tetrahedra exist. We observe empirically that Mg_2SiO_4 and $MgSiO_3$ are the only chemistries along the MgO-SiO_2 composition diagram that correspond to minerals. Under certain circumstances, some Si^{4+} may be in octahedral coordination and some Si^{4+} in tetrahedral coordination (inverse spinels), but the ratio of octahedra to tetrahedra remains unchanged. Starting with a pure $MgSiO_3$ composition, any attempt to add Si^{4+} to the enstatite structure in excess of a one to one ratio is rejected by the structure. Because extra Si^{4+} cannot be accommodated by the structure, it forms a separate mineral, quartz (SiO_2). We are saying, in effect, that enstatite and quartz are *immiscible*, that is, they display no mutual solid solubility toward each other. We must conclude that for all bulk chemistries defined by the regions *A*, *B*, and *C* in Figures 8.1 and 8.2, there are no corresponding single minerals. Two minerals in a proportion commensurate with the bulk composition will always form instead.

Returning now to our introductory statements about the liquid state, all bulk compositions between and including MgO and SiO_2 will eventually melt if heated to sufficiently high temperatures. The resulting liquids will all have different compositions along the binary diagram depending on the chemistry of the starting mixture, but each of them, regardless of composition, will be chemically homogeneous. At these elevated temperatures, the liquids can be mixed in *any* proportion

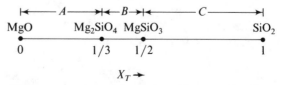

FIGURE 8.2 Positions of periclase, forsterite, enstatite, and quartz plotted in terms of fraction of cations in tetrahedral sites (X_T). Only values 0, 1/3, 1/2, and 1 are permissible.

to produce yet another homogeneous liquid, in much the same manner as two homogeneous gases mix together to form a single homogeneous gas of another composition. Liquids and gases that behave in this manner are said to be *miscible* in one another. Such behavior on the part of liquids and gases is due to the absence of long-range structure that would serve to constrain their compositional variation.

EFFECTS OF ADDITIONAL COMPONENTS

The previous example represents just one direction in an *n*-dimensional composition space. If we wish to investigate the effect of another component, for example, FeO, on minerals in the MgO-SiO₂ composition diagram, we can refer to the *ternary composition diagram* defined by MgO, SiO₂, and FeO (Figure 8.3). The position of FeO in composition space relative to MgO and SiO₂ is immaterial. The important fact is that straight lines exist between the three components, and these lines are coplanar. Each line defines a two-component system; the equilateral triangle in Figure 8.3 defines a three-component system.

The minerals wüstite (FeO), fayalite (Fe₂SiO₄), ferrosilite (FeSiO₃), and quartz (SiO₂) plot in the same relative positions with respect to FeO and SiO₂ as do minerals in the MgO-SiO₂ composition diagram. As in the latter diagram, no minerals

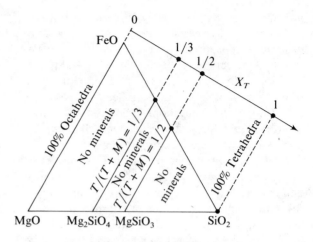

FIGURE 8.4   Relationships in Figure 8.3 expressed as the fraction of cations in tetrahedral sites X_T. Permissible values are 0, 1/3, 1/2, and 1. Areas between solid solutions delineate bulk compositions for which no corresponding minerals exist.

other than the ones shown are found in the system. The reason is again the crystal structures, in this case of fayalite and ferrosilite. Fayalite has the same space group symmetry as forsterite, although the size of the unit cell and all of its physical properties are markedly different. The compositional constraints imposed on fayalite with respect to its FeO to SiO₂ ratio are identical to those imposed on forsterite. Similar constraints also apply to ferrosilite and to enstatite. Both fayalite and forsterite have a one to one ratio of metal ions to Si. The similarity of the ionic radii of Fe^{2+} (0.068 nm) and Mg^{2+} (0.080 nm) in octahedral coordination and their identical charge enables these cations to mix in any proportion within the octahedral sites of the structure. Complete *solid solubility* or miscibility thus exists between fayalite and forsterite. This is known as the olivine *solid solution* series. Permissible variations in olivine chemistry within Figure 8.3 are constrained to the straight line between fayalite and forsterite. The line is unique insofar as every point on it corresponds to a crystalline structure that has twice as many octahedral, *M*, as tetrahedral, *T*, cation sites (Figure 8.4).

For the same reasons, substantial solid solubility also exists between enstatite and ferrosilite. Both end-members possess orthorhombic symmetry and have a one to one ratio of tetrahedra to octahedra. Like olivine, the orthopyroxenes in Figure 8.2 are a binary solid solution.

We can see that extensive regions of composition space exist in Figure 8.4 that do not correspond to compositions of any known minerals. If

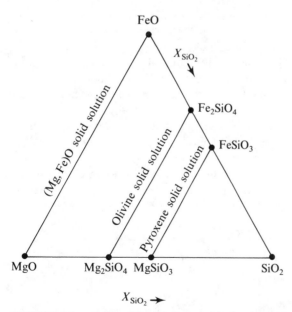

FIGURE 8.3   Compositional relationship of periclase (MgO) and quartz (SiO₂) extended to illustrate additional component wüstite (FeO). Triangle shape is arbitrary.

we were to expand our graphical representation to a volume by adding an additional component (Figure 8.5), we would discover large volumes within the tetrahedron, for example, between the Fe_2SiO_4-Mg_2SiO_4-Ca_2SiO_4 and the $FeSiO_3$-$FeSiO_3$-$CaSiO_3$ planes, that have no corresponding minerals. Similarly, no known minerals exist with compositions that fall in the volume between the Fe_2SiO_4-Mg_2SiO_4-Ca_2SiO_4 and FeO-MgO-CaO planes. The planes themselves correspond to those structurally permissible ratios of Si^{4+} tetrahedra and metal octahedra.

By analogy, unique compositional regions exist in an *n*-dimensional space that correspond to all known minerals and solid solutions. These regions are separated from one another by compositional fields in which minerals do not occur. The uniqueness of each region is due to the finite number of ways atoms and ions can be geometrically arranged with respect to each other and still comply with the rules of symmetry developed earlier.

The extent of solid solution in minerals has definite limits. If a substituting cation is too large or too small for a particular site, or if a difference in charge is not compensated by another substitution, then substitution may be energetically unfavorable. The extent of solid solution is also limited by the pressure and temperature of the mineral at the time of formation.

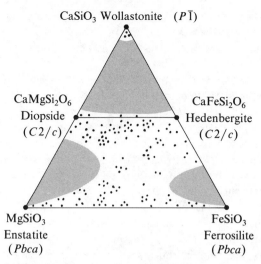

FIGURE 8.6 **The pyroxene "quadrilateral" with wollastonite ($CaSiO_3$) added for completeness. Dots represent plotted pyroxene analyses from numerous sources and from a wide range of geological conditions. Shaded regions represent gaps in solid solution.**

Important examples of these relationships are the natural and synthetic pyroxenes. Theoretically, all pyroxenes must plot in the $FeSiO_3$-$MgSiO_3$-$CaSiO_3$ plane of Figure 8.5, providing no metal cations other than Fe^{2+}, Mg^{2+}, and Ca^{2+} are there, and providing that the tetrahedra are exclusively occupied by Si^{4+}. In reality, many calcic pyroxenes have small amounts of other divalent cations, such as Mn^{2+}, in octahedral coordination, but these are ignored for the purposes of plotting in Figure 8.6. This diagram illustrates the variation in chemistry of pyroxenes or pyroxenoids, such as wollastonite ($CaSiO_3$), from a wide range of geological environments. Some are from high-temperature igneous environments; others are from lower temperature metamorphic environments. A range of pressures is also represented. A striking fact is apparent: The ternary diagram has large areas for which no known pyroxene compositions exist. Very few chemical analyses are more calcic than the binary solid solution between diopside ($CaMgSi_2O_6$) and hedenbergite ($CaFeSi_2O_6$), and some of these are calcic because other divalent cations in the chemical analyses have been ignored. This in effect increases the mole fraction of $CaSiO_3$ over the amount that is actually present had other components been considered. Chemical analyses of wollastonite from various geological localities exhibit very little solid solution with respect to Fe^{2+} and Mg^{2+}. Other conspicuous regions occur between diopside and enstatite and between hedenbergite and ferrosilite for which no analyses exist. One

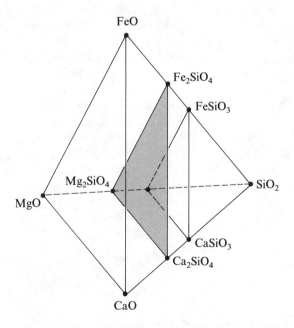

FIGURE 8.5 **Effect of a fourth component (CaO) on the solid solution relationships between oxides, olivines, and pyroxenes. Shaded triangle represents the plane for which $X_T = 1/3$. All Fe-Mg-Ca olivines must occupy this plane.**

might wonder if such vacant areas simply represent incomplete sampling on the part of the geologist, or whether true gaps in solid solubility are represented.

Whatever the reason for these gaps, it cannot be related to the ratio of tetrahedral to octahedral sites in the pyroxene structure, because that ratio is constant over the entirety of Figure 8.6. The answer lies in the distribution of metal cations between symmetrically different octahedral sites within the structures. The pyroxene structure has two such sites, referred to as *M1* and *M2* (Figure II.33). Detailed x-ray diffraction studies show that as the chemistry of orthopyroxenes along the enstatite-ferrosilite composition line changes toward the chemistry of pyroxenes along the diopside-hedenbergite composition line, the primary chemical substitution is Ca^{2+} for Mg^{2+} and Fe^{2+} in the *M2* octahedra. By examining the crystal structures of pyroxenes that have chemistries between ferrosilite and hedenbergite, we can understand exactly how *M2* changes shape as the proportion of Ca^{2+} to Mg^{2+} and Fe^{2+} increases. The individual *M2*—O bond lengths before substitution are all different and define an irregular octahedron. With the addition of Ca^{2+}, the octahedron expands, because the ionic radius of Ca^{2+} (0.120 nm) is substantially greater than that of either Fe^{2+} (0.068 nm) or Mg^{2+} (0.080 nm). The Si—O bonds in the tetrahedra and the *M1*—O bonds in the other octahedra are less affected, so *differential polyhedral distortion* occurs in much the same way as in thermal expansion and compressibility (Chapter 6). In the present case, however, the mechanism of *M2* octahedral expansion requires that the *shortest* bonds of the octahedron expand the most in order to accommodate the larger cation. The resulting strain (e.g., Figure 6.3a) accumulates as the amount of substituting Ca^{2+} increases until the orthorhombic structure is destabilized and a gap in solid solubility results.

The composition at which the miscibility gap first appears depends on temperature. In general, the extent of solid solubility is greater at higher temperature. This behavior can be understood by comparing the octahedral bond strains due to composition change with those due to thermal expansion. The longer bonds of the *M2* octahedron in clinopyroxene expand more during heating, because they are relatively weak (Figure 6.5). This is the opposite of the effect due to substitution. The overall effect is for increasing temperature to "undistort" octahedra that are already strained by cation substitution. The substituted composition is thereby stabilized at higher temperature.

CALCULATION OF CHEMICAL FORMULAS

Examples in the preceding sections illustrate the close relationship between the chemistry of a mineral and its structure. The simple integral subscripts in most mineral formulas convey the fact that the ratio of symmetrically nonequivalent sites is determined by the structure. Ionic substitutions in each site determine the exact chemistry of a mineral, but the ratio of the sites remains constant. For example, pure forsterite (Mg_2SiO_4), the Mg end-member of the olivine series, has two octahedral sites and one tetrahedral site per formula unit. The octahedral sites are fully occupied by Mg^{2+}, and the tetrahedral site is fully occupied by Si^{4+}. Ferrous iron (Fe^{2+}) readily substitutes for Mg^{2+} to yield olivine compositions between pure forsterite and pure fayalite (Fe_2SiO_4). In such cases, the substituting cations must sum to two. If the specific composition is unknown, the formula can be written as $(Mg, Fe)_2SiO_4$. If the ratio of octahedral cations is known, as it is, for example, in an olivine with a composition of 75% forsterite and 25% fayalite, the formula can be written $Mg_{1.50}Fe_{0.50}SiO_4$.

Most conventional chemical analyses performed in a laboratory represent the chemistry of a mineral in terms of an oxide set of components in units of weight. Such an analysis is called a *weight-percent analysis*, and although it fully describes the chemistry of a mineral, the numbers involved do not convey a clear relationship to structure. The purpose of formula calculation is to convert the components of the original analysis from oxides to elements, and to convert the units from weight (grams) to numbers of atoms.

Table 8.1 illustrates the procedure for a typical analysis of beryl. The most important oxides are Al_2O_3, SiO_2, and BeO, comprising 97.5% of the mineral by weight. The remaining 2.5 weight percent is accounted for by small amounts of additional oxides as shown. Dividing the weight percent oxide (second column) by the oxide molecular weight (third column) yields the molecular proportion of that oxide (fourth column) in the analysis. Viewed in slightly different terms, 18.58 grams of Al_2O_3 yields 0.1822 moles of Al_2O_3. As each mole of Al_2O_3 is equivalent to 2 moles of Al and 3 moles of oxygen, $0.1822 \ Al_2O_3 = 2(0.1822)$ Al $+ 3(0.1822)O = (0.3644)Al + (0.5466)O$. This procedure is repeated for each of the oxides, and the cation proportions and the oxygen proportions are tabulated in the fifth and sixth columns, respectively. We now know the relative proportions of ions in the mineral formula. Summing the

TABLE 8.1. *Procedure for Recalculating Mineral Analysis in Weight Percent Oxides to Standard Chemical Formula in Atomic Proportions*

Oxide	Weight Percent Oxide	Molecular Weight of Oxide	Mole Proportion of Oxide	Cation Proportion	Oxygen Proportion
Al_2O_3	18.58	101.96	0.1822	0.3644	0.5466
SiO_2	65.86	60.09	1.096	1.096	2.192
BeO	12.96	25.16	0.5151	0.5151	0.5151

Total weight = 97.50%   Total oxygen per formula = 3.254

Fe_2O_3	0.20
FeO	0.40
MgO	0.16
CaO	0.84
H_2O	0.90
	2.50%

Recalculated Analyses

1. $Al_{0.3644}Si_{1.096}Be_{0.5151}O_{3.254}$
2. $Al_{2.016}Si_{6.063}Be_{2.850}O_{18}$
3. $Al_2Si_6Be_3O_{18}$
4. $Al_{1.979}^{3+}Si_{5.951}^{4+}Be_{2.797}^{2+}Fe_{0.136}^{3+}Fe_{0.033}^{2+}Mg_{0.011}^{2+}Ca_{0.081}^{2+}O_{18} \cdot 0.27H_2O$
5. $(Al_{1.727}Fe_{0.136}^{3+}Fe_{0.033}^{2+}Mg_{0.011}Ca_{0.081})(Si_{5.951}Be_{0.049})(Be_{2.748}Al_{0.252})$
6. $(Al,Fe^{3+},Fe^{2+},Mg,Ca)_{1.99}(Si,Be)_{6.0}(Be,Al)_{3.0}O_{18} \cdot 0.27H_2O$

oxygens, we can write formula 1 in Table 8.1. The subscripts, however, require conversion to simple integers without changing the ionic proportions. At this point, if we suspect or know the mineral to be beryl, formula 1 in Table 8.1 can be normalized to an 18-oxygen basis by multiplying all subscripts by $18 \div 3.254 = 5.532$. This procedure yields formula 2 in the table.

The difference between formula 2 and ideal beryl (formula 3) is due in part to errors inherent in the chemical analysis, but the discrepancy is mostly due to the substituting cations that were not considered in the calculation. The structural formula for beryl (formula 3) is $Al_2Si_6Be_3O_{18}$, denoting six Si^{4+} tetrahedra and three Be^{2+} tetrahedra in the primary frame and two Al^{3+} octahedra in the secondary frame. Formula 2 shows that there is insufficient Be^{3+} to fill the available sites, and a little too much Si^{4+} and Al^{3+} to be accommodated in the remaining tetrahedral and octahedral sites. Formula 4 represents the entire beryl chemistry, again on the basis of 18 oxygens, but we assume that the water is molecular H_2O and occupies the large cages between the six-membered rings of the structure.

In formula 5 the cations are rearranged in the sites where ionic substitutions are most probable. The cations in the tetrahedral sites with 5.951 Si^{4+} are summed to six by adding 0.049 Be^{2+}. The

2.748 ions of Be^{2+} in the other tetrahedral sites combined with 0.252 Al^{3+} sum to three. The remaining Al^{3+} (1.727, formula 5) must be octahedrally coordinated with Fe^{3+}, Fe^{2+}, Mg^{2+}, and Ca^{2+}. The sum of the octahedral cations (1.99) is essentially the same as the structure predicts (formula 6).

Note that regardless of the assignment of cations to particular sites in the beryl structure, the formula remains electrically neutral. This is because the original weight-percent analysis was represented in terms of electrically neutral oxides, and no cations or oxygens were lost in the calculation.

ORDER AND ENTROPY

The simple integer subscripts of most mineral formulas imply a highly ordered state for each ion in the structure. In effect, each kind of cation or anion "knows" the structural site to occupy, because at that site the minimum bond strain is achieved. Ions in noncrystalline materials are much less constrained and consequently are less ordered. In both liquids and gases, ions are *disordered* with respect to each other and with respect to compositionally equivalent crystalline solids.

This sense of order, or lack of randomness, is embodied in the concept of entropy. All spontaneous processes in nature must be accompanied by an entropy change that is greater than or equal to zero. The entropy associated with such a process can never decrease. A simple example is the spontaneous mixing of two different gases initially separated by containers. Upon release to the atmosphere, the gases spontaneously mix to form a homogeneous mixture. The original state of the two gases was relatively ordered; their final state is relatively disordered. The process is quite irreversible. The mixture will not spontaneously segregate or "unmix" unless energy is provided from some external source. Unmixing could be accomplished by removing energy, thus lowering the temperature until one of the gases liquifies.

The fundamental relationship between entropy and the degree of disorder is given by

$$S = K \ln \Psi$$

where S is entropy and Ψ, for the purposes of our application here, is a measure of the randomness of atomic arrangements. The constant of proportionality that numerically relates these parameters is the Boltzmann constant (Table A1.5, Appendix 1). The number of possible different arrangements that gas molecules can have relative to each other is vastly greater than the possible arrangements between atoms or ions constrained by symmetry to occupy specified positions in a crystalline structure. The entropy of gases is consequently much greater than the entropy of crystalline solids. Liquids, in general, possess intermediate values of entropy, all other factors being equal, because of their intermediate state of atomic order.

The state of order of a substance depends on the energy available to it. The energy can be increased by simply heating the material. As the mineral ice (H_2O) is heated, the number of discrete energy levels available to each H_2O molecule increases, and the entropy gradually increases. When the temperature reaches the melting point, the thermal energy of molecular vibration exceeds the intermolecular bonding forces, and the crystalline structure breaks down. Further increase in temperature raises the entropy of water in a continuous manner until it boils. The gas is a high entropy state, and for that reason is the form of H_2O that is preferred at high temperature. Discontinuities occur in the variation of entropy with temperature when a major reorganization of atomic species takes place, as in the example of H_2O. We will examine this important aspect of entropy in a later section.

The effect of temperature on the compositional range of solid solutions is closely related to the concept of ordering. We learned earlier that the effect of increasing temperature is to extend the compositional range of crystalline solid solutions relative to the gaps in miscibility between solutions. The effect of temperature is represented in Figure 8.7 in which the range of possible chemistries between two hypothetical ideal endmembers A and B of a solid solution is permitted to vary. We will also assume that the structures of the end-members are different, but similar enough to permit solid solution under appropriate conditions. This can be accomplished by having two crystallographically distinct cation sites, as in the pyroxenes.

At absolute zero temperature, the entropy must be equal to zero as given by the third law of thermodynamics. The value of Ψ will then be unity since $\ln 1 = 0$. We interpret this to mean that the end-members A and B in our two-component system will be perfectly ordered. From a crystalchemical point of view, we can visualize each ion as knowing exactly in which site within the structure it prefers to be. Any attempt to place one cation in a structural position that is not preferred tends to distort the structure and thereby increases the energy of the crystal.

Consider now the effect of a rise in temperature on the extent to which one cation may occupy a site that was energetically unfavorable at lower temperature. The increased amplitudes of vibrations of all ions slightly relaxes the selectivity of the sites, and a small proportion of one cation may occupy the site of another and vice versa. Figure 8.7 represents this relationship as a *limited* solid solution between the two end-members at

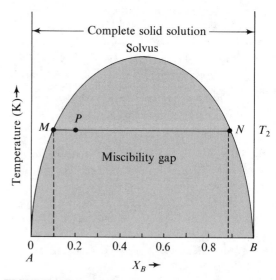

FIGURE 8.7 **Temperature-composition relationships for a hypothetical two-component system.**

temperature T_2. This argument can be repeated at successively higher temperatures with the result that progressively more solid solubility is allowed, and the gap in miscibility must eventually close. The higher temperatures are associated with a higher entropy for each mineral and with correspondingly higher states of disorder. Disorder in this context means that to an extent dependent on temperature, a certain proportion of the trillions of symmetrically equivalent cation sites in a single crystal will be occupied by two different cations.

At sufficiently high temperature, there may be complete solid solution between end-members A and B. The implication is that a physically homogeneous mineral may exist with any chemistry outside the shaded region in Figure 8.7. This is possible only because sufficient similarities in the radii and charge of the cations allow disorder within the structure at temperatures below melting. The solid solutions at high temperature in Figure 8.7 have relatively high entropy, but the energy of thermal vibrations is still far less than the bond energy that must be exceeded before melting takes place. Solid solution between enstatite ($MgSiO_3$) and quartz (SiO_2), or between enstatite and forsterite (Mg_2SiO_4), never occurs in detectable amounts, because the energy that would be required to generate solid solution is more than sufficient to melt the minerals.

What happens to bulk chemistries that are some proportion of A and B and at a temperature that places them in the shaded region of Figure 8.7? This is the miscibility gap within which the ordering processes are sufficiently strong to exclude the formation of a single, homogeneous mineral. Two minerals must form instead. Their compositions are given by the curved line in Figure 8.7 at the appropriate temperature. This line is called a *solvus* and gives the compositions of the two solid solutions at any temperature at which they *coexist* in equilibrium. The solvus is also a line of saturation in the sense that at any given temperature each mineral has the maximum amount possible of the soluble component dissolved in it. For example, the solid solution at point M has as much of component B dissolved in it as possible at temperature T_2. This corresponds to 10 mole percent (based on 100) of B. Any attempt to add more of component B (say another 10 mole percent) to the solid solution at M moves the bulk chemistry to point P within the miscibility gap. The additional component B is rejected by the saturated solid solution at M, so another saturated (in component A) solid solution at N must form.

Returning now to the distribution of pyroxene

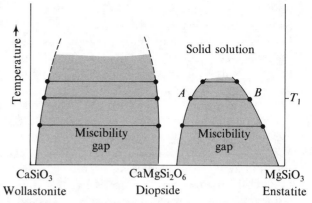

FIGURE 8.8 Idealized representation of gaps in solid miscibility as a function of temperature along the wollastonite-enstatite two-component system in Figure 8.6. At high temperature, solvi are intersected by melt processes. Shaded regions represent gaps in solid solutions as a function of temperature. Pressure is atmospheric.

analyses in Figure 8.6, we see how these principles apply. There is a solvus between diopside ($CaMgSi_2O_6$) and enstatite ($MgSiO_3$), and between diopside and wollastonite ($CaSiO_3$) as shown in Figure 8.8. Under normal geologic conditions, the solvi are intersected at high temperature by melting phenomena, and hence do not close. At low temperature, relatively little mutual solubility occurs between diopside and enstatite, and the width of the miscibility gap is correspondingly large. As temperature is increased, however, we observe a tendency for a less ordered, more random distribution of cations in the M sites. Progressively more Mg^{2+} ions occupy the $M2$ site at the expense of Ca^{2+}, which increases the compositional range of diopside toward enstatite. At elevated temperature, the probability that Ca^{2+} ions will occupy the $M2$ site in enstatite at the expense of Mg^{2+} is enhanced such that the compositional range of enstatite extends toward diopside. For example, at a given temperature T_1, Mg-bearing diopside A coexists with Ca-bearing enstatite B. The fact that such solid solutions are found in physical contact in rocks formed under equilibrium conditions is the best evidence we have that the miscibility gap exists. The contacts between the minerals are sharply defined, and no intermediate compositions exist between them.

EXSOLUTION

Exsolution is the process by which a homogeneous solid solution breaks down into two different minerals in response to a change in temperature or pressure or both. We commonly observe exsolution in natural feldspars, pyroxenes, amphiboles, and other mineral groups in which two or more cations may be disordered within a crystal structure. In metamorphic and igneous amphiboles, for example, exsolution lamellae (Figure 8.9) form along preferred crystallographic planes. This is strictly an *intracrystalline* phenomenon, as evidenced by the preservation of the original shape of the mineral as temperature is decreased. Figure 8.10 illustrates how these lamellae are formed. Beginning with a homogeneous solid solution at A, slow cooling takes place in the geological environment. Entropy decreases until the temperature T_1 is reached, at which a distinction can be made between the cation sites. At this temperature, two minerals become energetically more favorable than one because of the ordering tendency. Depending on the starting bulk composition, the more abundant mineral is referred to as the *host* and the less abundant forms the *lamellae*. Ideally, with a decrease in temperature both minerals change their compositions in a predictable manner in a direction that decreases the composi-

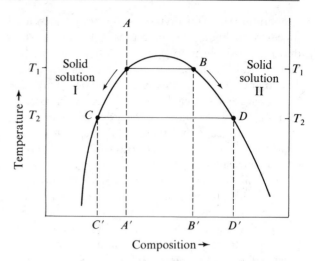

FIGURE 8.10   Schematic solvus illustrating progressive change of composition and proportion of host to lamellae as equilibrium is maintained with a lowering of temperature. Bulk composition at A remains constant throughout.

tion range of each and widens the miscibility gap between them. In Figure 8.10, solid solution I is the host phase at T_1. Its composition is given at A' on the composition coordinate. The lamellae formed at T_1 are at B and have a composition given at B'. As cooling proceeds to T_2, the host chemistry changes gradually to C', and the lamellae chemistry changes in the opposite direction to D'. The relative proportions of host to lamellae also decrease with decreasing temperature.

The extent to which the host and the lamellae remain in chemical communication depends on several factors. The mechanism of ion exchange within a crystal is solid state diffusion, and that process is greatly hindered at low temperature. Consequently, the chemistry of the host and of the lamellae cannot completely readjust as the cooling of the rock proceeds. The conditions of crystallization are thus recorded within a once homogeneous crystal.

The tendency for a disordered arrangement of ions at elevated temperatures may affect the chemistry of minerals that are normally very pure at lower temperatures. An example is quartz. Chemical analyses of quartz from low-temperature environments yield essentially 100% SiO_2. The structure has only one cation site, and it is tetrahedrally coordinated. In the absence of other cation sites, coupled substitutions involving three or more cations that would enable Al^{3+} to enter the structure are rare. Simple substitution by just two tetravalent cations avoids the coupling problem, and of the possibilities, Ti^{4+} seems the most likely. Its larger radius (0.061 nm), as com-

← 0.20 mm →

FIGURE 8.9   Photomicrograph of hornblende (Na $Ca_2(Fe,Mg)_5(OH)_2Si_7AlO_{22}$) crystals with cummingtonite (($(Fe, Mg)_5(OH)_2Si_8O_{22}$) exsolution lamellae. Specimen from the Ammonoosuc Volcanics near Keene, Vermont. Lamellae have formed on (001) by exsolution during cooling from metamorphic temperatures. Scale bar is 0.20 mm.

pared with Si (0.034 nm), prevents measurable substitution at low temperature. Quartz from high-temperature environments, however, shows detectable concentrations of foreign ions, and in particular Ti^{4+}. A careful optical or x-ray diffraction examination of such quartz frequently shows that the mineral rutile (TiO_2) is present as tiny needles dispersed throughout the quartz. A reasonable interpretation of these observations is that the quartz structure at high temperature can tolerate some substitution of Ti^{4+} in tetrahedral sites. Upon cooling, the ordering tendency becomes stronger, and the structure can no longer tolerate the titanium. Consequently, the Ti^{4+} is forced to unmix in the form of another mineral by exsolution in a pure quartz host. Some of the most beautiful museum specimens of rutilated quartz form in this manner.

ORDER AND SYMMETRY

Just as a tendency exists for high-temperature disordered forms of minerals to have high entropy relative to low-temperature ordered forms, so also a tendency exists for disordered forms to have high symmetry. Perhaps the simplest example is the comparison of any crystalline material and its compositionally equivalent melt. The highest symmetry any crystal can have is cubic, whereas the melt has infinite symmetry. An infinite number of rotation axes and mirror planes exist in a droplet of completely disordered liquid. The number of different ways the ions or atoms can be arranged in the liquid is obviously very large, and thus, so is its entropy. The solid crystal, however, is an ordered structure with a finite number of symmetry elements and a correspondingly lower entropy.

Yet another symmetry comparison is provided by the simple gas mixing model discussed earlier. Nitrogen and oxygen gas separated in their own identical containers have an ordered arrangement with low symmetry as a whole. The spontaneous increase in entropy upon release of the two gases to the atmosphere is accompanied by an increase in disorder and an increase in symmetry. If diffusion is complete, the resultant homogeneous gas mixture will have perfect symmetry.

The same tendency for high-temperature forms to have higher symmetry is observed among crystalline polymorphs. Several examples are listed in Table 8.2 along with their class symmetry. The polymorphs sanidine and microcline are particularly illustrative of the relationship between order and symmetry. Sanidine ($KSiAl_3O_8$) is the high-temperature disordered form of potassium feldspar. Its geologic occurrence is restricted to high-

temperature volcanic rocks that have quickly chilled on contact with the earth's atmosphere. X-ray diffraction studies show that high-temperature sanidine is monoclinic, having a $C2/m$ space group. Figure 8.11a is a stereoscopic view of the entire structure showing the two symmetrically distinct sites labeled *T1* and *T2*. Each of the tetrahedral sites occupies a general position and thus has eight equivalent positions in the unit cell. As there are 4 (KSi_3AlO_8) per unit cell, 4 Al^{3+} and 12 Si^{4+} must be distributed over two groups of eight equivalent sites.

This distribution can be achieved in several ways, some of which are consistent with the $C2/m$ symmetry and some of which are not. The simplest distribution is one in which all tetrahedra are identical. The actual observation in sanidine is that all *T1*—O and *T2*—O bond lengths are statistically the same (0.164 nm), from which we must conclude that the 8 *T1* sites and the 8 *T2* sites are occupied by identical cations. This conclusion might at first seem improbable in view of the 4 Al^{3+} and the 12 Si^{4+} that are distributed over the 16 tetrahedra.

The solution to the apparent inconsistency is to have a *disordered distribution* of Al^{3+} and Si^{4+}. In this arrangement, a tetrahedron is occupied by either Si^{4+} or Al^{3+}, but among the several trillion tetrahedra in a typical single crystal, 75% are silica tetrahedra and 25% are alumina tetrahedra. The structural refinement of a small single crystal always yields an average bond length, which in the case of *T*—O bonds in sanidine is between an ideal Si—O bond and an ideal Al—O bond. Figure 8.12 illustrates the relationship between average tetrahedral bond length and (Al, Si)-occupancy of the sites.

Returning to Figure 8.11, a disordered distribution of Al^{3+} and Si^{4+} on *T1* and *T2* is consistent with both the mirror plane and the twofold axis of rotation of the $C2/m$ space group. Upon cooling from high-temperature magmatic conditions, however, the Al^{3+} cations prefer to occupy the *T1* sites. That is, they become ordered into distinct sites of the structure. Figure 8.11b shows the effect of ordering on symmetry. The mirror plane is lost, as is the twofold axis, because the tetrahedra are no longer equivalent. The ordered arrangement is consistent with the lower symmetry of the triclinic space group $C\bar{1}$. Both low albite and ordered microcline have this symmetry.

At least one intermediate state of (Al, Si) ordering exists in which all of the Al^{3+} is in *T1* but is statistically disordered in that site. The *T2* site is occupied entirely by Si^{4+}, so the two *T* sites are symmetrically different. The cation distribution is

TABLE 8.2.   *Tabulation of Class Symmetry, Molar Volume, and Molar Entropy for Various Polymorphic Groups*

Chemistry	High-temperature Low-temperature Polymorphs	Crystal Class	Molar Volume $(10^{-5}\ \text{J/Pa})$*	Molar Entropy (J/K)**
H_2O	Steam	NC†	2478.92	188.83
	Water	NC	1.807	69.940
	Ice	$\frac{6}{m}\ \frac{2}{m}\ \frac{2}{m}$	1.963	44.685
SiO_2	SiO_2 glass	NC	2.727	47.400
	Low tridymite	222	2.653	43.932
	Low cristobalite	422	2.574	43.430
	Low quartz	32	2.269	41.460
	Coesite	$\frac{2}{m}$	2.064	40.376
	Stishovite	$\frac{4}{m}\ \frac{2}{m}\ \frac{2}{m}$	1.402	27.782
TiO_2	Rutile	$\frac{4}{m}\ \frac{2}{m}\ \frac{2}{m}$	1.882	50.290
	Anatase	$\frac{4}{m}\ \frac{2}{m}\ \frac{2}{m}$	2.052	49.915
Al_2SiO_5	Sillimanite	$\frac{2}{m}\ \frac{2}{m}\ \frac{2}{m}$	4.990	96.107
	Andalusite	$\frac{2}{m}\ \frac{2}{m}\ \frac{2}{m}$	5.153	93.220
	Kyanite	$\bar{1}$	4.409	83.764
KSi_3AlO_8	Glass	NC	11.650	261.600
	High sanidine	$\frac{2}{m}$	10.905	232.900
	Adularia	$\frac{2}{m}$ (?)	10.829	234.262
	Microcline	$\bar{1}$	10.872	214.200
$NaSi_3AlO_8$	Glass	NC	11.009	251.90
	High albite	$\frac{2}{m}$	10.043	226.400
	Low albite	$\bar{1}$	10.007	207.400
Mg_2SiO_4	Forsterite	$\frac{2}{m}\ \frac{2}{m}\ \frac{2}{m}$	4.379	95.186
	Modified spinel	$\frac{2}{m}\ \frac{2}{m}\ \frac{2}{m}$	4.048	80.500
	Spinel	$\frac{4}{m}\ \bar{3}\ \frac{2}{m}$	3.989	74.057
$CaSiO_3$	Pseudowollastonite	$\bar{1}$	4.008	86.446
	Wollastonite	$\bar{1}$	3.993	82.006
$Mg_2Al_4Si_5O_{18}$	Indialite	$\frac{6}{m}\ \frac{2}{m}\ \frac{2}{m}$	35.116	439.446
	Cordierite	$\frac{2}{m}\ \frac{2}{m}\ \frac{2}{m}$	23.332	407.200
$CaCO_3$	Calcite	$\bar{3}\ \frac{2}{m}$	3.693	91.710
	Aragonite	$\frac{2}{m}\ \frac{2}{m}\ \frac{2}{m}$	3.415	87.990

TABLE 8.2. (*continued*)

Chemistry	High-temperature Low-temperature Polymorphs	Crystal Class	Molar Volume (10^{-5}J/Pa)*	Molar Entropy (J/K)**
AlO(OH)	Boehmite	$\frac{2}{m}\frac{2}{m}\frac{2}{m}$	1.954	48.451
	Diaspore	$\frac{2}{m}\frac{2}{m}\frac{2}{m}$	1.776	35.271
ZnS	Wurtzite	$6mm$	2.385	58.840
	Sphalerite	$\bar{4}3m$	2.383	58.660
HgS	Metacinnabar	$\bar{4}3m$	3.017	96.232
	Cinnabar	32	2.842	82.509

SOURCE: Data from Robie, R. A., Hemingway, B. S., and Fisher, J. R., 1978. *U.S. Geological Survey Bulletin* No. 1452.
NOTE: Phases in each group are listed in order of decreasing temperature stability.
*10^{-5} J/Pa = 1 J/bar = 10 cm^3
**Measurements at 298 K
†Noncrystalline

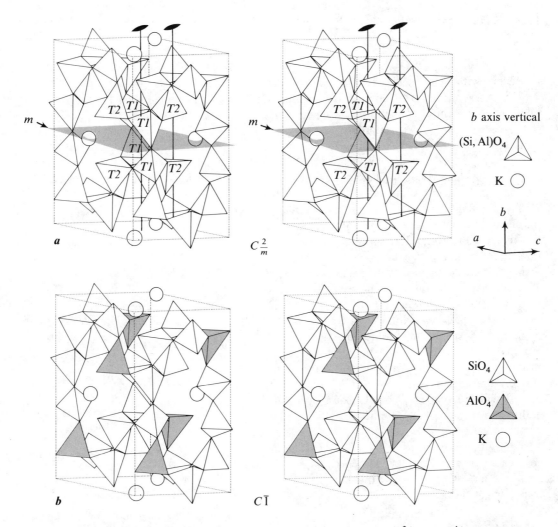

FIGURE 8.11 (a) Stereoscopic diagram of sanidine structure with mirror and twofold axes added. The Al^{3+} and Si^{4+} distribution is disordered. (b) Stereoscopic diagram of microcline. The Al^{3+} and Si^{4+} distribution is ordered.

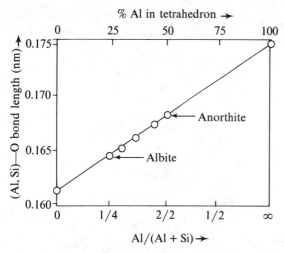

FIGURE 8.12   Variation of mean (Al, Si)—O bond lengths with percent of Al^{3+} in tetrahedral sites. One out of four sites is occupied by Al^{3+} in albite ($NaSi_3AlO_8$); two out of four are occupied by Al in anorthite ($CaSi_2Al_2O_8$). (After Smith, J. V., and Bailey, S. W., 1963. Second review of Al—O and Si—O tetrahedral distances. *Acta Crystallographica* 16, page 807, Figure 2.)

consistent with the $C2/m$ symmetry. This distribution and symmetry are found in orthoclase, an alkali feldspar typical of felsic intrusive rocks that have cooled more slowly than extrusive equivalents. Microcline, the fully ordered K-feldspar, is also typical of felsic intrusive rocks from environments in which slow cooling has taken place.

There are numerous exceptions to the generality that the high-temperature crystalline form of a particular chemistry has higher symmetry than the low-temperature form. The specific geometric arrangement of ions taken on by a mineral is that which yields the minimum energy state under a specified set of conditions. Those conditions include pressure and composition as well as temperature. All of these contribute to the entropy of a mineral, and since temperature may not always dominate, the symmetry of the stable structure may not always follow our reasoning based on temperature considerations alone. The molar volume is particularly important, because as a mineral expands, the volume contribution to entropy need not be expressed as a change in symmetry. Changes in proportion of bond type (e.g., ionic to covalent) can also be important, because the vibrational modes of the ions, and hence their entropy, can be affected.

PRESSURE EFFECTS ON ORDERING

We learned in Chapter 6 that an unconstrained

mineral *expands* on heating. The larger molar volume is associated with larger amplitudes of thermal vibration and higher entropy. The changes that accompany compression are exactly the opposite. A mineral generally responds to increased pressure by *reducing* its molar volume and becoming denser. At pressures of the earth's crust and mantle, ions in a crystalline structure are squeezed together to occupy space as efficiently as possible. Large anions, in particular, polarize more easily than cations, with the result that the radius ratio becomes larger and higher coordination is preferred. An example is stishovite, a high-pressure polymorph of SiO_2 in which Si^{4+} is in sixfold coordination rather than in the more common tetrahedral coordination (e.g., quartz). Another example is kyanite, the high-pressure polymorph of Al_2SiO_5 in which Al^{3+} is in octahedral coordination. The lower pressure polymorphs, andalusite and sillimanite, have some of their Al^{3+} in lower coordinated sites.

Because the effects of pressure are the opposite of the effects of temperature, we reason that a high-pressure environment would favor cation ordering. This conclusion is confirmed by the fact that entropy decreases with increasing pressure at constant temperature. Because there is "pressure" on the ions to fit in their sites, the ordering tendency at very high pressure is analogous to the ordering at very low temperature (Figure 8.8). A schematic solvus drawn at constant temperature (Figure 8.13) should then close toward higher

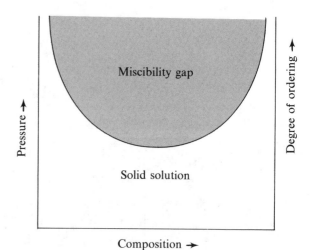

FIGURE 8.13   Qualitative effect of pressure on cation ordering in solid solutions. Effect of increasing pressure at constant temperature is to increase the ordering tendency and to promote less solid solution.

pressure—just the opposite effect of temperature at constant pressure.

Considerable care must be taken in the geological application of these relationships. Both pressure and temperature increase with depth in the earth, and we must therefore deal with both variables simultaneously. A homogeneous feldspar or pyroxene, for example, will generally exsolve if cooled slowly at constant pressure (depth). On the other hand, an exsolved assemblage will tend to homogenize as pressure is decreased at constant temperature. Both variables may also behave in such a manner that no change takes place.

MINERAL STABILITY

One of the ultimate objectives of mineralogical studies mentioned in Chapter 1 was to gain a better understanding of earth processes and history. To this end, the record of the geologic environment as recorded in minerals is of immense value. Once we understand the range of pressure and temperature over which a particular mineral will be stable in nature, we can then use the mineral as an indicator of conditions in the earth's crust. Our ability to read this record is impaired by the fact that the complete record is commonly not preserved. The geologic history of a rock may include several distinct events in its crystallization, and if only the last of these is recorded, then the previous history cannot be deciphered from an examination of the minerals alone. Because of this complexity, more than one independent approach to a problem in mineralogy may be required for its solution.

One approach is theoretical. It uses the basic principles of chemical thermodynamics in an attempt to predict the behavior of a mineral subject to different conditions in the earth's crust. A fundamental assumption of this approach is that minerals readjust their chemistry and structure in response to changes in the environment. For example, graphite is the stable form of elemental carbon over a wide range of pressure and temperature. At greatly elevated pressures corresponding to depths in the earth's upper mantle, graphite will react spontaneously to form diamond, the stable form of carbon under the new conditions. We can describe this situation by stating that diamond is the crystalline form of carbon at equilibrium in a high-pressure environment, whereas in lower-pressure environments, graphite is the form at equilibrium.

Under certain circumstances, minerals do not fully readjust to new conditions. We have evidence of this by the presence, at atmospheric pressure and temperature, of diamonds and other minerals formed under vastly different conditions. In fact, if this were not the case, mineralogy would be very dull indeed, for we would have only those minerals to study that are stable at the earth's surface.

The most important variables that control the stability of minerals are pressure, temperature, and chemistry. A study of the mutual relationships between these variables is contained within the framework of chemical thermodynamics. Although a complete treatment of the subject is beyond the scope of mineralogy, the basic relationships between these variables is of such fundamental importance to the understanding of mineral stability that we include them here.

EQUATIONS OF STATE

All substances that possess physical properties independent of time can be uniquely characterized by equations of state. The general form of such an equation for a mineral expresses the mutual relationships between a set of variables and the mathematical function describing that relationship. James Joule, in a series of experiments performed between 1843 and 1848, demonstrated that a given change in state of a substance, for example as measured by an increase in its temperature, could be brought about by two quite independent processes. First, a given volume of water placed in a perfect insulator at an initial temperature T_1 can be heated to a higher temperature T_2 simply by imparting energy to the water by work. In this case, the work done on the system is provided by a revolving paddle within the container connected through an insulated hole to a falling weight outside. Second, the same change in state can be duplicated by placing the same volume of water in a conducting container and permitting the water to warm up next to a hotter body. No energy is imparted to the water by work. If we examine the water after each experiment, no record whatsoever exists of the mechanism or path by which the new state was attained. In other words, the final state of the water is independent of the mechanism by which the change was brought about.

Joule recognized that the energy involved in each experiment has an effect on the internal state of the water. A change in that state can be described in terms of a change in *internal energy E*. Joule's contribution is embodied in the first law of thermodynamics:

$$dE = dq + dw$$

where the *dq* term represents an energy contribution in the form of heat when no work is done on the system, and the *dw* term represents an energy contribution in the form of work when no energy is transferred to the water in the form of heat. As in Joule's experiments, the algebraic sum of *dq* and *dw* is independent of path.

Given the usual expressions for heat and work,

$$dq = TdS$$

and

$$dw = -PdV$$

The first law takes the normal differential form

$$dE = TdS - PdV \qquad (8.1)$$

for a closed system, that is, a system in which the composition or chemistry is not free to change. This equation of state expresses the mutual relationship between five variables: internal energy *E*, temperature *T*, entropy *S*, pressure *P*, and volume *V*. The independent variables in Equation 8.1 are *S* and *V*, and depending on their values, *E* will be determined as the dependent variable. *T* and *P* are both defined and have knowable values for any designated values of *S* and *V*.

We can use Equation 8.1 to characterize the thermodynamic state of any mineral, liquid, or gas of fixed composition, provided that entropy and volume are suitable independent variables. Although useful for certain laboratory experiments, *S* and *V* do not in general represent the best choice of independent variables in nature. Pressure and temperature represent more realistic variable choices for minerals and rocks. Independent of its other physical properties, any given substance at depth in the earth's crust will experience the effects of the earth's geothermal gradient and the pressure exerted by the overlying lithosphere.

We therefore require an equation of state in which the roles of *S* and *T* and the roles of *P* and *V* in Equation 8.1 are reversed. The mathematical technique by which this is accomplished is a Legendre transformation. Without going into detail, it yields the equation

$$dG = -SdT + VdP \qquad (8.2)$$

for a closed system. The energy function *G* is called the Gibbs free energy. Changes in *G* now describe changes in the state of a mineral brought about by the independent variation of *P* and *T*. *S* and *V* will have defined and knowable values for any choice of *P* and *T*.

Note that Equations 8.1 and 8.2 are simply different ways of describing the same changes in the state of the system. In each case, only two independent variables are required to define the

state of the system. Given the chemical composition of a mineral and some arbitrarily fixed values of *P* and *T*, which, for example, might be realized at some depth in the earth's crust, we would find that all of the dependent variables (e.g., refractive index, density, volume) have fixed, measurable values. If we were to independently change *P* and *T* to some new set of values, all of the other variables would show a corresponding change in the manner predicted by Equation 8.2. If *P* and *T* are now independently returned to their original values, all of the other variables will have their original values as well. An inspection of the mineral and the values of all of its physical properties reveals nothing of its history.

EFFECT OF TEMPERATURE ON MINERAL STABILITY

The preceding statement illustrates a fundamental aspect of the thermodynamic approach to mineral stability. Properties of the internal state of a mineral, such as internal energy and Gibbs free energy, are independent of the path by which a given change was accomplished. Equation 8.2 shows that changes in *G* are effected *only* by changes in *P* and *T*. This being the case, then

$$G = f(P, T)$$

We may write an expression for the exact differential of *G*

$$dG = (\partial G/\partial P)_T dP + (\partial G/\partial T)_P dT \qquad (8.3)$$

which simply describes the total change in the state of the system in two parts, one due to the effect of pressure and the other due to the effect of temperature. We will investigate the latter term first.

By comparing the coefficient of *dT* in Equations 8.2 and 8.3, we make the identity

$$(\partial G/\partial T)_P = -S \qquad (8.4)$$

This equation states that the effect of a change in *T* on the Gibbs free energy of the system at constant *P* is given by the negative of entropy. An expression of this type may be written for any mineral, liquid, or gas that has an equation of state.

Figure 8.14 is diagrammatic of the energy-temperature relationships between two polymorphs of Al_2SiO_5. The curve *ACE* is Equation 8.4 evaluated for the low-temperature polymorph kyanite. Kyanite is also the lower entropy form, although the curvature of *ACE* is an indication that entropy increases as the temperature of kyanite increases. The curve *DCB* is Equation 8.4

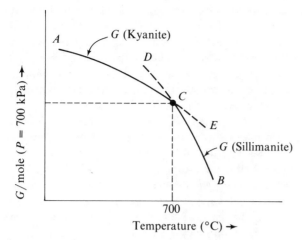

FIGURE 8.14 **Variation of molar Gibbs free energy (G) with temperature at constant pressure for kyanite (Al₂OSiO₄) and sillimanite (AlSiAlO₅). Approximate equilibrium temperature for coexistence at 700 kPa is 700°C.**

evaluated for sillimanite, the high-temperature and higher entropy form of Al_2SiO_5. The two curves intersect at point *C*, giving the temperature 700 °C at the stated pressure 700 kPa as the *P* and *T* at which the Gibbs free energy of the two minerals is equal. This means that neither mineral is favored energetically over the other. In other words, both are stable together at that particular *P* and *T*.

If we were to place a specimen of kyanite in the temperature range between *C* and *E* (dashed line), it would react to form sillimanite, because *all chemical reactions proceed in a direction that minimizes the Gibbs free energy*. In this example, a lower energy state exists along the curve *CB* for every corresponding point at the same temperature along *CE*. Thus, at the specified pressure, sillimanite will be the stable form of Al_2SiO_5 at temperatures above 700 °C.

For the same reasons, sillimanite along the dashed curve *CD* below 700 °C will spontaneously react to form kyanite, because all *G* values along *AC* are lower than corresponding values at the same *T* along *CD*. Kyanite is thus the stable form of Al_2SiO_5 at *T* < 700 °C at the stated pressure.

Note in Figure 8.14 the discontinuity in slope at point *C* between curves *AC* and *CB*. This represents a discontinuity in entropy at the temperature of transition and is due in part to the discontinuous change in structure in the transition from kyanite to sillimanite. Strong chemical bonds are broken during transition, and major changes in the coordination polyhedra occur. Sillimanite

has higher entropy and higher symmetry than kyanite. As we will see in a later section, the discontinuous change in entropy becomes important in evaluating the combined effects of pressure and temperature on the stability of minerals.

Because the entropy per mole of sillimanite is greater than that of kyanite, we can anticipate that for any polymorphic transition, the mineral with the higher entropy will have the higher temperature stability field. The relative entropy values of polymorphic transitions listed in Table 8.2 reveal this. In view of the relationships between entropy, temperature, and disorder discussed in preceding sections, the generality is not surprising.

EFFECT OF PRESSURE ON MINERAL STABILITY

Because minerals form at various depths in the earth's crust, they are subject to the pressure exerted on them by the weight of the overlying rock. To a first approximation, this pressure may be calculated by simply dividing the depth (km) by the average density (Mg/m³) of the overlying rock column. For example, a specimen of aragonite ($CaCO_3$) at a depth of 26 km is subject to a load pressure of 10,000 bars or 1 GPa if the average density is 2.7 Mg/m³. We make the basic assumption that the pressure is hydrostatic, that is, equal on all sides of the mineral as if it were at some depth in the ocean. Because water cannot sustain shear forces, the pressure must be uniform. Rocks and minerals, however, being solid, do have shear strength, and thus the hydrostatic model as applied to a mineral in the solid crust can be considered only an approximation.

We can evaluate the qualitative effect of pressure on the stability of minerals as follows: by inspection of Equations 8.2 and 8.3, the coefficients of *dP* are equated to give the relationship

$$(\partial G/\partial P)_T = +V \qquad (8.5)$$

for a specified mass. This equation states that the effect of "hydrostatic" pressure on the Gibbs free energy of a mineral at constant temperature is given by the volume of the mineral. Because the volume of a mineral, like the volume of a gas, will contract when pressure is applied, the energy of the mineral will change accordingly. This relationship is shown diagrammatically in Figure 8.15 for the polymorphic reaction between kyanite and sillimanite.

The slopes of the energy-pressure curves for kyanite (*DBE*) and sillimanite (*ABC*) must be positive for all pressures, since by definition volume is always a positive number. The slope of

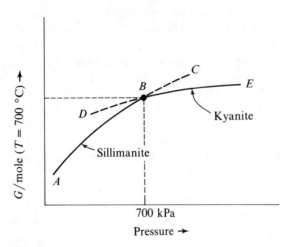

FIGURE 8.15 Variation of molar Gibbs free energy (G) with pressure at constant temperature for sillimanite and kyanite.

each curve, however, decreases as pressure increases due to the compression of the crystalline structure. As this process requires work on the mineral from the overlying rock column, energy is imparted to the mineral. At point *B*, the energies of kyanite and sillimanite are equal, and the minerals will therefore coexist in a stable association at that particular temperature and pressure.

If we were to place sillimanite in an environment of progressively increasing pressure, the mineral would remain stable over the range of pressures between *A* and *B*. An additional increase in pressure along the dashed curve *BC* would render sillimanite less stable than its polymorph kyanite, because for all pressures greater than the transition pressure at *B*, kyanite (curve *BE*) has a lower Gibbs free energy. The relationships illustrated in Figure 8.15 provide the basis for the general statement that for any mineral of fixed chemistry, the polymorph with the lower molar volume (the higher density) will be stable at higher pressure than the polymorph with higher molar volume. The examples given in Table 8.2 illustrate this statement.

THE COMBINED EFFECTS OF PRESSURE AND TEMPERATURE

Any mineral in the earth's crust will be subject to the combined effects of pressure and temperature. This is because both pressure and temperature increase uniformly with depth in response to the overlying load and the earth's geothermal gradient. The effects of these two variables are in opposite directions. Increasing temperature tends to increase the volume and the entropy of a min-

eral, whereas increasing pressure tends to decrease the values of these variables. The combined effect on the overall mineral stability is represented by the total Gibbs free energy in Equation 8.2. The difference in sign of the entropy and volume terms in the equation gives the contrasting effects of temperature and pressure, respectively.

The combined effects of temperature and pressure may be visualized graphically by combining Figures 8.14 and 8.15 along their common energy axis in the three-dimensional *G-P-T* diagram in Figure 8.16. An explicit evaluation of Equation 8.2 for all values of *P* and *T* defines a unique surface for each of the minerals. The values of *G*, *V*, and *S* will change from point to point on each surface, but for any specified choice of *P* and *T*, these variables will have fixed values. The intersection of the two surfaces defines the set of special conditions under which both polymorphs will coexist in stable equilibrium. The intersection is a line, commonly curved, along which *G*, *P*, and *T* are identical for the two minerals. These values will change along the line, but in such a way that any new set of values will be identical for each polymorph. The values for molar volume and molar entropy will be defined at each point on the line but will be different for each mineral. The equality of *P* and *T* when the two polymorphs coexist stably is, in fact, a necessary condition for chemical equilibrium in general. In the case when the two minerals have identical compositions, the Gibbs free energy values must also be equal for equilibrium to be attained.

The projection of each energy surface in Figure

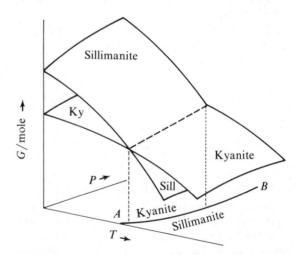

FIGURE 8.16 Intersecting free energy surfaces as a function of pressure and temperature. Line of intersection denotes the same free energy for both minerals in equilibrium.

8.16 onto the *PT* plane defines two regions that represent the range of pressure and temperature over which each of the polymorphs is stable. These are the *stability fields* of each mineral. The stability fields are separated by an equilibrium curve or reaction curve *AB*. The projection is accomplished in the direction parallel to the vertical *G* coordinate, so each point in the *PT* plane is characterized by a different Gibbs free energy.

The slope of the reaction curve in Figure 8.16 may be calculated if we know molar volumes and molar entropies of each polymorph as determined along the curve. The logical approach is to combine equations of state (Equation 8.2) for each polymorph in a single expression and then impose the constraints of chemical equilibrium.

$$dG_{Ky} = -S_{Ky}dT + V_{Ky}dP$$
$$dG_{Sill} = -S_{Sill}dT + V_{Sill}dP$$

We see from Figure 8.16 that any infinitesimal change of pressure and temperature consistent with the coexistence of kyanite and sillimanite must be matched by infinitesimal but identical changes of the Gibbs free energy of each polymorph. That is, $dG_{Ky} = dG_{Sill}$ everywhere on the reaction curve. For equal molar quantities of each polymorph, therefore,

$$-S_{Ky}dT + V_{Ky}dP = -S_{Sill}dT + V_{Sill}dP$$

or rearranging and combining terms,

$$(S_{Sill} - S_{Ky})dT = (V_{Sill} - V_{Ky})dP$$

and

$$\frac{dP}{dT} = \frac{S_{Sill} - S_{Ky}}{V_{Sill} - V_{Ky}} = \frac{\Delta S}{\Delta V}$$

Thus, the slope of the reaction curve in Figure 8.16 is given at any point by the ratio of the differences in entropies and volumes of the respective polymorphs as determined at that point. Even though entropy and volume will change for each mineral along *AB*, ΔS and ΔV remain constant in many reactions, particularly when the reactants and the products involve only solids. Reaction curves with little or no curvature result because of the small and nearly linear variation of compressibility and thermal expansion with pressure and temperature.

Chemical reactions characterized by large volume changes and relatively small entropy changes have flat slopes on a *P-T* diagram. Such reactions, like the one between calcite and aragonite, are especially sensitive to pressure changes. Reactions with relatively small volume changes have steep slopes on a *P-T* diagram and thus are especially sensitive to temperature changes.

HIGHER ORDER TRANSITIONS

The polymorphic reactions considered in the previous section are referred to as *first-order* transitions, because the first derivatives of the free energy (i.e., *S* and *V*) are discontinuous on the equilibrium curve. That the entropies and volumes of the polymorphs are different is intuitively clear if we compare any two polymorphs in equilibrium on the reaction curve under exactly the same *P* and *T*. The minerals are distinctly different, each having its unique structure. The molar volumes and entropies must therefore be different. In order for one polymorph to transform into the other, chemical bonds must be broken, and the ions reorganized into the new structure. For this reason, first-order transitions are commonly referred to as *reconstructive*.

Other kinds of polymorphic transitions exhibit no discontinuity in entropy or volume. Changes from an ordered to a disordered configuration of cations is one example. Structural states exhibit a complete variation from perfectly ordered (generally low temperature) to perfectly disordered (generally high temperature). Another type of polymorphic transition is exemplified by the low-high transition in quartz, which involves a gradual rotation of Si tetrahedra from a kinked arrangement having 32 point group symmetry to a more symmetrical 622 symmetry. No bonds are broken in the process (Figure 6.7b). The reaction is therefore referred to as *displacive*, because the ions or atoms show only small displacements.

Figure 8.17a illustrates the continuous increase of molar entropy with temperature across the stability fields of kyanite and sillimanite at a fixed pressure of 800 kPa. Figure 8.17b illustrates the variation of molar volume with pressure for the Al_2SiO_5 polymorphs at constant temperature. In both cases, the discontinuities in molar volume and entropy are typical of first-order transitions. Note how each mineral responds *continuously* to the changes in *P* or *T*. From a purely structural point of view, a given crystal structure can tolerate the distortion due to volume and entropy changes only up to a point. Beyond that point, a different structure can better accommodate further changes.

Figures 8.17c and 8.17d illustrate similar behavior for displacive transitions. Note the absence of discontinuities in either molar volume or entropy at the transition temperatures and pressures. The *slopes* of the entropy and volume curves are discontinuous, however. The term *second-order transition* is commonly used in this situation to denote that although the first derivative is contin-

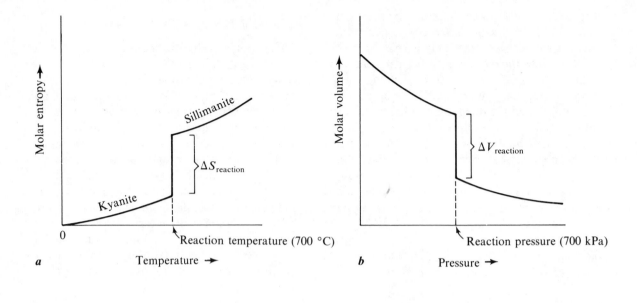

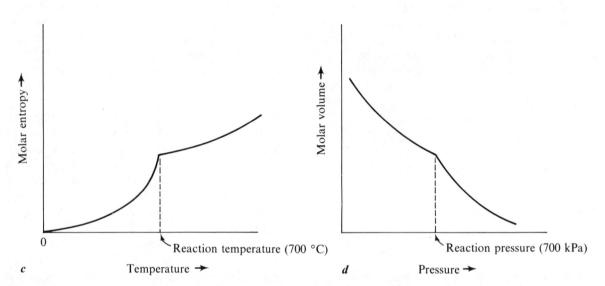

FIGURE 8.17 **(a) Variation of molar entropy S with temperature at constant pressure of 700 kPa. Entropy increases slowly in kyanite as temperature rises until sillimanite structure becomes stable. Discontinuity in entropy, ΔS, is characteristic of first-order transitions in which chemical bonds are broken. (b) Variation of molar** **volume with pressure at constant temperature is continuous within the stability field of each mineral, but discontinuous at points of transition. (c) Variation of molar entropy with temperature, and (d) variation of molar volume with pressure for a second-order transition.**

uous, the second derivatives of free energy with respect to pressure and temperature are not.

As might be expected, second-order transitions take place rapidly and reversibly. First-order transitions, in contrast, are sluggish and largely irre-versible because of the energy required to break chemical bonds. The existence of irreversible changes is fortunate, otherwise we would have no record of geologic history preserved in minerals.

CHAPTER NINE

MINERALS AND THEIR ASSOCIATIONS

Only rarely are minerals found isolated in nature. Almost always they are closely associated, if not in physical contact, with other minerals that have different crystal structures and chemistries. This fact helps us decipher the conditions of pressure and temperature under which rocks form, because in general, the range of possible conditions is smaller for two or more minerals that formed *together* than for any of the minerals by themselves. Mineral associations are also important for identification purposes. Frequently an intelligent guess can be made as to the identification of an unknown mineral simply by knowing the minerals with which it is found. For example, quartz and corundum or quartz and olivine rarely, if ever, occur together in nature, whereas quartz and feldspar almost always do. The reasons for these observations are found in the principles of free energy and composition space developed in the previous chapter.

ROCKS AS AGGREGATES OF MINERALS

The study of minerals as they occur together in rocks is formally within the study of petrology. This is not an arbitrary division between mineralogy and petrology. Minerals behave differently in association with each other from the way they behave when they are isolated. Everyone is familiar with the observation that the minerals ice (H_2O) and salt (NaCl) as a mechanical mixture

melt at a lower temperature than either mineral by itself. This is also true of all rocks, providing they consist of two or more different minerals. We thus speak of the "properties" of a particular rock as different from the properties of any of its constituent minerals.

An interesting and important fact about many rocks, especially those of igneous and metamorphic origin, is that the numbers of different minerals in each tend to be small. Basalt and its intrusive equivalent, gabbro, are volumetrically the most important rocks on earth. They consist almost entirely of plagioclase feldspar and clinopyroxene, with generally lesser amounts of olivine, orthopyroxene, and minor Fe oxide or Fe-Ti oxide. Rhyolite and granite, as well as most of the felsic rock types found in orogenic regions, consist of quartz, alkali feldspar, and plagioclase with lesser amounts of biotite, muscovite, and hornblende. The common metamorphic rocks such as biotite schist, amphibolite, and gneiss typically consist of small numbers of minerals, usually less than six and frequently only three or four.

This is less true for sedimentary rocks, especially those that have formed purely by mechanical erosion and are subject to rapid deposition. A typical sandstone formed under these conditions often contains most of the minerals found in the source region, and if the source regions have a variety of different rock types, the number of different minerals in the resulting sediment can be large. The minerals in such sedimentary rocks have very little interaction other than mechanical. Chemical interaction becomes important only after deposition, and usually not until after burial, when other factors such as load pressure and percolating solutions have had an effect.

Chemical interaction is the reason why any particular igneous or metamorphic rock will generally have a smaller number of different minerals than a sedimentary specimen. In nature, chemical interactions are in response to the environment and will always proceed in a direction that lowers the total free energy of the rock. In a later section, we will see exactly how this process tends to decrease the number of minerals. In Chapter 8, we learned how the stabilities of a mineral and its polymorphs are determined by their relative free energies. At any specified pressure and temperature, the mineral with the lowest value is the most stable. The same principle applies to rocks, but with the added complication that rock chemistries are much less restricted in their range than are the chemistries of minerals. These two factors, the tendency to minimize the total free energy and the less restricted, more variable rock chemistry, interact as the most important principles by which rocks form, and by which we can understand mineral associations.

THE FREE ENERGY OF ROCKS

The free energy of a rock is the sum of the free energies of its constituent minerals weighted according to relative abundance. For example, Figure 9.1 shows graphically the free energy of a rock that consists of equal amounts of the minerals forsterite (Mg_2SiO_4) and enstatite ($MgSiO_3$). If the proportions of the two are changed, two things happen. First, the composition of the rock changes toward the mineral that increased in amount, and second, the free energy of the rock changes in the same direction along the straight line connecting the energies of the minerals. Depending on proportions, the free energy of a forsterite-enstatite rock will thus range between that of a dunite (pure forsterite rock) and that of a pyroxenite (pure pyroxene rock).

Forsterite and enstatite are not the only minerals that lie along the Mg_2SiO_4-$MgSiO_3$ line in composition space. Both periclase (MgO) and quartz (SiO_2) do also. Figure 9.2 illustrates some alternative mineral associations for a rock with bulk composition fixed at point *P*. The bulk composition of a rock is a measure of total chemistry independent of the specific identity of the constituent minerals. From a chemical point of view, rock *P* could therefore consist of forsterite and enstatite in equal proportions (1 : 1), or periclase and enstatite (1 : 10), or forsterite and quartz (10 : 1), or periclase and quartz (1 : 1). The choice, so to speak, depends on the conditions under which the rock forms. In any case, the association that represents the lowest energy state will always be preferred and is called the *stable association*. In this example, under the conditions of 800 K temperature and 100 kPa pressure, olivine plus enstatite are stable together. The other chemical possibilities are not energetically feasible because the energies required are too high. These associations are referred to as *metastable*.

A mineral association that is stable under one set of pressure-temperature conditions may be metastable under another. We know from Chapter

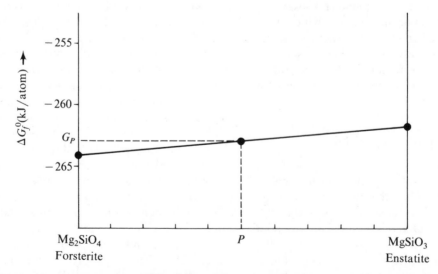

FIGURE 9.1   Free energy of formation ΔG_f^0 of forsterite and enstatite at 298 K and 100 kPa. Values are normalized to a per atom basis. Point *P* is the bulk composition of a rock with free energy given by G_P.

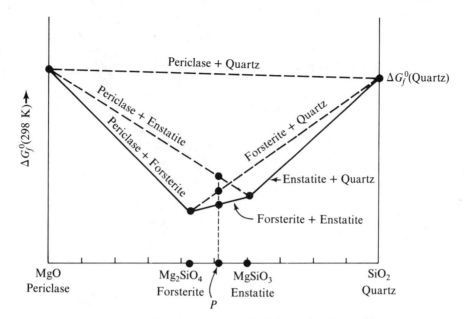

FIGURE 9.2 **Free energy of formation ΔG_f^0 of minerals in the MgO-SiO$_2$ system showing stable associations** (solid lines connecting energy values) and metastable associations (dashed lines) at 800 K and 100 kPa.

8 (Equation 8.4) that the free energy of each of the minerals in Figure 9.2 will decrease in proportion to its entropy as the temperature is raised. If all of the minerals had the same entropy, no relative change in stable associations would occur. The free energies would all decrease at the same rate. Because the entropies are different, however, the relative energy values in Figure 9.2 do change. By calculation, the higher entropy minerals like forsterite ultimately become stable with an SiO$_2$ polymorph, but not until a temperature is well over 2000 K. At that temperature, a real rock would have already melted, so we can safely say that the metastable associations in Figure 9.2 never occur under geologically reasonable temperatures.

A similar argument applies to the effect of pressure. As pressure increases, the free energies of each mineral in Figure 9.2 increase at a rate proportional to the molar volume (Equation 8.5). The ultimate result is that the densest minerals (lowest molar volume) will finally be stabilized at the highest pressures, because their energies are the lowest relative to less dense minerals under high-pressure conditions. The individual minerals commonly undergo polymorphic transitions, and in the case of quartz, with increasing pressure, coesite (SiO$_2$) and stishovite (SiO$_2$) appear. At very high pressure (40 GPa), periclase and stishovite appear to be the stable association relative to either forsterite or enstatite.

We can find numerous other examples in different chemical systems of incompatible minerals.

Quartz (SiO$_2$) and corundum (Al$_2$O$_3$) rarely occur together under normal crustal conditions, because a mineral of intermediate chemistry always can form between them. That mineral is andalusite (Al$_2$SiO$_5$) at low pressure, kyanite at high pressure, and sillimanite at high temperature. At very high pressures close to conditions in the earth's mantle, kyanite breaks down to form corundum and stishovite.

Another association of common minerals rarely observed in rocks is nepheline (NaAlSiO$_4$) and quartz. We should suspect immediately that minerals of intermediate composition exist that have relatively lower free energies, and in fact, they do. Both the minerals albite (NaSi$_3$AlO$_8$) and the pyroxene jadeite (NaAlSi$_2$O$_6$) occupy points in the region of composition space between nepheline and quartz (Figure 9.3). At relatively low pressures and temperatures (Figure 9.3a), albite and jadeite are stable together, as are albite and quartz. Under those conditions, jadeite plus quartz and albite plus nepheline are metastable associations. Under conditions of higher temperatures ($>$ 500 K), albite and nepheline are stable together, whereas the pairs albite plus jadeite and jadeite plus nepheline are not (Figure 9.3b). The stable association of albite and quartz is unaffected over this range of temperature, which is unfortunate from a petrologic point of view, because the association of these very common minerals in a rock tells us very little about the temperature of rock formation. Albite and nepheline are stable in igneous environments (nepheline

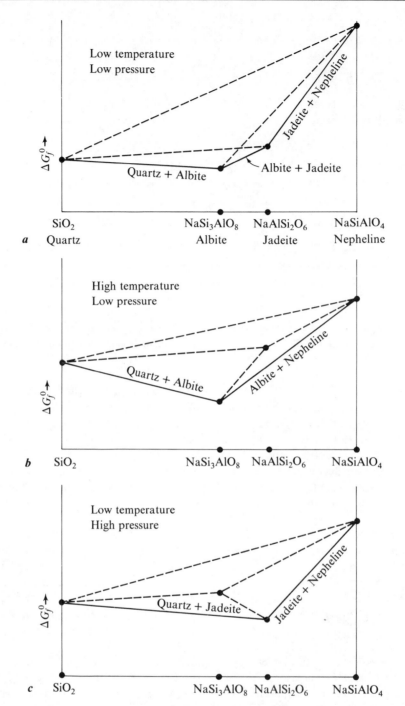

FIGURE 9.3 **Free energy-composition diagrams for the chemical system SiO₂-NaAlSiO₄ at various pressure and temperature conditions. (a) Low temperature, low pressure. Associations quartz + albite, albite + jadeite, and jadeite + nepheline are stable. Dashed assemblages** are metastable. **(b) High temperature, low pressure. Associations quartz + albite and albite + nepheline are stable. (c) Low temperature, high pressure. Associations quartz + jadeite and jadeite + nepheline are stable.**

syenites), as are albite and quartz (granites). Both pairs are known from metamorphic environments, and albite plus quartz are known to form as authigenic minerals in sedimentary rocks soon after burial.

Albite and quartz are not stable together over all pressures, however. At high pressure (Figure 9.3c), jadeite plus quartz are stabilized relative to albite, thus the associations albite plus quartz and albite plus jadeite that were stable at lower pressures are now metastable. These types of relationships are extremely helpful in interpreting the tectonic environment in which rocks form. For example, Figure 9.3 illustrates that a sandstone consisting of mostly quartz and lesser amounts of albite in a sedimentary environment can become in the high-pressure environment of a subduction zone a metamorphic rock consisting of jadeite plus quartz. Note how important the specific association is. Jadeite by itself or in association with albite or nepheline does *not* imply a high-pressure environment of formation (Figure 9.3a).

EFFECTS OF BULK COMPOSITION

Figures 9.2 and 9.3 illustrate the fact that under any particular set of conditions, the minerals we see in a rock are a result of the rock's bulk chemistry. A very siliceous (SiO₂-rich) igneous rock (Figure 9.3b) will consist of albite and quartz, and an SiO₂-poor rock will consist of albite and nepheline instead. An SiO₂-poor rock in Figure 9.3 consists of either periclase and forsterite, or more commonly, forsterite and enstatite. An SiO₂-rich rock in this system would consist of just enstatite plus quartz.

The effects of bulk composition can be extended to any chemical system provided that the mineral associations or assemblages are known for the conditions of interest. Diagrams that show these relationships are called *facies diagrams* in

metamorphic petrology, and isothermal-isobaric diagrams in igneous petrology. Identical diagrams are used for sedimentary rocks that have undergone diagenesis, the process by which minerals are formed in the sedimentary environment. The diagrams are all used for one common purpose— simply to show how the mineralogy of a rock varies with bulk composition under some specified pressure and temperature. High-temperature conditions over a broad range of pressures apply mainly to igneous rocks. Low-temperature and low-pressure conditions apply to sedimentary rocks, and the broad range in between is within the realm of metamorphic conditions.

The composition axis of Figure 9.2 is a facies diagram for the two-component system MgO-SiO₂ at 298 K and 100 kPa pressure. Like all facies diagrams, it represents the projection of stable minerals and their stable associations onto the composition axis of a free energy-composition diagram. The three parts of Figure 9.3 show the same chemical system, but each represents a different facies, for different mineral assemblages characterize each one. Although each diagram is drawn for a specific pressure and temperature, the mineral assemblages characteristic of a particular facies are usually stable over a restricted range of conditions.

Figure 9.4 is a facies diagram for the three-

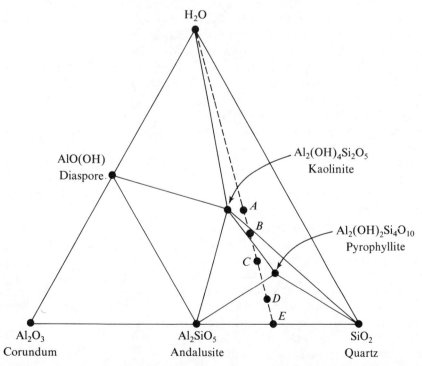

FIGURE 9.4 **Isothermal, isobaric facies diagram for the three-component system Al₂O₃-SiO₂-H₂O illustrating stable mineral assemblages at relatively low pressure** and temperature. Dashed line radial to H₂O shows effects of progressive dehydration on mineral assemblages.

component system Al_2O_3-SiO_2-H_2O under conditions of low temperature and pressure. It appears more complicated, but in principle, it does not differ from the two-component examples just considered. Simply imagine an energy coordinate normal to the paper along which each of the minerals has a particular free energy appropriate to the temperature and pressure of interest. Only those minerals with lowest energy are connected by lines, the projections of which are the lines on the facies diagram. The triangles in the diagram represent the various stable associations, such as andalusite, kaolinite, and pyrophyllite, of three minerals. Bulk compositions that plot interior to this triangle will consist only of these three minerals in proportions determined by the methods outlined in Chapter 8. Bulk compositions that plot along any of the lines consist only of the two minerals on that line. Bulk compositions that happen to coincide exactly with one of the mineral compositions are monomineralic. A pure quartzite, for example, plots at the point SiO_2.

Under conditions of even lower temperature than are represented in Figure 9.4, pyrophyllite becomes metastable relative to the assemblage kaolinite + andalusite + quartz. The new facies diagram for those conditions has pyrophyllite missing altogether so that any bulk composition plotting interior to the triangle defined by the minerals kaolinite, andalusite, and quartz consists just of those minerals, without pyrophyllite. Energetically, the essential difference between the two diagrams is whether the free energy of pyrophyllite lies above or below the triangle in an energy-composition diagram that has the free energies of kaolinite, andalusite, and quartz as its apices. If the free energy of pyrophyllite lies above, the mineral is metastable. If it lies below, pyrophyllite is stable, and the projection of that relationship onto the composition plane yields the facies diagram of Figure 9.4.

When any particular rock changes its mineralogy in response to changing temperature or pressure or both, but experiences no change in bulk composition, the process is referred to as *isochemical*. The mineralogy of any rock can also change under conditions of constant temperature and pressure, but the process is not isochemical. Frequently, the bulk composition will change as the temperature and pressure change. For example, an aluminous chert formed on the ocean floor and consisting of microcrystalline SiO_2 and kaolinite can undergo changes in mineralogy by simple dehydration. The original rock composition would plot at point *A* in Figure 9.4. As the sediment undergoes burial and perhaps an increase in tem-

perature within the range of conditions for which that facies is applicable, the rock dehydrates. By losing H_2O, the bulk chemistry moves directly away from the H_2O apex of the diagram. When all of the free (unbound) water is gone, the bulk composition has changed to the line between kaolinite and quartz, and thus must have that mineralogy. If more H_2O is lost, the bulk chemistry changes further to point *B* at which the rock has pyrophyllite as a new mineral in addition to kaolinite and quartz. Since pyrophyllite has less bound water than kaolinite, its prevalence over kaolinite reflects a more H_2O-deficient bulk chemistry. Continued dehydration to point *C* increases the amount of pyrophyllite and removes quartz as a mineral component. Dehydration to point *D* removes kaolinite completely, reintroduces quartz, and introduces andalusite for the first time. When no H_2O remains (point *E*), in principle the rock will contain only andalusite and quartz in approximately equal amounts. This mineralogy would form only in a metamorphic environment.

REACTIONS IN ROCKS

The only kind of chemical reaction that can take place in a one-component system involves two polymorphs. An example is the reaction between kyanite and sillimanite in Figure 8.16. Such reactions delineate the stabilities of the polymorphs and explain how an aluminous rock can contain kyanite over one range of conditions but sillimanite over another. In two-component systems, for example as illustrated in Figure 9.3, reactions must involve combinations of three minerals rather than two. Figure 9.5 shows two of the

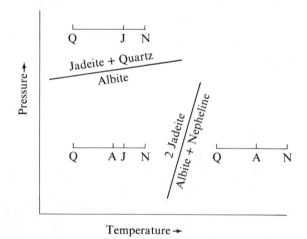

FIGURE 9.5   Pressure-temperature diagram involving the minerals quartz (Q), albite (A), jadeite (J), and nepheline (N) in a two-component system.

reactions in this chemical system, one involving the formation of jadeite from albite and nepheline, and the other involving the formation of albite from jadeite and quartz, according to the following equations:

$$2\,NaAlSi_2O_6 = NaSi_3AlO_8 + NaSiAlO_4 \qquad (9.1)$$

and

$$NaSi_3AlO_8 = NaAlSi_2O_6 + SiO_2 \qquad (9.2)$$

The two-component facies diagrams in the regions of pressure and temperature between these reactions consist of various combinations of *mineral pairs* generated by the reactions. The change from one facies to another involves continuous changes in the free energies of the minerals involved. Consider as an example the reaction in Equation 9.2 that relates the facies diagrams of Figures 9.3a and 9.3c. One of the three low-pressure assemblages is albite + quartz, an almost ubiquitous association in sandstones. Suppose now that such a sandstone is buried and eventually subjected to the very high pressures that might be encountered in a subduction zone environment. The gradual increase in pressure will cause each of the free energy values to rise according to its respective molar volume. For simplicity, suppose also that the effects due to temperature are unimportant. The relative values will remain the same over the range of pressure represented by the low-pressure facies diagram. In this range, the free energy of albite is always less than the chemically equivalent combination of jadeite and quartz. A pressure is reached, however, when this relationship is no longer true, and the free energy of albite is greater than that of jadeite and quartz. Albite is then metastable, and therefore does not appear on the high-pressure facies diagram. At any arbitrarily chosen temperature, the pressure at which the changeover takes place is uniquely determined, and in fact must correspond to some point on the reaction curve. There, all three minerals involved in the reaction are in chemical equilibrium with each other, whereas in the regions between the reactions, only pairs of minerals are in mutual equilibrium.

Chemical reactions in three-component systems must involve the mutual equilibrium of four minerals, rather than three (two-component system) or two (one-component system). The regions of pressure and temperature between the reaction curves will involve associations of three minerals for most bulk compositions and combinations of two or even one mineral for more specialized compositions. In general, if n is the number of chemical components, all reactions must involve the equilibrium of $n + 1$ mineral (or phase, if nonminerals such as a silicate melt or a gas are involved), and all facies diagrams must consist of combinations of n or fewer minerals. The maximum number of minerals that can coexist in equilibrium with each other is $n + 2$, and that happens only at very special temperatures and pressures referred to as *invariant points*. Examples of such points in one-component systems are the coexistence of ice, water, and steam in the H_2O system, and the coexistence of kyanite, andalusite, and sillimanite in the Al_2SiO_5 system. An example of a two-component system would be the coexistence of the four minerals quartz, albite, jadeite, and nepheline at the intersection of the two reaction curves in Figure 9.5.

THE PHASE RULE

The relationship between the number ϕ of minerals, or phases, in equilibrium with each other, the number of chemical components n, and the number of variables f that can operate independently to change the number of phases is called the *phase rule*. This relationship in its simplest and most elegant form was conceived by J. Willard Gibbs in 1892 and is expressed as follows:

$$f \leqslant n + 2 - \phi \qquad (9.3)$$

For our purposes, the only independent variables, f, of interest are pressure and temperature. If both are free to vary over a range of values, $f = 2$, and we have

$$2 \leqslant n + 2 - \phi \quad \text{or} \quad \phi \leqslant n \qquad (9.4)$$

This expression is often referred to as the *mineralogical phase rule*, a term coined by Victor M. Goldschmidt in 1918. The rule simply states that any rock that is subjected to some arbitrary choice of pressure and temperature will never have more minerals than the number of its chemical components. The inequality allows for the coexistence of fewer than ϕ minerals, which is exactly the observation in any of the facies diagrams discussed earlier. In three-component diagrams, combinations of three, two, and one minerals are allowed. In two-component diagrams, combinations of two and one minerals are allowed, and in one-component diagrams (e.g., Figure 8.20), only single minerals are allowed.

Note that the two variables P and T are independent of each other and are free to vary over a range of values. This is precisely the range of values over which any particular facies diagram will show no change. The term *divariant* (two-variation) *field* is often used to describe the total range, and the term *divariant assemblage* is used to describe the coexistence of *n* phases. Conditions that correspond to any point on a reaction curve or to an invariant point are explicitly *excluded*, because those do not represent values of temperature and pressure that can independently vary. If the requirement is made that the *P* and *T* conditions be on a reaction curve, then an independent choice of one variable such as temperature automatically determines the value of pressure. In the context of the phase rule, there is only one independent variable, and thus $f = 1$. The term *univariant* is used to describe this situation. Therefore, on any reaction curve,

$$f = 1 \leqslant n + 2 - \phi \quad \text{or} \quad \phi \leqslant n + 1$$

which states that the *maximum* number of minerals (or phases) that can coexist on a reaction curve is $n + 1$. Our observations on specific reactions show four phases in three-component systems, three phases in two-component systems, and two phases in one-component systems.

If $f = 0$, neither pressure nor temperature can be independently varied, and the coexistence of $n + 2$ phases at an invariant point is therefore permitted.

A slightly different view of the phase rule is to assume that the number of coexisting phases is predetermined. We then ask how many variables can be independently changed without disturbing that number. If a univariant ($\phi = n + 1$) situation is encountered, $f \leqslant 1$ and *at most one variable*, either *P* or *T*, can be changed without changing the number of phases. If a divariant ($\phi = n$) assemblage is in question, $f \leqslant 2$, and *at most two variables* can be independently changed. If fewer than two variables are changed, the number of coexisting phases is not affected.

It is important to remember that invariant mineral assemblages are very rare in nature, and univariant assemblages are unusual. Divariant assemblages are by far the most common assemblages and hence the appropriateness of the mineralogical phase rule. In nearly all rocks that have adjusted their mineralogies in response to their pressure-temperature environment, the number of minerals will be less than or equal to the number of chemical components. This is the fundamental reason why the number of different minerals in a given rock is generally small.

PROJECTIONS OF STABLE MINERAL ASSEMBLAGES

As the chemical complexity of rocks increases, so does the number of chemical components.

Most real rocks have several different minerals and therefore several components, usually more than can be readily represented on a piece of paper. To overcome this inconvenience, minerals that would normally plot in a tetrahedron or larger component space are *projected* into a two-dimensional plane. The resulting *projection* shows the essential relationships but with the missing dimensions left to the imagination. In all cases, a point of projection is required from which all minerals not already in the plane of interest are projected. One of the simplest kinds of projection is shown in Figure 9.6a in which SiO_2 is the point of projection, and the projection plane is defined by CaO, MgO, and Al_2O_3. We have four components, and divariant assemblages consist of various combinations of four minerals, each combination defining a subvolume within the CaO-MgO-Al_2O_3-SiO_2 tetrahedron. These assemblages can be more conveniently represented in a plane if we show only those assemblages that include quartz or other SiO_2 polymorphs as a phase. We can simply imagine SiO_2 as a common apex of several divariant assemblages, the bases of which are the three-phase triangles shown in the projection (Figure 9.6b). Phases that lie off the CaO-MgO-Al_2O_3 plane are projected onto the plane by effectively "sliding" the phase to the plane along a straight line radial to SiO_2. The point at which the phase hits the plane is the projected composition. Thus kyanite (Al_2SiO_5), for example, which lies halfway along the line from Al_2O_3 to SiO_2, projects to Al_2O_3. Corundum (Al_2O_3) does not appear as a phase on the facies diagram, because it does not occur in equilibrium with quartz. Similarly, both forsterite (Mg_2SiO_4) and enstatite ($MgSiO_3$) project from SiO_2 to the point MgO on the projection plane. Neither periclase (MgO) nor forsterite can coexist with quartz, so only enstatite appears on the facies diagram at that point. Diopside ($CaMgSi_2O_6$), which lies midway between $CaSiO_3$ and $MgSiO_3$ in the tetrahedron (Figure 9.6a), projects to a point that is midway between CaO and MgO.

A bulk rock composition that plots in the grossularite-wollastonite-diopside triangle in the resulting facies diagram will consist of the four-phase assemblage grossularite garnet ($Ca_3Al_2Si_3O_{12}$), wollastonite ($CaSiO_3$), diopside ($CaMgSi_2O_6$), and quartz (SiO_2). This is a moderately common assemblage in certain metamorphosed dolomites. A bulk composition that plots in the pyrope-diopside-enstatite triangle consists of the four-phase assemblage pyrope garnet, diopside, enstatite, and quartz. Such rocks are much less calcareous and in fact are more repre-

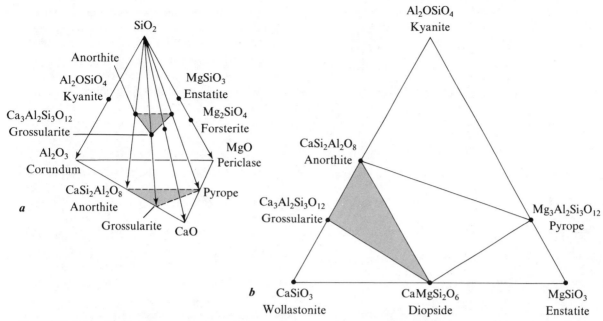

FIGURE 9.6 (a) **Four-component tetrahedron CaO-MgO-Al₂O₃-SiO₂ and associated mineral compositions. (b) Projection of minerals in (a) from SiO₂ onto the** **Al₂O₃-CaO-MgO plane. Rocks with four-phase assemblages in (a) are represented by triangles.**

sentative of the earth's deep mantle than of its crust.

It is important to remember that whenever a projection is used to represent mineral associations, the minerals through which the projection is made (e.g., SiO_2 in Figure 9.6a) are assumed to be present in the rock and in chemical equilibrium with the other minerals present. The range of pressure and temperature over which such a projection can be used is limited to the pressure-temperature stabilities of the various minerals, including those through which the projection is made.

Some minerals, like quartz, are so common in many rocks that using them as points of projection imparts little restriction on the range of rocks that can be represented. Gases or chemical components such as H_2O and CO_2, which may be present in certain minerals (amphiboles and carbonates) or were once present in the rock as a separate phase, can also be used as points of projection. For example, the mineral assemblages shown in Figure 9.4 can all be projected through H_2O onto the Al_2O_3-SiO_2 line to make a two-component system that is useful for hydrous mineral assemblages. Using such a chemical component as the projection point also leaves open another dimension for a new component, such as CO_2 or MgO, that may be of interest.

In this example and others like it, a distinction is made in projection between mineral assemblages that actually include the projected component as a free phase (mineral or gas) in equi-

librium with other minerals (e.g., diaspore + kaolinite + H_2O in Figure 9.4) and assemblages that include the component as part of the mineral chemistry only. An example of the latter would be the assemblage andalusite + quartz + pyrophyllite in Figure 9.4 in which H_2O as a separate phase is not stable but is present as a component in the hydrous mineral pyrophyllite. In these cases, the projection through H_2O can be interpreted loosely to mean that H_2O in the environment of formation was sufficient to stabilize the hydrous minerals shown in projection, but not enough H_2O was present to produce a free and separate gas or fluid phase.

Figure 9.7a uses the concept of multiple projections to illustrate conveniently some common assemblages in chemically complex rocks under metamorphic conditions. The assemblages shown are stable from approximately 400–500 °C over a range of low to moderate pressure. We can include both hydrous and calcareous minerals, because both H_2O and CO_2 are presumed present, under pressure, in sufficient quantity to stabilize the minerals shown. At higher temperatures or lower water pressure or both, the hydrous minerals such as tremolite and talc lose their water by chemical reactions, leaving the anhydrous minerals such as forsterite and diopside unaffected. Similarly, higher temperatures or lower pressure of CO_2 or both cause the chemical breakdown of the calcareous minerals like dolomite and magnesite.

At lower temperatures, the minerals forsterite,

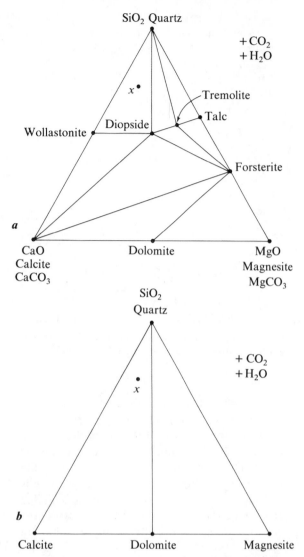

FIGURE 9.7   (a) Three-component system CaO-MgO-SiO₂ with sufficient CO₂ and H₂O to satisfy calcareous and hydrous minerals. Conditions around 500 °C and 300–500 kPa. (b) Same system under diagenetic conditions with stable low-temperature and low-pressure assemblages.

diopside, and wollastonite are no longer stable, and the facies diagram under those conditions has assemblages without those minerals. Under conditions of very low temperature and pressure in a sedimentary environment, the only mineral assemblages that are stable for this chemical system are shown in Figure 9.7b. These are the same rock chemistries shown in Figure 9.7a, but the specific mineral assemblages are quite different. A comparison of the figures illustrates exactly why the property of bulk chemistry has little to do with the conditions under which rocks form. The specific mineral assemblages that express the bulk chemistry give the clues to the pressures and temperatures of formation.

The implication is that any specific rock with a fixed and constant bulk chemistry (excepting H_2O and CO_2) will have quite different mineralogies depending on the pressure and temperature environment. The calcareous and dolomitic quartz sandstone (point x) in Figure 9.7b consists only of quartz, calcite, and dolomite. At 500 °C, the same rock (point x, Figure 9.7b) consists of quartz, wollastonite, and diopside.

In order to use mineral associations for identification purposes, we must learn as much as possible about the environment of formation. Frequently, the mineralogist recognizes several of the minerals present in a rock, and this recognition gives a general idea of the bulk rock chemistry. The known minerals and the geologic field relationships are almost always enough to tell the mineralogist whether a rock is sedimentary, metamorphic, or igneous. In later sections, we will discuss large groups of minerals that are restricted to each of these three broad categories.

EFFECTS OF SOLID AND LIQUID SOLUTION

We learned in Chapter 8 how the extent of solid solution in a mineral varies with pressure and temperature. The range of solid solution in composition space is broader at higher temperature, meaning that more of composition space is occupied by the mineral. This is also true of magmatic liquids (melts) in igneous rocks and of aqueous solutions in sedimentary rocks. In the presence of these common rock-forming materials, the usual facies diagrams become slightly more complicated. Instead of representing a mineral as a single point, we use regions to describe the extent of a solution at the stated pressure and temperature conditions. Such regions become larger at higher temperature because of the higher entropy and greater disorder among the atoms. The regions become smaller at lower temperature, as well as at higher pressure, because the entropy is lower, reflecting the greater order among the atoms.

Figure 9.8a is a facies diagram illustrating the effect of Fe^{2+}-Mg^{2+} solid solution in olivines and orthopyroxene at elevated temperature. Olivine under these conditions exhibits complete solid solution between forsterite and fayalite. Orthorhombic pyroxenes, however, are limited in their solid solution toward an ideal Fe end-member by the stable association of fayalite and quartz, found in metamorphosed iron formations and in certain granites. This diagram has the same importance for mineral associations as does Figure 9.4. Here

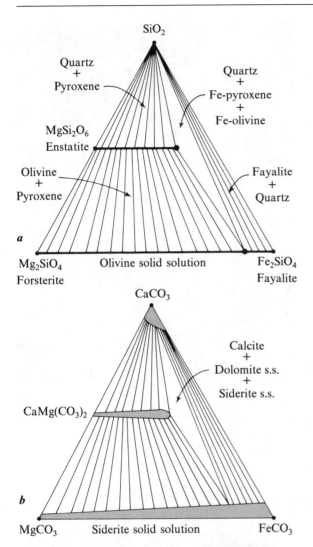

FIGURE 9.8   Effects of Fe²⁺-Mg²⁺ substitution on stability fields in isothermal, isobaric facies diagrams. (a) The Mg_2SiO_4-Fe_2SiO_4-SiO_2 system at moderate pressure and temperature. Solid solution series shown as heavy lines. (b) The $MgCO_3$-$FeCO_3$-$CaCO_3$ system at about 450 °C and low pressure. Solid solutions shown as shaded areas.

ingly, the pyroxenes found in fayalite granites are very Fe-rich and exhibit the strong pleochroism so common among Fe-rich silicates. Nor should we be surprised that pyroxenes associated with olivine from peridotites and basalts are very Mg-rich and usually quite colorless. These characteristics are simply a consequence of different bulk chemistries.

The same principles apply to any facies diagram involving solutions, whether they be solid solutions, magmatic liquids, gases, or low-temperature aqueous solutions. The compositions of the solutions will always reflect the bulk composition of the rock, providing equilibrium processes have operated. The other important variables for mineral associations are temperature and pressure, which control the extent of solution and the overall stability of minerals.

These three variables (composition, temperature, and pressure) enter into most rock classification. When a geologist speaks of a basalt, for example, a bulk chemistry enriched in Fe, Mg, and Ca, a high temperature, and a low pressure are implied. The mineralogy, especially the associations, will be closely related. When a geologist speaks of a pelitic schist, an aluminous bulk chemistry and a range of possible metamorphic temperatures and pressures are implied. A variety of probable minerals, all aluminous and stable under metamorphic conditions, immediately comes to mind. For this reason, a mineralogist who is accomplished at mineral identification is usually a good petrologist. Knowledge of the rocks can help solve identification problems with minerals and vice versa.

SEDIMENTARY ROCKS AND PROCESSES

Mineral associations in sedimentary rocks are probably more varied and less predictable than the associations in metamorphic and igneous rocks. In terrigenous rocks such as sandstones, processes of chemical equilibrium do not operate directly until the onset of recrystallization. These processes operate indirectly, however, during the chemical breakdown and weathering in the source regions. The relative chemical stability of minerals in a weathering environment is related to the number of strong Si—O bonds in the structure that must be broken during dissolution or reaction. Ferromagnesian silicates like pyroxene and olivine have a relatively low Si/O ratio and have little resistance to chemical attack. Consequently, these minerals are not common in sediments that have formed as a result of long weathering and

the minerals olivine and pyroxene, because they are solid solutions, are represented as lines rather than points. Virtually no solid solution exists between olivine and pyroxene, and between pyroxene and quartz, because of the energetics of exchanging cations between octahedral and tetrahedral sites. Olivine and pyroxene *together* are stable over a wide range of bulk composition as indicated by the family of lines, rather than by a single line as in Figure 9.4. Each of these lines connects the compositions of the two minerals when they are in mutual equilibrium. As the bulk chemistry becomes more Fe-rich, the compositions of olivine and pyroxene do also. Not surpris-

transport. Amphiboles (Si/O = 4:11) and micas (Si/O = 2:5) are more resistant to chemicals, but they are less resistant than quartz (SiO$_2$), one of the most important and ubiquitous minerals in sandstones because of its chemical resistance, low solubility, and lack of cleavage, which promotes mechanical breakdown. Quartz can survive many cycles of weathering, transport, and redeposition before consolidation as a sandstone, which frequently consists of nearly 100% rounded quartz grains. Olivine and pyroxene do form sand deposits in certain special environments such as Hawaii, but only because the source region is local and unique.

Feldspar (Si/O = 2:8 to 3:8) is generally more resistant to breakdown than the ferromagnesian silicates, but less resistant than quartz. It is also more abundant than other minerals in most source regions, which accounts for its presence in detrital sedimentary rocks. Feldspar, unlike quartz, cannot survive the sedimentary recycling because of its cleavage and susceptibility to chemical breakdown into clay minerals. The clays, because of their small size and low settling velocities, usually are deposited in relatively quiet water environments.

A great variety of other minerals with chemical and mechanical properties similar to quartz contribute to the composition of detrital sediments. All are generally of less importance than quartz in the source rocks. Examples commonly found in minor amounts in almost every detrital rock are zircon, sphene, apatite, and garnet.

The second broad class of sedimentary rocks concerns those of *biogenic* origin, including the limestones, cherts, and chalks that are formed from the skeletal remains of marine organisms. The resulting bulk chemistries are highly specialized in the sense that the sediment may consist entirely of SiO$_2$ or of CaCO$_3$, depending on the biogenic environment. If the environment provides for more than one organism, for example, calcareous foraminifera and siliceous radiolaria, the biogenic sediment will reflect that organism composition unless diagenetic processes operate to change the composition (e.g., dissolution of CaCO$_3$).

The third large category of sedimentary rocks is formed by *chemogenic* processes. These rocks are inorganic and are formed by chemical precipitation. All of the world's evaporite deposits are chemogenic, as are many dolomites, phosphate beds, ironstones, and cherts. Again, the processes that operate provide for an incredible chemical differentiation relative to the composition of most source rocks. This makes the task of mineral iden-

FIGURE 9.9   Scanning electron photomicrograph of authigenic quartz overgrowths on detrital quartz grains. (Courtesy of Andrew Hurst.)

tification easier, because the bulk chemistry is more restrictive.

After deposition, any of the rocks mentioned above can undergo changes due to *diagenesis*. This is the process by which a sedimentary rock responds to the chemical environment that exists in the sedimentary basin. Percolating pore solutions, increased pressure due to sedimentary overburden, and a possible increase in temperature due to subsidence, all operate to change the original mineralogy. Some minerals may dissolve totally or be removed by reaction, whereas others may grow larger (e.g., Figure 9.9), or may grow completely anew (e.g., Figure 9.10). These changes are subtle and often are not apparent until mineralogists make microscopic and x-ray studies. In most cases the rocks are still considered sedimentary after undergoing diagenesis. Only if the basic chemical processes intensify at higher temperature and pressure is the rock considered metamorphic.

SEDIMENTARY MINERALS AND THEIR ASSOCIATIONS

Sandstones

In principle, any mineral or any combination of minerals can occur in sandstones. The minerals

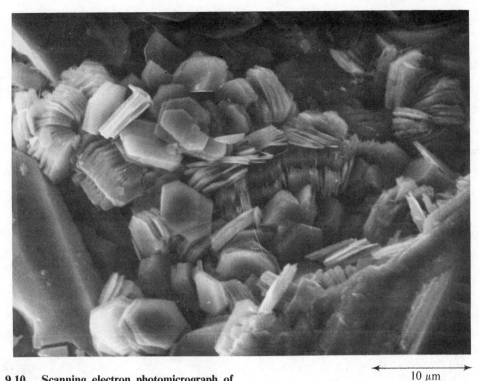

10 μm

FIGURE 9.10 Scanning electron photomicrograph of authigenic kaolinite in sandstone. (Courtesy of Andrew Hurst.)

may be of metamorphic, igneous, or sedimentary origin, or any combination of these. In practice, quartz and feldspar predominate because of factors mentioned earlier. Na-plagioclase and microcline are most common, because they are more resistant to breakdown and are frequently the most abundant feldspars in the source regions. Both have a higher Si/O ratio (3:8) than the more calcic plagioclases (2:8 for anorthite) because of the coupled substitution of $Na^{1+} + Si^{4+}$ for $Ca^{2+} + Al^{3+}$.

The various sheet silicates including the clay minerals (e.g., kaolinite, montmorillonite, illite, and glauconite) and the micas (biotite and muscovite) can be abundant in sandstones. Lithic fragments consisting of chert, chalcedony, or any other rock type resistant to weathering can also be important.

The minor and accessory minerals in sandstones are numerous. The most common ones are resistant to both chemical breakdown and mechanical abrasion. They include apatite, epidote, garnet, hornblende, ilmenite, magnetite, rutile, sphene, tourmaline, and zircon.

The authigenic minerals (those formed by diagenesis) in sandstones usually form by the interaction of detrital grains with percolating solutions. Euhedral overgrowths of quartz (Figure 9.9) and feldspar on previously rounded grains are com-

mon. Many other minerals such as kaolinite (Figure 9.10), a variety of zeolites, feldspar, and quartz grow for the first time, usually in conjunction with the dissolution or chemical breakdown of detrital grains. Authigenic anatase, for example, derives its Ti from the breakdown of either detrital biotite or ilmenite.

Detrital grains in sandstones are commonly held together by an authigenic cement, either quartz, calcite, barite, or iron oxide. The cement can have the same crystallographic orientation over areas of several square centimeters and can look like a single crystal in which is embedded smaller detrital grains. Barite-cemented sandstones, or barite "rosettes," are an example.

Siltstones and Shales

These rocks are finer grained than sandstones, having the same general mineralogy but in different proportions. This is because the clay-sized and mud-sized fractions have virtually all of the detrital clay minerals. They have hydrodynamic properties that differ from quartz and feldspar, and tend to be separated from them during transport. The dark color of many shales is imparted by a relatively small amount of carbonaceous material.

Many of the minor and accessory minerals in siltstones and shales tend to be derived from sources within the same marine or lagoonal environment. Carbonate, shell fragments, collophane, and glauconite occur along with virtually any of the accessory minerals mentioned earlier. Because of the fine grain size, most of these are not recognized until separated and concentrated into a "heavy fraction" in the laboratory.

Clay minerals, carbonates, chlorite, glauconite, silica, and pyrite can form as authigenic minerals.

Carbonates

Mineralogically, the carbonates are simple, consisting mainly of calcite (limestones), dolomite (the rock dolomite), or mixtures of the two. On weathered surfaces, the dolomite sometimes has a yellowish brown color distinct from the weathered bluish color of limestone. This is because of the solid solution of Fe^{2+} for Mg^{2+} in dolomite $(CaMg(CO_3)_2)$ but not in calcite. The subsequent oxidation of the Fe^{2+} during weathering imparts the yellowish brown color. On unweathered surfaces, the two carbonates may be indistinguishable except by prudent application of dilute HCl. The carbonates ankerite, siderite, and magnesite can also be present. Aragonite, an important constituent of modern calcareous organisms, is found in modern marine sediments, but it rapidly transforms to calcite. Primary aragonite is virtually unknown in older sedimentary rocks.

Carbonates, mainly calcite, are the major constituent of various surface deposits such as *caliche* and *travertine*. Both form as incrustations on older rocks in relatively arid regions where evaporation exceeds precipitation. Silica in the form of detrital quartz grains or as authigenic opal can be abundant in caliche. Formed by replacement, SiO_2 can constitute up to nearly 100% of the deposit, in which case the term *silcrete* is used.

Clay minerals, mainly kaolinite, illite, and montmorillonite, can be important detrital constituents in carbonates. They form as thin sedimentary laminations or are admixed with calcite and dolomite. Finely dispersed carbonaceous material imparts the dark color to many limestones but is not volumetrically important.

A surprising number of authigenic minerals can form in sedimentary carbonate rocks by the interaction of the primary carbonates and pore solutions. Both albite (Figure 9.11) and microcline form remarkably euhedral crystals in a pure carbonate matrix that resemble in form crystals that crystallize from an igenous rock. Quartz euhedra may form in the same way, often with a distinctive doubly terminated form. Pyrite, marcasite, hematite, and glauconite also form as authigenic minerals in the carbonate environment.

Authigenic clay minerals are relatively com-

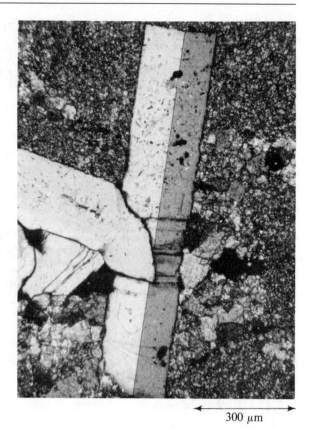

← 300 μm →

FIGURE 9.11   Photomicrograph of twinned authigenic albite in fine grained limestone.

mon in some carbonate rocks. Kaolinite, illite, montmorillonite, and even chlorite are important in caliche. The fibrous clays sepiolite and palygorskite are found in caliche and have economic importance because of their unusual sorptive properties. Gypsum and hematite are common accessory minerals in this environment.

Evaporites

Evaporites, consisting principally of the two minerals gypsum $(CaSO_4 \cdot 2H_2O)$ and anhydrite $(CaSO_4)$, form by precipitation from concentrated brines. Halite (NaCl), sylvite (KCl), and dolomite are also known and are locally abundant. Numerous other sulfate minerals, all relatively rare, can occur with any of the above. They include epsomite $(MgSO_4 \cdot 7H_2O)$, kieserite $(MgSO_4 \cdot H_2O)$, thenardite $(NaSO_4)$, glauberite $(NaCa(SO_4)_2)$, and polyhalite $(K_2MgCa_2(SO_4)_4 \cdot 2H_2O)$. All form by primary precipitation.

One of the most important authigenic minerals in sulfate evaporites is native sulfur. It forms by reduction of calcium sulfate minerals. The most extensive sulfur deposits, many of commercial importance, are found in limestone that directly overlies anhydrite or gypsum or both. Gypsum and anhydrite will replace each other by reaction with H_2O given the proper environment. So formed, they are authigenic in origin.

A wide variety of unusual evaporite minerals are found as precipitates from inland soda lakes. They include the borate minerals borax ($Na_2(H_2O)_8B_2B_2O_5(OH)_4$), kernite ($Na_2(H_2O)_3B_2B_2O_6(OH)_2$), ulexite ($NaCa(H_2O)_6B_2B_3O_7(OH)_4$), and colemanite ($Ca(H_2O)BB_2O_4(OH)_3$). Chemogenic nitrate minerals, mainly niter (KNO_3) and soda niter ($NaNO_3$), form in the same way.

Chert

Chert is composed almost entirely of microcrystalline or cryptocrystalline quartz. The original sediment in some cases appears to be amorphous silica, and we then consider the quartz to be authigenic. The SiO_2 may also be in the form of opal and chalcedony, recognized by its distinctive radiating texture.

We can frequently recognize the accessory minerals in chert by the color and properties they impart to the rock. Included hematite gives *jasper* its distinctive red color. Relatively small amounts of clay are responsible for the variety of chert called *porcellanite*. The gray to black color of *flint* is due to included organic matter. Small amounts of carbonate minerals may be present in chert, reflecting their common environment of formation. Occasionally, rock fragments containing ferromagnesian minerals (e.g., pyroxene) and magnetite will also be found, particularly if the chert forms near a submarine volcanic region such as a midocean ridge.

Phosphorites

Phosphatic sedimentary rocks are among the most important sources of phosphorus. The original sediments are generally of shallow water marine origin, but phosphatic nodules on the ocean floor at abyssal depths are known. The principal mineral in phosphorites is apatite, with a wide range of ionic substitution between OH^{1-}, Cl^{1-}, and F^{1-}. The important end-member compositions are $Ca_5F(PO_4)_3$ (fluoroapatite), $Ca_5Cl(PO_4)_3$ (chloroapatite), and $Ca_5(OH)(PO_4)_3$ (hydroxylapatite). Authigenic apatite is commonly a carbonate fluoroapatite in which $(CO_3)^{2-}$ substitutes for $(PO_4)^{3-}$. Amorphous phosphatic material is called *collophane*. Vivianite ($Fe_3(PO_4)_2 \cdot 8H_2O$) is a common authigenic mineral associated with both siderite and chamosite.

The accessory minerals in phosphorites are mainly glauconite, illite, quartz, and iron oxides. Both calcareous and siliceous fossil fragments can be common.

Ironstones

Fe-rich sedimentary rocks consist of two broad categories: the Precambrian banded *iron formations*, often known as *taconite*, and the Phanerozoic *ironstones*. The world's largest reserves of iron ore are found in these rocks. The principal Fe-bearing minerals are oxides, namely, goethite ($FeO(OH)$), hematite (Fe_2O_3), and magnetite (Fe_3O_4). In carbonate-rich iron formation and ironstones, siderite ($FeCO_3$) is the most important mineral. In cherty iron formation, the most important mineral is microcrystalline quartz. Iron silicates such as chamosite, greenalite, minnesotaite, and stilpnomelane, all fine-grained sheet silicates, are also found. Among the sulfides in iron formation, pyrite is the most common.

Some of these minerals may actually be authigenic. The fact that goethite, the principal iron oxide in Phanerozoic ironstones, is not known from the Precambrian iron formations suggests that the magnetite and hematite of the latter time period are authigenic. Most pyrite is authigenic, forming soon after deposition from amorphous FeS or from metastable iron sulfides such as mackinawite and greigite. Glauconite in the form of small (< 0.5 mm) bright green pellets is mostly authigenic. It also occurs as irregular fillings along cracks and as a replacement of carbonate.

Manganese-rich Rocks

As a result of deep-sea dredging over the last two decades and of direct bottom photography, we now know that vast areas of the modern ocean floors are covered with manganese nodules. Their origin is still problematic, but precipitation appears due to oxidation of Mn^{2+} by contact with oxygenated seawater. The principal mineral is manganite ($MnO(OH)$).

METAMORPHIC ROCKS AND PROCESSES

Metamorphic rocks are formed over the range of pressure and temperature conditions between sedimentary diagenesis and igneous processes. Although the pressure-temperature boundaries are completely gradational and thus our fixing of boundaries somewhat arbitrary, the textural changes in the rocks that accompany metamorphism are distinctive. These changes almost always include a gradual coarsening of grain size due to recrystallization, or a progressive development of a metamorphic fabric due to the preferred alignment of grains during growth, or both. In *regional metamorphism*, both recrystallization and development of fabric take place simultaneously because of the combined effects of temperature, pressure, and plastic deformation. In the sequence of rock names shale, slate, phyllite, schist, and gneiss—all are based primarily on textural development in response to progressive metamorphism. In *contact metamorphism*, high temperatures and steep thermal gradients prevail around igneous intrusions where plastic deformation is less important. Consequently, little development of fabric occurs other than the coarsening of grain size that accompanies the recrystallization. Texturally such rocks are referred to as *hornfelses.*

Unlike terrigenous sedimentary rocks prior to diagenesis, metamorphic rocks respond to the pressure and temperature environment by changes in mineralogy. The driving mechanism for all such changes is the irreversible process of attaining the minimum free energy state for the bulk chemistry of interest. The final mineral assemblage that results depends on the temperature and pressure, because the group of minerals representing the minimum energy state under one set of conditions will in general not be the stable group under a different set of conditions. This is the basis for the metamorphic *facies concept*. Relatively broad regions of pressure and temperature can be distinguished by the presence or absence of diagnostic minerals or mineral assemblages that have some-

what restricted stabilities. For example, the presence of hornblende in a metamorphic rock generally places the conditions of formation within the *amphibolite facies* (Figure 9.12), or broadly, between 450 and 650 °C and between 0.2 and 0.8 GPa. Under the lower temperature and pressure conditions of the *greenschist facies*, the same bulk chemistry no longer has hornblende, but has chlorite and actinolite instead. Under the higher temperature and pressure conditions of the *granulite facies*, the hornblende may react to form pyroxenes and garnet.

The term amphibolite facies is not restricted to metamorphic rocks that contain amphibole. It refers to a general range of conditions within which any rock will recrystallize. A kyanite schist with garnet and staurolite is considered to have formed in the amphibolite facies. A metamorphosed sandstone containing jadeite and lawsonite has formed within the pressure and temperature range of the *glaucophane schist facies*, so also has a schist containing glaucophane.

The boundaries between the various metamorphic facies (Figure 9.12) are not well defined, because the specific chemical reactions that delineate the stabilities of many minerals and their stable associations are not well understood. Moreover, many different reactions, at least one for each significantly different bulk composition,

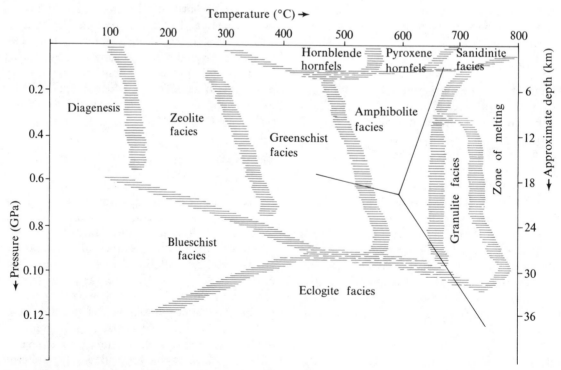

FIGURE 9.12 **General range of pressure and temperature conditions for various metamorphic facies. Granulite** **facies is well within zone of melting of quartzo-feldspathic rocks.**

could define more precisely a particular facies boundary. Even those would not have the same positions on a pressure-temperature diagram. The result is that two rocks of different composition that formed under identical metamorphic conditions could be referred to as representing different facies, depending on the specific reactions involved. For our general purposes, ignoring the exact locations of facies boundaries and simply using the diagnostic mineral assemblages as the criteria for identifying facies is more convenient and practical. For our purposes here, the specific associations are most important, because once a few of the more common minerals are identified in a particular rock, we have a reasonable idea of the bulk composition and general range of pressure and temperature. That information greatly restricts the identity of the minerals that remain in the rock.

In the simplest view, progressive metamorphism can be regarded as *isochemical*, that is, the bulk chemistry of the rock remains the same, and only the mineral assemblages change. This is certainly true of many of the chemical constituents that are mobile only to the extent of being reorganized into new mineral structures. Other constituents, in particular H_2O and CO_2, which are abundant in most sedimentary rocks, are extremely mobile and are commonly driven out of the rock during metamorphism, because many of the metamorphic reactions involve dehydration and decarbonation. The free H_2O and CO_2 thus released are invariably on the high-temperature side of reaction curves due to the high entropy and disorder of the free gas molecules relative to the solid phases. Once freed from the structural constraints of minerals, the gases diffuse in an intergranular manner to regions of lower pressure and temperature. In the process, the local diffusion required for the reconstitution of the crystalline phases is greatly enhanced, and the minerals tend toward equilibrium with each other.

Once the volatile constituents escape, the bulk chemistry is changed. Without H_2O, no micas or amphiboles can form in the rock, regardless of the pressure and temperature conditions. Without CO_2, carbonate minerals cannot form. For this reason, many of the highest grade metamorphic rocks contain only anhydrous minerals such as pyroxene, garnet, and sillimanite. Also for this reason, most metamorphic mineral assemblages are preserved for us to observe.

The pressure-temperature path followed by a rock to its final metamorphic environment can be reversed, as evidenced by the rocks at the earth's surface today. The mineralogy preserved at the peak of metamorphism responds to the trip back to the surface, but the reactions necessary to convert these minerals back to an assemblage stable under surface conditions cannot proceed. The volatile constituents necessary to form stable hydrous minerals and necessary to promote the kinetics of diffusion are no longer available.

METAMORPHIC MINERALS AND THEIR ASSOCIATIONS

Pelitic Schists

Most shales and many sandstones with significant clay contents will become pelitic schists upon metamorphism. Schists are recognized by the parallel alignment of micas, either muscovite or biotite or both, called *schistosity*. Both the necessary aluminum and potassium required by the micas are derived from the original clays. Because SiO_2, either detrital or biogenic, is an abundant constituent in environments where clay minerals accumulate, the bulk chemistry of these rocks always contains enough SiO_2 to more than satisfy the needs of the Fe-Mg silicates. Free quartz is consequently a ubiquitous mineral in pelitic schists. Plagioclase feldspar is rare, because the bulk chemistry is usually deprived in Ca and Na. In contrast, potash feldspar (microcline) is common.

The relatively restricted bulk chemistry of pelitic schists requires that the other minerals present be enriched in Al, K, Fe, Mg, and Si. In addition to muscovite, biotite, and quartz, the other common minerals in pelitic schists include sillimanite, andalusite, and kyanite (all Al_2SiO_5), garnet (($Fe, Mg)_3Al_2Si_3O_{12}$), chlorite, staurolite, chloritoid, cordierite, and microcline. All of these minerals are distinctive in hand specimens and are readily identified in the field. The various reactions that cause the first appearance of each of these during progressive metamorphism follow a classic sequence from chlorite to biotite to garnet to staurolite to kyanite to sillimanite with increasing temperature and pressure. Terms such as *chlorite zone* and *biotite zone* have evolved as a general indication of the particular pressure-temperature interval within this sequence. Within each interval, many but not all of the minerals mentioned will be stable in distinctive associations. Figures 9.13a and 9.13b represent just two possible facies, one at relatively low grade where chlorite and biotite are stable, and the other at relatively high grade where garnet and staurolite are stable. The diagrams are similar to Figure 9.8 and can be read and interpreted in the same way.

At the lowest grades of metamorphism, the first

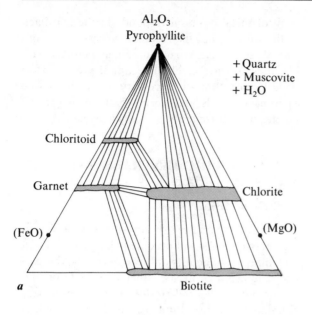

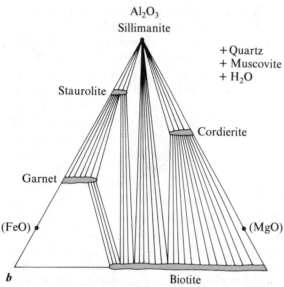

FIGURE 9.13   (a) Al$_2$O$_3$-MgO-FeO projection of low grade mineral assemblages in pelitic schists. Projection is through quartz and muscovite. (b) Projection of higher grade pelitic mineral assemblages. (After Thompson, James B., Jr., 1957. The graphical analysis of mineral assemblages in pelitic schists. *American Mineralogist* 42: 851.)

easily recognizable metamorphic minerals commonly appear as *porphyroblasts* or small "knots" in slates. The term *spotted slate* is used for such rocks. The porphyroblasts can be cordierite, garnet, biotite, chlorite, or andalusite, depending on the bulk composition and the exact pressure and temperature. Spotted slates typically occur in contact metamorphism, but are not restricted to that environment.

Calc-silicate Hornfelses and Marble

A complete gradation in bulk chemistry exists between pelitic schists and metamorphosed calcareous and magnesian sedimentary rocks. The addition of Ca and Mg (calcite and dolomite) in the protolith and the loss of Al (less clay) has a profound effect in determining which minerals will appear at various pressures and temperatures (Figure 9.14a). Several of these minerals, namely, wollastonite, akermanite, tillyite, merwinite, and larnite are found only in hornfelses (Figures 9.14b and 9.14c).

With the addition of H$_2$O, brucite, talc, tremolite, and other amphiboles become possible minerals in many of the assemblages shown in Figure 9.14. At the highest temperatures of contact metamorphism, both hydrous minerals and carbonates break down. Even calcite reacts to form CaO (lime) and CO$_2$ at high temperature and atmospheric pressure.

With the addition of FeO to the composition diagram of Figure 9.14a, the magnesian carbonates and silicates exhibit solid solution toward idealized Fe end-members. The MgO-SiO$_2$ edge of the CaO-MgO-SiO$_2$ triangle expands to Figure 9.8b. The CaO-MgO edge expands to Figure 9.8b in which the stability of calcite + dolomite solid solution + siderite solid solution is shown at about 450 °C.

The addition of Al$_2$O$_3$, perhaps due to original clay minerals, enables garnet, aluminous spinel, vesuvianite, zoisite, clinozoisite, epidote, and anorthite to form given the proper pressure and temperature conditions. If the metamorphic environment is somewhat oxidizing, the garnet will be mainly grossularite (Ca$_3$Al$_2$Si$_3$O$_{12}$) with solid solution toward andradite (Ca$_3$Fe$_2^{3+}$Si$_3$O$_{12}$). Vesuvianite, epidote, zoisite, and clinozoisite can also accommodate ferric iron in their structures.

During contact metamorphism, a magmatic gas phase at high temperature may contain chemical components such as fluorine, chlorine, sulfur, boron, and a large number of dissolved heavy metals. When these gases penetrate and react with limestone and dolomite, *skarns* are produced. The process by which the original sedimentary chemistry is changed is called *metasomatism*. Apatite, sphene, scapolite, tourmaline, fluorite, axinite, danburite (CaB$_2$Si$_2$O$_8$), and minerals of the humite group can form. If the rock has sufficient aluminum, topaz (Al$_2$(OH, F)$_2$SiO$_4$) and the lithium micas lepidolite and zinnwaldite may be present.

The ore minerals that develop in contact skarn deposits are as varied as the chemical diversity of

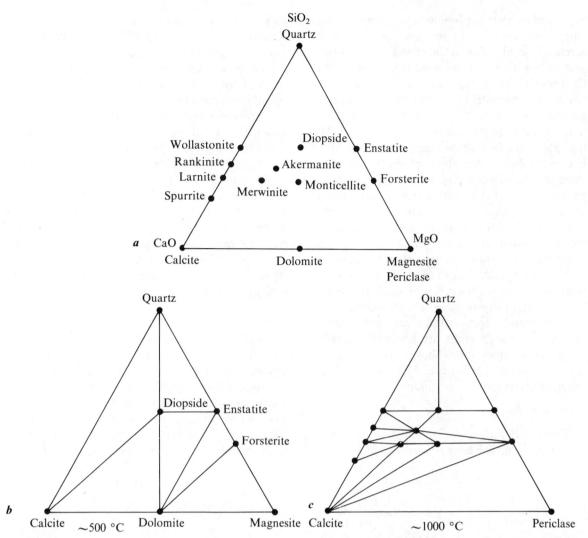

FIGURE 9.14 **(a) Some mineral compositions in the CaO-MgO-SiO$_2$ system with excess CO$_2$. (b) Facies** diagram at 500 °C and 100 kPa. (c) Facies diagram at 1000 °C and 100 kPa.

the magmatic gas. Sphalerite (ZnS), cassiterite (SnO$_2$), chalcopyrite (CuFeS$_2$), galena (PbS), bismuthinite (Bi$_2$S$_3$), molybdenite (MoS$_2$), scheelite (CaWO$_4$), cobaltite (CoAsS), arsenopyrite (FeAsS$_2$), and pyrite (FeS$_2$) are all known.

Greenstones and Greenschists

These rocks are the metamorphosed equivalents of mafic volcanic rocks, mainly of basaltic composition. Greenstones represent the lowest grade of metamorphism. Chlorite, epidote, albite, calcite, laumantite, prehnite, pumpellyite, and stilpnomelane are common with or without quartz. At slightly higher grade, actinolite becomes stable, and along with chlorite imparts the metamorphic fabric to greenschists. Unlike greenstones, the relict pillow structures, lava flow textures, and other features indicative of volcanic flow are less well preserved. Common accessory

minerals in greenstones and greenschists include sphene, calcite, magnetite, hematite, and quartz.

Amphibolites

With higher grades of metamorphism, greenschists become amphibolites. Actinolite, chlorite, and epidote are still stable, but now a blue green hornblende has appeared that imparts a distinctive dark color to the rock. Plagioclase of an intermediate composition is a major constituent, comprising up to 50% of the rock by volume. Biotite, garnet, quartz, ilmenite, and the colorless amphiboles cummingtonite and anthophyllite are common. Gedrite, the more aluminous and sodic variety of orthoamphibole, is usually found with both garnet and cordierite in rocks with sufficient Al$_2$O$_3$.

A great diversity of mineral assemblages is evident in amphibolites because of miscibility gaps within the amphiboles (as many as four amphiboles may coexist in some rocks!) and the extensive solid solution involving Fe^{2+}, Mg^{2+}, and Al^{3+} within each mineral group. Figure 9.15 illustrates many of the stable assemblages under typical amphibolite facies conditions. Again, a multiple projection (Figure 9.15a) is used, this time through quartz, magnetite, and plagioclase feldspar in the five-component system CaO-MgO-FeO-Al_2O_3-SiO_2. The plane of projection is defined by MgO-FeO-Al_2O_3, and like the muscovite projection in pelitic schists, some minerals plot in projection outside of the oxide triangle (Figure 9.15b). In this case, aluminous bulk chemistries are represented above the MgO-FeO line, and calcic bulk chemistries are represented below.

Note that under the range of pressure and temperature conditions represented by Figure 9.15, certain pairs of minerals are *incompatible*. Cordierite and cummingtonite, or cordierite and hornblende, for example, are not stable together for the bulk compositions shown. These combinations are known in nature, but they represent different metamorphic conditions.

Granulites

These rocks represent the highest grades of metamorphism. The temperatures are high enough to partially melt quartzo-feldspathic rocks, which accounts for the widespread occurrence of *migmatites* in granulite facies areas. Evidence of very high pressure is generally absent. (Kyanite is rare, for example, whereas both andalusite and sillimanite are common.) Possibly this is because of higher geothermal gradients during the Precambrian, the age of most granulites. These rocks are distinctive for their quartz and plagioclase, both of which can have a dark, resinous appearance.

Orthopyroxene is the distinctive mineral of granulites, usually forming with increasing temperature from the breakdown of the hydrous amphiboles. If water pressure is sufficient, the hydrous minerals biotite, muscovite, and amphibole may all be stable with orthopyroxene. Quartz is very common, as are garnet, cordierite, K-feldspar, clinopyroxene, and sillimanite.

Eclogites

These unusual rocks have the general bulk chemistry of basalts, and have been metamorphosed under conditions of very high pressure and a range of temperature. They are thought by many researchers to have originated in the earth's lower crust or upper mantle. Eclogitic rocks have only three abundant minerals—garnet, orthopyroxene, and clinopyroxene. The garnet is al-

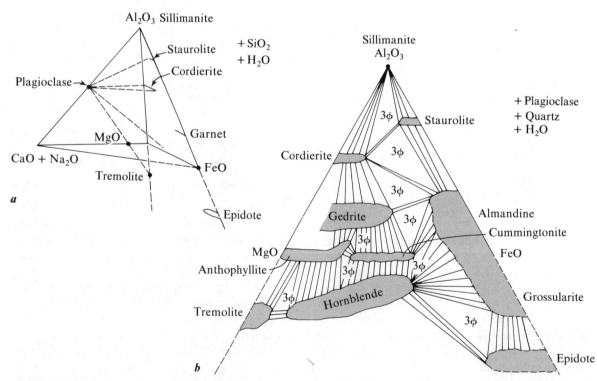

FIGURE 9.15 (a) **Five-component system CaO-Al_2O_3-MgO-FeO-SiO_2 with projection through plagioclase feldspar and quartz onto the Al_2O_3-MgO-FeO plane. (b) Mineral assemblages typical of the amphibolite facies.** The notation 3ϕ refers to three-phase assemblage. (After Stout, James H., 1972. **Phase petrology and mineral chemistry of coexisting amphiboles from Telemark, Norway.** *Journal of Petrology* **13:137.**)

ways rich in pyrope and may have a deep purple color. The clinopyroxene called *omphacite* is a solid solution between diopside ($CaMgSi_2O_6$) and jadeite ($NaAlSi_2O_6$). It has a distinctive bright green color. Together with the garnet and a light-colored orthopyroxene (enstatite), eclogites are among the most colorful rocks known. Common accessory minerals in eclogites include rutile, kyanite, hornblende, quartz, biotite, zoisite, spinel, and glaucophane.

Blueschists

Rocks that have been subjected to conditions of high pressure and relatively low temperature are said to have crystallized within the blueschist facies (Figure 9.12). Such conditions are associated with unusually low geothermal gradients as are found today only along the world's ocean trenches and in active subduction zones. When old rocks exhibit blueschist facies mineral assemblages, we can infer similar environments.

The blue color of a typical blueschist is imparted by glaucophane, the sodic amphibole. It is usually but not always associated with other sodic-rich minerals such as jadeite ($NaAlSi_2O_6$) and albite ($NaAlSi_3O_8$). More calcic bulk chemistries will have lawsonite, epidote, garnet, zoisite, pumpellyite, and the high-pressure $CaCO_3$ polymorph, aragonite. Quartz is common, as are biotite, stilpnomelane, epidote, rutile, and chlorite.

Metamorphosed Iron Formation

These Fe-rich and Si-rich rocks produce a variety of iron silicates upon metamorphism. They are usually well laminated and fine-grained, making visual mineral identification difficult. Hematite, magnetite, and quartz, with or without Fe-carbonates, are the most abundant minerals in the most common grades of metamorphism. At the lowest grades, chlorite, stilpnomelane, and minnesotaite (Fe-talc) are found. At higher grades, grunerite, hypersthene, hedenbergite, fayalite, and almandine become stable.

IGNEOUS ROCKS AND PROCESSES

The upper limit of rock metamorphism is usually considered as the onset of melting. This happens at temperatures as low as 550 °C for certain quartzo-feldspathic rocks enriched in H_2O and at temperatures as high as 1300 °C for anhydrous basalts and other rocks of similar chemistry. Under equilibrium conditions, all rocks share the property of melting over an interval of temperature that depends mainly on the specific bulk chemistry. All magmatic liquids, called *melts*,

share the property of crystallizing over a temperature interval upon cooling. In both cases, crystals and melt are in physical contact with each other in proportions that decrease with increasing temperature within that interval. Rocks that contain a quenched melt (glass) or that have evolved from a melt are called *igneous*.

Igneous melts form by two main processes; both involve the conversion of once solid rock into liquid. The first usually operates in subduction zones where oceanic lithosphere is subjected to high enough temperatures to promote melting. The sediments and underlying basalts of the oceanic lithosphere prior to subduction are well hydrated due to interaction with seawater. As pressure increases during subduction, the temperature of first melting actually *decreases* due to the higher potential solubility of water in igneous melts (Figure 9.16). As the pressure-temperature path of the rock passes across the reaction producing the first melt, and as the rock gradually becomes more "liquid," it also becomes more buoyant relative to the surrounding unmelted rock. The buoyancy forces drive the melt to higher levels of the crust to form a volcanic arc.

In contrast, the incredible volumes of basalt and related igneous rocks that make up the oceanic lithosphere develop from the solid earth mantle by the second igneous process of decompression melting (Figure 9.16). Because the earth's mantle contains relatively little water, its first melting curve has a positive slope on a pressure-temperature diagram. In extensional tectonic environments such as a midocean ridge, convecting mantle material passes through the melt curve as it rises rapidly to higher levels. By this process, melts are produced that, because of their buoyancy, separate to higher levels, leaving the more refractory residual minerals behind.

In both convergent and divergent environments, the melts may or may not reach the earth's surface before crystallizing. If the melt crystallizes at depth where cooling is slow, the crystals grow large, and the crystallization process is complete. A granite or granodiorite thus forms in the roots of an island arc, and a gabbro forms in the oceanic lithosphere. If the melt rises rapidly, however, only a few crystals may have time to form, and the remaining liquid will quench to a volcanic glass upon eruption at the surface. Rhyolite and andesite are the usual products in island arcs; basalts are the product on midocean ridges.

The diversity of igneous mineral assemblages is due mainly to the broad range of melt chemistries that develop first during the melting process and subsequently during crystallization. The composi-

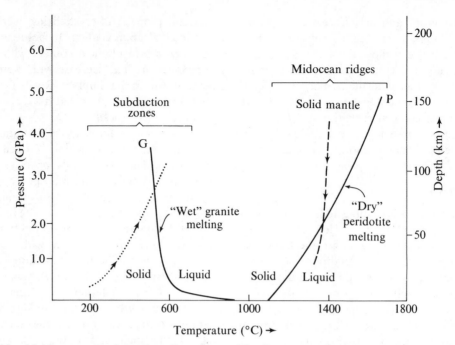

FIGURE 9.16 **Melting curves of granite, G, under water pressure and dry peridotite, P. Midocean basalts are produced by crossing curve P along dashed line.** **Rhyolites are produced by crossing curve G along dotted line.**

tion of the liquid at any time during either process depends on how much of the original rock has melted and on how much of the liquid so formed has crystallized. When the melt is removed by buoyancy forces from the partially melted host, the process is referred to as *fractional melting*. When crystals settle out of a crystallizing melt, the remaining liquid differs in chemistry from the original, and *fractional crystallization* has occurred. These two processes may operate repeatedly during the evolution of a particular igneous melt to produce the final chemistry responsible for the observed mineralogy.

Both the melt chemistry and the composition of minerals capable of solid solution change as an igneous rock cools. In the Fe-Mg silicates (e.g., olivines and pyroxenes) the Mg compositions always crystallize first at the highest temperatures. This causes a relative enrichment of Fe in the melt (Figure 9.17a). Removal of olivine of any composition by fractional crystallization causes the melt to be more siliceous, which will ultimately favor the crystallization of pyroxene and quartz. In the plagioclase feldspars, the Ca compositions always crystallize first (Figure 9.17b) at the highest temperature, causing the remaining melt to be enriched in Na and Si, and depleted in Ca and Al.

A common end product of these processes is the formation of *pegmatites* and their unusual and often exotic minerals. Fractional crystallization frequently enriches the melt in volatile compo-

nents such as F, Cl, and H_2O. These dissolved gases effectively lower the final crystallization temperature and thus give the minerals more time to grow. They also lower the melt viscosity, promoting diffusion and larger grain size. The results are spectacular in many pegmatites with single crystals up to several meters long.

Pegmatites are also enriched in those elements that, because of ionic radius and charge, are not removed by solid solution in earlier crystallized minerals. Li, Be, B, Sc, Y, Sn, U, W, Zr, Hf, Th, Nb, and Ta are some of those elements. When they combine with the enriched volatile components, the Li-micas, spodumene, beryl, tourmaline, and many rare and unusual minerals form.

Some pegmatites and their minerals form by solid-state replacement of preexisting minerals in much the same way as contact metasomatic skarns are formed. The gas phase of pegmatites is highly reactive, and may have high concentrations of dissolved salts that supply the chemical ingredients necessary to form new minerals. Spodumene, amblygonite, and lipidolite in Li-pegmatites are thought to form in this manner.

IGNEOUS MINERALS AND THEIR ASSOCIATIONS

Most classifications of igneous rocks are based on rock chemistry and mineralogy. Because of the great range of melt chemistries, a very large num-

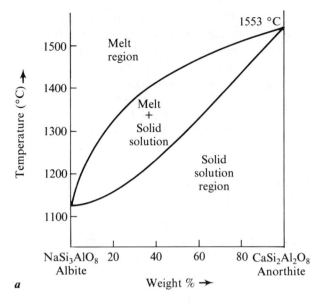

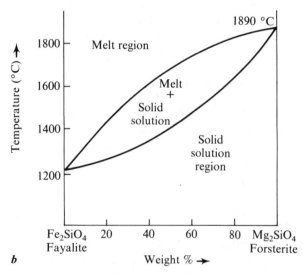

FIGURE 9.17   Melting relationships in solid solutions for (a) plagioclase feldspars and (b) olivine series.

ber of igneous rock names can be assigned only when the specific chemistries of the minerals and their relative proportions are known. The anorthite content of plagioclase feldspar is the most important igneous component in this regard. Determination of anorthite content helps us separate granites, monzonites, granodiorites, diorites, and gabbros from each other and also helps us distinguish their extrusive equivalents. Normally this cannot be done accurately without the aid of a petrographic microscope, x-ray diffraction, or direct analysis. By examining textures, we distinguish between intrusive, hypabyssal, and extrusive cooling histories, but textures are of limited value in mineral identification.

The simplest kind of igneous rock classification that is used implicitly by all petrologists is based on whether the rock is *felsic* (light colored) or *mafic* (dark colored). Most granites, rhyolites, and related rocks are felsic, because they consist mainly of K-feldspar, plagioclase, and quartz, with lesser amounts of darker minerals such as biotite and hornblende. Most gabbros and basalts are mafic, because they contain neither quartz nor alkali feldspar and have a higher proportion of dark minerals such as olivine, pyroxene, and hornblende.

The most general rule of thumb for the identification of igneous mineral assemblages is that minerals with similar chemistries are more likely to be found together. The Mg-rich olivines and pyroxenes are anhydrous and are usually associated with a Ca-rich plagioclase in gabbros and basalt. Biotites, muscovites, and amphiboles that are Fe-rich or alkali-rich or both are all hydrous and are invariably found with quartz, K-feldspar, and a Na-rich plagioclase.

Additional constraints on assemblages are the chemical reactions that prevent the stable coexistence of certain mineral pairs. In this context, the two most important reactions are:

1.   nepheline + quartz = albite

$$NaSiAlO_4 + 2SiO_2 = NaSi_3AlO_8$$

2.   forsterite + quartz = enstatite

$$Mg_2SiO_4 + SiO_2 = Mg_2Si_2O_6$$

Both reactions proceed spontaneously to the right under all igneous conditions, so nepheline + quartz and forsterite + quartz never occur together in equilibrium. Instead, reaction 1 produces albite + quartz (granite) and albite + nepheline (syenite), and reaction 2 produces olivine + pyroxene (gabbro) or pyroxene + quartz (diorite), depending on bulk composition. The presence or absence of quartz is the most important criterion and the easiest to determine in a hand specimen.

Quartz-bearing Felsic Igneous Rocks

Most granites, quartz monzonites, granodiorites, and their extrusive equivalents are in this category. All have K-feldspar and Na-rich plagioclase in a ratio that decreases from granite to granodiorite. Biotite, muscovite, and dark amphibole are all common.

Accessory minerals include apatite, zircon, magnetite, ilmenite, allanite, tourmaline, fluorite, and rarely, garnet. Very Fe-rich and Na-rich granites may have fayalite, riebeckite, and aegerine.

The extrusive rocks in this category have the

same bulk and mineral chemistries, but the structural states of some of the minerals reflect their rapid cooling. Sanidine appears instead of orthoclase or microcline, and either tridymite or cristobalite will be the high-temperature SiO_2 polymorph.

Quartz-absent Felsic Igneous Rocks

Bulk compositions that are not siliceous enough to crystallize albite + quartz (Figure 9.3b) will crystallize albite + nepheline instead. In syenites, K-feldspar, Na-plagioclase, and nepheline are common. Leucite ($KSiAlO_4$), like nepheline, never occurs with an SiO_2 phase. It is common in Si-undersaturated, K-rich volcanics.

The common mafic minerals in these rocks include aegerine, augite, sodic and Fe-rich amphibole (riebeckite and arfvedsonite), biotite, and fayalitic olivine. Magnetite, ilmenite, and apatite can be locally abundant. Occasionally, such rock will bear cancrinite, sodalite, nosean, hauyne, and corundum.

Some rocks in this category are felsic but are chemically more closely related to gabbros. Such are the anorthosites, consisting wholly of plagioclase feldspar of An_{50} to An_{90} composition. Small amounts of olivine, pyroxene, and Fe-Ti oxides may be present.

Quartz-bearing Mafic Igneous Rocks

This category includes many diorites, andesites, and even a few gabbros and basalts. Plagioclase of an intermediate composition is abundant, along with pyroxenes, which may include hypersthene, pigeonite, or augite or a combination of these. Biotite and hornblende are common. Olivine is absent. The usual accessory minerals are magnetite, ilmenite, picotite, and apatite. Quartz usually fills the interstitial areas between the feldspars and pyroxenes, showing that it crystallizes late in the cooling history.

Quartz-absent Mafic Igneous Rocks

In the absence of quartz, olivine of forsteritic composition is common with any combination of calcic plagioclase, enstatite (or hypersthene), augite, pigeonite (in extrusive rocks), and Fe-Ti oxides, mainly magnetite or ilmenite or both. In more oxidized environments, ferric iron is important, and can stabilize solid solutions in the pseudobrookite series and between ilmenite and hematite. Chromite, pyrite, and pyrrhotite may also be common.

Secondary minerals such as the various zeolites, chalcedony, and opal frequently are found as amygdule fillings in many basalts, andesites, and rhyolites. The chemical constituents are transported by a gaseous phase, often during cooling.

Olivine-bearing intrusive rocks such as gabbro and peridotite usually have some amounts of serpentine plus magnetite formed by reaction from olivine plus water. Dunites, rocks consisting of only forsteritic olivine, and minor Cr-rich spinel become serpentinites when sufficient water is available.

ORE MINERALS AND THEIR ASSOCIATIONS

Most of the metals used in our industrial society were originally concentrated through the action of hydrothermal solutions and subsequent precipitation as ore deposits. The majority of metals, such as Fe, Cu, Ni, and Zn, precipitate as sulfides, occasionally in a single episode but more frequently in several stages of mineralization. The resultant deposits may form at low temperatures in either sedimentary or metamorphic environments or at high temperatures in magmatic environments.

Sedimentary processes at low temperatures may concentrate ore minerals in a variety of ways. Dense minerals such as gold, platinum, and cassiterite that weather away from igneous rocks may concentrate by density segregation to form *placer deposits*. Other minerals such as diamond, ruby, and sapphire are preserved as detrital minerals in the sedimentary environment because of their great hardness and resistance to chemical degradation.

Weathering processes also concentrate relatively insoluble oxides by the chemical removal of the more soluble minerals. This process of concentration is particularly effective in tropical environments in which the warm, wet climate enhances the rate of chemical breakdown. Major iron deposits (laterite), aluminum deposits (bauxite), and deposits of nickel and cobalt form through this process of concentration.

Metamorphic processes concentrate ore minerals chiefly by the formation of *skarn deposits* developed in the country rock around igneous intrusions by the hydrothermal transport of metals and solutions from the intrusion. Regional metamorphism is responsible for most talc and asbestos deposits and in some instances for substantial quantities of graphite.

Igneous processes concentrate ore minerals in the early stages of crystallization by *magmatic segregation*. The relatively dense sulfide and oxide minerals settle to the bottom of a magma chamber, leaving the lighter silicates above. Magmatic

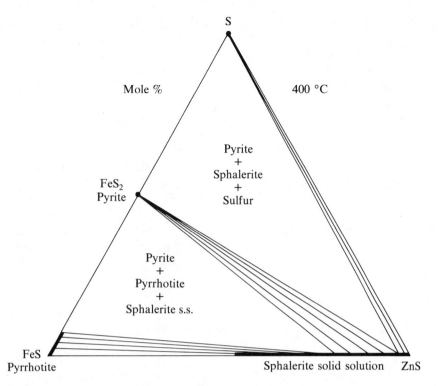

FIGURE 9.18 Ternary facies diagram for the system FeS-ZnS-S at 400 °C and 100 kPa. (Redrawn with permission after Craig, J. R., and Scott, S. D., 1974. *Sulfide Mineralogy*, Mineral. Soc. Amer. Short Course, p. CS-42.)

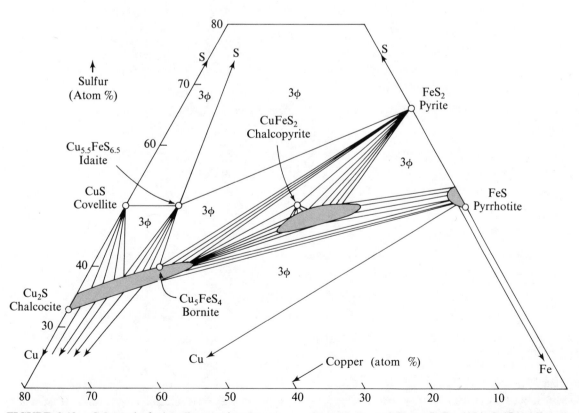

FIGURE 9.19 Schematic facies diagram for the system Cu-Fe-S at 400 °C and 100 kPa. (Redrawn after Craig, J. R., and Scott, S. D., 1974. *Sulfide Mineralogy*, Mineral. Soc. Amer. Short Course, p. CS-68.)

chromite, nickel, and magnetite deposits form in this way.

During the last stages of magmatic crystallization, volatile components such as H_2O, Cl_2, F_2, and CO_2, and a variety of metals such as gold and silver frequently concentrate, because none of these were incorporated in the structures of the earlier crystallized minerals. Such *hydrothermal solutions* may simply crystallize in place, or they may be squeezed by pressure into the fractures of the solidified part of the intrusive body or even into the country rock as ore-bearing veins. The solutions may also disseminate throughout the intrusive body in lesser concentrations but over a sufficiently large volume to justify their exploitation. An example is the *porphyry copper deposits*,

which are frequently extracted by openpit mining.

The mineral assemblages that form in ore deposits reflect both the environment of deposition and the overall bulk chemistry of the rock or solutions from which the deposits originated. Extensive experimental research has been conducted to better understand the variables that operate to produce ore deposits. Figure 9.18 is based on experimental work and illustrates some common mineral assemblages in the Fe-Zn-S system at 400 °C. Figures 9.19 and 9.20 illustrate common assemblages in the Cu-Fe-S and Ni-Fe-S systems, respectively, at 400 °C. These diagrams are read and interpreted in the same manner as the triangular facies diagrams that are discussed earlier in the chapter.

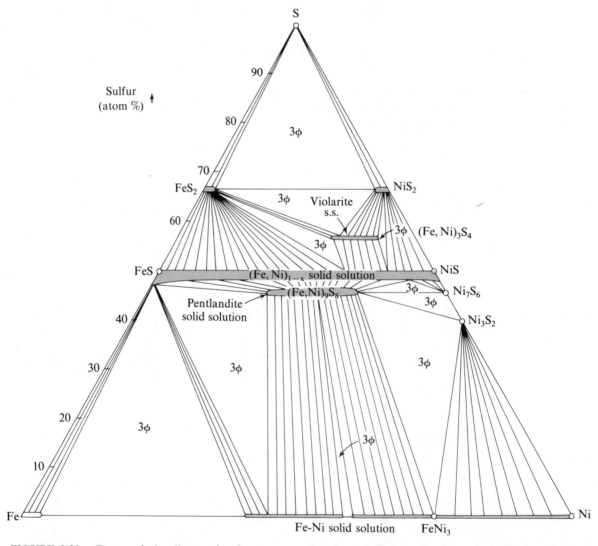

FIGURE 9.20 Ternary facies diagram for the system Fe-Ni-S at 400 °C and 100 kPa. (Redrawn after Craig, J. R., Naldrett, A. J., and Kullerud, G. 1968. 400 °C insothermal diagram. In *Carnegie Institution Yearbook* **66**, p. 441. By kind permission of Carnegie Institution of Washington, Washington, D.C.)

CHAPTER TEN

X-RAY MINERALOGY

X-rays are that part of the electromagnetic spectrum (Figure 4.29) with wavelengths generally between 10^{-1} and 10^{-6} nm. At the short end of this range are the *hard* x-rays used in medicine, which penetrate biological tissue without harm. In contrast, *soft* x-rays at the long end of this range (around 10^{-1} nm) can be very harmful, and all precautions must be taken to avoid direct exposure. Although not useful for medical purposes, the latter wavelengths are particularly well suited for diffraction studies of minerals.

Because x-rays, like visible light, have a wave-like nature, they can be diffracted. For this to happen, some rather special constraints are imposed upon the diffracting substance. As we will show later in this chapter, the distance between diffracting planes of the substance must be approximately equal to the wavelength of the incident radiation for diffraction to occur. For example, diffraction of optical wavelengths (350 to 700 nm) occurs in certain minerals (e.g., pyroxenes and feldspar) when the spacing of exsolution lamellae is about the same magnitude. The resulting "play of colors" can be observed on some polished building stone used all over the world. A similar light diffraction effect is responsible for the play of color in opal. If one is interested in lattice spacing in minerals, then the magnitude of the spacing is smaller, on the order of 0.01 to 4 nm. In order to see such spacings, the incident radiation must have correspondingly shorter wavelengths. X-rays satisfy this condition.

DIFFRACTION

When electromagnetic radiation of any wavelength encounters an obstacle, the radiation is scattered in all directions. In general, the scattered rays from many such obstacles will interfere with each other and be destroyed. In certain directions, however, the rays combine constructively to produce new rays. The phenomenon of cooperative scattering from a grating in three dimensions is called *diffraction*. Cooperative scattering from a continuous medium in two dimensions is called *reflection*. The terms are commonly interchanged.

Both diffraction and reflection of visible light were demonstrated as early as the 17th century. Sets of parallel slits, mirrors, prisms, and lenses were all used as the diffracting medium. If fine parallel grooves are engraved in a mirror, the closely spaced strips that remain act as scattering sources, as indicated in Figure 10.1a. The incident light is scattered radially from each source in a pattern like concentric ripples on a pond. The wavelength of the scattered light will be the same as the monochromatic incident light, as illustrated by the equal spacing of wave fronts. Only in certain directions will rays from adjacent strips combine in a *constructive* manner to produce a new ray having an enhanced intensity. The single condition for cooperative scattering, in this case reflection, is that all participating rays be *in phase*. This means that everywhere along a given wave front (*AD* in Figures 10.1a and 10.1b), the path

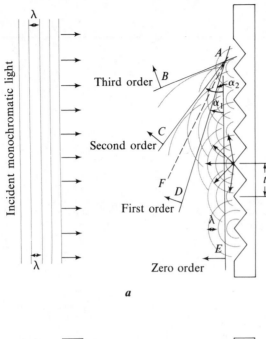

a

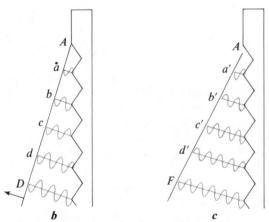

b   *c*

FIGURE 10.1   (a) Coherent scattering of monochromatic visible light of wavelength λ from a mirror grooved with a spacing *t*. Reflection is observed only at discrete values of α(α₁, α₂, . . .) that satisfy Equation 10.1. (b) Wave front *AD* corresponding to first-order diffraction at α₁. All rays along *AD* are *in phase*. (c) Line *AF* at a nondiffracting angle along which alternating rays are 180° *out of phase*.

difference between adjacent rays is an integer *n* number of wavelengths. Thus in Figure 10.1b, the ray at *a* has travelled one wavelength λ, the ray at *b* has travelled 2λ, and so on. When this happens, and it will only be at certain angles of reflection, all of the troughs (or crests) parallel to the wave front coincide and therefore reinforce each other to produce a coherent ray that propagates as a visible reflection. Figure 10.1c illustrates an angle at which the diffraction condition is not satisfied. Lines parallel to *AF* intersect some rays at a trough and other rays at a crest, so *destructive*

interference occurs. The rays are thus *out of phase*, and no reflection can be propagated.

Constructive interference occurs only when wave crests of adjacent rays lag behind by 1λ, 2λ, . . . *n*λ, where *n* is an integer that defines the *diffraction order*. The zero-order diffraction is when $n = 0$ (*AE*, Figure 10.1a), the first-order diffraction is when $n = 1$ (*AD*, Figure 10.1a), and so forth. The geometric relationship between the spacing of the grooves *t*, the wavelength λ, and the angle of the diffracted wave front to the mirror α was first defined by Friedrich M. Schwerd in 1835 as

$$\lambda = \frac{t}{n} \sin \alpha \quad \text{or} \quad n\lambda = t \sin \alpha \qquad (10.1)$$

This relationship indicates that for given values of λ and *t*, diffraction of a particular order (e.g., $n = 1$) occurs only for unique values of α. In the case of Figure 10.1, a series of bright, parallel lines separated by regions of darkness would be seen on a screen placed to the side of the incident light source. The equivalent experiment, but using slits rather than mirrors, was performed by Thomas Young in the early 1800s.

In 1895, Wilhelm Röntgen discovered x-rays, a feat worthy of the first Nobel Prize awarded in 1901. One of his early experiments was an attempt to produce diffraction with these new rays. His reasoning was that if these rays could be diffracted according to Equation 10.1, then their wavelike nature would be proved. Röntgen wrote in his third communication (March, 1896) that "I observed phenomena which looked very much like diffraction. But in each case a change in the experimental conditions . . . failed to confirm it." The problem was that the spacings in his grating were not small enough to diffract x-rays. Equation 10.1 shows that for first-order diffraction ($n = 1$), *t* must be equal to λ for a diffraction angle of 90 degrees. If the diffraction angle were 45 degrees, the value of *t* would be about 1.41λ. In other words, the spacing between diffracting components must be of the same order of magnitude as the wavelength. Visible light with a wavelength of 500 nm, for example, will not be diffracted by the same 0.5 nm spacings appropriate for x-ray diffraction.

In 1912, Max von Laue at Munich University had the idea that the lattice spacings of crystals might serve as a diffraction grating for x-rays. At that time scientists suspected, but still had not proved, that x-rays were wavelike in nature. In order to show that these mysterious rays could be diffracted, a finer grating had to be found. Paul Ewald suggested that crystals might provide such a grating. Two of Laue's students, Walter Friedrich and Paul Knipping, volunteered to do the experiment using a crystal of hydrated copper sulfate (Figure 10.2a). Sheets of lead with small holes were placed in front of an x-ray tube to direct the rays at the crystal. A sheet of photographic film encased in a black paper envelope was placed behind the crystal to record any diffraction effects. The developed film showed the direct beam (Figure 10.2b) that had passed straight through the crystal, and in addition, several weaker spots of variable intensity that looked like diffraction effects. The young experimenters discovered that when the crystal was tilted slightly, the spots changed position relative to the direct beam spot, thus proving that the "grating" was inside the crystal.

They soon tried the experiment with simpler crystals. The pretty blue crystals of hydrated copper sulfate were a poor first choice because of their low triclinic symmetry. Friedrich and Knipping discovered that cubic crystals yield symmetric arrangements of diffraction spots. When the incident beam was parallel to a fourfold axis, fourfold symmetry appeared on the photograph; when the beam was parallel to a threefold axis (cube diagonal), a threefold arrangement of spots appeared. This experiment proved that the diffracting elements were symmetrically arranged inside the crystal.

Confronted with Friedrich and Knipping's observations, Laue correctly concluded that the planar arrangement of atoms caused the diffraction. He further concluded that x-rays were electromagnetic waves like light waves but with a much shorter wavelength. Laue and his students had solved two big problems simultaneously, simply because the evidence for the solution of each depended on the other. X-rays are electromagnetic waves and atoms have planar arrangements in crystals.

X-RAY DIFFRACTION

In a short time, Laue derived the equations necessary to describe the experimental results. If an incident beam reaches a one-dimensional array of atoms at an angle α, the diffraction direction will be at an angle α' such that the path difference $p + r$ (Figure 10.3) between adjacent rays is an integer multiple of the wavelength $(n\lambda)$. That is,

$$p + r = n\lambda$$
$$a \cos \alpha + a \cos \alpha' = n\lambda$$

or

$$a(\cos \alpha + \cos \alpha') = n\lambda \qquad (10.2)$$

In a three-dimensional array of atoms, the same conditions must hold in each of the new directions:

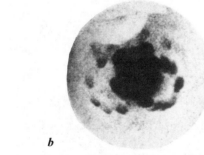

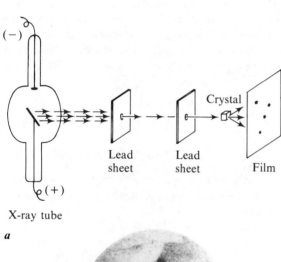

a

b

FIGURE 10.2   (a) Experimental setup used by Friedrich and Knipping to measure diffraction effects. (b) Observed diffraction effects from hydrous copper sulfate. (From Friedrich, W., Knipping, P., and von Laue, M., 1912. Interferenzen bei Röntgenstrahlen; Sitzungsberichte der mathmatisch-physikalen Klasse der K. B. Akademie der Wissenschaften zu München, vol. 2, figure 1. Reprinted in *Naturwissenschaften* 1952, vol. 16, p. 367. By kind permission of Springer-Verlag, New York, Inc.)

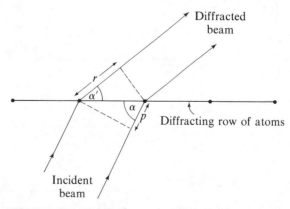

FIGURE 10.3   Conditions for diffraction from a one-dimensional array of atoms.

$$b(\cos \beta + \cos \beta') = n\lambda \qquad (10.3)$$

and

$$c(\cos \gamma + \cos \gamma') = n\lambda \qquad (10.4)$$

All three equations must be satisfied simultaneously to produce diffraction from a three-dimensional array of atoms. These equations are known today as the *Laue equations*.

A more easily visualized representation of the Laue equations was proposed by William Lawrence Bragg within a year of Laue's achievement. Bragg considered a two-dimensional case in which the diffraction condition was satisfied by lattice planes. Figure 10.4 shows that in order for all three lattice planes (1, 2, and 3) to diffract, the path difference between each of the waves (*ABC*, *DGI*, and *JLN*) must be an integer of the wavelength λ. That is,

$$FG + GH = n_1\lambda \quad \text{from plane 2}$$

and

$$KL + LM = n_2\lambda \quad \text{from plane 3}$$

In the case of plane 2,

$$\sin \theta = FG/d_{hkl} \quad \text{or} \quad FG = d_{hkl} \sin \theta$$

and

$$FG + GH = 2d_{hkl} \sin \theta = n\lambda$$

This equation is known as the *Bragg equation* and has been used extensively in diffraction studies since 1912 with but one minor modification. The integer *n*, which refers to the order of diffraction, is identified more directly with the diffracting plane to give

$$\lambda = \frac{2d_{hkl}}{n} \sin \theta = 2d_{nh\,nk\,nl} \sin \theta \qquad (10.5)$$

This form of the Bragg equation shows that the interplanar spacing *d* can also include the *d/n* spacing of semirational planes that will produce higher order diffraction. Planes such as d_{222} ($n = 2$) and d_{333} ($n = 3$) always have indices with a common denominator *n*. For example, the first-order diffraction of the *hkl* plane in Figure 10.5 is inclined to the rational lattice plane by the angle θ_1. The incident beam is labeled No. 1, and the distance *ABC* is equal to 1λ. Second-order diffraction from the same *hkl* plane takes place at θ_2. The incident beam is labeled No. 2, and the distance *A'B'C'* is equal to 2λ. This condition is equivalent to first-order diffraction by a semirational plane visualized halfway between the *hkl* planes with an interplanar spacing of d_{2h2k2l}. The distance *A"B"C"* is equal to 1λ.

The lattice of a crystal contains an infinite number of rational and semirational planes that are potential diffraction planes, but only those d_{hkl} planes will diffract that yield θ angles between zero and 90 degrees for a given wavelength of radiation. Consequently, the smallest possible *d* value (d_{min}) is the one that yields $\theta = 90$ degrees ($\sin \theta = 1$). That is,

$$\lambda = 2d_{hkl} \sin \theta = 2d_{min}$$

or

$$d_{min} = \frac{\lambda}{2} \qquad (10.6)$$

This relationship also means that the number of diffractions will depend on the wavelength used.

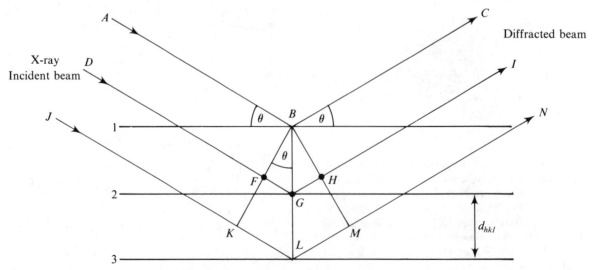

FIGURE 10.4 Diffraction of monochromatic x-rays from parallel lattice planes according to the Bragg equation.

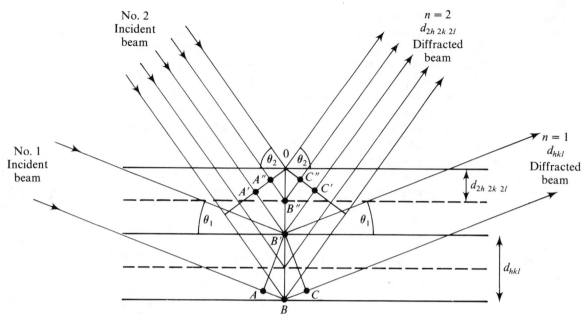

FIGURE 10.5 **Relationship between diffraction order *n* and semirational lattice planes *n*(*hkl*).**

Smaller values of λ give smaller $d_{\min}$ and thus larger numbers of diffractions.

The Bragg equation has a convenient graphical representation that helps clarify its real meaning. Note that Equation 10.5 has only three variables, namely, λ, θ, and the interplanar spacing d, usually referred to as the *d value*. If any two of the variables are known, the value of the third is uniquely determined. Figure 10.6 is a plot of $\sin\theta$ versus $1/\lambda$ that shows how each interplanar spacing d of a halite (NaCl) crystal is represented by a distinct curve. A different crystal would have a similar set of curves, but at different positions because of different lattice spacings. If the x-ray wavelength is constant, as it would be in a normal diffraction experiment, each of the interplanar spacings encountered along the line AB would satisfy the Bragg equation and would therefore diffract. By measuring the unique θ at which each spacing diffracts, the value of each d can be calculated. If a different x-ray wavelength is used, the same sequence of diffraction lines will be observed, but they will be either spread out or compressed, depending on whether the wavelength is longer or shorter.

The graphical relationship of the Bragg equation in Figure 10.6 exists regardless of how we choose to measure it. For some experimental purposes, θ may be held constant while we measure the different energies along line CD at which diffraction occurs. This is done with an energy dispersive detector. The diffraction spectrum that results looks much like a conventional spectrum,

except that energy rather than θ is the experimentally measured parameter.

THE NATURE OF X-RAYS

X-rays are generated by the sudden deceleration of fast-moving particles. In x-ray tubes, these particles are electrons emitted by a hot (2600–2800 K) tungsten wire filament. The free electrons are then accelerated by a positive potential, from 30 to 100 keV, between the filament and a metal target. When the electrons hit the target, several things happen. Most of the incident energy is

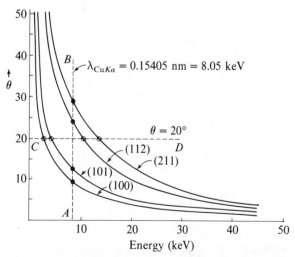

FIGURE 10.6 **Plot of θ versus reciprocal wavelength $1/\lambda$ for low quartz according to the Bragg equation.**

dissipated in the form of heat, which is the reason x-ray tubes have a cooling system. Some of the electrons, however, upon deceleration undergo a stepwise nuclear scattering by the target material. The energy emitted can have wavelengths in the x-ray region, providing the incident electrons have sufficient energy. If an incident electron lost all of its energy ΔE in a single collision with a target metal nucleus, the energy loss would be simply the voltage V times the electron charge e, or $\Delta E = eV$. The wavelength of the emitted energy is given by the frequency ν times Planck's constant h.

$$\Delta E = h\nu = \frac{hc}{\lambda} \quad \text{or} \quad \lambda = \frac{hc}{\Delta E}$$

As both h and c (speed of light in a vacuum) are constant, the above equation can be reduced to

$$\lambda(nm) = \frac{1.2398}{\Delta E(keV)} \qquad (10.7)$$

Because an electron normally decelerates by a sequence of collisions, a range of x-ray energies results. Taking the entire electron stream collectively, a *continuous spectrum* or so-called *white radiation* is produced. The shortest wavelength (highest energy) in the spectrum is given by Equation 10.7. As the potential on the x-ray tube is increased, from 20 keV to 30 keV, for example, the continuous spectrum expands to include shorter wavelengths, and its relative intensity increases (Figure 10.7).

As the potential is increased further, the incident electrons may have sufficient energy to eject one of the inner electrons of the target metal. When this happens, an electron from another energy level of the same atom will immediately fill the vacancy, emitting a *characteristic* wavelength with energy determined by the initial and final energy levels of the electron. The characteristic peak so generated appears as an energy spike superimposed on the continuous spectra (Figure 10.7). Some of the vacancies left by ejected K-shell electrons are filled by L_3 electrons, and others are filled by slightly less energetic L_2 electrons. The wavelengths of x-rays emitted by the process are thus slightly different. Electrons that fall into the K shell from L_3 are called $K\alpha_1$; those that fall into the K shell from L_2 are called $K\alpha_2$ (Figure 10.8). Less frequently, various electrons from the M shell will drop into a K-shell vacancy, which explains the origin of the $K\beta_1$ and $K\beta_2$ lines. The lines (e.g., K, L, M) are all a part of the *characteristic spectrum* of the target metal (Table 10.1).

The $K\alpha_1$ and $K\alpha_2$ peaks are the most useful for x-ray diffraction, because they are the most intense. When the Bragg equation (Equation 10.5) is used with a specified wavelength, for example, $K\alpha_1$, of the target metal, the x-rays are said to be *monochromatic*. To produce monochromatic radiation, the white radiation and the other peaks of the characteristic spectrum must be removed by crystal monochromators, or more commonly, by absorption filters. All materials absorb x-rays, but different materials have different levels of absorption depending on the wavelength. A typical example of an absorption curve is that of lead, shown in Figure 10.9a, where energy increases from right to left. Low x-ray energies, regulated by the x-ray tube voltage, are adequate to eject relatively loosely held M-shell electrons of the lead. The x-ray photons are absorbed in the process. The probability of an M-shell electron being ejected in this manner decreases rapidly as the x-ray wavelength decreases, because the collision probability with an x-ray photon decreases. As the x-ray energies increase, a value is reached that is high enough to knock out an L_3 electron. When this happens, the x-ray photons are suddenly and strongly absorbed, giving rise to a discontinuity in absorption, called the L_3 *absorption edge*. The magnitude of absorption then falls off as the wavelength decreases until energy increases to the value necessary to eject L_2 and, subsequently, L_1 electrons. As x-ray energies reach higher values, the most tightly held K-shell electrons will also be ejected, and the corresponding K absorption edge will be observed. The wavelengths of the absorp-

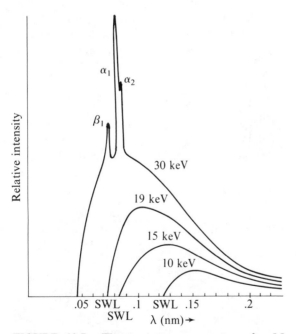

FIGURE 10.7   The x-ray energy spectrum for Mo metal at different accelerating voltages. The shortest attainable wavelength (SWL) is a function of voltage.

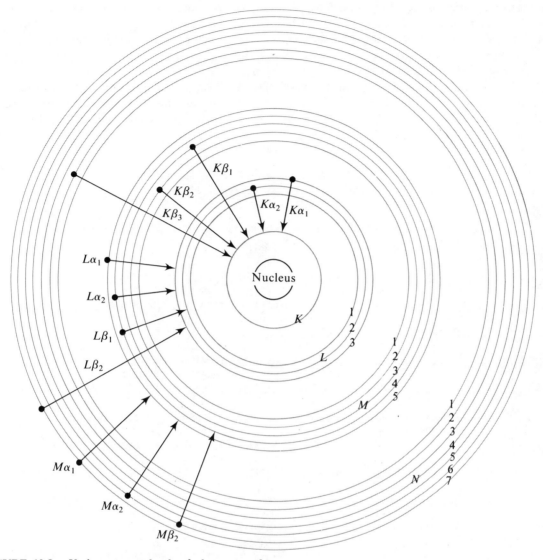

FIGURE 10.8 Various energy levels of electrons and their transitions that produce x-rays.

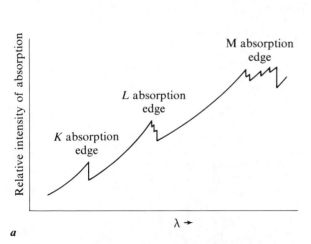

a

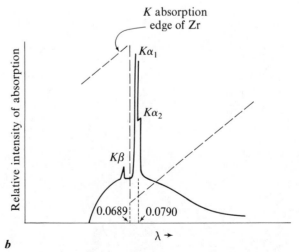

b

FIGURE 10.9 (a) Typical x-ray absorption behavior of the elements as a function of wavelength λ. Energy, ~1/λ, increases from right to left. (b) The use of a Zr metal filter to eliminate *Kβ* radiation from an Mo x-ray tube.

tion edges are characteristic of the absorbing elements and can be used for element analysis. The electrons ejected in the absorption process give rise to secondary or *fluorescent* radiation. Because the wavelengths of secondary electrons are diagnostic of the emitting element, they can be used for chemical analysis, and the method is known as *x-ray fluorescent analysis*.

Figure 10.9b illustrates the practical application of an absorption curve for filtering out unwanted x-ray wavelengths. Suppose one wishes to eliminate the K peak from the characteristic spectrum of molybdenum. A thin foil of zirconium metal placed in front of the x-ray tube will do the job. Its K absorption edge is at 0.0689 nm, which will allow the desirable K radiation of the molybdenum to pass but will absorb the $K\beta$ peak. Without the filter, the $K\beta$ peak is intense enough to gener-

TABLE 10.1.   *K Characteristic Line and Absorption Edge Wavelengths (nm) for the Lighter Elements*

| Element | | Emission lines (nm) | | | Absorption |
NO.	SYMBOL	$K\alpha_2$	$K\alpha_1$	$K\beta$	Edge, nm
11	Na		1.1909	1.1617	
12	Mg		0.98889	0.9558	0.95117
13	Al	0.833916	0.833669	0.7981	0.95117
14	Si	0.712773	0.712528	0.67681	0.67446
15	P		0.61549	0.58038	0.57866
16	S	0.537471	0.537196	0.50316	0.50182
17	Cl	0.473050	0.472760	0.44031	0.43969
18	A	0.419456	0.419162	(0.38848)	0.38707
19	K	0.374462	0.374122	0.34538	0.34364
20	Ca	0.336159	0.335825	0.30896	0.30701
21	Sc	0.303452	0.303114	0.27795	0.27573
22	Ti	0.275207	0.274844	0.215381	0.24973
23	V	0.250729	0.250348	0.228434	0.22690
24	Cr	0.229351	0.228962	0.208480	0.20701
25	Mn	0.210175	0.210175	0.191015	0.18964
26	Fe	0.193991	0.193597	0.175653	0.17433
27	Co	0.179278	0.178892	0.162075	0.16081
28	Ni	0.166169	0.165784	0.150010	0.14880
29	Cu	0.154433	0.154051	0.139217	0.13804
30	Zn	0.143894	0.143511	0.129522	0.12833
31	Ga	0.134394	0.134003	0.120784	0.11957
32	Ge	0.125797	0.125401	0.112890	0.11165
33	As	0.117981	0.117581	0.105726	0.10449
34	Se	0.110875	0.110471	0.099212	0.09797
35	Br	0.104376	0.103969	0.093273	0.091995
36	Kr	0.09841	0.09801	0.087845	0.086547
37	Rb	0.092963	0.092551	0.082863	0.081549
38	Sr	0.87938	0.087521	0.078288	0.076969
39	Y	0.083300	0.082879	0.074068	0.072762
40	Zr	0.079010	0.078588	0.070170	0.068877
41	Nb	0.074615	0.074615	0.066572	0.065291
42	Mo	0.071354	0.070926	0.063225	0.061977
43	Tc	(0.067927)	(0.067493)	(0.060141)	(0.05891)
44	Ru	0.064736	0.064304	0.057246	0.056047
45	Rh	0.061761	0.061324	0.054559	0.053378
46	Pd	0.058980	0.058541	0.052052	0.050915
47	Ag	0.056377	0.055936	0.049701	0.048582
48	Cd	0.053941	0.053498	0.047507	0.046409
49	In	0.051652	0.051209	0.045451	0.044388
50	Sn	0.049502	0.049056	0.043521	0.042468

ate a diffraction pattern of its own, which will be superimposed on the pattern generated by the $K\alpha$ peaks. As a rule of thumb, the appropriate filter for a particular target is the metal that has an absorption edge just below the $K\alpha$ wavelength of the target. Nickel, for example, with an absorption edge at $\lambda = 0.14880$ nm is the ideal filter for a copper x-ray tube (λ of $K\alpha_1 = 0.15405$ nm). These relationships are summarized in Table 10.1.

In practice, the $K\alpha_1$ and $K\alpha_2$ wavelengths are too close to be separated by normal filtering. A weighted average of the two is usually taken and called $K\alpha$ radiation, without a subscript. Because the $K\alpha_1$ peak is approximately twice the intensity of the $K\alpha_2$ peak, the weighted average for the wavelength is

$$\lambda K\alpha = \frac{2\lambda K\alpha_1 + \lambda K\alpha_2}{3} \qquad (10.8)$$

Another method for obtaining monochromatic radiation is to use a *crystal monochromator*. A flat crystal (Figure 10.10a) will diffract different characteristic rays in slightly different directions, because the diffraction angle θ depends on wavelength. With this geometry, a simple mechanical slit can be used to block out all but the desired wavelength. An alternative is to use a curved crystal monochromator (Figure 10.10b) and focus the divergent incident beam.

X-RAY POWDER DIFFRACTION

If a single crystal is placed in a monochromatic x-ray beam, the incident beam will be diffracted from those lattice planes that are inclined to the beam by such angles as satisfy the Bragg equation. If the crystal is stationary, only a few planes will satisfy this condition (Figure 10.11a). The

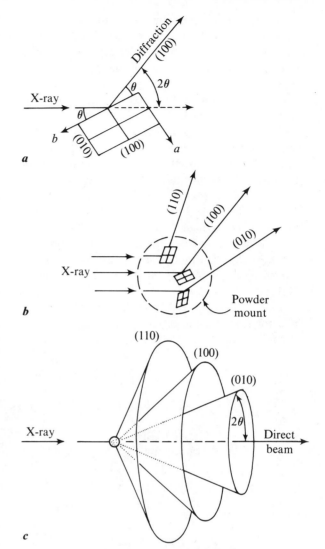

FIGURE 10.11 **Diffraction of monochromatic x-rays from (a) a single crystal and (b) an aggregate of small mineral fragments. (c) Diffraction cones produced by the powder method.**

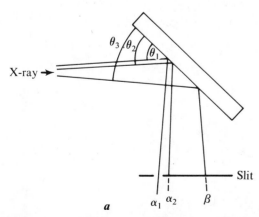

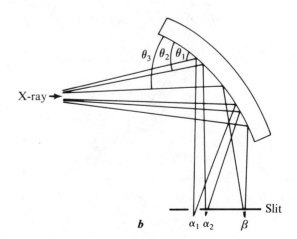

FIGURE 10.10 **The use of flat (a) and curved (b) crystal monochromators for producing monochromatic x-rays.**

experiment performed by Laue also involved a stationary crystal, but he observed more spots because of the range of wavelengths in the white radiation he used. By rotating the crystal without removing it from the x-ray beam, a greater number of planes will pass through a diffracting position (Figure 10.11b). Better yet, if the crystal is replaced by a fine-grained aggregate or powder, many crystals will always be simultaneously in diffracting positions. By rotating the powder continuously, all permissible lattice planes (with $d_{min} = \lambda/2 \sin 90$ degrees $= \lambda/2$) of the structure will diffract. The three-dimensional pattern of diffraction from a particular lattice plane is that of a cone, which has a solid angle twice the diffraction angle, or 2θ. Diffraction from all of the lattice planes defines a family of nested cones (Figure 10.11c).

In practice, we need only determine the value of θ for each cone in order to calculate, using the Bragg equation, the lattice spacing d for each lattice plane. The x-ray wavelength must also be known. Two methods are widely used for determining θ: (1) a powder camera that records a portion of each diffraction cone on photographic film, and (2) a powder diffractometer that records the diffraction electronically.

THE POWDER CAMERA

The most commonly used and most versatile powder camera is the Debye-Scherrer camera (Figure 10.12). A strip of flexible 35 mm film is placed around the interior of the cylindrical camera body. The specimen mount is positioned exactly in the center of the cylinder. The incident x-ray beam enters through a metal collimator that protrudes into the camera through a prepunched hole in the film. As most of the direct beam passes straight through the specimen, an exit collimator containing a leaded glass plug is used as a beam catcher.

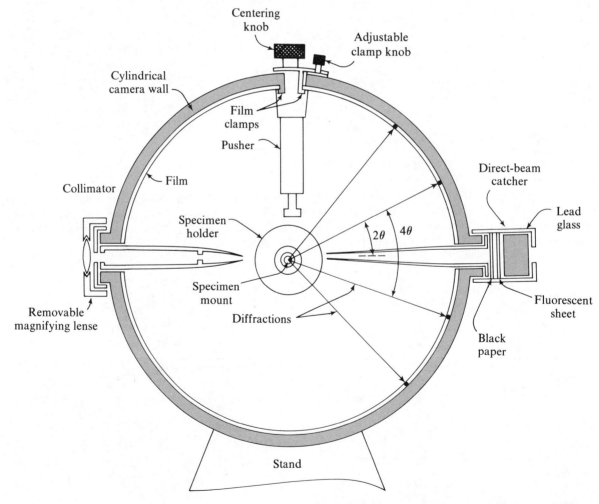

FIGURE 10.12   Debye-Scherrer powder camera (114.6 mm radius) in cross section.

Because of its position, the film intersects a portion of every diffraction cone from the material of interest. The diameter D of the camera is designed such that an integer value of millimeters measured on the film surface corresponds to an integer value of 2θ. That is,

$$\pi D = \frac{360}{2n} \quad \text{or} \quad D = \frac{360}{2\pi n} \qquad (10.9)$$

There are two common sizes of Debye-Scherrer cameras. The small size has a diameter of 57.26 mm and $n = 1$, which means that a 1 mm measurement on the film corresponds to 1 degree of 2θ (or 0.5 degree of θ). The larger size has a diameter of 114.5 mm and $n = 2$, which means that every 2 mm interval on the film corresponds to 1 degree of 2θ. The center of each of the two holes in the film (Figure 10.16) corresponds to $\theta = 0$ degrees (direct-beam catcher) and $\theta = 90$ degrees (collimator).

The usual collimator (Figure 10.12) is a tapered brass tube consisting of three internal slits. The first two are designed to reduce the incident x-rays to a relatively narrow and parallel beam. The third slit at the end of the tube absorbs the unwanted diffraction from the metal of the second slit. In some collimators, the first two slits are replaced with a single aluminum tube.

The beam catcher (Figure 10.12) looks externally like the collimator, but consists internally of a small piece of black paper to keep out room light, a fluorescent plate to convert x-rays to visible light, and a plug of transparent, leaded glass to permit direct observation of the direct beam as it impinges on the fluorescent plate. This arrangement is necessary for the proper alignment of the camera relative to the x-ray tube. When the specimen is centered between the collimator and the beam catcher, and when the beam image can be observed with room lights out, the beam is passing through the specimen as it should.

The first step in preparing a powder mount is grinding. The usual procedure is to crush the mineral fragments in a mortar and to grind them with a pestle until they will pass a 200 mesh sieve (mesh opening 0.074 mm). In most cases, overgrinding the mineral is practically impossible, so the sieving stage can be eliminated. If only a tiny amount of material is available, it can be immersed in a drop of clear fingernail polish on a glass slide and crushed by rolling a needle over it. Progress can be followed by using a microscope.

There are four common ways of forming the powder into a mount that can be loaded into the camera. (1) The *ball mount* (Figure 10.13a), designed for small amounts of specimen, is made by

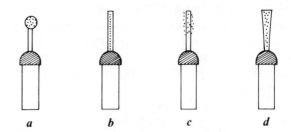

FIGURE 10.13 **The common types of powder mounts: (a) ball mount, (b) rod mount, (c) coated glass fiber mount, and (d) capillary tube.**

dipping a needle in clear fingernail polish (Duco cement and collodion also work well) and working the polish through the powder to form a small ball. This method has the advantage of providing a random orientation of even flaky or acicular mineral fragments. The ball is then glued on the end of a thin glass fiber by first dipping the fiber in a suitable cement and then quickly touching the fiber to the ball. The glass fiber is preattached to a brass pin, which is inserted in the camera with a pair of bent-nose pliers. (2) The *rod mount* (Figure 10.13b) requires more materials, because the powder is mixed with collodion or Duco cement and rolled between two glass slides until a thin (0.2 mm), cigar-shaped rod is formed. The rod can be glued directly to the brass pin. (3) A coated *glass fiber mount* (Figure 10.13c) is prepared by dipping the glass fiber, mounted on the brass pin, into glue or Vaseline, and then rolling the fiber in the powdered sample. Like method 1, the glass fiber mount has the advantage of being fast and simple. (4) In the *capillary tube mount* (Figure 10.13d) we use tubes made of boron glass with a small funnel at one end. This method is convenient when the mineral hydrates or dehydrates in air, or when only very small amounts of powder are available. One end of the tube is sealed by wax or glue, the powder is dropped into the funnel and gently tapped down, and then the unused part of the tube is broken away for use at another time.

Once prepared, the powder mount is loaded into the camera and centered. Centering is accomplished by rotating the mount while observing it through the collimator. If the mount rotates out of view, an adjustment (Figure 10.12) is made in the holder to bring it back into view. Repeated adjustments result in a centered mount.

The camera with its centered mount is then taken into a darkroom where the 35 mm filmstrip is loaded. Both the collimator and beam catcher are removed until the prepunched and cut film is pressed smoothly against the internal camera wall with the clamp knob. The two holes in the film must match the openings for the collimator and the beam catcher. When the collimator and beam stop are reinserted, the camera cover attached, and all film is put safely away, the lights can be turned on.

Exposure times depend on the type of specimen, the type and intensity of the x-rays, the type of film, and the size of the camera. The usual exposure time for the 114.5 mm powder camera is between 5 and 15 hours. The exposed film (Figure 10.14) is usually oriented with the $\theta = 0$-degree hole on the left. The left half of the film for which $\theta < 45$ degrees is the *front reflection* region and the right half ($\theta > 45$ degrees) is the *back reflection* region. Between 0 and 20 degrees of 2θ, some broad bands of darkening will usually occur due to x-ray scattering by air, glass, and the glue used in preparing the mount. If the specimen itself is amorphous, only these broad bands will appear on the film. If the specimen is crystalline, several dark, relatively well-defined arcs will also appear (Figure 10.14), representing the intersections of various diffraction cones with the film.

The number of diffraction lines (arcs) depends on the size of the unit cell and on the x-ray wavelength. The *smaller* the unit cell, the *larger* the number of lines. The *shorter* the wavelength of radiation, the *larger* the number of lines. The width of each diffraction line depends mainly on the size of the mount. The *smaller* the mount, the *sharper* the lines, and therefore the greater the accuracy of measurements. If broad lines still persist, they may be due to a low degree of crystallinity or to unusually high thermal vibration of the atoms. Discontinuous or dotted lines are caused by inadequate grinding or an inadequate quantity of powdered sample or both.

If the powder camera is run for many hours, the film may become overexposed and will have a dark background that often obliterates the diffraction lines. Film darkening can also be caused by secondary fluorescent radiation. This effect is especially pronounced if the sample contains elements that have K absorption edges (Table 10.1) just above the wavelength of radiation. For example, if we are using copper radiation ($\lambda = 0.1542$

nm), cobalt and iron with K absorption edges at $\lambda = 0.1608$ nm and $\lambda = 0.1743$ nm, respectively, will fluoresce and darken the film, especially in the back reflection region. If the darkening is too severe for the exposure time required, a different x-ray tube should be used.

Even when filtered radiation is used, both $K\alpha_1$ and $K\alpha_2$ wavelengths will be present in the incident beam. This effectively means that two diffraction patterns will be on the exposed film, one almost superimposed on the other. The difference in wavelength is so small (0.0003 to 0.0004 nm) that in the front reflection region the $K\alpha_1$ and $K\alpha_2$ diffractions appear as a single arc. For a Debye-Scherrer camera ($D = 114.5$ mm) and copper radiation, the separation at $\theta = 15$ degrees is only 0.08 mm, that is, less than the width of the arc. Because the location of arcs on the film is a function of $\sin\theta$, the separation of $K\alpha_1$ and $K\alpha_2$ becomes progressively greater as θ increases. At $\theta > 50$ degrees, the two arcs are separated by more than 0.5 mm and are therefore resolvable on the film if the sample is not too large.

The location of diffraction arcs on an exposed film can be measured with a ruler. More accurate measurements are made with a powder film measuring device like the one shown in Figure 10.15. The cross-hair is placed through the center of each arc, and readings are recorded for both arcs of a diffraction cone. The sum of each pair of measurements divided by two gives the location of the $\theta = 0$ degrees and the $\theta = 90$ degrees positions on the film for front and back reflection arcs, respectively. The difference between these two, in accordance with Equation 10.9, should be 90 mm for cameras with diameter $D = 57.26$ and 180 mm for cameras with $D = 114.5$ mm. Any difference between these values and the actual measurements is usually due to film shrinkage or expansion during development, to a radius error in the camera, or to measurement errors. We can correct for the difference error by dividing the difference by 180 and applying the resulting correction factor to all calculated θ values. For the usual Straumanis film mounting (Figures 10.12 and 10.14), these corrections cannot be made unless back reflection lines appear on the film. This is the case for the

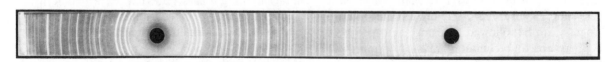

FIGURE 10.14   Typical x-ray powder film of a powdered quartz sample exposed to Cu $K\alpha$ radiation for 10 hours.

powder patterns of many minerals. These procedures are summarized in Figure 10.16.

Figure 10.16 illustrates that the difference in millimeters between two arcs of a cone in the front reflection region is equal to 4θ. In the back reflection region, the complement of 4θ is measured across the $\theta = 90$ degrees position, so that value must be subtracted from 360 degrees to yield the proper value of 4θ. From the calculated value of θ for each diffraction cone and the

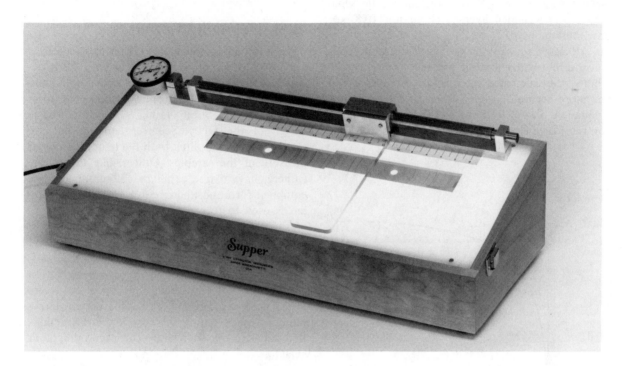

FIGURE 10.15   A standard x-ray powder film measuring device. (Courtesy of Charles Supper Co., Inc.)

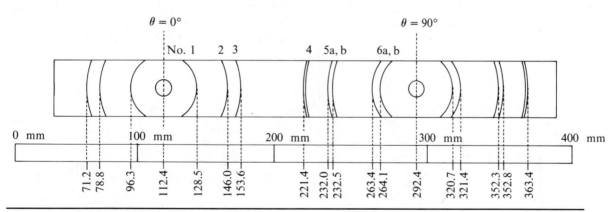

Arc No.	Measurements Left	Right	$\sum$ of meas.	Center	Diff. of meas.	$\theta(°)$	λ(nm)	d(nm)
1	128.5	96.3	224.8	112.4	32.20	8.50	0.15418	0.55
2	146.0	78.8	224.8	112.4	67.20	16.80	0.15418	0.267
3	153.6	71.2	224.8	112.4	82.40	20.60	0.15418	0.219
4	221.4	363.4	584.8	292.4	142.00	54.50	0.15418	0.09469
5a	232.0	352.8	584.8	292.4	120.80	59.80	0.15405	0.08912
b	232.5	352.3	584.8	292.4	119.80	60.05	0.15443	0.08912
6a	263.4	321.4	584.8	292.4	58.00	75.50	0.15405	0.07956
b	264.2	320.6	584.8	292.4	56.4	75.90	0.15443	0.07956

FIGURE 10.16   Procedures for measuring diffraction arcs on a powder photograph and for calculation of interplanar spacings. Radiation is Cu $K\alpha$; camera diameter = 114.6 mm.

known wavelength, we can use the Bragg equation (Equation 10.5) to calculate the value of *d* for each cone. When diffraction arcs due to $K\alpha_1$ and $K\alpha_2$ are resolvable, those wavelengths should be used in the calculations. In the low-angle region where $K\alpha_1$ and $K\alpha_2$ diffraction cannot be separated, the weighted average (Equation 10.8) for $K\alpha$ must be used.

Although we can measure each diffraction arc on the film with the same precision, or reproducibility, the calculated interplanar distances *d* will be more accurate at high values of θ than at low values. A plot of θ versus *d* (Figure 10.17) based on the Bragg equation shows why this is true. The

curve that results is characterized by constant λ. A measurement error of 2 degrees in θ at $\theta = 5$ degrees yields an error in *d* of 0.025 nm, whereas the same error in θ at $\theta = 35$ degrees yields an error in *d* of only 0.008 nm.

SPECIAL CAMERAS

In addition to the Debye-Scherrer cameras, a variety of other powder cameras have been designed for special applications. The Gandolfi camera is designed to take powder patterns from single crystals. Some cameras have the basic design of the Debye-Scherrer camera, but are equipped for controlling both the temperature and pressure of the samples. Others, like the micro-camera, have a fine capillary collimator and are equipped for atmospheric control. A similar camera is used to record powder diffraction of samples under extremely high pressure in the diamond anvil pressure cell.

Another category of special powder cameras is designed differently. A *flat film camera* has a planar film holder so that diffraction cones are intercepted as perfect circles. This design allows mineralogists to study the very low θ diffractions that occur in minerals with very large interplanar spacings.

Focusing cameras are designed to take advantage of the divergent nature of the incident beam and focus the diffracted rays on the film. A symmetrical camera that combines both front and back reflecting regions is illustrated in Figure 10.18a. The focusing capabilities of these cameras

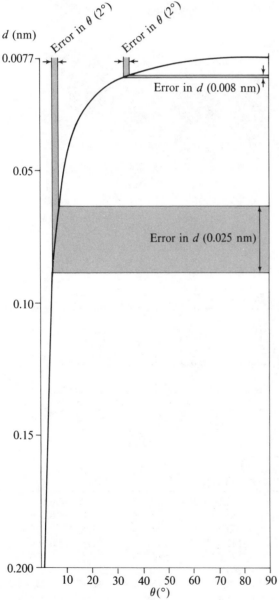

FIGURE 10.17   **Plot of diffraction angle θ versus interplanar spacing *d* based on the Bragg equation *d* = λ/2 sin θ. The curve represents constant wavelength.**

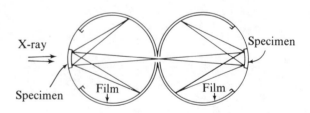

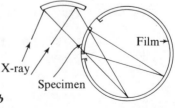

FIGURE 10.18   **(a) A pair of front-reflection and back-reflection focusing powder cameras. (b) An asymmetric focusing camera with curved crystal monochromator.**

can be enhanced by the use of crystal monochromators (Figure 10.18b). Both the flat film and focusing cameras can be equipped with multiple sample holders to simultaneously record the powder patterns of several samples including the standard.

POWDER DIFFRACTOMETERS

The powder diffractometer serves the same function as the powder camera, but the diffractometer records the diffraction effects by an electronic detector rather than on photographic film (Figure 10.19). The counter is mounted on a goniometer, which is rotated by a small motor through a 2θ range of almost 180 degrees. In most diffractometers, the specimen mount is replaced by a flat sample holder, which is rotated at one half the speed of the goniometer in order to preserve the relationship of θ and 2θ (Figures 10.11a, 10.19). The usual collimator is replaced by a linear slit or by soller slits, the latter consisting of sets of thin parallel plates. These insure that the incident beam does not diverge significantly before reaching the sample. Soller slits between the sample and the detector and two additional slits control the shape and width of the diffracted rays. If desirable, the diffractometer can be augmented with a curved crystal monochromator positioned in front of the detector. The monochromator allows only that diffraction due to $K\alpha$ radiation to be recorded.

The powdered mineral can be placed in a special holder supplied by the manufacturer, or it can be mixed with a suitable binder such as collodion and glued to a glass slide. The latter is similar to the "smear mount," for which the powder is spread evenly over part of a glass slide and held in place with a drop of acetone. As the acetone evaporates, the powder can be stirred to insure its even distribution over the slide. In all cases, the mineral should be ground to an even finer (< 400 mesh) powder to avoid orientation effects of platy and acicular fragments. Even then, additional precautions to avoid preferred orientation may be needed, such as fusing the powder with borax glass, regrinding, and remounting.

Diffracted x-rays that enter the detector cause a deflection of the pen of a strip chart recorder that is proportional to the diffraction intensity. Each diffraction is thus recorded on the paper chart (Figure 10.20) as a discrete peak. The sequence of peaks recorded corresponds to the sequence of diffraction arcs normally seen on a powder photograph of the same mineral. Peak position is routinely measured at the center of a line drawn across the peak at one half its height above background (Figure 10.20). For greater accuracy of 2θ determination, an internal standard (such as quartz) with well-known interplanar spacings can be mixed with the sample during preparation. Once the θ values for each diffraction are known, the d values are calculated with the Bragg equation.

Diffractometers have definite advantages over cameras in terms of speed, simplicity, and precision of measurement. When the detector scanning rate and recording speed are high, we can obtain a diffraction pattern suitable for most identification purposes in 15 minutes. If we use the slowest speeds, individual peaks are spread out and precise measurements can be made. One disadvan-

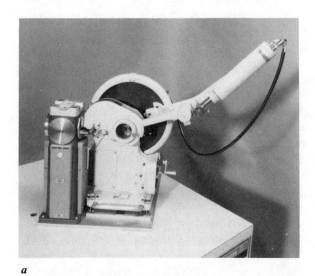

a

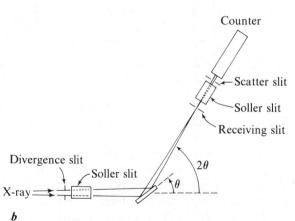

b

FIGURE 10.19 **(a) An x-ray diffractometer used for standard x-ray powder work. (Photograph courtesy of** **Phillips Electronic Instruments.) (b) Geometry of the x-ray diffractometer.**

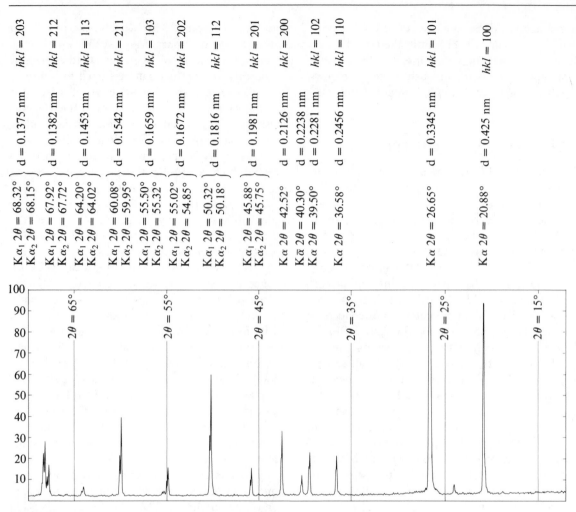

FIGURE 10.20   Strip chart recording of the powder diffraction spectrum of quartz. Cu *Kα* radiation.

tage of diffractometers is that they require a much larger sample than is needed for a powder camera. We can prepare a ball mount from a single grain, or even part of a grain, plucked from a petrographic thin section. Another disadvantage of diffractometers is the limited information provided on back reflections and on peaks that have weak intensities.

ENERGY DISPERSIVE POWDER DIFFRACTION

As mentioned earlier in the chapter, an energy dispersive diffraction spectrum looks very similar to the usual diffraction profile, except that energy rather than the angle of θ is the measured variable. The energy dispersive method has two major advantages. First, the geometry of the experiment is extremely simple, because there are no moving parts. The θ angle is fixed at some predetermined value, and both the sample and the solid state detector are stationary. This is necessary for some

experiments in which the sample is subjected to high static pressures and laser heating, and the specimen cannot be moved as in other powder methods. Some experimental devices impose another mechanical limitation, namely, that if a conventional film or diffractometer method is used, only a part of the usual range of θ can be scanned.

The second advantage is speed. All parts of the energy spectrum are recorded simultaneously by the energy dispersive detector and are stored in a multichannel analyzer. The usual output mode is a visual display on a cathode ray tube (CRT) where we can observe the photon energies from each diffraction accumulating over time. We frequently obtain a usable diffraction spectrum suitable for identification purposes within a single minute. The practical advantages of such short times are apparent in experimental situations when we have difficulty controlling temperature or pressure conditions or both on a sample for longer periods of time. With the energy dispersive detector we can also record rapidly the successive powder patterns of a sample subjected to a series of pressure or temperature conditions or both. In this way, we determine important physical properties such as thermal expansion and compressibility.

The basic relationship between photon energy (measured in keV) and wavelength (measured in nm) of an x-ray diffraction is stated in Equation 10.7. Substituting the Bragg equation (Equation 10.5) for wavelength λ,

$$d(\text{nm}) = \frac{1.2396}{2\,E(\text{keV})\sin\theta} \qquad (10.10)$$

$$= \frac{0.6199}{E(\text{keV})\sin\theta}$$

With this equation, the various photon energies of diffraction from each lattice plane can be easily converted to d values of the customary x-ray powder pattern. A simple graphical conversion of photon energies to interplanar spacings for different fixed values of θ is provided in Figure 10.21.

A disadvantage of the energy dispersive diffraction method is that it is relatively new, first introduced in the early 1960s, and is not as commonly available as other powder diffraction equipment. The method is costly, mainly because of the so-

phisticated electronics necessary to make accurate and high resolution measurements. Another disadvantage is the increased difficulty of obtaining high accuracy of lattice spacing determinations by energy dispersion because of the additional components, such as elemental fluorescence, that appear in the spectrum. Nonetheless, with state-of-the-art detectors with energy resolutions of nearly 140 eV, we can obtain d values accurate to a few parts in 10,000.

IDENTIFICATION OF POWDER PATTERNS

The most common use of x-ray powder diffraction patterns is for the identification of minerals. Since every mineral has a distinct chemical composition and structure, every mineral must have a distinct powder pattern. In practice, however, the differences between powder patterns of some minerals are so small that sophisticated computer methods of profile fitting may be required before we can make unambiguous identifications.

The locations of diffraction arcs on powder photographs and the diffraction peaks on diffractometer chart paper provide all the information necessary to calculate interplanar spacings.

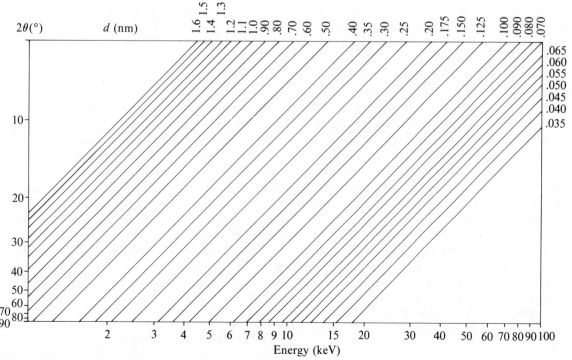

FIGURE 10.21   Graphical conversion of x-ray powder diffraction in terms of energy E to diffraction in terms of θ.

The spacings depend solely on the translational symmetry of the structure and therefore provide information only on the lattice. This is commonly sufficient to identify many minerals. Usually this information can and frequently does need to be supplemented by the relative *diffraction intensity* of each mineral, a property that depends both on structure and chemical composition and on many additional factors to be discussed. To a first approximation, however, diffraction intensity depends on volume. The larger the amount of powder, the greater the intensity. Consequently, the *relative* intensities *I* of the diffraction arcs and peaks are used along with the *d* values for routine mineral identification.

With practice, relative intensities can be visually estimated with reasonable accuracy. A more precise method is to use a photodensitometer for powder photographs and a planimeter for diffractometer peaks. Of particular interest is the *peak height intensity*, either the darkest portion of a diffraction arc on film or the highest point on a diffraction peak. Measurements made to these points are usually affected by numerous other factors that should be corrected for, if we are to achieve the best results. These factors include arc and peak broadening with decreasing particle size, instrumental peak broadening in diffractometers at high values of θ, and thermal vibrations among atoms of the sample. None of these factors significantly affects the total diffracted energy, so the area beneath a peak on a diffractometer chart remains proportional to the diffracted energy. Intensities measured in this manner are referred to as *integrated intensities*.

The relative intensities of diffraction arcs and diffractometer peaks are commonly expressed on a linear scale from zero (no intensity above background) to 100 (most intense). A complete record of a powder pattern should include all *d* values listed in order of decreasing magnitude, and all *I* values from zero to 100. Complete records of powder patterns can be lengthy, and parts of some patterns cannot always be obtained for all minerals because of factors mentioned previously. Often, only the three to eight strongest diffractions are required for positive identification, although all observed lines should be matched.

The most extensive and up-to-date record of powder patterns is maintained by the International Center for Diffraction Data. This record, earlier known as the ASTM file, is now known as the *X-ray Powder Diffraction File* (PDF) and is available at most university and industrial libraries. The basic file consists of a set of 3×5 in. cards, one for each mineral described. The complete powder pattern and other relevant chemical and crystallographic data along with references are given. In the top left-hand corner (Figure 10.22), the PDF set and sequence number are identified, and under that the *d* and *I* values of the three most intense peaks and the *d* and *I* of the lowest observed diffraction are listed. A special symbol in the upper right-hand corner indicates the relative reliability of the data. A star denotes high reliability. Under these symbols the chemical formula and mineral name are given. In the most modern laboratories, the entire PDF is stored on magnetic tape for computer search and retrieval.

Several search manuals are available to complement the PDF. The entries are grouped in ranges of *d* values of the strongest peak, and within each range, in order of decreasing *d* of the second and third strongest peaks. Books of the PDF cards divided into minerals and inorganic and organic compounds are also available. In addition, separate books are devoted to just the silicate minerals and the ore minerals.

The normal procedure for routine identification begins in the PDF search manual with the selection of those minerals that have the same three strongest diffraction lines as the unknown. These are narrowed down to a final choice by matching the *d* and *I* values of the remaining diffractions. The process can often be shortened by scanning the list for those common minerals one might expect based on other criteria such as association. When a match is found, any ambiguity is rare, although we must keep in mind that solid solution effects can cause small differences between the *d* and *I* values observed and those reported in the PDF catalog.

In the most modern x-ray diffraction laboratories, the precision of the *d* and *I* measurements and the accuracy of mineral identification are improved by more sophisticated electronic equipment and computerized controls and calculations. The simple measurement of the position and the intensity of the diffraction peaks is replaced by the calculated profile of peaks and the mathematical determination of their centroids and intensities. Unambiguous identification of a mineral is made by comparing the profile fitting of the unknown and the standard patterns.

ANALYSIS OF MIXED POWDER PATTERNS

We can often identify several minerals in a mixed sample by consecutively eliminating the diffractions of each from the total spectrum. This

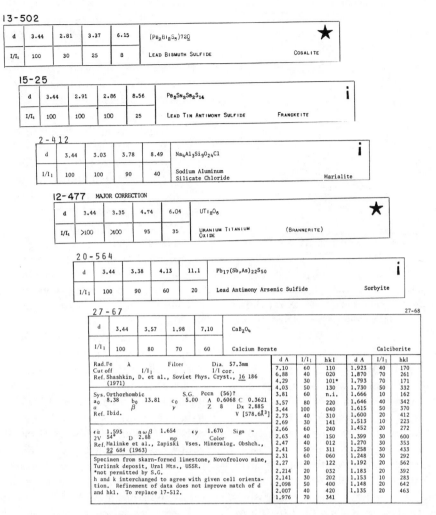

13-502

d	3.44	2.81	3.37	6.15	$(Pb_2Bi_2S_5)7\underline{2O}$	★
I/I_1	100	30	25	8	LEAD BISMUTH SULFIDE COSALITE	

15-25

d	3.44	2.91	2.86	8.56	$Pb_5Sn_3Sb_2S_{14}$	i
I/I_1	100	100	100	25	LEAD TIN ANTIMONY SULFIDE FRANCKEITE	

2-412

d	3.44	3.03	3.78	8.49	$Na_4Al_3Si_9O_{24}Cl$	i
I/I_1	100	100	90	40	Sodium Aluminum Silicate Chloride Marialite	

12-477 MAJOR CORRECTION

d	3.44	3.35	4.74	6.04	UTi_2O_6	★
I/I_1	>100	>100	95	35	URANIUM TITANIUM OXIDE (BRANNERITE)	

20-564

d	3.44	3.38	4.13	11.1	$Pb_{17}(Sb,As)_{22}S_{50}$	i
I/I_1	100	90	60	20	Lead Antimony Arsenic Sulfide Sorbyite	

27-67   27-68

d	3.44	3.57	1.98	7.10	CaB_2O_4	
I/I_1	100	80	70	60	Calcium Borate Calciborite	

Rad. Fe λ	Filter	Dia. 57.3mm	d Å	I/I_1	hkl	d Å	I/I_1	hkl
Cut off I/I_1		I/I cor.	7.10	60	110	1.923	40	170
Ref. Shashkin, D. et al., Soviet Phys. Cryst., 16 186			6.88	40	020	1.870	70	261
(1971)			4.29	30	101*	1.793	70	171
			4.03	50	130	1.730	50	332
Sys. Orthorhombic S.G. Pccn (56)?			3.81	60	n.i.	1.666	10	162
a_0 8.38 b_0 13.81 c_0 5.00 A 0.6068 C 0.3621			3.57	80	220	1.646	40	342
a β γ Z 8 Dx 2.885			3.44	100	040	1.615	40	370
Ref. Ibid.		V [578.6Å³]	2.73	40	310	1.600	20	412
			2.69	30	141	1.513	10	223
$\epsilon\alpha$ 1.595 $n\omega\beta$ 1.654 $\epsilon\gamma$ 1.670 Sign −			2.66	60	240	1.452	20	272
2V 54° D 2.88 mp Color			2.63	40	150	1.399	30	600
Ref. Malinke et al., Zapiski Vses. Mineralog. Obshch.,			2.47	40	012	1.270	30	353
92 684 (1963)			2.41	50	311	1.258	30	433
Specimen from skarn-formed limestone, Novofrolovo mine,			2.31	60	060	1.248	30	292
Turlinsk deposit, Ural Mts., USSR.			2.27	20	122	1.192	20	562
*not permitted by S.G.			2.214	20	032	1.183	20	392
h and k interchanged to agree with given cell orienta-			2.141	30	202	1.153	10	283
tion. Refinement of data does not improve match of d			2.098	50	400	1.148	20	642
and hkl. To replace 17-512.			2.007	40	420	1.135	20	463
			1.976	70	341			

FIGURE 10.22   A portion of the powder diffraction file (PDF) reprinted courtesy of the International Center for Diffraction Data. Note that *d* values (interplanar spacings) are given in Ångstroms (1Å = 0.1 nm).

procedure can be complicated by the fact that the individual spectra overlap, causing some diffractions to be broader and more intense than the *d* and *I* values on the appropriate PDF card indicate. Furthermore, weak diffractions in some minerals may be totally camouflaged by strong diffractions of others. Determining the *relative quantities* of the minerals present is a still more difficult task, because the diffraction intensities are not directly proportional to the amounts of the different minerals in the mixture.

One starting point for the quantitative analysis of a mixed powder pattern is to prepare a series of mixtures consisting of the minerals we know or suspect to be present. By varying the relative amounts of the minerals in each mixture, and then by determining their diffraction patterns under the same instrumental conditions, we can calibrate the intensity of one or more peaks for comparison with an unknown. The reliability of our comparison increases with the number of diffrac-

tions included in the calibration. In practice, the intensity of one or two peaks that show a variation in proportion to the amount present is used and compared with the intensity of the same peaks in the unknown mixture. This is a routine procedure when the constituent minerals, such as the various mixed-layer clays, are difficult or impossible to separate from each other.

We can use a similar calibration procedure to determine the effects of chemical substitutions on such solid solution series as the plagioclase feldspars and olivines. Figure 10.23 illustrates how the relative intensities and variations in *d* values of several different diffractions in the olivines vary with the amount of iron and magnesium substitution. In this case, all of the intensities are normalized against the intensity of one diffraction, I_{112}, which is the strongest in all of the patterns.

Another method of quantitative analysis is to calculate the absolute diffraction intensities for the powder patterns of all well-known crystal

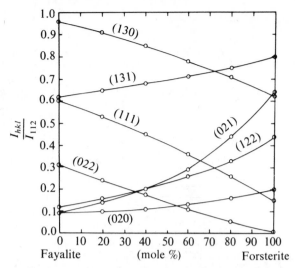

FIGURE 10.23 **Variation of the calculated intensity ratios with composition for several of the reflections in olivines. (After Jahanbagloo, I. C., 1969. X-ray diffraction study of olivine solid solution series.** *American Mineralogist* **54:250.)**

structures of interest. Done properly, these calculations take into account all of the various factors that influence intensity, such as x-ray absorption effects, multiplicity of diffraction peaks, Lorentz polarization, and thermal vibration effects. Fortunately, these calculated patterns are available for many minerals, including all of the entries in the PDF marked with a "C" in the upper right-hand corner. Instead of giving the absolute intensities, the relative intensities on a scale of 0 to 100 are calculated and listed to make direct comparison of patterns with experimental observations easier. A conversion factor is given, however, that allows the conversion of relative intensities to a scale on which the intensities are proportional to their relative amounts in a mixed pattern. This conversion factor is called the *normalization factor* (NF), or when divided by 100, it is called the *absolute scale factor* (ASF).

We use files of calculated patterns for direct quantitative analysis, because the observed intensities (properly corrected for the variables) can be converted to an absolute scale on which the total diffraction energies are proportional to the relative quantities of the component minerals. If we use films for quantitative analysis, overlapping diffraction lines should be omitted, and the absorption coefficient should be modified for the relative quantity of the component minerals in subsequent calculations. In analyzing a diffractometer pattern, the profile-fitting method offers special advantages, because the method separates overlapping peaks into components and thus verifies the presence of the peaks from different phases.

Given these data, we proceed with the quantitative analysis of a mixed pattern (Table 10.2) using the following steps:

1. Select only the well-resolved (no overlap) diffractions of each mineral, and sum their integrated intensities, $\sum I_{obs}$.
2. Sum the intensities, $\sum I_{std}$, of the same diffractions in the standard (preferably a calculated standard) pattern of each mineral.
3. Calculate the ratio of these sums, $\sum I_{obs} / \sum I_{std}$, for each mineral, and multiply it by the appropriate ASF times 100.
4. Divide the ratio for each mineral by the sum of these ratios for all of the minerals. Multiply by 100, and we have the volume percentage of each mineral in the mixture. We can convert these values to conventional weight percentages by multiplying by each formula weight and renormalizing to a weight percent total of 100.

Keep in mind that the reliability of these procedures is dependent on the quality of the crystal and its diffraction pattern. Natural mixtures are far more complicated to calibrate than synthetic mixtures prepared in the laboratory, because in natural mixtures we must deal with the combined effects of solid solution on each of the minerals, with variable particle size, with sample randomness, and with the presence of small amounts of undetected minerals. In spite of these difficulties, the analysis of mixed powder patterns may be our only recourse when the problem at hand involves, for example, very fine-grained sediments.

INDEXING POWDER PATTERNS

Each diffraction arc on a powder photograph and each diffraction peak on a diffractometer chart represents diffraction from a lattice plane that can be indexed. Given the lattice parameters, namely, a, b, c, α, β, and γ, the indices of any d may be determined by the general relationship

$$d_{hkl} = \sqrt{\frac{A}{B}} \qquad (10.11)$$

where

$$A = 1 - \cos^2\alpha - \cos^2\beta - \cos^2\gamma + 2\cos\alpha\cos\beta\cos\gamma$$

and

$$B = h^2\sin^2\alpha/a^2 + k^2\sin^2\beta/b^2 + l^2\sin^2\gamma/c^2$$
$$+ 2hk(\cos\alpha\cos\beta - \cos\gamma)/ab$$
$$+ 2hl(\cos\alpha\cos\gamma - \cos\beta)/ac$$
$$+ 2kl(\cos\beta\cos\gamma - \cos\alpha)/bc.$$

TABLE 10.2. *Procedure for Semiquantitative Analysis of Mixed Powder Patterns by Use of Calculated ASF Values of Separate Minerals*

Observed Pattern*		Halite**		Fluorite†		Rutile‡	
d (nm)	I	d (nm)	I	d (nm)	I	d (nm)	I
.315	50			.3153	94		
.282	100	.2821	100				
.249	17					.2487	50
.2296	3					.2297	8
.2188	5					.2188	25
.2055	3					.2054	10
.1994	54	.1994	55				
.1931	49			.1931	100		
.1688	18					.16874	60
.1647	16			.1647	35		
.1480	3					.14797	10
.1453	3					.14528	10
.1410	8	.1410	6				
.1261	18	.1261	11				
.1115	10			.1115	16		
.1051	4			.1051	7		
.0997	4	.09969	2				
.0940	8	.09401	3				
.0923	5			.09233	7		
$\sum I_{obs}$			192		134		52
$\sum I_{std}$			177		252		173
ASF			0.32		0.42		0.49
$\sum I_{obs}/\sum I_{std}$			1.08		0.53		0.30
$\sum I_{obs}/\sum I_{std} \times$ ASF			0.35		0.22		0.15
% volume			49		31		20
% weight			42		35		23
Supposed to be:							
% weight			50		30		20
By using I values of calculated pattern and two cycles of correction for absorption:							
% weight			48.9		28.8		22.3

*Limited to well-resolved diffractions (from powder photograph of mixture).
**From PDF card No. 5-628
†From PDF card No. 4-864
‡From PDF card No. 21-1276

This equation is simplified when all of the axial angles are 90 degrees, as is the case for isometric, tetragonal, and orthorhombic minerals, or when two angles are 90 degrees or 120 degrees, as in hexagonal minerals. Equation 10.11 describes the position of a lattice plane (*hkl*) in the *direct lattice*, that is, the normal array of lattice points as discussed in Chapter 2.

A simpler approach than Equation 10.11 is to view the lattice in a way that uses more directly the actual parameters measured in an x-ray experiment. The position of a plane in the direct lattice can be described in different ways, which we discussed in previous chapters. Direct intercepts and reciprocal intercepts (Chapter 2) work well for some purposes, the face normal (Chapter 3) works well for others and especially well here. Imagine the face normal (the pole) to a plane drawn out from the plane to a distance inversely proportional to *d*, the distance to the next plane parallel to it. In other words, every plane in the direct lattice can be represented by a point (at the end of each face normal) removed from the plane by the reciprocal of its interplanar spacing d. The resulting collection of points defines a three-dimensional array called the *reciprocal lattice*. Like the direct lattice, the reciprocal lattice represents translational symmetry. In addition, it represents all of the interplanar spacings as well as the relative inclination of those planes in much the same way as the external faces of a crystal are described by projection (Chapter 3). Each point in the reciprocal lattice represents a lattice plane. The reciprocal lattice consequently represents an array of translationally equivalent lattice planes.

We can better understand these concepts by converting a two-dimensional direct lattice into its reciprocal lattice (Figure 10.24a and Table 10.3). The direct lattice has unit translations a (0.28 nm) and b (0.42 nm) inclined to each other by the angle γ (75 degrees). Three lattice planes (100), (010), and (110) and their interplanar spacings d_{100} (0.27 nm), d_{010} (0.41 nm), and d_{110} (0.26 nm) are also shown. Now imagine each of the planes positioned at a common origin by normal translation. Perpendicular to each of these planes draw a line d^* proportional in length to the reciprocal of the d values given above (i.e., $d_{hkl}^* = 1/d_{hkl}$). To retain a convenient scale, the unit distance in the reciprocal lattice is reduced by one tenth of that in the direct lattice. For example, $d_{100}^* = 1/(0.27)(10) = 0.37$ RLU (reciprocal lattice units), $d_{010}^* = 0.25$ RLU, and $d_{110}^* = 0.39$ RLU. The end of each of these lines is a reciprocal lattice point. These points make up the reciprocal unit cell in Figure 10.24b and define the reciprocal unit translations a^* and b^*.

The resulting lattice (Figure 10.24b) represents each of the lattice planes and their interplanar spacings in a simple three-dimensional pattern. The planes and their d values are precisely the two geometric components of the lattice determined by diffraction. For convenience, unit length in the

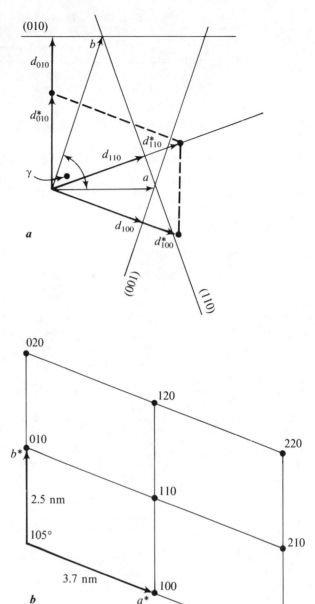

FIGURE 10.24　**(a) Reciprocal lattice with direct lattice vectors (d_{hkl}) and reciprocal lattice vectors (d_{hkl}^*). (b) Indexing of reciprocal lattice points. Scale: 1 nm = 1 cm.**

TABLE 10.3.　*Equations for Conversion of Direct Unit Cell Parameters to Reciprocal Values*

	Reciprocal Unit Translations	Reciprocal Lattice Angles
	$a^* = \dfrac{bc \sin \alpha}{V}$	$\cos \alpha^* = \dfrac{\cos \beta \cos \gamma - \cos \alpha}{\sin \beta \sin \gamma}$
	$b^* = \dfrac{ac \sin \beta}{V}$	$\cos \beta^* = \dfrac{\cos \alpha \cos \gamma - \cos \beta}{\sin \alpha \sin \gamma}$
	$c^* = \dfrac{ab \sin \gamma}{V}$	$\cos \gamma^* = \dfrac{\cos \alpha \cos \beta - \cos \gamma}{\sin \alpha \sin \beta}$
	$V = abc\sqrt{1 - \cos^2 \alpha - \cos^2 \beta - \cos^2 \gamma + 2 \cos \alpha \cos \beta \cos \gamma}$	

reciprocal lattice is defined as the wavelength of the radiation, λ, and hence

$$\frac{1}{d_{hkl}} \quad \text{becomes} \quad \frac{\lambda}{d_{hkl}} = \sigma_{hkl} \qquad (10.12)$$

Similarly, the Bragg equation (Equation 10.5) can be expressed in terms of the reciprocal lattice as

$$\sigma_{hkl} = 2 \sin \theta \qquad (10.13)$$

This expression (Equation 10.13) means, for example, that diffraction from the (010) lattice plane will occur when σ_{010} is equal to $2 \sin \theta$ as illustrated in Figure 10.25. That condition is satisfied when the reciprocal lattice point P_{010} is on the surface of the *sphere of reflection*, as all points in such position satisfy the condition. The radius of the sphere of reflection is 1 RLU and the distance AB is equal to 2 RLU. The angle between AP and AB is equal to θ, and that between CP and CB is equal to 2θ. The Bragg equation is satisfied for P_{010}, because $\sigma_{010} = 2 \sin \theta$ in the triangle APB.

This illustration of the Bragg equation and the reciprocal lattice (Figure 10.25) also offers a visual interpretation of the orientation of the crystal and of the direction of diffraction. The direction AP is

perpendicular to σ_{hkl} and therefore is parallel to the diffracting lattice plane. The direction CP then defines the direction of diffraction. The satisfaction of the Bragg equation, the orientation of the diffracting lattice plane, and the direction of diffraction are illustrated for P_{010} and $P_{3\bar{2}0}$ in Figure 10.25. The definition of $d_{\min}$ (page 240) in reciprocal space becomes the *limiting sphere*, with a radius of 2 RLU.

The relationship between the d_{hkl} values of the x-ray powder pattern and the σ_{hkl} values of the reciprocal lattice are illustrated in Figure 10.26. If the lattice of the crystal is known, the indexing of the diffractions becomes a simple routine procedure. The rather cumbersome direct space equation (Equation 10.11) expressing the value of d in terms of lattice parameters now becomes a relatively simple equation in reciprocal space:

$$\left(\frac{1}{d_{hkl}}\right)^2 = Q_{hkl} = h^2 a^{*2} + k^2 b^{*2} \qquad (10.14)$$

$$+ l^2 c^{*2} + 2klb^* c^* \cos \alpha^*$$
$$+ 2hla^* c^* \cos \beta^* + 2hka^* b^* \cos \gamma^*$$

For *isometric* minerals in which $a = b = c$ and

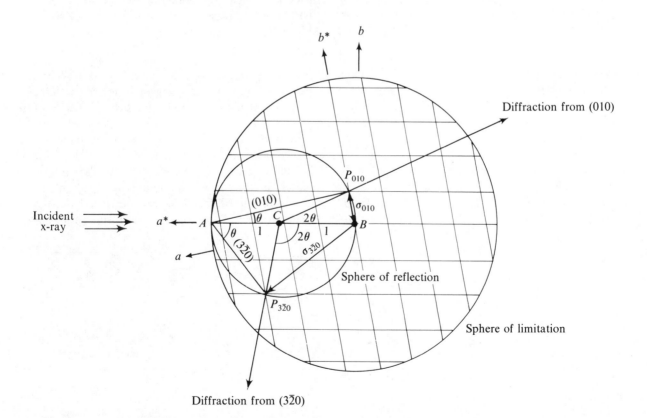

FIGURE 10.25 Geometry of the Bragg equation in reciprocal space. The position of the sphere of reflection (i.e., the incident angle of the x-ray beam) is appropriate in this diagram for reflection from the 010 and 320 reciprocal lattice points (i.e., lattice planes in direct space).

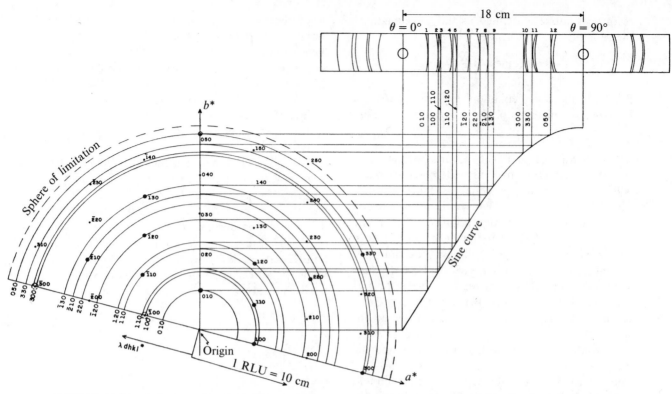

FIGURE 10.26   Indexing of an x-ray powder pattern with known lattice constants. Only (*hk*0) diffractions are shown for clarity. Reciprocal lattice points that diffract are circled. *a* = 0.28 nm, *b* = 0.42 nm, γ = 75°, Cu *Kα* radiation, 114.6 mm camera diameter. (After Zoltai, Ti-bor, and Navarro, E. F., 1981. Determinacion de indice en patrones de polvo. Introduccion de un nuevo metodo para determinar los indices en patrones isometricos, tetragonales y hexagonales desconocidos. *GEOS*, no. 26, Fig. 2, p. 48.)

α = β = γ = 90 degrees, Equation 10.14 becomes

$$Q_{hkl} = (h^2 + k^2 + l^2)a^{*2}$$

Taking the logarithm of both sides,

$$\log Q_{hkl} = \log(h^2 + k^2 + l^2) + \log a^{*2} \qquad (10.15)$$

In this equation, the first term on the right-hand side is independent of the value of *a**. That is, the first term is constant for a given (*hkl*) in all isometric powder patterns. Consequently, we can prepare a simple logarithmic scale (Figure 10.27) that will allow the indexing of isometric patterns and the calculation of *a*. In practice, the values of Q_{hkl} (i.e., $1/d_{hkl}^2$) are calculated for each observed diffraction and plotted on a strip of paper as shown in Figure 10.27b. The strip is slid along the isometric indexing scale (Figure 10.27a) until a match is found. In the matching position the indices of the diffractions can be read from the scale, and $\log a^{*2}$ is equal to the displacement of

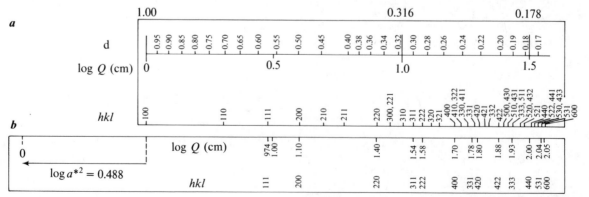

FIGURE 10.27   (a) The isometric (cubic) indexing scale and (b) procedure for indexing halite (NaCl). Log a^{*2} = 0.448 and *a* = 0.564 nm.

the 100 mark of the scale from the zero position of the strip.

Equation 10.14 for *tetragonal* crystals becomes

$$Q_{hkl} = (h^2 + k^2)a^{*2} + l^2 c^{*2}$$

or

$$\log Q_{hkl} = \log Q_{hk0} + \log Q_{00l} \qquad (10.16)$$

Note that in this equation the $hk0$ and $00l$ contributions are separated. We can obtain the indices for the d_{hk0} diffractions from the isometric indexing scale, provided that we consider only the $h00$ and $hk0$ marks. The $hk0$ indices and the value of a can be obtained from the $hk0$ marks of the isometric scale. Using Equation 10.16 and the Q values of the known $hk0$ diffractions, we can determine the magnitude of $\log c^*$.

Hexagonal powder patterns are indexed like tetragonal patterns, but we use an hexagonal $hk0$ scale constructed in accordance with the reciprocal lattice equation of hexagonal crystals:

$$Q_{hk0} = (h^2 + k^2 + hk)a^{*2}$$

$$\log Q_{hk0} = \log(h^2 + k^2 + hk) + \log a^{*2} \qquad (10.17)$$

Orthorhombic powder patterns that contain several $(h00)$, $(0k0)$, $(00l)$, $(hk0)$, $(0kl)$, and $(h0l)$ diffractions may be indexed by using the tetragonal scale three different times, once for each unit translation. Graphical indexing of *monoclinic* and *triclinic* powder patterns is impractical because of the larger numbers (4 and 6, respectively) of unknown lattice parameters. Analytical methods based on the general reciprocal lattice equation (Equation 10.14) must be used instead. One such method introduced by Tei-Ichi Ito assumes that the first two diffractions (the largest d values) are from the two longest interplanar spacings of the unit cell. The unknown lattice is assumed to have a primitive cell, in which case the two diffractions can be labeled d_{100} and d_{010}. Values of a^* and b^* are calculated by the relationships

$$Q_{110} = h^2 a^{*2} \quad \text{and} \quad Q_{010} = k^2 b^{*2}$$

If the value of Q_{110} is calculated from a^*, b^*, and γ^* of 90 degrees (Equation 10.14), no diffraction corresponding to that value will generally be found. Instead, diffractions corresponding to Q_{110} and $Q_{1\bar{1}0}$ for the real γ^* may be found at equal distances to both sides. The corresponding analytical expressions are

$$Q_{110}(\gamma^* \neq 90°) = a^{*2} + b^{*2} + 2a^*b^* \cos\gamma^*$$

$$Q_{110}(\gamma^* = 90°) = a^{*2} + b^{*2}$$

$$Q_{1\bar{1}0}(\gamma^* \neq 90°) = a^{*2} + b^{*2} - 2a^*b^* \cos\gamma^*$$

$$(10.18)$$

The difference between the right-hand sides of the nonorthogonal and orthogonal Q_{110} values is equal to $2a^*b^* \cos\gamma^*$, and therefore γ^* can be determined. If we find no Q_{110} and $Q_{1\bar{1}0}$ pairs, we can use any other d^*_{hk0} and d^*_{hk0} pairs.

The next step is to calculate c^* by assuming that the third largest d of the unknown pattern is d_{001}. If this d value happens to be coplanar with a^* and b^* ($d_{1\bar{1}0}$), the next largest d is chosen and c is calculated using Equation 10.14. At this point, only α^* and β^* remain unknown. They can be calculated from Q_{0kl} and Q_{h0l} types of diffractions in the same way we calculated γ^*.

When all six triclinic parameters are known, all diffractions can be indexed with Equation 10.14 by simply inserting different integer values for h, k, and l. The unit cell for which these determinations were made is *primitive*, and may or may not be the *reduced cell*. Several simple methods are available to determine the reduced (and multiple, if appropriate) cell from the primitive cell obtained by the Ito method, and to transform the indices to correspond with the new cell. Discussion of these methods, however, is beyond our scope.

Improved variations of the Ito method have been programmed for computers, and can be used efficiently and reliably for the indexing of powder patterns. Indeed, the widespread availability of computers has almost eliminated the use of graphical methods.

REFINEMENT OF LATTICE PARAMETERS

Once we have achieved optimal quality of the sample mount, determined the wavelength of radiation, and selected the instrumentation, we will obtain the most accurate lattice parameters by relying on the back-reflection data. For reasons given earlier (e.g., Figure 10.17), the uncertainty of a d value decreases as the diffraction angle increases. In fact, the error becomes vanishingly small at $\theta = 90$ degrees because of the form of the $\sin\theta$ term in the Bragg equation. Moreover, errors due to sample characteristics, such as absorption and eccentric position of sample, are highly dependent on θ and will be reduced to near zero at $\theta = 90$ degrees. This is why the back reflections (small d values) are so important and why care should be taken to record them if possible.

Accordingly, various methods have been introduced that allow the extrapolation of diffraction data from low θ regions to the 90-degree value at which the data are intrinsically more accurate. In 1945, Taylor, Sinclair, Nelson, and Riley intro-

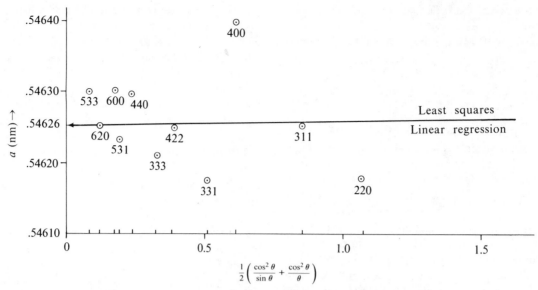

FIGURE 10.28   Extrapolation of diffraction data to $\theta = 90°$ **by the Nelson-Riley function to obtain precise lattice parameters. Example is fluorite with** $a =$ **0.54626 ± 0.00006 nm. Cu** $K\alpha$ **radiation. Diffractions with** $\theta < 18°$ **excluded.**

duced the most commonly used analytical device. This function is plotted for diffractions of fluorite (CaF_2) against its cubic cell edge a in Figure 10.28. The complicated form of the function is due to the empirical observation that the eccentricity error varies linearly with $\cos^2 \theta$, and the absorption error varies as a simple curve with $\cos^2 \theta$. When these errors are plotted against θ alone, no systematic trend is usually apparent, and no simple extrapolation to $\theta = 90$ degrees can be made. The Nelson-Riley function thus represents the "best" form we can use to make a *linear* extrapolation of the data. The scattering of the data points in Figure 10.28 is due to the actual uncertainties of each value and to the fact that the scale on the ordinate is greatly expanded. When sufficient data points are available, an analytical extrapolation, such as a least squares linear regression, can be used.

Today, most lattice parameter refinements are done by full matrix least squares refinements, which can be calculated speedily with computers. The statistical evaluations of the parameters, expressed in terms of the limits of reliability, are essential aspects of the refinement procedures. Low limits of reliability are not necessarily an indication of poor experimental procedures but may reflect instead the quality of the crystals used in the experiment.

SYMMETRIES OF POWDER PATTERNS

The usefulness of the powder method for determining symmetry is limited by two important factors. First, all lattice planes with the same interplanar spacing will diffract at the same angle and will therefore be represented by the same arc on a powder photograph. In a cubic crystal of $4/m\ \bar{3}\ 2/m$ symmetry, for example, {111} consists of eight symmetrically equivalent planes, namely, (111), ($\bar{1}$11), ($\bar{1}\bar{1}$1), ($\bar{1}\bar{1}\bar{1}$), (1$\bar{1}$1), (1$\bar{1}\bar{1}$), ($\bar{1}$1$\bar{1}$), and (11$\bar{1}$). In the 23 (cubic) point group, these planes have two different values of intensity, but their intensities cannot be distinguished in a powder photograph. Symmetries higher than the true symmetry of the crystal may consequently appear. The powder pattern of an orthorhombic crystal, for example, may be erroneously indexed as tetragonal if $a = b$. Similarly, no distinction can be made between the symmetries of $4/m$ and $4/m\ 2/m\ 2/m$ or other monaxial and multiaxial centric point groups.

By using single crystal and moving film methods discussed in the next section, we can eliminate the ambiguity in the symmetries of powder patterns due to overlap. Given even the true diffraction symmetry, the possible crystal symmetries are limited to the 11 centric point groups. The reason for this is inherent in all x-ray diffraction methods and is related to the so-called *phase problem*. All x-ray diffraction effects are centrosymmetric, that is, they possess a center of inversion regardless of whether the crystal has one or not. This fact is formalized as *Friedel's law*.

Despite these limitations, useful symmetry information can be obtained from powder photographs by the study of *systematic absences*, or *extinctions*, of certain types of diffractions. These extinctions are due to the presence of (1) unit cell centering, (2) glide planes, and (3) screw axes. When these symmetry elements are not present in the structure, all of the diffractions we would normally expect appear on a powder photograph. When the expected diffractions do not appear, we know that one or more of the three elements mentioned are present, and this knowledge helps us to establish or confirm the true crystal symmetry. For example, suppose that analysis of a powder pattern shows that $a = b$, and we index the diffractions as if the crystal were tetragonal. Absences in the pattern indicating the presence of a 4_1 or 4_3 screw axis would strongly support our choice, rather than a system of lower symmetry.

Systematic absences are generated when the translational component of lattice centering, glide planes, or screw axes causes a repetition of lattice planes at exactly half intervals of the normal interplanar spacing. For example, suppose plane A and plane B in Figure 10.29a are in a diffracting position. For this condition to exist, the planes must be tilted at just the appropriate angle, θ_1, to the incident beam, so the ray diffracted from plane B is exactly one wavelength behind the ray from plane A. If an additional diffraction plane, C, were to exist halfway between A and B, it, too,

would scatter the incident x-rays, but exactly 180 degrees ($\lambda/2$) out of phase relative to the scattered rays from A and B. Because the crests of one ray register with the troughs of each adjacent ray (along dashed line MN), the rays interfere destructively, and *no* diffraction is observed.

Extinction from this set of planes (e.g., $A, B, C, \ldots$) only occurs for odd orders of diffraction such as $n = 1, 3, 5$, and so forth. When n is even, as in Figure 10.29b where $n = 2$, diffraction occurs at one half of the interplanar spacing and at a different angle, θ_2. The diffracted ray from plane B is now 2λ behind that from A, and is naturally still in phase. Diffraction from an additional plane C is now an integer wavelength (1) behind that from A, and an integer wavelength (1) ahead of the ray from C. The rays all interfere constructively ($M'N'$), and diffraction is observed.

Figure 10.30a illustrates the effect of lattice centering in two dimensions. If a primitive unit cell is chosen for the lattice points, a_P and b_P are the unit translations that define d_{100} and d_{010}. Diffraction will be observed on a powder photograph corresponding to lattice planes with those indices. If, however, a centered multiple cell is chosen to describe the *same set* of lattice points, the lattice planes will index differently. The unit translations for the new choice of cell are a_m and b_m, and the new (100) and (010) planes are different from those defined on the basis of the primitive cell. Relative to the new cell, we now have an "additional" set of lattice planes, previously (110), that passes through the centered lattice points C and cuts a_m at exactly half the interplanar distance. These planes index as (200). Diffraction will occur for the set of lattice planes parallel to b_m ($h00$), but only when h is even.

The effects of glide planes (Figure 10.30b) requires similar cell centering. An a-axial glide in (001) will halve the a unit translation and will therefore generate "new" lattice planes interwoven with those indexed as (100). The "new" planes have (200) indices. Consequently, in the ($hk0$) and ($h00$) diffractions, the shortest d value is for $h = 2$, and diffractions with $h = 1$ are absent.

Screw axes (Figure 10.30c) have effects on powder patterns similar to glide planes. A 4_1 screw axis parallel to c produces lattice points at $c/4$ intervals in that direction. The shortest d values in (001)-type diffractions are consequently for $l = 4$. In this case, we would observe only diffractions corresponding to 004, 008, 0012, and so forth. If the screw axis were 2_1, lattice planes would exist at half intervals of c, and we would observe diffractions such as 002, 004, 006, and so forth.

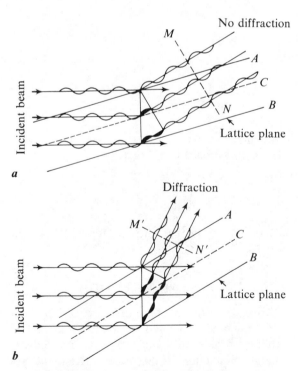

FIGURE 10.29 **(a) Systematic extinction of diffraction due to phase interference by lattice plane *C*. Scattered rays across line *MN* are out of phase for all odd values of *n*. (b) Observed diffraction for planes *A*, *B*, and *C* when *n* is an even integer. Scattered rays across line *MN* are in phase.**

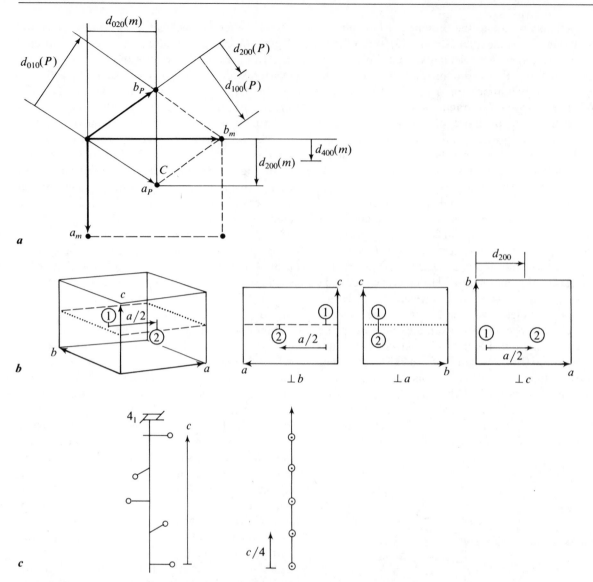

FIGURE 10.30 (a) A *c* projection of a *C*-centered unit cell and the absence of (100) and (010) lattice planes. (b) An *a* glide located in the (001) plane and the absence of (100) diffraction. (c) A 4_2 screw axis and effective translation of $c/4$ parallel to the *c* axis.

Table 10.4 summarizes the systematic absences for lattice centering, glide planes, and screw axes in various orientations. In using this table, always proceed first with the conditions for lattice centering, and keep in mind that some of the observed absences that are part of the more general conditions for lattice centering are also criteria for glide planes or for screw axes or both. For example, according to Table 10.4, the absence of $(h0l)$ diffraction for $h + k$ being odd indicates the presence of a diagonal glide n on (010). Diffractions such as (201) and (302) will thus be absent in the powder pattern. However, these same diffractions will also be absent if the unit cell is *C*-centered, because these absences are included in the more general condition that all (hkl) for odd $h + k$ be absent.

SINGLE CRYSTAL METHODS

Most of the shortcomings of the powder method can be eliminated by obtaining diffraction patterns from small, single crystals. The advantage of this method is that *each* lattice plane will yield a distinct diffraction. Several single crystal cameras and diffractometers are available, but we will mention only the most important types here. These include the Buerger precession camera (Figure 10.31a), which maintains parallelism between a selected lattice plane and the sheet of film and

TABLE 10.4. *Summary of Systematic Diffraction Absences due to Lattice Centering, Glide Planes, and Screw Axes*

Type of Absence	Index Type	Cause of Absence	Condition of Absence
Centered unit cell	hkl	A-centered cell	$k + l \neq 2n$
		B-centered cell	$h + l \neq 2n$
		C-centered cell	$h + k \neq 2n$
		I-centered cell	$h + k + l \neq 2n$
		F-centered cell	$h + k \neq 2n, h + l \neq 2n, k + l \neq 2n$
		R-centered cell	$-h + k + l \neq 3n$
Glide plane	$hk0$	a glide in (001)	$h \neq 2n$
		b glide in (001)	$k \neq 2n$
		n glide in (001)	$h + k \neq 2n$
		d glide* in (001)	$h + k \neq 4n$
	$h0l$	a glide in (010)	$h \neq 2n$
		c glide in (010)	$l \neq 2n$
		n glide in (010)	$h + l \neq 2n$
		d glide* in (010)	$h + l \neq 4n$
	$0kl$	b glide in (100)	$k \neq 2n$
		c glide in (100)	$l \neq 2n$
		n glide in (100)	$k + l \neq 2n$
		d glide* in (100)	$k + l \neq 4n$
	hhl**	c glide in (110)	$l \neq 2n$
		b glide* in (110)	$h \neq 2n$
		n glide* in (110)	$h + l \neq 2n$
		d glide* in (110)	$2h + l \neq 4n$
Screw axis	$h00$	2_1 along a axis	$h \neq 2n$
	$0k0$	2_1 along b axis	$k \neq 2n$
	$00l$	2_1 along c axis	$l \neq 2n$
		3_1 along c axis	$l \neq 3n$
		3_2 along c axis	$l \neq 3n$
		4_1 along c axis	$l \neq 4n$
		4_2 along c axis	$l \neq 2n$
		4_3 along c axis	$l \neq 4n$
		6_1 along c axis	$l \neq 6n$
		6_2 along c axis	$l \neq 3n$
		6_3 along c axis	$l \neq 2n$
		6_4 along c axis	$l \neq 3n$
		6_5 along c axis	$l \neq 6n$

NOTE: Rhombohedral crystals are considered in centered hexagonal system. The nonaxial screws in the isometric system are omitted.
*For certain centered lattices only.
**Limited to tetragonal and hexagonal systems.

will produce an enlarged, undistorted replica of the corresponding reciprocal lattice plane (Figure 10.31b). Each of these diffractions may be indexed, and the various σ_{hkl} values measured directly.

The Weissenberg camera produces a similar record of a reciprocal lattice plane by a synchronized rotation of the crystal and translation of the film. In both the Buerger and Weissenberg methods, the film is moving. Unlike a precession photograph, the diffraction spots on a Weissenberg photograph are distorted, but in a systematic way.

Both cameras are standard tools in modern crystallography.

Single crystal diffractometers are more sophisticated instruments than the two cameras. By simultaneously moving the crystal and an electronic detector, not only the position but the intensity of every diffraction can be determined. The most modern diffractometers are totally automated by computer; the researcher need only mount the crystal. The computer searches and finds each diffraction peak, indexes it, and measures its integrated intensity. If the apparatus is left on over-

a

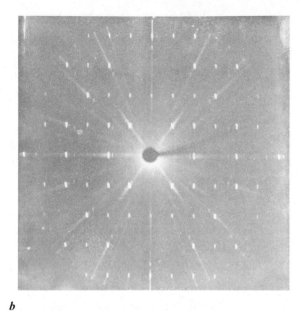

b

FIGURE 10.31 **(a) Buerger precession camera. (Courtesy of Charles Supper Co., Inc.) (b) Precession** photograph of the (*h*01) reciprocal lattice plane of wavellite.

night, 1000 or more diffractions can be routinely examined. The data are usually stored on a magnetic tape and then analyzed by various computer programs to produce the refined crystal structure.

THE CRYSTAL STRUCTURE

The direction of diffraction as given by the diffraction angle is dependent only on the lattice of the crystal. The intensity of the diffraction, however, is determined by both the scattering power and the arrangement of the atoms in the structure. When x-rays impinge on an atom, the electrons associated with that atom respond by oscillating with the same frequency as the x-rays. In a sense, each electron "scatters" a small percentage of the incident beam, and when all the scattered contributions are in phase, the coherent scattering known as diffraction occurs.

The *scattering power* of an atom is proportional to the number of its electrons, the wavelength of the radiation, and the angle of incidence. Atoms with high atomic numbers have large scattering powers; those with low atomic numbers have low scattering powers. The *intensity* of a diffracted ray is determined by the electron density in the lattice plane of interest, and that depends on the arrangement of atoms in the structure. When the diffraction intensity is high, the *amplitude* of the diffracted ray is large. The study of diffraction intensities thus tells us about crystal *structure*, that is, the actual arrangement of atoms.

The diffraction intensity is not simply the sum-

mation of all the diffractions from each lattice plane of a certain (*hkl*). Intensity depends also on the positions of atoms with respect to the diffracting planes. Figure 10.32 illustrates a simple structure in which heavy elements (large circles) are compounded with light elements (small circles). In Figure 10.32a, the (110) plane is in a diffracting position. Each plane consists of equal numbers of heavy and light elements, and even though the scattering powers of the two elements are different, the electron density of each plane is the same. The set of planes scatters coherently, and diffraction occurs. In Figure 10.32b, the (100) plane is in a diffracting position, but now the heavy atoms occupy one half of the planes, and the light atoms occupy the other half. Diffraction from the planes of light elements are out of phase with diffraction from the planes of heavy elements. Complete cancellation does not occur, because the scattering power of the light atoms is less. The observed intensity of the (110) planes is thus the sum of the two intensities.

It is a straightforward calculation to predict the intensity of a diffracted ray given the positions of the atoms in the unit cell, their scattering powers, the radiation wavelength, and other data. The task of solving an unknown structure is to deduce the atomic positions given only the diffraction intensities (and λ and so forth). This calculation is not the reverse of the intensity prediction. The problem is that from the intensities alone, we cannot know to what extent the contributing atoms, both light and heavy scatterers, are in phase. This is

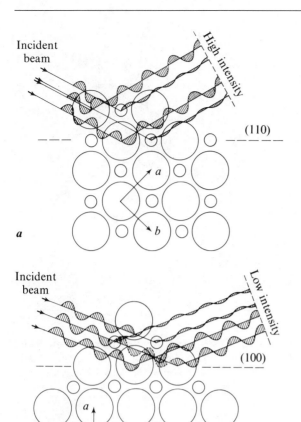

FIGURE 10.32 Variation of observed diffraction intensity depending on the relative positions of strongly scattering (large circles = high amplitude wave) and weakly scattering (small circles = low amplitude wave) atoms. (a) Relatively high intensity diffraction due to "in phase" positioning of strongly and weakly scattering atoms in the structure. (b) Relatively low intensity diffraction due to "out of phase" positioning in alternating (100) planes.

called the *crystallographic phase problem*. If its solution were known, the crystal structure could be easily and unambiguously solved. Instead, the solution often consists of making an educated guess at the structure, and then calculating what the diffraction pattern should be. If the calculated and observed intensities do not match, we make another guess and another until they do.

To see how this process actually works, consider the known structure of marcasite (FeS_2) shown in Figure 10.33. The origin of the lattice and its unit translations may be chosen in different ways. For our purposes, considering the lattice simultaneously located in six different positions, centered on each of six atoms in the unit cell ($Z = 6$), is convenient. For each choice, the atoms positioned at lattice points are separated by $n\lambda$

and therefore will scatter in phase at the appropriate diffraction angle. Because the interplanar spacing is obviously the same for the other five choices, atoms from those lattice points will also scatter in phase relative to each other at the same diffraction angle. The origins of each choice are *not* separated by $n\lambda$, however, so collectively the diffraction from the six sets of planes will not be in phase. The resulting intensity will depend on how the various diffracted rays interact, as illustrated in Figure 10.32.

For any particular choice of lattice, the phase difference between lattice planes will be 360 degrees, or 2π. The "in phase" condition for an ($hk0$) plane relative to the a and b unit translations is

$$\frac{a}{h} = 2 \quad \text{or} \quad a = 2h$$

and

$$\frac{b}{l} = 2 \quad \text{or} \quad b = 2l$$

The phase, or *phase angle P*, of the sulfur atom (S1, Figure 10.34) located at (x, y) is:

$$P_{S1} = ax + by = 2hx + 2ky = 2(hx + ky)$$

If we do this calculation for each of the six atoms, multiply by the appropriate scattering factors (Fe or S), and then sum, we obtain the *structure factor* F_{hk0}^2 for the marcasite structure. The calculated structure factor is a measure of the amplitude of the diffracted ray and is proportional to the square root of the predicted intensity. That is,

$$I \propto F_{hk0}^2 = \left[f_{Fe} \cos P_{Fe} + f_S (\cos P_{S1} + \cos P_{S2} \right.$$
$$\left. + \cos P_{S3} + \cos P_{S4}) \right]^2$$
$$+ \left[f_{Fe} \sin P_{Fe} + f_S (\sin P_{S1} + \sin P_{S2} \right.$$
$$\left. + \sin P_{S3} + \sin P_{S4}) \right]^2$$

where f_{Fe} and f_S are the atomic scattering factors, or the ratio of the amplitude of a ray from all of the electrons of an atom divided by the amplitude of a ray from a single free electron.

A graphical representation of the last equation is shown in Figure 10.34a for the (110) diffraction. The sine and cosine terms of the phase angles of each atom are added vectorially (Figure 10.34b) to yield the structure factor for (110). An identical procedure is followed for other (hkl) diffractions. In practice, the structure factor calculations are simplified by reducing the coordinates of the six atoms to just the one Fe and one S of the asymmetric portion of the unit cell. The phases of the other atoms are then calculated by generating equivalent positions of the space group. The gen-

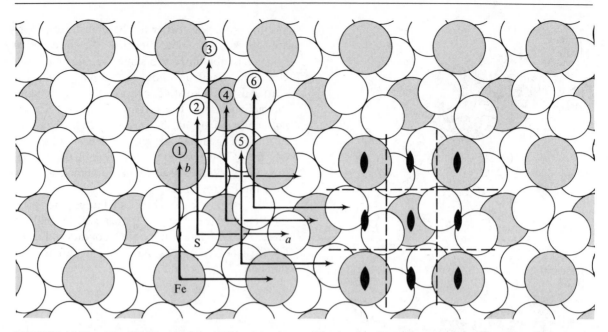

FIGURE 10.33 The [001] projection of the marcasite structure, its plane group symmetry, and six possible lattices with origins at atom centers. (After Buerger, Martin J., *Contemporary crystallography* (New York, 1970), Figure 5, p. 9. Copyright © 1970 and used with kind permission of McGraw-Hill Book Company.)

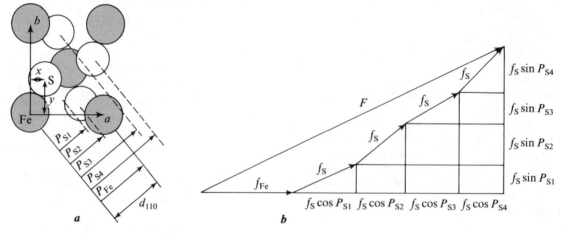

FIGURE 10.34 (a) The (111) lattice in the unit cell of marcasite showing the phases of Fe (positioned at 0) and S. (b) Vector sum F of the scattering powers of the six atoms in the d_{111} plane.

eralized equation for calculating the structure factor of an (hkl) plane for each atom j is

$$F_{hkl}^2 = \left[f_j \cos 2\pi (hx + ky + lz) \right]^2$$
$$+ \left[f_j \sin 2\pi (hx + ky + lz) \right]^2$$

If the crystal is centric, atoms at $(\bar{x}, \bar{y}, \bar{z})$ are symmetrically equivalent to atoms at (x, y, z). The intensities are also the same, but because the sine terms are opposite, cancellation occurs. For centric crystals, the calculation reduces to

$$F_{hkl} = 2 \left[f_j \cos 2\pi (hx + ky + lz) \right]$$

Before making these calculations, we must correct the observed intensities for systematic instru-

mental and sample-related distortions. These include polarization, the Lorentz factor (a distortion dependent on the instrument geometry), certain extinctions, multiplicity, and absorption. These corrected data are the set of diffraction intensities that will be matched by calculation when all of the atomic coordinates are properly assigned. Once these coordinates are known, the structure is solved. Interatomic distances are then calculated, which tells us something about the nature of chemical bonds. An accurate knowledge of interatomic distances is commonly used to determine the extent of chemical substitution on a particular site in the structure, for example, the exchange of Al and Si due to thermal disorder in feldspars.

CHAPTER ELEVEN

OPTICAL MINERALOGY

As we learned earlier, the unique combinations of crystalline structure and chemistry characteristic of every mineral determine its outward physical properties. Using our knowledge of properties such as cleavage and interfacial angles on a crystal, we can make intelligent statements concerning the symmetry of atomic arrangements at the unit cell scale. We could guess that other properties independent of a crystal's size will also reflect symmetry. The behavior of light as it is transmitted through a crystal is an important example. Knowledge of various optical properties easily determined with the aid of a petrographic microscope provides an accurate and unambiguous identification of all common rock-forming minerals. To this end, the crystal-chemical basis for the property of color in minerals, which was discussed at the end of Chapter 4, should be reviewed before continuing.

SAMPLE PREPARATION

In order to observe the optical properties of a mineral in transmitted light, the specimen must be thin enough for light to pass through it without significant reduction of intensity. The simplest way to prepare a mineral for petrographic observation is to crush a small portion in a mortar or between two glass slides. The crushed material is then gathered on one of the slides and a cover glass placed on top. The thickness of the crushed grains should be ideally between 0.10 and 0.15

mm. Next, place a single drop of immersion oil at the edge of the cover glass such that capillary action will draw the oil beneath the glass and immerse the grains. The sample is then ready for examination.

An alternative to oil immersion is the petrographic thin section. The thin section is prepared by first cutting off a small chip from a rock or mineral with a diamond saw. The flat surface of the chip is ground smooth with various abrasives on a circular wheel or lap, and then the chip is glued on a glass microscope slide and allowed to dry. Lastly, all but a 30 μm (0.03 mm) thickness is cut or ground away, leaving a very thin layer of mineral that will transmit light under the microscope. A cover glass is normally glued on top.

THE POLARIZING MICROSCOPE

The polarizing microscope (Figure 11.1) is the most useful instrument we have for determining the optical properties of minerals. To the trained eye, all of the common rock-forming minerals are familiar friends and can usually be identified on sight. The efficiency and speed of mineral identification is one of the most important advantages of the method. The principal disadvantage is that microscopic examination provides only indirect evidence of the chemistry and structure of minerals.

The primary function of the polarizing microscope is the same as that of any compound microscope. An objective lens on a rotating lens turret,

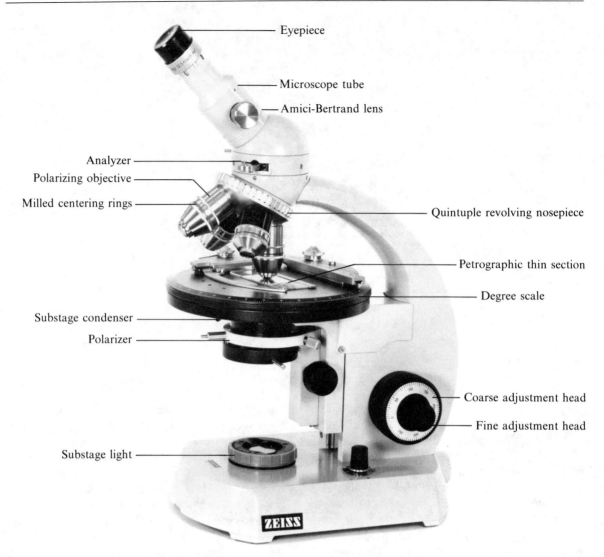

Eyepiece

Microscope tube

Amici-Bertrand lens

Analyzer

Polarizing objective

Milled centering rings

Quintuple revolving nosepiece

Petrographic thin section

Degree scale

Substage condenser

Polarizer

Coarse adjustment head

Fine adjustment head

Substage light

ZEISS

FIGURE 11.1   Components and accessories of a polarizing microscope. (Courtesy of Carl Zeiss, Inc.)

which is located directly above the object of interest, produces and magnifies an image that is further magnified by an ocular located at the observer's eye. The microscope has three or four objective lenses, commonly with magnifications of $3.5 \times$ (low power), $10 \times$ (medium power), and $40 \times$ (high power). The ocular will usually have either $8 \times$ or $10 \times$ magnification. The total magnification is the product of the ocular magnification and the objective magnification. For example, a $10 \times$ ocular in combination with the above objectives produces magnifications of $35 \times$, $100 \times$, and $400 \times$, respectively.

The polarizing microscope differs from an ordinary compound microscope in that it has a polarizer in the substage assemblage and an analyzer located in the vertical tube above the stage and sample. Both the polarizer and the analyzer (commonly referred to as the lower and upper nicol

after the inventor William Nicol) consist of synthetic Polaroid material oriented such that their polarizing directions are perpendicular to each other. Light originating from the substage illuminator is unpolarized, but after passing through the polarizer it vibrates exclusively in a specified direction within the vertical plane. If no sample is present on the microscope stage and if the analyzer is not inserted, the unaided eye cannot detect the polarized nature of the incident light. The intensity will be reduced because of absorption of the other components by the polarizer, but because the direction of preferred vibration of the analyzer is normal to that of the polarizer, insertion of the analyzer in the light path will result in complete absorption. We observe only darkness. The same effect is observed when a noncrystalline substance, such as glass, balsam, or a liquid, is placed on the stage and viewed with "crossed nicols." This is because such substances have no polarizing effect on the incident light. We use the polarizing microscope to greatest advantage in the study of minerals that do have a polarizing effect, because then certain components of light generated within the mineral will pass through the analyzer to be viewed.

RELATIONSHIPS BETWEEN OPTICAL PROPERTIES AND SYMMETRY

Before elaborating on the nature of light and its interaction with minerals, let us reconsider some general symmetry arguments developed in Chapter 6. We will first examine the optical effects produced by passing electromagnetic radiation (light) with a given symmetry through a crystal that has a generally different symmetry. This procedure is equivalent to examining the symmetry of an effect due to the interaction of a cause (light) and a mineral. By knowing the symmetry of the cause and by observing the symmetry of the effect, we deduce useful information concerning the symmetry of the crystal.

By analogy with the symmetry of thermal expansion discussed in Chapter 6, consider the symmetry of electromagnetic radiation emanating from an imaginary point source in a perfect vacuum. When the source is turned on, energy radiates in all directions. If we treat each ray as a vector, a sphere is generated after some arbitrary time. The situation is exactly analogous to the relationships shown in Figure 6.2a. Now consider the interaction of light with a cubic mineral having $4/m\overline{3}2/m$ point group symmetry. Symmetry arguments require that the *minimum* symmetry of the observable effect be the combination of those symmetry elements found both in the crystal and in the light. The observation made for all cubic minerals is that light travels with the same velocity in every direction. That is, we observe that cubic minerals have spherical symmetry. In practice, we do not measure the velocity directly. Instead, we find that the refractive index is the same for a given cubic mineral regardless of orientation. The important point at this stage is to recognize that we observe a higher symmetry effect than the crystal symmetry itself. Minerals for which light travels with the same velocity in all directions are said to be *optically isotropic*. This term is also used to describe all noncrystalline materials such as glass, air, and liquids that display these properties.

Most rock-forming minerals, however, are not isotropic, because they possess structural anisotropy in two directions. This causes the velocity of light to vary depending on the direction in which the light passes through the mineral. Consider, for example, the mineral zircon ($ZrSiO_4$), which has $4/m\,2/m\,2/m$ point group symmetry. Symmetry arguments applied to the interaction of light with this mineral require that the effect have at least $4/m\,2/m\,2/m$ symmetry, because each of the symmetry elements of this point group is obviously contained within the infinite symmetry of the cause. The visual observation is that zircon has two refractive indices, one for light that vibrates in the *ab* plane and another for light vibrating in a perpendicular direction parallel to the *c* axis. The result, shown in Figure 6.2b, is an ellipsoid of revolution with $\infty/m\,2/m\,2/m$ point group symmetry. Again, the effect has higher symmetry than the crystal. The same symmetry of effect is observed for all tetragonal, hexagonal, and trigonal minerals.

In contrast with zircon, forsterite (Mg_2SiO_4) with $2/m\,2/m\,2/m$ point symmetry is characterized by structural anisotropy in three mutually perpendicular directions rather than two as in the previous case. Consequently, forsterite has three different refractive indices corresponding to the three different velocities of light observed along each of the three crystallographic axes. The symmetry of this effect is orthorhombic ($2/m\,2/m\,2/m$), which corresponds to a triaxial ellipsoid. The same symmetry of effect is observed for all orthorhombic, monoclinic, and triclinic minerals. The relationship for the orthorhombic case is illustrated in Figure 6.2c.

By measuring refractive indices of minerals that crystallize in each of the 32 point groups, we discover that effect symmetries fall in the three categories summarized in Table 11.1. Recall from Chapter 6 that in terms of the crystal, the effect symmetries place only a maximum constraint on the possible crystal symmetry. That is, the *maxi-*

TABLE 11.1. *Summary of Relationships Between Cause Symmetry, Lattice Symmetry, and Effect Symmetry*

Cause Symmetry	Lattice Symmetry	Effect Symmetry
$\dfrac{\infty}{m}\dfrac{\infty}{m}\dfrac{\infty}{m}$	Isometric	$\dfrac{\infty}{m}\dfrac{\infty}{m}\dfrac{\infty}{m}$
$\dfrac{\infty}{m}\dfrac{\infty}{m}\dfrac{\infty}{m}$	Tetragonal Hexagonal Trigonal	$\dfrac{\infty}{m}\dfrac{2}{m}\dfrac{2}{m}$
$\dfrac{\infty}{m}\dfrac{\infty}{m}\dfrac{\infty}{m}$	Orthorhombic Monoclinic Triclinic	$\dfrac{2}{m}\dfrac{2}{m}\dfrac{2}{m}$

mum symmetry a crystal can have is the combined symmetry elements of both the cause and the effect. In principle, this means that if the effect symmetry is that of a sphere, the mineral can have the symmetry of any of the six crystal systems, because all of the systems have less symmetry than the observed effect. In practice, most minerals that are isotropic are in the isometric system, because those minerals must have spherical effect symmetry. In principle, if the effect symmetry is an ellipsoid of revolution, the crystal can have any symmetry except isometric. In practice, nearly all minerals with that effect symmetry are either hexagonal or tetragonal, because they must have at least $\infty/m\,2/m\,2/m$ symmetry. If the effect symmetry is that of a triaxial ellipsoid $(2/m\,2/m\,2/m)$, the crystal must be either orthorhombic, monoclinic, or triclinic. Visual observations in optical mineralogy are related to effect symmetry, and from this, the possible lattice symmetries or crystal systems are inferred.

THE BEHAVIOR OF LIGHT IN MINERALS

The interaction of electromagnetic radiation with minerals depends on the particular region of the spectra we consider. In the optical region (400–700 nm wavelength), minerals fall into two broad categories—those that are transparent to the range of optical frequencies and those that are not. The latter are generally referred to in optical mineralogy as the *opaque minerals*. They include some native metals (e.g., Cu, Au), most Fe and Cu sulfides (e.g., CuS, FeS_2), and several transition metal oxides (e.g., Fe_3O_4, $FeTiO_3$, $FeCr_2O_4$). Most of these minerals have energy gaps in their electronic structures that are much smaller than the range of energies (Figure 4.29) associated with the visible spectrum. The opaque minerals have metallic or partially metallic bonds characterized by an abundance of free electrons, and consequently, they absorb and reflect visible light. None of the constituent wavelengths of visible light are transmitted, and minerals like magnetite (Fe_3O_4) appear quite black, both in hand specimen and in petrographic thin section. In these cases, such absorption is only for energies in the visible region. Magnetite and other "opaque" minerals are transparent in other regions of the electromagnetic spectrum.

Optical studies of minerals in transmitted light are therefore restricted to minerals that are transparent to visible light energies. These include most of the rock-forming minerals, all of which are characterized by a predominance of ionic and covalent bonds. The lack of free electrons means good optical transparency, particularly in the visible region. When optical absorption does occur, it is usually selective at a particular wavelength or narrow range of wavelengths corresponding in energy to particular energy gaps of the ions present in the minerals (Chapter 4).

In addition to transporting energy and momentum, electromagnetic waves are reflected and refracted in certain circumstances. We can explain this behavior by using the modern concept of a propagating electromagnetic field. The wave model may be visualized as oscillating electric and magnetic fields that propagate through space normal to their vibration direction. The electric ($\mathbf{E}$) and magnetic ($\mathbf{H}$) vectors vibrate in all directions perpendicular to the propagation direction of the wave, so at any instant in time, a cross section of the wave will reveal electric and magnetic vectors in a radial pattern about the wave axis (Figure 11.2).

For the purposes of optical mineralogy, we need consider only the electric vector, because the interaction of a light wave with a mineral through which it passes is an electronic phenomenon involving only the outer electrons of each ion. The relationships in Figure 11.2 can therefore be considered to apply to the electric behavior of light alone. The important points are these. (1) The *vibration direction* of the wave is perpendicular to

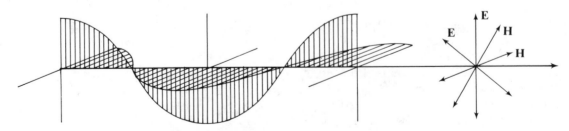

FIGURE 11.2   **Sinusoidal light waves consist of electric vector E and magnetic vector H at right angles to each other and at right angles to the propagation direction.**

the propagation direction. (2) The disturbance of electrons in the medium is perpendicular to the propagation direction, much like the motion we observe in a seagull at rest on the water's surface as successive waves pass by. Energy and momentum pass by, but the seagull only bobs up and down.

The optical properties of minerals in transmitted light depend on the specific manner in which the electric field of the incident light interacts with the atoms or ions in the structure. Under the influence of an external field, the normal electron distributions within the structure are distorted such that the net negative charge of each ion is no longer spherically symmetric around the nucleus. This behavior is called electron polarization and is discussed in Chapter 4. Each ion thus becomes an electric dipole (Figure 11.3) in the presence of visible light, and oscillates with the same range of frequencies ($\approx 10^{14}$ cycles per second) as the incident radiation. The electric dipole moment of a single ion, $\mathbf{p}$, is given by

$$\mathbf{p} = eS = \alpha \mathbf{E}_L \qquad (11.1)$$

where e is the electron charge, S is the displacement of positive and negative charges, α is the polarizability of the ion, and $\mathbf{E}_L$ is the local electric field near the dipole.

This equation simply states that the polarization of an ion is directly proportional to the strength of the local electric field around the ion. The constant of proportionality α is a measure of how easily the ion is polarized. Ease of polarization depends on how tightly bonded the outer electrons are. Anions are particularly susceptible to polarization, because their outer electrons are more weakly held to the central protons than are electrons in cations or neutral atoms. This phenomenon is purely electronic, for the mass of the nucleus of any ion is many orders of magnitude

greater than the mass of the outer electrons, and therefore cannot respond as readily to the incident frequencies of the electric field.

In a vacuum, the local field around a hypothetical isolated ion is the same as the external field, because there are no additional contributions from the surrounding medium. In a mineral, however, the local electric field also depends on the influence of the electric fields generated by each polarized ion in the coordination polyhedron. That is,

$$\mathbf{E}_L = \mathbf{E}_e + \mathbf{E}_d \qquad (11.2)$$

where $\mathbf{E}_e$ is the external electric field of the transmitted light, and $\mathbf{E}_d$ is the electric field due to surrounding dipoles.

It is important to evaluate $\mathbf{E}_L$, because the total polarization of a mineral, $\mathbf{P}$, is given by

$$\mathbf{P} = N_A \mathbf{p}_A + N_B \mathbf{p}_B + N_C \mathbf{p}_C + \cdots = (K - 1)\mathbf{E}_L \qquad (11.3)$$

where N_A is the number of A ions, N_B the number of B ions, and so forth, and $\mathbf{p}_A$ is the dipole moment of A, $\mathbf{p}_B$ the dipole moment of B, and so forth. K is the dielectric constant, which is related to the velocity of light by

$$K = \left(\frac{c}{V_m} \right)^2 \qquad (11.4)$$

where c is the velocity of light in a vacuum, and V_m is the velocity of light in a mineral. The terms in Equations 11.3 and 11.4 may be rearranged to give

$$V_m = c \left(\frac{\mathbf{E}_L}{\mathbf{E}_L + \mathbf{P}} \right)^{1/2} \qquad (11.5)$$

In a vacuum, the term in parentheses reduces to unity, because no medium exists to be polarized. In a mineral, $\mathbf{P}$ will always be positive, and hence the term in parentheses will always be less than

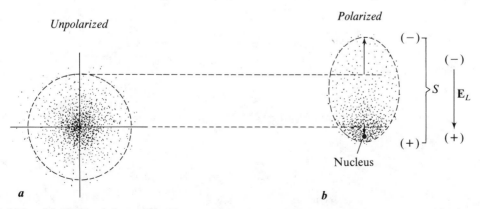

FIGURE 11.3 **(a) Idealized "normal" electron distribution in the absence of an external electric field. (b)** Idealized "polarized" distribution under the influence of external field E.

unity. This means that the velocity of light in a mineral, or in any substance for that matter, will always be less than the velocity of light in a vacuum.

Thus, to determine the velocity of light in a particular direction through a mineral, we must evaluate the total polarization **P**. As Equation 11.3 shows, total polarization depends on the number of dipoles per unit volume in that direction. Generally speaking, we expect polarization to be greatest, and hence the velocity least, in a direction through the mineral for which a high electron density exists. If the density of electrons is low, little polarization occurs, and the velocity is greater. These relationships give us a good qualitative basis for predicting how the velocity of light will differ from one mineral to another.

REFRACTIVE INDEX

In practice, we do not measure the velocity of light directly as it propagates through a mineral. Instead, we measure a parameter, the refractive index, n, which is related to velocity by

$$n = \frac{c_a}{V_m} \approx \frac{V_a}{V_m} = \frac{\text{constant}}{V_m} \qquad (11.6)$$

The refractive index of a vacuum is arbitrarily taken as unity. On this basis, the refractive index of air, c_a, is 1.00029 at 15 °C. Refractive indices of minerals are always positive numbers greater than unity. Clearly, a mineral through which light passes at a high velocity will have a low refractive index and vice versa. The common rock-forming minerals have refractive indices that range from about 1.3 to 2.1.

The fundamental relationship that enables mineralogists to measure refractive indices is Snell's law

$$\frac{V_a}{V_m} = \frac{n_m}{n_a} = \frac{\sin i}{\sin r} \qquad (11.7)$$

where the subscripts a and m refer to air and the mineral, respectively (Figure 11.4). The angles i and r refer to the angle of incidence and the angle of refraction, respectively. This equation means that when an incident beam of light passes from a medium of low refractive index, air in this example, to a medium of higher refractive index such as a mineral, the light is refracted or bent *into* the denser medium. Such behavior is readily observed with a petrographic microscope as illustrated in Figure 11.5. A common procedure for observing the refractive index of a mineral consists of immersing small grains of the mineral in a specially prepared oil[1] of known index and observing the direction of light refraction. Figure 11.5a shows a case in which the mineral has a higher index than the immersion oil, and consequently, the light rays are bent and concentrated around the inner margin of the grain. Figure 11.5b shows a case in which the immersion oil has the higher index, and the light rays are refracted into the oil rather than into the grain. The concentration of light is then along the outer margin of the grain. The region of light concentration is referred to as the *Becke line*. When we lower the stage of a polarizing microscope from a focused position, the Becke line will appear to move in the direction of the medium with the higher index of refraction (Figure 11.5c). If the stage of the microscope is raised from a focused position, the Becke line will appear to move into the medium with the lower index of refraction (Figure 11.5d). This procedure of raising and lowering the stage and observing the Becke line enables us to determine rapidly whether one mineral has a higher or lower refractive index than an adjacent mineral or the surrounding immersion oil. The absolute value of a mineral's refractive index is determined by first

[1] Sets of immersion oils are commercially available from firms such as R. P. Cargille Laboratories, Inc., 117 Liberty St., New York, N.Y.

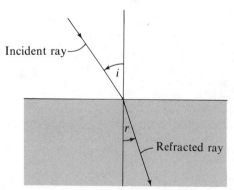

Air $\begin{cases} \text{Low density} \\ \text{Low index} \\ \text{High velocity} \end{cases}$

Mineral $\begin{cases} \text{High density} \\ \text{High index} \\ \text{Low velocity} \end{cases}$

FIGURE 11.4   Illustration of Snell's law, which gives the angle of refraction as a function of refractive index.

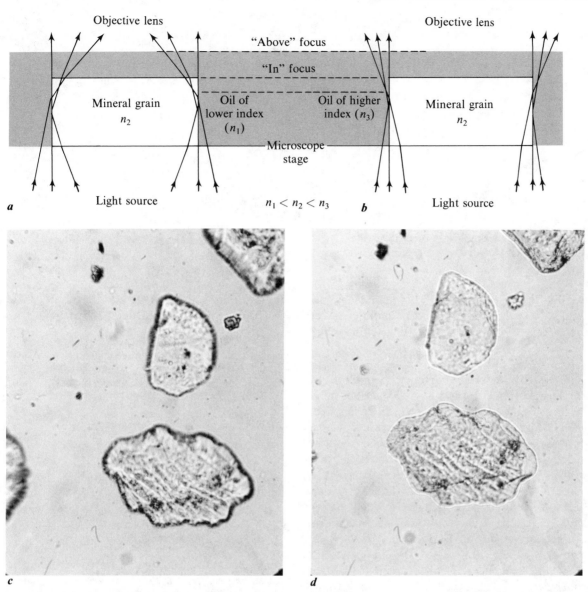

FIGURE 11.5 **Mineral grains of refractive index n_2 immersed in (a) oil of lower index n_1, and (b) oil of higher index n_3. (c) When immersion oil is of lower index than the mineral, lowering the stage relative to fixed objective causes the Becke line to move into the** mineral. **(d) When oil is of higher index than the mineral, lowering the stage causes the Becke line to move into immersion oil. Grains in (c) and (d) are quartz magnified 320 ×.**

judging the relative difference between the refractive index of an unknown mineral and an immersion oil of known refractive index, and then repeating the experiment with a different index until a "match" is achieved. This occurs when the index of the mineral is exactly the same as the oil, and the mineral outline is barely discernible.

Because the polarizability of electrons will vary from one atomic species to another, the velocity of light and therefore the refractive index becomes a measurable parameter that is quite sensitive to chemistry. As the geometric arrangement of the atoms relative to one another will affect the net

polarization, so the refractive index is a parameter that is also sensitive to structural changes. The accurate determination of the refractive indices of a mineral provides positive identification of any mineral, and extensive tables are available for this purpose.

The refractive index of a mineral in a particular direction through the mineral depends on ionic radius, valence, and coordination and atomic number. We may generalize these relationships by stating that minerals composed of atoms with high atomic numbers will have relatively large refractive indices. It is not surprising that as a mineral

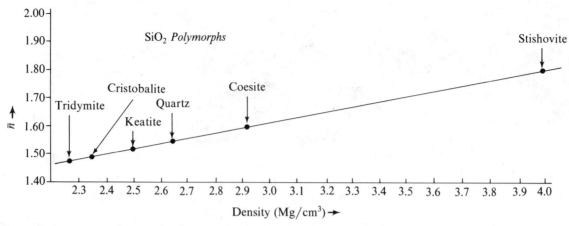

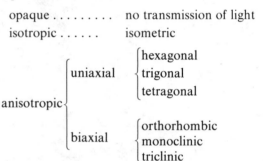

FIGURE 11.6   Relation between mean refractive index $\bar{n}$ and density of various structural forms of SiO_2.

expands upon heating, its refractive index decreases. Conversely, as a mineral contracts under the effects of pressure, its refractive index increases.

A few examples of the relationship between high refractive index and high density of electrons are illustrated in Figure 11.6 for the various polymorphs of SiO_2. A dense arrangement of the same chemical combinations of atoms produces a large decrease in the velocity of light and, therefore, an increase in the refractive index. The relationship between average atomic number and refractive index is illustrated by the olivine solid solution series in Figure 11.7. Both the refractive indices and the density increase in a uniform manner as the proportion of Fe ($Z = 26$) to Mg ($Z = 12$) increases in the series.

For the remainder of this chapter, we will be concerned with the optical properties of various groups of minerals. A broad classification of these properties is as follows:

opaque . . . .   no transmission of light

isotropic . . . . . .   isometric

$$\text{anisotropic} \begin{cases} \text{uniaxial} \begin{cases} \text{hexagonal} \\ \text{trigonal} \\ \text{tetragonal} \end{cases} \\ \\ \text{biaxial} \begin{cases} \text{orthorhombic} \\ \text{monoclinic} \\ \text{triclinic} \end{cases} \end{cases}$$

ISOTROPIC MINERALS

Minerals that transmit light in the visible region belong to two broad categories. As mentioned earlier, those minerals in which light propagates with the same velocity in all directions through the structure are termed *isotropic*. For reasons of crystal structure, optically isotropic minerals are restricted to the isometric system. All other minerals transmit light with velocities dependent on the structural anisotropy that is characteristic of the remaining crystal systems. These minerals are termed *anisotropic*.

What special feature of the isometric system requires light to travel with the same velocity in all directions of propagation? The actual observation for minerals such as halite (NaCl), fluorite (CaF_2), garnet ($Fe_3Al_2Si_3O_{12}$), and many others is that the characteristic refractive index is independent of the direction in which light passes through the structure. For example, halite has $n = 1.544$ for light transmitted normal to (100), (010), (001), (111), and so forth. The same observation is true for garnet, except that $n = 1.830$ for all faces and random sections cut through the mineral. The

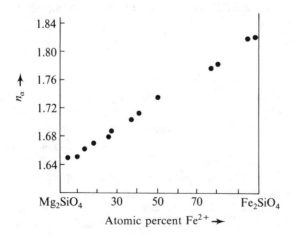

FIGURE 11.7   Variation of refractive index n_α in a specified direction with simple substitution of Fe^{2+} for Mg^{2+} in olivine.

reason for the difference between the *n* values of halite and garnet is that garnet is much denser ($\rho = 4.32$ Mg/m^3) than halite ($\rho = 2.17$ Mg/m^3), and therefore has a higher density of electrons. It follows from earlier discussions that light will travel faster through halite than through garnet, and thus the refractive index of halite is less than that of garnet.

Looking back at Equation 11.5, we see that in order for the velocity to be the same in all directions, the polarization must be the same in all directions. The polarization (Equation 11.3) depends not only on the density of the polarized ions or dipoles, but also on the strength of each dipole, which in turn is dependent on the local electric field. As we will see, the effect of neighboring dipoles on the polarization of a particular ion is highly dependent on the translational symmetry of the particular structure.

In order to proceed further, we must understand the nature of the electric field generated inside a mineral by an electric dipole. The magnitude of the electric intensity at any point in a crystal at a distance *r* from a single dipole may be calculated from the following expressions that are derived from Coulomb's law.

$$\mathbf{E}_r = \frac{1}{4\pi\epsilon_0} \frac{2eS\cos\theta}{r^3} \qquad (11.8)$$

and

$$\mathbf{E}_\theta = \frac{1}{4\pi\epsilon_0} \frac{eS\sin\theta}{r^3} \qquad (11.9)$$

where $\mathbf{E}_r$ is the component of $\mathbf{E}$ in the direction of *r*, $\mathbf{E}_\theta$ is the component of $\mathbf{E}$ perpendicular to *r*, and θ is the angle between the dipole axis and *r*. The quantity *eS* is the dipole moment of a single ion (Equation 11.1), and ϵ_0 is a constant. Figure 11.8a illustrates the relationships between these variables, and Figure 11.8b shows the total electric field about a dipole determined by calculating the vector sum of $\mathbf{E}_r$ and $\mathbf{E}_\theta$ at every point. The vector $\mathbf{E}_A$ is the same in both figures.

Consider now the polarizing effect of incident light on halite, which is isometric. Figure 11.9 shows a two-dimensional projection of the structure. Each ion, and in particular Cl^{1-}, is polarized by light waves that have in this case an electric vector oriented in an east-west direction parallel to one edge of the square lattice. The local electric field at each Cl^{1-} contributes to the total polarization (Equations 11.1 and 11.3) and depends on contributions made by neighboring Cl^{1-} dipoles (Equation 11.2). The vector sum of the electric fields of contributing dipoles is easily calculated. Assume in Figure 11.9 that a centrally coordinated Cl_c^{1-} is affected mainly by its four closest neighbors in this plane (Cl_a^{1-}, Cl_b^{1-}, Cl_d^{1-}, Cl_e^{1-}) and by two additional Cl^{1-} if the three-dimensional structure were considered. This

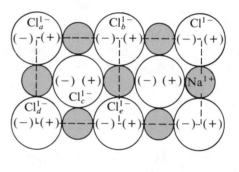

FIGURE 11.9 Two-dimensional projection of NaCl structure. Calculations determine the local electric field in the center of Cl_c^{1-} due to influence of dipole moments at Cl_a^{1-}, Cl_b^{1-}, Cl_d^{1-}, and Cl_e^{1-}.

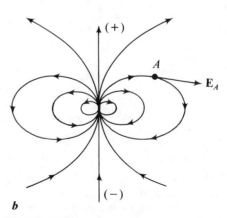

FIGURE 11.8 (a) Relationship between an electric dipole and the radial $\mathbf{E}_r$ and angular $\mathbf{E}_\theta$ components of its surrounding field. (b) Generalized electric field E around dipole in (a). The field at any point is tangent to the field lines.

model is a good approximation of Equation 11.8. Both show that the contributing effect of neighboring ions (dipoles) on the local electric field at Cl_c^{1-} falls off as the cube of the distance from Cl_c^{1-}. Let us also assume for arithmetic simplicity that the dipole moment is unity for all Cl^{1-} ions and that we can ignore the constant $1/4\pi\epsilon_0$, which appears in both E_r and E_θ. The calculation of E_r and E_θ at Cl_c^{1-} due to the presence of Cl_a^{1-}, Cl_b^{1-}, Cl_e^{1-}, and Cl_d^{1-} (Figure 11.10) begins with Equations 11.8 and 11.9.

$$E_{\theta(a)} = E_{\theta(e)} = \frac{\sin 45°}{r^3} = \frac{0.707}{r^3}$$

and

$$E_{r(a)} = E_{r(e)} = \frac{2\cos 45°}{r^3} = \frac{1.414}{r^3}$$

similarly,

$$E_{\theta(b)} = E_{\theta(d)} = \frac{0.707}{r^3}$$

and

$$E_{r(b)} = E_{r(d)} = \frac{1.414}{r^3}$$

From Figure 11.10, we see that $E_{r(a)}$ exactly cancels $E_{\theta(d)} + E_{\theta(b)}$, and $E_{r(d)}$ exactly cancels $E_{\theta(e)} + E_{\theta(a)}$. This leaves the vector sum of $E_{r(e)}$ and $E_{r(b)}$ as the local electric field at Cl_c. Its magnitude is given by

$$E_L = E_{r(e)} + E_{r(b)}$$

$$= \left(\left(\frac{1.414}{r^3} \right)^2 + \left(\frac{1.414}{3} \right)^2 \right)^{1/2} = \frac{2}{r^3}$$

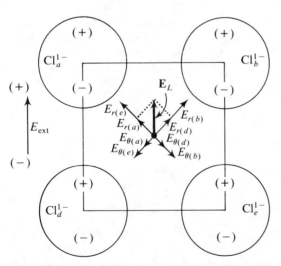

FIGURE 11.10   Arrangement of electric field components in the center of Cl_c^{1-} (Figure 11.9) due to contributions from four neighboring chlorine anions. The net local field is in the same direction as the external electric field generated by the transmitted light. External field direction is parallel to the square lattice edge.

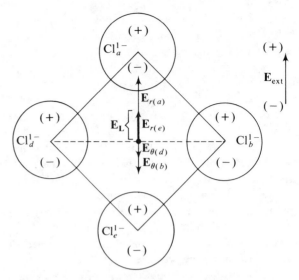

FIGURE 11.11   Arrangement of electric field components due to four neighboring dipoles such that external field direction is parallel to square lattice diagonal. Net local field has same magnitude as in Figure 11.10.

Because E_L is in the same direction as the external field E_{ext}, their sum gives the total polarization, **P**. From that, the velocity or refractive index can be calculated (Equation 11.5).

Now consider Figure 11.11 in which the electric vector of the incident field (E_{ext}) is oriented parallel to the diagonal of the square lattice rather than parallel to the lattice edge. In this case,

$$E_{\theta(a)} = E_{\theta(e)} = \frac{\sin 0°}{r^3} = 0$$

$$E_{r(a)} = E_{r(e)} = \frac{2\cos 0°}{r^3} = \frac{2}{r^3}$$

and

$$E_{\theta(b)} = E_{\theta(d)} = \frac{\sin 90°}{r^3} = \frac{1}{r^3}$$

$$E_{r(b)} = E_{r(d)} = \frac{2\cos 90°}{r^3} = 0$$

The vector $E_{r(a)}$ exactly cancels $E_{\theta(d)}$ and $E_{\theta(b)}$, leaving only $E_{r(e)}$ remaining. Its magnitude is $2/r^3$, which is exactly the same as calculated in Figure 11.10. The net polarization and hence the velocity of light is the same in these two directions through halite. The calculation for random directions of the incident field between the edge and diagonal orientations gives the same result.

We ignored the cations in this analysis, even though they are close neighbors to the Cl^{1-} of interest, because the polarizability of Na^{1+} in halite is relatively low due to the absence of easily deformable outer electrons. The Na^{1+} will still make a contribution, however small, with the same 4*mm* symmetry constraint. Since the symmetry must apply to the entirety of the plane around a specified point, it follows that any neighboring dipole effects, however distant from the point of calculation, will have cancellation of components as shown in Figures 11.10 and 11.11.

The cancellation of certain directional effects of neighboring dipoles therefore requires that the magnitudes of the components be the same and that their directions be opposing. Thus, the distance r to neighboring dipoles must be exactly the same, and the spatial relationship between the dipoles in a plane must have $4mm$ plane point symmetry. The symmetry requirement extends to include $3mm$ or $6mm$ plane point symmetry in the case of certain orientations of anisotropic minerals discussed in a later section.

In the context of this discussion, let us now consider how the local electric field behaves in a plane that has $2mm$ plane point symmetry such as would characterize sections through many anisotropic minerals. Figure 11.12a illustrates the case in which the external electric vector is parallel to the short edge of the lattice. The calculations are:

$$\mathbf{E}_{\theta(a)} = \mathbf{E}_{\theta(b)} = \mathbf{E}_{\theta(d)} = \mathbf{E}_{\theta(e)}$$

$$= \frac{\sin 56.31°}{r^3} = \frac{0.832}{r^3}$$

$$\mathbf{E}_{r(a)} = \mathbf{E}_{r(b)} = \mathbf{E}_{r(d)} = \mathbf{E}_{r(e)}$$

$$= \frac{2\cos 56.31°}{r^3} = \frac{1.109}{r^3}$$

$$\mathbf{E}_L = \frac{0.308}{r^3}$$

Figure 11.12b illustrates the opposite extreme with the external electric vector parallel to the long edge of the lattice. The calculations are

$$\mathbf{E}_{\theta(a)} = \mathbf{E}_{\theta(b)} = \mathbf{E}_{\theta(d)} = \mathbf{E}_{\theta(e)}$$

$$= \frac{\sin 33.69°}{r^3} = \frac{0.555}{r^3}$$

$$\mathbf{E}_{r(a)} = \mathbf{E}_{r(b)} = \mathbf{E}_{r(d)} = \mathbf{E}_{r(e)}$$

$$= \frac{2\cos 33.69°}{r^3} = \frac{1.664}{r^3}$$

$$\mathbf{E}_L = \frac{4.307}{r^3}$$

Clearly the magnitudes of the fields are different, which means that the velocity of light parallel to the long edge of the lattice will be different from the velocity of light parallel to the short edge. Such behavior is characteristic of all anisotropic minerals.

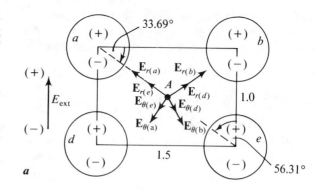

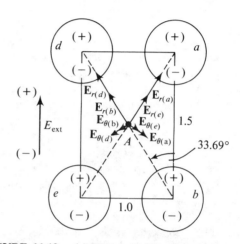

FIGURE 11.12 **(a) Electric field contributions to centrally located anion A by dipoles from four closest anion neighbors arranged in a $2mm$ plane lattice. Net local field at A is directed up with magnitude $0.308/r^3$. External electric vector is parallel to short edge of the lattice. (b) Electric field contributions to centrally coordinated anion A from four closest anion neighbors arranged in a $2mm$ plane lattice. Net local field at A is $4.307/r^3$. External electric vector is parallel to long edge of lattice.**

In summary of this section, the isotropic behavior of all isometric minerals is controlled by their relatively high symmetry. Gases and liquids have even higher symmetry (infinite) because of the random arrangement of atoms or molecules and are likewise isotropic. Because the point group symmetry represents the symmetry of space about any point in a crystal, a sphere of the type shown in Figure 6.2a can be thought to exist at every point in the crystal. If the equal radii represent the velocity of light, the sphere is referred to as an *isotropic ray velocity surface*. If the radii represent the refractive index, the sphere is referred to as an *isotropic indicatrix*. In either case, a random orientation of any isotropic mineral will display refractive indices with magnitudes proportional to the equal radii of the circle cut by that orientation through the indicatrix. Since all sections through the center of a sphere will give identical circles, all orientations of an isotropic mineral will display an identical refractive index.

UNIAXIAL MINERALS

In all minerals with symmetry other than isometric, the velocity of light will be faster in certain directions of the crystal than in others. The majority of rock-forming minerals are of this type and are referred to as *anisotropic*. Within this group are two categories, called *uniaxial* and *biaxial* depending on whether the mineral has one or two optic axes.

All uniaxial minerals must have either tetragonal, trigonal, or hexagonal symmetry. The effect symmetry due to transmitted light in these minerals is that of an ellipsoid of revolution. If we let the various radii of the ellipsoid be proportional to the refractive index, the ellipsoid is referred to as a *uniaxial indicatrix*. As Figure 6.2b shows, this indicatrix is characterized by a circular section perpendicular to the principal axis of rotation, whether it be a fourfold axis in the case of tetragonal minerals, or a threefold or sixfold axis as in trigonal and hexagonal minerals, respectively. The equal radii of the circular section represent the refractive index of light rays that travel at the same velocity. Therefore, all uniaxial minerals have a single isotropic orientation that is characterized by a constant velocity of light and a corresponding constant refractive index with magnitude designated by the symbol ω. Light rays that vibrate parallel to the plane of the isotropic section and have an ω refractive index are referred to as *ordinary rays*. Because the unique optic axis is normal to the isotropic section, it will correspond to the principal crystallographic axis of rotation of the system. An equivalent definition of the ordinary ray is any ray that vibrates perpendicular to the optic axis. The vibration direction is thus perpendicular to the direction of wave propagation through the mineral.

All other sections through the center of the ellipsoid of revolution are ellipses that have as their minimum principal axis the ω refractive index. The other principal axis varies in magnitude from ω as a minimum to a maximum value denoted by the symbol ϵ. The ϵ refractive index characterizes the *extraordinary ray*, defined as any light ray that vibrates parallel to the unique optic axis of uniaxial minerals. The ordinary and extraordinary rays have different refractive indices and hence different velocities as they propagate through a mineral.

The reason for such anisotropic behavior of minerals is that the degree of electron polarization varies with direction in the crystal structure. Let us examine the behavior of light as it passes through calcite ($CaCO_3$), a typical uniaxial mineral. The structure of calcite is shown diagrammatically in Figure 11.13. It consists of $(CO_3)^{2-}$ anionic groups coordinated by Ca^{2+} in a tetrahedral arrangement. A conspicuous anisotropism is due to the parallel arrangement of the $(CO_3)^{2-}$ planes that induces a strong polarization of the loosely held electrons. Because the three oxygen ions are symmetrically distributed around the central C^{4+}, the polarization of electrons in the $(CO_3)^{2-}$ plane is highly symmetric. Each oxygen becomes a dipole under the influence of the external electric field and produces a secondary field, which then contributes to the local field of neighboring oxygens.

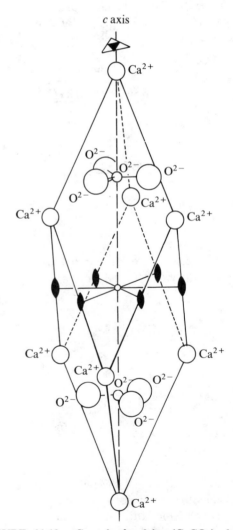

FIGURE 11.13   Crystal of calcite ($CaCO_3$) showing the pronounced structural anisotropy due to parallel alignment of $(CO_3)^{2-}$ planes normal to *c* axis.

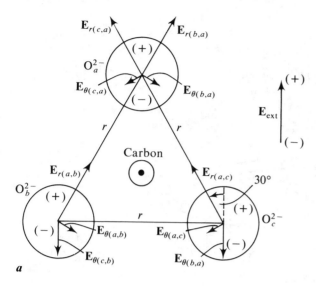

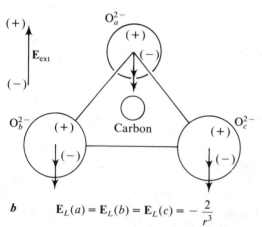

$$\mathbf{E}_L(a) = \mathbf{E}_L(b) = \mathbf{E}_L(c) = -\frac{2}{r^3}$$

b

FIGURE 11.14 **(a) Electric field contributions at oxygens due to dipole contributions from other two oxygens in $(CO_3)^{2-}$ anionic complex. External electric vector is parallel to $(CO_3)^{2-}$ plane and perpendicular to c axis of calcite. (b) Electric field contributions to oxygen in calcite due to dipole contributions from other two oxygens in the $(CO_3)^{2-}$ anionic complex. External electric vector is oriented normal to $(CO_3)^{2-}$ plane and parallel to c axis.**

Figure 11.14a illustrates the behavior of dipole-induced, local electric fields within calcite when the external electric field direction is parallel to the $(CO_3)^{2-}$ plane (perpendicular to the c axis). Because of the threefold symmetry, the east-west components of all vectors exactly cancel. The vector sum of the north-south components in the direction of the incident field (pointing up) is

$$\mathbf{E}_{\text{up}} = (\mathbf{E}_{r(b,a)} + \mathbf{E}_{r(c,a)} + \mathbf{E}_{r(a,b)} + \mathbf{E}_{r(a,c)})\cos 30°$$

$$= \frac{6}{r^3}$$

where $\mathbf{E}_{(b,a)}$ reads "the electric field at oxygen b due to the dipole at oxygen a." The vector sum in the opposite direction is

$$\mathbf{E}_{\text{down}} = (\mathbf{E}_{\theta(b,a)} + \mathbf{E}_{\theta(c,a)} + \mathbf{E}_{\theta(a,b)} + \mathbf{E}_{\theta(a,c)})\cos 60°$$

$$= \frac{3}{r^3}$$

The vector sum of these is clearly $3/r^3$ in the direction of the incident electric vector. In other words, the effect of neighboring dipoles is to enhance the polarization beyond the value it would have if contributing neighbors were absent.

Now consider the behavior of local fields when the external field is directed parallel to the c axis of calcite (Figure 11.14b). The vibration direction of the incident polarized light is perpendicular to the $(CO_3)^{2-}$ plane. The relevant components are:

$$\mathbf{E}_{\theta(b,c)} = \mathbf{E}_{\theta(c,b)} = \frac{\sin 90°}{r^3} = \frac{1}{r^3}$$

$$\mathbf{E}_{r(b,c)} = \mathbf{E}_{r(c,b)} = \frac{2\cos 90°}{r^3} = 0$$

$$\mathbf{E}_{\theta(b,a)} = \mathbf{E}_{\theta(c,a)} = \mathbf{E}_{\theta(a,b)} = \mathbf{E}_{\theta(a,c)}$$

$$= \frac{\sin 30°}{r^3} = \frac{0.5}{r^3}$$

$$\mathbf{E}_{r(b,a)} = \mathbf{E}_{r(c,a)} = \mathbf{E}_{r(a,b)} = \mathbf{E}_{r(a,c)}$$

$$= \frac{2\cos 30°}{r^3} = \frac{1.732}{r^3}$$

The conclusion we must draw from the analysis of Figure 11.14b is that each oxygen is polarized *less* in the presence of its closest neighbors than it would be if it were by itself. Consequently, we would expect light that vibrates in the $(CO_3)^{2-}$ plane (Figure 11.14a) to be more polarized than light that vibrates perpendicular to the plane (Figure 11.14b). The ordinary ray ω will therefore have a lower velocity and higher refractive index than the extraordinary ray ϵ. This is precisely the observation in calcite: $\omega = 1.658$ and $\epsilon = 1.486$.

These relationships in the case of calcite may be summarized with the uniaxial *ray velocity surface*, the dimensions of which are defined by the distance traveled by light rays in a specific time. Imagine in Figure 11.15 a hypothetical point light source located in the geometric center of a calcite crystal. Light emanating in all directions through the crystal will have electric vectors vibrating perpendicular to the direction of propagation. First, consider light advancing from left to right from the center of the crystal. It will have vibration directions parallel to the $(CO_3)^{2-}$ planes, as well as perpendicular to them. The light that vibrates in the $(CO_3)^{2-}$ plane has the ω refractive index,

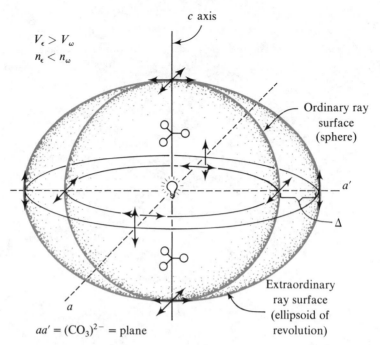

$$V_\epsilon > V_\omega$$
$$n_\epsilon < n_\omega$$

c axis

Ordinary ray surface (sphere)

a'

Δ

Extraordinary ray surface (ellipsoid of revolution)

a

$aa' = (CO_3)^{2-}$ = plane

FIGURE 11.15 **Uniaxial ray velocity surface for an optically negative mineral, calcite. Inner spherical surface is generated by the ordinary (ω) ray emanating from** **hypothetical light source inside the crystal. Outer surface is an ellipsoid of revolution generated by the extraordinary (ϵ) ray.**

and the light that vibrates perpendicular to the $(CO_3)^{2-}$ plane has the ϵ refractive index. Intermediate directions of vibration will be resolved by the crystal into vertical (parallel to *c* axis) and horizontal (normal to *c* axis) vector components. The component parallel to the $(CO_3)^{2-}$ plane encounters a high density of oxygen electrons, and therefore after some fixed period of time lags behind the component parallel to the *c* axis. Since the structure is equivalent by translation in the opposite direction, the same relationships must hold for light propagating from right to left, front to back, or along any radial line normal to the *c* axis.

Now consider the behavior of light advancing parallel to the *c* axis. All electric vectors will vibrate in the plane parallel to the $(CO_3)^{2-}$ groups, so the distance traveled after some period of time will be the same for all components. This must be true in exactly the opposite direction as well. Advancement of light perpendicular to the *c* axis and parallel to it are the extreme cases—either there is a maximum lag of one wave behind the other, or there is no lag at all.

The behavior of the ordinary ray with time, then, defines a spherical ray velocity surface identical to the surface for an isotropic mineral. The extraordinary ray, however, defines an ellipsoid of revolution that totally contains the spherical surface.

The two surfaces meet at only two points, cor-responding to the position of the crystallographic *c* axis. This direction is the unique *optic axis* common to all uniaxial minerals. For light propagating parallel to the optic axis, no distinction exists between the ordinary and extraordinary waves, for both travel at the same velocity. Any uniaxial mineral will therefore be isotropic when viewed directly down its optic axis. Observations made in all other directions will reveal anisotropic character.

Note that because the optic axis has only a directional property, translational symmetry requires that it be present everywhere in a crystal. That is why the ray velocity surfaces provide a meaningful model, for these relationships and their symmetry must necessarily prevail at all points in a particular mineral.

PLANE POLARIZED LIGHT

A characteristic of all anisotropic minerals is that transmitted light is constrained to vibrate through the mineral in certain mutually perpendicular directions. In the example of calcite, light associated with one direction travels with the ordinary ray velocity. The other ray travels with the extraordinary ray velocity. A convincing demonstration of these facts is provided by any large, clear cleavage rhombus of calcite (Figure 11.16a). A single dot placed on paper beneath the rhombus appears as two dots when viewed from above

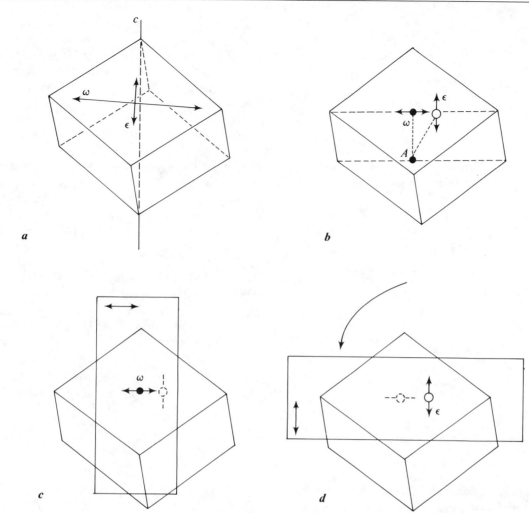

FIGURE 11.16 **(a)** A cleavage rhombus of calcite showing the positions (a) of the preferred vibration directions ω and ϵ as viewed normal to the {100} cleavage. **(b)** Double refraction causes a single dot at A beneath the rhombus to appear as two dots, one polarized at right angles to the other. **(c)** Positioning of Polaroid filter parallel to ϵ absorbs ω and transmits ω. **(d)** Rotation of Polaroid filter by 90° causes absorption of ω and transmission of ϵ.

(Figure 11.16b). One of the dots also appears deeper than the other, and in fact will rotate around the other when the crystal is rotated about a vertical axis. The fact that we see two images means that the two rays producing the images travel at different velocities, the faster ray being bent more than the slower one, as predicted by Snell's law. The outer dot, the faster one, is in fact the extraordinary ray ϵ, and is the dot that appears to be deeper.

Are the rays that produce the two images polarized at right angles to each other? We can demonstrate that they are by rotating a piece of synthetic Polaroid over the two dots and observing the dots alternately disappear and then reappear. When the vibration direction of the Polaroid filter is parallel to the long diagonal of the rhombus, it is also parallel to the ω vibration direction and perpendicular to the ϵ direction. Therefore, the ordinary image is transmitted and the extraordinary one is absorbed (Figure 11.16c). A 90-degree rotation of the Polaroid filter produces absorption of ω and transmission of ϵ (Figure 11.16d), which could happen only if the two rays were forced by the crystal structure to vibrate in mutually perpendicular directions. All anisotropic minerals have this property of *double refraction*. Calcite provides an extreme example of it.

A beam of light constrained to vibrate in a single plane is said to be *plane polarized*. Modern polarizing microscopes are based on this concept, using synthetic Polaroid material rather than natural crystals. Most Polaroid is now made of polyvinyl alcohol in which long-chain molecules are aligned in one direction. When positioned between the substage light source and the micro-

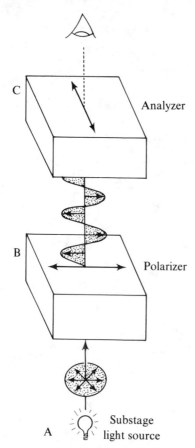

FIGURE 11.17   Schematic behavior of light rays as they pass from substage light source A, to polarizer B, and to analyzer C. Light is polarized to a plane at B and is totally absorbed at C, providing only isotropic materials are observed.

scope stage, the *polarizer*, as it is called, insures that the light reaching the stage is plane polarized. An identical Polaroid filter, called an *analyzer*, with a polarizing direction oriented at 90 degrees to that of the lower polarizer, is located between the objective lens and the ocular. The analyzer may be inserted or withdrawn at will. When inserted (Figure 11.17), it absorbs all light passed by the polarizer, provided there is no modification of that light by a mineral on the stage.

Isotropic minerals cause no observable modification of the plane-polarized light that passes through them. Therefore, the diagnostic test for any isotropic mineral in any orientation is to insert the analyzer, rotate the stage, and observe total darkness. No light can be transmitted.

OPTICAL PROPERTIES OF UNIAXIAL MINERALS

Now that the physical phenomena of light interaction has been described, let us consider sev-

eral optical properties that enable mineralogists to identify minerals based on this behavior. We use the uniaxial indicatrix, because it "indicates" the principal vibration directions and associated refractive indices in a form that is easily visualized. There are two types of indicatrices for uniaxial minerals. Optically ($+$) minerals like quartz (SiO_2) are represented by a prolate ellipsoid of revolution (Figure 11.18a), and optically ($-$) minerals like calcite ($CaCO_3$) are represented by an oblate ellipsoid of revolution (Figure 11.18b). For positive minerals, $n_\omega < n_\epsilon$. The distinctive feature of the indicatrices is their $\infty/m2/m2/m$ symmetry. Recall from Figure 6.2b that this effect symmetry in combination with the spherical symmetry of light is compatible only with hexagonal, trigonal, and tetragonal lattices. Note that each surface has a circular section normal to the c axis of all hexagonal, trigonal, and tetragonal minerals. The magnitude of the radius of the circular section is in proportion to the refractive index n_ω of the ordinary wave. All other sections that pass through the center of the indicatrix and at angles other than 90 degrees to the c axis are ellipses. The minor axis of each of these will be constant,

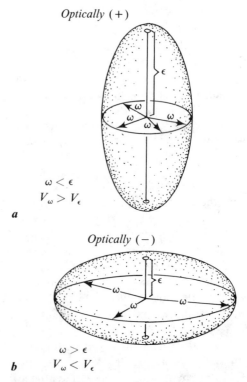

FIGURE 11.18   Uniaxial positive (a) and negative (b) indicatrices defined by the relative magnitudes of the ordinary (ω) and extraordinary (ϵ) rays. Both indicatrices have $\infty/m2/m2/m$ point group symmetry and are diagnostic of all hexagonal, trigonal, and tetragonal lattices.

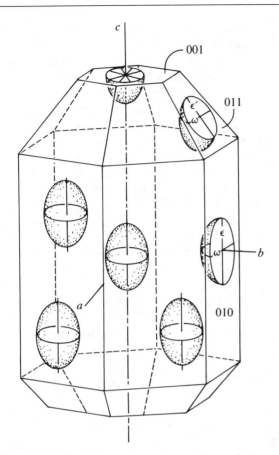

FIGURE 11.19 **Schematic representation of how translational symmetry provides identical point symmetry at every point in the mineral.**

equal to n_ω. The major axis of each will range in magnitude from n_ω to n_ϵ, depending on the angle of intersection with the c axis.

The utility of uniaxial indicatrices is illustrated in Figure 11.19. The indicatrix is imagined as situated at the geometric center of a crystal. Its point group symmetry $\infty/m2/m2/m$ is consistent with the symmetry of space around every point in the crystal, so we can imagine the indi-

catrix as existing at every point in the crystal. The indicatrix, optically (+) in this case, "indicates" the inherent vibration directions of the crystal and the relative magnitude of the refractive indices for all possible orientations of the crystal. This information becomes important when we are making observations on randomly oriented grains in an oil immersion mount or in a petrographic thin section. For example, if we observe a fragment of quartz in a direction parallel to its c axis, only the circular section of the indicatrix appears, and we can observe only the ordinary ray refractive index n_ω. That is, all light passing through the crystal in this direction will be vibrating normal to the c axis. Rotation of the microscope stage has no effect on the observed index. In contrast, if we observe a quartz fragment in a direction normal to the c axis, only the elliptical section through the indicatrix appears with its associated n_ω and n_ϵ refractive indices. The minimum index, n_ω for quartz, appears at one position of stage rotation; the maximum index n_ϵ appears with a rotation of 90 degrees. Intermediate orientations of the crystal and thus of the indicatrix will show n_ω and n_ϵ', intermediate values of the refractive index of light vibrating parallel to the c axis.

An important optical property of all anisotropic minerals is *extinction*. Observation is made with the analyzer inserted. When a grain is rotated such that its inherent vibration directions coincide with those of the polarizer and the analyzer, the grain will be in a position of extinction. Figure 11.20 illustrates that extinction will occur every 90 degrees of rotation, for only in those positions do the inherent vibration directions of the mineral coincide with those of the polarizer and the analyzer. In other positions of rotation, anisotropic minerals are illuminated. Maximum light intensity is attained at 45-degree intervals between positions of extinction. The ordinary and extraordi-

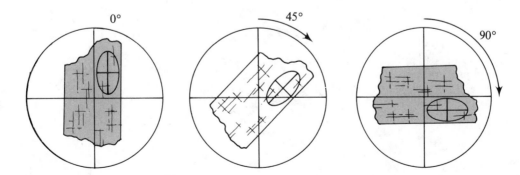

FIGURE 11.20 **Microscopic view with crossed polarizers of a mineral grain in three positions of rotation. Inherent vibration directions in the grain are at right** angles and parallel to cleavage. Parallel extinction occurs every 90° of rotation.

nary rays, vibrating at right angles and parallel to the inherent vibration directions of the mineral, combine upon emergence from the analyzer. The vibration plane of the resulting wave will be oriented at some angle to the vibration plane of the analyzer. In any case, the analyzer will resolve one component of the resultant wave for transmission. The other component will be absorbed by the analyzer.

Another characteristic of all uniaxial minerals (hexagonal, trigonal, and tetragonal) is that one of the two inherent vibration directions be parallel to the projection of the unique optic axis. The other vibration direction is perpendicular to the *c* axis. Uniaxial minerals will therefore display *parallel extinction* with respect to all prismatic or basal cleavages and crystal faces (Figure 11.20). If rhombohedral cleavage is developed, as in calcite, symmetry requires that the cleavage be symmetrical about the *c* axis and hence about the vibration directions. As shown in Figure 11.21, symmetrical extinction will result. Parallel extinction is also characteristic of orthorhombic minerals that are biaxial rather than uniaxial. The basic reason is the orthogonality of the crystallographic *c* axis with the plane defined by the remaining axes. That is, lattice symmetries $6/m2/m2/m$, $\bar{3}2/m$, $4/m2/m2/m$, and $2/m2/m2/m$ are all consistent with parallel extinction on prismatic and basal cleavages. The principal rotation axis in each point group coincides with the *c* axis, and the mirror plane normal to the principal rotation axis contains the *a* and *b* crystallographic axes.

Recognition of why extinction occurs is fundamental to the measurement of refractive index. A mineral that is in a position of extinction will transmit only that ray (ω or ϵ) that vibrates in the plane of the polarizer. In Figure 11.22a, the ordinary ray passes through calcite when the long diagonal of the rhombohedral cleavage is in the north-south position (the same orientation as the fixed polarizer beneath). With the analyzer withdrawn, the grain exhibits n_ω at every point. By trial and error, we find an immersion oil to exactly match the grain in that orientation. The grain is then rotated 90 degrees (Figure 11.22b) to bring the short diagonal and the extraordinary ray vibration direction into coincidence with the polarizer direction. In this position, only n_ϵ is observed. The difference in refractive indices is qualitatively expressed as a difference in *relief*. Relief is the contrast or degree of visibility of any transparent mineral relative to its surroundings. Calcite in Figure 11.22a is said to have high relief because of the great depth of the shadows around the grain boundaries. The orientation in Figure 11.22b has low relief due to the absence of contrast. Quartz ($n_\omega = 1.544$, $n_\epsilon = 1.553$) in contact with the epoxy adhesive ($n = 1.544$) of most petrographic thin sections exhibits little contrast and therefore has low relief. Garnet ($n = 1.830$), on the other hand, exhibits very high relief relative to epoxy.

Another important property of uniaxial minerals and of all anisotropic minerals in general is *birefringence*. This property is a measure of the difference in the two refractive indices observed for any particular grain orientation. For uniaxial minerals, birefringence, δ, ranges from a maximum observed for grains in a direction normal to the *c* axis to a minimum (zero) for grains oriented with their *c* axis parallel to the viewing direction. In the first case, the section through the indicatrix center (Figure 11.18) will have maximum ellipticity; in the second case, the section is a circle. Out of several hundred randomly oriented mineral grains on a glass slide, several will always be oriented with (001) parallel to the slide. Such orientations are isotropic, and with the analyzer

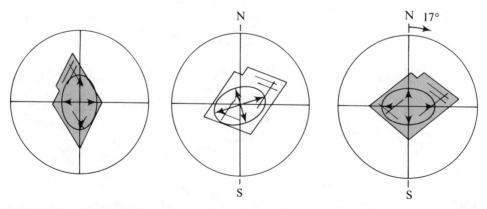

FIGURE 11.21   Mineral grain in three positions of rotation illustrating symmetrical extinction.

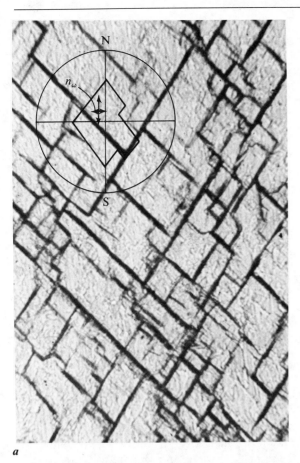

a

b

FIGURE 11.22 **Photomicrographs of a calcite grain in two mutually perpendicular positions of rotation. (a) The ordinary (ω) ray is parallel to the polarizer, and high** **relief is observed. (b) The extraordinary (ε) ray is parallel to the polarizer and low relief is observed. Analyzer removed, 128 × magnification.**

inserted, no light will be transmitted even with rotation of the stage. Many other grains that lie on (*hk*0) will show maximum birefringence. For these,

$$\delta = n_\epsilon - n_\omega = \frac{\Delta}{t}$$

where *t* is the grain thickness, and Δ is the *retardation*, or lag of the slow ray behind the fast ray after the light has passed through the grain (Figure 11.15). The graphical representation of this general relationship (the indices are different for biaxial minerals) for all the common minerals is illustrated by the Michel-Lévy chart (Figure 11.23). The maximum birefringence is a distinctive property of most minerals and thus is useful for identification purposes.

A closely related property is the algebraic sign of the birefringence, usually referred to as the *optic sign*. In the case of uniaxial minerals, if $n_\epsilon > n_\omega$, δ is positive. If $n_\epsilon < n_\omega$, δ is negative, and the mineral is said to be optically negative. Optically positive minerals have prolate indicatrices

(Figure 11.18), and optically negative minerals have oblate indicatrices. The actual procedure for determining optic sign is discussed in a later section.

How, then, does one actually determine birefringence? If we know the grain thickness (*t* = 30 μm for a standard petrographic thin section), we need only measure the retardation, and the problem is solved. Rather than measure Δ, a closely related phenomenon referred to as the characteristic *interference color* is evaluated instead. This is always done in crossed polarized light (analyzer inserted). The origin of the colors is related to the unequal transmission by the analyzer of the component wavelengths of white light. The extent of transmission depends on the retardation of each wavelength as the light advances through the mineral. Consider for a moment what happens to just one component, yellow light, for example, with a constant wavelength of 580 nm. If this ray propagates parallel to the unique optic axis of a uniaxial mineral, the vibration directions are within the isotropic section, and no retardation or birefrin-

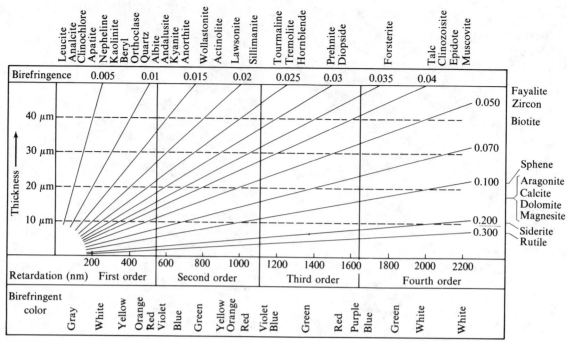

FIGURE 11.23   Michel-Lévy chart.

gence can occur (Figure 11.24a). This means that the plane polarized ray *CD* will be resolved by the crystal into two components, *OA* and *OB*, which travel through the crystal with the same velocity. Therefore, both components have the same refrac-

tive index and are actually indistinguishable. Because they remain "in phase," they will, after passing through any thickness *t*, recombine with the same *CD* vibration direction they started with. The analyzer will absorb all of this light.

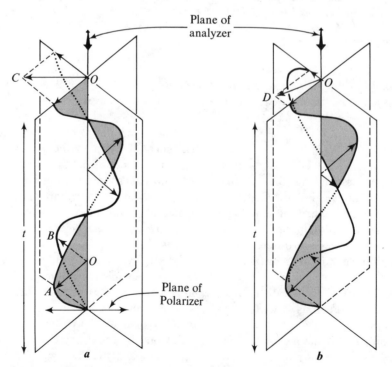

FIGURE 11.24   Effect of phase difference on transmission of light through the analyzer. (a) Two rays in phase recombine with the same orientation as the polarizer and are totally absorbed by the analyzer. Velocity and propa- gation are the same in each vibration direction. (b) Two rays out of phase recombine with a component in the analyzer direction, and hence will be transmitted.

Now consider what happens (Figure 11.24b) to the same ray when it passes through the mineral in a different direction. The fast component of the yellow light gets ahead of the slow component, and is generally "out of phase" with it after passing through some thickness t. When the components recombine, the resultant vibration direction is no longer parallel to that of the polarizer. The analyzer absorbs only part of the light, and the remainder passes through and is observed. In certain thicknesses, however, the fast ray will have advanced a full wavelength ahead of the slow ray, and the two will again be in phase. For these thicknesses (e.g., 580 nm, 1160 nm, 1740 nm in Figure 11.25), no light is transmitted. The condition for being in phase is simply that the retardation be an integer multiple of the wavelength in question. If this condition is not satisfied, light is transmitted through the analyzer.

This behavior for several of the component wavelengths of white light is summarized in Figure 11.25. Each row represents a specific wavelength and how the intensity of transmission varies with thickness. A wavelength of $\lambda = 700$ nm, for example, is interpreted by the human eye as the color red. At every integer multiple of 700 nm, extinction results. For values in between, red light of variable intensity is transmitted. Each component wavelength will behave in this manner but for different values of $n\lambda$. The resultant behavior that we actually observe with the petrographic microscope is shown in the last column of Figure 11.25. These are the interference colors that result when all wavelengths are combined for a fixed value of Δ minus the extinguished regions, and these are the characteristic colors of the Michel-Lévy chart.

An example of the use of such a chart follows:

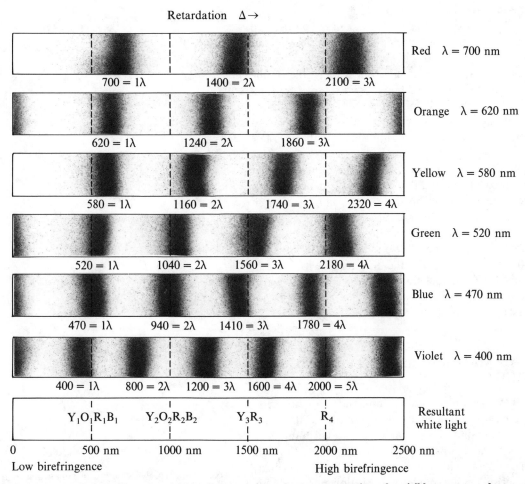

Retardation $\Delta \rightarrow$

Red $\quad \lambda = 700$ nm

$700 = 1\lambda \qquad 1400 = 2\lambda \qquad 2100 = 3\lambda$

Orange $\quad \lambda = 620$ nm

$620 = 1\lambda \qquad 1240 = 2\lambda \qquad 1860 = 3\lambda$

Yellow $\quad \lambda = 580$ nm

$580 = 1\lambda \qquad 1160 = 2\lambda \qquad 1740 = 3\lambda \qquad 2320 = 4\lambda$

Green $\quad \lambda = 520$ nm

$520 = 1\lambda \qquad 1040 = 2\lambda \qquad 1560 = 3\lambda \qquad 2180 = 4\lambda$

Blue $\quad \lambda = 470$ nm

$470 = 1\lambda \qquad 940 = 2\lambda \qquad 1410 = 3\lambda \qquad 1780 = 4\lambda$

Violet $\quad \lambda = 400$ nm

$400 = 1\lambda \quad 800 = 2\lambda \quad 1200 = 3\lambda \quad 1600 = 4\lambda \quad 2000 = 5\lambda$

$Y_1O_1R_1B_1 \qquad Y_2O_2R_2B_2 \qquad Y_3R_3 \qquad R_4$

Resultant white light

0 $\qquad$ 500 nm $\qquad$ 1000 nm $\qquad$ 1500 nm $\qquad$ 2000 nm $\qquad$ 2500 nm

Low birefringence $\qquad\qquad\qquad\qquad\qquad$ High birefringence

FIGURE 11.25 **Component wavelengths of white light are extinguished at every $n\lambda$ interval of retardation Δ. The wavelength is different for each component, so values of Δ for which extinction occurs are variable. The** sum of components gives the visible spectrum shown on bottom line. (After Phillips, Wm. Revell, 1971. Mineral optics, principles and techniques. San Francisco: W. H. Freeman and Company. Plate II.)

Suppose we observe that an unknown mineral to be identified displays a maximum birefringence color (analyzer is inserted) of bright blue. Lower birefringence colors will also be observed for most other grains if several hundred are crushed for simultaneous observation on a glass slide. Only a few of the grains, however, will have just the right orientation to give maximum birefringence. Reference to the Michel-Lévy chart shows the bright blue color corresponding to $\Delta = 600$ nm. If the mineral grains are sieved to a size interval close to 30 μm, or ground down to that thickness in the case of a petrographic thin section, then the maximum birefringence may be used as a property for identification. For $\Delta = 600$ nm and $t = 30$ μm = 30,000 nm, δ is determined by

$$\delta = \frac{\Delta}{t} = \frac{600}{30,000} = 0.020$$

Graphically, the diagonal line on the Michel-Lévy chart that passes through the intersection of $\Delta = 600$ nm, and $t = 30$ μm gives δ along the top edge of the chart, along with the common minerals characterized by that value of δ.

Another optical property that may be observed in many uniaxial minerals is *pleochroism*. This property is due to selective absorption of certain components of white light as a function of the vibration direction of light. For example, tourmaline is highly transparent to all wavelengths of white light vibrating parallel to its optic axis, but for light vibrating normal to the axis, all but a few wavelengths are absorbed. Those rays that pass impart a characteristic color to the mineral in that position of rotation that generally is dependent on the particular chemistry. For example, Fe-rich tourmaline is commonly pleochroic blue or green, whereas Mg-rich tourmaline is commonly pleochroic yellow. The color due to pleochroism is not to be confused with the interference color produced by unequal transmission of various wavelengths through the analyzer. The analyzer is not inserted when observing pleochroism.

UNIAXIAL MINERALS IN CONOSCOPIC LIGHT

Optical properties of uniaxial minerals discussed up to this point are determined with orthoscopic illumination. That is, a parallel bundle of rays from a substage light source passes through the mineral with the effects previously described. A second major mode of operation possible with the polarizing microscope uses conoscopic illumination. A converging lens in the substage assembly, a high-power objective lens, and an arrange-

ment for focusing on the back side of the objective lens are all the apparatus required. Focusing on the back side of the objective lens may be accomplished with a Bertrand lens in the body tube just before the ocular, or by removing the ocular and focusing with the unaided eye. Conceptually, however, the major difference that distinguishes conoscopic from orthoscopic illumination is the converging lens. It focuses an initially parallel bundle of light rays to a point at the geometric center of the mineral grain under examination (Figure 11.26). In so doing, we take

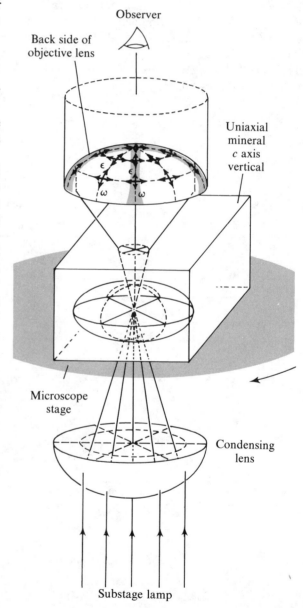

FIGURE 11.26 **Principle of conoscopic light illumination by use of a condensing lens. The pattern of intersecting vibration directions that appears on the back side of a high power objective lens has symmetry of the indicatrix of the mineral observed.**

full advantage of the indicatrix concept. A point light source now emanates rays in all directions through the grain. The objective lens intercepts all of the rays passing through a cone-shaped volume with the volume apex at the center of the crystal. A uniaxial indicatrix may also be imagined in the same location in order to analyze the various combinations of vibration directions associated with each of the now diverging rays. As an example, Figure 11.26 shows an optically negative uniaxial mineral with an optic axis perpendicular to the microscope stage. Upon rotation of the stage, the grain simply revolves around its optic axis. The resulting image obtained on the back side of the objective lens is called a centered optic axis *interference figure*. Its characteristic feature is a black cross centered in the field of view. The optic axis emerges from the center of the cross. The limbs of the cross are referred to as *isogyres*; their origin may be understood by studying Figure 11.27. The relative orientations of the ordinary and the extraordinary vibration directions are determined by appropriate sections through the indicatrix. The section normal to the optic axis is circular. When the analyzer is inserted, regions of the interference figure for which the vibration directions are aligned with the polarizer and the analyzer will be at extinction. We can visualize the entire image as made up of an infinite number of planar sections through the indicatrix distributed at every point in Figure 11.26. All sections other than the circular one are ellipses, each giving the vibration directions and the magnitude of the associated refractive indices for different ray paths through the crystal.

Mineral grains oriented for observation of the centered optic axis interference figures are easily recognized by their isotropic nature. Orientations slightly off the optic axis will show slight birefringence, and the center of the black cross, marking the emergence of the optic axis, will be oriented in an off-center position. Upon rotation of the stage, the center of the cross will rotate around the center of the field of view, and the isogyres will retain their north-south and east-west orientations.

Figures 11.27a and 11.27b illustrate centered uniaxial negative and positive interference figures, respectively. These differ in the relative magnitudes of n_ω and n_ϵ. In both cases, the ordinary wave vibration direction is tangential to the field of view, and the extraordinary wave direction is radial. The two wave types may be readily distinguished with the use of an auxiliary plate inserted between the objective lens and the Bertrand lens. The plate is commonly made of a crystal of quartz or gypsum oriented with its slow vibration direction (= high refractive index direction) perpendicular to the length of the plate. In the quartz plate, the thickness is such that light vibrating parallel to the slow direction is retarded by a constant 550 nm. The eye records this as a first-order red when the analyzer is in and only an isotropic mineral is on the stage. A normal interference figure observed in the absence of the accessory plate will exhibit minimal retardation in regions immediately adjacent to the optic axis. Typically, the retardation in those regions is on the order of 100 nm (white or light gray color). In regions farther removed from the optic axis, the diverging light rays emanating from the center of the mineral must travel through a greater thickness of crystal, and the retardation Δ will be correspondingly greater. Concentric rings, each characterized by a particular interference color, represent cones of

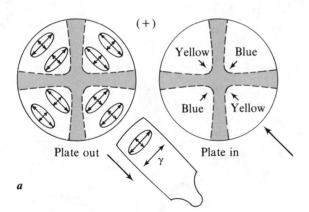

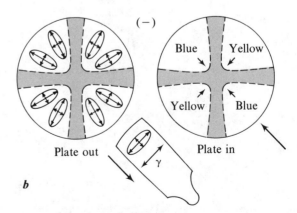

FIGURE 11.27 Use of an accessory quartz plate to determine optic sign in uniaxial minerals. The high refractive index (γ) direction of the plate is fixed perpendicular to the plate holder. (a) Insertion over ray path of positive minerals produces subtraction (fast ray on slow ray) and color pattern shown. (b) Insertion over ray path of negative minerals produces addition (fast ray on fast ray).

equal light retardation emanating from the center of the mineral. These are referred to as *isochromatic rings*.

When the accessory quartz plate is inserted, the net retardation seen by the eye is the combined retardation imposed by the mineral and the accessory plate. The specific combination will be additive or subtractive depending on the orientation of the inherent vibration directions of the mineral relative to that of the accessory plate. For example, consider in Figure 11.27a the effect of inserting a quartz plate over the interference figure of an optically negative, uniaxial mineral. Before entering the gypsum plate but after passing through the mineral, light vibrating parallel to the slow direction (n_ω in this case) of the crystal is 100 nm behind the fast ray (n_ϵ in this case) in the first-order gray regions next to the optic axis. Upon entering the quartz plate, the slow ray vibrates parallel to the slow ray vibration direction of the plate, and thus is retarded by an additional 550 nm relative to the fast ray. The net effect is additive, and a retardation of 650 nm reaches the eye as two blue spots located in the northwest and southeast quadrants next to the optic axis. In the northeast and southwest quadrants, the fast and slow vibration directions are rotated 90 degrees upon the ray's exit from the mineral. The fast ray in the mineral therefore becomes the slow ray in the quartz plate. The initial 100 nm retardation becomes − 450 nm after passage through the plate because of subtraction (100 nm − 550 nm = − 450 nm). The eye records this effect as two yellow spots situated next to the optic axis in the northeast and southwest quadrants.

Exactly the opposite effect is observed in the case of uniaxial positive minerals (Figure 11.27b). Insertion of the quartz plate results in subtraction in the northwest and southeast quadrants (yellow) and addition in the northeast and southwest quadrants. This is because the ordinary wave is now the slow direction rather than the fast direction, as in the preceding example. The distribution of blue and yellow interference next to the optic axis in both cases will be the same if the optic axis is off-center.

We can employ the same procedure to determine the *sign of elongation*. If the inherent vibration direction of the fast ray is parallel to the direction of elongation of the mineral, the mineral is said to be *length fast*. As the morphologic elongation is parallel to the *c* crystallographic axis in most uniaxial minerals, "length fast" corresponds with a negative sign and thus is sometimes called *negative elongation*. Minerals with slow ray vibration direction parallel to the elongation are termed *length slow*, or are said to have *positive elongation*.

OPTICAL PROPERTIES OF BIAXIAL MINERALS

We can now apply what we have learned about the anisotropic behavior of light in uniaxial minerals to the behavior of light in biaxial minerals. All orthorhombic, monoclinic, and triclinic minerals are biaxial. They are characterized by two optic axes (rather than one as in the uniaxial case), which are a consequence of the three principal vibration directions of all biaxial minerals. Light forced to vibrate in each of these directions travels

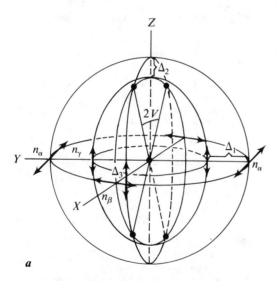

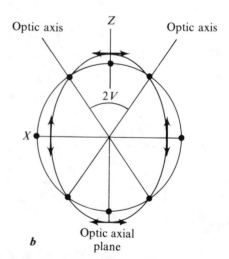

FIGURE 11.28   **(a) Ray velocity surface for a biaxial positive mineral with orthorhombic symmetry:** $X = a$, $Y = b$, **and** $Z = c$. **Retardation** Δ_1 **gives** $n_\gamma - n_\alpha$ **and** maximum birefringence on the (010) face. (b) Optic axial plane from (a) illustrating optic axes and retardations Δ_3 and Δ_2.

with a characteristic velocity dependent on the density of electrons and the polarizability of ions encountered by the propagating ray. These vibration directions are mutually perpendicular, and are called X (fast ray direction), Y (intermediate ray direction), and Z (slow ray direction). The associated refractive indices are referred to as n_α, n_β, and n_γ, respectively. In orthorhombic minerals, X, Y, and Z coincide with some combination of the unique a, b, and c crystallographic axes. In monoclinic and triclinic minerals, X, Y, and Z must remain mutually perpendicular even though a, b, and c in these systems need not be.

These relationships are readily visualized with the biaxial ray velocity surface (Figure 11.28) of a representative orthorhombic mineral. An imaginary point light source situated at the geometric center of the mineral emanates rays of light in all directions. First, consider light advancing parallel to the b axis, which has Y vibration direction in this case. The light is resolved by the mineral into two components vibrating perpendicular to each other and in the ac plane normal to the direction of propagation. One component with refractive index n_α and the other with refractive index n_γ vibrate parallel to X and Z, respectively. The former, n_α, refers to the fast ray. After some period of time, it will travel a greater distance than the ray n_γ vibrating perpendicular to it. The lag of one ray behind the other is the retardation, Δ_1, and as in the uniaxial case, may be related to the birefringence of the mineral in that orientation. Because of translation symmetry in the b direction, the same relationships must hold in the opposite direction as well.

Now consider light propagating parallel to the crystallographic c axis. The slow ray and its associated refractive index n_γ cannot be observed in this case. Instead, light vibrating parallel to X and Y will be observed. The fast ray will advance beyond the intermediate ray by a distance given by the retardation in this direction, Δ_2. Clearly, Δ_2 is less than Δ_1, so the birefringence of the mineral observed parallel to its c axis will be less than that observed parallel to its b axis.

Lastly, consider light that propagates parallel to X. The ray vibrating parallel to Y advances beyond the ray vibrating parallel to Z by an amount given by Δ_3. Because Δ_3 is related to the maximum birefringence the mineral may have in any orientation, it is a diagnostic optical property $(n_\gamma - n_\alpha)$ useful for identification purposes.

Of particular importance in Figure 11.28 is the fact that the surface defined by the intermediate wave velocity intersects that defined by the slow and fast wave velocities. At the point of intersec-

tion, all rays travel at the same velocity. This point and its symmetrically equivalent point, also in the XZ plane, are the two *optic axes*. Sections normal to the optic axes are isotropic and may be recognized on that basis. The plane containing the two optic axes is called the *optic axial plane*. These relationships for biaxial minerals can also be represented by a triaxial ellipsoid, as illustrated in Figure. 6.2. As in the uniaxial case, the length of the principal axes is in proportion to the refractive indices of light vibrating in those directions. In the case of biaxial minerals, however, three principal axes exist, given by the optical directions X, Y, and Z.

All sections through the indicatrix with but two exceptions will be ellipses. Their major and minor axes are perpendicular, and their magnitudes are in proportion to the characteristic refractive indices of light vibrating in those directions. The two exceptions are circular sections situated normal to the optic axes. The angle between the optic axes is called the *optic angle* and is abbreviated $2V$. Biaxial minerals are categorized as optically positive or negative, depending on whether the optic angle is bisected by the X optical direction $(-)$ or the Z optical direction $(+)$. We discuss procedures for making this distinction and methods for estimating $2V$ in the next section.

By analogy with uniaxial minerals, we see that various sections through the biaxial indicatrix will represent the mutual relationships between refractive indices for various orientations of any particular mineral. The $2/m\,2/m\,2/m$ point group symmetry of the indicatrix gives the optical symmetry of space around every point in the mineral, so it follows that the indicatrix as a geometric concept also exists at every point. Figure 11.29 illustrates these relationships for a representative orthorhombic mineral. If we were to proceed systematically with the determination of refractive indices n_α, n_β, and n_γ, certain orientations of the mineral would be more advantageous to examine than others. For example, (010) reveals two rays vibrating perpendicular to each other and having refractive indices n_α and n_γ. When the grain is rotated to a position of extinction, remove the analyzer and the entire grain will exhibit one index, for example, n_α. Rotate the stage 90 degrees, and n_γ may be determined. Note that in this last orientation, the mineral displays its maximum birefringence.

Similarly, we can determine n_β and n_γ by observations on (100). Observations on (001) reveal information about n_α and n_β. An orientation normal to the optic axis is recognized by its isotropism. It has but a single refractive index, n_β. Other orientations will result in elliptical sections

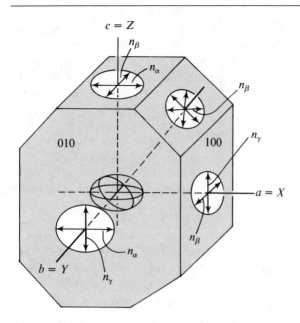

FIGURE 11.29 **Schematic representation of the role of translational symmetry in achieving identical point group symmetry at every point in a biaxial mineral.**

through the indicatrix. The difference between the length of the major and minor axes in each case will give the birefringence. In the routine examination of many mineral grains, either in thin section or in an oil immersion preparation, an appropriate choice of grains on the basis of birefringence will shorten the time required to determine the refractive indices. The darkest or least birefringent sections will be close to the optic axis, and thus will reveal n_β at all positions of rotation. Grains with the highest birefringence will show n_α and n_γ.

From Figure 11.29, we observe that orthorhombic minerals exhibit *parallel extinction* on most prismatic cleavages and crystal faces, because the optic directions X, Y, and Z are parallel to the crystallographic axes a, b, and c. This is not the case for monoclinic and triclinic minerals. For example, Figure 11.30a shows the optic orientation of common hornblende on a (010) cleavage fragment. Y and the b axis are coincident and normal to the page. The angle β between axes a and c is 105 degrees. Hornblende exhibits inclined extinction of 21 degrees in this orientation. Other prismatic orientations closer to (100) will show extinction angles of less than the maximum. All sections parallel to $b = Y$ will show parallel extinction, because the inherent vibration direction Z and the c axis lie in the same plane. The section showing the maximum extinction angle may be recognized as that showing maximum birefringence.

An examination of sections cut perpendicular to the c axis of hornblende reveals (110) cleavages

intersecting at an acute angle of 56 degrees and an obtuse angle of 124 degrees (Figure 11.30b). These cleavages are symmetrically arranged with respect to a and b, and therefore X and Y. Symmetrical extinction with respect to (110) cleavages results.

Following the procedure outlined for determining refractive indices, we can determine the pleochroism of biaxial minerals. The preferential absorption of certain wavelengths of white light depends on the particular vibration direction. As an example, the hornblende orientation in Figure 11.30b exhibits a yellow green color for light vibrating parallel to (010), but exhibits a green color when the grain is rotated such that (100) is parallel to the north-south direction of the polarizer. In contrast, the same mineral observed in a (010) orientation (Figure 11.30a) shows a blue green color when Z is parallel to the north-south vibration direction of the polarizer.

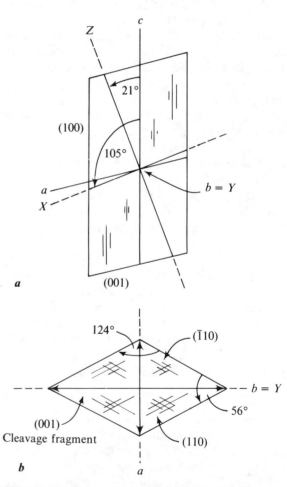

FIGURE 11.30 **(a) Photomicrograph of (010) cleavage plane of hornblende. The $2/m$ point group symmetry requires $b = Y$. Optical directions X and Z are constrained to plane ac (also the OAP). (b) Photomicrograph of (001) cleavage fragment showing {110} prismatic cleavage intersections. Optical directions X and Z are in plane ac (c is vertical to the page).**

BIAXIAL MINERALS IN CONOSCOPIC LIGHT

An understanding of the optical behavior of biaxial minerals in conoscopic illumination requires some additional nomenclature. Figure 11.31 illustrates the optic axial plane (OAP) in the plane of the page. Perpendicular to the optic axial plane is the Y optical direction. It follows that X and Z as well as the two optic axes must lie in the optic axial plane. The line bisecting the acute angle between the optic axes is defined as the *acute bisectrix*, and is abbreviated BXA. The line bisecting the obtuse angle between the optic axes is defined as the *obtuse bisectrix*, abbreviated BXO. Only two orientations of the bisectrices and the optical directions X and Z are possible. Figure 11.31a shows Z coincident with the BXA that defines optically positive biaxial minerals. Figure 11.31b shows X coincident with the BXA that defines optically negative biaxial minerals.

The purpose of the substage condensing system on the microscope is to focus an initially parallel bundle of light to a point in the mineral under examination. Figure 11.32 represents a typical bi-

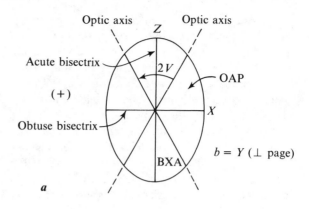

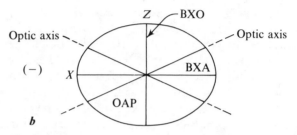

FIGURE 11.31 Schematic representation of optic axial planes (OAP) of (a) positive and (b) negative biaxial minerals. Optical direction Y coincides with crystallographic axis b, and both are normal to the page.

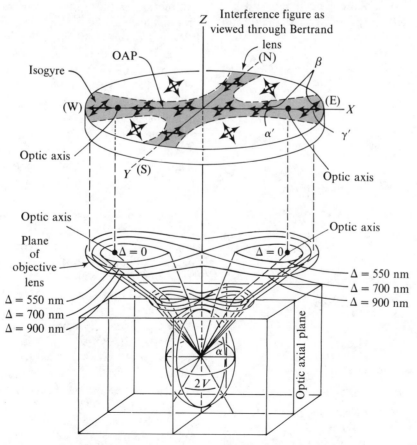

FIGURE 11.32 Schematic representation of an interference figure for a biaxial positive mineral in position of extinction. The isogyre arrangement resembles a uniaxial interference figure, but will break up into separate isogyres upon rotation of the stage.

axial positive mineral oriented with its BXA in a vertical position. Each light ray emanating from the center of the mineral will be associated with mutually perpendicular vibration directions. Several of these are shown in the field of view of the biaxial interference figure. When the mineral is in this position of rotation, a centered black cross similar in appearance to a centered uniaxial interference results. Unlike the uniaxial case, however, this cross will break up when the stage is rotated. The points of emergence of the two optic axes are sites of zero retardation, since the light rays propagating in these directions travel with the same velocity. Isochromatic rings representing cones of equal retardation emanating from the point light source are symmetrically arranged around the optic axes and the optic axial plane. Because the birefringence varies directly with the thickness of the mineral through which a ray passes, the birefringence associated with each of the isochromatic rings must increase away from the optic axes.

Figure 11.33 illustrates the change in appearance of the biaxial interference shown in Figure 11.32 with a 45-degree rotation of the microscope stage. The optic axial plane is now in a 45-degree position. The north-south and east-west positions

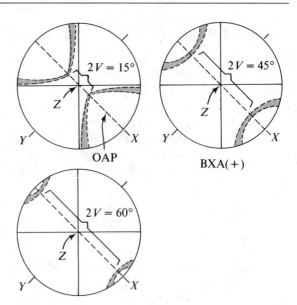

FIGURE 11.34   Interference figures of biaxial positive minerals with Z optical directions oriented normal to microscope stage. Magnitude of optic angle $2V$ may be qualitatively estimated.

of the inherent vibration directions mark areas of extinction, which again are symmetrically arranged around the optic axial plane. The Y direction must be perpendicular and in the plane of the page. Rotating the stage an additional 45 degrees will bring the isogyres back to the form shown in Figure 11.32. This will be the form of the centered BXA interference figure whenever the optic axial plane is in a north-south or east-west position of rotation.

Centered BXA interference figures in a 45-degree position of rotation are useful for determining both the optic angle $2V$ and the optic sign. The angular field of view of the interference figure on most microscopes is about 65 degrees. If the isogyres in the 45-degree position just reach the edge of the field of view, the $2V$ is near 65 degrees. Figure 11.34 illustrates the separation of optic axes for $2V$ values of 15, 45, and 60 degrees. Clearly a $2V$ of zero corresponds to a uniaxial interference figure.

The problem of determining optic sign reduces to whether the BXA is equivalent to X or Z. If the former, the mineral is negative; if the latter, the mineral is positive. This determination can easily be made by analyzing the vibration directions of Figures 11.33 and 11.35. If $Z = $ BXA, as in Figure 11.33, then the vibration direction in the optic axial plane must be X, since Y is always normal to the optic axial plane. The preferred vibration directions of the mineral in this orientation will be X and Y, with light vibrating parallel to X having

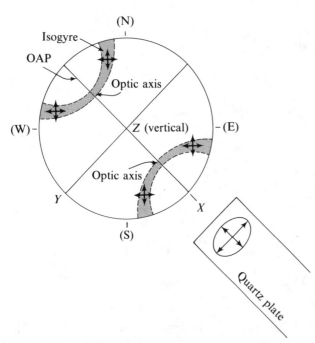

FIGURE 11.33   Positive biaxial interference figure in a position of rotation 45° from extinction. Isogyre pattern of Figure 11.32 has divided, and the OAP is in a 45° position. Isogyre pattern forms where vibration directions are coincident with the north-south polarizer and east-west analyzer.

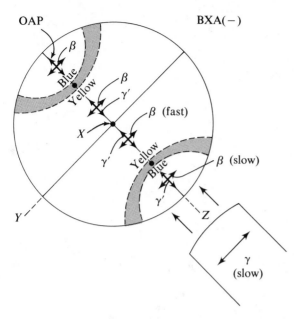

FIGURE 11.35   Biaxial negative interference figure showing relative positions and magnitudes of vibration directions. Insertion of accessory quartz plate produces subtraction to give a 550 nm retardation next to the optic axis on concave side of the isogyre.

the greater velocity and the lower refractive index. Insertion of the accessory quartz plate with its slow ray direction parallel to the slow (Y) vibration direction of the crystal produces addition (550 nm + 100 nm = first-order blue) on the convex side of the isogyres next to the optic axes. An examination of the ray velocity surface in Figure 11.28 shows that on the concave side of the iso-

gyre the relative velocities of light in the two vibration directions are reversed. On the concave side of the isogyre next to the optic axis, light vibrating parallel to Y is the faster ray. Insertion of the accessory plate places a slow direction parallel to a fast direction and subtraction (550 nm − 100 nm = first-order yellow) results. Figure 11.35 illustrates the opposite effect when $X = \text{BXA}$. In this case, the preferred vibration directions are parallel to Y and Z with Y being the faster direction on the convex side of the isogyre. Consequently, insertion of the gypsum plate will produce subtraction on the convex side and addition on the concave side. In the first-order gray regions adjacent to the optic axes, the diagnostic evidence for negative sign will be a blue spot on the concave side and a yellow spot on the convex side.

Centered optic axis figures are useful for the determination of $2V$ and optic sign. Such orientations are easily recognized by their lack of birefringence. Figure 11.36 illustrates how to estimate $2V$ from such figures. The optic sign determination is exactly the same as for a centered BXA figure, except that only one of the optic axes is visible. We can determine the orientation of the optic axial plane by the curvature of the isogyre. It is important that the OAP be consistently oriented in a northwest-southeast direction for the addition and subtraction effects as described to remain valid. If the OAP is oriented northeast-southwest, exactly the opposite effects will be observed.

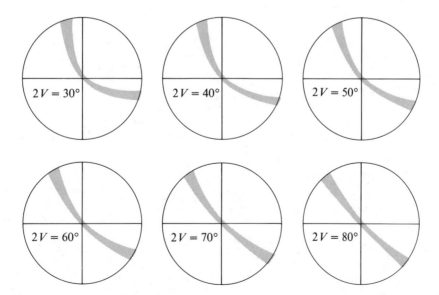

FIGURE 11.36   Representations of interference figures in an optic axis orientation. Such orientations are recognized in crossed polarized light by very low birefringence. Magnitude of the optic angle may be qualitatively estimated from such orientations.

PART II

DESCRIPTIVE MINERALOGY

The remainder of this text is devoted to the description of the structure, chemistry, and physical properties of most of the common or important minerals on earth. Our primary purpose is to describe the characteristics of minerals that are most likely to be encountered in an introductory mineralogy course. A secondary objective is to provide descriptions of minerals that illustrate concepts developed in Part I. Some of these minerals may be relatively rare and will not be encountered by most mineralogy students. A number of references listed at the end of Part II expand further on key topics.

The organization of mineral groups is as follows:

Silicate minerals
    Network silicates
    Layer silicates
    Chain silicates
    Ring silicates
    Isolated tetrahedral silicates
Native elements
    Metals
    Nonmetals
Sulfide minerals
    Simple sulfides
    Complex sulfides

Halide minerals
Oxide minerals
Hydroxide minerals
Carbonate and nitrate minerals
Borate minerals
Sulfate minerals
Chromate, tungstate, and molybdate
    minerals
Phosphate, arsenate, and vanadate
    minerals

The descriptive text for each mineral group begins with a summary of the general structure followed by specific sections on structure, chemistry, occurrence and associations, and distinguishing features for each mineral. The descriptive data for selected minerals in each group are presented in vertical tables organized in three categories. The first category refers to the chemistry and structure in the following format:

Formula: The standard chemical formula as discussed in Chapter 5, p. 118.
Coord: The coordination number for each symmetrically different ion in the chemical formula. For example, Si(4), O(2) in low quartz denotes fourfold coordination of oxygen around Si and twofold coordination of Si around oxygen.

System: The crystal system of the mineral.

a, b, c: Representative unit cell parameters in units of nanometers.

Z: The number of chemical formula units per unit cell.

Sp. group: The complete space group symbol.

Pt. group: The complete point group symbol. Refer to Chapter 2, p. 33 for discussion.

S; loops; RU: The sharing coefficient of the primary frame according to Equation 5.1, Chapter 5. "Loops" refers to the number of coordination polyhedra in closed loops of one-dimensional, two-dimensional, or three-dimensional structures. If such loops are absent, the repeat unit (RU) gives the number of polyhedra in the asymmetric unit. For example, the designation 6,8 for low quartz means that two different types of loops are in the structure, one with six tetrahedra and the other with eight tetrahedra.

SP plane: The symmetrical packing plane (hkl) designates the lattice plane parallel to anion sheets, which are arranged according to the packing symbol given below.

SP struct: The symmetrical packing structure as discussed in Chapter 5, p. 140. Continual reference should be made to Table 5.13. A hyphen in this row indicates that the packing of anions and void-filling cations in the structure is too complicated to be conveniently represented by symmetrical packing.

SPI: Symmetrical packing index calculated according to Equation 5.7, Chapter 5. The index expresses the percentage of volume occupied by anions in the structure. Refer to p. 148 for discussion.

The second category of properties refers to crystal form, cleavage, twinning, and important optical properties.

Cleavage: The cleavage orientation and prominence (pf = perfect, gd = good, pr = poor).

Twinning: The twin law by name, the twin axis [uvw], and composition plane (hkl), or twin plane (hkl).

RI: Representative refractive indices. Refer to Chapter 11, p. 268 for discussion.

δ: Birefringence as calculated from refractive indices.

$2V$: Representative optic angle for biaxial minerals. "None" refers to uniaxial and opaque minerals.

Sign: Refer to Chapter 11, p. 281 for discussion. Positive ($+$) sign refers to Z = BXA; negative ($-$) sign refers to X = BXA.

The third category of properties refers to standard physical properties.

H: Representative hardness of the mineral on the Mohs scale from one (softest) to ten (hardest). Refer to Chapter 1, p. 15 for discussion.

D: Representative density for the structure and chemistry given. Units are megagrams per cubic meter (Mg/m^3), which are equivalent to grams per cm^3.

Color: The usual color of commonly encountered hand specimens. Refer to Chapter 1, p. 13 and Chapter 4, p. 102 for discussions.

Streak: The usual color of the powdered mineral. Refer to Chapter 1, p. 13 for discussion.

Luster: Luster of commonly encountered hand specimens.

Fracture: Fracture of commonly encountered hand specimens.

Habit: A statement of the most common appearance in hand specimens. Compare with representative crystal form at top of table.

Minerals written in all capital letters at the top of table columns are illustrated with stereo pair drawings.

THE SILICATE MINERALS

The silicates are volumetrically the most important of all minerals, comprising over 90% of the earth's crust and mantle. Feldspars and quartz are most abundant and widespread in the crust; olivine and pyroxenes are the most abundant minerals in the upper part of the mantle. All silicate minerals have a primary frame consisting principally of silica polyhedra linked together in various ways. Nearly all of the silica polyhedra are tetrahedra; a few are octahedra. The ways in which the tetrahedra link together in the structure provide the basis for classification as discussed in Chapter 5.

The following descriptive sections on silicate minerals begin with the three-dimensional network structures so characteristic of the most common minerals on earth, namely, feldspar and quartz. We then proceed to silicates such as the micas and clays that have two-dimensional frame structures. The one-dimensional structures (e.g., pyroxenes and amphiboles) are discussed next, followed by minerals with zero-dimensional frames. The latter include important minerals such as olivine and garnet. This sequence is the same as the sequence used for the classification of tetrahedral frame structures in Chapter 5, beginning with the most completely developed or polymerized frames and proceeding systematically to conclude with those having less extensive tetrahedral linkages.

The Network Silicates

The groups of minerals included within the network silicates and discussed here are as follows: silica minerals, feldspars, feldspathoids, scapolites, other network silicates, and zeolites.

In all of these silicates, known also as the tectosilicates, the linkage of tetrahedra constitutes a continuous three-dimensional network that extends throughout the structures. The many substructures differ in detail, some having distinctive features that are characteristic of a particular mineral. Examples are the four-membered tetrahedral loops in feldspars, the six-membered loops in beryl and cordierite, and the linked chains of four-membered tetrahedra units in the fibrous zeolites such as natrolite.

Although many silicates possess distinctive geometric groups within their structures, the principal criterion for placing a mineral within the network structures is that the *connectivity* of tetrahedra be three-dimensional. On this basis, minerals such as beryl and cordierite, both of which have been classified in the past as ring silicates because of their distinctive six-membered loops, are considered here as network silicates. Tourmaline, on the other hand, possesses similar six-membered "loops," but because the "loops" are not connected to one another in three dimensions, they are isolated rings and we do not treat tourmaline as a network silicate.

Division of the network silicates into the various categories is based mainly on structural differences. Distinctive physical properties commonly result from these differences. One such case is the zeolites. The usual criterion for classifying a particular network silicate as a zeolite is its ability to exchange cations and to gain or lose water without changing its structure. The mineral analcime has such properties, but its structure is more like that of feldspathoids then of typical zeolites. The mineral sodalite is commonly referred to as a zeolite. Its structure has some similarities to the feldspars, but the channels connecting the tetrahedral cages are larger than in the feldspathoids. In this text, we treat analcime as a feldspathoid and sodalite as a zeolite. The wisdom of these decisions can be debated, but the important issue is the relationship between structure and physical properties. The classification itself is only an attempt to organize this relationship.

Nearly all of the network silicates have sharing coefficients (Chapter 5) of exactly 2.00. Pellyite ($K_2CaSi_2(Fe, Mg)_6O_{17}$; $S = 1.94$), barylite ($Si_2Be_4O_7$; $S = 2.375$), and willemite ($SiZn_2O_4$; $S = 3.0$) are among the few exceptions, but in each of these unusual cases a divalent cation is in tetrahedral coordination with Si. Stishovite (SiO_2, $S = 3.00$), the very high-pressure polymorph of silica, has no cations other than Si, but in this case the coordination is octahedral, and the octahedra share edges.

The Silica Minerals

The silica minerals all have the chemistry SiO_2 and are second only to feldspars in abundance in the earth's crust. They occur in several polymorphs (Figure II.1), depending on the temperature and pressure of formation. Quartz is the most common of these, being an essential constituent of many sedimentary, metamorphic, and igneous rocks.

Structure. The general structure of all of the SiO_2 polymorphs except stishovite consists of a three-dimensional network of silica tetrahedra, each linked to four other tetrahedra by sharing of apical oxygens (Figure II.2). Differences in the basic tetrahedral frame distinguish the SiO_2 polymorphs from one another. The three-dimensional

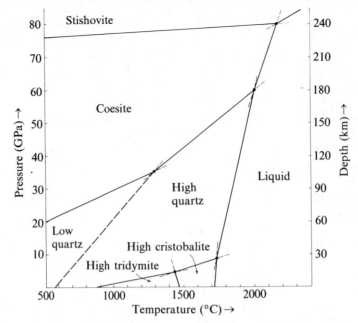

FIGURE II.1   Pressure-temperature phase diagram for SiO₂ polymorphs. Long dashes represent possible sec- **ond-order transition. Short dashes represent metastable extensions of univariant reaction curves.**

Quartz
SiO₂

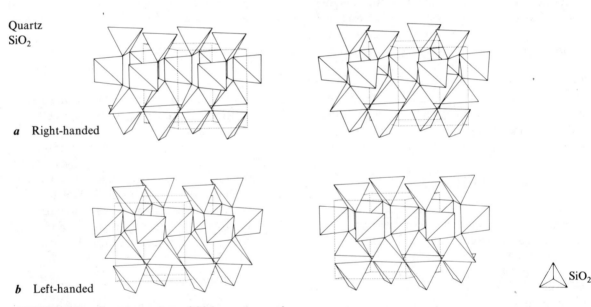

a Right-handed

b Left-handed

FIGURE II.2   Crystal structure of high quartz, *c* axis vertical. (a) right-handed, (b) left-handed.

structure of quartz (Figure II.2) reveals both six-membered and eight-membered loops of tetrahedra in a relatively compact arrangement. Figures 6.7a and 6.7b are (001) projections of low and high quartz illustrating these loops and how they are more expanded and more symmetrical in the high-temperature polymorph. The structure of tridymite (Figure II.3), an even higher temperature polymorph, has two sets of six-membered loops that are arranged less compactly than the loops in either low or high quartz. Consequently, the den-

sity and refractive indices of tridymite are less than those of quartz. Cristobalite (Figure II.4) has only a single set of six-membered loops in an arrangement that actually yields a higher density ($\rho = 2.33$ Mg/m³) than the lower temperature tridymite ($\rho = 2.28$ Mg/m³). Cristobalite must therefore be stable at higher pressure than tridymite when both are at the same temperature. The reaction curve between phases on a pressure-temperature diagram (Figure II.1) expresses this fact with a negative slope.

High tridymite
SiO$_2$

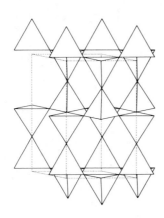
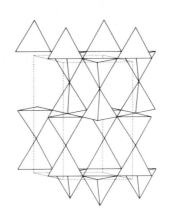

SiO$_2$

FIGURE II.3   Crystal structure of high tridymite, *c* axis vertical.

High cristobalite
SiO$_2$

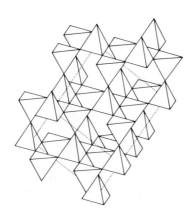
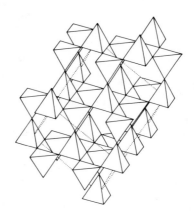

SiO$_2$

FIGURE II.4   Crystal structure of high cristobalite, [111] direction vertical.

The structures of both tridymite and cristobalite can also be described in terms of symmetrical packing (Chapter 5). The oxygen sheets in each case, however, are incomplete. Only 75% (A sheet) and 25% (B sheet) of alternating symmetrical sites are occupied by oxygen. In these terms, tridymite is an *AB*-type structure, and cristobalite is an *ABC*-type structure. The triangular bases of each tetrahedra in the six-membered loops of both tridymite and cristobalite define the A sheet of oxygens in the (001) and (111) planes, respectively. In both structures, adjacent loop tetrahedra point alternately up and down, giving the B sheet only 25% occupancy by oxygen. In tridymite (Figure II.3), "up-pointing" tetrahedral apices of the *B* sheet are shared with "down-pointing" apices of tetrahedra that have triangular bases defining an overlying *A* sheet. In cristobalite (Figure II.4), however, the equivalent triangular bases are rotated 60 degrees to define a C sheet. This rotation increases the distance between nonlinking oxygens in the *A* and *C* sheets of cristobalite, which lessens

the repulsive forces relative to tridymite. Consequently, the tetrahedral layer in cristobalite is slightly thinner than in tridymite, and this accounts for the slightly higher density of cristobalite.

The crystal structure of coesite (Figure II.5) is considerably more compact than other tetrahedral frame polymorphs, and has correspondingly greater density and higher refractive indices. Four-membered loops of tetrahedra dominate the structure in a pattern that cannot be conveniently represented by a symmetrical packing model. Stishovite is the highest pressure polymorph, having the highest density, highest refractive indices, and highest packing index. The structure is a common type, usually referred to as the *rutile structure type*. The three-dimensional network of silica octahedra is characterized by edge-sharing single chains. This structure may also be described in terms of symmetrical packing with oxygen in *AB* stacking sheets and alternate octahedral voids occupied by Si.

**THE SILICA
MINERALS**

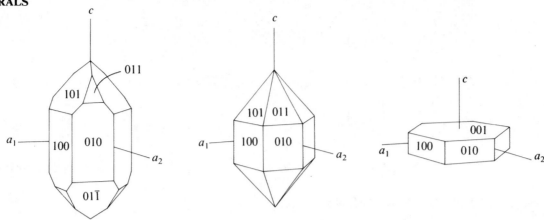

	LOW QUARTZ	High Quartz	LOW TRIDYMITE
Formula	SiO_2	SiO_2	SiO_2
Coord.	Si(4)	Si(4)	Si(4)
System	Trigonal	Hexagonal	Orthorhombic
$a =$	0.4913	0.4999	0.99
$b =$			1.71
$c =$	0.5405	0.5457	1.63
$Z =$	3	3	64
Sp. group	$R3_121$ (right)	$P6_222$ (right)	$P222$
	$R3_221$ (left)	$P6_422$ (left)	
Pt. group	32	622	222
S; loops	2.00; 6, 8	2.00; 6, 8	2.00; 6, 6
SP plane			(001)
SP struct.			$A_{75}(-\frac{1}{4}-)B_{25}(-\frac{1}{4}-)$;
			$A_{75}(--\frac{1}{4})B_{25}(--\frac{1}{4})$;
SPI $=$	49	47	42
Parting			{120}, (001)
Twinning	Dauphiné [001]	{102}, {302}	{110}
	Comp. plane	{201}, {112}	
	{$hk0$} or irregular		
	Brazil {110}		
	Japan {112}		
RI	$n_\omega = 1.544$	$n_\omega = 1.540$	$n_\alpha = 1.478$
	$n_\epsilon = 1.533$	$n_\epsilon = 1.533$	$n_\beta = 1.479$
			$n_\gamma = 1.481$
$\delta =$	0.009	0.007	0.003
$2V =$			70°
Sign	(+)	(+)	(+)
Transp.	Transp. to transl.	Transp. to transl.	Transp. to transl.
$H =$	7	7	6–7
$D =$	2.65	2.53	2.28
Color	Colorless, variable	Colorless	Colorless
Streak	White	White	White
Luster	Vitreous	Vitreous	Vitreous
Fracture	Conchoidal	Conchoidal	Conchoidal
Habit	Prismatic, massive	Stubby bipyramidal	Wedge shaped
Remarks			pseudomorph
			Structural variety: opal

fibrous massive: Chalcedony
pyroelectric and piezoelectric

THE SILICA MINERALS

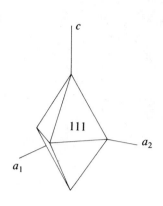
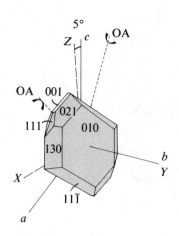
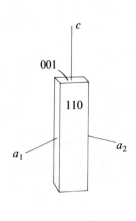

	LOW CRISTOBALITE	COESITE	Stishovite
Formula	SiO_2	SiO_2	SiO_2
Coord.	Si(4)	Si(4)	Si(6)
System	Tetragonal	Monoclinic	Tetragonal
$a =$	0.497	0.717	0.418
$b =$		1.233	
$c =$	0.693	0.717	0.266
$\beta =$		120.0°	
$Z =$	4	16	2
Sp. group	$P4_32_12$	$C2/c$	$P4_2/m\,2_1/n\,2/m$
Pt. group	422	$2/m$	$4/m\,2/m\,2/m$
S; loops	2.00; 6	2.00; 4, 6, 8, 9	3.00; 2, 4
SP plane	{112}		{110}
SP struct.	$A_{75}(\frac{1}{4}--)B_{25}(\frac{1}{4}--)C_{75}(--\frac{1}{4})$; $A_{25}(--\frac{1}{4})B_{75}(-\frac{1}{4}-)C_{25}(-\frac{1}{4}-)$;		$A(--\frac{1}{2})B(--\frac{1}{2})$;
SPI $=$	43	52	70
Cleavage			{110}
Twinning	Spinel {111}	(100), {021}	{011}
RI	$n_\omega = 1.489$ $n_\epsilon = 1.482$	$n_\alpha = 1.59$ $n_\beta = 1.60$ $n_\gamma = 1.60$	$n_\omega = 1.799$ $n_\epsilon = 1.826$
$\delta =$	0.007	0.01	0.027
$2V =$		64°	
Sign	$(-)$	$(+)$	$(+)$
Transp.	Transparent	Transparent	Transparent
$H =$	6–7	7–8	
$D =$	2.33	2.93	4.30
Color	Colorless	Colorless	Colorless
Streak	White	White	White
Luster	Vitreous	Vitreous	Vitreous
Fracture	Conchoidal	Conchoidal	
Habit	Cubic, coarse aggr.	Tabular	Prismatic
Remarks	Structural variety: opal	High pressure	Extreme pressure Rutile structure

Coesite
SiO_2

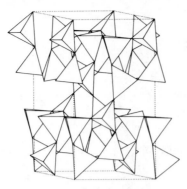
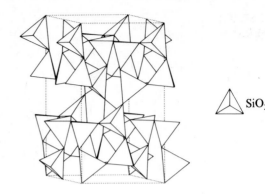

SiO_2

FIGURE II.5   Crystal structure of coesite, *b* axis vertical.

The silica minerals provide one of nature's best examples of *reconstructive* and *displacive* transformations. The reconstructive changes require a new structural arrangement of ions and hence the breaking and reforming of chemical bonds. All of the phase boundaries shown as solid lines in Figure II.1 are of this type. Because bonds must be broken, the transitions are sluggish, and polymorphs that are stable at relatively high pressure and temperature are commonly preserved under normal conditions. In contrast are the displacive transformations between the low and high forms of quartz, low and high tridymite, and low and high cristobalite. These structural changes involve a change in symmetry but no breaking of bonds. The transformations are rapid, and consequently, only the low-temperature forms are preserved.

Several varieties of cryptocrystalline quartz have important aggregate structures. Chalcedony is the most common, consisting of submicroscopic grains, rods, or fibers of quartz. Banded chalcedony is called *agate. Jasper, chert, onyx,* and *carnelian* are varieties of chalcedony. *Opal* is hydrated

silica and is composed of small, partially crystallized spheres (a few hundred nm) in a close-packed arrangement. In precious opal, this arrangement constitutes a three-dimensional grating that diffracts light and creates a fiery luster.

Several of the silica mineral structures can be taken as a basis for deriving other network structures and corresponding chemical formulas. For example, the crystal structure of feldspar can be derived from the coesite structure by substituting a lower valence Al^{3+} cation for every third Si^{4+}, and balancing the negative charge of the tetrahedral frame by adding monovalent Na^{1+} or K^{1+}. This is not to say that coesite and feldspar are isostructural. The coesite structure has no void space large enough to accommodate such monovalent cations. If, however, the coesite structure were distorted to such an extent that alkalies could enter the structure, a feldsparlike structure would result. Table II.1 gives a list of the more common silicates that can be considered as derivatives of the various SiO_2 structures.

Chemistry. The silica minerals are essentially

TABLE II.1.   *Silica Minerals and Their Structural Derivatives*

Silica Mineral	Tetrahedral Substitution per one Si in SiO_2	Extra Cation per Tetrahedral Substitution	Derivative Silicate	
Coesite	$\frac{1}{4}Al^{3+}$	$\frac{1}{4}K^{1+}$	Microcline	KSi_3AlO_8
Coesite	$\frac{1}{4}Al^{3+}$	$\frac{1}{4}Na^{1+}$	Albite	$NaSi_3AlO_8$
Coesite	$\frac{1}{2}Al^{3+}$	$\frac{1}{4}Ca^{2+}$	Anorthite	$CaSi_2Al_2O_8$
Tridymite	$\frac{1}{2}Al^{3+}$	$\frac{1}{2}K^{1+}$	Kalsilite	$KSiAlO_4$
Tridymite	$\frac{1}{2}Al^{3+}$	$\frac{3}{8}Na^{1+} + \frac{1}{8}K^{1+}$	Nepheline	$KNa_3Si_4Al_4O_{16}$
Cristobalite	$\frac{1}{2}Al^{1+}$	$\frac{1}{2}Na^{1+}$	Carnegieite	$NaSiAlO_4$
High quartz	$\frac{1}{2}Al^{3+}$	$\frac{1}{2}Li^{1+}$	Eucryptite	$LiSiAlO_4$
Keatite*	$\frac{1}{3}Al^{3+}$	$\frac{1}{3}Li^{1+}$	High spodumene	$LiSi_2AlO_6$

*Does not occur naturally.

pure SiO_2 because of the severe constraints imposed by the structure on the type and extent of solid solution. Only the more open structures of tridymite and cristobalite tolerate small but notable cation substitution, mainly Al^{3+} for tetrahedral Si^{4+}. Charge balance is maintained by monovalent alkalies such as Na^{1+} and K^{1+} occupying the large structural cavities in these high-temperature forms. The other polymorphs are much denser and have smaller cavities that are less receptive to foreign constituents.

Occurrence and Associations. Quartz is exceedingly widespread in detrital sedimentary rocks because of its resistance to chemical weathering and mechanical disintegration. It also forms as authigenic crystals and as overgrowths on detrital grains. Metamorphic rocks that are derived from siliceous sediments are quartz-rich. Such sediments are usually clay-rich, and therefore alumina-rich minerals such as kyanite, sillimanite, andalusite, staurolite, and cordierite are common associates of quartz. Among the igneous rocks, granites and related intrusives such as pegmatites are usually enriched in quartz. Alkali feldspars are nearly always present. Quartz is never associated with nepheline because of their reaction (Figure 9.3) to form albite. Similarly, quartz is never associated with corundum (Al_2O_3) because of their reaction to form the Al_2SiO_5 polymorphs. For this reason, quartz is also not found in syenites and other silica-undersaturated igneous rocks. Quartz and other SiO_2 polymorphs are never found with forsterite because of their reaction to form enstatite (Figure 9.2) nor is quartz found in most basalts, gabbros, and other mafic igneous rocks.

Extrusive or volcanic equivalents of granites and related rocks are the principal rocks in which tridymite and cristobalite occur. Rapid cooling from high temperatures preserves these minerals before they have time to revert to low quartz, the most stable polymorph under room pressure and temperature. The high-temperature alkali feldspar, called sanidine, is a common association along with biotite, hornblende, and Fe-Ti oxides.

Coesite and stishovite, the high-pressure polymorphs, are rare in nature. They are known from meteor impact craters and from a few localities where deep-seated mantle material has been brought to the earth's surface by kimberlites. Both polymorphs have been experimentally synthesized at high pressure with the results shown in Figure II.1.

The silica minerals are important for industrial purposes. Quartz is a basic constituent of glasses and fluxes; it is a common abrasive material and its crystals are used for oscillators and piezoelectric devices. Tridymite and cristobalite are common in high-temperature refractory bricks. Colored varieties of quartz such as amethyst and citrine provide a number of precious and semiprecious gemstones (Table 4.9).

Because of its chemical and physical durability, quartz is one of the most common constituents of particulate air and water pollution. Concentration of quartz as a pollutant due to construction, mining, and manufacturing processes can be extensive and can cause silicosis.

Distinguishing Features. Low quartz is the most frequently encountered silica mineral. In hand specimen, large single crystals commonly develop as hexagonal prisms and dipyramids (Figure II.6) with distinctive striations and etch pits. The forms may be either right-handed ($P3_121$) or left-handed ($P3_221$), but this difference can only be detected morphologically by the presence of hkl-type faces such as {111} and {511}. Twinning is commonly observed in hand specimens, the most common types being the Dauphiné (twin axis [001]), the Brazil (twin plane {110}), and combinations of these. When large, well-developed crystals are not present, we cannot usually see twinning. Instead, quartz is distinguished by its vitreous luster, lack of cleavage, and its relative hardness. Quartz is usually colorless when in fine-grained occurrences. Because of its resistance to weathering, it appears unaltered and darker colored compared to feldspars with which it is commonly associated.

In thin section, quartz is distinguished by its low relief, low birefringence, lack of color, and lack of alteration. The presence of undulatory extinction, the absence of visible twinning, and uniaxial positive character also aid in identification.

Tridymite and cristobalite are usually so fine-grained that positive identification can be conveniently made only with a petrographic microscope or x-ray diffraction. The low birefringence and typical occurrence as wedge-shaped (tridymite) crystals in vesicle and cavity linings of volcanic rocks is characteristic.

The Feldspar Minerals

The feldspars are the single most abundant mineral group in the earth's crust. They are the major constituent of virtually all igneous rocks, and are present and usually abundant in most metamorphic and sedimentary rocks. The structural variations among the feldspars reflect the variations in pressure and temperature environment of formation of the rocks in which they occur. The compositional range of feldspars expresses the bulk chemistry (Chapter 9) of the

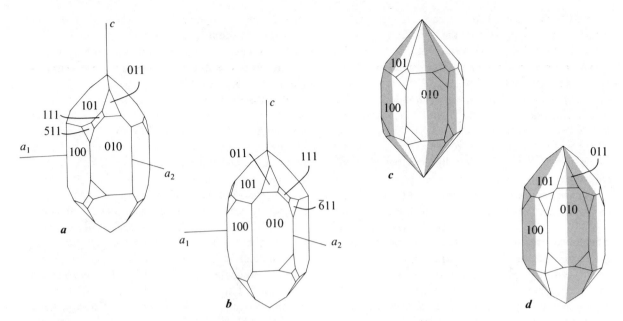

FIGURE II.6   **(a) Left-handed and (b) right-handed quartz crystals. (c) Dauphiné twinning, twin axis [001], and (d) Brazil twinning, twin plane {110}.**

rocks in which they occur, and hence feldspar compositions are an important criterion for igneous rock classification.

Structure. All feldspars are characterized by a common tetrahedral frame, which can be derived from the structure of coesite. Figure II.7 compares the structures of coesite and feldspar as viewed down their *b* axes. In both cases, a lower tetrahedral layer consists of chains of four-membered loops of tetrahedra that run parallel to the *c* axis. The lower layer in each instance is linked to a similar overlying tetrahedral layer by the sharing of apical oxygens. In coesite (Figure II.7a), the linkage between the two layers is the same except that half of the overlying loop tetrahedra have edges parallel to *b*. This arrangement is more compact than the loop arrangement in feldspar and decreases the volume of the cavity beneath the upper loops. In feldspar (Figure II.7b), the bases of the loop tetrahedra are in the *ac* plane, an arrangement that provides a substantial cavity directly below each of the upper layer loops.

The most important difference between the feldspar and coesite structures, however, is the manner in which the double layers repeat. In feldspar, a simple mirror in the *ac* plane beneath the lower tetrahedral layer effectively doubles the volume of the cavity referred to earlier, and allows the structure to accommodate large cations such as K^{1+}, Na^{1+}, and Ca^{2+} (Figure II.7c). In coesite, the double layer repeats by a *c* glide. We can

imagine this operation in two parts, first as a simple mirror that doubles a somewhat smaller cavity than in feldspar, and then as a translation of *c*/2 that effectively closes the cavity to the extent that no additional cations can be accommodated (Figure II.7d).

The interstitial voids in the feldspar framework are most commonly occupied by K^{1+}, Na^{1+}, and Ca^{2+}. Feldspars with K^{1+} and Na^{1+} are called *alkali feldspars* (Figures II.8, II.9, and II.10) and form a complete solid solution at high temperature. All members possess monoclinic symmetry at high temperature. In order for the mirror plane to exist as shown in Figure II.7c, the $(Al, Si)O_4$ tetrahedra must remain symmetrically equivalent, which requires either a disordered arrangement of Al^{3+} and Si^{4+} on tetrahedral sites, as in high-temperature sanidine (Figure II.8), or a symmetrically ordered arrangement of sites as in orthoclase (Figure II.9). Both the twofold axis and the mirror plane are missing in the fully ordered microcline structure (Figure II.10). The K^{1+} in alkali feldspars is coordinated by nine nearest oxygen neighbors, whereas the Na^{1+} has sixfold or sevenfold coordination.

Feldspars with Na^{1+} and Ca^{2+} in the interstitial voids are called *plagioclase* (Figure II.11). Charge neutrality requires that each divalent cation be associated with an additional Al^{3+} substitution in the tetrahedral sites as illustrated by the SiO_2 derivative structures in Table II.1. Plagioclase over the composition range from pure albite

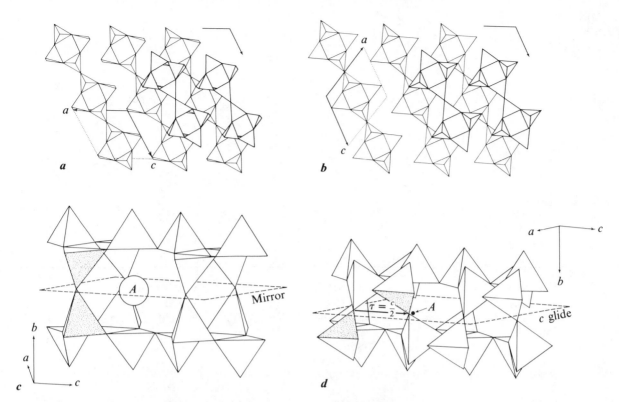

FIGURE II.7 **Tetrahedral layers in (010) plane of (a) coesite and (b) feldspar, (c) mirror plane and alkali site** in feldspar, (d) glide plane and absence of alkali site in coesite.

($NaSi_3AlO_8$) to nearly pure anorthite ($CaSi_2Al_2O_8$) has a disordered arrangement of Al^{3+} and Si^{4+} at high temperature, and thus has the same monoclinic structure as the high-temperature alkali feldspars.

As both the alkali feldspars and plagioclase cool from high-temperature environments, ordering of Al^{3+} and Si^{4+} in tetrahedral sites causes a change in symmetry. Tetrahedra previously related by the mirror plane of the monoclinic point group are no longer equivalent, and the structure becomes triclinic (see Figure 8.11). The completely disordered K-feldspar with a random distribution of Al^{3+} and Si^{4+} on tetrahedral sites is called high sanidine and has monoclinic symmetry. With slow cooling to about 800 °C (Figure II.12), ordering of Al^{3+} into one of the two tetrahedral sites occurs, and the K-feldspar called orthoclase is formed. The structure is still monoclinic $C2/m$, because the ordering is only partial. Upon further cooling to around 600 °C, ordering of Al^{3+} becomes more complete and ordered microcline (called maximum microcline when completely ordered) is stabilized (Figure II.12). The low-temperature K-feldspar adularia is probably microcline twinned on a submicroscopic scale.

The completely disordered monoclinic albite is called monalbite and is stable only above about 1000 °C. Below 1000 °C, it transforms to a triclinic polymorph high albite (also called analbite). The symmetry change is not due to Al-Si ordering but rather to the effect of decreasing temperature on the coordination of the alkali cation. As K^{1+} substitutes for Na^{1+} in the high albite structure, the transformation temperature lowers until the reaction curve intercepts the alkali feldspar solvus. Only when temperature lowers to around 700 °C does the Al-Si ordering in pure $NaSi_3AlO_8$ cause a symmetry change to triclinic low albite. The monalbite-high albite transition appears to be displacive. The other transitions are represented in Figure II.12 as reconstructive, although the evidence is scanty.

The low-temperature anorthite (Figure II.11) is triclinic, and because of the Al/Si ratio of 2:2, alternating Al^{3+} and Si^{4+} causes the c axis to double to 1.4 nm compared with 0.7 nm for albite. Even at high temperature, no Al—O—Al linkages exist, and the structure appears ordered. Because the ordering patterns of albite and anorthite are unique and different, it is likely that no truly homogeneous, ordered solid solutions exist between them. When some of the Ca^{2+} in the anorthite structure is replaced by Na^{1+}, the replacement must be accompanied by a coupled substitution of Si^{4+} for Al^{3+}, and consequently, the per-

THE FELDSPAR MINERALS

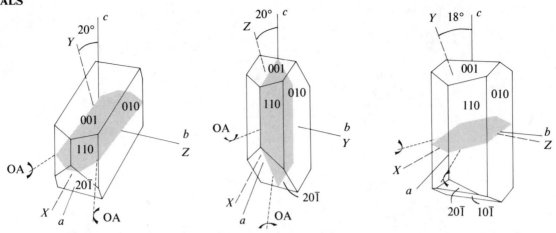

	ORTHOCLASE	SANIDINE	MICROCLINE
Formula	KSi_3AlO_8	KSi_3AlO_8	KSi_3AlO_8
Coord.	K(10), Si(4), Al(4)	K(10), Si(4), Al(4)	K(10), Si(4), Al(4)
System	Monoclinic	Monoclinic	Triclinic
$a =$	0.856	0.856	0.858
$b =$	1.299	1.303	1.296
$c =$	0.719	0.717	0.721
$\alpha =$			89.7°
$\beta =$	116.01°	116.58°	115.97°
$\gamma =$			90.87°
$Z =$	4	4	4
Sp. group	$C2/m$	$C2/m$	$P\bar{1}$
Pt. group	$2/m$	$2/m$	$\bar{1}$
S; loops	2.00; 4, 6, 8, 10	2.00; 4, 6, 8, 10	2.00; 4, 6, 8, 10
SPI =	42	42	43
Cleavage	pf(001), gd(010)	pf(001), gd(010)	pf(001), gd(010)
	pr{110}	pr{110}	pr{110}
Twinning	Carlsbad [001]	Carlsbad [001]	Albite [010]*
	Comp. plane (010)	Comp. plane (010)	Comp. plane (010)
	Baveno [021]		Pericline [010]
	Comp. plane {021}		Comp. plane (h0*l*)
	Manebach [001]*		
	Comp. plane (001)		
RI	$n_\alpha = 1.521$	$n_\alpha = 1.521$	$n_\alpha = 1.518$
	$n_\beta = 1.525$	$n_\beta = 1.525$	$n_\beta = 1.524$
	$n_\gamma = 1.528$	$n_\gamma = 1.528$	$n_\gamma = 1.528$
$\delta =$	0.007	0.007	0.010
$2V =$	60–65°	80–85°	77–84°
Sign	(−)	(−)	(−)
Transp.	Translucent	Transp. to transl.	Translucent
H =	6	6	6
D =	2.56	2.56	2.56
Color	White, turbid	White, variable	White, green
Streak	White	White	White
Luster	Pearly, vitreous	Vitreous	Pearly, vitreous
Fracture	Uneven	Uneven	Uneven
Habit	Prismatic	Prismatic	Prismatic
Remarks	Partial Si/Al order	Disordered	Ordered

Isomorphic series: *alkali feldspars*

*Reciprocal translation, perpendicular to planes of equivalent *hkl* indices

THE FELDSPAR MINERALS

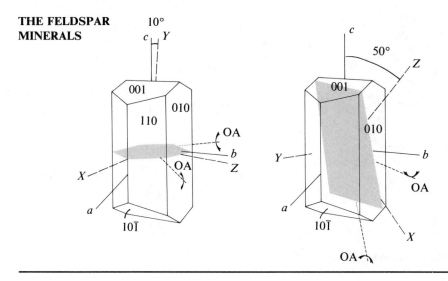

	Albite	ANORTHITE
Formula	$NaSi_3AlO_8$	$CaSi_2Al_2O_8$
Coord.	Na(7), Si(4), Al(4)	Ca(7), Si(4), Al(4)
System	Triclinic	Triclinic
$a =$	0.814	0.817
$b =$	1.279	1.288
$c =$	0.716	1.416
$\alpha =$	93.17°	93.33°
$\beta =$	115.85°	115.60°
$\gamma =$	87.65°	91.22°
$Z =$	4	8
Sp. group	$C\bar{1}$	$P\bar{1}$
Pt. group	$\bar{1}$	$\bar{1}$
S; loops	2.00; 4, 6, 8, 10	2.00; 4, 6, 8, 10
SPI =	45	48
Cleavage	pf(001), gd(010), pr{110}	pf(001), gd(010) pr{110}
Twinning	Albite [010]* Comp. plane (010) Pericline [010] Comp. plane ($h0l$)	Albite [010]* Comp. plane (010) Pericline [010] Comp. plane ($h0l$)
RI	$n_\alpha = 1.527$ $n_\beta = 1.531$ $n_\gamma = 1.538$	$n_\alpha = 1.577$ $n_\beta = 1.585$ $n_\gamma = 1.590$
$\delta =$	0.011	0.013
$2V =$	77°	78°
Sign	(+)	(−)
Transp.	Translucent	Translucent
H =	$6–6\frac{1}{2}$	$6–6\frac{1}{2}$
D =	2.62	2.76
Color	White, gray, green	White, gray
Streak	White	White
Luster	Pearly, vitreous	Pearly, vitreous
Fracture	Uneven	Uneven
Habit	Prismatic, tabular	Prismatic, tabular
Remarks	Polysynthetic twinning causes striation on (010) faces	

End-members of the *plagioclase* series

*Reciprocal translation, perpendicular to plane of other two axes

Sanidine
KSi₃AlO₈

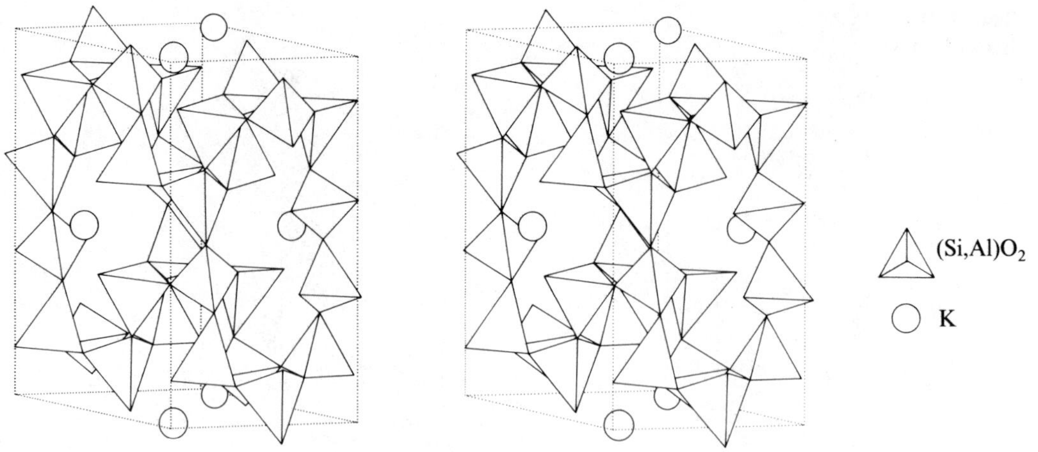

(Si,Al)O₂

K

FIGURE II.8   Crystal structure of sanidine, *b* axis vertical.

Orthoclase
KSi₃AlO₈

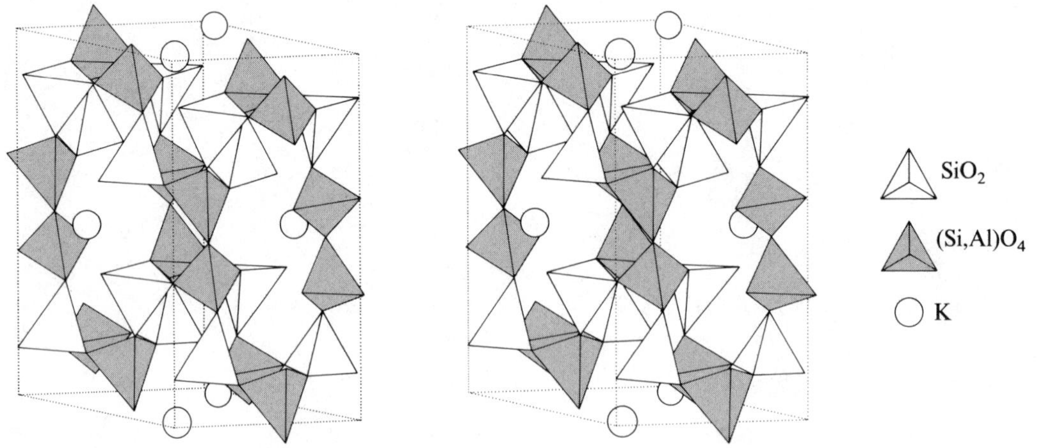

SiO₂

(Si,Al)O₄

K

FIGURE II.9   Crystal structure of orthoclase, *b* axis vertical.

Microcline
KSi₃AlO₈

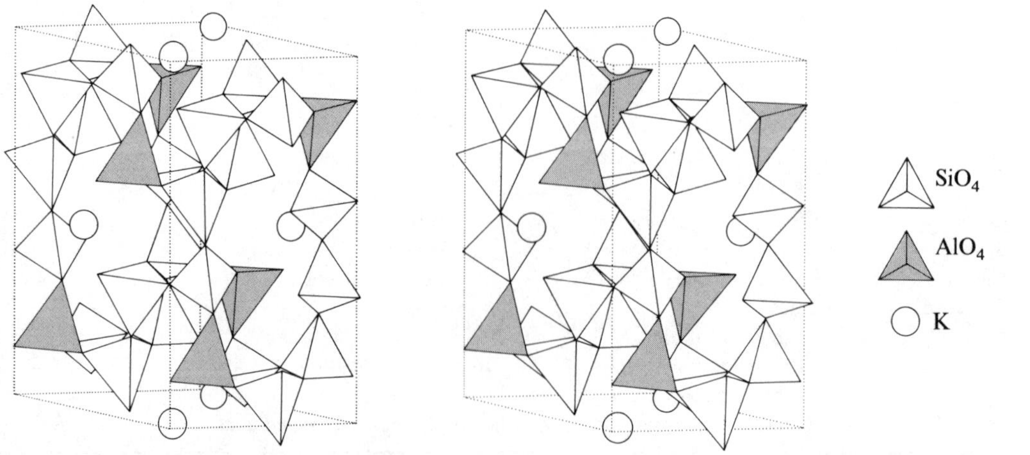

SiO₄

AlO₄

K

FIGURE II.10   Crystal structure of maximum microcline, *b* axis vertical.

Anorthite
CaSi$_2$Al$_2$O$_8$

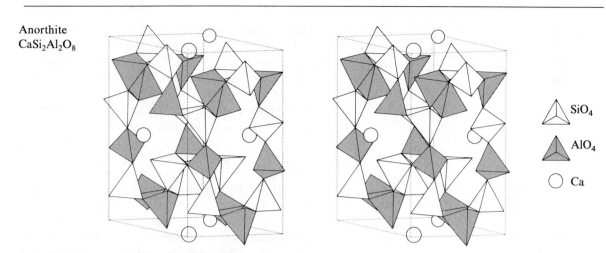

SiO$_4$

AlO$_4$

Ca

FIGURE II.11 Crystal structure of anorthite, *b* axis vertical shaded tetrahedra are 50% Si and 50% Al. Only half of the unit cell is shown along the *c* axis.

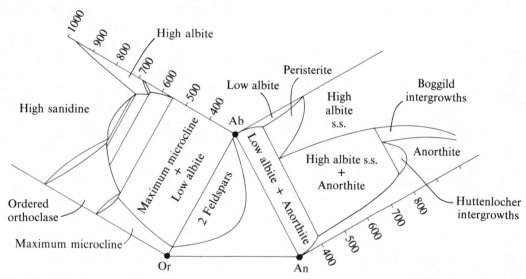

FIGURE II.12 Temperature-composition diagram for ternary feldspars.

fectly alternating sequence of Al and Si in the anorthite structure is disrupted. These facts help explain the great variety of exsolution and domain structures of intermediate composition in plagioclase.

Contact, penetration, and polysynthetic twins (Figure II.13) are common in feldspars and are closely related to structure. In orthoclase and sanidine, the Carlsbad contact twin is one of the most common types, typically with only two individuals. The twin axis is [001], and the composition plane is (010). Twinning according to the albite law with a (010) composition plane and twinning according to the pericline law with a [010] twin axis are common in all triclinic feldspars. The "scotch plaid" or "grid" twinning of microcline is a combination of both albite and pericline twins. Such twins do not occur in the

higher temperature monoclinic feldspars, because the twin axis then coincides with the twofold axis of the untwinned structure (Figure II.8).

The alkali feldspars near albite in composition most commonly exhibit albite and pericline twins, as do members of the plagioclase feldspars. The albite twins are usually polysynthetic and form conspicuous striations parallel to (010) that can be observed with the unaided eye or with a hand lens. Other twin laws in the feldspars that are less common or difficult to identify are the Baveno law with a (021) twin plane and the Manebach law with a (001) twin plane.

Chemistry. The alkali feldspars consist of two chemical end-members, KSi$_3$AlO$_8$ (K-feldspar) and NaSi$_3$AlO$_8$ (Na-feldspar or albite), which are related by a simple K^{1+}-Na^{1+} substitution. The Al/Si ratio remains constant at 1:3. Solid solution

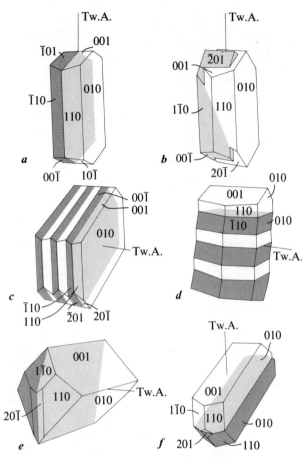

FIGURE II.13   Common twin laws in feldspars. (a) Carlsbad (contact), (b) Carlsbad (interpenetrant), (c) albite, (d) pericline, (e) Baveno, and (f) Manebach.

between the end-members is complete at high temperature (Figure II.12), but because of cation ordering, gaps in miscibility appear at lower temperatures. Cooling through the alkali feldspar solvus produces intracrystalline intergrowths of K-rich and Na-rich feldspars. If the K-feldspar end-member is more abundant in the original solid solution, the exsolution texture is called *perthite*. The K-rich feldspar is the host, and the Na-rich feldspar forms the less abundant lenses and lamellae within it. If the Na end-member is the more abundant in the original solid solution, the texture is called *antiperthite*, because the K-rich feldspar forms by exsolution in a more abundant Na-feldspar host.

The alkali feldspars exhibit solid solution toward a barium end-member called celsian ($BaSi_2Al_2O_8$) by the coupled substitution (Na^{1+}, K^{1+}) + Si^{4+} = Ba^{2+} + Al^{3+}. The mineral hyalophane is an intermediate end-member in a probably complete solid solution series between the alkali feldspars and celsian.

The plagioclase feldspars consist of two chemi-

cal end-members, $NaSi_3AlO_8$ (albite) and $CaSi_2Al_2O_8$ (anorthite) related by the coupled substitution $Na^{1+} + Si^{4+} = Ca^{2+} + Al^{3+}$. A completely homogeneous solid solution probably does not exist between these end-members at any temperature because of the distinct Al-Si ordering patterns of each. Exsolution textures are rampant at lower temperatures and give rise to intracrystalline intergrowths of two feldspars. *Peristerite* is a slightly iridescent variety, forming between 2 and 20% anorthite, abbreviated An_2 and An_{20} (Figure II.12). *Boggild intergrowths* in the range An_{45}-An_{60} are responsible for the refraction of optical wavelengths that cause the visible iridescence, or schiller effect, in intermediate composition plagioclase. *Huttenlocher intergrowths* form in the composition range An_{60} to An_{85}. These exsolution features are not usually visible to the unaided eye and can be reliably identified only by x-ray diffraction techniques.

The alkali and the plagioclase feldspars have a common end-member, namely, albite. Together, the two feldspar solid solutions define the ternary feldspars (Figure II.12). The extent of solid solution within the ternary system depends on both temperature and pressure, but in general is much more restricted between the KSi_3AlO_8 and $CaSi_2Al_2O_8$ end-members. More extensive solubility exists nearer the albite end-member, and those compositions go by a number of different names. K-albite in Figure II.12 is usually called *anorthoclase*.

Occurrence and Associations. Feldspars are present and usually abundant in nearly every igneous and metamorphic rock. They are also common in many sedimentary rocks, but because of their good cleavage and susceptibility to chemical weathering, they are not often found in the more mature quartz-rich sandstones. Both K-feldspar and plagioclase are commonly found together with quartz in the more siliceous range of igneous rocks that includes granites and granodiorites. The plagioclase in these occurrences is invariably enriched in the albite component simply because Na, K, and Si tend to concentrate together through igneous crystallization and crystal fractionation. For the same reason, the alkali feldspars are commonly associated with other potassic minerals such as muscovite and biotite. In the less siliceous igneous rocks such as basalts and gabbros, quartz and K-feldspar are usually absent, and the plagioclase composition is calcic.

The feldspars of sedimentary and metamorphic rocks reflect the composition of the source rocks from which they are derived. Calcic plagioclase, sodic plagioclase, K-feldspar, and quartz can all occur together when the source rocks are of mixed types such as gabbro and granite. Calcic plagioclase is more susceptible to chemical weathering then the more sodic plagioclase, so it is less abundant. Most metamorphic rocks reflect the original chemistry of the sediments from which they were derived. Clay-rich sediments are abundant in K, Na, and Si, and upon recrystallization under metamorphic pressures and temperatures, alkali feldspars and Na-rich plagioclase along with quartz are commonly formed. Feldspars, especially Na-feldspar and K-feldspar, are commonly formed as authigenic minerals in sedimentary rocks, often with striking euhedral forms.

Among the feldspar polymorphs, the sequence of sanidine to orthoclase to microcline as a function of the cooling rate from an initial high temperature is well known. Sanidine preserves the high-temperature disordered arrangement of Al and Si in tetrahedral sites (Figure II.8). Microcline has an ordered arrangement (Figure II.10). Orthoclase appears to represent an intermediate structural state (Figure II.9). This reasoning is consistent with geological occurrence. Sanidine is restricted to relatively fine-grained extrusive volcanic rocks; microcline is typical of coarse-grained plutonic rocks that have cooled slowly at depth.

Feldspars are important industrial minerals, having widespread use in porcelain production and in glass making. They are also used as a flux in iron smelting. A few opalescent or colored feldspars such as moonstone, peristerite, and amazonite (green) have value as semiprecious gems.

Feldspars are the most abundant constituent of dust produced either naturally or artificially. Because of the natural cleavage of feldspars, a large percentage of the dust fragments are elongate and fiberlike.

Distinguishing Features. In hand specimen, plagioclase can usually be distinguished from microcline and orthoclase by its whitish, weathered appearance, elongate form, and the presence of polysynthetic twinning, which can be seen with a hand lens. The K-feldspars, on the other hand, generally lack observable twinning except for a single Carlsbad twin (Figure II.13a) that divides single crystals into two halves parallel to the *c* axis. On fresh surfaces, K-feldspars may be pink compared to colorless plagioclase, and perthitic intergrowths can usually be observed with a hand lens.

In thin section, the "grid" or "tartan" twinning (Figure 3.21a) of microcline and its low relief (less than the mounting medium) are distinctive. Higher relief and polysynthetic twinning (Figure 3.21b) are typical of plagioclase. Measurement of extinction angle, optic angle, and sign are adequate to determine the anorthite content of plagioclase to within 5 mole percent.

The Feldspathoid Minerals

The structures of all minerals within the feldspathoid group consist of a three-dimensional network of tetrahedra that are occupied by Si^{4+} and Al^{3+} in ratios between 1:1 (e.g., nepheline, $KNa_3Si_4Al_4O_{16}$) and 4:1 (e.g, petalite, $LiSi_4AlO_{10}$). Compared to the alkali feldspars in which one fourth of the tetrahedra are occupied by Al^{3+}, nepheline and leucite are less siliceous than feldspars and always form in silica-undersaturated (quartz-free) environments.

The important feldspathoid minerals are leucite and analcime, which have structures comparable to feldspar, and nepheline, which is a structural derivative of high tridymite (Table II.1).

Structure. The structural frame of leucite, analcime (Figure II.14), pollucite, paracelsian, and

Analcime
$NaSi_2AlO_6 \cdot 6H_2O$

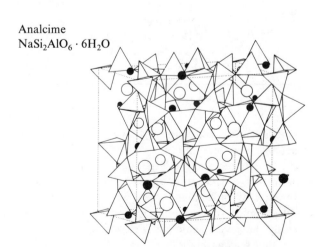
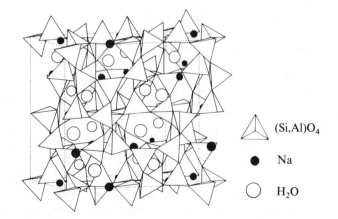

△ $(Si,Al)O_4$

● Na

○ H_2O

FIGURE II.14   Crystal structure of analcime.

THE FELDSPATHOID MINERALS

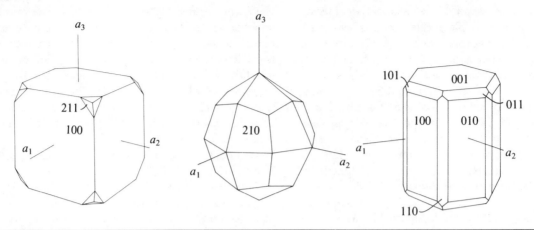

	ANALCIME	Leucite	NEPHELINE
Formula	$NaSi_2AlO_6 \cdot H_2O$	KSi_2AlO_6	$KNa_3Si_4Al_4O_{16}$
Coord.	Na(6), Si(4), Al(4)	K(12), Si(4), Al(4)	Na(8), K(9), Si(4), Al(4)
System	Isometric	Tetragonal	Hexagonal
$a =$	1.371	1.304	1.001
$c =$		1.385	0.841
$Z =$	16	16	2
Sp. group	$I4_1/a\bar{3}2/d$	$I4_1/a$	$P6_3$
Pt. group	$4/m\bar{3}2/m$	$4/m$	6
S; loops	2.00; 4, 6, 8	2.00; 4, 6, 8	2.00; 6, 6
SPI $=$	38	39	43
Cleavage	pr{100}	pr{100}, pr(001)	pr{100}, pr(001)
Twinning	{100}, {110}	{100}, {112}	{100}, {112}, {335}
RI	$n = 1.482$	$n_\omega = 1.508$	$n_\omega = 1.540$
		$n_\epsilon = 1.509$	$n_\epsilon = 1.536$
$\delta =$		0.001	0.004
Sign		(+)	(−)
Transp.	Transp. to transl.	Transp. to transl.	Transp. to transl.
$H =$	$5\frac{1}{2}$	$5\frac{1}{2}$–6	$5\frac{1}{2}$–6
$D =$	2.26	2.48	2.60
Color	White, gray, pink	White, gray	Colorless, turbid
Streak	White	White	White
Luster	Vitreous	Vitreous	Vitreous
Fracture	Uneven	Conchoidal	Subconchoidal
Habit	Trapezohedral	Trapezohedral	Prismatic
Remarks	Similar to zeolites	Pseudoisometric	Derivative of tridymite struct. K ⇌ Na: kalsilite

Members of the *feldspathoid* group

danburite consists of connected four-membered tetrahedral loops characteristic of feldspars and coesite. These zeolites also have relatively open six-membered loops that define channels through the structure. The channels account for the less compact frame of the zeolites compared to the feldspars. Thus, the feldspathoid structure, unlike that of feldspar, has some intercommunication between the cavities. Two types of cavities are associated with the channels; the smaller cavity is partially occupied by monovalent cations (e.g., K^{1+} in leucite), and the larger cavity by molecular water (e.g., H_2O in analcime). Because of the presence of molecular water, we have classified feldspathoids such as analcime as zeolites. Structurally, however, analcime is closer to leucite and the other feldspathoids.

High-temperature (> 625 °C) leucite is cubic,

but as it cools, the primary frame appears to collapse around the K^{1+} with a resultant symmetry change to tetragonal. The transformation is like the suspected displacive transformation between monalbite and high albite in the alkali feldspars.

The other important feldspathoid is nepheline, $(Na, K)SiAlO_4$, which is a derivative of the high-temperature structure of tridymite (Table II.1) by substitution of one half of the Si^{4+} by Al^{3+} and charge balancing with Na^{1+} and K^{1+}. Like tridymite, nepheline has six-membered tetrahedral loops and the same symmetrical packing index. The loops in nepheline, however, are distorted such that the structure has two symmetrically different alkali sites instead of a single, empty site as in tridymite. One site is relatively large and accommodates K^{1+} in an irregular eightfold to ninefold coordination. The other site is smaller and better accommodates Na^{1+}. The two sites are in a ratio of 1:3 and account for the ideal nepheline formula of $KNa_3Si_4Al_4O_{16}$.

The structure of nepheline is given in Figure II.15, and is idealized in order to demonstrate its similarity to the structure of high tridymite.

Chemistry. The chemical formulas of leucite, analcime, and pollucite reflect the structure of the primary feldspathoid frame and the intraframe cavities. In leucite, relatively large K^{1+} cations occupy the cavities with limited substitution (~ 10 mole percent) toward a Na^{1+} end-member. No molecular water is generally present. In analcime, smaller Na^{1+} cations occupy the smaller cavities along the equal amounts of molecular H_2O. In the larger cavities, H_2O is apparently excluded in the presence of K^{1+}. In pollucite, the simple substitution of Na^{1+} for Cs^{1+} allows for the presence of variable amounts of water, suggesting the substitution $Cs^{1+} = Na^{1+} + H_2O$. Beyond the primary

frame, only limited substitution of Al^{3+} for Si^{4+} occurs, although in analcime some correlation is apparent between Si^{4+} and water content.

Occurrence and Associations. Most feldspathoids are restricted to rocks in which quartz is not present. This restriction is because the very broad stability field of feldspars precludes the stable coexistence of feldspathoids and quartz in the ideal Na-Al-Si-O system. Feldspathoids such as nepheline are nearly always associated with the alkali feldspars, especially in SiO_2-undersaturated igneous rocks such as syenites. Leucite is characteristic of potassic volcanic rocks, and is usually associated with sanidine, a high-temperature K-feldspar. It is not known from plutonic igneous rocks or from metamorphic rocks.

Analcime is a widespread feldspathoid found in many igneous rocks, either as a primary magmatic phase in silica-undersaturated lavas or as a secondary phase in hydrothermal veins and vesicle linings. In the latter occurrence, the common associations are with prehnite, zeolites, and other hydrous minerals that have a low-temperature origin. As a primary mineral in lavas, analcime is associated with leucite and olivine. Analcime also forms as an authigenic mineral in sandstones and volcanic tuffs.

The Li-rich feldspathoids petalite and eucryptite (a derivative of the quartz structure), and the Cs-feldspathoid called pollucite are generally found in pegmatites associated with tourmaline, spodumene, and alkali feldspars.

Distinguishing Features. In hand specimens, nepheline and other feldspathoids are distinguished by their equant form, their ubiquitous association with alkali feldspars, and their lack of association with quartz. These minerals typically "weather out" to leave pockmarked, weathered surfaces. In thin section, very low birefringence,

Nepheline
$KNa_3Si_4Al_4O_{16}$

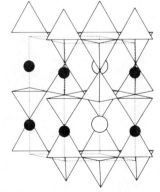

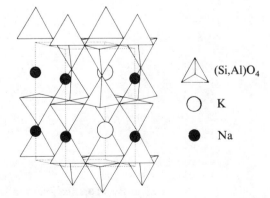

$\triangle$ $(Si,Al)O_4$

$\bigcirc$ K

$\bullet$ Na

FIGURE II.15   Crystal structure of nepheline, *c* axis vertical.

low relief, and uniaxial character are distinctive of nepheline and leucite. Leucite may be distinguished from nepheline and analcime by the presence of "grid" twinning similar to that observed in microcline.

The Scapolite Minerals

Structure. The scapolite minerals have a structure like the feldspars in that they have a three-dimensional network of four-membered loops (Figure II.16). Unlike the feldspars, however, scapolites have two symmetrically different loops and cavities within the tetrahedral network. In these characteristics they resemble nepheline. The larger cavities are occupied by Cl^{1-}, F^{1-}, and $(OH)^{1-}$, or by the anionic groups $(CO_3)^{2-}$ and $(SO_4)^{4-}$. The smaller cavities are occupied by Na^{1+}, K^{1+}, and Ca^{2+} in amounts necessary to provide charge balance.

Chemistry. The two common end-members of the scapolite series are called marialite ($Na_4ClSi_9Al_3O_{24}$) and meionite ($Ca_4CO_3Si_6Al_6O_{24}$).

THE SCAPOLITE MINERALS

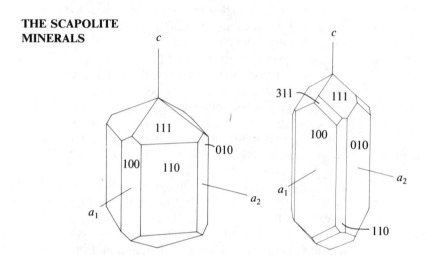

	MARIALITE	Meionite
Formula	$Na_4ClSi_9Al_3O_{24}$	$Ca_4CO_3Si_6Al_6O_{24}$
Coord.	Na(6), Si(4), Al(4)	Ca(6), Si(4), Al(4)
System	Tetragonal	Tetragonal
$a =$	1.211	1.223
$c =$	0.756	0.768
$Z =$	2	2
Sp. group	$I4/m$	$I4/m$
Pt. group	$4/m$	$4/m$
S; loops	2.00; 4, 5, 8	2.00; 4, 5, 8
SPI =	45	45
Cleavage	gd{100}, pr{110}	gd{100}, pr{110}
RI	$n_\omega = 1.540$	$n_\omega = 1.595$
	$n_\epsilon = 1.536$	$n_\epsilon = 1.558$
$\delta =$	0.004	0.037
Sign	$(-)$	$(-)$
Transp.	Transp. to transl.	Transp. to transl.
H =	5–6	5–6
D =	2.55	2.76
Color	Colorless, variable	Colorless, variable
Streak	White	White
Luster	Vitreous	Vitreous
Fracture	Conchoidal	Conchoidal
Habit	Prismatic	Prismatic
Remarks	May be fluorescent orange yellow in UV light	
	Members of the *scapolite* group	

Marialite
$Na_4ClSi_9Al_3O_{24}$

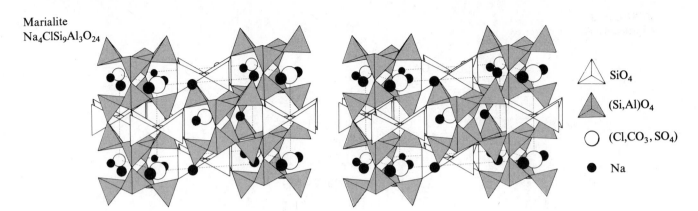

FIGURE II.16　**Crystal structure of marialite, *c* axis vertical.**

The main cation substitution is $Na^{1+} + Si^{4+} = Ca^{2+} + Al^{3+}$, as in the plagioclase feldspars. An anion substitution between $(Cl)^{1-}$ and $(CO_3)^{2-}$ also occurs and is conveniently represented by $4Na^{1+} + 3Si^{4+} + Cl^{1-} = 4Ca^{2+} + 3Al^{3+} + (CO_3)^{2-}$. Hydroxyl $(OH)^{1-}$ and sulfate $(SO_4)^{4-}$ may also substitute, depending on the bulk chemistry of the environment of formation. Intermediate compositions are referred to as calcian marialite and sodian meionite.

Occurrence and Associations. Scapolite minerals are known principally from Ca-rich metamorphic rocks, particularly marbles and skarns in which Cl, F, CO_3, or SO_4 are present. Scapolite in gneisses and amphibolites is typically calcic and is associated with calcic plagioclase, calcic clinopyroxene, hornblende, apatite, and sphene. Scapolites are not known from sedimentary rocks and only rarely from igneous rocks.

Distinguishing Features. Scapolite commonly occurs in hand specimens as yellowish to white stubby grains in association with plagioclase. Pris-

matic cleavage can be well developed but typically is obscured by cracks and parting. In thin section, scapolite is colorless and may resemble quartz and feldspar. The uniaxial character distinguishes scapolite from feldspar; the negative sign and somewhat higher birefringence (calcic varieties) distinguish it from quartz.

Beryl and Cordierite

Structure. Both beryl and cordierite have essentially the same structure consisting of a three-dimensional network of tetrahedra that have distinctive six-membered loops (Figure II.17). Within these loops are broad channels that are connected to each other by other tetrahedra that form four-membered loops. Beryl has traditionally been classified as a ring silicate because of its distinctive six-membered loops or rings. The three-dimensional connectivity of the tetrahedral frame, however, distinguishes these minerals from true ring silicates such as tourmaline.

The six-membered loops in cordierite have an

Beryl
$Al_2Si_6Be_3O_{18}$

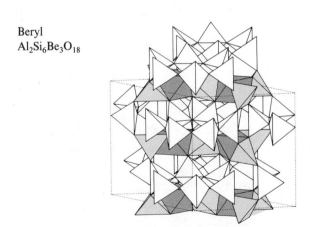
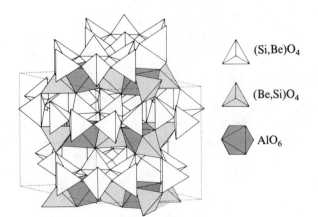

FIGURE II.17　**Crystal structure of beryl, *c* axis vertical.**

BERYL AND CORDIERITE

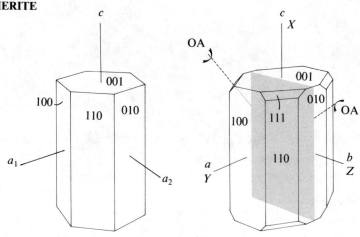

	BERYL	Cordierite
Formula	$Al_2Si_6Be_3O_{18}$	$Mg_2Si_5Al_4O_{18}$
Coord.	Al(6), Si(4), Be(4)	Mg(6), Si(4), Al(4)
System	Hexagonal	Orthorhombic
$a =$	0.923	1.713
$b =$		0.980
$c =$	0.919	0.935
$Z =$	2	4
Sp. group	$P6/m\,2/c\,2/c$	$C2/c\,2/c\,2/m$
Pt. group	$6/m\,2/m\,2/m$	$2/m\,2/m\,2/m$
S; loops	2.00; 4, 6, 8	2.00; 4, 6, 8
SPI =	50	48
Cleavage	pr(001)	gd(010), pr(100), pr(100)
Twinning	{311}, {110}	{110}, {130}
RI	$n_\omega = 1.568$	$n_\alpha = 1.54$
	$n_\epsilon = 1.562$	$n_\beta = 1.55$
		$n_\gamma = 1.56$
$\delta =$	0.006	0.02
$2V =$		65–105°
Sign	(−)	(+ or −)
Transp.	Transp. to transl.	Transp. to transl.
Pleochr.	ω = green	X = pale orange
	ϵ = yellow green	Y = blue violet
		Z = pale blue violet
H =	$7\frac{1}{2}$–8	7
D =	2.7–2.9	2.5–2.8
Color	Colorless, variable	Gray blue
Streak	White	White
Luster	Vitreous	Vitreous
Fracture	Even	Even
Habit	Prismatic	Prismatic
Remarks	Isostr. w. cordierite	Pseudohexagonal

ordered distribution of two Al^{3+} and four Si^{4+} cations that imparts an orthorhombic symmetry to the structure. The disordered distribution makes all six tetrahedra equivalent, and consequently, the loops have $6/m$ symmetry. The high-temperature, hexagonal polymorph of cordierite, called indialite, has this structure. In ideally ordered beryl, all six loop tetrahedra are Si, and the structure is hexagonal.

Chemistry. The connecting tetrahedra are occupied by Be and Si in natural beryl, and by Si and Al in cordierite. Connecting octahedra in beryl are occupied by Al^{3+}. In cordierite, they are occupied by Fe^{2+}, Mg^{2+}, and Mn^{2+}. The open channels are the only sites available for molecular water (H_2O), which is nearly always present in beryl and cordierite. Other gases, notably CO_2, He, and Ar may also be present. Both Na^+ and K^+ occupy the channels of cordierite, possibly coordinating with oxygens that bridge adjacent Al and Si in the six-membered rings.

The important substitutions in cordierite are $Fe^{2+} \rightleftharpoons Mg^{2+}$ in the octahedral sites and Fe^{3+} for Al^{3+}. In beryl and cordierite, simple substitution of the monovalent Na, Li, and Cs ions for each other is common.

Occurrence and Associations. Beryl is a common accessory mineral in many igneous rocks and will almost always form if any appreciable Be is in the environment. Beryl is concentrated in granitic pegmatites where it is an important ore mineral for Be. It also is found as a detrital mineral in sediments and in metamorphic rocks.

Beryl is an important gem mineral, the deep green variety (emerald) being the best known. The blue variety is called aquamarine; the pink variety is called morganite.

Cordierite is most commonly found in metamorphic rocks rich in alumina. Typical associations are with the aluminosilicates (e.g., andalusite), garnet, biotite, muscovite, and quartz. Aluminous orthoamphibole is a common associate of cordierite in more mafic bulk compositions. Cordierite also occurs in association with quartz in granitic pegmatites, and more rarely in plutonic rocks.

Cordierite has a unique historical importance— Viking navigators used it as a polarizer to locate the sun's position on cloudy or overcast days.

Distinguishing Features. Beryl is easily recognized by its hexagonal form and prismatic faces. Cordierite is more difficult to recognize, especially on fresh surfaces, due to its similarity to quartz. Large crystals may have a bluish tint. On weathered surfaces, cordierite is commonly altered to a microcrystalline aggregate of chlorite and sericite, sometimes called pinite, which can have a rusty appearance. Cordierite's equant form and association with other aluminous minerals are useful for identification purposes.

In thin section, relatively low birefringence is typical of cordierite. The presence of pinite alteration along cleavages and fractures, and distinctive yellow pleochroic halos around included zircons are diagnostic. Twinning may also be present, occasionally lamellar but more commonly cyclic with twin lamellae at angles of 30, 60, or 120 degrees.

The Zeolite Minerals

The zeolite minerals constitute a large and important group within the network silicates. More than 40 different natural species are found in a broad range of geological environments, and at least 100 synthetic species have been made in laboratories for commercial applications. Like all network silicates, the zeolites have a primary frame made up of SiO_4 and AlO_4 tetrahedra that link together by corner sharing to form a connected, three-dimensional network. Like the feldspars and feldspathoids, the negative charge on the zeolite frame is balanced by cations that occupy the intraframe cavities.

The most important structural differences between the zeolites and the other network silicates are the dimensions of the intraframe cavities and the connecting channels between them. Recall that in the feldspar structure the cavities are relatively small, and the cations occupying them are so strongly bonded to the primary frame that their substitution by other cations having different valences requires a change in the Al/Si ratio. The cavities are not connected and are occupied only by monovalent and divalent cations. The feldspathoid frame is more expanded than the feldspar structure, and some connectivity exists between the intraframe cavities. Some of the cavities are occupied by cations, and other cavities are large enough to accommodate molecular water. In comparison with the feldspathoid frame, the primary frame of the zeolites is even more expanded, containing larger cavities that are connected by broad channels. Molecular water may pass in and out of the zeolite structure through the channels without disruption of the primary frame.

Upon heating, molecular water is readily driven out of the zeolite structure. The anhydrous material can then absorb other kinds of molecules, providing they are not larger than the channels through which they must pass. This special aspect of the structure makes zeolites useful as molecular sieves.

THE ZEOLITE MINERALS

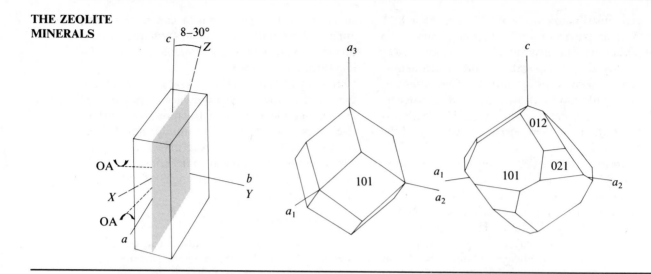

	Laumontite	SODALITE	Chabazite
Formula	$CaSi_4Al_2O_{12} \cdot 4H_2O$	$Na_2Si_3Al_3O_{24} \cdot Na_2Cl$	$CaSi_4Al_2O_{12} \cdot 6H_2O$
Coord.	Ca(6), Si(4), Al(4)	Na(7), Si(4), Al(4)	Ca(7), Si(4), Al(4)
System	Monoclinic	Isometric	Trigonal
$a =$	1.475	0.887	1.317
$b =$	1.310		
$c =$	0.755		1.506
$\beta =$	111.5°		
$Z =$	4	2	6
Sp. group	Cm	$P\bar{4}3m$	$R\bar{3}2/m$
Pt. group	m	$\bar{4}3m$	$\bar{3}2/m$
S; loops	2.00; 4, 6, 6, 10	2.00; 4, 6, 12	2.00; 4, 6, 8, 12
SPI =	33	40	30
Cleavage	pf(010), pf{110}	gd{110}	pr{101}
Twinning	(100)	{111}	{100}, [001]
			Comp. plane {101}
RI	$n_\alpha = 1.51$	$n = 1.485$	$n_\omega = 1.484$
	$n_\beta = 1.52$		$n_\epsilon = 1.481$
	$n_\gamma = 1.52$		
$\delta =$	0.01		0.003
$2V =$	25–45°		
Sign	(−)		(−)
Transp.	Transp. to transl.	Transp. to transl.	Transp. to transl.
$H =$	3–4	$5\frac{1}{2}$–6	4–5
$D =$	2.3	2.3	2.1
Color	White	Blue, white	Colorless, red
Streak	White	White	White
Luster	Vitreous	Vitreous	Vitreous
Fracture	Uneven	Conchoidal	Uneven
Habit	Prismatic	Dodecahedral	Rhombohedral
Remarks	Pyroelectric		Penetr. twin: phacolite

Members of the *zeolite* group

THE ZEOLITE MINERALS

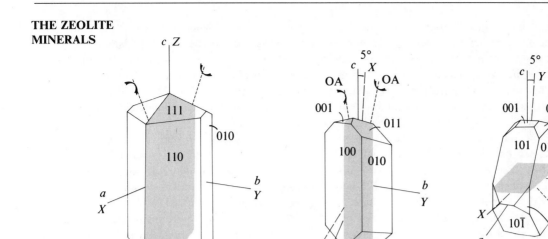

	NATROLITE	STILBITE	Heulandite
Formula	$Na_2Si_3Al_2O_{10} \cdot 2H_2O$	$(Ca, Na)Si_7Al_2O_{18} \cdot 7H_2O$	$(Ca, Na)Si_7Al_2O_{18} \cdot 6H_2O$
Coord.	Na(6), Si(4), Al(4)	Ca(6), Na(6), Si(4), Al(4)	Ca(6), Na(6), Si(4), Al(4)
System	Orthorhombic	Monoclinic	Monoclinic
$a =$	1.830	1.364	1.773
$b =$	1.863	1.824	1.782
$c =$	0.660	1.127	0.743
$\beta =$		129.16°	116.3°
$Z =$	8	4	4
Sp. group	$Fd2d$	$C2/m$	Cm
Pt. group	$m2m$	$2/m$	m
S; loops	2.00; 4, 8, 9	2.00; 4, 5, 6, 8	2.00; 4, 5, 6, 8
SPI =	34	31	32
Cleavage	pf{110}	pf(010), pr(001)	pf(010)
	Parting (010)	pr(101)	
RI	$n_\alpha = 1.48$	$n_\alpha = 1.49$	$n_\alpha = 1.49$
	$n_\beta = 1.48$	$n_\beta = 1.50$	$n_\beta = 1.50$
	$n_\gamma = 1.49$	$n_\gamma = 1.50$	$n_\gamma = 1.50$
$\delta =$	0.012	0.010	0.005
$2V =$	38–62°	30–50°	35°
Sign	(+)	(−)	(+)
Transp.	Transp. to transl.	Transp. to transl.	Transp. to transl.
$H =$	5–5½	3½–4	3½–4
$D =$	2.23	2.15	2.15
Color	Colorless, gray	Gray	White, variable
Streak	White	Gray	White
Luster	Vitreous	Pearly	Vitreous
Fracture	Uneven	Subconchoidal	Subconchoidal
Habit	Acicular, fibrous	Prismatic, striated	Platy
Remarks	"Fibrous zeolites" incl. thomsonite, eddingtonite, erionite (Fibr. var.: "stone wool")	Crystals curved May be fibrous	Na ⇌ K: clinoptilolite most common zeolite

Members of the *zeolite* group

Structure. The zeolite minerals are distinguished from one another by different configurations of tetrahedra within the three-dimensional network and by the size and shape of the resulting channels. The basic configurations are single loops of 4, 5, 6, 8, 10, and 12 tetrahedra and double loops of 4, 6, and 8 tetrahedra. The loops combine to form larger and more complicated polyhedral cages characteristic of many of the natural zeolites. The truncated octahedron (14-hedron), the truncated cuboctahedron (26-hedron), and other types are known. Table II.2 divides the zeolite structures into six groups, and summarizes the structures based on their ring configurations and higher *n*-hedra. Note that analcime, discussed earlier as a feldspathoid, is included in the phillipsite group of the zeolites because of its physical properties. Its fraction of void space (column 5, Table II.2) is smaller than any of the zeolites, as is the diameter of the smallest channel aperture.

Zeolite minerals within the *phillipsite group* have structures based on parallel loops of four and eight tetrahedra that link together to form a two-dimensional channel system. The *sodalite group* structures (Figure II.18) consist of loops of six tetrahedra linked by loops of four tetrahedra to form complex polyhedral cages (Table II.2).

TABLE II.2. *Classification of Zeolite Minerals*

Zeolite Mineral	Formula	Rings 4	6	8	10	*n*-hedron *n*	Void Fraction	Minimum Channel Dimension (nm)
Phillipsite Group								
(Analcime)	$NaSi_2AlO_6 \cdot H_2O$	×	×				0.18	0.26
Phillipsite	$KCaSi_5Al_3O_{16} \cdot 6H_2O$	×		×			0.31	0.28 × 0.48
Harmotome	$BaSi_6Al_2O_{16} \cdot 6H_2O$	×		×			0.31	0.42 × 0.44
Laumontite	$CaSi_4Al_2O_{12} \cdot 4H_2O$	×	×		×		0.34	0.46 × 0.63
Gismondine	$CaSi_2Al_2O_8 \cdot 8H_2O$	×		×			0.46	0.31 × 0.44
Paulingite	complex	×		×			0.49	0.39
Yugawaralite	$CaSi_6Al_2O_{16} \cdot 4H_2O$	×					0.27	0.28 × 0.36
Sodalite Group								
Sodalite	$Na_3Si_3Al_3O_{12} \cdot NaCl$	×	×			14	0.35	0.22
Cancrinite	$(Na_3Ca)_2CO_3Si_3Al_3O_{12} \cdot 2H_2O$	×	×			11		
Erionite	complex	×	×		(×)	11, 23	0.35	0.36–0.52
Offretite	complex	×	×			11, 14	0.40	0.36–0.52
Levyne	$CaSi_4Al_2O_{12} \cdot 6H_2O$	×	×			17	0.40	0.32–0.51
Chabazite Group								
Chabazite	$CaSi_4Al_2O_{12} \cdot 6H_2O$		×	×		20	0.47	0.37–0.42
Faujasite	$(Na, Ca)Si_4Al_2O_{12} \cdot 8H_2O$		×	×		14, 26	0.47	0.74
Gmelinite	$(Na, Ca)Si_4Al_2O_{12} \cdot 6H_2O$		×	×		14	0.44	0.36–0.39
Natrolite Group								
Natrolite	$Na_2Si_3Al_2O_{10} \cdot 2H_2O$						0.23	0.26 × 0.39
Scolecite	$CaSi_3Al_2O_{10} \cdot 3H_2O$						0.31	0.26 × 0.39
Mesolite	$NaCaSi_4Al_3O_{14} \cdot 4H_2O$						0.30	0.26 × 0.39
Thomsonite	$NaCa_2Si_5Al_5O_{20} \cdot 6H_2O$						0.32	0.26 × 0.39
Gonnardite	$Na_2CaSi_6Al_4O_{20} \cdot 7H_2O$						0.31	0.26 × 0.39
Edingtonite	$BaSi_3Al_2O_{10} \cdot 4H_2O$						0.36	0.35 × 0.39
Heulandite Group								
Heulandite	$(Ca, Na)Si_7Al_2O_{18} \cdot 6H_2O$			×	×			
Brewsterite	$(Sr, Ba, Ca)Si_6Al_2O_{16} \cdot 5H_2O$			×				
Clinoptilolite	$(Na, K)Si_7Al_2O_{18} \cdot 6H_2O$				×			
Stilbite	$NaCa_2Si_{13}Al_5O_{36} \cdot 14H_2O$							
Mordenite Group								
Mordenite	$NaSi_5AlO_{12} \cdot 6H_2O$						0.28	0.29 × 0.57
Dachiardite	$Na_5Si_{19}Al_5O_{24} \cdot 12H_2O$						0.32	0.36 × 0.48
Epistilbite	$CaSi_6Al_2O_{16} \cdot 6H_2O$						0.25	0.37 × 0.44
Ferrierite	complex						0.28	0.34 × 0.48
Bikitaite	$LiSi_2AlO_6 \cdot H_2O$						0.23	0.32 × 0.49

After Breck, D. R. *Zeolite molecular sieves* (New York: John Wiley & Sons, Inc., 1974), pp. 48–50, by kind permission of the publisher.

Structures in the *chabazite group* have distinctive double rings of six tetrahedra that define hexagonal prisms. The prisms link complex polyhedral cages into an open network having a void fraction of nearly 50%.

Zeolites of the *natrolite group* differ from others by having linked chains of four-membered loops, as illustrated in the structural model of thomsonite (Figure II.19). This chainlike structure gives rise to the distinctive acicular habit of minerals in this group. The natrolite group is sometimes referred to as the fibrous zeolites.

The zeolites within the *heulandite group* have four-membered and five-membered loops arranged in parallel layers instead of chains. This structure is illustrated in the model of stilbite (Figure II.20). The parallel layers of loops account for the platelike and tabular form of this group.

Minerals in the *mordenite group* of zeolites all have a special configuration of five-membered

Sodalite
$Na_4ClSi_3Al_3O_{12}$

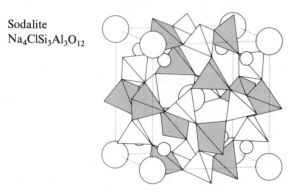
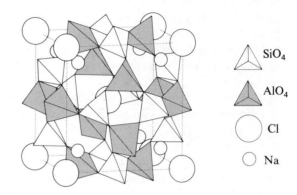

FIGURE II.18   Crystal structure of sodalite.

Thomsonite
$NaCa_2Si_5Al_5O_{20} \cdot 6H_2O$

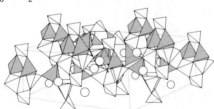
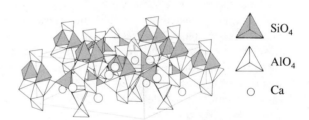

FIGURE II.19   Crystal structure of thomsonite, *c* axis vertical.

Stilbite
$NaCa_2Si_7Al_2O_{18} \cdot 7H_2O$

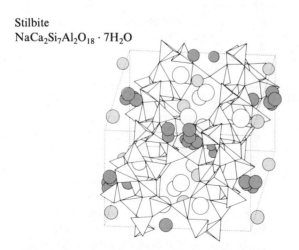
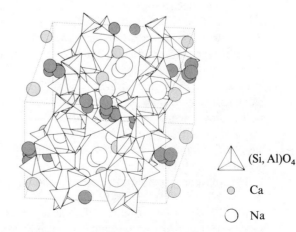

FIGURE II.20   Crystal structure of stilbite, *b* axis vertical.

loops that form chains cross-linked to each other by four-membered loops. The specific nature of cross-linking varies, but the elongated habit typical of the natrolite group is not developed.

Chemistry. All zeolite minerals have a tetrahedral cation to oxygen ratio of 1:2, and within the tetrahedra, Al^{3+} may never be in excess of Si^{4+}. Depending on the ratio ($\leqslant 1$) of Al^{3+} to Si^{4+} in tetrahedral sites, various amounts of Na, Ca, K, and Ba may occupy the intraframe cavities. The coupled substitution of $Ca^{2+} + Al^{3+} = Na^{1+} + Si^{4+}$ characteristic of other network silicates such as the plagioclase feldspars does not apply to zeolites. Unlike the feldspars, the zeolites have so much intraframe volume that cations need not be exchanged on a one to one basis. For example, natrolite can be derived from scolecite (Table II.2) by substituting $2\ Na^{1+}$ for Ca^{2+}. The only requirement is that charge neutrality be maintained.

Occurrence and Associations. Zeolites are recognized today as among the most abundant and widespread authigenic minerals in sedimentary rocks. Zeolites are also a major constituent of volcanic tuffs and volcano-clastic sediments formed by reaction of volcanic glass and conate water trapped during sedimentation, especially in saline lakes. In this type of occurrence, clinoptilolite, chabazite, erionite, mordenite, and phillipsite are the zeolites most commonly found. Zeolites also form by reaction of volcanic glass with percolating meteoric water and can occur in deposits up to several hundred meters thick. Clinoptilolite and mordenite are the common zeolites in this occurrence, comprising up to 90% of the rocks in some cases. Such deposits are of commercial value and are exploited.

Clinoptilolite and phillipsite are the most abundant zeolites in deep-sea sediments, sometimes comprising as much as 80% of the sediment. Analcime, erionite, and laumontite are found as well. The typical occurrence is in fine-grained pelagic sediments in which these authigenic zeolites occur as euhedral crystals in the sediment matrix.

Zeolites occur abundantly in low-grade metamorphic rocks, especially as vug and vesicle fillings in altered volcanics. Fine euhedral specimens are frequently found. In deeply weathered volcanics, the volcanic rock has eroded away, leaving the zeolite free from its host. Thomsonite from the North Shore of Lake Superior and other localities forms in this manner.

The unique structural features of zeolites are responsible for their extensive commercial use as molecular sieves and as ion-exchange material in water softeners. As sieves, zeolites are used for the removal of water and CO_2 from gaseous hydrocarbons and petroleum, for oil-spill cleanup, and for wastewater and effluent treatment. The ion-exchange properties in water softeners allow hard water, initially high in dissolved Ca^{2+}, to become soft by exchanging its Ca^{2+} with $2\ Na^{1+}$ supplied by the natural or synthetic zeolite.

Distinguishing Features. Many of the common zeolite minerals are found in cavity linings of altered volcanic rocks associated with other zeolites and calcite. Crystals are generally small and colorless, frequently with a vitreous or pearly luster. The sheaflike aggregates of stilbite and the radial growth pattern of thomsonite are distinctive. As an alteration product of volcanic glass, zeolites are typically white and punky, and like clay minerals may momentarily stick to one's tongue.

In thin section the zeolites are distinguished by their very low relief, very low birefringence, and characteristic association.

The Layer Silicates

The most important mineral groups within the layer silicates (also known as phyllosilicates) are the *micas* and the *clays*. The most common micas are biotite and muscovite, both of which are found in igneous, metamorphic, and sedimentary rocks. The most common clay minerals are kaolinite, illite, and montmorillonite. They are not found in igneous and metamorphic rocks, but constitute up to 40% of most sediments, sedimentary rocks, and certain hydrothermal deposits.

The tetrahedral frame of all layer silicates is a *two-dimensional network* in which three of the four apical oxygens of every tetrahedron are shared with other tetrahedra (Figure 5.8). The resulting tetrahedral layer usually has a sharing coefficient of 1.75 compared with 2.00 for most network silicates. Because of the strong Si—O bonds with each layer, all layer silicates, including micas and clays, possess perfect basal cleavage and a distinctive platy crystallization habit. A direct relationship exists between the crystal morphology and structure, as we can see by comparing Figures II.21 through II.24 with the crystal form diagrams of typical layer silicates.

The term "clay" has two meanings. The first usage refers to minerals that possess a claylike crystal structure, and the term is used without consideration of grain size. The second usage refers to any mineral with dimensions less than 2 μm. Feldspars and quartz, for example, can be considered as clay-sized particles in many sedimentary rocks, but they are not true clays in the structural sense. All layer silicates may be considered true clays if they are in the clay-size range.

THE LAYER SILICATES

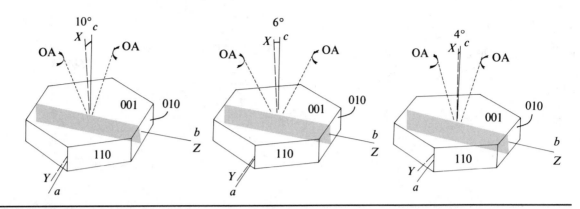

	KAOLINITE (*1Tc*)	PYROPHYLLITE (*1Tc*)	MUSCOVITE (*2M₁*)
Formula	$Al_4(OH)_8Si_4O_{10}$	$Al_2(OH)_2Si_4O_{10}$	$KAl_2(OH)_2Si_3AlO_{10}$
Coord.	Al(6), Si(4)	Al(6), Si(4)	K(6), Al(6), Si(4), Al(4)
System	Triclinic	Triclinic	Monoclinic
$a =$	0.515	0.516	0.519
$b =$	0.892	0.896	0.904
$c =$	0.738	0.935	2.008
$\alpha =$	91.8°	90.03°	
$\beta =$	104.8°	100.37°	95.5°
$\gamma =$	90.0°	89.75°	
$Z =$	1	2	4
Sp. group	$P\bar{1}$	$P\bar{1}$	$C2/c$
Pt. group	$\bar{1}$	$\bar{1}$	$2/m$
S; loops	1.75; 6	1.75; 6	1.75; 6
SP plane	(001)	(001)	(001)
SP struct.	$AoB + C =$;	$AoB + CoA =$;	$AoB + CoAc$
	$CoA + B =$;	$CoA + BoC =$;	$BoC + AoBc$;
	$BoC + A =$;	$BoC + AoB =$;	$CoA + BoCc$
SPI =	45	51	58
Cleavage	pf(001)	pf(001)	pf(001)
Twinning			[310]
			Comp. plane (001)
RI	$n_\alpha = 1.556$	$n_\alpha = 1.553$	$n_\alpha = 1.565$
	$n_\beta = 1.563$	$n_\beta = 1.588$	$n_\beta = 1.596$
	$n_\gamma = 1.565$	$n_\gamma = 1.600$	$n_\gamma = 1.600$
$\delta =$	0.007	0.047	0.035
$2V =$	40°	52–62°	30–40°
Sign	(−)	(−)	(−)
Transp.	Translucent	Translucent	Transparent
H =	$2–2\frac{1}{2}$	$1\frac{1}{2}$	$2\frac{1}{2}$
D =	2.6	2.8	2.8
Color	White	White	White, gray brown
Streak	White	White	White
Luster	Dull	Pearly	Vitreous
Fracture	Flexible	Flexible	Elastic
Habit	Platy, massive	Platy, foliated	Lamellar
Remarks	Dickite = *2M₁*	Occasionally fibrous	"White mica"
	Nacrite = *2M₂*		

THE LAYER SILICATES

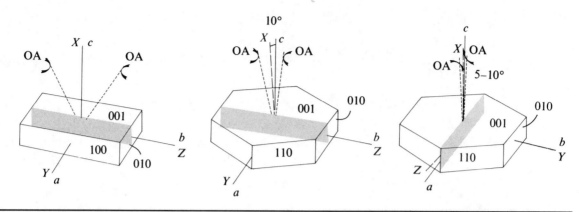

	Antigorite (*2M₁*)	Talc (*2M₁*)	Phlogopite (*1M*)
Formula	$Mg_6(OH)_8Si_4O_{10}$	$Mg_3(OH)_2Si_4O_{10}$	$KMg_3(OH)_2Si_3AlO_{10}$
Coord.	Mg(6), Si(4)	Mg(6), Si(4)	K(12), Mg(6), Si(4), Al(4)
System	Monoclinic	Monoclinic	Monoclinic
$a =$	0.532	0.529	0.531
$b =$	0.950	0.910	0.919
$c =$	1.49	1.881	1.015
$\beta =$	101.9°	100.00°	95.18°
$Z =$	2	4	2
Sp. group	*C2/m*	*Cc*	*C2/m*
Pt. group	*2/m*	*m*	*2/m*
S; loops	2.00; 6	2.00; 6	2.00; 6
SP plane	(001)	(001)	(001)
SP struct.	*AoB + C =*	*AoB + CoA =*	*AoB + CsBc*;
	CoA + B =;	*CoA + BoC =*;	*BoC + AsCc*;
	BoC + A =		*CoA + BsAc*;
SPI =	48	55	62
Cleavage	pf(001)	pf(001)	pf(001)
Twinning			[310]
			Comp. plane (001)
RI	$n_\alpha = 1.56$	$n_\alpha = 1.54$	$n_\alpha = 1.56$
	$n_\beta = 1.57$	$n_\beta = 1.58$	$n_\beta = 1.60$
	$n_\gamma = 1.57$	$n_\gamma = 1.58$	$n_\gamma = 1.60$
$\delta =$	0.007	0.05	0.04
$2V =$	20–60°	6–30°	0–20°
Sign	(−)	(−)	(−)
Transp.	Translucent	Translucent	Transparent
H =	3–4	1	$2\frac{1}{2}$
D =	2.6	2.8	2.8
Color	Green yellow	White, gray	Yellow brown
Streak	White	White	White
Luster	Resinous, silky	Resinous, silky	Pearly
Fracture	Flexible	Flexible	Elastic
Habit	Platy, massive	Platy, massive	Platy, foliated
Remarks	Similar: lizardite	Massive: soapstone	"Brown mica"
	Fibrous: chrysotile		

THE LAYER SILICATES

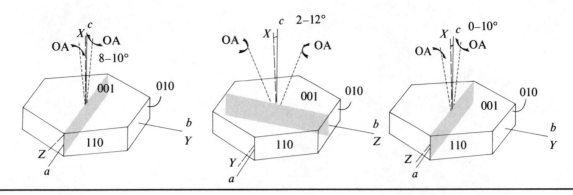

	Biotite (*1M*)	Margarite (*2M₁*)	CHLORITE (*1M*)
Formula	$K(Fe, Mg)_3(OH)_2Si_3AlO_{10}$	$CaAl_2(OH)_2Si_2Al_2O_{10}$	$(Mg, Fe, Al)_6(OH)_8(Si, Al)_4O_{10}$
Coord.	K(6), Fe(6), Mg(6), Si(4), Al(4)	Ca(6), Al(6), Si(4), Al(4)	Mg(6), Fe(6), Si(4), Al(4)
System	Monoclinic	Monoclinic	Monoclinic
$a =$	0.533	0.514	0.537
$b =$	0.931	0.900	0.930
$c =$	1.016	0.981	1.425
$\beta =$	99.3°	100.8°	97.4°
$Z =$	2	4	2
Sp. group	$C2/m$	$C2/c$	$C2/m$
Pt. group	$2/m$	$2/m$	$2/m$
S; loops	2.00; 6	2.00; 6	2.00; 6
SP plane	(001)	(001)	(001)
SP struct.	$AoB + CoAc;$	$AoB + CoAc$	$AoB + CoA = A - C =;$
	$BoC + AoBc;$	$BoC + AoBc;$	$CoA + BoC = C - B =;$
	$CoA + BoCc;$	$CoA + BoCc$	$BoC + AoB = B - A =;$
		$AoB + CoAc;$	
		$BoC + AoBc$	
		$CoA + BoAc;$	
SPI =	61	59	52
Cleavage	pf(001)	pf(001)	pf(001)
Twinning	[310]		[310]
	Comp. plane (001)		Comp. plane (001)
RI	$n_\alpha = 1.57$	$n_\alpha = 1.635$	$n_\alpha = 1.56–1.60$
	$n_\beta = 1.60$	$n_\beta = 1.645$	$n_\beta = 1.57–1.61$
	$n_\gamma = 1.61$	$n_\gamma = 1.648$	$n_\gamma = 1.58–1.61$
$\delta =$	0.04	0.013	0.006–0.020
$2V =$	0–32°	45°	0–40°
Sign	(−)	(−)	(−)
Transp.	Transp. to transl.	Transp. to transl.	Transp. to transl.
Pleochr.	X = pale brown, green		
	Y = dark brown, green		
	Z = dark brown, green		
H =	$2\frac{1}{2}$–3	$3\frac{1}{2}$–$4\frac{1}{2}$	2–3
D =	3.0	3.1	3.0
Color	Black, brown	Gray yellow	Green, variable
Streak	White	White	White, green
Luster	Pearly	Vitreous	Vitreous
Fracture	Elastic	Uneven	Flexible
Habit	Foliated	Foliated	Foliated, scaly
Remarks	"Black mica"	"Brittle mica"	Data given for prochlorite

THE LAYER
SILICATES

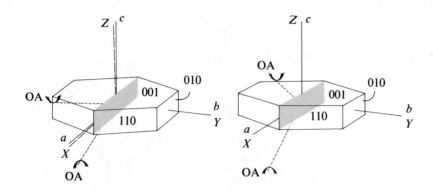

	PALYGORSKITE	Sepiolite
Formula	$Mg_5(OH)_2(H_2O)_4Si_8O_{20} \cdot 4H_2O$	$Mg_4(OH)_2(H_2O)_2Si_6O_{15} \cdot 4H_2O$
Coord.	Mg(6), Si(4)	Mg(6), Si(4)
System	Monoclinic	Orthorhombic
$a =$	0.52	0.528
$b =$	1.80	2.68
$c =$	1.34	1.34
$\beta =$	90–93°	
$Z =$	2	2
Sp. group	$A2/m$	$P2_1/n\,2/c\,2/n$
Pt. group	$2/m$	$2/m\,2/m\,2/m$
S; loops	2.00; 6, 6	2.00; 6, 6
SP plane	(100)	(100)
SP struct.	$A^*oBoC^* + A^*oBoC^* +$;	$A^*oBoC^* + A^*oBoC^* +$;
SPI =	16	16
Cleavage	gd{110}	gd{110}
RI	$n_\alpha = 1.52$	$n_\alpha = 1.51$
	$n_\beta = 1.53$	$n_\beta = 1.52$
	$n_\gamma = 1.55$	$n_\gamma = 1.53$
$\delta =$	0.03	0.02
$2V =$	0–60°	20–70°
Sign	(−)	(−)
Transp.	Transl. to transl.	Transp. to transl.
H =	$2-2\frac{1}{2}$	$2-2\frac{1}{2}$
D =	2.2	2
Color	White, gray	White
Streak	White	White
Luster	Vitreous, dull	Dull
Fracture	Subconchoidal	Subconchoidal
Habit	Fibrous, platy	Platy, fibrous
Remarks	Double subchains	Triple subchains
	Local name: attapulgite	Massive variety: meerschaum

$^*B$ and C sheets are $7/12^{th}$ filled in palygorskite and $5/9^{th}$ in sepiolite

THE LAYER SILICATES

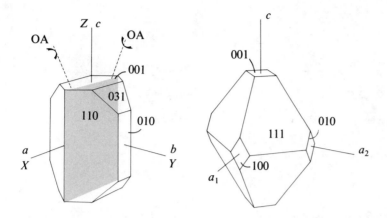

	PREHNITE	Apophyllite
Formula	$Ca_2Al(OH)_2Si_3AlO_{10}$	$KCa_4FSi_8O_{20} \cdot 8H_2O$
Coord.	Ca(7), Si(4), Al(4)	K(8), Ca(7), Si(4)
System	Orthorhombic	Tetragonal
$a =$	0.465	0.896
$b =$	0.548	
$c =$	1.849	1.578
$Z =$	2	2
Sp. group	$P2/n\,2_1/c\,2/m$	$P4/m\,2_1/n\,2/c$
Pt. group	$2/m\,2/m\,2/m$	$4/m\,2/m\,2/m$
S; loops	1.75; 6	1.75; 4, 8
SPI =	47	31
Cleavage	gd(001)	pf(001), pr 110
RI	$n_\alpha = 1.625$	$n_\omega = 1.535$
	$n_\beta = 1.635$	$n_\epsilon = 1.537$
	$n_\gamma = 1.655$	
$\delta =$	0.03	0.002
$2V =$	65–70°	
Sign	(+)	(+)
Transp.	Transp. to transl.	Transp. to transl.
H =	$6-6\frac{1}{2}$	$4\frac{1}{2}-5$
D =	2.9	2.3
Color	Pale green	White, gray
Streak	White	White
Luster	Vitreous	Pearly
Fracture	Uneven	Uneven
Habit	Tabular, prismatic globular, reniform	Short prismatic, tabular Occasionally pseudocubic
Remarks	Fades on exposure	Tetr. layer similar to feldspar sublayer

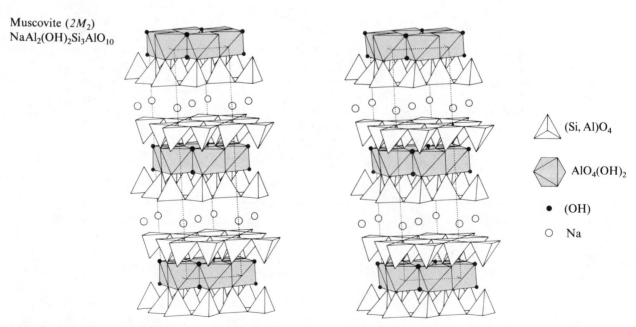

Kaolinite (1M)
Al$_4$(OH)$_8$Si$_4$O$_{10}$

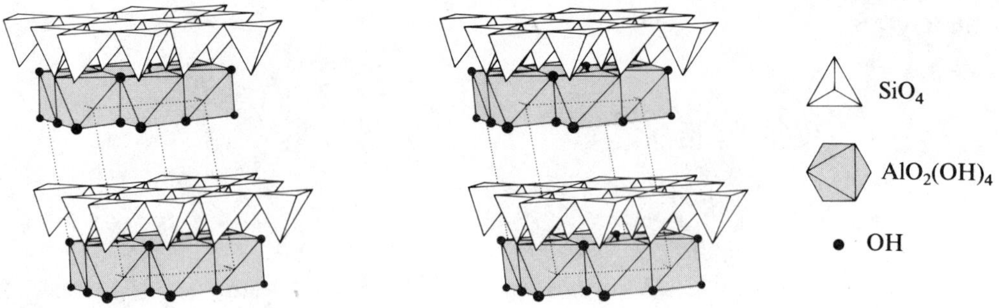

△ SiO$_4$

⬡ AlO$_2$(OH)$_4$

● OH

FIGURE II.21   **Crystal structure of 1M kaolinite, c axis vertical.**

Pyrophyllite (2M$_2$)
Al$_2$(OH)$_2$Si$_4$O$_{10}$

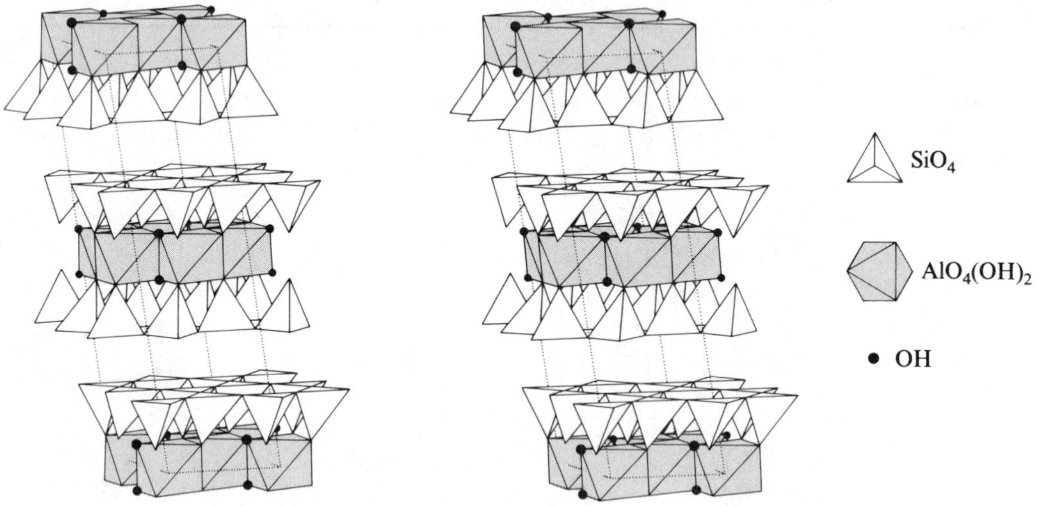

△ SiO$_4$

⬡ AlO$_4$(OH)$_2$

● OH

FIGURE II.22   **Crystal structure of 2M$_2$ pyrophyllite, c axis vertical.**

Muscovite (2M$_2$)
NaAl$_2$(OH)$_2$Si$_3$AlO$_{10}$

△ (Si, Al)O$_4$

⬡ AlO$_4$(OH)$_2$

● (OH)

○ Na

FIGURE II.23   **Crystal structure of 2M$_2$ muscovite, c axis vertical.**

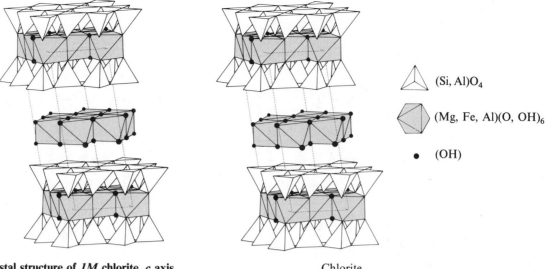

FIGURE II.24 **Crystal structure of *1M* chlorite, *c* axis vertical.**

Chlorite
$(Mg, Fe, Al)_6(OH)_8(Si, Al)_4O_{10}$

$(Si, Al)O_4$

$(Mg, Fe, Al)(O, OH)_6$

(OH)

Structure. The basic structure of the common layer silicates consists of tetrahedral layers alternating with parallel layers of octahedrally coordinated cations. Two types of alternation occur: one referred to by the symbol *TM* to denote single tetrahedral, *T*, and octahedral layers, *M*, joined together by sharing oxygens, and the other referred to as *TMT* to denote a structural unit that consists of a single octahedral layer sandwiched between two tetrahedral layers (Table II.3). Kaolinite and other Al-rich *TM* layer silicates are sometimes referred to as "kandites," and the Fe-rich serpentines are referred to as "septechlorites."

The *TM* and *TMT* structures are further divided on the basis of the cation valence within the octahedral layer. There are two valence possibilities. If the cations are divalent, such as Mg^{2+} and Fe^{2+}, all octahedral sites in the octahedral layer must be occupied to provide charge balance with the tetrahedral frame, hydroxyl $(OH)^{1-}$ ions and other cations. The structure of kaolinite (Figure II.21) shows that each apical oxygen of each tetrahedron is shared with the apical oxygens of three octahedra in an adjoining layer. If all three octahedra are occupied by divalent cations, the layer is referred to as a *trioctahedral* layer, and the mica or clay is referred to in like manner. If the cations in the octahedral layer are trivalent, such as Al^{3+}, only two of three octahedral sites are occupied, and the layer is referred to as *dioctahedral*. Muscovite $(KAl_2(OH)_2Si_3AlO_{10})$ is a dioctahedral mica; biotite $(KFe_3(OH)_2Si_3AlO_{10})$ is a trioctahedral mica. The trioctahedral layers are also called *brucite* layers, and the dioctahedral layers are called

gibbsite layers because of the respective layer resemblances to the structures of these minerals.

The manner in which the *TM* and *TMT* units are connected in a layered sequence further distinguishes the layer silicates and determines some of their distinctive physical properties. The *TM* units of kaolinite, serpentines, and related minerals (first row, Table II.3) are joined by relatively weak hydrogen bonds. The *TMT* units of pyrophyllite (Figure II.22) and talc are also neutral, and bonded together by van der Waals forces (Chapter 4). All of these minerals are easily cleaved parallel to the layers.

TMT units are joined in additional ways, which we symbolize as $TMT + H_2O + C$ (smectites, or expandable clays), $TMT + C$ (micas), and $TMT + M$ (chlorites). In the first case, loosely bound cations, *C*, and molecular water, H_2O, are present between *TMT* units in which some substitution of Al^{3+} for Si^{4+} occurs. The generic name *smectite* applies to a group of *TMT* layer silicates that includes a number of fine-grained clay minerals such as montmorillonite, beidellite, and saponite. These minerals differ from other clays in their cation exchange properties. Because of unbalanced substitutions (e.g., $Mg^{2+} \rightleftharpoons Al^{3+}$), the *T* or *M* layer or both layers gain a negative charge, which is balanced by various interlayer cations. In the case of micas, one out of four tetrahedra is occupied by Al^{3+}, and the excess negative charge is balanced by a monovalent cation, usually K^{1+} or Na^{1+} positioned between the *TMT* units (Figure II.23). In the so-called "brittle micas," margarite and clintonite, two of the four tetrahedra are occupied by Al^{3+} cations, and the charge

TABLE II.3. *Classification of Dioctahedral and Trioctahedral Layer Silicates*

Layer units:	Dioctahedral		Trioctahedral				Layer height (nm)
+	M = Al	M = Al + Mg(Fe)	M = Mg + Al	M = Mg	M = Mg + Fe	M = Fe	
M, T	kandites			serpentines		septechlorites	
	KAOLINITE — dickite, nacrite		amesite	ANTIGORITE — lizardite		greenalite, cronstedtite	0.7
M, T (+ H_2O)		halloysite	hydroamesite	hydroantigorite		parahalloysite	0.9
T, M, T	PYROPHYLLITE			TALC (Talcs)		minnesotaite*	0.9
T, M, T (C_n + H_2O; C = Na, Ca)		beidellite; montmorillonite, nontronite (Fe) (Smectites)		saponite			1.4
T, M, T (M_n + H_2O)						vermiculite	1.4
T, M, T, +C (C = K, Na, Ca; K, Na, Ca)	MUSCOVITE, paragonite; MARGARITE; glauconite; phengite (Fe), celadonite (Fe)		xanthophyllite	PHLOGOPITE (Micas)		annite; anandite (C = Ba)	1.0
M, T (+ H_2O)	hydromuscovite, hydroparagonite			hydrophlogopite		hydrobiotite; stilpnomelane*	1.1
T, M, T, +M, T, M, T	cookeite		clinochlore, penninite, prochlorite (Chlorites)			chamosite	1.4

*Modified structure.

is compensated by the divalent cation Ca^{2+}. As we might predict, the bonding between *TMT* units in margarite and clintonite is stronger than in other micas, with a consequent decrease in cleavage and flexibility. In the case of chlorites (Figure II.24), the *TMT* units are joined by another octahedral layer *M*. In vermiculites, the extra *M* layer is only partially developed and is mixed with water molecules.

In detail the layer silicate structures are more complicated than Table II.3 would lead us to believe. When the first structural studies of these minerals were made, the tetrahedral layer was assumed to have ideal hexagonal symmetry (Figure II.25b). After further studies, mineralogists found the actual symmetry of the layer to be lower, approaching the trigonal symmetry of the symmetrically packed model in Figure II.25a and II.25c in which the tetrahedra are rotated 30 degrees with respect to the tetrahedra in Figure II.25b. In the symmetrically packed model, all of the oxygens in the bottom sheet of the tetrahedral layer are corners of the basal tetrahedral faces. Two thirds of the anions in the top sheet are oxygens at the free apices of tetrahedra that link to the octahedral layer. The other one third of the anions are hydroxyl and are not considered part of the tetrahedral layer. These $(OH)^{1-}$ anions occupy the apices of the missing tetrahedra (marked with a dot in Figure II.25a) and are in fact corners of octahedra in the overlying *M* layer.

In terms of the unit translations in the common layer silicates, the dimensions of the symmetrically packed tetrahedral layer are $a = 0.456$ nm and $b = 0.795$ nm, as illustrated in Figure II.26a. These dimensions measure the separation of translationally equivalent, apical oxygens that must be shared with octahedra of an overlying *M* layer. In a dioctahedral *M* layer, the separation of translationally equivalent oxygens (Figure II.26b) corresponds to $a = 0.468$ nm and $b = 0.810$ nm, and

hence there is a mismatch between the dioctahedral and tetrahedral layers. In a trioctahedral *M* layer (Figure II.26c), the corresponding dimensions are $a = 0.514$ nm and $b = 0.891$ nm, and again, a mismatch results. Because the mismatch cannot exist in actual minerals, it was once thought that the *T* layer was bent in such a way that the distance between apical oxygens was stretched enough to provide a perfect registry with the corresponding oxygens in the *M* layer. We now know that the mismatch is compensated for by rotation and tilt of the tetrahedra in the *T* layer. The required rotation α (or $\bar{\alpha}$, Figure II.25) is given by

$$\alpha = 30 - \bar{\alpha}$$
$$= \cos^{-1}\left(\sqrt{3/4}\ \frac{\text{octahedral edge length}}{\text{tetrahedral edge length}}\right)$$

Figure II.26d shows that a 3-degree rotation is required for perfect registry of a *T* layer with a dioctahedral (Al) layer, and Figure II.26e shows that an 18-degree rotation is required for perfect registry of a *T* layer with a trioctahedral (Mg) layer.

In a relatively rare silicate mineral called catoptrite $(Mn_5Sb_2Si_2Mn_8Al_4O_{28})$, the anions are all oxygens, and all of the tetrahedral sites are occupied. The tetrahedral layer of micas and clays can be derived from the catoptrite structure in the same way as the dioctahedral layer of mica is derived from the trioctahedral layer, namely, by removing one out of three cations. In this sense, the catoptrite structure is the direct prototype of talc (*TMT*), and indirectly, the prototype of all clays and micas.

The structures of the layer silicates are further complicated by a large number of possible polytypes. The various polytypes of the layer silicates are discussed under the heading "biopyriboles," because the concepts controlling these polytypes are common to all minerals in this group.

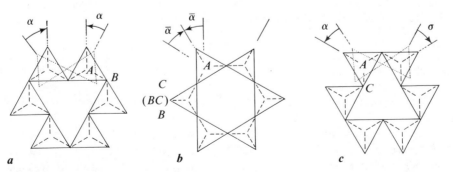

FIGURE II.25 (a) Ideal hexagonal rings in tetrahedral layer (E sheet). (b) Ditrigonal arrangement of rings by positive rotation α. (c) Ditrigonal arrangement of rings by negative rotation α.

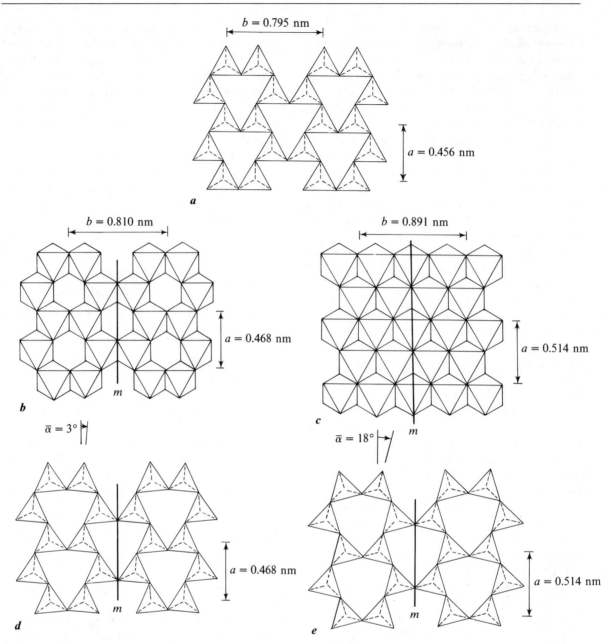

FIGURE II.26   **In a *TM* layer unit, a tetrahedral and an octahedral layer are joined by sharing corners. (a) Ideal symmetrical tetrahedral layer with orthogonal *a* and *b* unit translations, (b) ideal dioctahedral layer (gibbsite or Al layer), and (c) trioctahedral layer (brucite or Mg layer). The misfit between the tetrahedral and two octa-** hedral layers is adjusted by rotating tetrahedra in the tetrahedral layer. (d) A 3° rotation makes the tetrahedral layer fit the dioctahedral layers and (e) an approximately 18° rotation is required for registry of the trioctahedral layer. Layer mirrors are shown with *m*.

The crystal structures of palygorskite (Figure II.27) and sepiolite are similar to the micas and clays except that alternating bands of tetrahedra point up and down in both tetrahedral layers of the *TMT* unit. Both of these minerals are true layer silicates because of the two-dimensional extent of the tetrahedral frame, but they also provide an important structural transition to the multiple chain silicates discussed in the following sec-

tion. Figure 5.8 in Chapter 5 summarizes these structural relationships. Palygorskite and sepiolite are probably the only silicates that normally have a fibrous crystallization habit.

Prehnite (Figure II.28) and apophyllite are two relatively common layer silicates with structures that are different from those of the clays and micas. The tetrahedral layer in prehnite is undulating and contains six-membered loops of un-

Palygorskite
Mg$_5$(OH)$_2$Si$_8$O$_{20}$ · 4H$_2$O

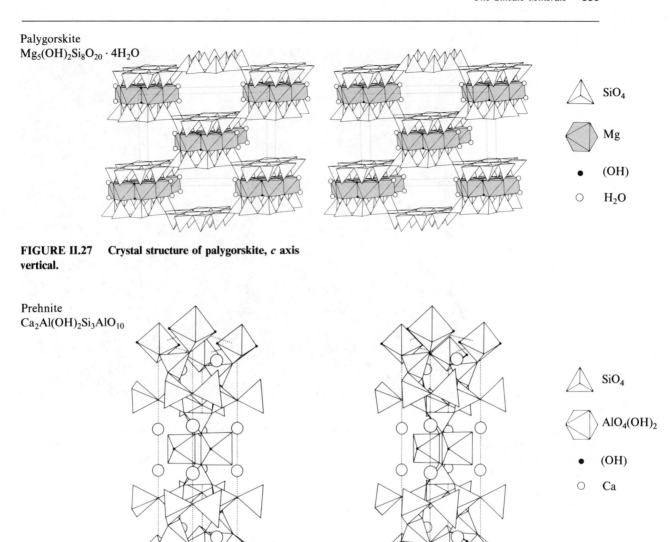

△ SiO$_4$

⬡ Mg

● (OH)

○ H$_2$O

FIGURE II.27 **Crystal structure of palygorskite, *c* axis vertical.**

Prehnite
Ca$_2$Al(OH)$_2$Si$_3$AlO$_{10}$

△ SiO$_4$

⬡ AlO$_4$(OH)$_2$

● (OH)

○ Ca

FIGURE II.28 **Crystal structure of prehnite, *c* axis vertical.**

usual symmetry connected by Al octahedra and ninefold coordinated Ca. The tetrahedral layer in apophyllite resembles the four-membered and six-membered loops in the (100) section of the feldspar frame. The layers are connected by K^{1+} and Ca^{2+} cations in eightfold and sevenfold coordination, respectively.

Some layer silicates possess a unique crystallization habit known as *asbestiform*. The *TM* or *TMT* layers are curled up and form tubular or scroll-like structures (Figure II.29). These crystals are extremely thin, long, strong, and flexible, and are referred to as "fibers" based on their resemblance to organic fibers. The best known example is *chrysotile*, an asbestiform Mg-serpentine. Other examples are tubular halloysite and a highly fi-

brous variety of talc sometimes called *agalite*. Asbestiform palygorskite in matted sheets is known as *mountain leather*. Massive fibrous sepiolite has the varietal name *meerschaum*. The fibrous structures of palygorskite and sepiolite are not fully understood. Apparently they do not have the chrysotile type of tubular structure. Massive fibrous occurrences of serpentine are known as *bowenite* and make excellent carving stones.

Chemistry. Table II.3 and Figure II.30 summarize the major cation substitutions within the structures of the layer silicates. Complete substitution occurs between Fe^{2+} and Mg^{2+} in the trioctahedral layer, which allows for a complete solid solution between the trioctahedral micas biotite and phlogopite, between talc and minne-

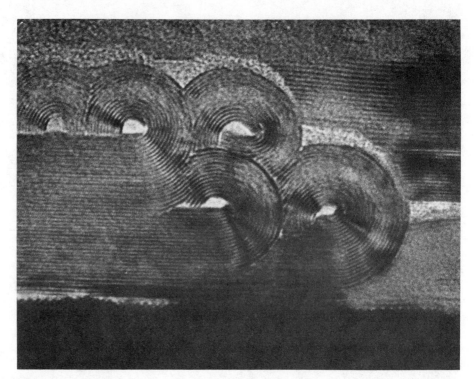

a

b

20 nm

FIGURE II.29   A high resolution transmission electron microscope (HRTEM) photograph of serpentine (S) and talc (T) showing curved atomic layers characteristic of chrysotile. (Courtesy of D. R. Veblen, Johns Hopkins University. From Veblen, D. R., and Buseck, P. R., 1979. Serpentine minerals: intergrowths and new combination structures. *Science* **206:4425, pp. 1398–1400, Fig. 2b. Copyright December 1979 by the American Association for the Advancement of Science.)**

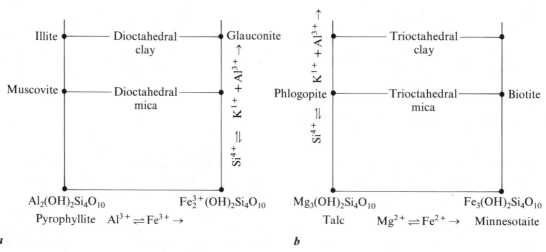

a

b

FIGURE II.30   A portion of composition space for (a) dioctahedral layer silicates and (b) trioctahedral layer silicates.

sotaite, and between antigorite and ferroantigorite. There is also a complete solid solution between daphnite (Fe^{2+} chlorite) and the Mg end-member of the chlorites. In the trioctahedral *TMT* + *C* structures, the "*C*" cation is usually

K^{1+}, Na^{1+}, or Ca^{2+} with appropriate Al-Si substitution in the tetrahedral layers.

A number of trioctahedral smectites have chemistries derived from the talc structure. Hectorite results by substitution of Li^{1+} for Mg^{2+} in octahedral layers with charge balance supplied by Ca^{2+} or Na^{1+} in interlayer positions. Saponite results from $Al^{3+} \rightleftharpoons Si^{4+}$ substitution in tetrahedral layers with the same charge balance mechanism as in hectorite.

Within the dioctahedral layer silicates, substitution of Fe^{3+} for Al^{3+} in the dioctahedral layers produces glauconite from illite within the clays, nontronite from beidellite among the smectites, and Fe^{3+}-bearing muscovite, paragonite, and margarite. A common substitution involving the "C" cation in $TMT + C$ structures is Na^{1+} for K^{1+}. Muscovite and paragonite are related in this manner, and in fact, a gap in solid solubility between these end-members is structurally analogous to the solvus in the alkali feldspars. Similarly miscibility gaps may exist between illite and Na-beidellite, and between glauconite and Na-nontronite. As in the alkali feldspars, such miscibility gaps imply that solid solutions on opposing limbs of the solvus may be found together in an equilibrium association.

Among the dioctahedral layer silicates, the chemical relationship between pyrophyllite, muscovite, and illite may be viewed in terms of the successive application of the coupled $K^{1+} + Al^{3+} \rightleftharpoons Si^{4+}$ substitution in the sequence

$$Al_2(OH)_2Si_4O_{10} \rightarrow KAl_2(OH)_2Si_3AlO_{10}$$
$$\text{(pyrophyllite)} \qquad \text{(muscovite)}$$

$$\rightarrow K_{1.5}Al_2(OH)_2Si_{2.5}Al_{1.5}O_{10}$$
$$\text{(illite)}$$

The structures of these minerals are quite different, so little solid solubility exists between them. Glauconite may be derived chemically from illite by the substitution of Fe^{3+} for Al^{3+}.

The generalized chemistries of the dioctahedral smectites, namely, montmorillonite and beidellite, can also be derived from the chemistry of pyrophyllite. By substituting Mg^{2+} for Al^{3+} in the dioctahedral layer and charge balancing with Ca^{2+} or Na^{1+} or both in interlayer sites, montmorillonite results. By substituting Al^{3+} for Si^{4+} in the tetrahedral layer and charge balancing again with Ca^{2+} or Na^{1+} or both in interlayer positions, beidellite results. By substituting Fe^{3+} for Al^{3+} in beidellite, the simplified chemistry of nontronite results.

Only limited solid solution exists between the trioctahedral and dioctahedral layer silicates. Consequently, both types of layer silicates occur together frequently in many igneous and metamorphic rocks. Biotite and muscovite are commonly found together if the bulk chemistry of the rock is appropriate. Layer silicates with chemistries between the ideal dioctahedral and trioctahedral end-members exist mostly for poorly crystallized, disordered polytypes. Some of the mixed-layer clays are in this category and are extremely difficult to characterize without careful x-ray diffraction studies.

Occurrence and Associations. Because all of the layer silicates are hydrous aluminosilicates, their occurrence and associations are severely limited by bulk chemistry. For the most part, they form only in hydrous, alumina-rich environments, and consequently, the minerals with which they are associated have similar chemistries. The weathering products of many igneous and metamorphic rocks are mostly clay, derived in large part from feldspars. Such clays find their way into sedimentary environments where they either remain unaltered or react to form authigenic clays, feldspars, and a host of other minerals. The chemical environment is enriched in Al and Si, and in K and Na, along with various amounts of Fe and Mg. Upon metamorphism of such sediments, coarse-grained muscovite and biotite frequently form in association with other alumina-rich silicates such as kyanite, sillimanite, andalusite, garnet, staurolite, and many others (Figure 9.13). Because of the alkalies present in many layer silicates, the alkali feldspars are a common associate. Most felsic igneous rocks will have biotite, muscovite, or both in association with an albitic plagioclase and a K-feldspar. Because of the water in most layer silicates, other hydrous silicates such as amphiboles also form.

The Fe/Mg ratio in the chemical environment has a profound effect on the composition of Fe-Mg solid solutions within the layer silicates. For example, phlogopite and talc are found only in mafic to ultramafic igneous rocks and in metamorphosed dolomites and other Mg-enriched rocks. Minnesotaite, the Fe end-member, is found only in very Fe-rich environments such as iron formations. The same relationships are true for other layer silicate solid solutions. The mineral chemistry reflects the chemical environment of formation.

The layer silicates have a wide range of commercial uses. Muscovite and phlogopite are well known for their thermal and electrical insulating properties. Talc is used extensively for powder in the cosmetics industry. Clay minerals, especially kaolinite, are used extensively in ceramics and as a filler for a number of food and paper products. Palygorskite (variety attapulgite) is frequently used as an absorbent. Meerschaum is a favored pipe material. Chrysotile, the asbestiform serpentine, is widely used in fabrics, filters, cement and tile composites, and other building materials because of its strength, flexibility, and fireproof qualities.

Distinguishing Features. All of the layer silicates possess a perfect basal cleavage and platy crystallization habit that distinguishes them from most other minerals. The trioctahedral layer silicates are generally dark colored (e.g., biotite) because of ferrous and ferric iron in the trioctahedral layer, although pure Mg compositions may be colorless. The dioctahedral layer silicates are generally colorless (e.g., muscovite) unless a transition metal such as Fe^{3+} is substituting for Al^{3+} in the dioctahedral layer.

In thin section, the layer silicates large enough to observe as discrete grains show a distinctive texture with crossed polarizers that resembles the surface of bird's eye maple, a wood frequently used in fine furniture. Parallel extinction, small optic angle, and strong dichroism in the colored layer silicates are usually diagnostic.

Positive identification of the clay-sized layer silicates is best made with special x-ray diffraction techniques.

The Chain Silicates

The most common and important chain silicates, also referred to as the inosilicates, are the pyroxenes, pyroxenoids, and the amphiboles. Their general structures consist of alternating tetrahedral chain and octahedral band layers that are parallel to the (100) plane. Each of the tetrahedra in the pyroxenes and pyroxenoids links with two others to form infinite single chains (Figure 5.7) that run parallel to the *c* axis. In the amphiboles, the tetrahedra form infinite double chains (Figure 5.7) that also run parallel to the *c* axis. In both pyroxenes and amphiboles, the Si—O—Si bonds in the chain directions are the strongest in the structure and control the dominant (110) and $(1\bar{1}0)$ cleavages and prismatic crystallization habit.

Within the chain silicate group, many other types of chains exist. The simplest of all is the single chain of synthetic iscorite ($Fe_6O_4SiFeO_6$), which is known only from industrial slags. It has a repeat unit of just one tetrahedron and can be viewed as a structural prototype for the other chain silicates. In addition to the most common single and double chains, there are a variety of multiple-chain silicates, several of which are discussed in Chapter 5.

The Pyroxenes

The pyroxenes are important minerals in most mafic to ultramafic igneous rocks and in high-grade metamorphic rocks such as granulites and eclogites. Their general chemical and structural division is summarized in the composition diagram of Figure II.31. This representation is referred to as the pyroxene quadrilateral because of the four compositions used to describe many natural pyroxene solid solutions. Pyroxenes along the base of the diagram are Fe-Mg solid solutions and have orthorhombic symmetry. They are usually referred to as the *orthopyroxenes*. Pyroxenes be-

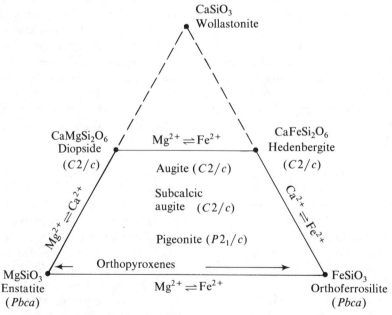

FIGURE II.31 The pyroxene quadrilateral composition diagram in the system $MgSiO_3$-$FeSiO_3$-$CaSiO_3$.

tween diopside and hedenbergite form a complete Fe-Mg solid solution as well but have monoclinic symmetry. These are referred to as the *clinopyroxenes*. At the top of the composition diagram is wollastonite, a pyroxenoid with triclinic symmetry. The details of the equilibrium relationships within this diagram are discussed in Chapter 8.

Structure. The basic structural components of the pyroxenes are the single tetrahedral chains and the double octahedral bands. At one time, the tetrahedral chain was believed to be straight as shown in Figure II.32b. This configuration is referred to as the E-chain and is similar to a single tetrahedral chain segment of the layer silicates shown in Figure II.25b. Mineralogists later discovered that very few pyroxenes have this type of chain. Instead, nearly all pyroxene chains have symmetries intermediate to those shown in Fig-

ures II.32b and II.32c, which are referred to as "kinked" chains and are similar to the single tetrahedral chain segments of the layer silicates shown in Figures II.25a and II.25c.

The tetrahedral chains in the pyroxenes (e.g., Figure II.33) run parallel to the *c* axis, and the tetrahedra in alternating chains point up and down as shown in Figure II.34a. Their free apices are connected to the double octahedral bands, which also run parallel to the *c* axis (Figure II.34b) by sharing oxygen ions. Each octahedral band will thus have a tetrahedral chain linked to it on both sides, resembling an I-beam (Figure II.34c). The complete pyroxene structure consists of a staggered pattern of I-beams as shown in Figure II.34d. Adjacent I-beams are connected by sharing the free basal tetrahedral corners with the free corners of the *M2* octahedra.

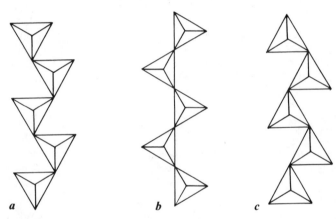

FIGURE II.32 **Major variations of the tetrahedral chains of pyroxenes. (a) and (b) are the ideal symmetrically packed chains, and (c) is the E-chain.**

Enstatite
$Mg_2Si_2O_6$

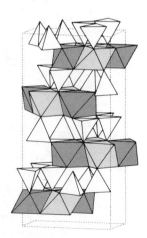

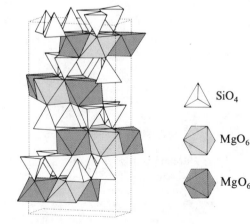

SiO₄

MgO₆

MgO₆

FIGURE II.33 **Crystal structure of enstatite, *a* axis vertical.**

THE PYROXENES

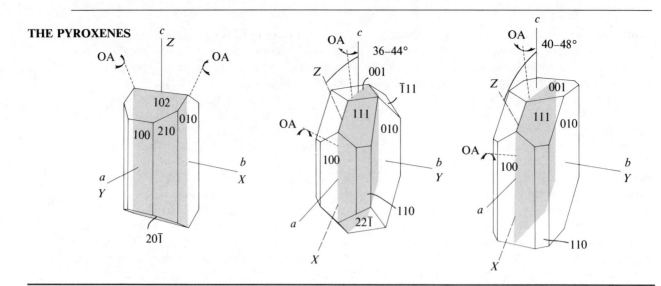

	ENSTATITE	DIOPSIDE	Hedenbergite
Formula	$Mg_2Si_2O_6$	$CaMgSi_2O_6$	$CaFeSi_2O_6$
Coord.	Mg(6 or 8), Si(4)	Ca(8), Mg(6), Si(4)	Ca(8), Fe(6), Si(4)
System	Orthorhombic	Monoclinic	Monoclinic
$a =$	1.822	0.97	0.985
$b =$	0.881	0.89	0.902
$c =$	0.521	0.525	0.526
$\beta =$		105.83°	104.33°
$Z =$	4	4	4
Sp. group	$P2_1/b\,2_1/c\,2_1/a$	$C2/c$	$C2/c$
Pt. group	$2/m\,2/m\,2/m$	$2/m$	$2/m$
S; RU	1.50; 2	1.50; 2	1.50; 2
SP plane	(100)	(100)	(100)
SP struct.	$AooB/C - BooA -$ $CooB/A + BooC +;$	$AooB + CooA +;$ $BooC + AooB +;$ $CooA + BooC +;$	$AooB + CooA +;$ $BooC + AooB +;$ $CooA + BooC +;$
SPI =	66	73	72
Cleavage	pf{210}	pf{110}	pf{110}
Twinning	{101} rare	(100), (001)	(100), (001)
Pt. group	(100), (010)	(001), (100)	(001), (100)
RI	$n_\alpha = 1.657$ $n_\beta = 1.659$ $n_\gamma = 1.665$	$n_\alpha = 1.665$ $n_\beta = 1.672$ $n_\gamma = 1.695$	$n_\alpha = 1.727$ $n_\beta = 1.735$ $n_\gamma = 1.756$
$\delta =$	0.008	0.030	0.029
$2V =$	54°	56–62°	58–64°
Sign	(+)	(+)	(+)
Transp.	Translucent	Transp. to transl.	Transp. to transl.
Pleochr.			X = yellow green Y = yellow green Z = green blue
H =	5–6	$5\frac{1}{2}$–$6\frac{1}{2}$	$5\frac{1}{2}$–$6\frac{1}{2}$
D =	3.2–3.5	3.2–3.5	3.3–3.6
Color	Gray, green	Green, variable	Green, variable
Streak	White, gray	White, gray	White, gray
Luster	Vitreous, pearly	Vitreous	Vitreous
Fracture	Uneven	Uneven	Uneven
Habit	Prismatic, acicular	Prismatic	Prismatic
Remarks	(Fe, Mg): hypersthene	(Mg, Fe): augite	(Mn, Fe): johannsenite

Members of the *pyroxene* group

THE PYROXENES

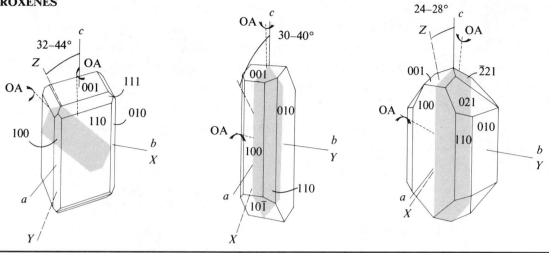

	Pigeonite	Jadeite	Spodumene
Formula	(Ca, Mg) (Mg, Fe)Si_2O_6	$NaAlSi_2O_6$	$LiAlSi_2O_6$
Coord.	Ca(8), Mg(8), Mg(6), Fe(6), Si(4)	Na(8), Al(6), Si(4)	Li(6), Al(6), Si(4)
System	Monoclinic	Monoclinic	Monoclinic
$a =$	0.973	0.950	0.952
$b =$	0.895	0.861	0.832
$c =$	0.526	0.524	0.525
$\beta =$	108.55°	107.43°	110.46°
$Z =$	4	4	4
Sp. group	$C2/c$	$C2/c$	$C2/c$
Pt. group	$2/m$	$2/m$	$2/m$
S; RU	1.50; 2	1.50; 2	1.50; 2
SP plane	(100)	(100)	(100)
SP struct.	$AooB + CooA +$; $BooC + AooB +$; $CooA + BooC +$;	$AooB + CooA +$; $BooC + AooB +$; $CooA + BooC +$;	$AssB + AssB +$;
SPI =	74	74	66
Cleavage	pf{110}	pf{110}	pf{110}
Parting	(100), (010), (001)		(100), (010)
Twinning	(100), (001)	(100), (001)	(100)
RI	$n_\alpha = 1.69$ $n_\beta = 1.69$ $n_\gamma = 1.72$	$n_\alpha = 1.65$ $n_\beta = 1.66$ $n_\gamma = 1.67$	$n_\alpha = 1.65$ $n_\beta = 1.66$ $n_\gamma = 1.67$
$\delta =$	0.025	0.02	0.02
$2V =$	38–44°	70–75°	60–80°
Sign	(+)	(+)	(+)
Transp.	Transp. to transl.	Translucent	Transp. to transl.
Pleochr.	X = pale yellow Y = pale brown Z = pale green	X = pale green Y = colorless Z = pale yellow	
H =	6	$6\frac{1}{2}$	$6\frac{1}{2}$–7
D =	3.40	3.30	3.15
Color	Brown, green	Green, variable	Colorless, variable
Streak	White	White	White
Luster	Vitreous	Vitreous, greasy	Vitreous
Fracture	Uneven	Uneven	Uneven
Habit	Prismatic	Prismatic, tabular	Prismatic
Remarks	Commonly zoned	One of the jade minerals	Colored varieties pleochroic

Members of the *pyroxene* group

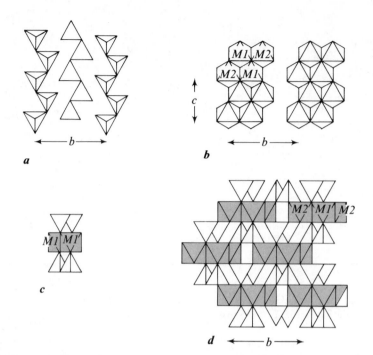

FIGURE II.34 **Basic components of the pyroxene structure illustrated by the structure of spodumene. (a) Characteristic tetrahedral chains, (b) octahedral chains** **(with identification of *M1* and *M2* octahedra), (c) a single *TM* band, and (d) view of chains along *c* axis.**

The oxygens in the ideal pyroxene structure, composed of chains like those shown in Figures II.32a and II.32b, are in hexagonal sheets that can be stacked in various sequences. The details of the different permissible stacking sequences and the consequent structure types will be discussed together with the layer silicates under the heading "biopyriboles."

We should expect considerable departure from the ideal symmetrically packed model in the actual pyroxene structures. In the common layer silicates, the degree of tetrahedral rotation is essentially a function of the relative sizes of the cations in tetrahedral and octahedral positions (p. 329), and all structures can have practically any degree of tetrahedral rotation. The linkage of the octahedral and tetrahedral chains in the pyroxene and amphibole structures is similar to that in the layer silicates insofar as the linkage of the up-pointing and down-pointing tetrahedral corners is concerned. In chain silicates, however, the distance between the unshared basal tetrahedral corners and the appropriate octahedral edges must match, a condition achieved by tetrahedral rotation. That rotation necessitates a coupled distortion of the *M2* (or *M4* in amphiboles) octahedra because of the shift in the position of the oxygen shared between the tetrahedra and the octahedra at the edge of the two chains. The distortion of these octahedra (*M2* or *M4*) is different for differ-

ent stacking sequences. The ideal models of the *A + B* octahedral chains followed by the oxygen sheets of the tetrahedral layer in *C* and in *A* position are illustrated, respectively, in Figures II.35a and II.35c. In both cases, tetrahedral rotation is achieved by shifting a row of oxygens in the *B* sheet, as shown by the arrows in Figures II.35b and II.35d.

In forward alphabetical stacking order, as in *A + BC* (Figures II.35a and II.35b), the *M2* octahedron increases in size with increasing tetrahedral rotation, and the shifting oxygen approaches a second *M2* cation to form a new bond. In reverse alphabetical stacking, as in *A + BA* (Figures II.35c and II.35d), the *M2* octahedron becomes smaller with increasing tetrahedral rotation, and the distance between that oxygen and the neighboring *M2* cation increases. Consequently, no new bond is formed. At the 30 degree maximum value of $\bar{\alpha}$, however, the shifting oxygen reaches the bond distance with a neighboring *M2* cation. That is, at $\bar{\alpha} = 30$ degrees the distortion patterns of both *A + BC* and *A + BA* stackings are alike (Figure II.35e).

If the oxygen sheets on both sides of the octahedral layer are in forward alphabetical order, as in *CA + BC*, the sixfold coordination of the *M2* cation changes to eightfold. If one oxygen sheet is in forward and the other in reverse alphabetical order, as in *CA + BA* or *BA + BC*, the coordina-

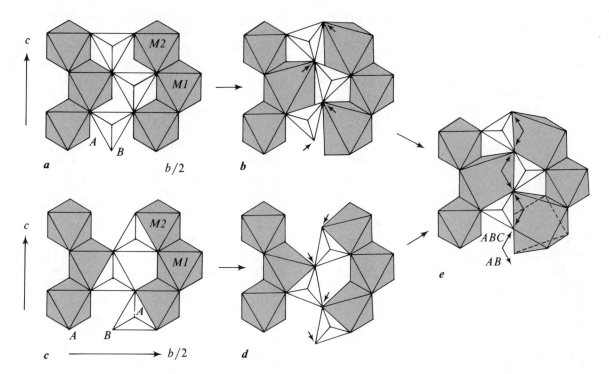

FIGURE II.35 Schematic illustration of the coordinated effect of tetrahedral rotation and distortion of the *M2* octahedral site. (a) Ideal model of the *AB* octahedral layer followed by a *C* oxygen sheet, and (b) the relative distortion of the *M2* octahedra at around 15° tetrahedral rotation. (c) The ideal model of the *A* + *B* octahedral layer followed by an *A* oxygen sheet, and (d) the distortion of the *M2* octahedra at 15° tetrahedral rotation. (e) At full rotation of the tetrahedra ($\bar{\alpha} = 30°, \alpha = 0$) both models are identical and contain the E-chain.

tion of the *M2* site becomes sevenfold. If both sheets are in reverse alphabetical order, as in *BA* + *BA*, the coordination remains sixfold until $\bar{\alpha} = 30$ degrees.

Consequently, the different stacking sequences of the pyroxene (and amphibole) structure are not just simple polytypes, but represent different structural models suitable for different cation combinations. For example, large cation pyroxenes such as Ca and Na (radii around 0.10 nm) have the oxygen sheets in forward alphabetical order as in diopside ($CaMgSi_2O_6$) shown in Figure II.36 and jadeite ($NaAlSi_2O_6$). The smaller cation pyroxenes such as Li and Al (radii 0.074 and 0.053 nm, respectively) have the oxygen sheets in reverse alphabetical order as in spodumene ($LiAlSi_2O_6$).

Chemistry. The possible chemistries of pyroxenes are constrained by the physical limits of each cation site in the structure. Within these constraints, widespread solid solution exists between ideal end-member compositions. Any chemical

Diopside
$CaMgSi_2O_6$

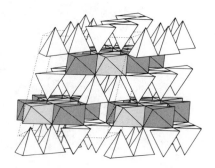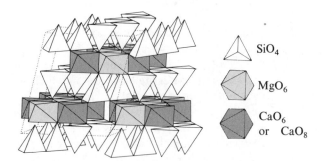

FIGURE II.36 Crystal structure of diopside, *a* axis vertical.

variation from one end-member to another must be accomplished by substitutions that maintain a continuous state of electrical neutrality. For example, most of the common pyroxenes can be represented in the pyroxene quadrilateral (Figure II.31 and Figure 8.6). Ferrous iron (Fe^{2+}) and Mg^{2+} substitute readily because of their identical charge and similar ionic radii. Starting with enstatite ($Mg_2Si_2O_6$) as a reference, ferrosilite ($Fe_2Si_2O_6$) and all solid solutions between can be generated by the $Fe^{2+} \rightleftharpoons Mg^{2+}$ substitution. Ferrosilite is the physical limit of that substitution in the octahedral sites. Starting with diopside ($CaMgSi_2O_6$), all solid solutions toward hedenbergite are generated by the same $Fe^{2+} \rightleftharpoons Mg^{2+}$ substitution, but hedenbergite is as far as the substitution can be carried. As we learned in Chapter 8, a substitution of Ca^{2+} for Mg^{2+} in enstatite will generate solid solutions toward diopside, and if carried to the physical limits, wollastonite ($CaSiO_3$) results. We do not mean to imply that all pyroxene solid solutions generated in this way are necessarily stable. Severe distortions commonly develop, and stability is then determined by the temperature and temperature of formation in each case.

Numerous pyroxenes lie in the interior part of the quadrilateral composition space. Augite, subcalcic augite, and pigeonite are the most common. Their chemistries can be derived from any of the end-member pyroxenes by a combination of $Fe^{2+} \rightleftharpoons Mg^{2+}$ and $Mg^{2+} \rightleftharpoons Ca^{2+}$ substitutions. Many pyroxenes that are called augite have substantial Al^{3+} and other cations and thus cannot be strictly represented in the quadrilateral plane. Additional substitutions are required, and hence additional dimensions to the pyroxene composition space (Chapter 8). In practice, a chemical analysis of a real pyroxene is recalculated in terms of ideal end-members, and then the most important of these are chosen for graphical representation.

Among the more common chemical substitutions that represent additional dimensions to the pyroxene composition space are $Na^{1+} + Al^{3+} \rightleftharpoons 2(Mg^{2+})$ and $Mg^{2+} + Si^{4+} \rightleftharpoons 2(Al^{3+})$. If enstatite is used as a reference composition, the first of these substitutions leads to the pyroxene jadeite ($NaAlSi_2O_6$). The second substitution leads to the ideal tschermakite ($MgAlSiAlO_6$) in which Al^{3+} is substituted for tetrahedral Si^{4+}, and Al^{3+} in octahedral coordination is substituted for Mg^{2+}. If ferric (Fe^{3+}) iron substitutes for Al^{3+} in the jadeite substitution, end-member aegerine ($NaFe^{3+}Si_2O_6$) results. If Li^{1+} substitutes for Na^{1+} instead, solid solutions toward spodumene ($LiAlSi_2O_6$) are generated. Natural pyroxene analyses, once recalculated, may have amounts of several end-member components.

Occurrence and Associations. Minerals in the pyroxene group are widespread in igneous and metamorphic rocks. Because of their susceptibility to both chemical and mechanical (due to cleavage) weathering, they are rare in sedimentary rocks. Their conditions of formation are almost exclusively at elevated temperature or elevated pressure or both. They are thus common only at the higher grades of metamorphism. Both orthopyroxenes and clinopyroxenes in the pyroxene quadrilateral can crystallize directly from Mg-rich and Fe-rich melts at high temperature. They are commonly associated with olivine or plagioclase or both in gabbros and basalts. Orthopyroxenes rich in enstatite and clinopyroxenes rich in diopside are important constituents of the earth's mantle as evidenced by their presence in ultramafic nodules from lavas that originate at mantle depths.

Solid solutions between diopside and hedenbergite are commonly associated with other Ca-rich minerals such as wollastonite and grossularite in metamorphosed siliceous dolomites.

Distinguishing Features. The pyroxenes in hand specimen are distinguished by their stubby pris-

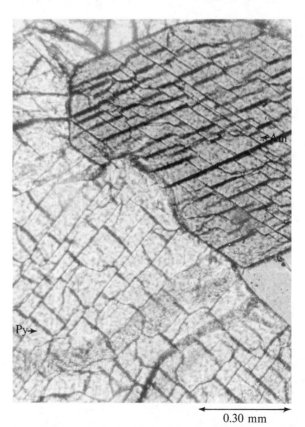

0.30 mm

FIGURE II.37   Intersections of prismatic cleavages in a pyroxene (Py) and an amphibole (Am).

matic form and prismatic cleavage intersection at almost 90 degrees. Clinopyroxenes such as augite and diopside-hedenbergite solid solutions are generally black, whereas the orthopyroxenes tend to be brown to bronze in color. Pure diopside and pure enstatite are colorless. Jadeite is apple green to white depending on iron content, and is one of the toughest known materials when occurring as *jade*, a microcrystalline massive variety.

In thin section, the pyroxenes are distinguished by their moderate relief and prismatic cleavage intersections at nearly 90 degrees (Figure II.37). Calcic clinopyroxenes have moderate birefringence, whereas the orthopyroxenes generally have lower, first-order gray birefringence. Jadeite and omphacite, a solid solution between jadeite and diopside, usually display anomalous "berlin blue" birefringence and frequently have a radial form.

The Amphiboles

The amphiboles are important minerals in a variety of plutonic rocks that span the composition range from granite to gabbro. They are also a major mineral in many metamorphic rocks, especially those derived from mafic igneous rocks and impure dolomites. The general chemistry and structure of amphiboles is summarized in the composition diagram of Figure II.38, called the amphibole quadrilateral. Like the pyroxenes, the amphiboles along the base of the diagram have orthorhombic symmetry and are frequently referred to as orthoamphiboles (Figure II.39). The amphiboles between tremolite and ferroactinolite are referred to as the calcic amphiboles and are both structurally and compositionally similar to the calcic clinopyroxenes (Figure II.31).

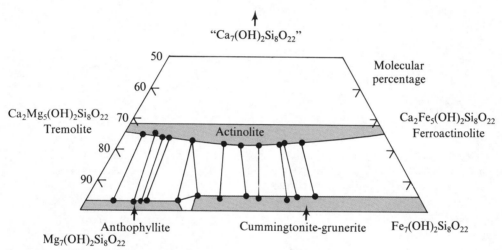

FIGURE II.38 Amphibole quadrilateral composition diagram. Compare with Figure II.31.

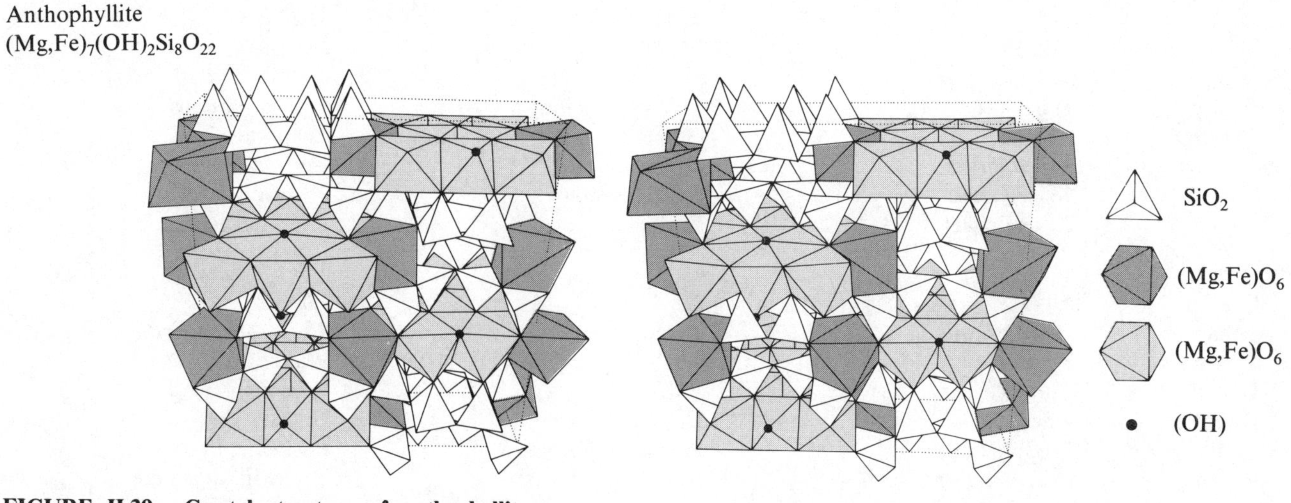

Anthophyllite
$(Mg,Fe)_7(OH)_2Si_8O_{22}$

FIGURE II.39 Crystal structure of anthophyllite, *a* axis vertical.

THE AMPHIBOLES

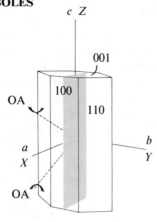

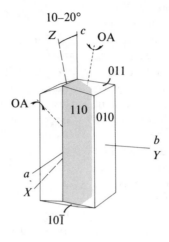

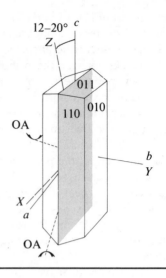

	ANTHOPHYLLITE	Cummingtonite	Tremolite
Formula	$(Mg, Fe)_7(OH)_2Si_8O_{22}$	$Mg_7(OH)_2Si_8O_{22}$	$Ca_2Mg_5(OH)_2Si_8O_{22}$
Coord.	Mg(6 or 7), Fe(6 or 7), Si(4)	Mg(6 or 8), Si(4)	Ca(8), Mg(6), Si(4)
System	Orthorhombic	Monoclinic	Monoclinic
$a =$	1.856	0.951	0.986
$b =$	1.801	1.819	1.811
$c =$	0.528	0.533	0.534
$\beta =$		101.83°	105.00°
$Z =$	4	2	2
Sp. group	$P2/n2/m2/a$	$C2/m$	$C2/m$
Pt. group	$2/m2/m2/m$	$2/m$	$2/m$
S; RU (loops)	1.625; 2 (6)	1.625; 2 (6)	1.625; 2 (6)
SP plane	(100)	(100)	(100)
SP struct.	$AooB/C - BooA -$ $CooB/A + BooC +$;	$AooB + CooA +$; $BooC + AooB +$; $CooA + BooC +$;	$AooB + CooA +$; $BooC + AooB +$; $CooA + BooC +$;
SPI =	60	57	64
Cleavage	pf{210}, pr(100)	pf{110}	pf{110}
Parting	(001)		(100)
Twinning		(100)	(100), (001)
RI	$n_\alpha = 1.60$ $n_\beta = 1.62$ $n_\gamma = 1.63$	$n_\alpha = 1.644$ $n_\beta = 1.657$ $n_\gamma = 1.674$	$n_\alpha = 1.608$ $n_\beta = 1.618$ $n_\gamma = 1.630$
$\delta =$	0.03	0.030	0.022
$2V =$	65–90°	80–90°	85°
Sign	(−)	(+)	(−)
Transp.	Transp. to transl.	Transp. to transl.	Transp. to transl.
Pleochr.	Gedrite: X = colorless Y = colorless Z = pale violet	Grunerite: X = pale yellow Y = pale brown Z = pale green	Actinolite: X = pale yellow Y = pale green Z = blue green
H =	$5\frac{1}{2}$–6	6	5–6
D =	2.9–3.2	2.9–3.2	3.0–3.3
Color	Gray, brown	White, green	White, green
Streak	White	White	White
Luster	Vitreous	Vitreous	Vitreous
Fracture	Uneven	Uneven	Uneven
Habit	Prismatic, fibrous	Prismatic, fibrous	Prismatic, fibrous
Remarks	(Na, Al): gedrite	Fe⇌Mg: grunerite	Fe⇌Mg: actinolite

Members of the *amphibole* group

THE AMPHIBOLES

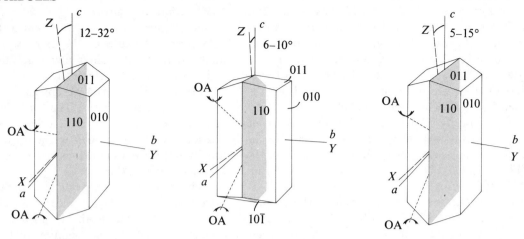

	HORNBLENDE	Glaucophane	Kaersutite
Formula	$NaCa_2(Mg, Fe, Al)_5(OH)_2$ $(Si, Al)_8O_{22}$	$Na_2Mg_3Al_2(OH)_2Si_8O_{22}$	$Ca_2Na(Mg, Fe)_4Ti(OH)_2$ $Si_6Al_2O_{22}$
Coord.	Ca(8), Na(8), Mg(6), Si(4), Al(4)	Na(8), Mg(6), Al(6), Si(4)	Na(8), Mg(6), Si(4), Al(4)
System	Monoclinic	Monoclinic	Monoclinic
$a =$	0.897	0.958	0.993
$b =$	1.801	1.780	1.825
$c =$	0.533	0.530	0.534
$\beta =$	105.75°	103.76°	104.9°
$Z =$	2	2	2
Sp. group	$C2/m$	$C2/m$	$C2/m$
Pt. group	$2/m$	$2/m$	$2/m$
S; RU (loops)	1.625; 2 (6)	1.625; 2 (6)	1.625; 2 (6)
SP plane	(100)	(100)	(100)
SP struct.	$AooBaCooAa$; $BooCaAooBa$; $CooAaBooCa$;	$AooB + CooA +$; $BooC + AooB +$; $CooA + BooC +$;	$AooBaCooAa$; $BooCaAooBa$; $CooAaBooCa$;
SPI =	65	59	64
Cleavage	pf{110}	pf{110}	pf{110}
Parting	(100), (001)		(100), (001)
Twinning	(100)	(100)	(100)
RI	$n_\alpha = 1.65$ $n_\beta = 1.66$ $n_\gamma = 1.67$	$n_\alpha = 1.66$ $n_\beta = 1.67$ $n_\gamma = 1.65$	$n_\alpha = 1.68$ $n_\beta = 1.71$ $n_\gamma = 1.73$
$\delta =$	0.02	0.01–0.02	0.02–0.08
$2V =$	50–80°	0–50°	68–82°
Sign	(−)	(−)	(−)
Transp.	Translucent	Transp. to transl.	Translucent
Pleochr.	X = light brown Y = light brown Z = dark brown	X = blue green Y = yellow green Z = brown green	X = yellow Y = red brown Z = dark brown
H =	5–6	5–6	6
D =	3.0–3.5	3.1–3.2	3.2–3.3
Color	Green, black	Gray, blue	Brown, black
Streak	White	White	White
Luster	Vitreous	Vitreous	Vitreous
Fracture	Uneven	Uneven	Uneven
Habit	Prismatic	Prismatic, fibrous	Prismatic
Remarks	High Fe^{3+}/Fe^{2+} ratio: basaltic hornblende	$Fe \rightleftharpoons (Mg, Al)$: riebeckite	No Ti: barkevikite

Members of the *amphibole* group

Structure. The amphiboles differ structurally from pyroxenes in the width of their tetrahedral and octahedral chains. The tetrahedral chain is doubled (Figure II.40a) in the amphiboles, and the octahedral chains are alternately three and four octahedra wide (Figure II.40b). The corresponding I-beam representation (Figure II.40c) is therefore wider, and the structure has four crystallographically distinct octahedral sites instead of two as in the pyroxenes. The octahedral sites in the amphiboles are labeled *M1*, *M2*, *M3*, and *M4* (Figure II.40d). The *M4* site is most similar to the *M2* site in pyroxenes and accommodates Ca^{2+}, as does the *M2* site in the pyroxenes. The amphiboles have two each of the *M1*, *M2*, and *M4* sites, but only one *M3* site to give a total of seven octahedral cations in a standard formula unit. When the *M4* site is fully occupied by Ca, the formula $Ca_2(Mg, Fe)_5(OH)_2Si_8O_{22}$ results. This formula is analogous to the calcic pyroxene formula $CaMgSi_2O_6$ that results when the *M2* site is fully occupied by Ca^{2+}.

Unlike the pyroxenes, the amphiboles have octahedral corners (marked with dots in Figure II.40d) that are not shared with tetrahedra. Instead, the anion at these corners is usually $(OH)^{1-}$, but may also be F^{1-} or Cl^{1-} depending on the chemical environment in which the amphibole forms. Because of the extra width of the double tetrahedral chain, a row of additional octahedral sites, usually referred to as "*A*" sites (and so labeled in Figure II.40c), appears between the regular octahedral chains. The "*A*" sites accommodate Na^{1+} and K^{1+} to the extent necessary to provide overall charge balance, and therefore are usually not fully occupied, as is the case in hornblende (Figure II.41).

In other structural respects, the amphiboles are extremely similar to the pyroxenes. The effects of tetrahedral rotation and octahedral distortion are similar, and the enumeration of stacking sequences and possible polytypes follows identical reasoning.

A consequence of the broader tetrahedral and octahedral chain widths in amphiboles is a smaller intersection of {110} cleavage. In pyroxenes, the narrower I-beams account for prismatic cleavage angles near 90 degrees. Cleavage between the broader I-beams in amphiboles is about 56 degrees and 124 degrees (Figure II.37).

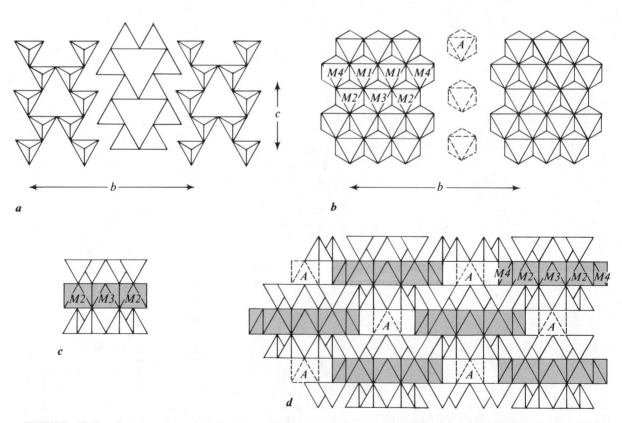

FIGURE II.40   Basic components of the amphibole structure. (a) Tetrahedral double chains, (b) three and four octahedra-wide chains with the conventional *M1*, *M2*, *M3*, *M4* labeling and the octahedra of the *A* site (dotted), (c) a single *TMT* I-beam, and (d) view of an amphibole structure along the *c* axis.

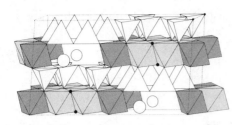

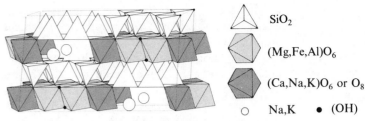

FIGURE II.41   Crystal structure of hornblende, *a* axis vertical.

Hornblende
$(Ca,Na,K)_{2-3}(Mg,Fe,Al)_5(OH)_2Si_8O_{22}$

Chemistry. The amphiboles differ chemically from the pyroxenes in two major respects. As *hydrous* silicates, the amphiboles are stable only in hydrous environments where water is available to be incorporated in the structure as $(OH)^{1-}$. In rather special compositional circumstances, the hydroxyl may be replaced by Cl^{1-} or F^{1-} or both. The second major compositional difference originates in the "*A*" site of amphiboles where K^{1+} usually resides. The pyroxenes do not have an equivalent site that will accommodate potassium.

The ionic substitutions that control the compositional variations in pyroxenes are the same in amphiboles. An amphibole composition space, consisting of ideal end-members that have compositions determined by the physical limits of cation substitution, is represented in Figure II.42. In parentheses are the equivalent end-members for the pyroxenes, if a pyroxene end-member is physically possible. Several of the end-members have exact counterparts in nature, but others such

as the ideal tschermakite end-member are only approached by real compositions. Note that not all of the cation substitutions in Figure II.42 are independent. Some can be expressed as linear combinations of others.

Because of the large number of dimensions in the amphibole composition space, amphibole analyses span a remarkable range of chemistries. Recalculation of amphibole analyses usually requires several end-member components, and only occasionally is the three-component treatment represented in Figure II.38 considered complete.

Occurrence and Associations. Because of the wide range of structurally permissible cation substitutions, amphiboles can crystallize in both igneous and metamorphic rocks that have bulk chemistries spanning a broad range. In igneous rocks, the common amphibole is calcic and is generally called hornblende. Most hornblendes consist of major amounts of the tremolite, tschermakite, edenite, and pargasite end-members. Only rarely does the noncalcic amphibole cummingtonite oc-

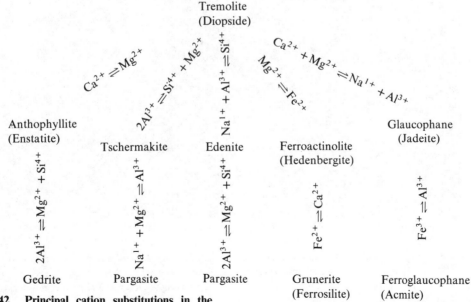

FIGURE II.42   Principal cation substitutions in the amphibole composition space. Pyroxene equivalents in parentheses.

cur in igneous rocks, and then it occurs only in dacites and rhyolites. Orthorhombic amphiboles are unknown from igneous rocks.

In metamorphic rocks, many different amphiboles are stable, and they frequently are found in association with each other. Figure 9.15 illustrates some of the common metamorphic associations as a function of bulk chemistry. Tremolite, anthophyllite, and hornblende are thus stable together only in rocks relatively enriched in MgO, and gedrite, cummingtonite, and garnet are found only in rocks relatively enriched in Al and Fe. All of these amphiboles are commonly associated with plagioclase feldspar and quartz, as well as biotite, chlorite, and opaque oxide minerals. Gedrite, the sodic and aluminous orthoamphibole, is commonly associated with garnet, cordierite, and occasionally with staurolite.

Another common metamorphic occurrence of amphiboles is in skarns and calc-silicate rocks. Tremolite and solid solutions toward ferroactinolite are frequently associated with other calcic minerals such as wollastonite and grossular.

Although the common crystallization habit of amphiboles is prismatic or acicular, most of the amphiboles are also known to crystallize in the asbestiform habit. Next to chrysotile, amphiboles constitute the most important commercial source of asbestos. The fiber structure of chrysotile is conspicuously different (Figure II.29) from the structure of normal serpentine crystals, but the asbestiform and acicular amphibole crystals do not have a comparable difference in structure. This could account for the extreme strength and flexibility of their fibers. High-resolution electron microscopy reveals that asbestiform crystals contain more structural defects, such as the intermittent triple chains illustrated in Figure II.43, than common acicular crystals. These defects are probably due to the rapid growth of the fibers, but they do not explain the fibers' unusual mechanical properties. Recent research indicates that the surface structure of the amphibole asbestos fibers is stronger than the normal structure of amphibole crystals and that the surface structure may be responsible for the high strength and flexibility of the fibers. The high strength of synthetic fibers, known as "whiskers," is similarly attributed to the extreme strength of the material's unique and essentially defect-free surface structures.

The asbestiform variety of riebeckite is known as *crocidolite* or blue asbestos. *Amosite* is a trade name of asbestiform varieties of primarily cummingtonite-grunerite, and occasionally of actinolite-tremolite or anthophyllite. The finely fibrous and massive variety of actinolite-tremolite is known as *nephrite*.

Distinguishing Features. The prismatic form, vitreous luster, and cleavage intersections of 56 and 124 degrees are usually adequate to identify amphiboles in hand specimens. Common hornblende is nearly always black. The colorless amphiboles

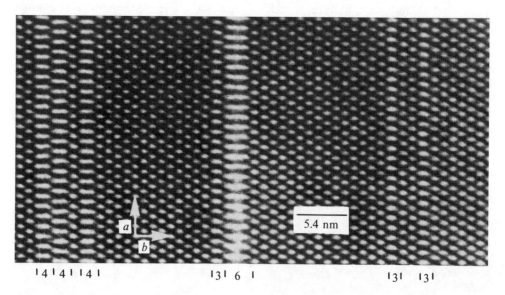

FIGURE II.43 **High resolution transmission electron microscope (HRTEM) images of a disordered sequence of double ("2") and triple ("3") chain sequences. (Courtesy of D. R. Veblen, Johns Hopkins University. From Veblen, D. R., Buseck, P. R., and Burnham, C. W.,** 1977. Asbestiform chain silicates: new minerals and structural groups. *Science* 198:4315, pp. 359–365, Figure 7. Copyright October 1977 by the American Association for the Advancement of Science.)

such as cummingtonite and tremolite can easily be misidentified or even overlooked if fine-grained. The orthorhombic amphiboles are usually a clove brown color that distinguishes them from other amphiboles.

In thin section, the monoclinic amphiboles are distinguished from the orthorhombic amphiboles by their inclined extinction. Within the monoclinic amphiboles, hornblende is distinguished from other amphiboles by its color and pleochroism.

The Pyroxenoids

The important minerals within the pyroxenoid group are wollastonite, rhodonite, and pectolite. Like the pyroxenes, the pyroxenoid frame is a single chain, but the individual tetrahedra are twisted relative to those in the pyroxene chain. Figure 5.6 compares the different single chain configurations and illustrates why the repeat distance in the chains differs. In wollastonite, the repeat unit has three tetrahedra with a repeat distance of about 0.73 nm, compared to pyroxenes with repeat units of two tetrahedra and a repeat distance of 0.52 nm (Figure II.44). The other pyroxenoids have similar structures but with larger repeat units and larger dimensions. Rhodonite has five tetrahedra in the repeat unit; pyroxmangite has seven tetrahedra in the repeat unit.

Structure. All of the minerals within the pyroxenoid group have triclinic symmetry (Figure II.45). The general relationships between symme-

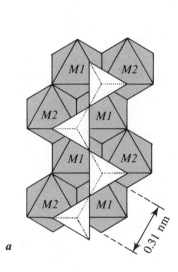

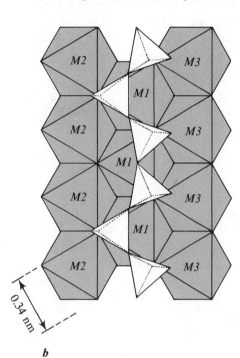

FIGURE II.44 Comparison of octahedral dimensions of (a) pyroxene and (b) wollastonite idealized structures. Differences are due to geometry of the tetrahedral chain.

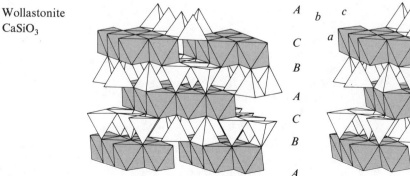

Wollastonite
CaSiO$_3$

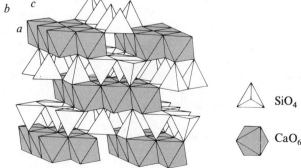

FIGURE II.45 Crystal structure of wollastonite, ($\bar{1}$01) horizontal.

THE PYROXENOIDS

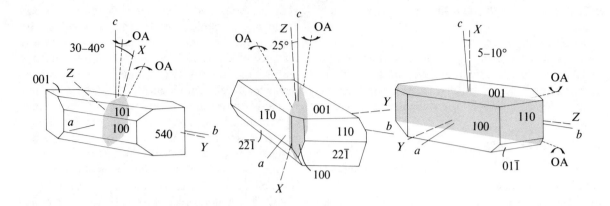

	WOLLASTONITE	Rhodonite	Pectolite
Formula	CaSiO$_3$	(Mn, Ca)SiO$_3$	Ca$_2$NaSi$_3$O$_8$(OH)
Coord.	Ca(6), Si(4)	Mn(6), Ca(6), Si(4)	Ca(6), Na(6), Si(4)
System	Triclinic	Triclinic	Triclinic
$a =$	0.794	0.768	0.799
$b =$	0.732	1.182	0.704
$c =$	0.707	0.671	0.702
$\alpha =$	90.03°	92.35°	90.05°
$\beta =$	95.37°	93.95°	95.28°
$\gamma =$	103.43°	105.67°	102.47°
$Z =$	4	2	2
Sp. group	P$\bar{1}$	P$\bar{1}$	P$\bar{1}$
Pt. group	$\bar{1}$	$\bar{1}$	$\bar{1}$
S; RU	1.50; 3	1.50; 3	1.50; 5
SPI =	65	63	63
Cleavage	pf(100), pf(001)	pf(110), pf(110)	pf(100), pf(001)
Twinning	[010] comp. plane (100)	(010) rare	[010] comp. plane (100)
RI	$n_\alpha = 1.620$	$n_\alpha = 1.717$	$n_\alpha = 1.59$
	$n_\beta = 1.632$	$n_\beta = 1.720$	$n_\beta = 1.61$
	$n_\gamma = 1.634$	$n_\gamma = 1.730$	$n_\gamma = 1.63$
$\delta =$	0.014	0.013	0.04
$2V =$	39°	63–76°	35–63°
Sign	(−)	(+)	(+)
Transp.	Transp. to transl.	Transp. to transl.	Translucent
Pleochr.		X = pale orange	
		Y = pink	
		Z = yellow orange	
H =	4½–5	5½–6	4½–5
D =	3.1	3.5–3.7	2.9
Color	White	Pink, red	White
Streak	White	White	White
Luster	Silky	Vitreous	Silky
Fracture	Uneven	Conchoidal	Uneven
Habit	Prismatic, needlelike	Tabular, massive	Radiating,
Remarks	May be fibrous	(Ca, Mn): bustamite	Fibrous
			Fe, Mg increases RI

try and chemistry within the single chain silicates are summarized in Figure 8.6. Recall that in the calcic clinopyroxenes, the *M2* octahedral site is filled with Ca^{2+}, and the smaller *M1* site is occupied by smaller Fe^{2+} and Mg^{2+} cations. In wollastonite, all of the octahedral sites are occupied by Ca^{2+}, which requires that the normal pyroxene chain be modified. Figure II.44a illustrates the normal pyroxene in which the maximum length of the octahedral edge is 0.31 nm. The dimension of this octahedron, *M1*, is constrained by the configuration of the single chain, because both the octahedra and tetrahedra share the same oxygens. The *M1* octahedron can only accommodate cations with radii that do not exceed 0.09 nm, and the radius of Ca^{2+} is 0.11 nm. The only way the larger cation can be accommodated is if the octahedral edge becomes longer, and that can happen only if the tetrahedral chain configuration changes (Figure II.44b).

Chemistry. There is less extensive solid solubility among the pyroxenoid minerals than within the pyroxenes. Wollastonite is nearly pure $CaSiO_3$, showing only small amounts of MgO and FeO in typical analyses. Significant amounts of Fe^{2+} and Mg^{2+} in the larger octahedra of the pyroxenoid structure will cause sufficient distortion to stabilize a pyroxene such as diopside or hedenbergite. This is the basic reason for the gap in miscibility between the pyroxenes and pyroxenoids shown in Figure 8.8.

Wollastonite does exhibit solid solubility toward rhodonite because of a $Ca^{2+} \rightleftharpoons Mn^{2+}$ substitution. Rhodonite can have up to about 20% $CaSiO_3$ in its structure, as well as smaller amounts of Fe^{2+} and other divalent cations. The mineral bustamite, another pyroxenoid with the general formula $(Mn, Ca, Fe)SiO_3$, may exhibit complete solid solubility with wollastonite but not with rhodonite. Pyroxmangite $((Mn, Fe)SiO_3)$ has only small amounts of CaO and MgO.

Occurrence and Associations. Wollastonite is one of the characteristic minerals of metamorphosed limestones and dolomites, forming from calcite plus quartz at relatively high temperature. Its prismatic habit, white color, and common association with diopside and garnet are distinctive. Wollastonite is occasionally found in certain Si-undersaturated igneous rocks.

Rhodonite and bustamite are typically found in manganese ore bodies, usually as a product of metamorphism. Pyroxmangite has a similar occurrence, and like rhodonite and bustamite is associated with other Mn-rich minerals such as spessartine, tephroite, and rhodochrosite.

Pectolite resembles the zeolites, and like the zeolites is typically found as cavity fillings and as linings along joint planes in mafic igneous rocks. Pectolite usually occurs as radiating, fibrous aggregates of acicular crystals.

Distinguishing Features. The pyroxenoids in general have a well-developed prismatic cleavage and tabular crystal form. Wollastonite is easily mistaken for tremolite but can be distinguished by its two perfect cleavages at an angle of about 84 degrees compared with about 56 degrees for tremolite. Rhodonite is typically pink in color. Pyroxmangite is also pink but is frequently coated with black or brown Mn-oxides.

Biopyriboles

The similarities between the crystal structures of the major layer silicates (clays and micas) and the chain silicates (pyroxenes and amphiboles) have long been recognized. In 1970, James B. Thompson formalized the structural relationship by identifying two structural units or "modules" that could be used to construct all such silicates. According to this view, pyroxene structures are composed of I-beam modules (Figure II.33), and mica structures are composed of layer modules. The two modules are similar in that both contain a band of octahedra sandwiched between two chains of tetrahedra, and when the modules are combined in certain ways, all other minerals in the layer silicate and chain silicate groups can be constructed. The amphibole structure, for example, is composed of a pyroxene, P, and a mica, M, module. Thompson revived Johannsen's 1911 expression *biopyriboles* as the collective name for these minerals (BIOtite, PYRoxenes, and amphIBOLES), and he predicted possible combinations of these modules other than those currently known. When the first triple-chain silicate was discovered, it was named jimthompsonite $((Mg, Fe)_{10}(OH)_4Si_{12}AlO_{32})$.

We can recognize four types of layer modules, the dioctahedral and trioctahedral modules of talcs and micas, symbolized as Di, Tri, Di*C*, and Tri*C*, which are illustrated in Figure II.46a. The compositions of the asymmetric units of these modules are:

$$\text{Di module} = M_2A_2Si_4O_{10}$$
$$\text{Tri module} = M_3A_2Si_4O_{10}$$
$$\text{Di}C \text{ module} = CM_2A_2Si_4O_{10}$$
$$\text{Tri}C \text{ module} = CM_3A_2Si_4O_{10}$$

where *M* is the octahedral cation, *C* is the *A*-site cation, and *A* is the anion not bonded to Si within the module. The *A* anion is oxygen when bonded to Si and is (OH) when bonded to more than one *M* cation. One half of each *A* anion is an H_2O

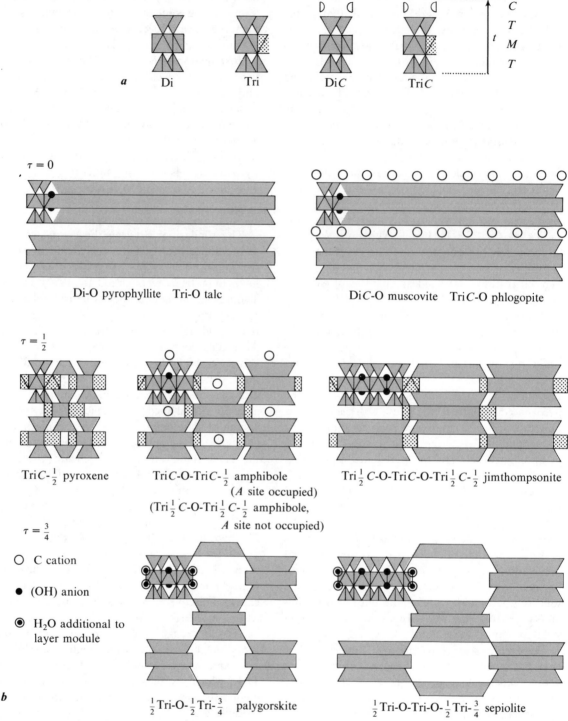

FIGURE II.46 (a) The four types of layer modules and (b) their combinations forming the important biopyriboles. (After Zoltai, Tibor, 1981. Mineralogical Society of America, *Reviews in Mineralogy,* vol. 9a, pp. 237–278.)

molecule when the anion is bonded to only one *M* cation.

These modules can be linked to form a variety of complete chain and layer silicate structures with or without a vertical displacement between the modular units. The major types of biopyribole minerals are illustrated in Figure II.46b.

The pyroxenes are single Tri*C* modules displaced by *t*/2. One half of the cations in the *M*2 octahedra are provided by the Tri module itself, and the other half are provided by the C cations. All of the A anions are bonded to Si tetrahedra and therefore are oxygens according to the relationship

$$\text{Tri}\,C = CM_3A_2\text{Si}_4O_{10} = \text{Mg}_4\text{Si}_4O_{12} = \text{Mg}_2\text{Si}_2O_6$$

The structure of amphiboles is composed of either (1) a pair of $\text{Tri}\,\tfrac{1}{2}C$ modules, where "$\tfrac{1}{2}$" indicates that only half of the C sites are used (those on the outside of the double module) or (2) two complete $\text{Tri}\,C$ modules. Both types of double module are displaced by $t/2$. The former module type produces the amphibole structure without the A sites between the I-beams, according to the relationship

$$\text{Tri}\tfrac{1}{2}C\,\text{Tri}\tfrac{1}{2}C = 2(\tfrac{1}{2}CM_3A_2\text{Si}_4O_{10})$$
$$= CM_6A_4\text{Si}_4\text{Si}_8O_{20}$$

The latter module type produces A sites between the I-beams according to the relationship

$$\text{Tri}\,C\,\text{Tri}\,C = 2(CM_3A_2\text{Si}_4O_{10}) = C_2M_6A_4\text{Si}_8O_{20}$$

All of the C cations in (1) and one of the two C cations in (2) provide the other half of the cations for the $M4$ octahedra. The extra C cations in (2) constitute the A sites. One half of the A anions (bonded to Si) are oxygen, and the other half, (bonded only to M cations) are (OH). The chemical formulas of the two amphiboles are therefore (1) $M_7(\text{OH})_2\text{Si}_8O_{22}$ and (2) $CM_7(\text{OH})_2\text{Si}_8O_{22}$.

The palygorskite structure is constructed with a $3t/4$ displacement between a pair of $\tfrac{1}{2}\text{Tri}$ modules, where "1/2" indicates that only half of the extra trioctahedral sites are occupied, that is, those sites between the two linked modules. One half of the A anions (bonded to three M cations) are (OH), and the remaining one half (bonded to only one M cation) are H_2O molecules as shown by the relationship

$$\tfrac{1}{2}\text{Tri}\tfrac{1}{2}\text{Tri} = M_5A_4\text{Si}_8O_{20}$$

$$= \text{Mg}_5(\text{OH})_2(\text{H}_2\text{O})_4\text{Si}_8O_{20}(\text{plus}\cdot4\text{H}_2\text{O})$$

Polytypism and Isotypism

The simple TM and TMT structures of the layer silicates are complicated by variations in the stacking of the layer units. In some instances, the layer units are randomly stacked, which is revealed in the x-ray powder pattern by the absence of the (hkl) reflections. In other cases, the stacking of the layer units is regular. Many regular stackings are possible, and many are known in the layer silicates. The various stacking patterns of the layer units are referred to as *polytypes*. In 1956, Smith and Yoder proposed a classification system in which the various polytypes are distinguished

by the different displacements between the adjacent layer units. The pattern of displacement can be represented by the projection of "stagger vectors" shown in Figure II.47. The polytypes are identified by a symbol consisting of (1) a number that refers to the number of layer units contained in a unit cell, (2) a capital letter that identifies the point group symmetry of the structure, and (3) a subscript that distinguishes between various polytypes having the same point group symmetry. Since 1956, many other polytypes have been described and observed.

In 1970, Papike and Ross outlined some of the many variations possible in the I-beam diagrams of amphibole structures and proposed a simple scheme for their description based on the earlier work of Thompson. In their scheme, the octahedral bands in the I-beam are labeled "$+$" if the top triangular face of the octahedra point in the direction of the c axis, or "$-$" if the faces point in the opposite direction. The orientation of the tetrahedral chains can be of two types. If the basal triangular faces of the tetrahedra point in the same direction as the adjacent octahedral faces, the chain is referred to as an "S" chain (Similar). If the two faces point in the opposite direction, the chain is referred to as an "O" chain (Opposite). Figure II.48 illustrates the I-beam diagram of the cummingtonite structure.

Both the major layer and chain silicate structures can also be described in terms of symmetrical packing, using the symbols described in Chapter 5. Some examples are:

Kaolinite	$A\left(-\tfrac{2}{3}-\right)B\left(\tfrac{2}{3}--\right)C\left(---\right),$
($1M$):	$C\left(\tfrac{2}{3}--\right)A\left(--\tfrac{2}{3}\right)B\left(---\right),$
	$B\left(--\tfrac{2}{3}\right)C\left(-\tfrac{2}{3}-\right)A\left(---\right);$
Muscovite	$A\left(-\tfrac{2}{3}-\right)B\left(\tfrac{2}{3}--\right)C\left(--\tfrac{2}{3}\right)$
($3T$):	$A\left(--\tfrac{1}{3}\right)B\left(--\tfrac{2}{3}\right)C\left(-\tfrac{2}{3}-\right)$
	$A\left(\tfrac{2}{3}--\right)B\left(\tfrac{1}{3}--\right)C\left(\tfrac{2}{3}--\right)$
	$A\left(--\tfrac{2}{3}\right)B\left(-\tfrac{2}{3}-\right)C\left(-\tfrac{1}{3}-\right);$
Palygorskite:	$A\left(-\tfrac{1}{3}-\right)B^*\left(\tfrac{5}{12}--\right)C^*\left(--\tfrac{1}{3}\right)$
	$A\left(-\tfrac{1}{3}-\right)B^*\left(\tfrac{5}{12}--\right)C^*\left(--\tfrac{1}{3}\right);$
	(* only 7/12th of anion sites filled)
Spodumene:	$A\left(--\tfrac{2}{3}\right)B\left(\tfrac{1}{3}\tfrac{1}{3}-\right)A\left(--\tfrac{2}{3}\right)B\left(\tfrac{1}{3}\tfrac{1}{3}-\right);$

Unfortunately, these symbols become lengthy for extensive polytypes and thus are too cumbersome for routine use. A simplified description of the biopyribole structure retains the stacking symbols of the oxygen sheets, but the fractional expressions of void occupancy are replaced by the "$+$" or "$-$" symbols for the octahedral chains

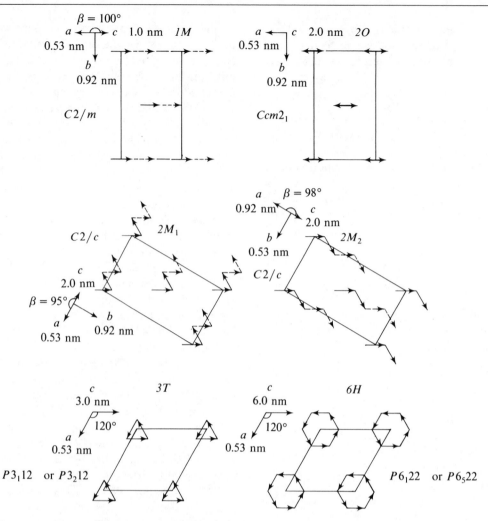

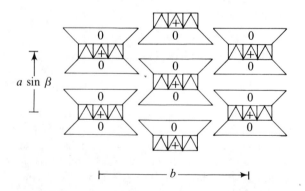

FIGURE II.47   The six important mica polytypes. (Reprinted courtesy of the Mineralogical Society from Smith, J. V., and Yoder, H. S. "Experimental and Theo- retical Studies of the Mica Polymorphs," *Mineralogical Magazine*, vol. 31 (1956):213.)

and the "*o*" and "*s*" symbols for the tetrahedral chains. These symbols are a direct consequence of the stacking order of the anion sheets and as such are redundant, but they do augment the descriptive value of the system. The symmetrical packing symbols entered in the columns of properties for the biopyriboles consist of the following parts:

1. The octahedral band or layer is marked as "+" or "−". If the two sheet symbols of the octahedral layer are in alphabetical order, the layer is "+". If they are in reverse alphabetical order, the layer is "−".

2. The tetrahedral layers are "*o*" if the stacking symbol of the basal tetrahedral sheet is different from the stacking symbol of the second anion sheet of the octahedral layer (the sheet that is not common to the tetrahedral and octahedral layers). If the two sheet symbols are identical, the tetrahedral layer is labeled "*s*".

FIGURE II.48   I-beam diagram of the cummingtonite structure. (After Thompson, J. B., 1970. Geometrical possibilities for amphibole structures: model biopyriboles. *American Mineralogist* 55:292–293, and Papike, J. J., and Ross, M., 1970. Gedrites: crystal structures and intracrystalline cation distributions. *American Mineralogist* 55:1945–1972.

3. Special features are marked with special symbols: vacant layers are identified by "=", layers containing the *C* cations in micas are labeled "*c*" without the "+" and "−" designation, and octahedral layers containing *A* sites in amphiboles are labeled "*a*".

4. The portion of the stacking symbol that corresponds to one structural unit cell is identified by ";".

The symbols of the layer and chain silicates can thus be rewritten as:

Kaolinite (*1M*):
$$AoB + C = ; CoA + B = ; BoC + A = ;$$
Muscovite (*3T*):
$$AoB + CoAcBoC + AoBcCoA + BoCc;$$
Palygorskite: $A*oBoC* + A*oBoC* +;$
Spodumene: $A + BssA + Bss;$

These symbols illustrate, for example, that a layer is vacant in kaolinite and that the layer of the extra *C* cation is filled in muscovite. The symbols also illustrate that palygorskite has a double tetrahedral layer, because the tetrahedra point up in one layer and down in the other connected tetrahedral layer. In addition, the symbols tell us that the tetrahedral chains in the pyroxenes alternately point up and down within a single layer. The two "*o*" or "*s*" symbols illustrate the relationship of the tetrahedral chains to the adjacent octahedral bands. The palygorskite symbol also expresses the mineral's structural similarity to both the layer and the chain silicates, a property that characterizes this mineral.

The uppercase letters in these symbols identify the stacking of the anion sheets. The permissible stacking sequences are numerous but not unlimited. Table 5.12 gives the unique and permissible stacking sequences of monatomic sheets. The same concept can be extended to derive the possible stacking orders of the biopyribole structures. In the application of Table 5.6, however, we must consider certain restrictions concerning the different biopyribole structures. For example, in the polytypes of kaolinite, two identical letter symbols can be adjacent across a vacant layer, a condition not permissible in the stacking of the sheets in monatomic and in most other symmetrically packed structures.

Although the suggested symbols for biopyriboles give an adequate picture of the structure, the system is still incomplete because the labels do not identify the relative positions of the vacant tetrahedral and octahedral sites. The system can be extended to include a distinction between the three possible locations of the vacant sites. One of the three possible locations is identified by no

superscript, and the other two locations are identified by superscripts (') and ("). With these additional symbols, we can derive and identify all of the possible polytypes. For example, the symbol of the kaolinite structure given above is one of the nine possible one-layer, monoclinic polytypes that use the same stacking sequence of the anions:

$$AoB + C = ; CoA + 'B = ; BoC + ''A = ;$$
$$AoB + C = ; CoA + 'B = ; Bo'C + A = ;$$
$$AoB + C = ; CoA + 'B = ; Bo''C + 'A = ;$$
$$AoB + C = ; Co'A + ''B = ; BoC + ''A = ;$$
$$AoB + C = ; Co'A + ''B = ; Bo'C + A = ;$$
$$AoB + C = ; Co'A + ''B = ; Bo''C + 'A = ;$$
$$AoB + C = ; Co''A + B = ; BoC + ''A = ;$$
$$AoB + C = ; Co''A + B = ; Bo'C + A = ;$$
$$AoB + C = ; Co''A + B = ; Bo''C + 'A = ;$$

As we pointed out in our discussion of the pyroxenes and amphiboles, the different "polytypes" of these structures are not valid polytypes. The various stackings of the anion sheets provide different environments for the *M2* (and *M4*) octahedra, allow different types of distortion, and may even change the coordination number of the cations in these sites. Because of these structural modifications and consequent compositional preferences, the stacking patterns of the different pyroxene (and amphibole) structures cannot be referred to as polytypes. The structures are more appropriately identified as *isotypes*, a term that expresses the structural relationship but includes minor structural and chemical variations. Isotypes thus have some of the structural characteristics of polytypes and some of the compositional characteristics of isomorphs.

Note that in some pyroxene and amphibole structures a stacking irregularity, known as *parity violation*, occurs. The oxygens in these sheets are in two different orientations. For example, in the structure of enstatite, the oxygens in the top sheet of the first tetrahedral layer are in *C* orientation for the down-pointing tetrahedra and in *A* orientation for the up-pointing tetrahedra. This modification of the stacking sequence is expressed in the symbol of enstatite by using both letters with a slash: C/A.

The Ring Silicates

Tourmaline is the only common mineral that is a true ring silicate as defined in this text. Its structure (Figure II.49) consists of isolated six-membered tetrahedral rings that are not connected to one another by other tetrahedra, as they are in the structures of beryl and cordierite discussed earlier. Dioptase ($Cu(OH)_2SiO_2$) and be-

**THE RING
SILICATES**

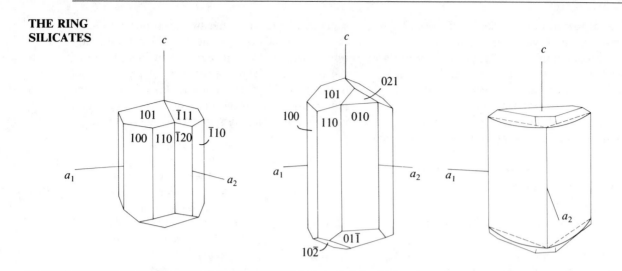

	DRAVITE	Schorl	Elbaite
Formula	$NaMg_3Al_6(OH)_4B_3O_9Si_6O_{18}$	$NaFe_3Al_6(OH)_4B_3O_9Si_6O_{18}$	$NaLi_3Al_6(OH)_4B_3O_9Si_6O_{18}$
Coord.	Na(6), Mg(6), Al(6), B(3), Si(4)	Na(6), Fe(6), Al(6), B(3), Si(4)	Na(6), Li(6), Al(6), B(3), Si(4)
System	Trigonal	Trigonal	Trigonal
$a =$	1.594	1.603	1.584
$c =$	0.722	0.715	0.710
$Z =$	3	3	3
Sp. group	$R3m$	$R3m$	$R3m$
Pt. group	$3m$	$3m$	$3m$
S; ring	1.50; 6	1.50; 6	1.50; 6
SPI =	56	57	56
Cleavage	pr{101}, pr{110}	pr{101}, pr{110}	pr{101}, pr{110}
Twinning	{101}	{101}	{101}
RI	$n_\omega = 1.650$	$n_\omega = 1.668$	$n_\omega = 1.646$
	$n_\epsilon = 1.628$	$n_\epsilon = 1.639$	$n_\epsilon = 1.625$
$\delta =$	0.022	0.029	0.021
Sign	(−)	(−)	(−)
Transp.	Translucent	Transp. to transl.	Translucent
Pleochr.	= light brown	= dark blue	
	= pale yellow	= green brown	
$H =$	$7–7\frac{1}{2}$	$7–7\frac{1}{2}$	$7–7\frac{1}{2}$
$D =$	3.02	3.27	2.9
Color	Brown, yellow	Black, blue	Blue, variable
Streak	White	White	White
Luster	Resinous	Vitreous	Resinous
Fracture	Subconchoidal	Subconchoidal	Subconchoidal
Habit	Prismatic	Prismatic	Prismatic
Remarks	Ditrigonal cross section, striated prismatic faces		
	Exceptionally good polarizers and are pyroelectric		
	Members of the *tourmaline* group		

nitoite ($BaTiSi_3O_9$) also have true ring structures but are relatively rare minerals.

Structure. In the crystal structure of tourmaline, each silica tetrahedron shares two of its apical oxygens with neighboring tetrahedra to form a six-membered ring having the chemistry $(Si_6O_{18})^{12-}$. The rings are connected to borate ($BO_3)^{1-}$ groups, and to both oxygen and hydroxyl octahedra of Al^{3+}, Mg^{2+}, and other cations. The borate groups are distributed about a threefold axis that extends through the six-membered rings parallel to the c axis. Cations such as Ca^{2+}, Na^{1+}, and K^{1+} coordinate to the borate groups and occupy special positions on the threefold axis.

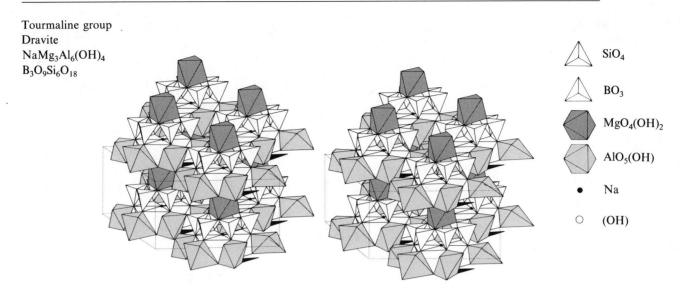

Tourmaline group
Dravite
$NaMg_3Al_6(OH)_4$
$B_3O_9Si_6O_{18}$

SiO_4

BO_3

$MgO_4(OH)_2$

$AlO_5(OH)$

• Na

○ (OH)

FIGURE II.49 **Crystal structure of a tourmaline,** *c* **axis vertical.**

Chemistry. The compositional variations in tourmaline are extensive, and account for the many different colored varieties and the diversity of names. Varieties with mostly Fe^{2+} in octahedral sites are called *schorl* and are usually black. Varieties with Mg^{2+} in the octahedral site are called *dravite*. A complete solid solution seems to exist between the Fe^{2+} and the Mg^{2+} end-members. If the octahedral site is occupied mainly by Al^{3+} and Li^{1+}, the name *elbaite* is used. Mn^{2+}, Cr^{3+}, Fe^{3+}, and other transition metals can also occupy this site, and they impart a wide range of color depending on the specific substitution. Compositional zoning is common in tourmaline, resulting in spectacular color variation within single crystals.

Occurrence and Associations. Tourmaline is a widespread accessory mineral in many granites and can be abundant in pegmatites. As one of the few minerals in which boron may reside in appreciable amounts, tourmaline nearly always crystallizes if boron is present. Common associations are quartz, muscovite, and alkali feldspars. The Li-bearing tourmalines are typical of many pegmatites and are associated with beryl, Li-mica, fluorite, and apatite. Tourmaline is also found in metamorphic rocks and commonly as detrital grains in sediments.

Distinguishing Features. Tourmaline has a distinctive habit, forming prominent trigonal prisms that are vertically striated. The luster is vitreous, and the fracture is conchoidal, much like quartz. Basal sections typically have the form of rounded triangles. In thin section, strong dichroism, high relief, and uniaxial character are diagnostic. Tourmaline is known as an excellent polarizer.

Isolated Tetrahedral Silicates

The last category of silicates is a large group of important minerals that have primary frames consisting of either isolated tetrahedra or small groups of isolated tetrahedra. In this sense, the frames have a zero-dimensional extent. Sillimanite is one major exception, but we include it with the isolated tetrahedral silicates because of its association and polymorphism with andalusite and kyanite. The latter are both isolated tetrahedral silicates and have sharing coefficients of 1.00. Sillimanite, however, is structurally a chain silicate, because alternating Al and Si tetrahedra form infinite double chains with a sharing coefficient of 1.75.

The groups of minerals included in the isolated tetrahedral silicates and discussed in this section are the silicates with (1) single tetrahedral structures, (2) double tetrahedral structures, and (3) double and single tetrahedral structures.

The Garnet Group

Garnets are found most commonly in pelitic schists and gneisses from regionally metamorphosed terrain, and to a lesser extent in igneous rocks ranging in composition from granite to peridotite. They also occur as detrital grains in sedimentary rocks.

THE GARNET GROUP

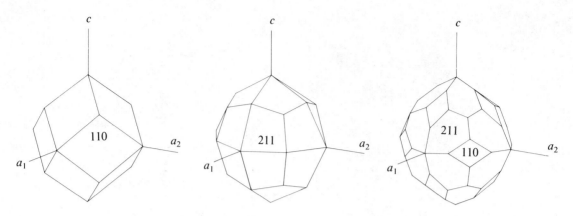

	Pyrope	Almandine	Spessartine
Formula	$Mg_3Al_2Si_3O_{12}$	$Fe_3Al_2Si_3O_{12}$	$Mn_3Al_2Si_3O_{12}$
Coord.	Mg(8), Al(6), Si(4)	Fe(8), Al(6), Si(4)	Mn(8), Al(6), Si(4)
System	Isometric	Isometric	Isometric
$a =$	1.146	1.153	1.162
$Z =$	8	8	8
Sp. group	$I4_1/a\bar{3}2/d$	$I4_1/a\bar{3}2/d$	$I4_1/a\bar{3}2/d$
Pt. group	$4/m\bar{3}2/m$	$4/m\bar{3}2/m$	$4/m\bar{3}2/m$
S	1.00	1.00	1.00
SPI =	63	62	61
Parting	{110}	{110}	{110}
RI	$n = 1.71$	$n = 1.83$	$n = 1.80$
Transp.	Transp. to transl.	Transp. to transl.	Transparent
$H =$	$6\frac{1}{2}-7\frac{1}{2}$	$6\frac{1}{2}-7\frac{1}{2}$	$6\frac{1}{2}-7\frac{1}{2}$
$D =$	3.54	4.33	4.19
Color	Red, black	Red, brown, black	Red, violet
Streak	White	White	White
Luster	Resinous	Resinous	Resinous
Fracture	Subconchoidal	Subconchoidal	Subconchoidal
Habit	Dodecahedral	Dodecahedral	Dodecahedral
Remarks		Members of the *garnet* group	

Structure. The garnet structure (Figure II.50) consists of isolated silica tetrahedra that share apical oxygens with two types of coordination polyhedra. One is a slightly distorted octahedron that accommodates trivalent cations, mainly Al^{3+} and Fe^{3+}, and the other is a distorted dodecahedron that houses divalent cations in eightfold coordination. Relative to other silicates, garnets have a large symmetrical packing index, because one half of the octahedral edges are shared with adjacent dodecahedra, and there is no tetrahedral polymerization.

An unusual but important garnet for mantle mineralogy is majorite. One fourth of the Si^{4+} is in octahedral coordination, and like stishovite, majorite is stable only in high-pressure environments.

Chemistry. Subdivision of the garnet group is based on ideal end-member compositions determined by the physical limits of cation substitution in the octahedral and dodecahedral sites. The more common end-members are listed beneath the physical properties table for garnets. Complete solid solution exists between pyrope, almandine, and spessartine because of the mutual substitution of Mg^{2+}, Fe^{2+}, and Mn^{2+} in the dodecahedral site. Complete solid solution appears to exist between grossular, andradite, and uvarovite because of the unlimited substitution of Al^{3+}, Fe^{3+}, and Cr^{3+} within the octahedral site. No such solubility

THE GARNET GROUP

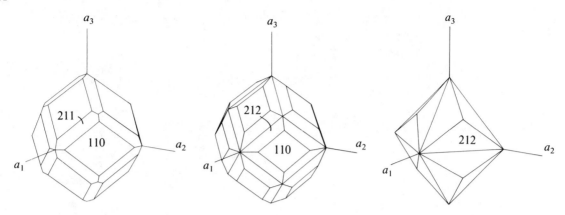

	GROSSULAR	Andradite	Uvarovite
Formula	$Ca_3Al_2Si_3O_{12}$	$Ca_3Fe_2Si_3O_{12}$	$Ca_3Cr_2Si_3O_{12}$
Coord.	Ca(8), Al(6), Si(4)	Ca(8), Fe(6), Si(4)	Ca(8), Cr(6), Si(4)
System	Isometric	Isometric	Isometric
$a =$	1.185	1.205	1.200
$Z =$	8	8	8
Sp. group	$I4_1/a\bar{3}2/d$	$I4_1/a\bar{3}2/d$	$I4_1/a\bar{3}2/d$
Pt. group	$4/m\bar{3}2/m$	$4/m\bar{3}2/m$	$4/m\bar{3}2/m$
$S =$	1.00	1.00	1.00
SPI =	59	63	62
Parting	{110}	{110}	{110}
RI	$n = 1.75$	$n = 1.87$	$n = 1.85$
Transp.	Transparent	Transp. to transl.	Transp. to transl.
$H =$	$6\frac{1}{2}-7\frac{1}{2}$	$6\frac{1}{2}-7\frac{1}{2}$	$7\frac{1}{2}$
$D =$	3.56	3.86	3.80
Color	Colorless, red, yellow	Yellow, brown, green	Emerald green
Streak	White	White	White
Luster	Resinous	Resinous	Resinous
Fracture	Subconchoidal	Subconchoidal	Subconchoidal
Habit	Dodecahedral	Dodecahedral	Dodecahedral
Remarks		Frequently show uniaxial or biaxial optics	
		Members of the *garnet* group	

appears to exist between these two groups, however, suggesting a gap in miscibility. Exsolution has been theoretically predicted, and exsolution in garnets of intermediate composition has been observed.

A group of silica-deficient garnets known as the hydrogarnets is becoming an increasingly important topic of study as more occurrences are found. Despite the name, little or no molecular water is in the structure. Instead, hydrogen enters tetrahedral sites by the substitution $Si^{4+} \rightleftharpoons 4H^+$. The limits of this substitution are not as yet known.

Occurrence and Associations. Garnets are widespread in a variety of pelitic schists and gneisses where they are typically associated with many other aluminous minerals such as the micas, kyanite, sillimanite, and staurolite (Figure 9.13). In more mafic rocks such as amphibolites and peridotites, the garnets are more pyrope-rich and are associated with hornblende and pyroxenes (Figure 9.15b). Garnets from metamorphosed dolomites and skarn deposits are generally enriched in grossular and andradite because of the Ca^{2+} and Fe^{3+} in the rock chemistry. Common associations with grossular include diopside and wollastonite.

Garnet group
Grossular
Ca₃Al₂Si₃O₁₂

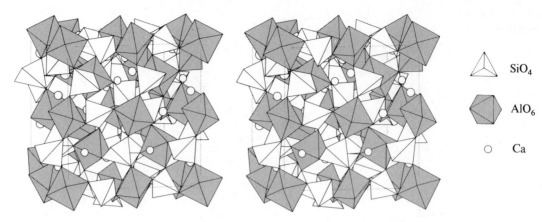

SiO₄

AlO₆

○ Ca

FIGURE II.50 **Crystal structure of grossular, a garnet.**

For andradite, the association with hedenbergite and magnetite is well known.

Hydrogarnets are known principally from metamorphosed marls and limestones and from hydrothermally altered igneous rocks.

Distinguishing Features. In hand specimen, garnet typically displays an equant, dodecahedral form and is commonly deep red to brownish black in color. In thin section, its high relief, lack of color, and isotropic character are diagnostic. Birefringent garnets, possibly having lower than cubic symmetry, are known from certain skarn deposits.

The Olivine Group

Olivine is an abundant mineral in most mafic igneous rocks and is the most abundant mineral in the earth's upper mantle. The general structure (Figure II.51) consists of isolated Si tetrahedra linked by divalent cations in octahedral coordination. Forsterite (Mg_2SiO_4) and fayalite (Fe_2SiO_4) are the principal end-member compositions. All end-members possess orthorhombic symmetry.

Structure. Olivine may be viewed as having either a polyhedral frame structure or a symmetrically packed structure (Chapter 5, p. 109). Its secondary frame (Figure II.51) shows zigzag, double chains of edge-sharing octahedra that run

parallel to the *c* axis and are connected by corner sharing in an octahedral network. These subchains control the (100) and (010) cleavages and the dominance of (*hk*0) crystal faces.

As a symmetrically packed structure, the oxygens in olivine are in hexagonal sheets in the (100) plane. The packing symbol indicates an *AB* stacking sequence and that both the tetrahedral and octahedral voids are only partially occupied. The pattern of partial void occupancy produces the orthorhombic symmetry of olivines in spite of the hexagonal symmetry of the oxygen stacking. Two symmetrically nonequivalent octahedra result, one referred to as *M1*, which shares 6 of its 12 edges with four octahedra and two tetrahedra, and the other referred to as *M2*, which shares only 3 edges with two octahedra and a tetrahedron.

Chemistry. Several chemical end-members within the olivines express the physical limits of cation substitution. The most important of these are forsterite (Mg_2SiO_4) and fayalite (Fe_2SiO_4) between which complete solid solution exists. Continuous solid solution also exists between fayalite and tephroite (Mn_2SiO_4), and between forsterite and tephroite. Solid solubility of Mg-Fe-Mn compositions toward the Ca_2SiO_4 end-member is less extensive, and one or more gaps in

Olivine group
Forsterite
Mg₂SiO₄

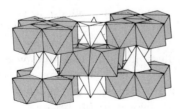

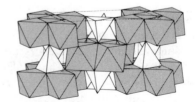

SiO₄

MgO₆

FIGURE II.51 **Crystal structure of forsterite, *a* axis vertical.**

THE OLIVINE GROUP

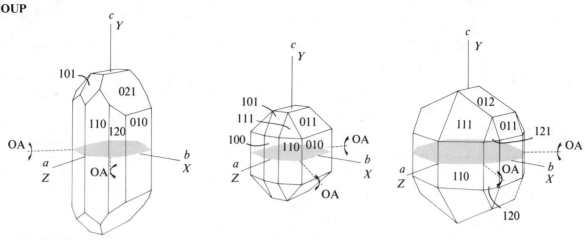

	FORSTERITE	Fayalite	Monticellite
Formula	Mg_2SiO_4	Fe_2SiO_4	$CaMgSiO_4$
Coord.	Mg(6), Si(4)	Fe(6), Si(4)	Ca(6), Mg(6), Si(4)
System	Orthorhombic	Orthorhombic	Orthorhombic
$a =$	0.478	0.481	0.482
$b =$	1.028	1.061	1.108
$c =$	0.600	0.611	0.638
$Z =$	4	4	4
Sp. group	$P2_1/b\,2/n\,2/m$	$P2_1/b\,2/n\,2/m$	$P2_1/b\,2/n\,2/m$
Pt. group	$2/m\,2/m\,2/m$	$2/m\,2/m\,2/m$	$2/m\,2/m\,2/m$
S	1.00	1.00	1.00
SP plane	(100)	(100)	(100)
SP struct.	$A(\frac{1}{8}\frac{1}{8}\frac{1}{2})B(\frac{1}{8}\frac{1}{8}\frac{1}{2})$;	$A(\frac{1}{8}\frac{1}{8}\frac{1}{2})B(\frac{1}{8}\frac{1}{8}\frac{1}{2})$;	$A(\frac{1}{8}\frac{1}{8}\frac{1}{2})B(\frac{1}{8}\frac{1}{8}\frac{1}{2})$;
SPI =	65	65	63
Cleavage	gd(010), pr(100)	gd(010), pr(100)	gd(010), pr(100)
Twinning	(100), {011}, {012}	(100), {011}, 012	{031}
RI	$n_\alpha = 1.635$	$n_\alpha = 1.827$	$n_\alpha = 1.645$
	$n_\beta = 1.651$	$n_\beta = 1.877$	$n_\beta = 1.655$
	$n_\gamma = 1.670$	$n_\gamma = 1.880$	$n_\gamma = 1.665$
$\delta =$	0.035	0.053	0.020
$2V =$	85–90°	47–54°	72–82°
Sign	(+)	(−)	(−)
Transp.	Transp. to transl.	Transp. to transl.	Transp. to transl.
$H =$	$6\frac{1}{2}$	7	$5\frac{1}{2}$
$D =$	3.2	3.4	3.15
Color	Colorless, green	Green, yellow	Colorless, green
Streak	White	White yellow	White
Luster	Vitreous	Vitreous	Vitreous
Fracture	Conchoidal	Conchoidal	Conchoidal
Habit	Tabular, prismatic	Tabular, prismatic	Tabular, prismatic
Remarks			Distorted olivine struct.

End-members of the *olivine* group

miscibility may exist. Complete solubility appears to exist between Fe_2SiO_4 and $CaFeSiO_4$, and between $CaFeSiO_4$ and $CaMgSiO_4$, but not between Mg_2SiO_4 and $CaMgSiO_4$. This behavior may be a result of the ordering of Ca in the *M2* site in a manner analogous to the pyroxenes discussed in Chapter 8. The most calcic olivine appears to be $Ca(Mg, Fe, Mn)SiO_4$, suggesting that the distortion caused by placing Ca^{2+} in the *M1* site is sufficient to destabilize the olivine structure. The mineral larnite (Ca_2SiO_4) does not have the olivine structure. In fact, the compositional gap between monticellite ($CaMgSiO_4$) and larnite may be structurally analogous to the compositional gap between diopside ($CaMgSi_2O_6$) and wollastonite ($Ca_2Si_2O_6$).

The current practice is to reference the compositions of olivines in terms of the four end-members forsterite (Fo), fayalite (Fa), tephroite (Te), and larnite (La). The first two are by far the most common. A typical forsteritic olivine from a gabbro would be symbolized as Fo_{85}, for example.

Occurrence and Associations. Olivines are found most commonly in mafic to ultramafic igneous rocks where they are associated with Mg-rich pyroxenes, calcic plagioclase, and Fe-Ti oxides. Olivines in these associations have compositions in the range Fo_{60}–Fo_{90}. They are never associated with quartz because of the reaction producing pyroxene (Chapter 9, p. 227). More Fe-rich olivines are found with quartz in certain granites and quartz syenites where they are associated with other Fe-rich silicates such as hedenbergite, aegerine, and arfvedsonite.

Metamorphic olivines are less common, but are important minerals in certain impure marbles and metamorphosed ultramafic rocks. Unlike most igneous occurrences, metamorphic olivine commonly has an elongate crystal habit.

Distinguishing Features. Olivines are distinguished in hand specimens by their generally equant form, green color, and association with pyroxenes and plagioclase in mafic igneous rocks. More Fe-rich compositions are light green to yellow in color. In thin section, the high relief, moderate birefringence, high optic angle, and lack of cleavage are fairly distinctive. Alteration to serpentine plus magnetite along olivine grain boundaries is common.

The Humite Group

The four important minerals within the humite group are norbergite, chondrodite, humite, and clinohumite. All are structurally similar to olivine in that they have the same *AB* stacking sequence of hexagonal oxygen sheets in which one half of the octahedral sites are occupied (Figure II.52). Unlike olivine, however, some of the oxygen is replaced by $(OH)^{1-}$ or F^{1-} or both with charge balance provided by divalent Mg^{2+} in the octahedral layer.

Structure. As in the olivine structure, the anions are in an *AB* stacking sequence and the structures are dominated by edge-sharing octahedral subchains enclosing single tetrahedra. The octahedral subchain direction is parallel to the *c* axis and appears to control the prismatic cleavage and crystal habit.

The humite minerals are distinguished structurally from each other by the variations in the pattern of the octahedral subchains and by the occupancy of tetrahedral sites. For comparison, one eighth of the tetrahedral sites are occupied in olivine, as indicated by its packing symbol. Only one twelfth of the tetrahedral sites are occupied in norbergite, one tenth in chondrodite, and one ninth in clinohumite.

Chemistry. The chemistry of the humite minerals is commonly represented by the formula $nMg_2SiO_4 \cdot Mg(OH, F)_2$ where $n = 1, 2, 3,$ and 4, respectively, for norbergite, chondrodite, humite, and clinohumite. Octahedral cation substitutions in the olivinelike part of the formula resemble the olivine series to a much lesser extent, except for Mn^{2+} substitution for which complete solid solution may exist. The octahedral site in the brucite-like part of the formula is smaller than other octahedra in the structure and may be occupied by Ti^{4+}, especially in chondrodite and clinohumite.

Occurrence and Associations. Humites are for the most part restricted in their occurrence to metamorphosed limestones and dolomites and to skarn deposits. Common associations are with dolomite, phlogopite, and Mg-spinel, and with forsterite or monticellite in skarns.

Norbergite
$Mg_3(OH)_2SiO_4$

 SiO_4

 MgO_6

FIGURE II.52 **Crystal structure of norbergite, *b* axis vertical.**

THE HUMITE GROUP

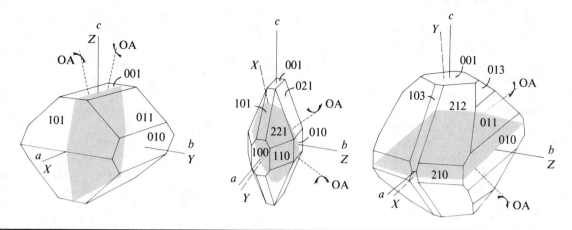

	NORBERGITE	Chondrodite	Clinohumite
Formula	$Mg_3(OH, F)_2SiO_4$	$Mg_5(OH, F)_2Si_2O_4$	$Mg_9(OH, F)_2Si_2O_8$
Coord.	Mg(6), Si(4)	Mg(6), Si(4)	Mg(6), Si(4)
System	Orthorhombic	Monoclinic	Monoclinic
$a =$	0.470	0.473	0.475
$b =$	1.022	1.027	1.027
$c =$	0.872	0.787	1.368
$\beta =$		109.1°	100.8°
$Z =$	4	2	2
Sp. group	$P2_1/b\,2_1/n\,2_1/m$	$P2_1/b$	$P2_1/b$
Pt. group	$2/m\,2/m\,2/m$	$2/m$	$2/m$
S	1.00	1.00	1.00
SP plane	(010)	(010)	(010)
SP struct.	$A(\frac{1}{12}\frac{1}{12}\frac{1}{2})B(\frac{1}{12}\frac{1}{12}\frac{1}{2})$	$A(\frac{1}{10}\frac{1}{10}\frac{1}{2})B(\frac{1}{10}\frac{1}{10}\frac{1}{2})$	$A(\frac{1}{9}\frac{1}{9}\frac{1}{2})B(\frac{1}{9}\frac{1}{9}\frac{1}{2})$
	$A(\frac{1}{12}\frac{1}{12}\frac{1}{2})B(\frac{1}{12}\frac{1}{12}\frac{1}{2})$;	$A(\frac{1}{10}\frac{1}{10}\frac{1}{2})B(\frac{1}{10}\frac{1}{10}\frac{1}{2})$;	$A(\frac{1}{9}\frac{1}{9}\frac{1}{2})B(\frac{1}{9}\frac{1}{9}\frac{1}{2})$;
SPI =	64	62	64
Cleavage		pr(100)	pr(100)
Twinning	(001)	(001)	(001)
RI	$n_\alpha = 1.561$	$n_\alpha = 1.60$	$n_\alpha = 1.63$
	$n_\beta = 1.570$	$n_\beta = 1.62$	$n_\beta = 1.64$
	$n_\gamma = 1.587$	$n_\gamma = 1.63$	$n_\gamma = 1.59$
$\delta =$	0.026	0.03	0.03–0.04
$2V$	44–50°	60–90°	73–76°
Sign	(+)	(+)	(+)
Transp.	Transp. to transl.	Translucent	Translucent
Pleochr.		X = brown yellow	X = brown yellow
		Y = medium yellow	Y = medium yellow
		Z = pale yellow	Z = pale yellow
H =	$6\frac{1}{2}$	$6\frac{1}{2}$	6
D =	3.16	3.16–3.26	3.21–3.35
Color	White, yellow	White, yellow	White, yellow
Streak	White	White	White
Luster	Vitreous	Vitreous	Vitreous
Fracture	Subconchoidal	Subconchoidal	Subconchoidal
Habit	Tabular	Equant	Equant
Remarks		Members of the *humite* group	

Titaniferous clinohumite and chondrodite are of particular interest to petrologists because of their high-pressure stability and hence their role as hydrous minerals in the earth's mantle.

Distinguishing Features. Chondrodite is the most common of the humite minerals and can be distinguished in hand specimen by its yellow to brown color, prismatic form, and pseudo-hexagonal cross sections. The occurrence of chondrodite in metamorphosed calc-silicate rocks is also distinctive. In thin section, its deep yellow to tan pleochroism is similar only to that of staurolite from which it is easily distinguished by association.

The Aluminosilicate Group

In the aluminosilicate group of isolated tetrahedral silicates, we include andalusite and kyanite for structural reasons, and sillimanite because of chemistry and polymorphism. The three polymorphs are among the most important minerals in

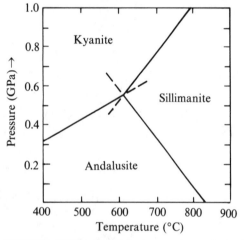

FIGURE II.53 Pressure-temperature phase diagram for the Al$_2$SiO$_5$ polymorphs.

pelitic schists, serving as indicators of relative pressure and temperature in metamorphic rocks. The phase diagram for the Al$_2$SiO$_5$ system is represented in Figure II.53.

Structure. All three Al$_2$SiO$_5$ polymorphs have all silicon in tetrahedral coordination, and one of two Al^{3+} cations in octahedral coordination. A major difference in the structures is the coordination of the remaining Al^{3+}, being four-coordinated with oxygen in sillimanite, five-coordinated in andalusite, and six-coordinated in kyanite. This difference in coordination is expressed in the respective chemical formulas of these minerals. This sequence of coordination corresponds to the order of increasing density of the three polymorphs and to their stability with increased pressure, kyanite being the most stable. The symmetrical packing indices of the three minerals are also consistent with these observations.

The structure of kyanite (Figure II.53), the high-pressure polymorph, can be viewed as complex octahedral layers consisting of subchains of edge-shared octahedra that run parallel to [001]. Both the perfect {100} cleavage and the good {010} cleavage of kyanite are parallel to the subchains, and the cleavages are undoubtedly controlled to some extent by them. Moreover, the well-known differential hardness of kyanite (4–5 parallel to *c* on {100}, 6–7 parallel to *b* on {100}) is an expression of the structural anisotropy, the "harder" direction being across the octahedral subchains.

We can also view the structure of kyanite as symmetrically packed oxygen sheets in an *ABC* stacking sequence parallel to (011), and also to (110), in which 10% of the tetrahedral voids are occupied by Si^{4+}, and 40% of the octahedral voids are occupied by Al^{3+}.

The structure of andalusite (Figure II.54) dif-

Andalusite
AlOSiO$_4$

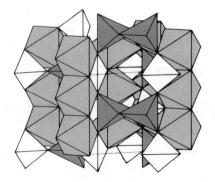
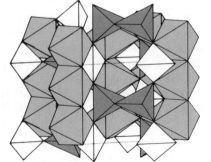

 SiO$_4$

 AlO$_5$

 AlO$_6$

FIGURE II.54 Crystal structure of andalusite, *c* axis vertical.

THE ALUMINOSILCATE GROUP

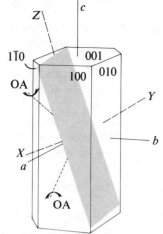
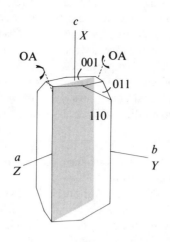
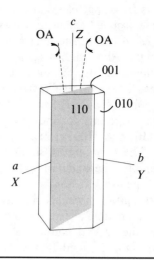

	KYANITE	ANDALUSITE	SILLIMANITE
Formula	Al_2OSiO_4	$AlAlOSiO_4$	$AlSiAlO_5$
Coord.	Al(6), Si(4)	Al(5), Al(6), Si(4)	Al(6), Si(4), Al(4)
System	Triclinic	Orthorhombic	Orthorhombic
$a =$	0.710	0.778	0.744
$b =$	0.774	0.792	0.760
$c =$	0.557	0.557	0.575
$\alpha =$	90.08°		
$\beta =$	101.03°		
$\gamma =$	105.73°		
$Z =$	4	4	4
Sp. group	$P\bar{1}$	$P2_1/n\,2_1/n\,2/m$	$P2_1/b\,2/m\,2/n$
Pt. group	$\bar{1}$	$2/m\,2/m\,2/m$	$2/m\,2/m\,2/m$
S; loops	1.00	1.00	1.50; 4
SP plane	(011)		
SP struct.	$A(-\,-\tfrac{3}{5})B(\tfrac{1}{5}\tfrac{1}{5}\tfrac{1}{5})C(-\tfrac{3}{5}-)$; $A(\tfrac{1}{5}\tfrac{1}{5}\tfrac{1}{5})B(\tfrac{3}{5}-\,-)C(\tfrac{1}{5}\tfrac{1}{5}\tfrac{1}{5})$;		
SPI =	68	58	63
Cleavage	pf(100), gd(010)	gd{110}, pr(100)	pf(010)
Parting	(001)		
Twinning	(100)	(101) rare	
RI	$n_\alpha = 1.712$	$n_\alpha = 1.632$	$n_\alpha = 1.658$
	$n_\beta = 1.720$	$n_\beta = 1.640$	$n_\beta = 1.662$
	$n_\gamma = 1.728$	$n_\gamma = 1.642$	$n_\gamma = 1.680$
$\delta =$	0.016	0.010	0.022
$2V =$	82–83°	75–85°	20–30°
Sign	(−)	(−)	(+)
Transp.	Transp. to transl.	Transp. to transl.	Transp. to transl.
Pleochr.	X = colorless	X = rose	
	Y = colorless	Y = colorless	
	Z = light blue	Z = colorless	
H =	5–7	$7\tfrac{1}{2}$	6–7
D =	3.60	3.18	3.23
Color	Blue, white	Brown, red	White, brown
Streak	White	White	White
Luster	Vitreous, pearly	Vitreous	Vitreous
Fracture	Uneven	Subconchoidal	Uneven
Habit	Blady, tabular	Prismatic, massive	Prismatic
Remarks	Subfibrous	Carbonaceous inclusions: chiastolite	Fibrous: fibrolite
	High pressure	Low pressure	Intermed. pressure

fers from that of kyanite (Figure II.55) by having true, isolated chains of edge-shared octahedra parallel to [001] instead of subchains. Only one half of the Al^{3+} cations occupy these octahedra; the other half occupy distorted trigonal bipyramids that have an irregular fivefold coordination. As in kyanite (Figure II.55), the octahedral chain direction accounts for the {110} prismatic habit and {110} and {100} cleavages.

The structure of sillimanite (Figure II.56) resembles the andalusite structure in that one half of the Al^{3+} cations occupy edge-sharing octahedra that form chains parallel to [001]. In addition, in sillimanite the remaining Al^{3+} cations occupy tetrahedra that share three of four apical oxygens with the isolated Si tetrahedra. The Al and Si tetrahedra alternate to form a double chain parallel to [001]. The octahedral and tetrahedral chains together impart a strong structural anisotropy and explain the prismatic habit and good {010} cleavage. A fibrous variety of sillimanite, referred to as *fibrolite* when fine-grained, is elongate in the chain direction.

Chemistry. With few exceptions, none of the aluminosilicates shows a significant departure from pure Al_2SiO_5 chemistry. Only Fe^{3+}, Mn^{3+}, and Cr^{3+} have been demonstrated to substitute for Al^{3+}, and then only in small amounts. The substitution of Mn^{3+} for Al^{3+} in andalusite can be substantial because of the affinity of high-spin Mn^{3+} for the distorted octahedral site. Such andalusites may have a pink color and slight pleochroism.

Occurrence and Associations. The aluminosilicate polymorphs are found almost exclusively in pelitic schists and gneisses as the product of metamorphosed argillaceous sediments. Andalusite is the stable form at the lowest grades of metamorphism, followed by sillimanite as temperature increases, or by kyanite as both temperature and pressure increase. The most common associations are with quartz, muscovite, garnet, staurolite,

Kyanite
Al_2OSiO_4

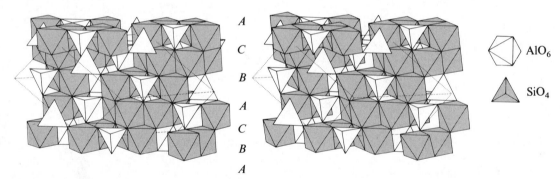

FIGURE II.55   Crystal structure of kyanite, oxygen sheet in (011), [011] vertical.

Sillimanite
$AlSiAlO_5$

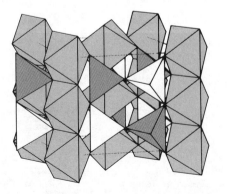

FIGURE II.56   Crystal structure of sillimanite, oxygen sheet in (130).

biotite, and a number of other aluminous minerals (Figure 9.13). Kyanite is found in certain eclogites associated with garnet, jadeitic pyroxene, and zoisite. Kyanite is also known from amphibolites in association with cordierite and gedrite.

Distinguishing Features. Kyanite typically occurs as blue, tabular to bladed crystals in mica schists. In contrast, sillimanite is colorless and forms long, slender crystals, commonly with a pearly luster. Andalusite forms white, square prisms and occasionally may be found as *chiastolite*, the varietal name applied when dark-colored carbonaceous inclusions form a cruciform pattern. Both andalusite and sillimanite commonly alter to muscovite.

In thin section, the high relief and distinctive cleavage of kyanite is diagnostic. Sillimanite has lower relief, a distinctive rhombus-shaped cross section, and a low optic angle. Andalusite has low birefringence and moderately high relief.

Staurolite

Staurolite is a metamorphic mineral that is found almost exclusively in medium-grade schists. In general, it has an isolated tetrahedral structure related to that of kyanite.

Structure. The oxygen sheets in the staurolite structure (Figure II.57) are in an *ABC* stacking sequence parallel to the (130) and the (101) planes. One eighth of the tetrahedral sites are occupied by Si^{4+} and by Fe^{2+} cations. Staurolite is one of the few silicates in which ferrous iron occupies a tetrahedral site. A similar and also unusual feature of this structure is the octahedral coordination of the hydrogen cation. The octahedral frame is dominated by edge-sharing chains, like those in the octahedral networks of olivine and humites and in the complex layer of kyanite. The weakest plane in the octahedral network is (010), which coincides with the only cleavage plane of staurolite.

Chemistry. The most important cation substitution in staurolite is the replacement of Fe^{2+} by Mg^{2+}, possibly within the tetrahedral site. The substitution is limited in most natural staurolites to about 30% of the ideal Mg end-members, which possibly reflects the fact that Mg^{2+} seldom enters four-coordinated sites. The $Fe^{2+} \rightleftharpoons Mg^{2+}$ substitution in the tetrahedral site is as yet unproven. Additional Fe^{2+} along with Al^{3+} cations may occupy the octahedral sites of the monolayer, and thus the Mg^{2+} substitution is restricted to that site.

One interesting facet of staurolite chemistry is the unusual affinity of this mineral for zinc. Any Zn^{2+} in the bulk chemistry of the original sediment seems to be always taken up by staurolite, and very probably in the tetrahedral site because of the close similarity in the radii of Fe^{2+} and Zn^{2+}.

Occurrence and Associations. With few exceptions, staurolite is found only in pelitic schists and related metamorphic rocks relatively enriched in Al^{3+} and Fe^{2+}. Associated minerals are the aluminosilicate polymorphs, quartz, muscovite, biotite, chloritoid, and garnet (Figure 9.13). Staurolite is occasionally found associated with gedrite and garnet and with plagioclase in aluminous amphibolites.

Distinguishing Features. Staurolite crystals are black and prismatic, and commonly have a distinctive cruciform twin on either {031} or {231}; the two members intersect at 90 and 60 degrees, respectively (Figure II.58). In thin section, staurolite has colorless to yellow pleochroism that can be mistaken only for pleochroic humite.

Chloritoid

Chloritoid is a common associate of staurolite in pelitic schists, and like staurolite, is restricted to metamorphic rocks that are relatively enriched in Fe and Al. As the name implies, chloritoid bears

Staurolite
$Al_9HSi_4Fe_2O_{24}$

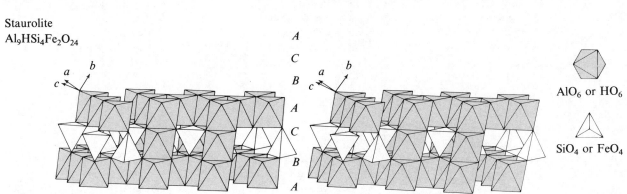

FIGURE II.57 Crystal structure of staurolite, *c* axis vertical.

STAUROLITE

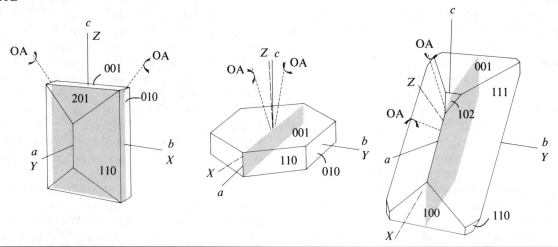

	STAUROLITE	Chloritoid (*2M*)	Titanite (sphene)
Formula	$Al_9HSi_4Fe_2O_{24}$	$(Fe, Mg)Al_2O(OH)_2SiO_4$	$CaTiOSiO_4$
Coord.	Al(6), H(6), Si(4), Fe(4)	Fe(6), Mg(6), Al(6), Si(4)	Ca(7), Ti(6), Si(4)
System	Monoclinic	Monoclinic	Monoclinic
$a =$	0.782	0.952	0.656
$b =$	1.652	0.547	0.872
$c =$	0.563	1.819	0.744
$\beta =$	90.0°	101.65°	119.72°
$Z =$	2	8	4
Sp. group	$C2/m$	$C2/c$	$C2/c$
Pt. group	$2/m$	$2/m$	$2/m$
S	1.00	1.00	1.00
SP plane	(130)		(011)
SP struct.	$A(--\frac{2}{3})B(\frac{1}{4}\frac{1}{4}\frac{1}{4})C(-\frac{2}{3}-)$ $A(\frac{1}{4}\frac{1}{4}\frac{1}{4})B(\frac{2}{3}--)C(\frac{1}{4}\frac{1}{4}\frac{1}{4})$;		
SPI =	67	62	55
Cleavage	pf(001)	gd(001)	pf{110}, pr(100)
Twinning	90° cross {031} 60° cross {231}	(100), {221}	(100), {221} lamellar
RI	$n_\alpha = 1.740$ $n_\beta = 1.744$ $n_\gamma = 1.753$	$n_\alpha = 1.715$ $n_\beta = 1.720$ $n_\gamma = 1.725$	$n_\alpha = 1.86$ $n_\beta = 1.93$ $n_\gamma = 2.10$
$\delta =$	0.013	0.010	0.15
$2V =$	80–88°	45–65°	23–50°
Sign	(+)	(+)	(+)
Transp.	Translucent	Transparent	Transp. to transl.
Pleochr.	$X =$ colorless $Y =$ pale yellow $Z =$ yellow	$X =$ yellow green $Y =$ indigo blue $Z =$ colorless	$X =$ colorless $Y =$ colorless $Z =$ pale brown
$H =$	$7–7\frac{1}{2}$	$6\frac{1}{2}$	5
$D =$	3.75	3.50	3.50
Color	Black, brown	Dark green	Gray, variable
Streak	White, gray	Gray	White
Luster	Resinous, vitreous	Pearly	Adamantine
Fracture	Subconchoidal	Uneven	Uneven
Habit	Prismatic, tabular	Platy, foliated	Tabular
Remarks	Structure similar to kyanite	Tetrahedral layers between different size Fe and Al octahedral layers	Wedge-shaped crystals

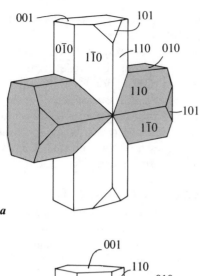

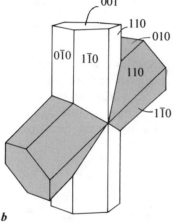

FIGURE II.58   Cruciform twinning in staurolite. (a) twin plane {031} and (b) twin plane {231}.

some resemblance to chlorite, which is a layer silicate. In fact, the structure of chloritoid is dominated by octahedral layers. This accounts for its perfect basal cleavage, resembling the cleavage of chlorite and mica, but unlike these minerals, the chloritoid layers are bound together by isolated Si tetrahedra.

Structure. Two types of layers are in the chloritoid structure. Trioctahedral layers are occupied by Fe^{2+}, Mg^{2+}, Mn^{2+}, and Al^{3+}, and unusual octahedral layers are occupied by Al^{3+}. In the latter, three fourths of the octahedral voids are empty, in contrast to the two thirds occupancy in the dioctahedral layers of clays and micas. The isolated Si tetrahedra are linked to both sides of the Al octahedral layers and have their basal faces in the position of the triangular faces of the unfilled octahedral voids. The free corners of the tetrahedra are on the outside of the Al octahedral layer and are connected to the trioctahedral layer. The two octahedral layers differ in their dimensions because of the smaller ionic radii of the Al^{3+} cations compared with the radii of Fe^{2+} and

Mg^{2+}. Consequently, the edges of the Al octahedra are cosine 30 degrees times the edges of the (Fe, Mg) octahedra, and a 30-degree rotation must occur between the two octahedra in order to compensate for the difference in sizes.

Both monoclinic and triclinic chloritoids are known. Because of the pseudohexagonal geometry of the octahedral and trioctahedral layers, the possibilities for chloritoid isotypes are similar to the isotype possibilities of the true layer silicates.

Chemistry. The principal cation substitution in chloritoid is $Mg^{2+} \rightleftharpoons Fe^{2+}$ within the octahedral layer. Natural chloritoids exhibit up to about 40 mole percent solubility toward an ideal Mg end-member, and up to about 10–15% solubility toward the Mn end-member. Small amounts of Fe^{3+} may also substitute for Al^{3+}.

Occurrences and Associations. Chloritoid is fairly common in the lower grades of metamorphism where it is associated with biotite, muscovite, and quartz in fine-grained schists. In schists of medium-grade metamorphism, chloritoid may be associated with garnet, staurolite, and kyanite.

Distinguishing Features. Because of its perfect basal cleavage and physical similarity to the layer silicates, chloritoid may be difficult to distinguish from biotite and chlorite in hand specimen. In thin section, the distinctive high relief, indigo blue pleochroism, and common association with muscovite and quartz are diagnostic.

Titanite (Sphene)

Titanite is a common accessory mineral in most felsic plutonic rocks and in a variety of metamorphic rocks in which it may be the dominant Ti-bearing mineral.

Structure. As in most isolated tetrahedral silicates, the structure of titanite is dominated by the secondary octahedral frame. Parallel chains of corner-shared Ti octahedra run parallel to the [110] direction and are linked together by isolated Si tetrahedra. Octahedral voids between the chains are occupied by Ca^{2+} in sevenfold coordination.

Chemistry. The ionic substitutions that affect the chemistry of titanite are relatively simple, chiefly small amounts of Al^{3+}, Fe^{3+}, and Sn^{2+} in the octahedral sites for Ti^{4+}, and a variety of rare earth elements (e.g., Nb^{5+} and Ta^{5+}) substituting for Ca^{2+} in sevenfold octahedral voids. We also have some evidence of $(OH)^{1-}$ and F^{1-} substitution for oxygen.

Occurrence and Associations. Titanite is a widespread accessory mineral in many igneous and metamorphic rocks. Notable occurrences are in nepheline syenites in association with apatite and alkali feldspars, and in certain marbles associated with calcite, dolomite, and calc-silicates such as grossular and diopside.

Distinguishing Features. Titanite is distinguished in hand specimens by its wedge-shaped crystals and clove brown color. In thin section, its rhombic form in cross section, high relief, and birefringence are diagnostic.

Topaz

Topaz is a F-rich aluminosilicate found commonly in hydrothermal veins associated with felsic igneous rocks and pegmatites. Its principal use is as a gemstone.

Structure. The topaz structure (Figure II.59) is characterized by an unusual "double hexagonal" stacking of hexagonal sheets of O^{2-}, OH^{1-}, and F^{1-} in an *ABCB* sequence. The layers are parallel to (010), and all have the same content of edge-sharing double octahedra that alternate with isolated tetrahedra to form chains running parallel to the c axis. Four of the six anions about each Al^{3+} are oxygens that are shared with Si tetrahedra.

The remaining two anions (OH, F) share with other double octahedra.

Chemistry. The only significant compositional variation in topaz is the substitution of up to 30 mole percent $(OH)^{1-}$ for F^{1-}.

Occurrence and Associations. Topaz typically forms during the latest stages of the solidification of siliceous igneous rocks, when F^{1-} and OH^{1-} ions are sufficiently concentrated. Common associates are fluorite, tourmaline, cassiterite, apatite, quartz, and muscovite.

Distinguishing Features. In hand specimens, topaz frequently occurs as clear or pale blue stubby crystals with wedge-shaped terminations. In thin section, its high relief, low birefringence, and prismatic form are fairly distinctive.

Zircon

Zircon ($ZrSiO_4$) is a widespread accessory mineral in most igneous and metamorphic rocks and is commonly found as detrital grains in sediments. The zircon structure (Figure II.60) consists of isolated Si tetrahedra connected by a network of

Topaz

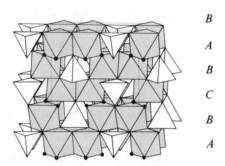 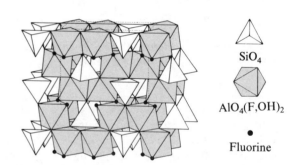

FIGURE II.59   Crystal structure of topaz, *a* axis vertical.

Zircon
$ZrSiO_4$

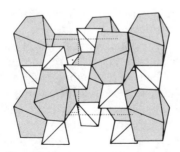 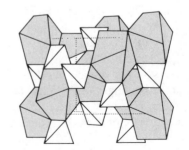

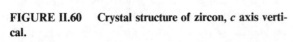

FIGURE II.60   Crystal structure of zircon, *c* axis vertical.

TOPAZ

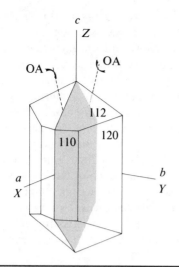

ZIRCON

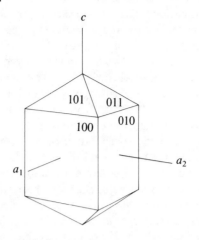

	TOPAZ
Formula	$Al_2(F, OH)_2SiO_4$
Coord.	Al(6), Si(4)
System	Orthorhombic
$a =$	0.465
$b =$	0.880
$c =$	0.840
$Z =$	4
Sp. group	$P2_1/b\,2/n\,2/m$
Pt. group	$2/m\,2/m\,2/m$
S	1.00
SP plane	(010)
SP struct.	$A(\frac{1}{6}-\frac{1}{3})B(\frac{1}{3}-\frac{1}{6})C(\frac{1}{3}-\frac{1}{6})B(\frac{1}{6}-\frac{1}{3})$;
SPI	66
Cleavage	pf(001)
RI	$n_\alpha = 1.61$
	$n_\beta = 1.61$
	$n_\gamma = 1.62$
$\delta =$	0.01
$2V =$	48–65°
Sign	(+)
Transp.	Transp. to transl.
$H =$	8
$D =$	3.5–3.6
Color	Colorless, variable
Streak	White
Luster	Vitreous
Fracture	Subconchoidal
Habit	Prismatic
Remarks	Occ. fluorescent (U.V.) acicular: pycnite

	ZIRCON
Formula	$ZrSiO_4$
Coord.	Zr(4)(+2)*, Si(4)
System	Tetragonal
$a =$	0.659
$c =$	0.599
$Z =$	4
Sp. group	$I4_1/a\,2/m\,2/d$
Pt. group	$4/m\,2/m\,2/m$
S	1.00
SP plane	(101)
SP struct.	$A(\frac{1}{4}\frac{1}{4}-)B(--\frac{1}{2})$ $A(\frac{1}{4}\frac{1}{4}-)B(--\frac{1}{2})$;
SPI =	63
Cleavage	pr(100), pr{101}
Twinning	{111}
RI	$n_\omega = 1.99$
	$n_\epsilon = 1.93$
$\delta =$	0.06
Sign	(+)
Transp.	Transp. to transl.
$H =$	$7\frac{1}{2}$
$D =$	4.68
Color	Brown, green
Streak	White
Luster	Adamantine
Fracture	Conchoidal
Habit	Prismatic
Remarks	Ca shifted toward tetrahedral layers, increases CN to 8
	Contains Hf
	*Close second neighbor bonds

unusually distorted cubes with Zr^{4+} in sixfold coordination. The Si tetrahedra are in regular symmetrically packed layers occupying one quarter of both types of tetrahedral void. The Zr^{4+} is located between two tetrahedral layers near the center of the octahedral voids. The open arrangement of the tetrahedra gives room for the larger Zr^{4+} cations to shift from the octahedral void center to a higher coordination position. This adjustment of the coordination of the larger cations in a symmetrically packed tetrahedral frame is rather common. For example, among the sulfates we observe such adjustments in barite and anhydrite, among the tungstates in scheelite, and among the phosphates in xenotime, vivianite, and erythrite, which have comparable structures. Note, however, that the pattern of the large cation displacement is different in most of these structures.

Zircons contain a variety of additional components in small amounts, notably Hf, which seems to reside almost exclusively in this mineral. Other elements that are sometimes present are U, Th, Fe, Sn, Nb, Y, and Ta. Because of the content of radioactive elements, radiation can damage the crystal structure. Zircon and other minerals so affected are said to be *metamict*.

Zircon is recognized in hand specimens by its square prismatic form, vitreous luster, and brown to yellow color. In thin section, the high relief, double refraction, and doubly terminated prismatic form are distinctive.

Double Tetrahedral Structures

Lawsonite $(CaAl_2(OH)_2Si_2O_7 \cdot H_2O)$ and tilleyite $(Ca_5(CO_3)_2Si_2O_7)$ are among the few silicates that have a truly double tetrahedral structure. In lawsonite, pairs of Si tetrahedra share apical oxygens and are connected in a relatively open secondary frame by edge-sharing Al octahedra that run parallel to the *b* axis. The edge-sharing octahedra are somewhat similar to the octahedral chains in sillimanite (Figure II.56).

Lawsonite is restricted in occurrence to glaucophane schists where it typically occurs as white, blocky crystals in a finer grained matrix of dark blue glaucophane, chlorite, epidote, garnet, and pumpellyite. Tilleyite and rankinite $(Ca_2Si_2O_7)$, also a double tetrahedral structure, are restricted in occurrence to calc-silicate rocks affected by high-temperature contact metamorphism.

Double and Single Tetrahedral Structures

Vesuvianite (Idocrase)

Vesuvianite $(Ca_{10}(Mg, Fe)_2(OH)_4Al_4(SiO_4)_5$ $(Si_2O_7)_2)$ is usually formed in impure limestones and dolomites subjected to contact metamorphism. The vesuvianite structure (Figure II.61) consists of independent Si tetrahedra and Si_2O_7 double Si tetrahedra connected in a network of (Mg, Fe) octahedra and Al octahedra that bond to both oxygen and hydroxyl. The holes in the network are occupied by Ca^{2+}.

Vesuvianite shows substantial substitution of Mn^{2+} for Mg^{2+}, Fe^{2+} or Na^{1+} for Ca^{2+}, and Al^{3+} for Ti^{4+} in such a manner that charge balance is maintained. Vesuvianite is commonly associated with other calcic minerals of contact metamorphism such as grossular, wollastonite, and diopside. The typical form in hand specimens is as brown tetragonal prisms.

Vesuvianite
(idocrase)
$Ca_{10}(Mg,Fe)_2Al_4(OH)_4Si_5O_{20}Si_4O_{14}$

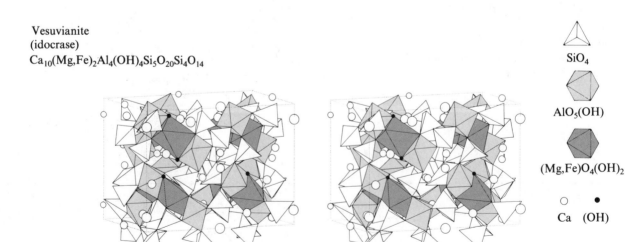

SiO_4

AlO_5(OH)

(Mg,Fe)O_4(OH)_2

○ ●
Ca (OH)

FIGURE II.61   Crystal structure of vesuvianite, *c* axis vertical.

DOUBLE TETRAHEDRAL STRUCTURES

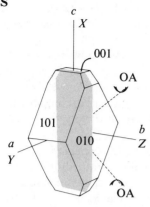

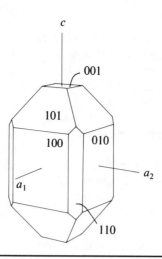

	Lawsonite	VESUVIANITE (idocrase)
Formula	$CaAl_2(OH)_2Si_2O_7 \cdot H_2O$	$Ca_{10}(Mg, Fe)_2Al_4(OH)_4(SiO_4)_5(Si_2O_7)_2$
Coord.	Ca(8), Al(6), Si(4)	Ca(8), Mg(6), Fe(6), Al(6), Si(4)
System	Orthorhombic	Tetragonal
$a =$	0.890	1.566
$b =$	0.576	
$c =$	1.333	1.185
$Z =$	4	4
Sp. group	$C2/c\,2/m\,2/m$	$P4/n\,2/n\,2/c$
Pt. group	$2/m\,2/m\,2/m$	$4/m\,2/m\,2/m$
S; group	1.25; pairs	1.11; singles and pairs
SP plane	(100)	
SP struct.	$A(--\frac{1}{2})B(-\frac{1}{4}\frac{1}{4})C_{25}(-\frac{1}{4}\frac{1}{4})$	
	$B(--\frac{1}{2})A(\frac{1}{4}-\frac{1}{4})C_{25}(\frac{1}{4}-\frac{1}{4})$;	
SPI =	53	50
Cleavage	pf(100), pf(010)	pr{100}
Twinning	{110}	
RI	$n_\alpha = 1.665$	$n_\omega = 1.706$
	$n_\beta = 1.674$	$n_\epsilon = 1.701$
	$n_\gamma = 1.685$	
$\delta =$	0.020	0.005
$2V =$	76–86°	
Sign	(+)	(−)
Transp.	Transparent	Transparent
H =	7–8	$6\frac{1}{2}$
D =	3.1	3.4
Color	Gray, white	Green, brown, yellow
Streak	White	White
Luster	Vitreous, greasy	Vitreous
Fracture	Uneven	Subconchoidal
Habit	Tabular, prismatic	Prismatic
Remarks	SP struct. distorted	Biaxial varieties are known
	Similar struct. ilvaite:	
	$CaFe_2O(OH)Si_2O_7$	

The Epidote Minerals

The minerals of the epidote group include zoisite, which is orthorhombic, and clinozoisite, epidote, piemontite, and allanite, all of which are monoclinic. With the exception of allanite, which is a characteristic accessory mineral in granites and syenites, the epidote minerals are found almost exclusively in metamorphic rocks relatively enriched in Ca.

Structure. The structure of the epidote minerals (Figure II.62) is similar to that of titanite. Edge-sharing chains of Al octahedra run parallel to the *b* axis with Fe^{3+} octahedra attached to these chains, also by shared edges. The chains are connected to each other by both isolated and double Si tetrahedra. The large cavities between the chains are occupied by Ca^{2+} in an irregular, eightfold coordination. In clinozoisite, the Fe octahedra are occupied by Al^{3+}.

The hydrogen in the epidote group minerals is not bound as $(OH)^{1-}$, but rather by hydrogen bonds between adjacent oxygens.

Chemistry. The chemistry of orthorhombic zoisite has little variation. The main substitution is Mn^{2+} for Ca^{2+}, producing the pink variety of zoisite called *thulite*. The minor substitution of Al^{3+} by Fe^{3+} also occurs. In contrast, the monoclinic epidote minerals exhibit extensive solid solution, chiefly between clinozoisite ($Ca_2Al_3HSi_3O_{13}$) and epidote ($Ca_2Fe_3HSi_3O_{13}$). Substitution of Mn, Pb, and Sr for Ca in the eightfold site is also known.

Allanite analyses show a substitution of ferrous iron for Al^{3+} with a balancing substitution of trivariant Ce, La, and Y for Ca^{2+}. The common presence of radioactive components, chiefly Th and U, can lead to the physical breakdown of the structure.

Occurrence and Associations. Epidote minerals are fairly common in medium-grade amphibolites and gneisses where they are associated with hornblende, garnet, plagioclase, and calcite. In lower grade rocks, epidote and clinozoisite occur with actinolite, chlorite, quartz, and plagioclase.

Distinguishing Features. Epidote is recognized in hand specimens by its prismatic to acicular form and a distinctive pistachio green to yellow green color. Radiating clusters of acicular needles on joint surfaces are common, as are striations parallel to the *b* axis. In thin section, the moderate relief equant form and anomalous "berlin blue" birefringence are characteristic. Epidote, because of its ferric iron content, has a greenish yellow pleochroism; clinozoisite and zoisite are colorless.

Epidote
$Ca_2Al_2FeHO_2SiO_4Si_2O_7$

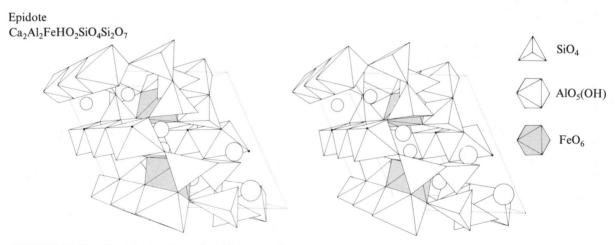

SiO₄

AlO₅(OH)

FeO₆

FIGURE II.62   Crystal structure of epidote, *c* axis vertical.

THE NATIVE ELEMENTS

About 24 elements are known to occur in their native state in rocks. These elements are referred to as "native" because they occur in their most reduced state, that is, uncombined with anions of, for example, oxygen or sulfur. Compared to other minerals, the native elements are rare but nonetheless important. They are valued as a major source of certain precious metals (gold and silver), for their use in industry, and as gemstones (diamond).

THE EPIDOTE MINERALS

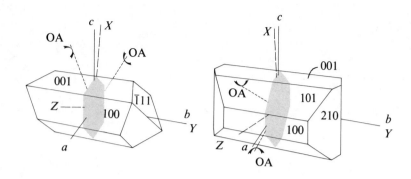

	EPIDOTE	Clinozoisite
Formula	$Ca_2Al_2FeHO_2SiO_4Si_2O_7$	$CaAl_3HO_2SiO_4Si_2O_7$
Coord.	Ca(6 or 9), Fe(6), Si(4)	Ca(7), Al(6), Si(4)
System	Monoclinic	Monoclinic
$a =$	0.898	0.894
$b =$	0.564	0.561
$c =$	1.022	1.023
$\beta =$	115.4°	115.0°
$Z =$	2	2
Sp. group	$P2_1/m$	$P2_1/m$
Pt. group	$2/m$	$2/m$
S; group	1.167; singles, pairs	1.167; singles, pairs
SPI =	56	52
Cleavage	pf(001), pr(100)	pf(001)
Twinning	(100)	(100)
RI	$n_\alpha = 1.71$–1.75	$n_\alpha = 1.67$–172
	$n_\beta = 1.72$–1.78	$n_\beta = 1.67$–1.72
	$n_\gamma = 1.73$–1.80	$n_\gamma = 1.69$–1.73
$\delta =$	0.01–0.05	0.005–0.015
$2V =$	90–115°	14–90°
Sign	(−)	(+)
Transp.	Transp. to transl.	Transp. to transl.
Pleochr.	X = colorless	
	Y = green yellow	
	Z = colorless	
H =	$6\frac{1}{2}$	$6\frac{1}{2}$
D =	3.4–3.5	3.1–3.4
Color	Yellow, green	Green, yellow
Streak	White	White
Luster	Vitreous	Vitreous
Fracture	Uneven	Uneven
Habit	Prismatic	Prismatic
Remarks	Also known as pistacite	Isostr. w. epidote
	Mn: piemontite	Orthorhombic: zoisite
	Ce, La, Fe: allanite	Fe, Mn: thulite
	H bonds instead of (OH) anions	

**THE NATIVE
ELEMENTS**

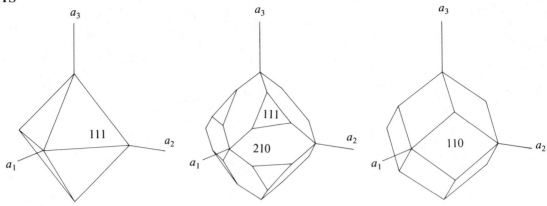

	Gold	Silver	Platinum
Formula	Au	Ag	Pt
Coord.	Au(12)	Ag(12)	Pt(12)
System	Isometric	Isometric	Isometric
$a =$	0.40783	0.40856	0.39237
$Z =$	4	4	4
Sp. group	$F4/m\bar{3}2/m$	$F4/m\bar{3}2/m$	$F4/m\bar{3}2/m$
Pt. group	$4/m\bar{3}2/m$	$4/m\bar{3}2/m$	$4/m\bar{3}2/m$
CP plane	{111}	{111}	{111}
CP struct.	ABC	ABC	ABC
SPI =	74	74	74
Twinning	{111}	{111}	{111}
RI	$n = 0.368$	$n = 0.181$	$n = 4.28$
Refl.	85%	94%	70%
H =	$2\frac{1}{2}$–3	$2\frac{1}{2}$–3	4–$4\frac{1}{2}$
D =	15.6–19.3	10.1–10.5	16.5–18.0
Color	Gold yellow	Silver white	Gray silver
Luster	Metallic	Metallic	Metallic
Fracture	Hackly	Hackly	Hackly
Habit	Octahedral, dendritic	Octahedral, filiform	Cubic
Remarks		Usually wiry	May be magnetic

Isostructural metals, malleable, and ductile

The generally accepted classification of the native elements is based on properties that are derived from the metallic bond (Chapter 4, p. 95). Those elements that are characterized by metallic luster, opaqueness to visible light, high electrical and thermal conductivity, and malleability are referred to as the *native metals*. They all have pure metallic bonds in which the outer electrons are largely delocalized. Gold, silver, platinum, copper, and iron are the best known elements in the native metals category. In contrast, elements that possess highly localized electrons form covalent bonds and are characterized by nonmetallic properties such as poor electrical conductivity and nonmetallic luster. These elements are referred to as the *native nonmetals* and include graphite, diamond, and sulfur as the only important minerals. Elements that possess properties between the metallic and nonmetallic extremes are referred to as the *native semimetals*. This category includes native arsenic, antimony, and bismuth.

Division within each of the above categories is based on structure type as follows:

1. Metals: ABC (or SD): Au, Ag, Cu, Pb, Pt, Pd, Ni, Ir, Rh, In
AB: Os, Zn
$\sqrt{4/3}$ SD: low Fe, Ta

2. Semimetals: $nABC$ (distorted): As, Sb, Bi, Se, Te

THE NATIVE ELEMENTS

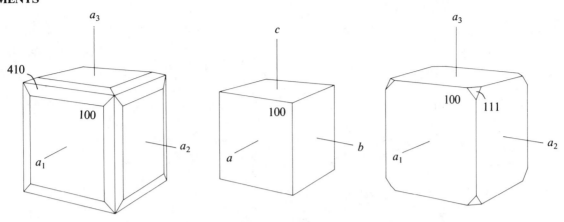

	Copper	Low Iron	Zinc
Formula	Cu	Fe	Zn
Coord.	Cu(12)	Fe(7)(+4)*	Zn(12)
System	Isometric	Isometric	Hexagonal
$a =$	0.36153	0.28664	0.2664
$c =$			0.4945
$Z =$	4	2	2
Sp. group	$F4/m\bar{3}2/m$	$I4/m\bar{3}2/m$	$P6_3/m2/m2/c$
Pt. group	$4/m\bar{3}2/m$	$4/m\bar{3}2/m$	$6/m2/m2/m$
CP plane	{111}	{100}	(001)
CP struct.	*ABC*	$\sqrt{4/3}$ *SD*	*AB*
SPI =	74	68	74
Cleavage		pf{100}	
Twinning	{111}	{111}	
RI	$n = 0.641$	$n = 2.36$	$n = 1.96$
Refl.	81%	56%	73%
H =	$2\frac{1}{2}$–3	4–5	2
D =	8.7–8.9	7.3–7.8	7.1
Color	Copper red	Steel gray	White, gray
Luster	Metallic	Metallic	Metallic
Fracture	Hackly	Hackly	Hackly
Habit	Cubic, dendritic	Irreg. grains	Irregular
Remarks	Isostr. w. gold	Magnetic	Malleable
	Malleable, ductile	Malleable	
		Close second neighbor bonds	

3. Nonmetals: $\sqrt{8/3}$ *AABBCC*: diamond
$\sqrt{3}$ *ABAC*: *2H* graphite
$\sqrt{3}$ *ABACBC*: *3R* graphite
Molecular rings: low-temperature sulfur

Structure. The details of each of the metal structure types are discussed in Chapter 5 in which we tabulate the structures of all of the elements (Table 5.10). Briefly, the *ABC* stacking, referred to as cubic closest-packed (CCP), has atoms in twelvefold coordination and has a packing efficiency of 74.05% (Figure 5.14). These structures can be represented equally well by an *SD* stacking (Figure 5.20). The *AB* stacking is referred to as hexagonal closest-packed (HCP), has atoms in twelvefold coordination, and like CCP structures, has a packing efficiency of 74.05% (Figure 5.15). The $\sqrt{4/3}$ *SD* stacking is referred to as body-centered cubic (BCC), has atoms in eightfold coordination (with four close second neighbors), and a packing efficiency of only 68.02% (Figure 5.19).

THE NATIVE ELEMENTS

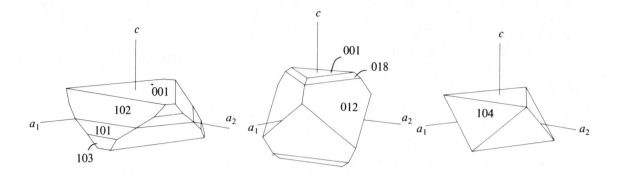

	ARSENIC	Antimony	Bismuth
Formula	As	Sb	Bi
Coord.	As(3) (+3)*	Sb(3) (+3)*	Bi(3) (+3)*
System	Trigonal	Trigonal	Trigonal
$a =$	0.376	0.431	0.455
$c =$	1.055	1.127	1.186
$Z =$	6	6	6
Sp. group	$R\bar{3}2/m$	$R\bar{3}2/m$	$R\bar{3}2/m$
Pt. group	$\bar{3}2/m$	$\bar{3}2/m$	$\bar{3}2/m$
CP plane	(001)	(001)	(001)
CP struct.	*ABC* distorted	*ABC* distorted	*ABC* distorted
SPI =	43	42	40
Cleavage	pf(001)	pf(001)	pf(001)
RI		$n = 1.70–1.80$	$n = 2.26$
Refl.	62%	75%	68%
Pleochr.	ω = lighter	ω = lighter	
	ϵ = darker	ϵ = darker	
H =	1–2	$3–3\frac{1}{2}$	$2\frac{1}{2}–3$
D =	5.7	6.7	9.8
Color	Tin white	Tin white	Silver
Luster	Metallic	Metallic	Metallic
Fracture	Uneven	Uneven	Uneven
Habit	Rhombohedral	Lamellar	Reticulated
Remarks		Brittle, isostructural *semimetals*	
		*Close second neighbor bonds	

The crystal structures of the semimetals resemble those of the *ABC* closest-packed pattern, but because of the low symmetry of the orbital overlaps, the structures are distorted from isometric to rhombohedral. The distortion is especially high in the [111] direction, and may be as high as 25% to 40%. The structure of arsenic is illustrated in Figure II.63.

The structures of the nonmetals are quite variable. Diamond (Figure II.64) has an unusual structure in which the carbon sheets ($n = \sqrt{8/3}$) are in the {111} plane of the isometric unit cell. Because of its isometric symmetry, the diamond structure is related to the *ABC* type, but has an atomic arrangement described by the $\sqrt{8/3}$ *AABBCC* type. The strongly covalent C—C bonds (0.154 nm) in diamond form a three-dimensional network that makes diamond the hardest known substance as well as the best known thermal conductor. The density of bonds across the {111} planes is lower than in other directions, which explains the {111} cleavage.

THE NATIVE ELEMENTS

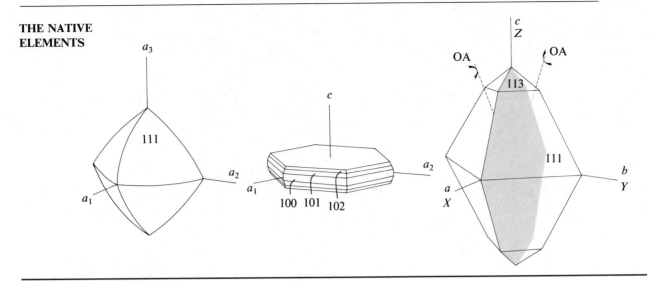

	DIAMOND	GRAPHITE (3R)	SULFUR
Formula	C	C	S
Coord.	C(4)	C(3)	S(2) (8/molecule)
System	Isometric	Hexagonal (rhombohedral)	Orthorhombic
$a =$	0.35668	0.246	1.044
$b =$			1.284
$c =$		1.006	2.437
$Z =$	8	6	128
Sp. group	$F4/d\bar{3}2/m$	$R\bar{3}2/m$	$F2/d2/d2/d$
Pt. group	$4/m\bar{3}2/m$	$\bar{3}2/m$	$2/m2/m2/m$
CP plane	{111}	(001)	
CP struct.	$\sqrt{8/3}\,AABBCC$	$\sqrt{3}\,ABACBC$	
SPI =	30	19	67
Cleavage	pf{111}	pf(001)	pr{101}, pr{110}
Twinning	{111}		
RI	$n = 2.419$	$n = 1.93\text{–}2.07$	$n_\alpha = 1.958$
			$n_\beta = 2.038$
			$n_\gamma = 2.245$
$\delta =$		0.01	0.29
$2V =$			69°
Sign		(−)	(+)
Transp.	Transparent	Opaque	Transp. to transl.
Refl.		$\omega = 5\%,\ \epsilon = 25\%$	
Pleochr.			X = brown
			Y = yellow brown
			Z = yellow
H =	10	1–2	$1\frac{1}{2}$–$2\frac{1}{2}$
D =	3.5	2.1–2.2	2.1
Color	Colorless	Lead gray	Bright yellow
Streak	White	Black	White
Luster	Adamantine	Submetallic	Resinous
Fracture	Conchoidal	Flexible, elastic	Conchoidal
Habit	Octahedral	Lamellar	Pyramidal, tabular
Remarks	Black: carbonado	Usually *2H* polytype	

Arsenic
As

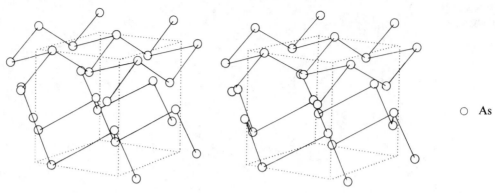

 As

FIGURE II.63   **Crystal structure of arsenic, *c* axis vertical.**

Diamond
C

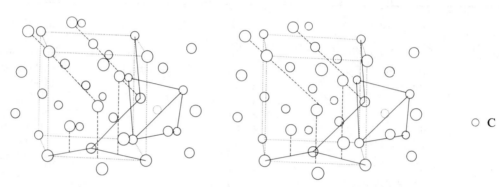

○ C

FIGURE II.64   **Crystal structure of diamond.**

Diamond has an hexagonal polymorph known as lonsdaleite, an extremely rare mineral found in certain meteorites. The carbon stacking of lonsdaleite (Chapter 5, p. 136) is $\sqrt{8/3}$ *AABB*, and is related to $\sqrt{8/3}$ *AABBCC* of diamond in the same way as *AB*-type hexagonal structures are related to *ABC*-type isometric structures. In both polymorphs, the carbon atoms are in fourfold coordination.

The principal use of diamond as a precious gem stems from its very high refractive index (2.42) and the consequent dispersion of light, which accounts for the gem's brilliance. In industry, diamonds are used extensively as abrasives and for impregnating saws and steel drill bits.

Graphite is the lower pressure polymorph (Figure II.65) of carbon. Unlike the other nonmetals, it has a gray metallic appearance. The strong C—C bonds (0.142 nm) in graphite (Figure II.66) are localized in the (001) plane, which imparts a characteristic flakey cleavage. The graphite structure consists of hexagonal open sheets within which the carbon atoms are separated by $n = \sqrt{3}$ rather than unity, as is the case for closest-

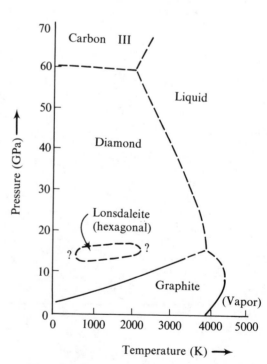

FIGURE II.65   **Pressure-temperature phase diagram for the carbon system.**

Graphite (*2H*)

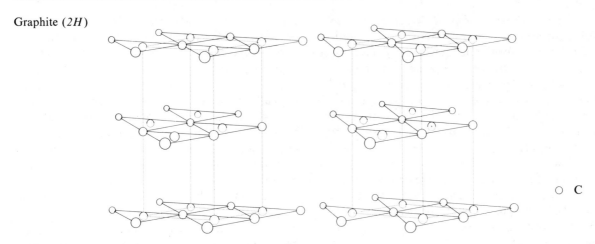

○ C

FIGURE II.66 **Crystal structure of graphite, *c* axis vertical.**

packing. The value of *n* is sufficiently large that the atoms of the overlying carbon sheet will just fit into the triangular voids of the underlying sheet. Alternate sheets are thus in identical positions, with the intervening sheet translated by $1/3a_1 + 2/3a_2$. The distance between adjacent sheets is 0.34 nm, implying the effect of relatively weak van der Waals forces (Chapter 4), which explains the low hardness of graphite, $1\frac{1}{2}$ on the Mohs scale. Because of its low hardness, graphite is used extensively as a refractory material and as a lubricant. It is also used as "lead" in pencils.

Sulfur is known to crystallize in several polymorphs, but only the low-temperature orthorhombic form is common in nature. Figure II.67 illustrates the structure in which units of eight strongly bonded sulfur atoms form eight-membered rings held together by relatively weak molecular bonds. At atmospheric pressure, orthorhombic sulfur converts to a monoclinic polymorph at about 96 °C, and that phase melts at about 120 °C. The principal use of sulfur is for the production of sulfuric acid and for explosives and fertilizers.

Chemistry. Extensive solid solution exists between several of the native metals, especially between Au and Ag, Fe and Ni, Pt and Fe, and between other native metals having similar atomic radii. Because the elements are electrically neutral, solid solution is determined largely by size considerations and to a lesser extent by bonding requirements. Figure A5.2, Appendix 5, compares the metallic radii of the elements. Gold and silver have identical radii ($r = 0.144$ nm), as do nickel and iron ($r = 0.124$ nm). The elements Pt, Pd, Ir, Os, Ru, and Rh have radii within less than 1% of each other, and hence readily substitute for each other within the metal structures.

A number of native metal alloys are considered as separate minerals and have specific names. Examples are kamacite (Fe, Ni) with about 4–8% Ni and taenite (Fe, Ni) with Ni in excess of 24%. Both of these minerals are found in association with native Fe in iron meteorites. Other examples of apparently ordered metal alloys are potarite (PdHg) and moschellandsbergite (Ag_2Hg_3).

The semimetals As, Sb, and Bi are isostruc-

Sulfur
S_8

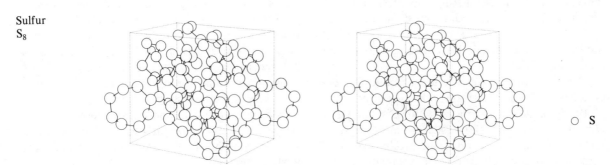

○ S

FIGURE II.67 **Crystal structure of sulfur, *c* axis vertical.**

tural, but because of their differing radii (Table A5.1, Appendix 5), only limited solid solution exists between them. The mineral stibarsen (SbAs) is an ordered alloy with hexagonal symmetry.

Among the nonmetals, diamond exhibits only trace amounts of boron and nitrogen, both of which have covalent radii close to that of carbon (Table A5.1, Appendix 5). Native sulfur frequently has small amounts of Se and Te in solid solution.

A number of native metal carbides, nitrides, and phosphides have been identified, but in general they are quite rare. Examples are cohenite $(Fe, Ni, Co)_3C$, moissanite (SiC), osbornite (TiN), and schreibersite $(Fe, Ni)_3P$.

Occurrence and Associations. The most common occurrence of native gold is in hydrothermal quartz veins, typically in metamorphic slates and schists. Pyrite, known as "fool's gold," is a common association along with other sulfide minerals. Chemical weathering liberates the native gold, enabling it to enter the sedimentary cycle. The gold is then usually concentrated in placer deposits because of its high density.

Native silver is also a product of hydrothermal solutions, and like gold is frequently associated with Fe sulfides. Silver also occurs as fine wire and in arborescent (fernlike) forms with calcite, quartz, and sulfides. In a few localities, silver occurs with native copper as vesicle fillings in basalts or greenstones, and is associated with quartz, calcite, prehnite, and zeolites.

Native copper is usually found with cuprite (Cu_2O), malachite $(Cu_2(OH)_2CO_3)$, and azurite $(Cu_3(OH)C_2O_6)$ in ore deposits associated with basalts or greenstones. The ultimate source of Cu appears to be basalt, but the deposition by solution may be either within the basalt as fracture or vesicle fillings or in adjacent sediments as veins or coatings.

Native platinum occurs principally in ultramafic igneous rocks such as dunites where it is associated with olivine, chromite, pyroxenes, and magnetite. Upon weathering, the Pt is released from its host and is commonly concentrated in placer deposits. The second most important source of the Pt metals is in solid solution in Ni and Cu sulfides.

Native iron is a rare mineral, found in meteorites and also in basalts, where it appears to form by reduction associated with assimilated carbonaceous material. Native nickel-iron is found in meteorites and is an apparent product of low-temperature serpentinization of ultramafic rocks. Native nickel is also known from the latter association. Native lead is quite rare in nature, asso-

ciated with Mn oxides and hydroxides in altered granites. Very rarely, native indium is found with native Pb in this association.

Among the native semimetals, bismuth is perhaps the most important as a principal ore. It is typically found with silver, cobalt, and nickel ores in hydrothermal veins. Antimony has a similar hydrothermal origin and is associated with native arsenic and antimony sulfides. Native arsenic occurs in hydrothermal veins with silver, nickel, and cobalt ores.

The nonmetallic native elements have diverse origins. Diamond is found only in rocks that originate in the earth's mantle at relatively high pressure (Figure II.65). The most common occurrence is in kimberlite pipes, pluglike intrusive bodies of generally mafic composition. These rocks undergo extensive weathering, which frees the chemically inert diamonds from their host. Consequently, diamond-bearing gravels and alluvium are major sources of this valuable mineral.

Graphite, on the other hand, has a totally different origin as a usually minor constituent of marbles, schists, and some gneisses. In most of these rocks, the graphite is the result of metamorphism of original organic material. More rarely, graphite is found in hydrothermal veins, in some silica-undersaturated igneous rocks, and in iron meteorites.

Native sulfur is most commonly found as encrustations around volcanic fumaroles where it forms by precipitation from hot gases. Large commercial deposits are associated with calcite, anhydrite, and gypsum in salt domes.

Distinguishing Features. With few exceptions, the native metals are recognized by their high density, metallic luster, softness, and hackly fracture. Given these properties, color is generally a reliable criterion for distinguishing the metals from each other. Gold has a rich yellow color that becomes more pale with silver content. Gold may be distinguished from pyrite, chalcopyrite, and other "gold-looking" sulfides by its malleability. The sulfides are distinctly brittle in comparison.

Native silver is brilliant white on freshly exposed surfaces, but otherwise is tarnished by a gray or black surface layer. The intricate wirelike habit can also be distinctive. On fresh surfaces, copper has a diagnostic copper red color that is not likely to be confused with any other mineral. Green staining is common on weathered surfaces. Its associations with bright green malachite, bright blue azurite, and encrusted cuprite crystals are fairly common.

Platinum is usually bluish gray in color, but may be dark gray to black with iron content. Its hardness (4–$4\frac{1}{2}$) is significantly greater than that of Ag. Native iron is steel gray and strongly magnetic.

The native semimetals have a submetallic luster and are distinctly brittle in comparison with the native metals. Arsenic is bright white on fresh surfaces but has a dark gray tarnish on weathered surfaces. The typical occurrence is as nodular or reniform masses. Bismuth has a metallic luster and a pinkish white color on fresh surfaces that readily tarnishes to a gray, somewhat iridescent film. Antimony is brilliant white but commonly has a gray antimony oxide coating or encrustation.

The only important nonmetals are diamond, graphite, and sulfur. Diamond is distinguished by its great hardness, adamantine luster, and occasional crystal form of an octahedron with slightly curved faces. Color is not a good criterion for diamond identification, as appearance may range from perfectly colorless, transparent crystals to less perfect, gray varieties called *bort* to black varieties called *carbonado*.

Graphite is distinguished by its submetallic luster, softness, and flakey crystallization habit. It also has a distinct greasy feeling indicative of its value as a lubricant. Sulfur has a pale to deep yellow color, is usually clear to translucent with a somewhat resinous luster, and is distinctly brittle.

THE SULFIDE MINERALS

The sulfides are an important group of minerals. To form a sulfide, one or more metals are usually combined with sulfur. This mineral group has great economic importance as the major source of most of our metals. Although the sulfides are much less abundant than the silicates, they include a disproportionately large number of distinct minerals because of their broad range of structure and chemistry. Unlike the silicates in which chemical bonding is predominantly ionic, most of the bonding in the sulfide minerals is covalent or metallic or both. The electron configuration of sulfur allows for several valences (Figure 4.9), and numerous possibilities exist within the *d* orbitals for hybridization. An important consequence of the sulfur bonding characteristics is the large number of possible coordination polyhedra of metals with sulfur, resulting in many different sulfide structures. The atomic separation of metals is frequently small, which allows for metallic bonds. Most sulfides consequently have metallic luster, but unlike the true metals, the sulfides are brittle.

The sulfide minerals as a structural group en-compass not only the sulfides, which have sulfur as the major nonmetal, but also those minerals that have arsenic (the arsenides), selenium (the selenides), tellurium (the tellurides), antimony, or bismuth as the major anion. The sulfide group has two conventional divisions, the metal *sulfides* in which transition metals coordinate with sulfur, and the *sulfosalts* in which the semimetals As and Sb occupy some of the metal sites. This division is somewhat artificial because of the close structural similarities between the sulfides and sulfosalts. An alternative distinction that we adopt here is to distinguish between *simple sulfides* and *complex sulfides* according to structural complexity. On this basis, the simple sulfides include most of the metal sulfides previously mentioned and some of the sulfosalts, and the complex sulfides include most of the conventional sulfosalts and some of the metal sulfides.

With few exceptions, the structures of *simple* sulfides have sulfur or other nonmetals in symmetrical packing with metal cations occupying the voids. Distinctions between the simple sulfides are based on whether the tetrahedral voids, the octahedral voids, or both types of voids are occupied. If two metals or a metal and a nonmetal were to occupy only the close-packed sites within the sheets and the voids were empty, the structure would be that of a metal alloy rather than a sulfide. Most of the *complex* sulfide structures consist of clusters, bands, or layers of symmetrically packed, simple sulfide units that are held together by highly directional, molecularlike bonds. In the following section, the sulfide minerals are discussed under the headings of simple and complex sulfides, and are classified by their coordination polyhedra.

The Tetrahedral Sulfides

Within this category are several common and important sulfide minerals including sphalerite (ZnS), chalcopyrite ($CuFeS_2$), bornite (Cu_5FeS_4), cubanite ($CuFe_2S_2$), and enargite (Cu_3AsS_4). These minerals have great economic value as major sources of Zn (sphalerite) and copper.

Structure. In all of the tetrahedral sulfides, sulfur has either an *AB* or *ABC* stacking. The metal atoms occupy only tetrahedral voids with the octahedral voids remaining vacant. The metal atom usually occupies the center of the tetrahedral void as in sphalerite (Figure II.68) and chalcopyrite (Figure II.69). In some sulfides such as bornite, however, the metal atoms are shifted toward the centers of tetrahedral faces, so that on an average, one fourth of a metal atom is near the center of each face. Strong directional bonds with sulfur

THE TETRAHEDRAL SULFIDES

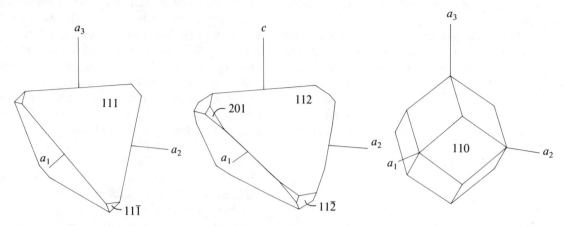

	SPHALERITE	CHALCOPYRITE	HIGH BORNITE
Formula	ZnS	$CuFeS_2$	Cu_5FeS_4
Coord.	Zn(4)	Cu(4), Fe(4)	Cu(4), Fe(4)
System	Isometric	Tetragonal	Isometric
$a =$	0.543	0.525	0.550
$c =$		1.032	
$Z =$	4	4	1
Sp. group	$F\bar{4}3m$	$I\bar{4}2d$	$F4/m\bar{3}2/m$
Pt. group	$\bar{4}3m$	$\bar{4}2m$	$4/m\bar{3}2/m$
S; loops	3.00; 3	3.00; 3	4.00; 2
SP plane	{111}	{112}	{111}
SP struct.	$A(-1-)B(--1)C(1--)$;	$A(-1-)B(--1)C(1--)$ $A(-1-)B(--1)C(1--)$;	$A(\frac{3}{4}\frac{3}{4}-)B(-\frac{3}{4}\frac{3}{4})C(\frac{3}{4}-\frac{3}{4})$;
Cleavage	pf{110}	pr{011}	pr{111}
Twinning	{111}	{112}	{111}
RI	$n = 2.42$		
Transp.	Transp. to opaque	Opaque	Opaque
Refl.	19%	41%	19%
$H =$	$3\frac{1}{2}$–4	$3\frac{1}{2}$–4	3
$D =$	4.0	4.2	6
Color	Orange red	Brass yellow	Bronze
Streak	Brown, yellow	Green black	Gray
Luster	Resinous, metallic	Metallic	Metallic
Fracture	Conchoidal	Uneven	Conchoidal
Habit	Tetrahedral	Tetr. bisphenoidal	Cubic
Remarks	Usually zoned	(S, Sn) stannite	Tarnishes blue
	HgTe: coloradoite	Ga $\rightleftharpoons$ Fe: gallite	
	HgSe: tiemannite		

apparently stabilize the structure with the metals closer to the face centers. In high-temperature bornite, the metal atoms occupy any of six possible positions on each tetrahedral face, giving a total of 24 different subsites per tetrahedron.

The tetrahedral sulfides differ in percentage of void occupancy. The symmetrical packing symbols indicate this. In bornite (Cu_5FeS_4), the Cu and Fe atoms are randomly distributed and oc-cupy 75% of the up-pointing and down-pointing tetrahedral voids, whereas in sphalerite (ZnS) and chalcopyrite ($CuFeS_2$), only one half of the tetrahedral voids are occupied. In sphalerite (Figure II.68) and chalcopyrite (Figure II.69), the tetrahedra share apical sulfurs; in bornite (Figure II.70), the tetrahedra share edges. In sulvanite, (Cu_4VS_4), Cu occupies 75% of one type of tetrahedral void, and vanadium occupies 25% of the other

THE TETRAHEDRAL SULFIDES

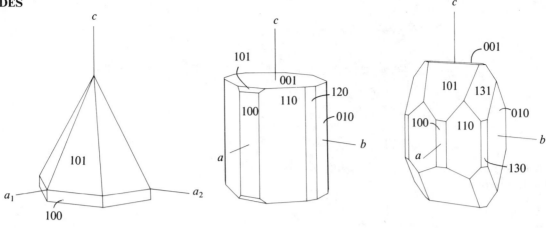

	WURTZITE	ENARGITE	CUBANITE
Formula	ZnS	Cu_3AsS_4	$CuFe_2S_3$
Coord.	Zn(4)	Cu(4), As(4)	Cu(4), Fe(4)
System	Hexagonal	Orthorhombic	Orthorhombic
$a =$	0.385	0.647	0.646
$b =$		0.744	1.112
$c =$	0.629	0.619	0.623
$Z =$	2	1	4
Sp. group	6_3mc	$Pn2,m$	$P2_1/c\,2_1/m\,2_1/n$
Pt. group	$6mm$	$m2m$	$2/m\,2/m\,2/m$
S; loops	3.00; 3	3.00; 3	3.333; 2, 3
SP plane	(001)	(001)	(100)
SP struct.	$A(-1-)B(1--)$;	$A(-1-)B(1--)$;	$A(\frac{1}{2}\frac{1}{2}-)B(\frac{1}{2}\frac{1}{2}-)$;
Cleavage	gd(100), gd(001)	pf{110}, gd(100), gd(010), pr(001)	gd{110}
Twinning		{320}	{110}
RI	$n_\omega = 2.356$		
	$n_\epsilon = 2.378$		
$\delta =$	0.022		
Sign	(+)		
Refl.	19%	22–25%	41%
$H =$	$3\frac{1}{2}$–4	3	$3\frac{1}{2}$
$D =$	4.0	4.5	4.1
Color	Orange red	Bronze	Gray, black
Streak	Brown, yellow	Dark gray	Gray
Luster	Resinous	Metallic	Metallic
Fracture	Conchoidal	Uneven	Subconchoidal
Habit	Pyramidal	Tabular	Lamellar
Remarks	Cd$\rightleftharpoons$Fe: greenockite	Pseudohexagonal	Magnetic
			Ag$\rightleftharpoons$Cu: argentopyrite

type as expressed by the packing symbol $A(\frac{1}{4}\frac{3}{4}-)$ $B(-\frac{1}{4}\frac{3}{4})C(\frac{3}{4}-\frac{1}{4})$. One consequence of this void-filling pattern is that the Cu tetrahedra share edges with the V tetrahedra.

Wurtzite is the stable polymorph of sphalerite above 1020 °C, although it commonly forms at much lower temperatures due to sulfur deficiency in the structure. Wurtzite (Figure II.71) is hexagonal with sulfur atoms occupying symmetrical sheets in an AB sequence. Zinc occupies one half of the tetrahedral voids. The many different polytypes of wurtzite appear to differ in structure, composition, and stability with respect to temperature and sulfur concentration.

Sphalerite
ZnS

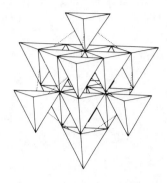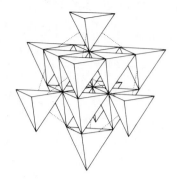

 ZnS$_4$

FIGURE II.68 Crystal structure of sphalerite, [111] direction vertical.

Chalcopyrite
CuFeS$_2$

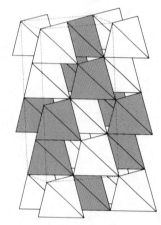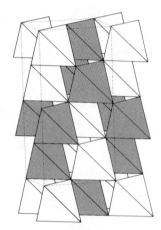

 CuS$_4$

 FeS$_4$

FIGURE II.69 Crystal structure of chalcopyrite, *c* axis vertical.

Bornite
Cu$_5$FeS$_4$

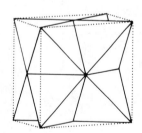 (Cu,Fe)S$_4$

FIGURE II.70 Crystal structure of bornite.

Wurtzite
ZnS

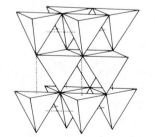

 ZnS$_4$

FIGURE II.71 Crystal structure of wurtzite, *c* axis vertical.

The structure of enargite (Figure II.72) is related to wurtzite as chalcopyrite is to sphalerite. The structure of cubanite (Figure II.73) is related to wurtzite as sulvanite is related to sphalerite, except that both types of tetrahedral voids are partially filled.

Chemistry. The chemistry of the tetrahedral sulfide minerals is controlled by many factors such as bulk chemistry, solid solubility as a function of temperature, and in some cases, nonstoichiometry. Among the tetrahedral sulfides, the last factor is important in sphalerite. A large body of data shows that Zn and S deviate from equal proportions, probably because of metal vacancies within the ideally filled tetrahedral voids. The composition range is small, but it does have a large effect on the inversion temperature to hexagonal ZnS (wurtzite) at about 1020 °C. An important solid solution referred to simply as the "i.s.s." (intermediate solid solution) phase includes a large range of compositions in the Cu-Fe-S system (Figure 9.19). The i.s.s. phase has a sphalerite structure and is slightly sulfur deficient. In contrast, chalcopyrite has an ordered tetragonal structure that allows little deviation from the ideal $CuFeS_2$ composition.

Sphalerite exhibits extensive solid solution toward ideal FeS up to 40–45 mole percent at elevated temperature (Figure 9.18). Dark colored or even black sphalerite is usually Fe-rich, although darker color is known to be imparted by metal vacancies. In addition to Fe^{2+}, Mn^{2+} and Cd^{2+} may enter the sphalerite structure by substitution for Zn^{2+}. Observed exsolution of chalcopyrite in sphalerite implies mutual solubility between these phases at elevated temperature. Similar evidence for enhanced solid solubility at high temperature is the exsolution of sphalerite and cubanite in chalcopyrite, and chalcopyrite exsolution from bornite. Figure 9.19 shows extensive solid solution between bornite and digenite.

Occurrence and Associations. Figures 9.18 and 9.19 summarize many of the common associations of sulfide minerals in ore deposits. Among the tetrahedral sulfides, sphalerite is frequently associated with pyrite or pyrrhotite or both, and chalcopyrite is frequently associated with pyrite, bornite, or pyrrhotite, depending on bulk chemistry and temperature. Sphalerite is often found with galena in lead-zinc deposits in association with chalcopyrite, calcite, and dolomite.

Sphalerite is the major ore of zinc. Chalcopyrite

Enargite
Cu_3AsS_4

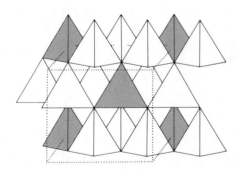
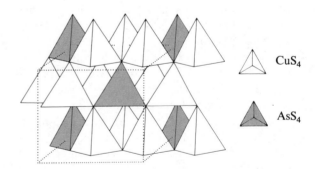

FIGURE II.72 **Crystal structure of enargite, *c* axis vertical.**

Cubanite
$CuFe_2S_3$

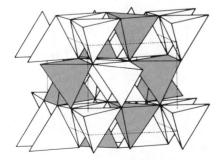
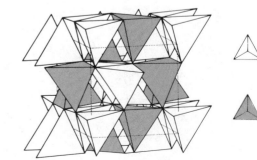

FIGURE II.73 **Crystal structure of cubanite, *b* axis vertical.**

is one of the most important copper ores, and when associated with pentlandite ((Fe, Ni)$_9$S$_8$), it is frequently mined with nickel. Bornite and cubanite are also mined for Cu in other deposits.

Distinguishing Features. Sphalerite is readily distinguished from other sulfides by its resinous luster and perfect cleavage, although massive sphalerite may have metallic luster and is sometimes mistaken for massive galena. Chalcopyrite has a distinctive brass yellow color and a greenish black streak. Bornite is distinguished by its purple tarnish, and enargite is recognized by its perfect (110) cleavage and striated tabular crystals.

The Octahedral Sulfides

Galena (PbS), pyrrhotite (Fe$_{1-x}$S), and niccolite (NiAs) are the most common octahedral sulfides. These minerals are important commercial sources of lead, iron, and nickel, respectively.

Structure. The most common sulfur stacking sequences in the octahedral sulfides are *AB* and *ABC*. The metal atoms occupy only the octahedral voids; the tetrahedral voids remain empty. In most of the octahedral sulfides, the octahedral void is fully occupied, as indicated by the appropriate packing symbols. In some of the tellurides (e.g., krennerite), the octahedral void is only partially occupied. In most cases, the metal atoms occupy the centers of voids. As in the tetrahedral sulfides, however, the metal atoms may shift their positions toward octahedral faces due to strong directional bonds.

The galena structure is a fairly common one, found also in halite (NaCl), sylvite (KCl), and periclase (MgO). This structure is frequently referred to as the *rocksalt* structure type.

The structures of niccolite and pyrrhotite (Figure II.74) are the *AB* stacking equivalent of galena. We see from these structures that in *AB* stacking the occupied octahedra must share faces, an arrangement rarely encountered in the silicate

minerals in which ionic bonds are important. Polytypes within pyrrhotite are common, and most of them appear to have a distinct chemistry within the Fe$_{1-x}$S composition range. For example, the *4M* polytype is Fe$_7$S$_8$, the *5H* polytype is Fe$_9$S$_{10}$, and the *6H* polytype is Fe$_{11}$S$_{12}$. Pure FeS is known as troilite. This structure type is usually referred to as the *nickel arsenide structure*.

Chemistry. Galena usually contains Ag in solid solution or as mineral impurities. In some cases, galena is mined as a silver ore. Nearly complete solubility appears to exist between galena and its isostructural selenide, clausthalite (PbSe). The chemistry of pyrrhotite is complicated by nonstoichiometry, namely, the deviation of the Fe/S ratio to values less than unity. The deficiency in Fe is due to vacancies in up to 20% of the normally occupied octahedral sites. Complete solid solution exists between pyrrhotite and Ni$_{1-x}$S, with substantial amounts of Cu and Co also possible (Figure 9.20).

Niccolite exhibits slight solubility toward Fe^{2+} and Co^{2+} by substitution for Ni^{2+} in octahedral sites, and toward Se by substitution for S in the close-packed sites.

Occurrences and Associations. Galena is one of the most common sulfides in hydrothermal sulfide ore bodies where it is associated with pyrite, chalcopyrite, and sphalerite. In low-temperature lead-zinc deposits, galena is commonly associated with chalcopyrite, sphalerite, marcasite, calcite, dolomite, and quartz. Pyrrhotite is usually associated with Ni-sulfides and chalcopyrite. Niccolite is found with other Ni-sulfides, pyrrhotite, chalcopyrite, and other arsenides in sulfide ore bodies. Figure 9.20 summarizes many of the common sulfide assemblages.

Distinguishing Features. Galena is distinguished from other sulfides by its cubic form, silver gray color, and metallic luster. Pyrrhotite has a bronze yellow color, commonly with an iridescent dark

Pyrrhotite
Fe$_{1-x}$S

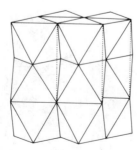 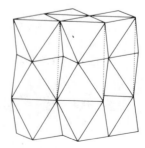

FeS$_6$

FIGURE II.74 **Crystal structure of pyrrhotite, *c* axis vertical.**

THE OCTAHEDRAL SULFIDES

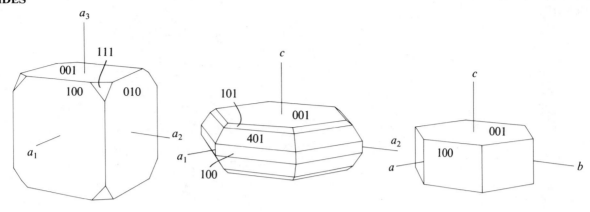

	Galena	PYRRHOTITE (troilite)	Niccolite
Formula	PbS	FeS	NiAs
Coord.	Pb(6)	Fe(6)	Ni(6)
System	Isometric	Hexagonal	Hexagonal
$a =$	0.594	0.569	0.358
$c =$		1.175	0.511
$Z =$	4	6	2
Sp. group	$F4/m\bar{3}2/m$	$P\bar{6}2c$	$P6_3/m2/m2/c$
Pt. group	$4/m\bar{3}2/m$	$\bar{6}2m$	$6/m2/m2/m$
S; loops	6.00; 2	6.00; 2	6.00; 2
SP plane	(111)	(001)	(001)
SP struct.	$A(--1)B(1--)C(-1-)$;	$A(--1)B(--1)$	$A(--1)B(--1)$
		$A(--1)B(--1)$;	$A(--1)B(--1)$;
Cleavage	pf{100}	pr(001)	pr(001)
Twinning	{111}, {114}		
RI	$n = 4.30$		
Refl.	38%	37%	56%
Pleochr.		ω = red brown	ω = brown rose
		ϵ = brown	ϵ = yellow rose
$H =$	$2-2\frac{1}{2}$	$3\frac{1}{2}-4\frac{1}{2}$	$5-5\frac{1}{2}$
$D =$	7.6	4.6	4.6
Color	Lead gray	Bronze	Copper red
Streak	Lead gray	Gray	Brown
Luster	Metallic	Metallic	Metallic
Fracture	Subconchoidal	Uneven	Uneven
Habit	Cubic, octahedral	Tabular	Massive
Remarks	May be fibrous	May be magnetic	Reniform
	Fe usually < 1	Pt$\rightleftharpoons$Pb: niggliite	Sb$\rightleftharpoons$As: breithauptite
	Mn$\rightleftharpoons$Pb: alabandite		

brown tarnish, and is usually magnetic. Niccolite has no cleavage and a very distinctive metallic pink hue.

Mixed Tetrahedral and Octahedral Sulfides

Pentlandite ((Fe, Ni)$_9$S$_8$), the principal ore of nickel, is the only abundant sulfide in which portions of both tetrahedral and octahedral voids are occupied by metal atoms.

Structure. The structure of pentlandite (Figure II.75) is based on the *ABC* stacking of hexagonal sheets of sulfur located in the {111} lattice plane. Both Ni and Fe metals occupy tetrahedral and octahedral voids. Alternate layers have different void occupancy: one layer has one fourth of both the tetrahedral and octahedral voids filled, and the other has three fourths of the tetrahedral voids filled. Because of the two different layers, com-

MIXED TETRAHEDRAL AND OCTAHEDAL SULFIDES

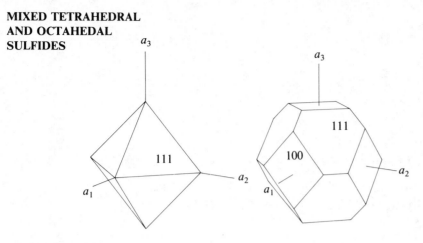

	PENTLANDITE	Violarite
Formula	$(Ni, Fe)_9S_8$	$FeNi_2S_4$
Coord.	Ni(6), Fe(6), Fe(4)	Ni(6), Fe(4)
System	Isometric	Isometric
$a =$	1.005	0.953
$Z =$	4	8
Sp. group	$F4/m\bar{3}2/m$	$F4/d\bar{3}2/m$
Pt. group	$4/m\bar{3}2/m$	$4/m\bar{3}2/m$
S; loops	2.25; 2, 6	1.00
SP plane	(111)	(111)
SP struct.	$A(\frac{1}{4}\frac{1}{4}\frac{1}{4})B(-\frac{3}{4}\frac{3}{4})C(\frac{1}{4}\frac{1}{4}\frac{1}{4})$	$A(--\frac{3}{4})B(\frac{1}{4}\frac{1}{4}\frac{1}{4})C-\frac{3}{4}-)$
	$A(\frac{3}{4}\frac{3}{4}-)B(\frac{1}{4}\frac{1}{4}\frac{1}{4})C(\frac{3}{4}-\frac{3}{4})$;	$A(\frac{1}{4}\frac{1}{4}\frac{1}{4})B(\frac{3}{4}--)C(\frac{1}{4}\frac{1}{4}\frac{1}{4})$;
Cleavage	pf{100}, gd{111}	pf{111}
Twinning		{111}
Refl.	51%	45%
$H =$	$3\frac{1}{2}$–4	$4\frac{1}{2}$–$5\frac{1}{2}$
$D =$	5.0	4.5–4.8
Color	Light bronze	Violet gray
Streak	Light bronze	Gray
Luster	Metallic	Metallic
Fracture	Uneven	Uneven
Habit	Massive	Octahedral
Remarks	Commonly associated with pyrrhotite	

pleting a stacking sequence requires an even number of sheets, that is, two sets of sequences. The ratio of the number of filled tetrahedra to filled octahedra is 8 : 1. Eight tetrahedra form edge-shared clusters, and these in turn share corners to complete a three-dimensional network.

The structure of violarite is similar to that of pentlandite. The sheets of sulfur atoms are in the {111} plane and are stacked in the *ABC* sequence. Again, two layers alternate: one has one fourth of both tetrahedra and octahedra occupied, and the other has three fourths of the octahedra filled. The ratio of the number of filled tetrahedra to filled octahedra is thus decreased to 1 : 2. The tetrahedra are occupied by Fe and the octahedra by Ni. The tetrahedra are isolated; the octahedra form an edge-sharing network. This structure is commonly known as the *spinel structure*.

Chemistry. Figure 9.20 illustrates the extent of $Fe^{2+} \rightleftharpoons Ni^{2+}$ substitution in pentlandite at 400 °C. The gap in miscibility between $(Fe, Ni)_9S_8$ and $(Fe, Ni)_{1-x}S$ is due to the structural differences between these two solid solutions. At about 610 °C, pentlandite forms by reaction with nearly pure pyrrhotite and Ni_3S_2. Below 556 °C, Ni_3S_2 inverts to a stable mineral called heazlewoodite.

Pentlandite
(Fe,Ni)$_9$S$_8$

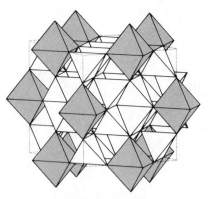

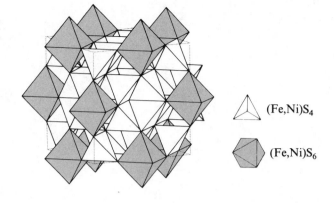

△ (Fe,Ni)S$_4$

⬡ (Fe,Ni)S$_6$

FIGURE II.75   Crystal structure of pentlandite.

Natural pentlandites may have up to 40 weight percent Co^{2+} in solid solution.

Violarite exhibits complete solubility with isostructural polydymite (Ni$_3$S$_4$) and siegenite (Co$_3$S$_4$) at moderate temperatures.

Occurrence and Associations. Pentlandite is the most important ore of Ni and is a relatively common mineral in massive sulfide deposits associated with mafic igneous intrusives. It is frequently associated with pyrite, chalcopyrite, and pyrrhotite. Exsolution of pentlandite within pyrrhotite is relatively common. Figure 9.20 summarizes some of the other associations of pentlandite as a function of bulk chemistry in the Fe-Ni-S system at 400 °C. Violarite is not shown, because it usually forms at lower temperature through weathering processes.

Distinguishing Features. Pentlandite looks much like pyrrhotite in hand specimen but has a yellow bronze streak and a good parting on {111}. Pentlandite is also nonmagnetic, whereas pyrrhotite generally is magnetic.

Sulfides With Unusual Coordination

Several important sulfide minerals have structures that contain metals in unusual coordination with sulfur. This feature is generally attributed to the variable character and degree of covalent bonding. The important sulfides in this group are molybdenite (MoS$_2$), millerite (NiS), cinnabar (HgS), covellite (CuS), chalcocite (Cu$_2$S), and argentite (Ag$_2$S).

Structure. The crystal structures of each of these sulfides is characterized by coordination polyhedra not found in the sulfide structures considered earlier. The structures are consequently more complicated and continual reference should be made to the appropriate structure diagrams. The structure of molybdenite (Figure II.76) has Mo coordinated to sulfur in trigonal prisms that are positioned directly over the threefold axis in the *A* and *B* symmetrically packed positions (Table 5.10). Tungstenite (WS$_2$) is isostructural also.

Molybdenite
MoS$_2$

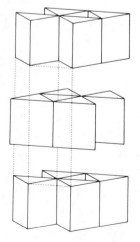

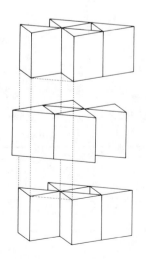

 MoS$_6$

FIGURE II.76   Crystal structure of molybdenite, *c* axis vertical.

SULFIDES WITH UNUSUAL COORDINATION

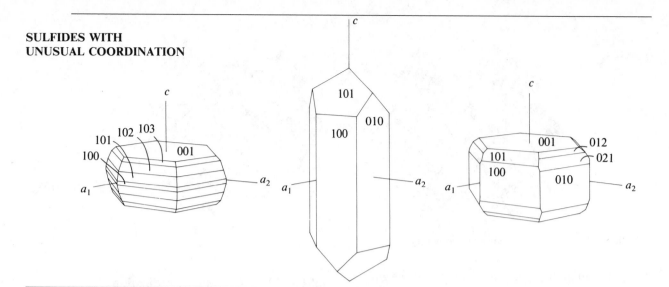

	MOLYBDENITE (*2H*)	MILLERITE	CINNABAR
Formula	MoS_2	NiS	HgS
Coord.	Mo(6) (trig. prism)	Ni(5) (tetr. pyram.)	Hg(6) (rhomb. domes)
System	Hexagonal	Trigonal	Hexagonal
$a =$	0.316	0.962	0.415
$c =$	1.232	0.315	0.950
$Z =$	2	9	3
Sp. group	$P6_3/m\,2/m\,2/m$	$R3m$	$P3_121$
Pt. group	$6/m\,2/m\,2/m$	$3m$	32
S; loops	3.00; 2	3.00; 3, 6	6.00; 2
SP plane	(001)	(001)*	(001)
SP struct.	$A(1-)A(---)$	$A(1-)$	$A(--1)D(--1)E(--1);$
	$B(1-)B(---);$		
Cleavage	pf(001)	pf{101}, pf{012}	pf{100}
Twinning			(001), {121} penetr.
RI	$n_\omega = 4.33$		$n_\omega = 2.90$
	$n_\epsilon = 2.03$		$n_\epsilon = 3.25$
$\delta =$	2.3		0.35
Sign	$(-)$		$(+)$
Refl.	$\omega = 33\%, \epsilon = 18\%$	54%	25%
Pleochr.		ω = yellow brown	ω = pale yellow
		ϵ = light yellow	ϵ = pink
$H =$	$1–1\frac{1}{2}$	$3–3\frac{1}{2}$	$2–2\frac{1}{2}$
$D =$	4.7	5.5	8.1
Color	Silver	Brass yellow	Red
Streak	Green gray	Green gray	Scarlet
Luster	Metallic	Metallic	Adamantine
Fracture	Flexible	Uneven	Subconchoidal
Habit	Foliated, flexible	Filiform	Tabular
Remarks	Several polytypes	Fibrous	Occasional coating
		*Corrugated sheet	

SULFIDES WITH UNUSUAL COORDINATION

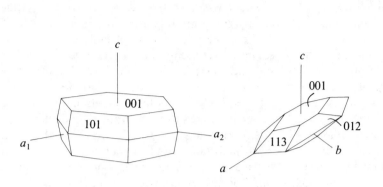
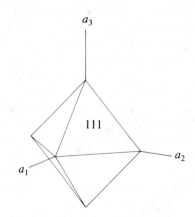

	COVELLITE	Chalcocite	Argentite
Formula	CuS	Cu_2S	Ag_2S
Coord.	Cu(3 and 4)	Cu(3 and 4)	Ag(4)
System	Hexagonal	Monoclinic	Isometric
$a =$	0.380	1.525	0.489
$b =$		1.188	
$c =$	1.636	1.349	
$\beta =$		116.35°	
$Z =$	6	48	2
Sp. group	$P6_3/m\,2/m\,2/c$	$P2_1/c$	$I4/m\,\bar{3}\,2/m$
Pt. group	$6/m\,2/m\,2/m$	$2/m$	$4/m\,\bar{3}\,2/m$
S; loops	2.75; 3, 4 (IV)	4.00; 3 (IV)	3.00; 2, 4
SP plane	(001)	(102)	(001)
SP struct.	$A--(-1-)B-1(-1-)$	$A-\frac{2}{3}(\frac{2}{3}\frac{2}{3}-)B-\frac{2}{3}(\frac{2}{3}\frac{2}{3}-)$	$\sqrt{4/3}\ S(\frac{1}{3}\frac{1}{3})\frac{1}{3}\frac{1}{3}\frac{1}{3}\frac{1}{3}$
	$A--(---)A--(--1)$	$A-\frac{2}{3}(\frac{2}{3}\frac{2}{3}-)B-\frac{2}{3}(\frac{2}{3}\frac{2}{3}-);$	$D(\frac{1}{3}\frac{1}{3})\frac{1}{3}\frac{1}{3}\frac{1}{3}\frac{1}{3}$
	$C-1(--1)A--(---);$		
Cleavage	pf(001)	pr{110}	pr{100}
Twinning		{110}, {032}, {112}	{111}
RI	$n_\omega = 1.60$		
	$n_\epsilon = 1.45$		
$\delta =$	0.15		
Sign	(+)		
Refl.	$\omega = 15\%, \epsilon = 24\%$	32%	33%
Pleochr.	ω = darker blue		
	ϵ = lighter blue		
$H =$	$1\frac{1}{2}$–2	$2\frac{1}{2}$–3	2–$2\frac{1}{2}$
$D =$	4.6	5.8	7.1
Color	Blue	Blue white	Lead gray
Streak	Dark gray	Dark gray	Shiny black
Luster	Submetallic	Metallic	Metallic
Fracture	Conchoidal	Conchoidal	Subconchoidal
Habit	Platy	Tabular	Octahedral, filiform
Remarks	Tarnish purple	Cu content = 1.96: djurelite	< 180 °C: acanthite
	Se ⇌ S: klockmannite		

The structure of millerite (Figure II.77) is based on an *AA* stacking. The hexagonal sheets undulate and the voids of the trigonal prisms are changed to tetragonal pyramids. Consequently, the coordination of Ni in millerite is fivefold.

The structure of cinnabar, HgS (Figure II.78), is based on an *ADE* stacking of sulfur sheets, which produces voids with the shape of tetragonal disphenoids. Mercury occupies these voids. The structure of covellite, CuS (Figure II.79), is a combination of double tetrahedral layers and layers of copper in triangular coordination with sulfur. The layered structure accounts for the perfect {001} cleavage of covellite. The structure of chalcocite is like that of covellite in that both have Cu atoms in triangular coordination with sulfur.

Chemistry. The chemistry of these sulfides shows little variation from the ideal formulas, possibly because the unusual coordination polyhedra characteristic of this group are unique to the bonding characteristics of the particular metals involved. A notable exception is millerite (NiS),

which may exhibit metal vacancies analogous to those in pyrrhotite.

Occurrence and Associations. All of the sulfides in this group are important ore minerals. Molybdenite is the principal ore of molybdenum, and is commonly concentrated in quartz veins with fluorite and topaz in association with granites. Associated ores are scheelite, Fe-sulfides, and Cu-sulfides. Millerite is a low-temperature sulfide that frequently forms as an alteration product of other nickel minerals in Ni-Co-Ag ores. Cinnabar is the most important ore of mercury, forming in near-surface environments from solutions associated with hot springs and volcanic activity. Pyrite, marcasite, and Cu-sulfides are commonly associated ores.

Covellite and chalcocite are frequently found together in association with bornite, pyrite, and chalcopyrite. Chalcocite is one of the most important copper ores, and is concentrated by precipitation from percolating solutions in near-surface environments or appears as disseminated grains in

Millerite
NiS

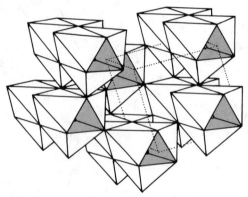 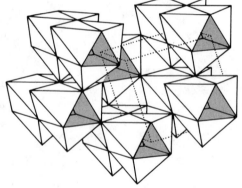

NiS$_5$

FIGURE II.77   Crystal structure of millerite, *c* axis vertical. Basal faces of the pyramids are shaded.

Cinnabar
HgS

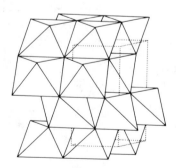 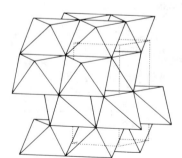

HgS$_6$

FIGURE II.78   Crystal structure of cinnabar, *c* axis vertical.

Covellite
CuS

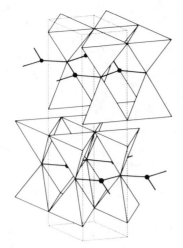
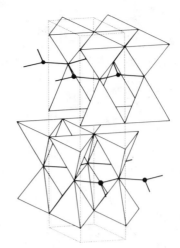
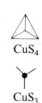

FIGURE II.79   Crystal structure of covellite, *c* axis vertical.

porphyritic igneous rocks. The latter occurrence is known as a *porphyry copper*.

Distinguishing Features. Molybdenite is easily distinguished by its softness, greasy feeling, and bright metallic luster. Molybdenite can only be mistaken for graphite, which has a black rather than greenish streak. Millerite is pale brass yellow in color and may form velvety or fine hairlike coatings on other minerals. Cinnabar is unmistakable with its vermillion red color and adamantine luster.

Covellite has a distinctive indigo blue color and micaceous cleavage in contrast with its common associate chalcocite, which has a lead gray color.

Complex Sulfides

In a number of important sulfide minerals, the bonding characteristics are so directional that the resulting structures can be quite complex. In some cases like pyrite, the covalent bonding between pairs of sulfur atoms is reasonably symmetric and

a symmetrically packed model is approached. In other structures such as that of stibnite, symmetrically packed subunits derived from simpler sulfides such as galena can be identified only within the larger structure. In a few cases such as realgar and orpiment, the bonding characteristics are entirely molecular, and the symmetrical stacking representation is no longer useful.

Structure. The structures of the various sulfides in this group are sufficiently complex that the structural models must be carefully studied if the structures are to be understood. The structures of pyrite (Figure II.80), ullmannite (Figure II.81), and cobaltite (Figure II.82) may be viewed as symmetrically packed in an *ABC* stacking if the S_2 molecules or the midpoints of each strong sulfur-sulfur (Sb—S and As—S) bond are positioned in the symmetrical sheet. The resulting octahedral voids are occupied by iron. This idealized structure is similar to that of galena. Because of the ordered distribution of the two different

Pyrite
FeS_2

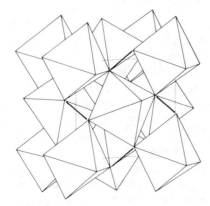
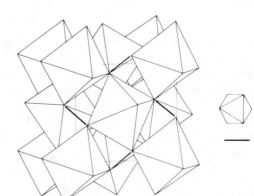

FIGURE II.80   Crystal structure of pyrite.

COMPLEX SULFIDES

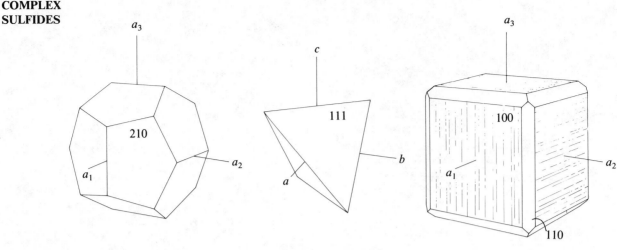

	PYRITE	ULLMANNITE	COBALTITE
Formula	FeS_2	$NiSbS$	$CoAsS$
Coord.	Fe(6)	Ni(6)	Co(6)
System	Isometric	Isometric	Orthorhombic
$a =$	0.542	0.588	0.558
$b =$			0.558
$c =$			0.558
$Z =$	4	4	4
Sp. group	$P2_1/a\bar{3}$	$P2_13$	$Pa2c$
Pt. group	$2/m\bar{3}$	23	$m2m$
SP plane	$\{111\}$	$\{111\}$	$\{111\}$
SP struct.	$A(--\frac{1}{2})B(\frac{1}{2}--)C(-\frac{1}{2}-);$	$A(--\frac{1}{2})B(\frac{1}{2}--)C(-\frac{1}{2}-);$	$A(--\frac{1}{2})B(\frac{1}{2}--)C(-\frac{1}{2}-);$
Cleavage	pr$\{100\}$	pr$\{100\}$	gd$\{100\}$
Twinning	Iron cross $\{110\}$		
Refl.	54%	42%	53%
$H =$	$6-6\frac{1}{2}$	$5-5\frac{1}{2}$	$5\frac{1}{2}$
$D =$	5.1	6.7	6.3
Color	Brass yellow	Steel gray	Tin white
Streak	Green gray	Gray	Gray black
Luster	Metallic	Metallic	Metallic
Fracture	Subconchoidal	Uneven	Uneven
Habit	Cubic, pyritohedral	Tetrahedral	Cuboctahedral
Remarks	Faces striated	Usually massive	Usually massive
	Pyroelectric	As$\rightleftharpoons$Sb: gersdorffite	Faces striated
	$PtAs_2$: sperrylite		

anions in ullmannite and cobaltite, the symmetries of these structures are lower than that of pyrite. The point group of pyrite is $2/m3$. The point groups of ullmannite and cobaltite are 23 and $m2m$, respectively. The structures of marcasite (Figure II.83), a polymorph of pyrite, and of arsenopyrite (Figure II.84) are based on an AB stacking of sulfur-sulfur (and As—S) pairs, and again, one half of the octahedral sites are occupied, resembling the structure of pyrrhotite. In skutterudite (Figure II.85), the covalent bond is between four As atoms. The interpretation of this structure in terms of symmetrical packing is complicated.

The structure of stibnite consists of bands of four incomplete Sb octahedra connected by covalent bonds to form a network (Figure II.86). The structure of realgar (Figure II.87) consists of distinct molecules of four As and four S atoms. The molecular unit in the orpiment structure is a layer that controls the (010) cleavage.

Occurrence and Associations. Pyrite is the most abundant and widespread of the sulfides, occurring over a broad range of temperatures and in a

Ullmannite
NiSbS

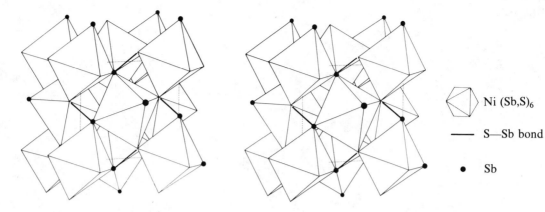

Ni (Sb,S)$_6$

— S—Sb bond

• Sb

FIGURE II.81   Crystal structure of ullmanite.

Cobaltite
CoAsS

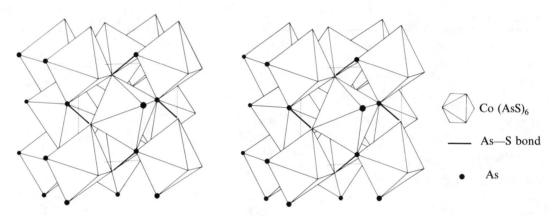

Co (AsS)$_6$

—— As—S bond

• As

FIGURE II.82   Crystal structure of cobaltite.

Marcasite
FeS$_2$

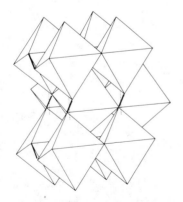
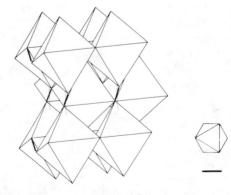

FeS$_6$

— S—S bond

FIGURE II.83   Crystal structure of marcasite, *b* axis vertical.

great variety of rocks. It is generally not mined for its iron, but rather for the gold and copper frequently associated with it. Marcasite is much less common, found typically as low-temperature concretions in limestone and shale. Arsenopyrite is an important ore of arsenic and is found chiefly in high-temperature veins associated with tin and tungsten ores. Pyrite, chalcopyrite, sphalerite, and galena are commonly associated with arsenopyrite.

COMPLEX SULFIDES

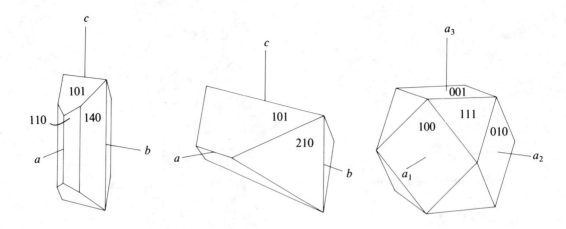

	MARCASITE	ARSENOPYRITE	SKUTTERUDITE
Formula	FeS_2	$FeAsS$	$(Co, Ni)As_3$
Coord.	Fe(6)	Fe(6)	Co(6), Ni(6)
System	Orthorhombic	Monoclinic	Isometric
$a =$	0.444	0.951	0.820
$b =$	0.541	0.565	
$c =$	0.338	0.642	
$\beta =$		90.00°	
$Z =$	2	8	8
Sp. group	$P2_1/n\,2_1/n\,2/m$	$B2_1/d$	$I2/m\bar{3}$
Pt. group	$2/m\,2/m\,2/m$	$2/m$	$2/m\bar{3}$
SP plane	(100)	(310)	{111}
SP struct.	$A(--\frac{1}{2})B(--\frac{1}{2})$;	$A(--\frac{1}{2})B(--\frac{1}{2})$ $A(--\frac{1}{2})B(--\frac{1}{2})$;	$A_{75}(--\frac{1}{4})B_{75}(\frac{1}{4}--)C_{75}(-\frac{1}{4}-)$;
Cleavage	pr{101}	gd(101)	pr{100}, pr{111}
Twinning	{101}	(100), (001), (101), {012}	{112}
Refl.	46%	52%	54%
Pleochr.	X = light brown Y = yellow Z = beige	X = yellow red Y = red yellow Z = white	
$H =$	$6-6\frac{1}{2}$	$5\frac{1}{2}-6$	$5\frac{1}{2}-6$
$D =$	4.9	6.1	6.1–6.8
Color	White green	Silver	Tin white
Streak	Gray black	Black	Black
Luster	Metallic	Metallic	Metallic
Fracture	Uneven	Uneven	Uneven
Habit	Cockscomb aggregate Also tabular	Prismatic	Cubic
Remarks	{$h0l$} faces curved $As \rightleftharpoons S$: loellingite	Prism. striation a, b, c same as pyrite $Sb \rightleftharpoons As$: gudmundite	May be skeletal

COMPLEX SULFIDES

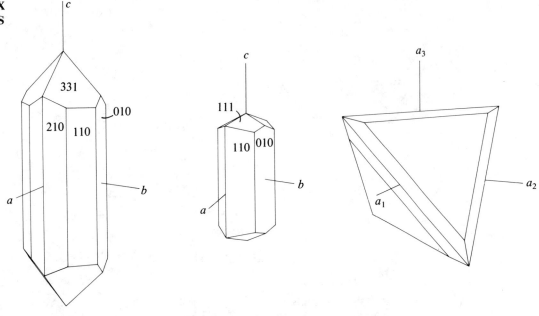

	STIBNITE	Boulangerite	Tetrahedrite
Formula	Sb_2S_3	$Pb_5Sb_4S_{11}$	$Cu_{12}Sb_4S_{13}$
Coord.	Sb(7)	Pb(7), Sb(3)	Cu(4 and 3)
System	Orthorhombic	Monoclinic	Isometric
$a =$	1.122	2.156	1.034
$b =$	1.130	2.351	
$c =$	0.384	0.809	
$\beta =$		100.8°	
$Z =$	4	8	2
Sp. group	$P2_1/b\ 2_1/n\ 2_1/m$	$P2/a$	$I\bar{4}3m$
Pt. group	$2/m\ 2/m\ 2/m$	$2/m$	$\bar{4}3m$
Cleavage	pf(010)	gd(100)	
Twinning	{130}		{111}
RI	$n_\alpha = 3.194$		$n = 2.72$
	$n_\beta = 4.046$		
	$n_\gamma = 4.303$		
$\delta =$	1.11		
$2V =$	26°		
Sign	(−)		
Refl.	25–38%	35%	24%
Pleochr.	X = light gray	X = green gray	
	Y = darker gray	Y = light gray	
	Z = bright white	Z = light gray	
H =	2	$2\frac{1}{2}$–3	3–4
D =	4.6	6.0–6.2	4.5–5.1
Color	Lead gray	Purple gray	Silver
Streak	Lead gray	Gray	Brown, black
Luster	Metallic	Metallic	Metallic
Fracture	Subconchoidal	Conchoidal	Subconchoidal
Habit	Prismatic	Prismatic, tabular	Tetrahedral
Remarks	Crystals bent and striated	Similar composition: jamesonite	Sb ⇌ As: tennatite
			(Cu, Ag): freibergite

Arsenopyrite
FeAsS

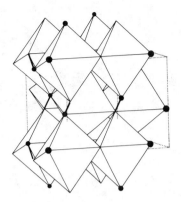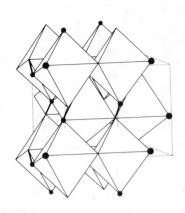

Fe(AsS)$_6$

—— As—S bonds

● As

FIGURE II.84 **Crystal structure of arsenopyrite, *b* axis vertical.**

Skutterudite
(Co,Ni)As$_3$

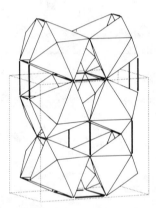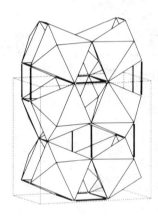

(Co,Ni)As$_3$

—— As—As bonds

FIGURE II.85 **Crystal structure of skutterudite, *c* axis vertical.**

Stibnite
Sb$_2$S$_3$

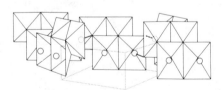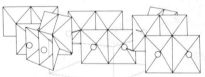

S$_5$

○ Sb

FIGURE II.86 **Crystal structure of stibnite, *c* axis vertical.**

Realgar
AsS

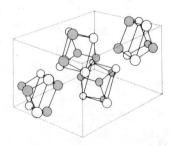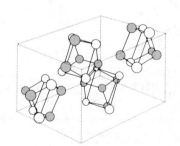

○ S

◍ As

FIGURE II.87 **Crystal structure of realgar, *c* axis vertical.**

**COMPLEX
SULFIDES**

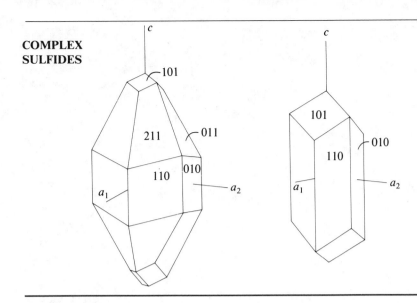

	Proustite	Pyrargyrite
Formula	Ag_3AsS_3	Ag_3SbS_3
Coord.	Ag(2), As(3)	Ag(2), Sb(3)
System	Trigonal	Trigonal
$a =$	1.076	1.106
$c =$	0.866	0.873
$Z =$	6	6
Sp. group	$R3c$	$R3c$
Pt. group	$3m$	$3m$
Cleavage	gd{101}	gd{101}
Twinning	{101}, {104}	{101}, {104}
RI	$n_\omega = 2.98$	$n_\omega = 3.08$
	$n_\epsilon = 2.71$	$n_\epsilon = 2.88$
$\delta =$	0.17	0.20
Sign	(−)	(−)
Refl.	$\omega = 26\%, \epsilon = 22\%$	$\omega = 36\%, \epsilon = 35\%$
Pleochr.	$\omega =$ white yellow	$\omega =$ lighter gray
	$\epsilon =$ gray	$\epsilon =$ darker gray
$H =$	$2–2\frac{1}{2}$	2
$D =$	5.57	5.85
Color	Ruby red	Ruby red
Streak	Red	Red purple
Luster	Adamantine	Adamantine
Fracture	Subconchoidal	Subconchoidal
Habit	Prismatic, rhombohedral	Prismatic
Remarks	Often referred to as the "ruby silvers"	

Stibnite is the principal ore of antimony. It is usually concentrated in low-temperature hydrothermal veins in association with galena, sphalerite, and both realgar and orpiment. The latter two minerals often appear together in lead and silver deposits with stibnite and other sulfide minerals.

Distinguishing Features. Reference should be made to the physical data tables for a direct comparison of the numerous properties of these sulfides. Pyrite is pale brass yellow in color and forms distinctive cubes that are usually striated. Marcasite is lighter colored than pyrite and forms radiating "cockscomb" masses in limestone. Arsenopyrite has a silver white color. Realgar is recognized by its reddish orange color. Orpiment is bright yellow.

COMPLEX SULFIDES

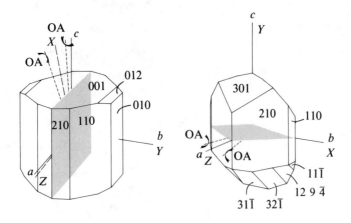

	REALGAR	Orpiment
Formula	As_4S_4	As_2S_3
Coord.	AsS molecules	Molecular layers
System	Monoclinic	Monoclinic
$a =$	0.929	1.149
$b =$	1.353	0.959
$c =$	0.657	0.425
$\beta =$	106.55°	90.45°
$Z =$	4	4
Sp. group	$P2_1/n$	$P2_1/n$
Pt. group	$2/m$	$2/m$
Cleavage	gd(010)	pf(010)
Twinning	(100)	
RI	$n_\alpha = 2.538$	$n_\alpha = 2.40$
	$n_\beta = 2.864$	$n_\beta = 2.81$
	$n_\gamma = 2.704$	$n_\gamma = 3.02$
$\delta =$	0.166	0.62
Sign	$(-)$	$(+)$
$2V =$	41°	76°
Transp.	Transp. to opaque	Transp. to opaque
Refl.	22–27%	22–27%
$H =$	$1\frac{1}{2}$–2	$1\frac{1}{2}$–2
$D =$	3.56	3.49
Color	Red orange	Yellow orange
Streak	Orange red	Yellow
Luster	Resinous	Resinous
Fracture	Conchoidal	Even
Habit	Short prismatic	Foliated, curved
Remarks	Almost always occur together	

THE HALIDE MINERALS

The halides consist of minerals in which a halogen element is an essential anion. All chlorides, fluorides, bromides, and iodides are included. Perhaps 90 halides are known, but collectively they constitute far less than 0.1% of the earth's crust. Only halite (NaCl) and fluorite (CaF_2) are common.

Structure. The structures in the various halide minerals depend to a large extent on the identity of the cation (or cations) and on the bonding to the halogen anion. Na^{1+} (halite) and K^{1+} (sylvite) form perfect ionic bonds with Cl^{1-} to yield a highly symmetric structure, namely, an *ABC* packing of chlorine with all of the octahedral voids occupied by monovalent cations (Figure II.88).

THE HALIDE MINERALS

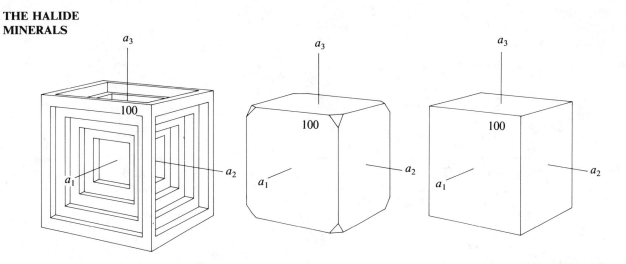

	HALITE	Sylvite	Chlorargyrite
Formula	NaCl	KCl	AgCl
Coord.	Na(6)	K(6)	Ag(6)
System	Isometric	Isometric	Isometric
$a =$	0.56404	0.629	0.555
$Z =$	4	4	4
Sp. group	$F4/m\,\bar{3}\,2/m$	$F4/m\,\bar{3}\,2/m$	$F4/m\,\bar{3}\,2/m$
Pt. group	$4/m\,\bar{3}\,2/m$	$4/m\,\bar{3}\,2/m$	$4/m\,\bar{3}\,2/m$
S; loops	6.00; 2	6.00; 2	6.00; 2
SP plane	{111}	{111}	{111}
SP struct.	$A(--1)B(1--)C(-1-)$	$A(--1)B(1--)C(-1-)$	$A(--1)B(1--)C(-1-)$
	$A(--1)B(1--)C(-1-)$;	$A(--1)B(1--)C(-1-)$;	$A(--1)B(1--)C(-1-)$;
SPI =	74	74	74
Cleavage	pf{100}	pf{100}	pr{100}
RI	$n = 1.5446$	$n = 1.490$	$n = 2.071$
Transp.	Transp. to transl.	Transp. to transl.	Transp. to transl.
H =	$2\frac{1}{2}$	2	$2\frac{1}{2}$
D =	2.16	1.99	5.55
Color	Colorless, variable	Colorless, variable	Colorless, gray
Streak	White	White	White
Luster	Vitreous	Vitreous	Resinous
Fracture	Conchoidal	Uneven	Subconchoidal
Habit	Cubic	Cubic	Cubic
Remarks	Saline taste	Bitter taste	Sectile
	Table salt	Br $\rightleftharpoons$ Cl: bromargyrite	Violet tarnish

Halite
NaCl

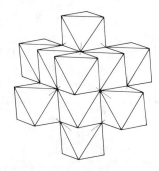 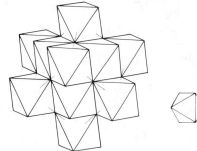

FIGURE II.88   Crystal structure of halite.

THE HALIDE MINERALS

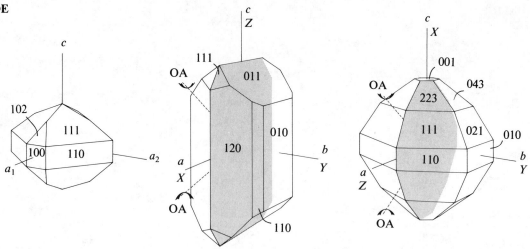

	CALOMEL	Atacamite	Carnallite
Formula	HgCl	$Cu_2Cl(OH)_3$	$KMgCl_3 \cdot 6H_2O$
Coord.	Hg(5) (tetr. pyram.)	Cu(6)	Mg(6), K(6)
System	Tetragonal	Orthorhombic	Orthorhombic
$a =$	0.446	0.602	0.956
$b =$		0.915	1.605
$c =$	1.091	0.685	2.256
$Z =$	2	4	12
Sp. group	$I4/m\,2/m\,2/m$	$P2_1/n\,2/a\,2/m$	$P2/b\,2/a\,2/n$
Pt. group	$4/m\,2/m\,2/m$	$2/m\,2/m\,2/m$	$2/m\,2/m\,2/m$
S; loops	5.00; 2	3.00; 3, 6	2.00; 4
SP plane	(001)	101	?
SP struct.	$S1(--)D1(--)$	$A(--\frac{2}{3})B(\frac{2}{3}--)C(-\frac{2}{3}-);$	$A_{75}(--\frac{1}{4})B_{75}(\frac{1}{4}--)C_{75}(-\frac{1}{4}-)$
	$E1(--)F1(--);$		$A_{75}(--\frac{1}{4})B_{75}(\frac{1}{4}--)C_{75}(-\frac{1}{4}-);$
SPI =	62	45	49
Cleavage	gd{100}	pf(010), gd{101}	
Twinning	{110}	{110} rare	
RI	$n_\omega = 1.973$	$n_\alpha = 1.831$	$n_\alpha = 1.467$
	$n_\epsilon = 2.656$	$n_\beta = 1.861$	$n_\beta = 1.474$
		$n_\gamma = 1.880$	$n_\gamma = 1.496$
$\delta =$	0.683	0.049	0.029
$2V =$		75°	70°
Sign	(+)	(−)	(+)
Transp.	Translucent	Transp. to transl.	Transp. to transl.
H =	1–2	$3–3\frac{1}{2}$	$2\frac{1}{2}$
D =	6.48	3.76	1.602
Color	White, yellow gray	Green	Colorless
Streak	White	Green	White, red
Luster	Adamantine	Adamantine	Vitreous
Fracture	Conchoidal	Conchoidal	Conchoidal
Habit	Tabular, pyramidal,	Acicular, prismatic	Pyramidal
		Striated prism faces	Pseudohexagonal
Remarks	Coating on crystals	Occasionally fibrous	Isostr. w. perovskite
			Bitter taste

THE HALIDE MINERALS

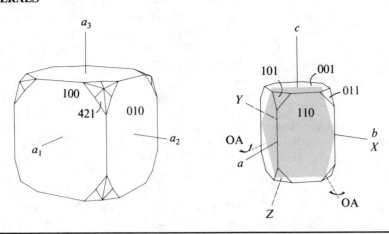

	FLUORITE	Cryolite
Formula	CaF_2	Na_2AlNaF_6
Coord.	Ca(8)	Na(12), Al(6), Na(6)
System	Isometric	Monoclinic
$a =$	0.5463	0.546
$b =$		0.560
$c =$		0.780
$\beta =$		90.18°
$Z =$	4	2
Sp. group	$F4/m\,\bar{3}\,2/m$	$P2_1/n$
Pt. group	$4/m\,\bar{3}\,2/m$	$2/m$
S; loops	4.00; 6	2.00; 4
SP plane	$\{100\}$	$\{111\}$
SP struct.	$S(\frac{1}{2})S(\frac{1}{2})$;	$A_{75}(--\frac{1}{4})B_{75}(\frac{1}{4}--)C_{75}(-\frac{1}{4}-)$
		$A_{75}(--\frac{1}{4})B_{75}(\frac{1}{4}--)C_{75}(-\frac{1}{4}-)$;
SPI =	60	60
Cleavage	pf$\{111\}$	
Twinning	$\{111\}$	(001), $\{110\}$, (101)
RI	$n = 1.434$	$n_\alpha = 1.3385$
		$n_\beta = 1.3389$
		$n_\gamma = 1.3396$
$\delta =$		0.0011
$2V =$		43°
Sign		(+)
Transp.	Transparent	Transp. to transl.
$H =$	4	$2\frac{1}{2}$
$D =$	3.18	2.97
Color	Colorless, variable	Colorless
Streak	White	White
Luster	Vitreous	Vitreous
Fracture	Conchoidal, splintery	Uneven
Habit	Cubic, octahedral	Pseudocubic, lamellar
Remarks	Fluorescent (U.V.)	Distorted perovskite structure
	Color zones common	

As the cations become heavier and less electropositive, the bonding to halogen anions becomes less ionic, and the structures become less symmetric. Calomel (HgCl) has a unique structure consisting of an *SDDS* stacking (Figure II.89) of Cl^{1-}, which creates voids having the form of tetragonal pyramids. Mercury occupies these voids in fivefold coordination with Cl^{1-}.

The carnallite structure is complicated and not well understood. Available data indicate that Mg^{2+} has octahedral coordination with six H_2O molecules and that K^{1+} is in octahedral coordination with six Cl^{1-} anions.

Fluorite has an *SS* packing of F^{1-}, which creates cubic voids (Figure II.90), and half of these are occupied by Ca^{2+}. Each Ca^{2+} thus has eight flourine neighbors, and each F^{1-} has four

calcium neighbors. In the case of CsCl, all cubic voids are occupied, and the cation coordination around the anion is eightfold. Although Cs-halides are not known in nature, the structure may be important deep within the earth's mantle.

Chemistry. Halite exhibits solid solution toward sylvite (KCl) in the same manner as the solid solutions of Na-feldspars and K-feldspars. At elevated temperature, the solvus between NaCl and KCl closes, and complete stability exists between them. At low temperature, the two minerals are nearly pure NaCl and KCl.

Mutual anion substitution of Cl^{1-}, Br^{1-}, and I^{1-} gives rise to bromargyrite (AgBr) and iodargyrite (AgI). Complete solid solution exists between cerargyrite (AgCl) and bromargyrite.

Occurrence and Associations. The alkali halides

Calomel
$KMgCl_3 \cdot 6H_2O$

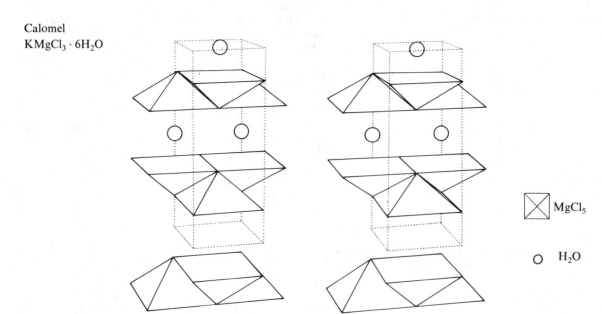

$\boxtimes$ MgCl$_5$

$\bigcirc$ H$_2$O

FIGURE II.89 **Crystal structure of calomel, *c* axis vertical.**

Fluorite
CaF_2

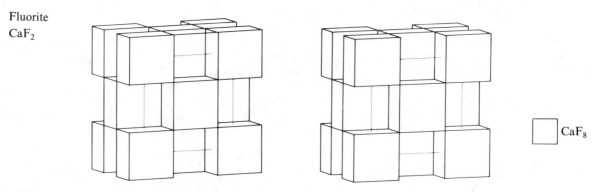

$\square$ CaF$_8$

FIGURE II.90 **Crystal structure of fluorite.**

occur chiefly in sedimentary rocks where they form by precipitation from seawater or alkaline brines. During the formation of marine evaporite deposits, halite is among the first minerals to precipitate along with gypsum. With further evaporation, ion concentrations increase until sylvite, carnallite, anhydrite, polyhalite, and a number of other salts become stable.

Calomel is a secondary mineral formed in the oxidized parts of mercury ore deposits and is usually associated with cinnabar, pyrite, quartz, and calcite. Cerargyrite also occurs as a secondary mineral, but in hydrothermal veins associated with argentite (AgS), pyrargyrite (Ag_3SbS_3), native silver, limonite, and manganese oxides.

Fluorite is frequently found in lead-silver veins associated with barite, quartz, and dolomite, as a common accessory mineral in the hydrothermal alteration of granites, and less frequently as a cement in sandstones associated with calcite, dolomite, and gypsum. Atacamite forms as an alteration product of copper ores, mainly in arid regions where it is associated with malachite, cuprite, and gypsum.

Distinguishing Features. Halite may be distinguished by its cubic form, perfect {100} cleavage, and saline taste. It may also form distinctive hopper-shaped crystals that have stepped, concave faces. Sylvite is similar to halite but has a more bitter taste. Carnallite also has a bitter taste but is commonly reddish in color due to hematite inclusions.

Fluorite is distinguished by its perfect octahedral {111} cleavage, cubic form, and vitreous luster. It is harder than calcite and unaffected by dilute HCl. The color of fluorite is highly variable from colorless to purple, blue, green, yellow, and brown.

Calomel commonly has a pale yellow color. It is usually massive but occasionally occurs as small crystals with perfect cleavage. Atacamite has a distinctive dark green color and is associated with other copper minerals.

THE OXIDE MINERALS

The minerals in this relatively large group consist of one or more metal cations bonded to either oxygen or hydroxyl anions. The structural properties that result are generally different from those found in mineral groups such as the silicates, sulfates, or carbonates, for example, in which an anionic group dominates the structure. The SiO_2 minerals, although oxides in the strict chemical sense, are thus more appropriately discussed with the silicates.

The usual classification of oxides and hydroxides is based on the ratio of cations to oxygen or hydroxyl in the standard chemical formula. In this scheme, the simple oxides are X_2O, XO, and X_2O_3 types (X is the metal cation), and the multiple oxides are of the XY_2O_4 type (Y denotes a metal different from X). In essence, this classification is based on the valence of the metal cation (or cations) and conveys little essential information regarding structural systematics. As an example, periclase (MgO) is an XO-type oxide, and corundum (Al_2O_3) is an X_2O_3-type, but in each the cations (Mg^{2+} and Al^{3+}) occupy octahedral voids in a symmetrically packed arrangement of oxygens.

We prefer a classification of oxides and hydroxides that follows the classification system of the sulfides discussed earlier. Accordingly, we divide the oxides and hydroxides into the following subgroups based on their coordination polyhedra.

Tetrahedral Oxides

Zincite (ZnO) is the one well-known oxide in which the metal cation occupies only the tetrahedral void. The zincite structure has an AB stacking of oxygens in which the tetrahedral voids are fully occupied. Zincite is isostructural with wurtzite (ZnS, see Figure II.71) and with bromellite (BeO).

Zincite exhibits some solid solution toward Mn^{2+}, which probably imparts the reddish color. Pure ZnO is white. Zincite is a rare mineral, found only in abundance in zinc deposits at Franklin, New Jersey. The typical occurrence is as rounded blebs with calcite, franklinite, and willemite.

Octahedral Oxides

Numerous important oxides have only the octahedral voids occupied by cations. Among them are hematite, an important iron ore, and ilmenite, a major source of titanium metal. Cassiterite and pyrolusite are the most important ores of tin and manganese, respectively.

Structure. The structure of periclase (MgO) is based on an ABC packing of oxygens in which the octahedral voids are fully occupied by Mg^{2+} cations. This arrangement gives sixfold coordination for both Mg^{2+} and O^{2-}, and in fact, the periclase structure is identical to that of halite (Figure II.88), as we note from their identical packing symbols.

Corundum (Al_2O_3) and hematite (Fe_2O_3) are isostructural with an AB stacking of oxygens in which two thirds of the octahedral voids are occupied by the same cation (Figure II.91). The octahedra are slightly distorted in these structures in

TETRAHEDRAL OXIDES

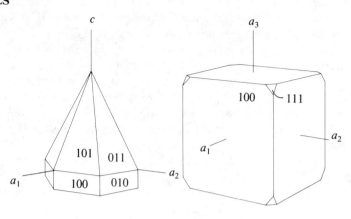

	Zincite	Periclase
Formula	ZnO	MgO
Coord.	Zn(4)	Mg(6)
System	Hexagonal	Isometric
$a =$	0.325	0.421
$c =$	0.519	
$Z =$	2	4
Sp. group	$P6_3mc$	$F4/m\bar{3}2/m$
Pt. group	$6mm$	$4/m\bar{3}2/m$
S; loops	3.00; 3	6.00; 2
SP plane	(001)	{111}
SP struct.	$A(-1-)B(1--)$	$A(--1)B(1--)C(-1-)$
		$A(--1)B(1--)C(-1-)$;
SPI =	74	74
Cleavage	pf(001)	pf{100}, pr{111}
Parting	{110}	
Twinning		{111}
RI	$n_\omega = 2.013$	$n = 1.736$
	$n_\epsilon = 2.029$	
$\delta =$	0.016	
Sign	(+)	
Transp.	Transp. to transl.	Transp. to transl.
H =	$4-4\frac{1}{2}$	$5\frac{1}{2}$
D =	5.4–5.7	3.56
Color	Red, orange yellow	Colorless, gray
Streak	Orange yellow	Orange yellow
Luster	Subadamantine	Vitreous
Fracture	Subconchoidal	Uneven
Habit	Hex. pyramidal	Octahedral
Remarks	Isostr. w. wurtzite	Isostr. w. halite
	$Cu \rightleftharpoons Zn$: tenorite	$Fe \rightleftharpoons Mg$: wüstite

contrast to the regular octahedra in halite. The reason can be explained (1) by the face sharing between some octahedra, which causes the O—O distances to decrease on the shared faces, and (2) by the expansion of the octahedra toward the vacant octahedral sites. Ilmenite has a nearly identical structure (Figure II.92), except that Fe^{2+} and Ti^{4+} cations are ordered in alternating octahedral layers, each of which is two thirds occupied. The AB stacking sequence is the same as in corundum and hematite, but the point group symmetry is reduced from $\bar{3}2/m$ to $\bar{3}$.

OCTAHEDRAL OXIDES

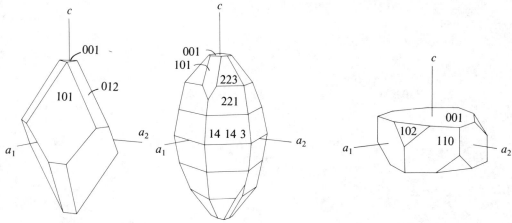

	HEMATITE	Corundum	ILMENITE
Formula	Fe_2O_3	Al_2O_3	$FeTiO_3$
Coord.	Fe(6)	Al(6)	Fe(6), Ti(6)
System	Trigonal (rhombohedral)	Trigonal (rhombohedral)	Trigonal (rhombohedral)
$a =$	0.504	0.495	0.508
$c =$	1.376	1.378	1.408
$Z =$	6	6	6
Sp. group	$R\bar{3}2/c$	$R\bar{3}2/c$	$R\bar{3}$
Pt. group	$\bar{3}2/m$	$\bar{3}2/m$	$\bar{3}$
S; loops	3.00; 2, 3, 6	3.00; 2, 3, 6	3.00; 2, 3, 6
SP plane	(001)	(001)	(001)
SP struct.	$A(--\frac{2}{3})B(--\frac{2}{3})$	$A(--\frac{2}{3})B(--\frac{2}{3})$	$A(--\frac{2}{3})B(--\frac{2}{3})*$
	$A(--\frac{2}{3})B(--\frac{2}{3})$	$A(--\frac{2}{3})B(--\frac{2}{3})$	$A(--\frac{2}{3})B(--\frac{2}{3})*$
	$A(--\frac{2}{3})B(--\frac{2}{3})$;	$A(--\frac{2}{3})B(--\frac{2}{3})$;	$A(--\frac{2}{3})B(--\frac{2}{3})$; *
SPI =	71	70	78
Parting	(001), {101}	{100}, {101}	{100}, {101}
Twinning	(001), {101}	{101}, {101}	(001), {101}
RI	$n_\omega = 3.22$	$n_\omega = 1.768$	$n_\omega = {>}\,2.7$
	$n_\epsilon = 2.96$	$n_\epsilon = 1.760$	$n_\epsilon = {>}\,2.7$
$\delta =$	0.28	0.008	
Sign	$(-)$	$(-)$	$(-)$
Transp.	Transl. to opaque	Transp. to transl.	Opaque
Refl.	28%		20%
Pleochr.	White to gray blue in reflected light	$\omega =$ dark blue $\epsilon =$ light blue, green	
$H =$	$5\frac{1}{2}$–$6\frac{1}{2}$	9	5–6
$D =$	4.9–5.3	3.9–4.1	4.5–5
Color	Steel gray, red	Brown, variable	Iron black
Streak	Red	White	Black, brown
Luster	Submetallic	Adamantine	Metallic
Fracture	Subconchoidal	Uneven	Subconchoidal
Habit	Botryoidal, tabular	Botryoidal, foliated	Prismatic, tabular
Remarks	Flakes flexible	Blue: sapphire	Slightly magnetic
	Platy: red ocher	Red: ruby	Derivative of the
	Metallic: specularite	Isostr. w. hematite	hematite structure
			*Occupied by Ti

**OCTAHEDRAL
OXIDES**

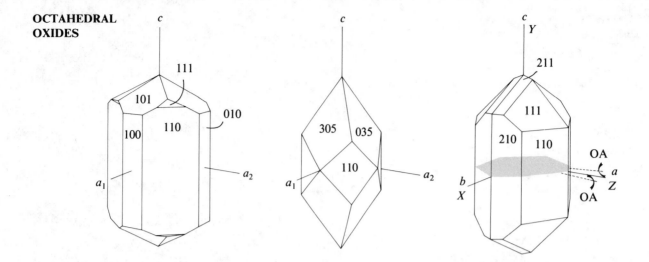

	RUTILE	ANATASE	BROOKITE
Formula	TiO_2	TiO_2	TiO_2
Coord.	Ti(6)	Ti(6)	Ti(6)
System	Tetragonal	Tetragonal	Orthorhombic
$a =$	0.459	0.368	0.918
$b =$			0.545
$c =$	0.296	0.951	0.515
$Z =$	2	4	8
Sp. group	$P4_2/m\ 2_1/n\ 2/m$	$I4_1/m\ 2/a\ 2/d$	$P2_1/b\ 2_1/c\ 2_1/a$
Pt. group	$4/m\ 2/m\ 2/m$	$4/m\ 2/m\ 2/m$	$2/m\ 2/m\ 2/m$
S; loops	3.00; 2, 3, 4	3.00; 2, 3, 4	3.00; 2, 3, 4
SP plane	{110}	111	(001)
SP struct.	$A(--\frac{1}{2})B(--\frac{1}{2})$ $A(--\frac{1}{2})B(--\frac{1}{2})$;	$A(--\frac{1}{2})B(\frac{1}{2}--)C(-\frac{1}{2}-)$;	$A(--\frac{1}{2})B(\frac{1}{2}--)$ $C(\frac{1}{2}--)B(--\frac{1}{2})$;
SPI =	72	65	69
Cleavage	gd{100}, gd{110}, pr{111}	pf{001}, pf{111}	pr{110}, pr(001)
Twinning	{101}, {301}	{112}	
RI	$n_\omega = 2.90$ $n_\epsilon = 2.61$	$n_\omega = 2.561$ $n_\epsilon = 2.488$	$n_\alpha = 2.583$ $n_\beta = 2.585$ $n_\gamma = 2.72$
$\delta =$	0.29	0.073	0.14
$2V =$			0–30°
Sign	(+)	(−)	(+)
Transp.	Transp. to transl.	Transp. to transl.	Transp. to transl.
H =	$6–6\frac{1}{2}$	$5\frac{1}{2}–6$	$5–5\frac{1}{2}$
D =	4.24	3.90	4.14
Color	Red brown	Orange brown	Brown, black
Streak	Light brown	Light yellow	White, gray
Luster	Submetallic, adamantine	Adamantine	Submetallic, adamantine
Fracture	Subconchoidal	Subconchoidal	Subconchoidal
Habit	Prismatic, acicular	Dipyramidal, tabular	Dipyramidal, prismatic
Remarks	Inclusion in "star sapphire" and quartz		SP structure distorted

OCTAHEDRAL OXIDES

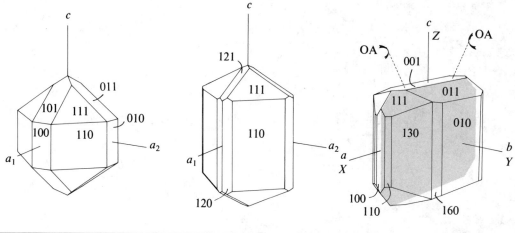

	Cassiterite	Pyrolusite	COLUMBITE
Formula	SnO_2	MnO_2	$(Fe, Mn)(Nb, Ta)_2O_6$
Coord.	Sn(6)	Mn(6)	Fe(6), Mn(6), Nb(6), Ta(6)
System	Tetragonal	Tetragonal	Orthorhombic
$a =$	0.474	0.440	0.510
$b =$			1.427
$c =$	0.319	0.287	0.574
$Z =$	4	2	2
Sp. group	$P4_2/m\ 2_1/n\ 2/m$	$P4_2/m\ 2_1/n\ 2/m$	$P2_1/b\ 2/c\ 2_1/n$
Pt. group	$4/m\ 2/m\ 2/m$	$4/m\ 2/m\ 2/m$	$2/m\ 2/m\ 2/m$
S; loops	3.00; 2, 3, 4	3.00; 2, 3, 4	3.00; 2, 3, 4
SP plane	{110}	{110}	(010)
SP struct.	$A(--\frac{1}{2})B(--\frac{1}{2})$	$A(--\frac{1}{2})B(--\frac{1}{2})$	$A(--\frac{1}{2})B(--\frac{1}{2})$
	$A(--\frac{1}{2})B(--\frac{1}{2})$;	$A(--\frac{1}{2})B(--\frac{1}{2})$;	$A(--\frac{1}{2})*B(--\frac{1}{2})$
			$A(--\frac{1}{2})B(--\frac{1}{2})$; *
SPI $=$	71	73	68
Cleavage	gd{100}, pr{111}	pf{110}	gd(010)
Twinning	{011}		{101}
RI	$n_\omega = 2.006$		$n_\alpha = 2.44$
	$n_\epsilon = 2.097$		$n_\beta = 2.32$
			$n_\gamma = 2.38$
$\delta =$	0.091		0.12
$2V =$			75°
Sign	(+)		(+)
Transp.	Transp. to transl.	Opaque	Transl. to opaque
Refl.		30–50%	17%
$H =$	6–7	$2\frac{1}{2}$–$6\frac{1}{2}$	5
$D =$	7.00	4.5–5.0	6.0
Color	Red brown, black	Black	Black, brown
Streak	Gray, brown	Black	Brown
Luster	Adamantine	Metallic	Submetallic
Fracture	Subconchoidal	Uneven	Subconchoidal
Habit	Acicular, needlelike	Dendritic, prismatic	Prismatic
Remarks	Fibrous variety: "wood tin"	Good crystals: polianite	*Occupied by Nb, Ta cations Derivative of the rutile structure

Isostructural with rutile

Hematite
Fe_2O_3

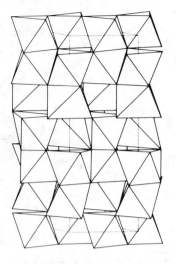
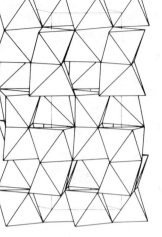

FeO$_6$

FIGURE II.91   Crystal structure of hematite, *c* axis vertical.

Ilmenite
$FeTiO_3$

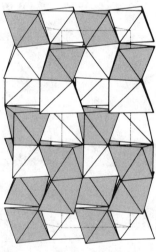
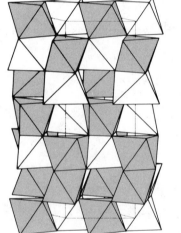

FeO$_6$

TiO$_6$

FIGURE II.92   Crystal structure of ilmenite, *c* axis vertical.

The structure of rutile (TiO_2) is based on an *AB* stacking of oxygens in which one half of the octahedral voids are occupied by Ti^{4+}. Rutile (Figure II.93) is isostructural with stishovite, the high-pressure polymorph of SiO_2 in which Si^{4+} has sixfold coordination in edge-shared octahedra. Both cassiterite and pyrolusite are isostructural with rutile.

Rutile has two polymorphs, anatase (Figure II.94) and brookite (Figure II.95). The polymorphic structures also contain hexagonal sheets of oxygens but with different stacking sequences as indicated by their stacking symbols. Although one half of the octahedral voids are occupied in all three structures, the pattern of occupancy differs.

In rutile, two edges of every octahedron are shared, in brookite three edges are shared, and in anatase four edges of every octahedron are shared.

The structure of columbite and its isostructural minerals is based on an *AB* stacking sequence of oxygens in which one half of the octahedral voids are occupied (Figure II.96). The stacking sequence is the same as in rutile and stishovite, and their structures are similar insofar as the networks consist of edge-sharing single subchains. Because of the two different cations in columbite, the symmetry of the structure is reduced to orthorhombic.

Wolframite (($Fe, Mn)WO_4$) is practically isostructural with columbite, except that the tung-

Rutile
TiO$_2$

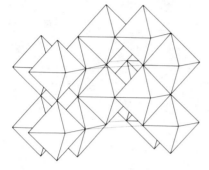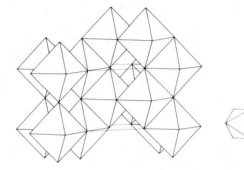

TiO$_6$

FIGURE II.93 **Crystal structure of rutile, *c* axis vertical.**

Anatase
TiO$_2$

TiO$_6$

FIGURE II.94 **Crystal structure of anatase, *c* axis vertical.**

Brookite
TiO$_2$

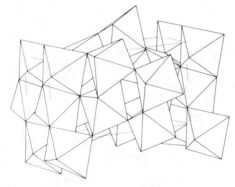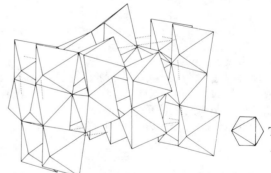

TiO$_6$

FIGURE II.95 **Crystal structure of brookite, *c* axis vertical.**

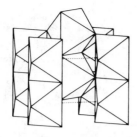

sten octahedron is irregular, having four shortest bond lengths. The structure thus approaches the tetrahedral coordination typical of the tungstates. Consequently, we treat the structure of wolframite in more detail under the category of the tungstate minerals (p. 444).

Chemistry. Periclase (MgO) exhibits substantial solid solubility toward ideal wüstite (FeO) as a result of the Mg^{2+} ⇌ Fe^2 substitution. Wüstite appears to be nearly always cation deficient, meaning that not all of the octahedral voids are occupied as implied in the ideal structure. The

same nonstoichiometry is found in the octahedral iron sulfide, pyrrhotite.

Corundum shows little solubility toward other components, but its physical properties are dramatically affected by substitutions. Very small amounts of Cr^{3+} substituting for Al^{3+} account for the deep red color of ruby because of the crystal field effect (Chapter 4, p. 106). Small amounts of Fe^{2+} and Ti^{4+} in corundum produce sapphire.

Hematite exhibits complete solid solubility toward ilmenite at temperatures above approximately 800 °C. Below 800 °C, a solvus exists. At

Columbite
$(Fe,Mn)(Nb,Ta)_2O_6$

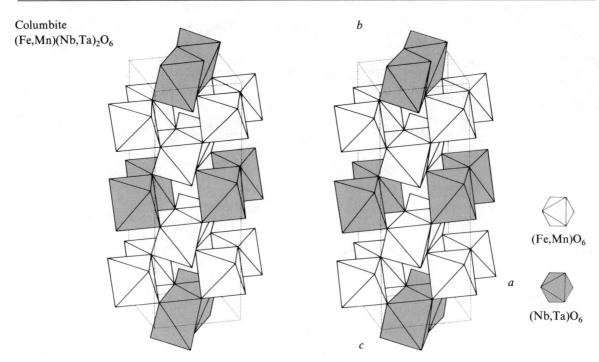

$(Fe,Mn)O_6$

$(Nb,Ta)O_6$

FIGURE II.96   Crystal structure of columbite, *b* axis vertical.

600 °C, ferrian ilmenite (Hem_{20}) coexists with titanhematite ($\sim Hem_{80}$), usually as exsolution of one in the other. The entire solid solution series shows significant contents of the pyrophanite ($MnTiO_3$) and geikielite ($MgTiO_3$) end-members, depending on bulk chemistry.

Rutile and its polymorphs are essentially pure, showing solubility chiefly toward Fe_2O_3 and rarely toward Nb_2O_5, Cr_2O_3, and V_2O_3. Columbite and tantalite exhibit $Fe^{2+} \rightleftharpoons Mn^{2+}$ substitution in one of the octahedral sites and $Ta^{5+} \rightleftharpoons Nb^{5+}$ substitution in the other.

Occurrence and Associations. Periclase is a relatively rare mineral found in metamorphosed dolomites in association with calcite and forsterite at relatively high temperatures. Corundum is also found in metamorphosed impure limestones, but more commonly is found as an accessory mineral in silica-undersaturated igneous rocks such as nepheline syenites and certain feldspathic pegmatites.

Hematite is a widespread weathering product of other Fe-bearing minerals. Much of the world's iron production comes from sedimentary iron formations associated with chert, and from laterite deposits formed by weathering and oxidation in tropical regions. Hematite is also a common cement in sandstones, especially in redbeds and associated clastic sediments.

Anatase is the low-temperature polymorph of TiO$_2$ found typically in hydrothermal veins, low-grade schists, and in miarolitic cavities in granite associated with quartz, adularia, and chlorite. Rutile forms instead of anatase in higher grade metamorphic rocks, where it is associated with ilmenite and magnetite, or with magnetite and hematite. Rutile is a common mineral in eclogites (Figure 9.12), which form at high pressure in association with diopside, enstatite, and garnet.

Cassiterite is most commonly found in high-temperature hydrothermal veins in granites and pegmatites. It is the most common ore of tin and is frequently found in placer deposits. Common associations are tourmaline, apatite, and fluorite. Associated ores are wolframite and molybdenite. Pyrolusite is found generally in low-temperature sedimentary environments where it precipitates from solution as nodules on the seafloor or as coatings or layers in bogs and lake bottoms.

Columbite is a relatively rare mineral found in granitic pegmatites as well-developed crystals associated with tourmaline, apatite, beryl, muscovite, and alkali feldspars.

Distinguishing Features. Periclase is distinguished by its vitreous luster, gray white color, and occurrence as small, rounded grains in high-grade marbles. Corundum resembles periclase in hand specimens but differs in its great hardness and prismatic or barrel-shaped crystals. Rutile commonly is found as black, vertically striated needles within single crystals of quartz and corundom (star sapphire). Rutile may also have distinctive knee-shaped twins not found in either of its polymorphs.

Cassiterite also develops knee-shaped twins, but more typically, it occurs as reddish brown to black, short, prismatic crystals. Fibrous concretionary masses of cassiterite are known as "wood tin." Pyrolusite is generally black and has a dull to earthy luster, especially in the common powdery form that soils one's fingers.

Mixed Tetrahedral and Octahedral Oxides

An important group of oxides are those in which both tetrahedral and octahedral voids are partially occupied by metal cations. Spinel ($MgAl_2O_4$), magnetite (Fe_3O_4), and a number of isostructural oxides are in this group, and all have the *spinel structure type*. Both tetrahedral and octahedral voids cannot be fully occupied in this group, because that would necessitate face-sharing of occupied voids. Electrostatic repulsion of the cations and the resulting high-energy state prevents this from happening.

Structure. Figure II.97 illustrates the basic spinel structure in which oxygens are in closed-packed sheets stacked parallel to {111} in an *ABC* sequence. Comparing the structure with the symmetrical stacking symbol, we see that tetrahedra occupy alternate layers with half as many octahedra, and the intervening layers consist of octahedra only. Because occupied tetrahedra and octahedra tend not to share faces, the ratio of tetrahedral to octahedral sites in the spinel structure type is $1:2$.

The general formula for "normal" spinel minerals is XY_2O_4. X and Y are cations of different valences in the ratio $X/Y = 1:2$. Since the structure has twice as many Y cations as X cations, the "normal" distribution is for the X cation to occupy the tetrahedral site and the Y cations to occupy the two octahedral sites. Examples are chromite ($FeCr_2O_4$) and hercynite ($FeAl_2O_4$). Several spinel minerals, however, have an inverted cation distribution with all of the X cations and half of the Y cations filling half of the octahedral sites, and the remaining half of the Y cations filling the tetrahedral site. These spinels are referred to as *inverse spinels* and have the general formula $Y[XY]O_4$ in which the bracket denotes cations in octahedral coordination. Examples of inverse spinels are magnetite (Fe_3O_4 or $Fe^{3+}[Fe^{2+}Fe^{3+}]O_4$), spinel ($Al^{3+}[Mg^{2+}Al^{3+}]O_4$), and ulvospinel ($Fe^{2+}[Fe^{2+}Ti^{4+}]O_4$). Most natural spinels have cation distributions between these two extremes.

Chemistry. With the spinel structures, extensive solid solution exists between the end-members listed in Table II.4. There is continuous solid

TABLE II.4. *Classification of Normal and Inverse Spinel Structures*

Mineral Name	Formula	Structure
Chromite	$Fe^{2+}[Cr^{3+}]_2O_4$	N
Magnesiochromite	$Mg^{2+}[Cr^{3+}]_2O_4$	N
Hercynite	$Fe^{2+}[Al^{3+}]_2O_4$	N
Galaxite	$Mn^{2+}[Al^{3+}]_2O_4$	N
Franklinite	$Zn^{2+}[Fe^{3+}]_2O_4$	N
"Silicate Spinel"	$Si^{4+}[Mg^{2+}]_2O_4$	N
Magnetite	$Fe^{3+}[Fe^{2+}Fe^{3+}]O_4$	I
Magnesioferrite	$Fe^{3+}[Mg^{2+}Fe^{3+}]O_4$	I
Jacobsite	$Fe^{3+}[Mn^{2+}Fe^{3+}]O_4$	I
Spinel	$Al^{3+}[Mg^{2+}Al^{3+}]O_4$	I
Ulvospinel	$Fe^{2+}[Fe^{2+}Ti^{4+}]O_4$	I
"Silicate Spinel"	$Mg^{2+}[Si^{4+}Mg^{2+}]O_4$	I

NOTE: N = normal. I = inverse.

Spinel
$MgAl_2O_4$

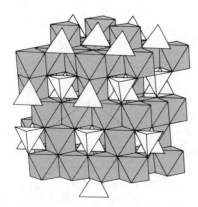
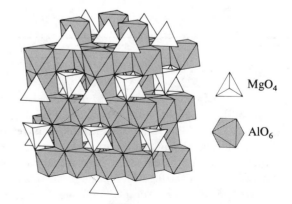

MgO$_4$

AlO$_6$

FIGURE II.97 Crystal structure of spinel.

**MIXED TETRAHEDRAL
AND OCTAHEDRAL
OXIDES**

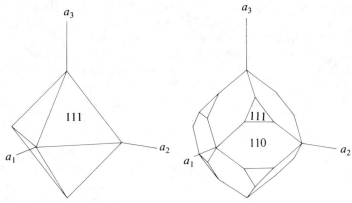

	SPINEL	Magnetite
Formula	$MgAl_2O_4$	$FeFe_2O_4$
Coord.	Mg(4), Al(6)	Fe(4 and 6)
System	Isometric	Isometric
$a =$	0.809	0.8397
$Z =$	8	8
Sp. group	$F4/d\,\bar{3}\,2/m$	$F4/d\,\bar{3}\,2/m$
Pt. group	$4/m\,\bar{3}\,2/m$	$4/m\,\bar{3}\,2/m$
S	1.00	1.00
SP plane	$\{111\}$	$\{111\}$
SP struct.	$A(--\frac{3}{4})B(\frac{1}{4}\frac{1}{4}\frac{1}{4})C(-\frac{3}{4}-)$	$A(--\frac{3}{4})B(\frac{1}{4}\frac{1}{4}\frac{1}{4})C(-\frac{3}{4}-)$
	$A(\frac{1}{4}\frac{1}{4}\frac{1}{4})B(\frac{3}{4}--)C(\frac{1}{4}\frac{1}{4}\frac{1}{4})$;	$A(\frac{1}{4}\frac{1}{4}\frac{1}{4})B(\frac{3}{4}--)C(\frac{1}{4}\frac{1}{4}\frac{1}{4})$;
SPI =	70	73
Cleavage	pr$\{111\}$	
Twinning	$\{111\}$	$\{111\}$
RI	$n = 1.74$	
Transp.	Transp. to transl.	Opaque
Refl.		22%
H =	$7\frac{1}{2}$–8	$5\frac{1}{2}$–$6\frac{1}{2}$
D =	3.5–4.0	5.20
Color	Red, variable	Black
Streak	White	Black
Luster	Vitreous	Metallic
Fracture	Conchoidal	Subconchoidal
Habit	Octahedral	Octahedral
Remarks		Highly magnetic
		$\{110\}$ striated

Members of the *spinel* group

solution between magnetite and ulvospinel by virtue of the coupled substitution $2Fe^{3+} \rightleftharpoons Ti^{4+} + Fe^{2+}$. This is the same substitution that relates hematite and ilmenite in a continuous solid solution. The continuous series between spinel and hercynite is due to the $Mg^{2+} \rightleftharpoons Fe^{2+}$ substitution.

Cation deficient spinels, maghemite and titanomaghemite, are formed by progressive oxidation of magnetite and any of the titanomagnetites in the solid solution series toward ulvospinel.

Occurrence and Associations. The spinel minerals are common accessories in igneous and metamorphic rocks, and are found as detrital grains in most clastic sediments. Magnetite is the most abundant of the group and can form important iron ore deposits. Spinel is found in high-grade

**MIXED TETRAHEDRAL
AND OCTAHEDRAL
OXIDES**

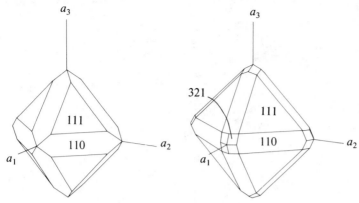

	Chromite	Franklinite
Formula	$FeCr_2O_4$	$ZnFe_2O_4$
Coord.	Fe(4), Cr(6)	Zn(4), Fe(6)
System	Isometric	Isometric
$a =$	0.837	0.843
$Z =$	8	8
Sp. group	$F4/d\,\bar{3}\,2/m$	$F4/d\,\bar{3}\,2/m$
Pt. group	$4/m\,\bar{3}\,2/m$	$4/m\,\bar{3}\,2/m$
S	1.00	1.00
SP plane	$\{111\}$	$\{111\}$
SP struct.	$A(--\frac{3}{4})B(\frac{1}{4}\frac{1}{4}\frac{1}{4})C(-\frac{3}{4}-)$	$A(--\frac{3}{4})B(\frac{1}{4}\frac{1}{4}\frac{1}{4})C(-\frac{3}{4}-)$
	$A(\frac{1}{4}\frac{1}{4}\frac{1}{4})B(\frac{3}{4}--)C(\frac{1}{4}\frac{1}{4}\frac{1}{4})$;	$A(\frac{1}{4}\frac{1}{4}\frac{1}{4})B(\frac{3}{4}--)C(\frac{1}{4}\frac{1}{4}\frac{1}{4})$;
SPI =	69	62
Parting		$\{111\}$
Twinning	$\{111\}$	
RI	$n = 2.16$	$n = 2.36$
Refl.	13%	15%
H =	$5\frac{1}{2}$	$5\frac{1}{2}$–$6\frac{1}{2}$
D =	5.10	5.32
Color	Black, brown	Black
Streak	Brown	Black, red brown
Luster	Metallic	Metallic
Fracture	Conchoidal	Conchoidal
Habit	Octahedral	Octahedral
Remarks	Occ. magnetic	

Members of the *spinel* group

metamorphic rocks and in mantle peridotites. Chromite occurs chiefly in ultramafic igneous rocks, and like magnetite, forms important ore deposits.

With few exceptions, franklinite is found only with zincite and willemite in limestone at the Franklin, New Jersey, zinc deposits.

Several silicate spinels (e.g., Mg_2SiO_4 or $SiMg_2O_4$) have been synthesized under high pressure and are believed to represent stable phases in the earth's lower mantle.

Distinguishing Features. Spinel may be distinguished by its vitreous luster, octahedral form, and white streak. Magnetite is black, has a metallic luster, yields a black streak, and is magnetic. Franklinite is black, slightly magnetic, but yields a reddish brown streak. Chromite has a submetallic luster and is commonly massive.

**MIXED TETRAHEDRAL
AND OCTAHEDRAL
OXIDES**

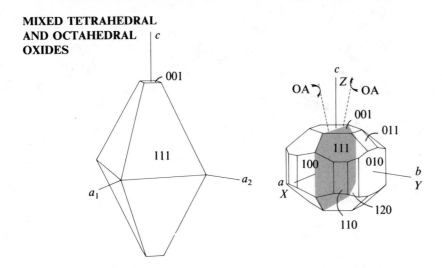

	Hausmannite	Chrysoberyl
Formula	$MnMn_2O_4$	Al_2BeO_4
Coord.	Mn(4 and 6)	Al(6), Be(4)
System	Tetragonal	Orthorhombic
$a =$	0.576	0.424
$b =$		0.939
$c =$	0.944	0.547
$Z =$	4	4
Sp. group	$I4_1/a\,2/m\,2/d$	$P2_1/b\,2_1/n\,2_1/m$
Pt. group	$4/m\,2/m\,2/m$	$2/m\,2/m\,2/m$
S	1.00	1.00
SP plane	$\{112\}$	(100)
SP struct.	$A(--\frac{3}{4})B(\frac{1}{4}\frac{1}{4}\frac{1}{4})C(-\frac{3}{4}-)$	$A(\frac{1}{8}\frac{1}{8}\frac{1}{2})B(\frac{1}{8}\frac{1}{8}\frac{1}{2});$
	$A(\frac{1}{4}\frac{1}{4}\frac{1}{4})B(\frac{3}{4}--)C(\frac{1}{4}\frac{1}{4}\frac{1}{4});$	
SPI =	64	69
Cleavage	pf(001)	gd$\{011\}$, pr(010)
Twinning	$\{112\}$, $\{101\}$	$\{031\}$
RI	$n_\omega = 2.46$	$n_\alpha = 1.747$
	$n_\epsilon = 2.15$	$n_\beta = 1.748$
		$n_\gamma = 1.757$
$\delta =$	0.31	0.010
$2V =$		45°
Sign	$(-)$	$(+)$
Transp.	Transl. to opaque	Transp. to transl.
Refl.	17%	
Pleochr.		X = light pink
		Y = pale yellow
		Z = pale green
$H =$	$5-5\frac{1}{2}$	$8\frac{1}{2}$
$D =$	4.856	3.7–3.8
Color	Brown, black	Green, yellow
Streak	Light brown	White
Luster	Submetallic	Vitreous
Fracture	Uneven	Subconchoidal
Habit	Pseudo-octahedral	Tabular
Remarks	Distorted spinel struct.	Isostr. w. olivine
		(001) striated

**CUBIC
OXIDES**

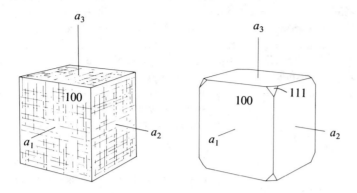

	PEROVSKITE	Uraninite
Formula	$CaTiO_3$	UO_2
Coord.	Ti(6), Ca(12)	U(2)(+6)*
System	Orthorhombic	Isometric
$a =$	0.537	0.54682
$b =$	0.764	
$c =$	0.544	
$Z =$	4	4
Sp. group	$P2_1/c\ 2_1/m\ 2_1/n$	$F4/m\ \bar{3}\ 2/m$
Pt. group	$2/m\ 2/m\ 2/m$	$4/m\ \bar{3}\ 2/m$
S; loops	2.00; 4	
SP plane	$\{111\}$	(100)
SP struct.	$A_{75}(--\frac{1}{4})B_{75}(\frac{1}{4}--)C_{75}(-\frac{1}{4}-);$	$S(\frac{1}{2})S(\frac{1}{2});$ (distorted)
SPI =	69	55
Cleavage	pr$\{100\}$	
Twinning	$\{111\}$	
RI	$n_\alpha = 2.34$	
	$n_\beta = 2.34$	
	$n_\gamma = 2.34$	
$\delta =$	0.002	
$2V =$	90°	
Sign	$(-)$	
Transp.	Transp. to opaque	Opaque
Refl.	16%	14%
H =	$5\frac{1}{2}$	5–6
D =	4.0	7–9.5
Color	Brown, black	Black
Streak	Light brown	Brown, black
Luster	Adamantine, metallic	Metallic
Fracture	Subconchoidal	Conchoidal
Habit	Cubic, reniform	Cubic, massive: pitchblende
Remarks	Faces striated	$Th \rightleftharpoons U$: thorianite
	Contains Ce	*Close second neighbor bonds

Cubic Oxides

Uraninite (UO_2) and thorianite (ThO_2) are the only important oxides that have oxygen in an *SS* stacking sequence with metals occupying the cu-bic voids. The structure is the same as that of fluorite (Figure II.90).

A complete solid solution exists between UO_2 and ThO_2, and significant amounts of Pb, Ce, and Ra may substitute in the structure. Lead and helium exist in the structure by virtue of the radioactive decay of ^{235}U and ^{232}Th. Uraninite is an important source of uranium and may be distinguished by its black, pitchy luster from which the name *pitchblende* is derived. Uraninite occurs in high-temperature hydrothermal veins commonly associated with sulfides of Sn, Fe, Cu, and As. It may also occur as single crystals in granitic pegmatites and as detrital grains in clastic sedimentary rocks.

Mixed Octahedral and Cuboctahedral Oxides

In perovskite ($CaTiO_3$), the titanium cations occupy one quarter of the octahedral voids in an *ABC* sequence of symmetrically packed oxygens in which 75% of each sheet is occupied (Figure II.98). One quarter of the octahedral voids are filled with Ti^{4+}; the tetrahedral voids are vacant. This octahedral frame is similar to that of halite (Figure II.88) with alternating octahedra removed. The octahedra share corners in an arrangement

that creates large cuboctahedral voids where calcium is located.

Although perovskite is essentially pure $CaTiO_3$, small amounts of other constituents can be accommodated by the structure. The most important of these are the rare earths, notably Ce but also La, Y, Ta, and Nb. Perovskite occurs in some metamorphosed impure limestones associated with wollastonite, spinel, and larnite, and as an accessory mineral in nepheline syenites. Perovskite also occurs in kimberlite pipes associated with diamonds.

Oxides With Unusual Coordination

Cuprite (Cu_2O) and psilomelane ($Mn_5(Ba, H_2O)O_{10}$) are oxides with unusual coordination. The structure of cuprite has Cu coordinated to only two oxygens (Figure II.99), and each oxygen has four Cu atoms surrounding it. Two independent networks of four-membered CuO_2 loops result.

Perovskite
$CaTiO_3$

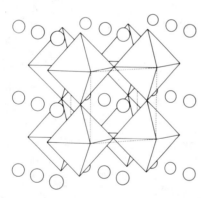
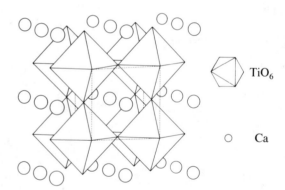

TiO$_6$

Ca

FIGURE II.98   Crystal structure of perovskite, *c* axis vertical.

Cuprite
Cu_2O

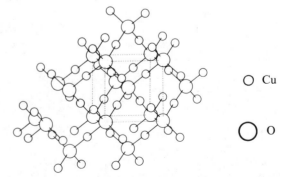

Cu

O

FIGURE II.99   Crystal structure of cuprite.

OXIDES WITH UNUSUAL COORDINATION

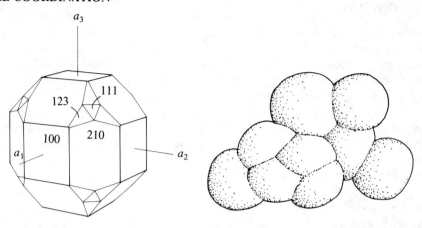

	CUPRITE	Romanechite (psilomelane)
Formula	Cu_2O	$Ba(OH)_4Mn^{2+} Mn_8^{4+}O_{16}$
Coord.	Cu(2)	Ba(10), Mn(6)
System	Isometric	Monoclinic
$a =$	0.427	0.956
$b =$		0.288
$c =$		1.385
$\beta =$		90.50
$Z =$	2	2
Sp. group	$P4_2/n\,\bar{3}\,2/m$	$A2/m$
Pt. group	$4/m\,\bar{3}\,2/m$	$2/m$
S; loops	4.00; 6 (Cn = 2)	3.00; 2, 4, 10
SP plane	{100}	(100)
SP struct.	$\sqrt{4/3}\,S(11)D(11);$	$A(--\frac{2}{3})B(--\frac{1}{3})$
		$A(--\frac{1}{3})C(--\frac{1}{3});$
SPI =	33	73
Cleavage	pr{111}	
RI	$n = 2.849$	
$\delta =$		High
Refl.	23%	22–24%
H =	$3\frac{1}{2}$–4	5–6
D =	5.9–6.1	3.5–4.7
Color	Red	Black
Streak	Brown red	Brown black
Luster	Submetallic	Submetallic
Fracture	Conchoidal	Conchoidal
Habit	Cubic, octahedral	Reniform, botryoidal, dendritic
Remarks	"Copper ruby"	Usually associated w. cryptomelane,
	Fibrous: chalcotrichite	$K(Mn^{4+}, Mn^{2+})_8O_{16}$ and hollandite, $Ba(Mn^{4+}, Mn^{2+})_8O_{16}$
	Cu located on void edges	

Cuprite is typically found as shiny red octahedra or dodecahedra associated with malachite and azurite in the oxidized zone of copper veins and ore deposits. Cuprite may also be massive or earthy and in some localities is an important copper ore.

Psilomelane has an unusual structure consisting of large tunnels bounded by four double chains of octahedra that run parallel to the b axis. Oxygen, Ba^{2+}, and H_2O occupy the tunnel sites.

Psilomelane commonly occurs with pyrolusite and other manganese ore minerals.

THE HYDROXIDE MINERALS

Minerals in the hydroxide group have hydroxyl $(OH)^{1-}$ as their essential anion. All of the important minerals in this group such as gibbsite $(Al(OH)_3)$ and brucite $(Mg(OH)_2)$ have hydroxyl anions in either an *AB* or *ABC* sequence of symmetrical sheets. Only the hexagonal voids are occupied; the tetrahedral voids are always vacant.

The dioctahedral and trioctahedral layers found in the layer silicate structures are the mineralogical equivalents of the hydroxide minerals gibbsite and brucite. Because each layer unit is electrically neutral, the layers are held together in the respective structures by weak residual bonds. Both gibbsite and brucite therefore have perfect {001} cleavage.

Structure. The brucite structure (Figure II.100)

is based on an *AB* stacking sequence of hydroxyl ions in which all of the octahedral voids are occupied by Mg^{2+} in the *AB* layers and the *BA* layers are vacant. Because all three octahedra around each $(OH)^{1-}$ in a layer are occupied, the layer is said to be *trioctahedral*. In gibbsite (Figure II.101), hydroxyl ions are in an *ABBA* stacking sequence. In the *AB* and *BA* layers, two thirds of the octahedral voids are occupied by Al^{3+}, and the *AA* and *BB* layers are vacant. Because only two of the three octahedral voids around each $(OH)^{1-}$ are filled, the layer is said to be *dioctahedral*.

In manganite $(MnO(OH))$, both oxygen and (OH) are in symmetrical sheets, and the structure is very similar to that of rutile (Figure II.93). All of these structures are based on an *AB* stacking sequence in which one half of the octahedral voids are occupied and form edge-shared subchains.

Brucite
$Mg(OH)_2$

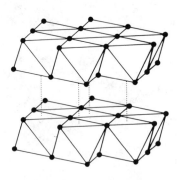 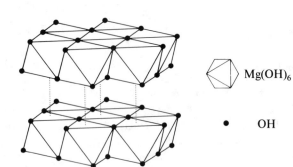

FIGURE II.100   Crystal structure of brucite, *c* axis vertical.

Gibbsite
$Al(OH)_2$

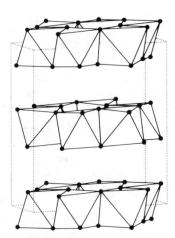 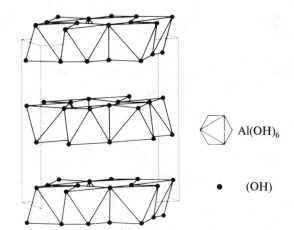

FIGURE II.101   Crystal structure of gibbsite, *c* axis vertical.

THE HYDROXIDE MINERALS

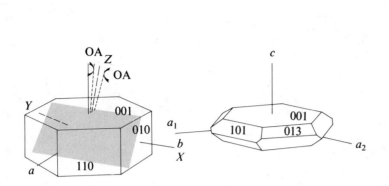

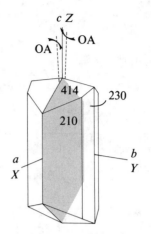

	GIBBSITE	BRUCITE	Manganite
Formula	$Al(OH)_3$	$Mg(OH)_2$	$MnO(OH)$
Coord.	$Al(6)$	$Mg(6)$	$Mn(6)$
System	Monoclinic	Trigonal	Monoclinic
$a =$	0.8641	0.3147	0.884
$b =$	0.507		0.523
$c =$	0.9719	0.4769	0.574
$\beta =$	94.57°		90.0°
$Z =$	8	1	8
Sp. group	$P2_1/n$	$P\bar{3}\,2/m\,1$	$B2_1/d$
Pt. group	$2/m$	$32/m$	$2/m$
S; loops	2.00; 2, 6	3.00; 2	3.00; 2, 3, 4
SP plane	(001)	(001)	(100)
SP struct.	$A(--\frac{2}{3})B(---)$ $B(--\frac{2}{3})A(---)$;	$A(--1)B(---)$;	$A(--\frac{1}{2})B(--\frac{1}{2})$;
SPI =	39	38	59
Cleavage	pf(001)	pf(001)	pf(010), gd{110}
Twinning	{310}, (001)		{011}
RI	$n_\alpha = 1.57$ $n_\beta = 1.57$ $n_\gamma = 1.59$	$n_\omega = 1.57$ $n_\epsilon = 1.58$	$n_\alpha = 2.24$ $n_\beta = 2.24$ $n_\gamma = 2.53$
$\delta =$	0.02	0.02	0.29
$2V =$	0–40°		Small
Sign	(+)	(+)	(+)
Transp.	Transp. to transl.	Transp. to transl.	Opaque
Refl.			15%
H =	$2\frac{1}{2}$–$3\frac{1}{2}$	$2\frac{1}{2}$	4
D =	2.40	2.4–2.5	4.2–4.4
Color	White, gray	White, green	Black, gray
Streak	White	White	Red brown
Luster	Pearly, vitreous	Pearly, vitreous	Submetallic
Fracture	Uneven, tough	Sectile	Sectile
Habit	Tabular, foliated	Tabular, foliated	Prismatic
Remarks	Pseudohexagonal Major comp. in bauxite	Fibrous: nemalite (Occ. w. biaxial optics)	Isostr. w. rutile (Distorted)

THE HYDROXIDE MINERALS

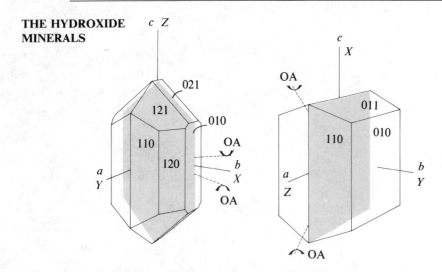

	GOETHITE	Diaspore
Formula	FeO(OH)	AlO(OH)
Coord.	Fe(6)	Al(6)
System	Orthorhombic	Orthorhombic
$a =$	0.465	0.442
$b =$	1.002	0.940
$c =$	0.304	0.284
$Z =$	4	4
Sp. group	$P2_1/b\ 2_1/n\ 2_1/m$	$P2_1/b\ 2_1/n\ 2_1/m$
Pt. group	$2/m\ 2/m\ 2/m$	$2/m\ 2/m\ 2/m$
S; loops	3.00; 2, 3, 6	3.00; 2, 3, 6
SP plane	(001)	(001)
SP struct.	$A(--\frac{1}{2})B(--\frac{1}{2})$;	$A(--\frac{1}{2})B(--\frac{1}{2})$;
SPI =	56	57
Cleavage	pf(010)	pf(010), pr 210
RI	$n_\alpha = 2.26\text{–}2.27$	$n_\alpha = 1.68\text{–}1.71$
	$n_\beta = 2.39\text{–}2.41$	$n_\beta = 1.71\text{–}1.72$
	$n_\gamma = 2.40\text{–}2.52$	$n_\gamma = 1.73\text{–}1.75$
$\delta =$	0.15	0.04
$2V =$	0–27°	85°
Sign	(−)	(+)
Transp.	Transl. to opaque	Translucent
Refl.	14%	
Pleochr.	X = yellow	
	Y = yellow brown	
	Z = orange to olive	
H =	5–5½	6½–7
D =	4.3	3.2–3.5
Color	Yellow, brown	Colorless, yellow, brown
Streak	Brown yellow	White yellow
Luster	Subadamantine	Vitreous, pearly
Fracture	Uneven	Conchoidal
Habit	Prismatic, fibrous, reniform	Prismatic, platy, massive
Remarks	Mn ⇌ Fe: groutite	Isostr. w. goethite
	Double octahedral chains	Major comp. in bauxite

THE HYDROXIDE MINERALS

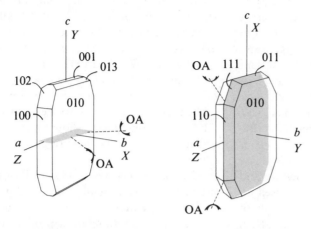

	LEPIDOCROCITE	Boehmite
Formula	FeO(OH)	AlO(OH)
Coord.	Fe(6)	Al(6)
System	Orthorhombic	Orthorhombic
$a =$	0.388	0.369
$b =$	1.254	1.22
$c =$	0.307	0.286
$Z =$	4	4
Sp. group	$A2_1/m\,2/a\,2/m$	$A2/m\,2/a\,2/m$
Pt. group	$2/m\,2/m\,2/m$	$2/m\,2/m\,2/m$
S; loops	2.67; 2	2.67; 2
SP plane	$\{101\}$	$\{101\}$
SP struct.	$A(--\tfrac{1}{2})B(\tfrac{1}{2}--)C(-\tfrac{1}{2}-);$	$A(--\tfrac{1}{2})B(\tfrac{1}{2}--)C(-\tfrac{1}{2}-);$
SPI =	52	51
Cleavage	pf(010), gd(001)	pf(010)
RI	$n_\alpha = 1.94$	$n_\alpha = 1.64\text{--}1.65$
	$n_\beta = 2.20$	$n_\beta = 1.65\text{--}1.66$
	$n_\gamma = 2.51$	$n_\gamma = 1.65\text{--}1.67$
$\delta =$	0.57	0.02
$2V =$	83°	80°
Sign	$(-)$	$(+)$
Transp.	Translucent	Transp. to transl.
Pleochr.	X = yellow	
	Y = orange red	
	Z = red orange	
H =	5	$3\tfrac{1}{2}$–4
D =	4.0	3.1
Color	Red brown	White
Streak	Orange	White
Luster	Submetallic	Vitreous
Habit	Blady, tabular	Flaky, nodular
Remarks	Major component in limonite	Isostruc. w. lepidocrocite
	Zigzag octahedral layers	Major component in bauxite

The tetragonal symmetry of rutile is lowered to monoclinic in manganite for reasons not fully understood, but probably related to the ordering of oxygen and (OH). Goethite (Figure II.102) and diaspore are isostructural, and have oxygen and (OH) in the *AB* stacking sequence with one half of the octahedral voids occupied by either Fe^{3+} or Al^{3+}. Unlike manganite, which has single subchains of octahedra running parallel to the *c* axis, goethite and diaspore have double subchains running in that direction.

Lepidocrocite (FeO(OH)) and boehmite (AlO(OH)) are both polymorphs of goethite and diaspore, respectively. Their structure is based on an *ABC* stacking sequence in which one half of the octahedral voids are occupied but in a pattern of corrugated sheets (Figure II.103) in the (010) plane. Consequently, both of these minerals possess a perfect {010} cleavage.

We have some commonly used names in mineralogy that refer to mechanical aggregates of various hydroxide minerals. *Bauxite* refers to a mixture of gibbsite, boehmite, and diaspore, and is usually an oolitic or pisolitic aggregate of colloi-dal-sized grains. Bauxite is the most important ore of aluminum. *Limonite* is a mixture of hydrated lepidocrocite, goethite, and occasionally hematite. *Wad* is a mixture of hydrated manganese oxides and hydroxides, and can be an ore of manganese.

Chemistry. The hydroxide minerals have a simple chemistry. Brucite may contain small amounts of Fe^{2+} and Mn^{2+} substituting for Mg^{2+}, and gibbsite usually shows some Fe^{3+} substituting for Al^{3+}. The same substitutions operate in goethite and diaspore and in their respective polymorphs (dimorphs).

Occurrence and Associations. All of the hydroxide minerals are generally restricted in occurrence to low-temperature, hydrous environments. Brucite is a common alteration product of periclase, but also is found with calcite as the reaction product of the breakdown of dolomite to produce brucite marbles. Gibbsite frequently forms as an alteration product of corundum, but is best known as a constituent of bauxite. Gibbsite is common in soils and is an abundant clay mineral on the ocean bottom in equatorial latitudes.

Manganite occurs as a low-temperature vein

Goethite
FeO(OH)

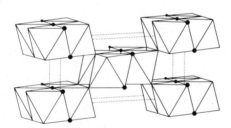 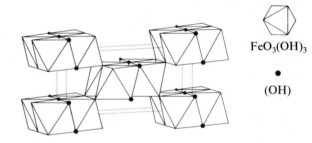

FeO₃(OH)₃

(OH)

FIGURE II.102 **Crystal structure of goethite, *a* axis vertical.**

Lepidocrocite
FeO(OH)

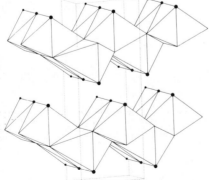 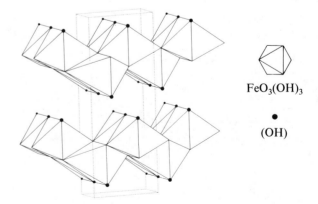

FeO₃(OH)₃

(OH)

FIGURE II.103 **Crystal structure of lepidocrocite, *b* axis vertical.**

mineral associated with carbonates—usually calcite, siderite, and barite. It is also associated with pyrolusite and psilomelane in deposits formed through the action of surface water.

Goethite and lepidocrocite both form as a weathering product of Fe-bearing minerals under relatively oxidizing conditions that produce ferric iron. Diaspore and boehmite are found with other oxides and hydroxides of aluminum in most bauxite deposits formed by weathering in tropical latitudes. Diaspore is also common with corundum in emery deposits that form by the metamorphism of bauxite.

Distinctive Features. The very fine-grained occurrences of most of the hydroxide minerals make their positive identification difficult without the use of x-ray diffraction. Brucite is soft and has a pearly luster on its perfect (001) cleavage. Unlike the micas, cleavage plates of brucite have no elasticity. Manganite is black with a submetallic luster and frequently occurs as columnar crystals with longitudinal striations. Goethite is usually massive or earthy but yields a yellow brown streak. Brown to yellow varieties are used as yellow ochre, a common pigment. Lepidocrocite commonly occurs as reddish brown, scaly aggregates that yield a dull orange streak.

THE CARBONATE AND NITRATE MINERALS

The carbonates and nitrates include those minerals that have either the $(CO_3)^{2-}$ or $(NO_3)^{1-}$ anionic group as an essential part of their structures. A large number of different carbonate minerals exist (Table II.5), but only a few are common. Calcite ($CaCO_3$) and dolomite ($CaMg(CO_3)_2$) are the predominant minerals of limestone and dolomite. There is a much smaller number of nitrate minerals; nitratite (soda niter) and niter (saltpeter) are the only relatively common ones.

As mentioned in Chapter 1, the physical properties associated with carbonates as a mineral group are more closely related to their common $(CO_3)^{2-}$ anionic complex than to the particular cation or cations in each member. The cation C^{4+} has a relatively high positive valence and a relatively small radius, which imparts a strong covalent character to each of the three surrounding oxygens. These oxygens do not define a simple symmetrical sheet with respect to the oxygens of coplanar CO_3 units, because the separation of oxygens between units is much greater than within units. Consequently, the crystal structure models for the carbonate minerals have the entire CO_3 unit centered on the symmetrically packed sites, a procedure that allows us to treat this group of minerals as symmetrically packed structures for visual illustration only. The octahedral voids in this mode of representation have carbon atoms, not oxygen, at their apices. It is thus important to remember during the following discussion that any cation occupying the octahedral "void" in this model is not bonded to the carbon atoms, but rather to the oxygens immediately around them. The coordination of Ca^{2+} to oxygen is therefore not six, as is normally the case for the octahedra in the crystal models.

Structure. By adopting this mode of representation, we can consider calcite (Figure II.104) as a derivative of the NaCl structure, providing we make no distinction between the Ca and Mg sites. Both structures have the same symmetrical packing symbol. Their structural equivalents can be visualized by replacing the chlorine in halite with the carbonate radical, and the sodium in halite with calcium.

Dolomite has the same symmetrical stacking sequence as calcite and its isostructural minerals,

---

TABLE II.5. *Classification of Carbonate Minerals*

Calcite Structure Type		Dolomite Structure Type	
Calcite	$CaCO_3$	Dolomite	$CaMg(CO_3)_2$
Magnesite	$MgCO_3$	Ankerite	$CaFe(CO_3)_2$
Siderite	$FeCO_3$	Kutnahorite	$CaMn(CO_3)_2$
Rhodochrosite	$MnCO_3$		
Smithsonite	$ZnCO_3$	Aragonite Structure Type	
Otavite	$CdCO_3$	Aragonite	$CaCO_3$
Sphaerocobaltite	$CoCO_3$	Witherite	$BaCO_3$
		Strontianite	$SrCO_3$
Other Structures		Cerussite	$PbCO_3$
Malachite	$Cu_2(OH)_2CO_3$		
Azurite	$Cu_3(OH)_2(CO_3)_2$		

**THE CARBONATE
AND NITRATE
MINERALS**

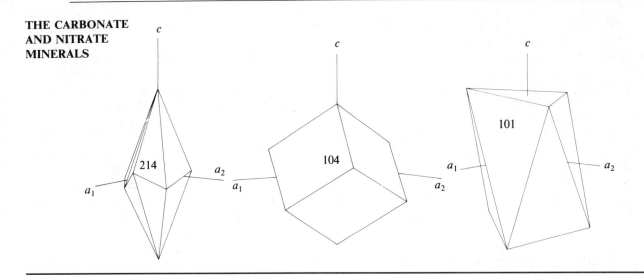

	CALCITE	Magnesite	Siderite
Formula	$CaCO_3$	$MgCO_3$	$FeCO_3$
Coord.	Ca(6), C(3)	Mg(6), C(3)	Fe(6), C(3)
System	Trigonal (rhombohedral)	Trigonal (rhombohedral)	Trigonal (rhombohedral)
$a =$	0.499	0.459	0.472
$c =$	1.704	1.487	1.546
$Z =$	6	6	6
Sp. group	$R\bar{3}2/c$	$R\bar{3}2/c$	$R\bar{3}2/c$
Pt. group	$\bar{3}2/m$	$\bar{3}2/m$	$\bar{3}2/m$
SP plane	(001)	(001)	(001)
SP struct.	$A(--1)B(1--)C(-1-)$;	$A(--1)B(1--)C(-1-)$;	$A(--1)B(1--)C(-1-)$;
SPI =	51	57	54
Cleavage	pf{104}*	pf{104}*	pf{104}*
Twinning	(001), {104}, {018}		{018}
RI	$n_\omega = 1.658$	$n_\omega = 1.700$	$n_\omega = 1.875$
	$n_\epsilon = 1.486$	$n_\epsilon = 1.509$	$n_\epsilon = 1.633$
$\delta =$	0.172	0.191	0.242
Sign	$(-)$	$(-)$	$(-)$
Transp.	Transp. to transl.	Transp. to transl.	Translucent
H =	3	4	4
D =	2.71	3.00	3.96
Color	Colorless, variable	White	Yellow brown
Streak	White	White, gray	White
Luster	Vitreous	Porcelaneous	Vitreous
Fracture	Conchoidal	Conchoidal	Subconchoidal
Habit	Prismatic, large variety of forms	Massive, rhombohedral	Rhombohedral faces occ. curved
Remarks	CO_3 in A, B, C	Isostr. w. calcite	Isostr. w. calcite

Also fibrous and stalactitic, and frequently fluorescent
*In morphological unit cell of calcite ($a = 0.997$ nm, $c = 0.852$) cleavage is {101}

but the distribution of Ca^{2+} and Mg^{2+} is ordered in alternating layers (Figure II.105). The effect of ordering is to reduce the symmetry, removing the twofold axes of calcite as well as the c glide plane.

Aragonite, the orthorhombic polymorph of calcite, has a higher density and closer packing than calcite. The packing sequence (Figure II.106) is a highly distorted $A(-\frac{1}{2}-)A(-\frac{1}{2}-)$ in which calcium increases its coordination relative to that in calcite. The higher coordination in aragonite is more favorable to accommodating larger divalent cations such as Ba (witherite), Pb (cerussite), and Sr (strontianite). In the case of nitrates, K^{1+} favors the aragonite structure, whereas the smaller Na^{1+} favors the calcite structure.

Two hydrous copper carbonates, malachite

**THE CARBONATE
AND NITRATE
MINERALS**

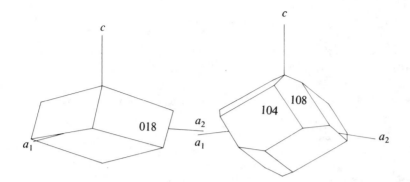

	Rhodochrosite	Smithsonite
Formula	$MnCO_3$	$ZnCO_3$
Coord.	Mn(6), C(3)	Zn(6), C(3)
System	Trigonal (rhombohedral)	Trigonal (rhombohedral)
$a =$	0.474	0.461
$c =$	1.551	1.488
$Z =$	6	6
Sp. group	$R\bar{3}2/m$	$R\bar{3}2/m$
Pt. group	$\bar{3}2/m$	$\bar{3}2/m$
SP plane	(001)	(001)
SP struct.	$A(--1)B(1--)C(-1-)$;	$A(--1)B(1--)C(-1-)$;
SPI =	50	40
Cleavage	pf{104}*	pf{104}*
RI	$n_\omega = 1.816$	$n_\omega = 1.850$
	$n_\epsilon = 1.597$	$n_\epsilon = 1.625$
$\delta =$	0.219	0.225
Sign	$(-)$	$(-)$
Transp.	Transp. to transl.	Transp. to transl.
H =	$3\frac{1}{2}$–4	4–$4\frac{1}{2}$
D =	3.70	4.43
Color	Pink	Gray, green
Streak	White	White
Luster	Pearly, vitreous	Vitreous, pearly
Fracture	Uneven	Subconchoidal
Habit	Massive	Massive
Remarks	Also reniform, botryoidal, stalactitic	
	Rarely in curved rhombohedral crystals	
	Isostructural with calcite	
	*In morphological unit cell cleavage is {101}	

(Figure II.107) and azurite, are well known but have relatively complicated structures.

Chemistry. The three kinds of structures within the carbonate group control to some extent the possible substitutions. In calcite and its isostructural carbonates (Table II.5), substitution is widespread among the divalent cations. Calcites associated with dolomite can have several percent Mg^{2+} substituting for Ca^{2+}. A solvus between calcite and dolomite is strongly skewed toward dolomite at high temperature, indicating up to 50 mole percent solubility of dolomite in magnesium calcite but little solution in the opposite direction. Dolomite exhibits extensive solution toward ankerite, and a complete solid solution exists between magnesite and siderite.

Calcite
CaCO₃

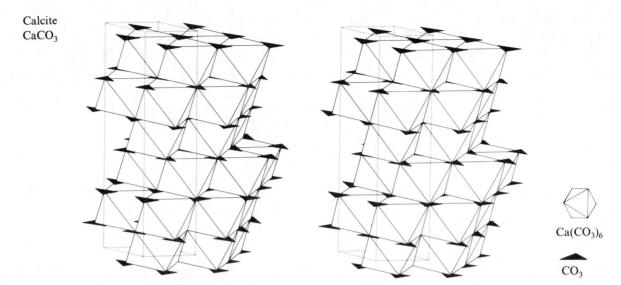

Ca(CO₃)₆

CO₃

FIGURE II.104   Crystal structure of calcite, c axis vertical.

Dolomite
CaMg(CO₃)₂

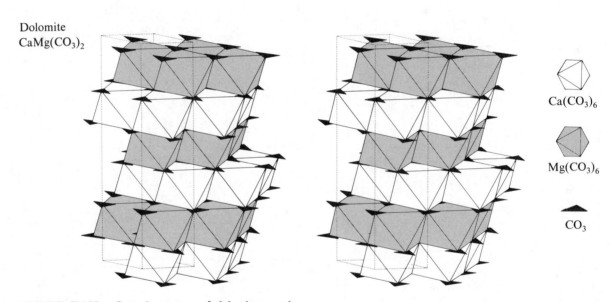

Ca(CO₃)₆

Mg(CO₃)₆

CO₃

FIGURE II.105   Crystal structure of dolomite, c axis vertical.

Aragonite
CaCO₃

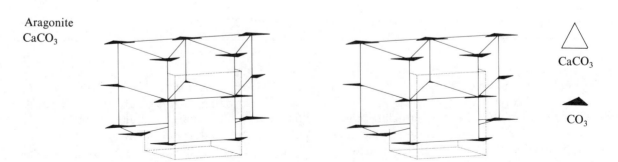

CaCO₃

CO₃

FIGURE II.106   Crystal structure of aragonite, c axis vertical.

THE CARBONATE
AND NITRATE
MINERALS

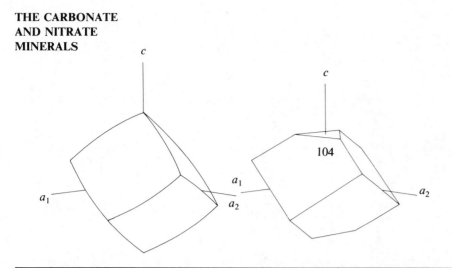

	DOLOMITE	Ankerite
Formula	$CaMg(CO_3)_2$	$CaFe(CO_3)_2$
Coord.	Ca(6), Mg(6), C(3)	Ca(6), Fe(6), C(3)
System	Trigonal (rhombohedral)	Trigonal (rhombohedral)
$a =$	0.484	0.482
$c =$	1.596	1.614
$Z =$	3	3
Sp. group	$R\bar{3}$	$R\bar{3}$
Pt. group	$\bar{3}$	$\bar{3}$
SP plane	(001)	(001)
SP struct.	$A(--1)B(1--)C(-1-)$;	$A(--1)B(1--)C(-1-)$;
SPI =	53	57
Cleavage	pf{104}*	pf{104}*
Twinning	(001), {100}	(001), {100}
	{110}, {012}	
RI	$n_\omega = 1.679$	$n_\omega = 1.750$
	$n_\epsilon = 1.500$	$n_\epsilon = 1.548$
$\delta =$	0.179	0.202
Sign	$(-)$	$(-)$
Transp.	Transp. to transl.	Transp. to transl.
H =	$3\frac{1}{2}-4$	$3\frac{1}{2}-4$
D =	2.85	3.10
Color	White, pink	White, yellow brown
Streak	White	White
Luster	Vitreous	Vitreous
Fracture	Subconchoidal	Subconchoidal
Habit	Rhombohedral with curves faces	
	Botryoidal, globular, stalactitic	
Remarks	Derivatives of the calcite structure	

*In morphological unit cell cleavage is {101}

Most aragonites are relatively pure, showing solid solution only toward calcite or magnesite at elevated temperature, and substitution of Sr^{2+} for Ca^{2+}. The larger divalent cations have a tendency toward more extensive solid solution of Ca^{2+} in the aragonite structure than in the calcite or dolomite structures.

In the nitrates, only very limited solid solution exists between $NaNO_3$ and KNO_3 because of their different structures.

**THE CARBONATE
AND NITRATE
MINERALS**

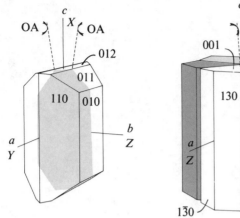

	ARAGONITE	Witherite
Formula	$CaCO_3$	$BaCO_3$
Coord.	Ca(6), C(3)	Ba(6), C(3)
System	Orthorhombic	Orthorhombic
$a =$	0.495	0.526
$b =$	0.796	0.885
$c =$	0.573	0.655
$Z =$	4	4
Sp. group	$P2_1/m\ 2_1/c\ 2_1/n$	$P2_1/m\ 2_1/c\ 2_1/n$
Pt. group	$2/m\ 2/m\ 2/m$	$2/m\ 2/m\ 2/m$
SP plane	(001)	(001)
SP struct.	$A(\frac{1}{2})A(\frac{1}{2})$;	$A(\frac{1}{2})A(\frac{1}{2})$;
SPI =	57	53
Cleavage	gd(010), gd{110}	gd(010), pr{110}, pr{012}
Twinning	{110}	{110}
RI	$n_\alpha = 1.530$	$n_\alpha = 1.529$
	$n_\beta = 1.681$	$n_\beta = 1.676$
	$n_\gamma = 1.685$	$n_\gamma = 1.677$
$\delta =$	0.155	0.148
$2V =$	18°	16°
Sign	(−)	(−)
Transp.	Transp. to transl.	Transp. to transl.
H =	$3\frac{1}{2}$–4	3–$3\frac{1}{2}$
D =	2.94	4.29
Color	Colorless, white	Colorless, gray
Streak	White	White
Luster	Vitreous	Resinous
Fracture	Subconchoidal	Uneven
Habit	Acicular	Mammillary
	Reniform, globular or fibrous	Globular and coarse, fibrous
Remarks		Isostr. w. aragonite

CO_3 plane A is corrugated along the a axis
{110} twinning produces pseudohexagonal plates

THE CARBONATE AND NITRATE MINERALS

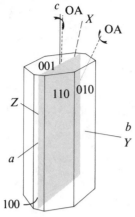
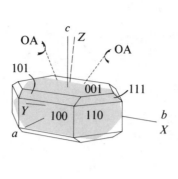

	MALACHITE	Azurite
Formula	$Cu_2(OH)_2CO_3$	$Cu_3(OH)(CO_3)_2$
Coord.	Cu(4) (+2)*, C(3)	Cu(5), C(3)
System	Monoclinic	Monoclinic
$a =$	0.948	0.497
$b =$	1.203	0.584
$c =$	0.321	1.029
$\beta =$	98.0°	92.4°
$Z =$	4	2
Sp. group	$P2_1/a$	$P2_1/c$
Pt. group	$2/m$	$2/m$
SPI =	48	46
Cleavage	pf(201)	pf{011}, gd(100)
Twinning	(100)	
RI	$n_\alpha = 1.655$	$n_\alpha = 1.730$
	$n_\beta = 1.875$	$n_\beta = 1.756$
	$n_\gamma = 1.909$	$n_\gamma = 1.836$
$\delta =$	0.254	0.106
$2V =$	43°	68°
Sign	(−)	(+)
Transp.	Transp. to transl.	Transp. to transl.
Pleochr.	X = colorless	X = light blue
	Y = yellow green	Y = medium blue
	Z = green	Z = dark blue
H =	$3\frac{1}{2}$–4	$3\frac{1}{2}$–4
D =	3.7–4.0	3.77
Color	Grass green	Azure blue
Streak	Pale green	Blue
Luster	Adamantine	Vitreous, adamantine, dull, earthy
Fracture	Subconchoidal	Conchoidal
Habit	Massive, botryoidal, massive fibrous	Tabular, massive
Remarks	These two minerals usually occur together	
	*Close second neighbor bonds	

**THE CARBONATE
AND NITRATE
MINERALS**

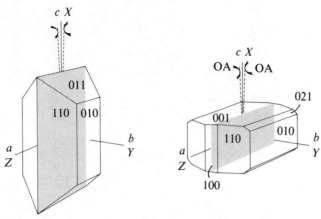

	Strontianite	Cerussite
Formula	$SrCO_3$	$PbCO_3$
Coord.	Sr(6), C(3)	Pb(6), C(3)
System	Orthorhombic	Orthorhombic
$a =$	0.513	0.515
$b =$	0.842	0.847
$c =$	0.609	0.611
$Z =$	4	4
Sp. group	$P2_1/m\ 2_1/c\ 2_1/n$	$P2_1/m\ 2_1/c\ 2_1/n$
Pt. group	$2/m\ 2/m\ 2/m$	$2/m\ 2/m\ 2/m$
SP plane	(001)	(001)
SP struct.	$A(\frac{1}{2})A(\frac{1}{2})$;	$A(\frac{1}{2})A(\frac{1}{2})$;
SPI =	53	54
Cleavage	gd{110}, pr(010)	gd{110}, pr{021}
Twinning	{110}	{110}, {130}
RI	$n_\alpha = 1.520$	$n_\alpha = 1.804$
	$n_\beta = 1.667$	$n_\beta = 2.076$
	$n_\gamma = 1.668$	$n_\gamma = 2.078$
$\delta =$	0.148	0.274
$2V =$	7°	9°
Sign	(−)	(−)
Transp.	Transp. to transl.	Transp. to transl.
H =	$3\frac{1}{2}$	$3–3\frac{1}{2}$
D =	3.72	6.55
Color	Pale green, yellow	Colorless, gray
Streak	White	White
Luster	Vitreous	Adamantine
Fracture	Uneven	Conchoidal
Habit	Acicular, spadelike	Tabular, clustered
Remarks	Both can be massive, compact, or fibrous	
	Isostructural with aragonite	

**THE CARBONATE
AND NITRATE
MINERALS**

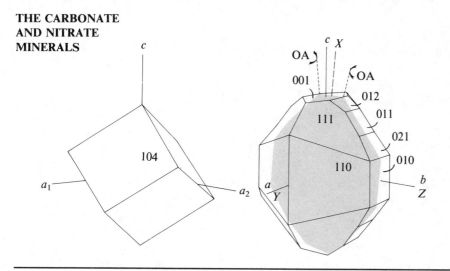

	Nitratite (soda niter)	Niter (saltpeter)
Formula	$NaNO_3$	KNO_3
Coord.	Na(6), N(3)	K(6), N(3)
System	Trigonal (rhombohedral)	Orthorhombic
$a =$	0.507	0.543
$b =$		0.919
$c =$	1.682	0.646
$Z =$	6	4
Sp. group	$R\bar{3}2/c$	$P2_1/c\ 2_1/m\ 2_1/n$
Pt. group	$\bar{3}2/m$	$2/m\ 2/m\ 2/m$
SP plane	(001)	(001)
SP struct.	$A(--1)B(1--)C(-1-)$;	$A(\tfrac{1}{2})A(\tfrac{1}{2})$;
SPI =	37	57
Cleavage	pf{104}*	pf{011}
Twinning	(001)	{110}
RI	$n_\omega = 1.587$	$n_\alpha = 1.333$
	$n_\epsilon = 1.336$	$n_\beta = 1.505$
		$n_\gamma = 1.505$
$\delta =$	0.251	0.172
$2V =$		7°
Sign	$(-)$	$(-)$
Transp.	Transp. to transl.	Translucent
H =	1–2	2
D =	2.29	2.10
Color	Colorless	White
Streak	White	White
Luster	Vitreous	Vitreous
Fracture	Conchoidal	Uneven
Habit	Rhombohedral	Acicular
Remarks	*(See cleavage under calcite)	{110} twinning yields pseudohexagonal prisms
	Isostr. w. calcite	Isostr. w. aragonite
	(Distorted)	

Malachite
Cu$_2$(OH)$_2$CO$_3$

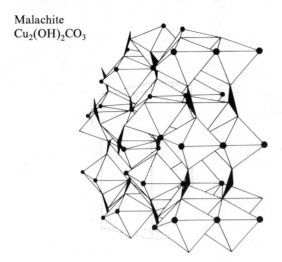 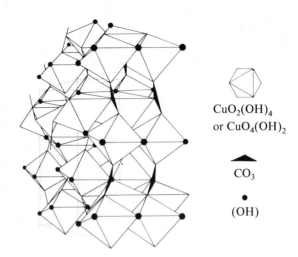

CuO$_2$(OH)$_4$
or CuO$_4$(OH)$_2$

CO$_3$

(OH)

FIGURE II.107   **Crystal structure of malachite, *b* axis vertical.**

Occurrence and Associations. Calcite and dolomite are easily the most abundant carbonate minerals. Calcite is the principal mineral in limestones and is a primary precipitate from seawater as well as the major constituent of calcareous cave deposits and travertine. The bulk of all calcareous testes and skeletal material deposited in marine environments is composed of calcite and aragonite together. These are the minerals responsible for limestone and chalk.

Aragonite that forms in organisms is apparently metastable with respect to calcite under the same conditions. Aragonite is stable under relatively high pressure. The mineral is then about 8% denser than calcite (Table 8.2). Aragonite is the usual carbonate in glaucophane schists, commonly associated with jadeite, lawsonite, and quartz.

The occurrence and associations of the remaining carbonates are strongly dependent on the bulk chemistry of the environment. Rhodochrosite forms only in Mn-rich rocks, commonly in hydrothermal veins associated with ores of lead and silver. Smithsonite is usually found in zinc deposits associated with sphalerite, calcite, cerussite, and limonite. Witherite is found with galena and barite. Cerussite is also associated with galena as well as sphalerite, anglesite, and limonite in secondary lead deposits.

The hydrous copper carbonates, malachite and azurite, are found almost exclusively in the oxidized portions of copper veins, usually in limestone. Both are popular semiprecious gemstones. Associations are cuprite, native copper, and iron oxides.

The nitrate minerals are easily soluble in water

and hence are found only in relatively arid climates. Soda niter is a constituent of soils and usually forms as encrustations associated with gypsum, halite, and other salts.

Distinguishing Features. All of the rhombohedral carbonates are distinguished by a perfect rhombohedral cleavage and vitreous luster. In general, all of the carbonates are relatively soft with hardnesses between 3 and 5 on the Mohs scale. In addition, calcite, smithsonite, aragonite, strontianite, and the hydrous copper carbonates all effervesce in dilute, cold HCl. Dolomite effervesces readily only when powdered.

The carbonates can generally be distinguished from each other by their color, although exceptions occur. Calcite is generally colorless or white, siderite and smithsonite are brown, and rhodochrosite is pink. Dolomite frequently has a rusty brown weathering surface because of its siderite component, and may also exhibit curved or saddle-shaped cleavage faces. Malachite is bright green; azurite is azure blue.

The nitrate minerals are distinguished by their solubility in water and a peculiar cooling taste when touched to one's tongue.

THE BORATE MINERALS

Borates are structurally similar to the carbonates and nitrates in that boron strongly bonds to oxygen to form either a (BO$_3$)$^{3-}$ or (BO$_4$)$^{5-}$ anionic group. Boron can enter into either triangular or tetrahedral coordination with oxygen. Well over 100 borate minerals are known, but only a few are relatively common. They are divided into anhydrous and hydrous groups. The only com-

THE BORATE MINERALS

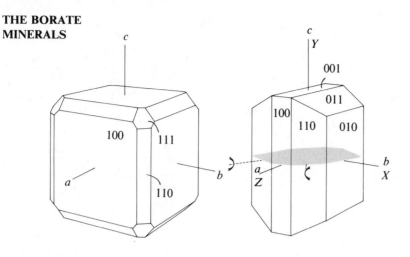

	Boracite	Sinhalite
Formula	$Mg_6Cl_2B_6B_8O_{25}$	$MgAlBO_4$
Coord.	Mg(6), B(3 or 4)	Mg(6), Al(6), B(4)
System	Orthorhombic	Orthorhombic
$a =$	0.854	0.4328
$b =$	0.854	0.9878
$c =$	1.207	0.5675
$Z =$	2	4
Sp. group	$Pc2a$	$P2_1/m\ 2_1/c\ 2_1/n$
Pt. group	$m2m$	$2/m\ 2/m\ 2/m$
S; loops	2.00; 3, 8	1.00
SP plane		(100)
SP struct.		$A(\frac{1}{8}\frac{1}{8}\frac{1}{2})B(\frac{1}{8}\frac{1}{8}\frac{1}{2})$;
SPI =	61	62
Cleavage	pr 111	gd(010)
RI	$n_\alpha = 1.662$	$n_\alpha = 1.67$
	$n_\beta = 1.647$	$n_\beta = 1.70$
	$n_\gamma = 1.673$	$n_\gamma = 1.71$
$\delta =$	0.011	0.04
$2V =$	82°	55°
Sign	(+)	(+)
Transp.	Transp. to transl.	Transp. to transl.
$H =$	7	$6\frac{1}{2}$–7
$D =$	2.95	3.42
Color	White, yellow	Green blue
Streak	White	White, yellow
Luster	Vitreous	Vitreous
Fracture	Conchoidal	Conchoidal
Habit	Pseudocubic	Prismatic
Remarks	Pyroelectric	Isostr. w. olivine

mon anhydrous borate is boracite $Mg_6Cl_2B_6B_8O_{25}$. Borax ($Na_2(H_2O)_3B_2B_2O_6(OH)_2$) is the most widespread and probably best known of the hydrous borates.

Structure. The borax structure (Figure II.108) consists of isolated groups of double boron tetrahedra and double boron triangles connected to each other by single chains of Na octahedra.

THE BORATE MINERALS

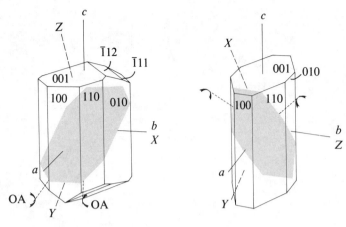

	BORAX	Kernite
Formula	$Na_2(H_2O)_8B_2B_2O_5(OH)_4$	$Na_2(H_2O)_3B_2B_2O_6(OH)_2$
Coord.	Na(6), B(3 or 4)	Na(5), B(3 or 4)
System	Monoclinic	Monoclinic
$a =$	1.184	1.568
$b =$	1.063	0.909
$c =$	1.232	0.702
$\beta =$	106.58°	108.87°
$Z =$	4	4
Sp. group	$C2/c$	$P2/a$
Pt. group	$2/m$	$2/m$
S; RU (loops)	1.714; 6 (4) chain	1.857; 2 (3) chain
SPI =	24	36
Cleavage	pf(100), pf{110}	pf(100), pf(001)
	pr(010)	pr{101}
RI	$n_\alpha = 1.447$	$n_\alpha = 1.454$
	$n_\beta = 1.469$	$n_\beta = 1.472$
	$n_\gamma = 1.472$	$n_\gamma = 1.488$
$\delta =$	0.025	0.034
$2V =$	40°	80°
Sign	(−)	(−)
Transp.	Translucent	Transparent
H =	$2–2\frac{1}{2}$	3
D =	1.7–1.9	1.90
Color	White, gray	Colorless, white
Streak	White	White
Luster	Vitreous, resinous	Vitreous, pearly
Fracture	Conchoidal	Uneven
Habit	Prismatic	Coarse aggregates, cleavable massive
Remarks	Sweet alkaline taste	

THE BORATE MINERALS

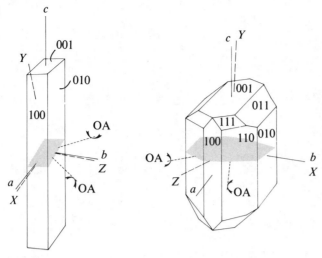

	Ulexite (natroborocalcite)	Colemanite
Formula	$NaCa(H_2O)_6B_2B_3O_7(OH)_4$	$Ca(H_2O)BB_2O_4(OH)_3$
Coord.	Na(6), Ca(9), B(3 or 4)	Ca(7), B(3 or 4)
System	Triclinic	Monoclinic
$a =$	0.873	0.874
$b =$	1.275	1.126
$c =$	0.670	0.610
$\alpha =$	90.27°	
$\beta =$	109.13°	110.12°
$\gamma =$	105.12°	
$Z =$	2	4
Sp. group	$P\bar{1}$	$P2_1/a$
Pt. group	$\bar{1}$	$2/m$
S; RU	1.67; 5 groups	1.73; 4 chain
SPI =	26	41
Cleavage	pf(010)	pf(010)
RI	$n_\alpha = 1.491$	$n_\alpha = 1.586$
	$n_\beta = 1.505$	$n_\beta = 1.592$
	$n_\gamma = 1.520$	$n_\gamma = 1.614$
$\delta =$	0.029	0.028
$2V =$	73°	56°
Sign	(+)	(+)
Transp.	Transp. to transl.	Transp. to transl.
$H =$	$2\frac{1}{2}$	$4–4\frac{1}{2}$
$D =$	1.96	2.42
Color	White	Colorless, white
Streak	White	White
Luster	Silky	Vitreous
Fracture	Uneven	Subconchoidal
Habit	Acicular, fibrous	Short prismatic, cleavable massive
Remarks	Also known as "boronatrocalcite"	

Borax
$Na_2(H_2O)_3B_2B_2O_6(OH)_2$

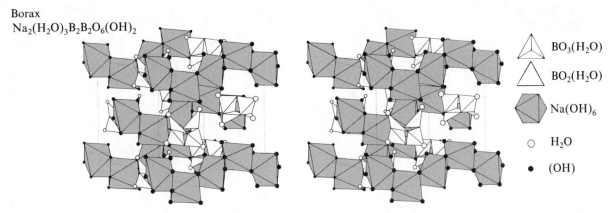

△	$BO_3(H_2O)$
△	$BO_2(H_2O)$
⬡	$Na(OH)_6$
○	H_2O
●	(OH)

FIGURE II.108   Crystal structure of borax, *b* axis vertical.

Because the octahedral corners are H_2O molecules, the single octahedral chains are weakly bonded to the ribbonlike strips of boron polyhedra. Borax consequently has perfect prismatic {100} cleavage and prismatic crystal form.

Occurrence and Associations. All of the borate minerals form in relatively arid regions from brines that develop by evaporation in enclosed basins. Common associations are with halite and gypsum.

Distinguishing Features. Borax has a vitreous luster, perfect {100} cleavage, and dissolves readily in water.

THE SULFATE MINERALS

The sulfate minerals all have the $(SO_4)^{2-}$ anionic group as an essential part of their structure. Like the carbonates, nitrates, and borates, the sulfates owe their common physical properties such as cleavage to the anionic group and its arrangement in the structure, rather than to the diverse properties of the other cations that may enter the structure.

The sulfates are generally divided into two groups, the anhydrous sulfates and the hydrous sulfates as follows:

> Anhydrous Sulfates
>   Anhydrite ($CaSO_4$)
>   Barite ($BaSO_4$)
>   Celestite ($SrSO_4$)
>   Anglesite ($PbSO_4$)
>
> Hydrous Sulfates
>   Gypsum ($CaSO_4 \cdot 2H_2O$)
>   Chalcanthite ($CuSO_4 \cdot 5H_2O$)
>   Epsomite ($MgSO_4 \cdot 7H_2O$)

> Antlerite ($Cu_3(OH)_4SO_4$)
> Alunite ($KAl_3(OH)_6(SO_4)_2$)

Structure. All of the anhydrous sulfates have structures consisting of isolated $(SO_4)^{2-}$ tetrahedra connected by various types of cation polyhedra. Of the two basic types of cation polyhedra, one provides eightfold coordination for cations such as Ca^{2+} in anhydrite (Figure II.109), and the other provides twelvefold coordination for larger cations such as Ba^{2+}, Sr^{2+}, and Pb^{2+}. Barite ($BaSO_4$) has a structure resembling anhydrite, with Ba in sevenfold instead of eightfold coordination. Celestite and anglesite are isostructural with barite. Other sulfates such as glauberite ($Na_2Ca(SO_4)_2$) and glasserite ($K_3Na(SO_4)_2$) have monovalent and divalent cations occupying two different kinds of eight-coordinated sites in an ordered distribution.

There are two types of "hydrous" sulfates, those with hydroxyl in their structures and those that contain molecular H_2O. Antlerite (Figure II.110) is an example of the hydroxyl type in which SO_4 tetrahedra are connected by triple chains of Cu octahedra that share hydroxyl ions. In antlerite as well as in related brochantite ($Cu_4(OH)_6SO_4$), alunite ($KAl_3(OH)_6(SO_4)_2$), and jarosite ($KFe_3(OH)_6(SO_4)_2$), the divalent or trivalent cation is in sixfold coordination. In alunite and jarosite, K^{1+} has twelvefold coordination.

The hydrous sulfates that contain molecular H_2O include gypsum ($CaSO_4 \cdot 2H_2O$) and the related sulfates chalcanthite ($CuSO_4 \cdot 5H_2O$), epsomite ($MgSO_4 \cdot 7H_2O$), and melanterite ($FeSO_4 \cdot 7H_2O$). In gypsum (Figure II.111), isolated SO_4 tetrahedra are connected by eight-coordinated polyhedra that consist of Ca^{2+} bonded to six oxygens and two water molecules. The

THE SULFATE MINERALS

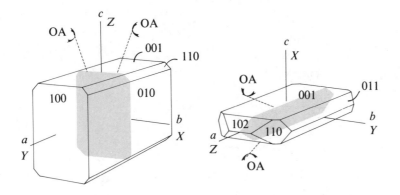

	ANHYDRITE	Barite
Formula	CaSO$_4$	BaSO$_4$
Coord.	Ca(6) (+2)*, S(4)	Ba(6) (+6)*, S(4)
System	Orthorhombic	Orthorhombic
$a =$	0.622	0.887
$b =$	0.697	0.545
$c =$	0.696	0.714
$Z =$	4	4
Sp. group	$C2/c\ 2/m\ 2_1/m$	$P2_1/n\ 2_1/m\ 2_1/a$
Pt. group	$2/m\ 2/m\ 2/m$	$2/m\ 2/m\ 2/m$
S	1.00	1.00
SP plane	(011)	(101)
SP struct.	$A(\frac{1}{4}\frac{1}{4}-)B(--\frac{1}{2})$	$A(\frac{1}{4}\frac{1}{4}-)B(--\frac{1}{2})$
	$A(\frac{1}{4}\frac{1}{4}-)B(--\frac{1}{2})$	$A(\frac{1}{4}\frac{1}{4}-)B(--\frac{1}{2})$;
	$A(\frac{1}{4}\frac{1}{4}-)B(--\frac{1}{2})$;	
SPI $=$	70	60
Cleavage	gd(001), gd(010), pr(100)	pf(001), gd{210} pr(011)
Twinning	{011}	
RI	$n_\alpha = 1.570$	$n_\alpha = 1.636$
	$n_\beta = 1.575$	$n_\beta = 1.637$
	$n_\gamma = 1.614$	$n_\gamma = 1.648$
$\delta =$	0.044	0.012
$2V =$	44°	37°
Sign	(+)	(+)
Transp.	Transparent	Transp. to transl.
H $=$	$3\frac{1}{2}$	$3–3\frac{1}{2}$
D $=$	2.98	4.5
Color	Colorless	White
Streak	White	White
Luster	Vitreous, pearly	Vitreous, pearly
Fracture	Uneven, splintery	Uneven
Habit	Tabular, massive, lamellar, fibrous	Tabular, prismatic, lamellar, fibrous
Remarks	Twinning often penetrating	Cockscomb aggregates: "desert roses"

*Close second neighbor bonds

isolated SO$_4$ tetrahedra and the neighboring Ca^{2+} cations that share oxygens define planar units within the structure. These units are separated from each other by H$_2$O layers parallel to {010}, and this is the structural feature that accounts for the perfect {010} cleavage of gypsum.

THE SULFATE MINERALS

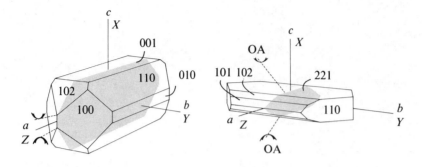

	Celestite	Anglesite
Formula	$SrSO_4$	$PbSO_4$
Coord.	Sr(6) (+6)*, S(4)	Pb(6) (+6)*, S(4)
System	Orthorhombic	Orthorhombic
$a =$	0.838	0.847
$b =$	0.537	0.539
$c =$	0.685	0.694
$Z =$	4	4
Sp. group	$P2_1/n\,2_1/m\,2_1/a$	$P2_1/n\,2_1/m\,2_1/a$
Pt. group	$2/m\,2/m\,2/m$	$2/m\,2/m\,2/m$
S	1.00	1.00
SP plane	(101)	(101)
SP struct.	$A(\frac{1}{4}\frac{1}{4}-)B(--\frac{1}{2})$	$A(\frac{1}{4}\frac{1}{4}-)B(--\frac{1}{2})$
	$A(\frac{1}{4}\frac{1}{4}-)B(--\frac{1}{2})$;	$A(\frac{1}{4}\frac{1}{4}-)B(--\frac{1}{2})$;
SPI =	63	60
Cleavage	pf(001), gd{210}, pr{011}	pf(001), gd{210}, pr{011}
RI	$n_\alpha = 1.622$	$n_\alpha = 1.877$
	$n_\beta = 1.624$	$n_\beta = 1.883$
	$n_\gamma = 1.631$	$n_\gamma = 1.894$
$\delta =$	0.009	0.017
$2V =$	51°	75°
Sign	(+)	(+)
Transp.	Transp. to transl.	Translucent
$H =$	$3-3\frac{1}{2}$	$2\frac{1}{2}-3$
$D =$	3.97	6.38
Color	Colorless, blue	White
Streak	White	White
Luster	Vitreous, pearly	Adamantine
Fracture	Uneven	Conchoidal
Habit	Tabular, radiated fibrous	Tabular, prismatic, massive, granular
Remarks	Isostructural with barite	
	*Close second neighbor bonds	

THE SULFATE MINERALS

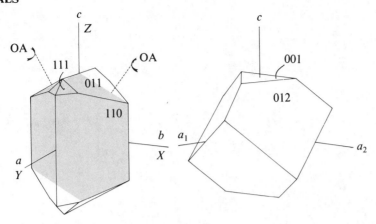

	ANTLERITE	Alunite
Formula	$Cu_3(OH)_4SO_4$	$KAl_3(OH)_6(SO_4)_2$
Coord.	Cu(6), S(4)	K(12), Cu(6), S(4)
System	Orthorhombic	Trigonal
$a =$	0.824	0.697
$b =$	1.199	
$c =$	0.603	1.738
$Z =$	4	3
Sp. group	$P2_1/a\,2_1/a\,2_1/m$	$R3m$
Pt. group	$2/m\,2/m\,2/m$	$3m$
S	1.00	1.00
SP plane	(120)	
SP struct.	$A(\frac{1}{8}\frac{1}{8}-)B(\frac{5}{8}--)C(\frac{1}{8}-\frac{1}{8})$	
	$A(--\frac{5}{8})B(-\frac{1}{8}\frac{1}{8})C(-\frac{5}{8}-);$	
SPI $=$	54	71
Cleavage	pf(010)	gd(001), pr(101)
RI	$n_\alpha = 1.726$	$n_\omega = 1.572$
	$n_\beta = 1.738$	$n_\epsilon = 1.592$
	$n_\gamma = 1.789$	
$\delta =$	0.063	0.020
$2V =$	53°	
Sign	$(+)$	$(+)$
Transp.	Transp. to transl.	Transp. to transl.
Pleochr.	X = yellow green	
	$Y = Z$ = blue green	
$H =$	$3\frac{1}{2}$–4	$3\frac{1}{2}$–4
$D =$	3.9	2.6–2.9
Color	White, gray	White, gray
Streak	Green, gray	White, gray
Luster	Vitreous	Vitreous
Fracture	Uneven	Conchoidal
Habit	Prismatic, tabular	Rhombohedral, fibrous
Remarks	Pyroelectric	Na $\rightleftharpoons$ K: natroalunite
		Pyroelectric

THE SULFATE MINERALS

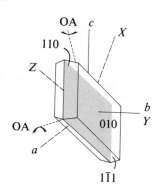

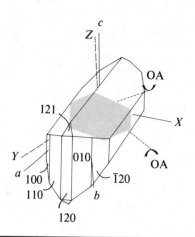

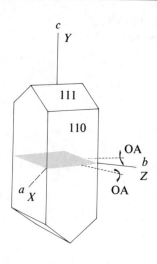

	GYPSUM	Chalcanthite	Epsomite
Formula	$CaSO_4 \cdot 2H_2O$	$CuSO_4 \cdot 5H_2O$	$MgSO_4 \cdot 7H_2O$
Coord.	Ca(8), S(4)	Cu(6), S(4)	Mg(6), S(4)
System	Monoclinic	Triclinic	Orthorhombic
$a =$	0.568	0.612	1.196
$b =$	1.5518	1.069	1.205
$c =$	0.629	0.596	0.688
$\alpha =$		97.58°	
$\beta =$	113.83°	107.17°	
$\gamma =$		77.33°	
$Z =$	4	2	4
Sp. group	$A2/n$	$P\bar{1}$	$P2_12_12_1$
Pt. group	$2/m$	$\bar{1}$	222
S	1.00	1.00	1.00
SPI $=$	44	35	30
Cleavage	pf(010), pf(100), gd{011}	gd(110), gd(110), pr(111)	pf(010), gd{011}
Twinning	(100)		
RI	$n_\alpha = 1.520$	$n_\alpha = 1.514$	$n_\alpha = 1.433$
	$n_\beta = 1.523$	$n_\beta = 1.537$	$n_\beta = 1.455$
	$n_\gamma = 1.529$	$n_\gamma = 1.543$	$n_\gamma = 1.461$
$\delta =$	0.009	0.029	0.028
$2V =$	58°	56°	52°
Sign	(+)	(−)	(+)
Transp.	Transp. to transl.	Translucent	Transp. to transl.
H $=$	2	$2\frac{1}{2}$	2–$2\frac{1}{2}$
D $=$	2.32	2.30	1.68
Color	Colorless, variable	Sky blue	Colorless
Streak	White	White	White
Luster	Vitreous, pearly	Vitreous	Vitreous
Fracture	Conchoidal	Conchoidal	Conchoidal
Habit	Tabular, variable	Tabular, fibrous	Botryoidal, prismatic
		Massive, reniform	
Remarks	Clear crystals: selenite	Metallic taste	Fibrous encrustation
	Compact: alabaster		Bitter taste
	Fibrous: satin spar		

Anhydrite
CaSO$_4$

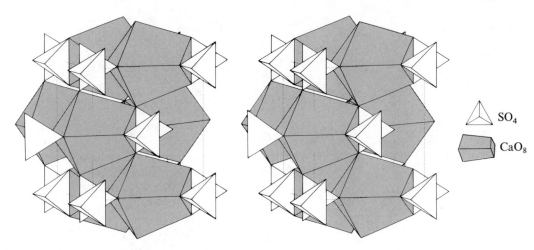

FIGURE II.109 **Crystal structure of anhydrite, *c* axis vertical.**

Antlerite
Cu$_3$(OH)$_4$SO$_4$

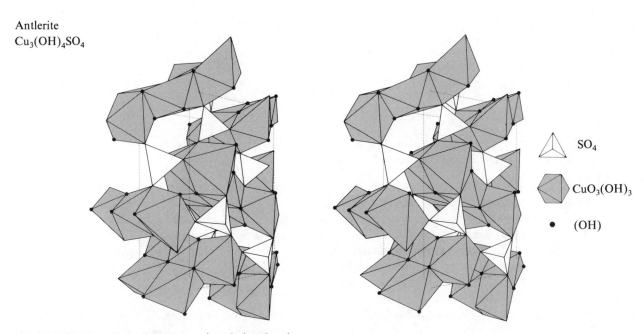

FIGURE II.110 **Crystal structure of antlerite, *b* axis vertical.**

Chemistry. Cation substitution is extensive within the 12-coordinated sites of the barite-type sulfates, especially between Ba^{2+} and Sr^{2+} to form a complete solid solution between barite and celestite. Solid solution is less extensive between barite and anglesite. Mutual solubility is relatively little between sulfates with the barite-type structure and those with the anhydrite-type structure because of the difference in the coordination of the cation sites. There is some indication of substitution of anionic groups between the sulfates, phosphates, arsenates, and vanadates.

Occurrence and Associations. Because the sulfates are oxidized relative to the sulfides, they are commonly found associated with the oxidized zones of sulfide ore deposits. Barite typically occurs around hydrothermal veins in which lead, silver, and copper ores are located, but may also occur as an authigenic cement in certain sandstones. The "barite roses" consist of radiating tabular crystals formed in this authigenic mode. Celestite forms a solid solution with barite, but in its pure form is found as cavity linings associated with calcite, dolomite, or fluorite. Anglesite is

Gypsum
CaSO$_4 \cdot$ 2H$_2$O

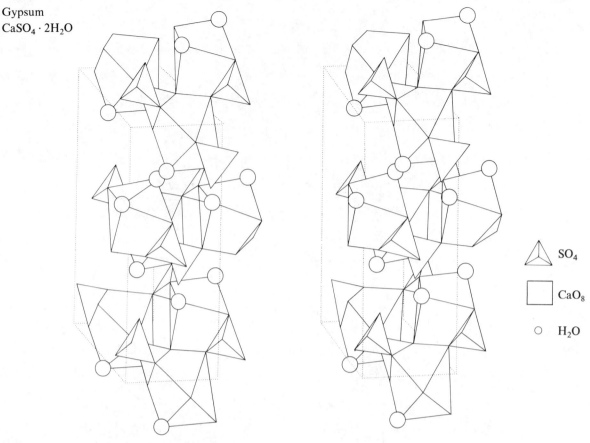

SO$_4$

CaO$_8$

H$_2$O

FIGURE II.111   Crystal structure of gypsum, *b* axis vertical. Ca has distorted cubic coordination.

almost always found in the oxidized zone of galena deposits where it is associated with galena, cerussite, sphalerite, and iron oxides.

Anhydrite is frequently associated with gypsum as veins or fracture fillings in limestone and salt deposits. Both minerals can precipitate during evaporation of saline brines to form evaporite deposits, although most anhydrite appears to be a secondary mineral formed by the dehydration of gypsum.

The other hydrous sulfates have similar associations, antlerite with the oxidized portions of copper deposits and jarosite as coatings on oxidized iron ores. Alunite is generally found in association with hot springs and volcanic fumaroles.

Distinguishing Features. The sulfates can be distinguished as a group by their vitreous luster, perfect cleavage, and relative softness. Gypsum, especially, has perfect {010} cleavage and a hardness on the Mohs scale of only 2. Swallowtail twins on {100} are fairly common. Anhydrite is similar to gypsum and calcite in appearance, but has a greater hardness and three cleavages at right angles. Barite usually forms as tabular prisms with perfect {001} cleavage; its most distinctive feature

is its high density (4.5 Mg/m³). Anglesite is also quite dense, but unlike barite, is almost always associated with galena.

THE CHROMATE, TUNGSTATE, AND MOLYBDATE MINERALS

This group of minerals is structurally like the sulfates in that the metal atom of the anionic group has fourfold coordination with oxygen. Consequently, the structures of the principal minerals in this group consist of isolated tetrahedra connected in various ways depending on the radii and bonding characteristics of the other cations in the structures. Most of these cations enter the structures in eightfold coordination, as in the case of Ca^{2+} in scheelite (CaWO$_4$).

Structure. The major minerals under this heading are scheelite (CaWO$_4$, Figure II.112), wulfenite (PbMoO$_4$), and crocoite (PbCrO$_4$). The first two minerals are isostructural, consisting of isolated tetrahedra joined together by eight coordinated cations. The structure of crocoite is similar to that of anhydrite. The Ca coordination in scheelite is very similar to that of Ca in gypsum.

THE CHROMATE, TUNGSTATE, AND MOLYBDATE MINERALS

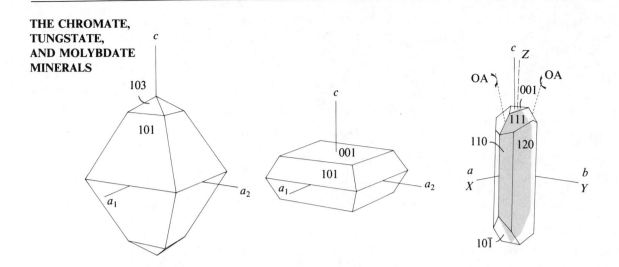

	SCHEELITE	Wulfenite	Crocoite
Formula	$CaWO_4$	$PbMoO_4$	$PbCrO_4$
Coord.	Ca(4) (+4)*, W(4)	Pb(6) (+2)*, Mo(4)	Pb(6) (+3)*, Cr(4)
System	Tetragonal	Tetragonal	Monoclinic
$a =$	0.525	0.542	0.711
$b =$			0.741
$c =$	1.140	1.210	0.681
$\beta =$			102.55
$Z =$	4	4	4
Sp. group	$I4_1/a$	$I4_1/a$	$P2_1/n$
Pt. group	$4/m$	$4/m$	$2/m$
S	1.00	1.00	1.00
SP plane	$\{112\}$	$\{112\}$	$\{101\}$
SP struct.	$A(\frac{1}{8}\frac{1}{8}\frac{1}{4})B(\frac{1}{4}\frac{1}{8}\frac{1}{8})C(\frac{1}{8}\frac{1}{4}\frac{1}{8})$ $A(\frac{1}{8}\frac{1}{8}\frac{1}{4})B(\frac{1}{4}\frac{1}{8}\frac{1}{8})C(\frac{1}{8}\frac{1}{4}\frac{1}{8})$;	$A(\frac{1}{8}\frac{1}{8}\frac{1}{4})B(\frac{1}{4}\frac{1}{8}\frac{1}{8})C(\frac{1}{8}\frac{1}{4}\frac{1}{8})$ $A(\frac{1}{8}\frac{1}{8}\frac{1}{4})B(\frac{1}{4}\frac{1}{8}\frac{1}{8})C(\frac{1}{8}\frac{1}{4}\frac{1}{8})$;	$A(\frac{1}{4}\frac{1}{4}-)B(--\frac{1}{2})$ $A(\frac{1}{4}\frac{1}{4}-)B(--\frac{1}{2})$;
SPI =	69	73	65
Cleavage	gd$\{101\}$, pr$\{112\}$	gd(001), pr(001)	pf$\{110\}$, pr(001)
Twinning	$\{110\}$ occ. penetr.	(001) rare	
RI	$n_\omega = 1.920$ $n_\epsilon = 1.934$	$n_\omega = 2.404$ $n_\epsilon = 2.283$	$n_\alpha = 2.31$ $n_\beta = 2.37$ $n_\gamma = 2.66$
$\delta =$	0.014	0.121	0.35
$2V =$			54°
Sign	(+)	(−)	(+)
Transp.	Transp. to transl.	Transp. to transl.	Translucent
$H =$	$4\frac{1}{2}$–5	3	$2\frac{1}{2}$–3
$D =$	6.11	6.7–7.0	6.0
Color	White, yellow, brown, variable	Yellow, brown, green	Red orange
Streak	White	White	Orange
Luster	Subadamantine	Adamantine, silvery	Adamantine
Fracture	Uneven	Subconchoidal	Subconchoidal
Habit	Bipyramidal, tabular	Tabular, bipyramidal	Acicular
Remarks	Fluorescent (U.V.)	Fragile, zoned Isostr. w. scheelite	$\{110\}$ striated $\|c$ Sectile, wk. pleochr. Pb has diff. CN from CN in wulfenite

*Close second neighbor bonds

THE CHROMATE, TUNGSTATE, AND MOLYBDATE MINERALS

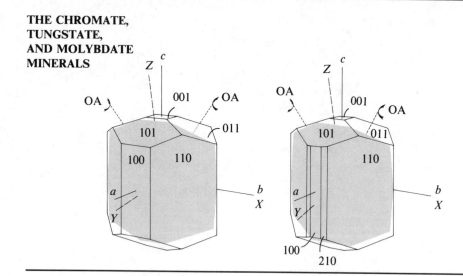

	Ferberite	Huebnerite
Formula	$FeWO_4$	$MnWO_4$
Coord.	Fe(6), W(4) (+2)*	Mn(6), W(4) (+2)*
System	Monoclinic	Monoclinic
$a =$	0.471	0.485
$b =$	0.570	0.577
$c =$	0.494	0.498
$\beta =$	90.0°	90.88°
$Z =$	2	2
Sp. group	$P2/c$	$P2/c$
Pt. group	$2/m$	$2/m$
S; loops	3.00; 2, 3, 4 (CN = 6)	3.00; 2, 3, 4 (CN = 6)
SP plane	(010)	(010)
SP struct.	$A(--\frac{1}{2})B(--\frac{1}{2})$;	$A(--\frac{1}{2})B(--\frac{1}{2})$;
SPI =	44	41
Cleavage	pf(010)	pf(010)
Parting	(100)	(100)
Twinning	[001] comp. pl. (100)	{001} comp. pl. (100)
	{023} rare	{023} rare
RI	$n_\alpha = 2.31$	$n_\alpha = 2.17$
	$n_\beta = 2.40$	$n_\beta = 2.22$
	$n_\gamma = 2.46$	$n_\gamma = 2.30$
$\delta =$	0.15	0.13
$2V =$	79°	73°
Sign	(+)	(+)
Refl.	16%	15%
H =	4–4½	4–4½
D =	7.60	7.25
Color	Black	Red brown
Streak	Black	Red brown
Luster	Submetallic	Submetallic
Fracture	Uneven	Uneven
Habit	Short prismatic	Long prismatic
Remarks	Isomorphic series	

Occasionally magnetic, (100) striated $\parallel c$
*Close second neighbor bonds
End-members of the wolframite series

Scheelite
$CaWO_4$

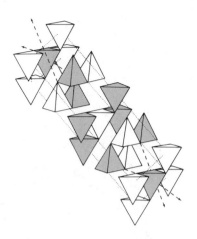

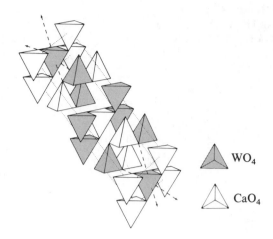

WO₄

CaO₄

FIGURE II.112 **Crystal structure of scheelite, [112] direction vertical. Dashed lines are close second neighbor bonds of Ca.**

The structures of wolframite ($(Fe, Mn)WO_4$) and its end-members ferberite and huebnerite are unique among the tungstates. The coordination around W consists of four closest oxygens with two additional oxygens at a slightly greater distance. The Fe^{2+} and Mn^{2+} cations are also in sixfold coordination. Wolframite is almost isostructural with columbite (Figure II.96).

Chemistry. The substitution of Pb^{2+} for Ca^{2+} in scheelite produces the lead tungstate known as stolzite ($PbWO_4$). A partial solid solution exists between these end-members, and a partial solid solution also exists between scheelite and powellite, the calcium molybdate ($CaMoO_4$). Within the wolframite-type minerals, mutual substitution between Fe^{2+}, Mn^{2+}, Mg^{2+}, and Ni^{2+} in the octahedral sites is well known, with complete solid solution between ferberite ($FeWO_4$) and huebnerite ($MnWO_4$).

Occurrence and Associations. Wolframite is the major ore of tungsten, usually found in high-temperature quartz veins and pegmatites associated with granitic intrusives. Common associations are with scheelite, cassiterite, pyrite, and galena. Scheelite has a similar occurrence and associations, and is also an important tungsten ore. Crocoite and wulfenite are relatively, rare minerals found in the oxidized portions of lead veins.

Distinguishing Features. Wulfenite is distinguished by its softness (H = 3), waxy orange appearance, and association with other lead minerals. Crocoite forms distinctive orange red prismatic crystals. Scheelite can be difficult to identify in hand specimens but usually is white to yellow in color, has a nonmetallic luster, and is unusually dense for a nonmetallic mineral. Fluorescence

(Chapter 4, p. 106) under a short-wavelength ultraviolet light is diagnostic.

Wolframite is generally black and typically forms bladed crystals with one good cleavage {010}.

THE PHOSPHATE, ARSENATE, AND VANADATE MINERALS

The minerals under this heading all have either phosphorus, arsenic, or vanadium tetrahedrally coordinated to oxygen in discreet, anionic groups that control many of the structural and chemical similarities of the phosphates, arsenates, and vanadates. As with the sulfates, we group these minerals in three categories according to (1) whether they are anhydrous, (2) whether they contain an essential hydroxyl or halogens, and (3) whether they contain molecular water.

Structure. The structures of most of the common phosphate, arsenate, and vanadate minerals consist of $(PO_4)^{3-}$, $(AsO_4)^{3-}$, or $(VO_4)^{3-}$ anionic groups as isolated tetrahedra connected by various types of cation polyhedra. Among the anhydrous phosphates, xenotime is isostructural with zircon (Figure II.60). The structure of monazite also has isolated PO₄ tetrahedra, but they are connected by 12 coordinated Ce, La, Y, or Th cations that are randomly distributed.

Among the hydroxyl-bearing and halogen-bearing minerals in this group, apatite is easily the most abundant. Its structure (Figure II.113) consists of isolated PO₄ tetrahedra connected by a network of distorted octahedra and twisted trigonal prisms, both of which accommodate Ca^{2+}. The coordination of Ca^{2+} around the F^{1-} anions describes an equilateral triangle in the (001) plane.

THE PHOSPHATE, ARSENATE, AND VANADATE MINERALS

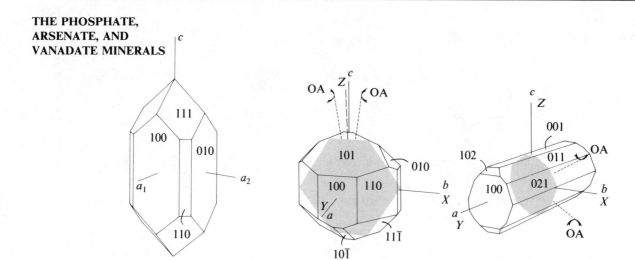

	Xenotime	Monazite	Triphylite
Formula	YPO_4	$(Ce, La, Th)PO_4$	$Li(Fe, Mn)PO_4$
Coord.	Y(6), (+2)*, P(4)	Ce(8), P(4)	Li(8), Fe(8), P(4)
System	Tetragonal	Monoclinic	Orthorhombic
$a =$	0.689	0.679	0.601
$b =$		0.701	0.468
$c =$	0.604	0.646	1.036
$\beta =$		104.4°	
$Z =$	4	4	4
Sp. group	$I4_1/a\ 2/m\ 2/d$	$P2_1/n$	$P2_1/m\ 2_1/c\ 2_1/n$
Pt. group	$4/m\ 2/m\ 2/m$	$2/m$	$2/m\ 2m\ 2/m$
S	1.00	1.00	1.00
SP plane	$\{101\}$	$\{111\}$	(100)
	$A(\frac{1}{4}\frac{1}{4}-)B(--\frac{1}{2})$	$A(\frac{1}{4}\frac{1}{4}-)B(--\frac{1}{2})$	$A(\frac{1}{8}\frac{1}{8}\frac{1}{2})B(\frac{1}{8}\frac{1}{8}\frac{1}{2})$
	$A(\frac{1}{4}\frac{1}{4}-)B(--\frac{1}{2})$;	$A(\frac{1}{4}\frac{1}{4}-)B(--\frac{1}{2})$;	$A(\frac{1}{8}\frac{1}{8}\frac{1}{2})B(\frac{1}{8}\frac{1}{8}\frac{1}{2})$;
SPI =	47	58	56
Cleavage	pf(100)	pf(001), gd(100)	pf(001), gd(010)
Parting		(001), {110}	
RI	$n_\omega = 1.816$	$n_\alpha = 1.785–1.800$	$n_\alpha = 1.68$
	$n_\epsilon = 1.721$	$n_\beta = 1.786–1.801$	$n_\beta = 1.68$
		$n_\gamma = 1.838–1.850$	$n_\gamma = 1.69$
$\delta =$	0.095	0.05	0.01
$2V =$		10–20°	0–56°
Sign	(+)	(+)	(−)
Transp.	Translucent	Translucent	Transp. to transl.
H =	$4\frac{1}{2}$	$5\frac{1}{2}$	$5–5\frac{1}{2}$
D =	4.5	4.9–5.2	3.5–5.5
Color	Yellow brown	Red, brown	Blue green, variable
Streak	Pale brown	White	White, gray
Luster	Resinous	Subresinous	Vitreous, resinous
Fracture	Uneven, splintery	Subconchoidal	Subconchoidal
Habit	Prismatic	Tabular, prismatic	Coarse massive
Remarks	Dipyramidal	Usually cont. Y, Di	(Mn, Fe): lithiophilite
	Isostr. w. zircon	Isostr. w. crocoite	Na ⇌ Li: natrophilite
	*Close second neighbor bonds		Distorted olivine struct.

**THE PHOSPHATE,
ARSENATE, AND
VANADATE MINERALS**

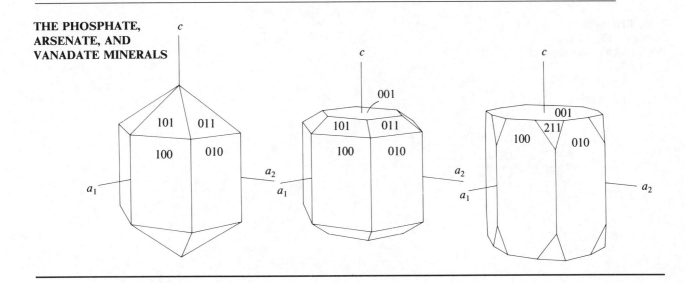

	APATITE	Pyromorphite	Vanadinite
Formula	$Ca_5(OH, F, Cl)(PO_4)_3$	$Pb_5Cl(PO_4)_3$	$Pb_5Cl(VO_4)_3$
Coord.	Ca(6), P(4)	Pb(6), P(4)	Pb(6), V(4)
System	Hexagonal	Hexagonal	Hexagonal
$a =$	0.938	0.997	1.033
$c =$	0.686	0.732	0.735
$Z =$	2	2	2
Sp. group	$P6_3/m$	$P6_3/m$	$P6_3/m$
Pt. group	$6/m$	$6/m$	$6/m$
S	1.00	1.00	1.00
SPI =	50	66	69
Cleavage	gd(001), pr{100}	pr{100}, pr{101}	
RI	$n_\omega = 1.633$	$n_\omega = 2.058$	$n_\omega = 2.416$
	$n_\epsilon = 1.630$	$n_\epsilon = 2.048$	$n_\epsilon = 2.350$
$\delta =$	0.003	0.010	0.066
Sign	$(-)$	$(-)$	$(-)$
Transp.	Transp. to transl.	Transp. to transl.	Translucent
H =	5	$3\frac{1}{2}$–4	3
D =	3.2	7.0	6.9
Color	Green, yellow, variable	Green, yellow, variable	Orange, red
Streak	White	White, yellow	White, yellow
Luster	Subresinous	Resinous	Resinous
Fracture	Conchoidal	Subconchoidal	Subconchoidal
Habit	Prismatic, tabular, occ. fibrous	Prismatic (barrel-shaped), occ. fibrous	Prismatic Hollow prisms
Remarks	(OH), F, Cl cont. variable	$V\langle\rightleftarrows\rangle As$: mimetite Isostr. w. apatite	Encrustation Isostr. w. apatite

Apatite
$Ca_5(F,OH)(PO_4)_3$

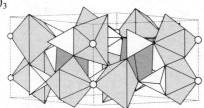 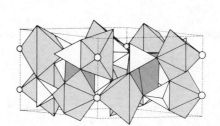 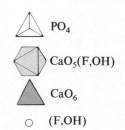

△ PO_4

⬡ $CaO_5(F,OH)$

▲ CaO_6

○ (F,OH)

FIGURE II.113   Crystal structure of apatite, *c* axis vertical.

THE PHOSPHATE, ARSENATE, AND VANADATE MINERALS

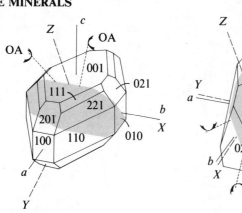
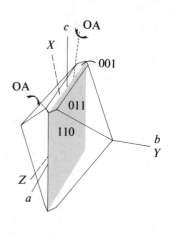

	AUGELITE	Amblygonite	Lazulite
Formula	$Al_2(OH)_3PO_4$	$(Li, Na)(F, OH)PO_4$	$(Fe, Mg)Al_2(OH)_2(PO_4)_2$
Coord.	Al(5 and 6), P(4)	Li(6), Na(6), P(4)	Fe(6), Mg(6), Al(6), P(4)
System	Monoclinic	Triclinic	Monoclinic
$a =$	1.312	0.519	0.716
$b =$	0.799	0.712	0.726
$c =$	0.507	0.504	0.724
$\alpha =$		112.02°	
$\beta =$	112.25°	97.82°	120.67°
$\gamma =$		68.12°	
$Z =$	4	2	2
Sp. group	$C2/m$	$P\bar{1}$	$P2_1/c$
Pt. group	$2/m$	$\bar{1}$	$2/m$
S	1.00	1.00	1.00
SPI =	43	52	51
Cleavage	gd(101)	pf(100), gd(110), pr(011)	pr(011)
Twinning		(111)	
RI	$n_\alpha = 1.574$	$n_\alpha = 1.59$	$n_\alpha = 1.612$
	$n_\beta = 1.588$	$n_\beta = 1.60$	$n_\beta = 1.634$
	$n_\gamma = 1.576$	$n_\gamma = 1.62$	$n_\gamma = 1.643$
$\delta =$	0.014	0.03	0.031
$2V =$	51°	52–90°	70°
Sign	(+)	(−)	(−)
Transp.	Transp. to transl.	Transp. to transl.	Translucent
Pleochr.			X = colorless
			$Y = Z$ = blue
H =	5	6	6
D =	2.7	3.0	3.0
Color	Colorless, white	White, green	Azure blue
Streak	White	White	White
Luster	Vitreous	Vitreous, greasy	Vitreous
Fracture	Uneven	Subconchoidal	Uneven
Habit	Tabular, massive	Equant	Massive, prismatic
Remarks		RI increases w. (OH)	(Fe, Ni): scorzalite

THE PHOSPHATE, ARSENATE, AND VANADATE MINERALS

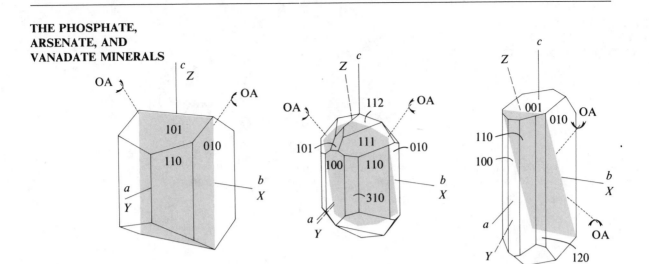

	WAVELLITE	Vivianite	Erythrite
Formula	$Al_3(OH)_3(PO_4)_3 \cdot 5H_2O$	$Fe_3(PO_4)_3 \cdot 8H_2O$	$Co_3(AsO_4)_2 \cdot 8H_2O$
Coord.	Al(6), P(4)	Fe(6), P(4)	Co(6), As(4)
System	Orthorhombic	Monoclinic	Monoclinic
$a =$	0.962	1.008	1.026
$b =$	1.736	1.343	1.337
$c =$	0.699	0.470	0.474
$\beta =$		104.5°	105.1°
$Z =$	4	2	2
Sp. group	$P2_1/c\,2_1/m\,2_1/n$	$C2/m$	$C2/m$
Pt. group	$2/m\,2/m\,2/m$	$2/m$	$2/m$
S	1.00	1.00	1.00
SP plane		(100)	(100)
SP struct.		$A(--\tfrac{3}{8})B(-\tfrac{1}{4}\tfrac{1}{4})C(-\tfrac{3}{8}-)$	$A(--\tfrac{3}{8})B(-\tfrac{1}{4}\tfrac{1}{4})C(-\tfrac{3}{8}-)$
		$A(\tfrac{1}{4}\tfrac{1}{4}-)B(\tfrac{3}{8}--)C(\tfrac{1}{4}-\tfrac{1}{4})$;	$A(\tfrac{1}{4}\tfrac{1}{4}-)B(\tfrac{3}{8}--)C(\tfrac{1}{4}-\tfrac{1}{4})$;
SPI =	35	38	39
Cleavage	pf{101}, pf(010)	pf(010), pr(100)	pf(010)
RI	$n_\alpha = 1.525$	$n_\alpha = 1.579$	$n_\alpha = 1.626$
	$n_\beta = 1.534$	$n_\beta = 1.603$	$n_\beta = 1.661$
	$n_\gamma = 1.552$	$n_\gamma = 1.633$	$n_\gamma = 1.699$
$\delta =$	0.027	0.054	0.073
$2V =$	72°	83°	90°
Sign	(+)	(+)	(−)
Transp.	Translucent	Transp. to transl.	Translucent
Pleochr.		X = blue	X = pink
		$Y = Z$ = green	Y = violet
			Z = red
H =	3–4	2	1–2
D =	2.36	2.58	3.06
Color	White, yellow, green	Colorless, green	Purple red
Streak	White	White	Pale purple
Luster	Vitreous	Pearly, vitreous	Adamantine
Fracture	Subconchoidal	Sectile	Sectile
Habit	Globular, radiating, spherulitic	Reniform, fibrous	Reniform, fibrous
Remarks	Crystals tabular	Flexible lamellae	Isostr. w. vivianite
			"Cobalt bloom"
			$Ni\langle\rightleftharpoons\rangle Co$: annabergite

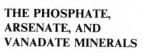

THE PHOSPHATE, ARSENATE, AND VANADATE MINERALS

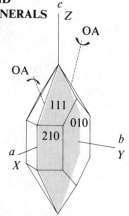

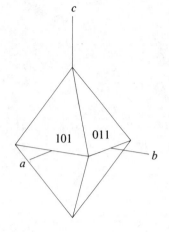

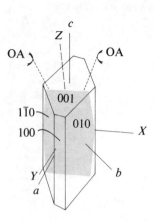

	Scorodite	Variscite	Turquoise
Formula	$FeAsO_4 \cdot 4H_2O$	$AlPO_4 \cdot 2H_2O$	$CuAl_6(OH)_2(PO_4)_4 \cdot 4H_2O$
Coord.	Fe(6), As(4)	Al(6), P(4)	Cu(6), Al(6), P(4)
System	Orthorhombic	Orthorhombic	Triclinic
$a =$	1.043	0.987	0.748
$b =$	0.896	0.957	0.995
$c =$	1.015	0.852	0.769
$\alpha =$			111.65°
$\beta =$			115.38°
$\gamma =$			69.43°
$Z =$	8	8	1
Sp. group	$P2_1/c\ 2_1/a\ 2_1/b$	$P2_1/c\ 2_1/a\ 2_1/b$	$P\bar{1}$
Pt. group	$2/m\ 2/m\ 2/m$	$2/m\ 2/m\ 2/m$	$\bar{1}$
S; group	1.25; pairs	1.00	1.00
SPI $=$	44	39	35
Cleavage	gd{120}, pr(100), pr(010)	gd(010)	pf(001), gd(010)
RI	$n_\alpha = 1.784$	$n_\alpha = 1.55–1.56$	$n_\alpha = 1.61$
	$n_\beta = 1.796$	$n_\beta = 1.57–1.58$	$n_\beta = 1.62$
	$n_\gamma = 1.814$	$n_\gamma = 1.58–1.59$	$n_\gamma = 1.65$
$\delta =$	0.030	0.03	0.04
$2V =$	54°	48–54°	40°
Sign	(+)	(−)	(+)
Transp.	Transp. to transl.	Transp. to transl.	Translucent
$H =$	3–4	4	6
$D =$	3.2	2.5	2.7
Color	Green, brown	Green, yellow	Blue, green
Streak	White	White	White, green
Luster	Resinous	Resinous	Resinous, waxy
Fracture	Uneven	Subconchoidal, splintery	Subconchoidal, brittle
Habit	Prismatic, dipyramidal	Prismatic	Reniform
Remarks	Occ. earthy	Isostr. w. scorodite Fe⇌Al: strengite	

THE PHOSPHATE, ARSENATE, AND VANADATE MINERALS

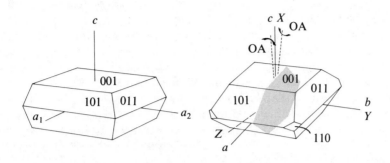

	Autunite	Carnotite
Formula	$CaU_2O_4(PO_4)_2 \cdot 10H_2O$	$K_2U_2O_4(VO_4)_2 \cdot 3H_2O$
Coord.	Ca(6), U(2) (+4)*, P(4)	V(5), V(4), K(9), U(2) (+5)*
System	Tetragonal	Monoclinic
$a =$	0.700	1.047
$b =$		0.841
$c =$	2.067	0.691
$\beta =$		103.67°
$Z =$	4	1
Sp. group	$I4/m\ 2/m\ 2/m$	$P2_1/a$
Pt. group	$4/m\ 2/m\ 2/m$	$2/m$
S; group	1.00	1.40; 2 (CN = V)
SPI =	32	30
Cleavage	pf(001), gd(100), gd(010), gd{110}	pf(001)
RI	$n_\omega = 1.577$ $n_\epsilon = 1.553$	$n_\alpha = 1.75$ $n_\beta = 1.92$ $n_\gamma = 1.95$
$\delta =$	0.024	0.20
$2V =$		38–44°
Sign	(−)	(−)
Transp.	Transp. to transl.	Translucent
Pleochr.	ω = dark yellow ϵ = pale yellow	
H =	$2–2\frac{1}{2}$	2
D =	3.15	4.5
Color	Yellow	Yellow, green
Streak	Yellow	Yellow
Luster	Adamantine	Dull, earthy
Fracture	Uneven	
Habit	Tabular, foliated	Fine powder
Remarks	Cu $\rightleftharpoons$ Ca: torbernite Fluorescent (U.V.)	Crystals rare

*Close second neighbor bonds

The minerals pyromorphite, mimetite, and vanadinite are all isostructural with apatite.

The structure of augelite (Figure II.114) consists of isolated PO_4 tetrahedra connected by two types of isolated polyhedral units, both of which accommodate Al^{3+}. One unit is a pair of Al octahedra, and the other is a pair of polyhedra in which Al^{3+} has fivefold coordination. The structure of amblygonite also has pairs of five-coordinated polyhedra, but they are occupied by Li^{1+}. The Li polyhedra are connected to chains of corner-sharing octahedra that form a zigzag pattern along the *b* axis. Each PO_4 tetrahedron in the amblygonite structure is connected to an Al octahedron.

A variety of structures within the group of phosphates, arsenates, and vanadates contain essential molecular water. The structure of wavellite (Figure II.115) is fairly representative, consisting of single chains of corner-sharing Al octahedra connected by isolated PO_4 tetrahedra. Molecular water occupies the larger cavities in the structure. Two important uranyl phosphates, autunite and carnotite, are also in this group.

Chemistry. Both xenotime and monazite are well known for their contents of rare earth elements, especially Ce, La, U, and Th substituting in the relatively large cation site. The rare earth elements Nd, Pr, and Dy can be sufficiently concentrated along with Ce in monazite to make mining of monazite sands a profitable venture.

The structure of apatite (Figure II.113) allows for complete solid solution between fluorapatite ($Ca_5FP_3O_{12}$), chlorapatite ($Ca_5ClP_3O_{12}$), and

Augelite
$Al_2(OH)_3PO_4$

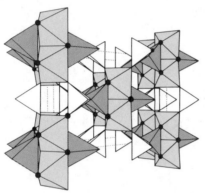
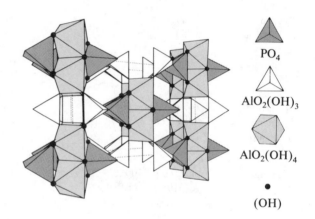

FIGURE II.114   Crystal structure of augellite, *b* axis vertical.

Wavellite
$Al_3(OH)_3(PO_4)_2 \cdot 5H_2O$

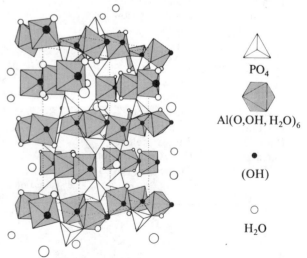

FIGURE II.115   Crystal structure of wavellite, *b* axis vertical.

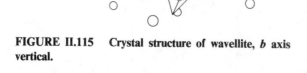

hydroxylapatite ($Ca_5OHP_3O_{12}$) by virtue of the mutual substitution of F^{1-}, Cl^{1-}, and $(OH)^{1-}$. In addition, the anionic groups $(CO_3)^{2-}$ and $(SO_4)^{2-}$ may substitute for $(PO_4)^{3-}$ in apatite, in which case charge balance is maintained by a coupled substitution of Ca^{2+} by Na^{1+}. Complete solid solution also exists between pyromorphite and mimetite by substitution of $(PO_4)^{3-}$ and $(AsO_4)^{3-}$. Although the two latter minerals are isostructural with vanadinite (and with apatite), the substitution of $(VO_4)^{3-}$ for $(PO_4)^{3-}$ and $(AsO_4)^{3-}$ anionic groups appears not to be complete.

Occurrence and Associations. Xenotime and monazite are both found as common accessory minerals in granites, syenites, and certain pegmatites. Because of their resistance to weathering, both minerals are common constituents of alluvial sands in which they may form commercial deposits. Apatite has a similar occurrence but is more widespread in the more mafic igneous rocks and in most metamorphic rocks. It is also very widespread in sedimentary rocks, especially in phosphorite deposits interbedded with sandstones and shales and as an essential component of fossil bones. The term *collophane* is used loosely to denote sedimentary phosphates of a very fine-grained aggregate of hydroxylapatite, carbonate-phosphate, and other apatites of variable composition.

The lead minerals pyromorphite, mimetite, and vanadinite almost always occur as secondary minerals in lead deposits. They are frequently associated with other lead minerals such as cerussite and anglesite. A common secondary mineral in the oxidized zone of cobalt deposits is erythrite ($Co_3(AsO_4)_2 \cdot 8H_2O$), known also as *cobalt bloom* for its crimson red color, and used as a prospecting guide to cobalt deposits. Annabergite ($Ni_3(AsO_4)_2 \cdot 8H_2O$) is called *nickel bloom* because of its green color and its usefulness as a guide to oxidized nickel deposits.

Turquoise is a secondary mineral formed by the alteration of aluminous igneous rocks and usually is found in veins associated with chalcedony and limonite. The principal use of turquoise is as a semiprecious decorative stone.

Wavellite (Figure II.115) occurs as radiating aggregates on the joint faces of phosphatic shales and slates as well as in sedimentary phosphate deposits.

Autunite and its copper analogue torbernite are secondary minerals in the oxidized zone of uraninite veins and are distinguished by their bright lemon yellow and green colors, respectively. Carnotite also has a bright lemon yellow color and is a secondary mineral formed by the oxidation of uranium deposits.

ADDITIONAL READINGS

Breck, D. R., 1974. *Zeolite molecular sieves*. New York: John Wiley and Sons.

Grim, R., 1968. *Clay mineralogy*. 2nd ed. New York: McGraw-Hill.

Mineralogical Society of America Reviews in Mineralogy. Washington, D.C.: Mineralogical Society of America:

Volume 1, 1974. *Sulfide Mineralogy* edited by P. H. Ribbe.

Volume 2, 1975. *Feldspar Mineralogy* edited by P. H. Ribbe.

Volume 3, 1976. *Oxide Minerals* edited by D. R. Rumble, III.

Volume 4, 1977. *Mineralogy and Geology of Natural Zeolites* edited by F. A. Mumpton.

Volume 5, 1978. *Orthosilicates* edited by P. H. Ribbe.

Volume 6, 1979. *Marine Minerals* edited by R. G. Burns.

Volume 7, 1980. *Pyroxenes* edited by C. T. Prewitt.

Volume 8, 1981. *Kinetics of Geochemical Processes* edited by A. C. Lasaga and R. J. Kirkpatrick.

Volumes 9A, 9B, 1982. *Amphiboles and Other Hydrous Pyriboles* edited by D. R. Veblen.

Smith, J. V., 1974. *Feldspar minerals*. 2 vols. New York: Springer-Verlag.

APPENDIX 1

UNITS AND CONSTANTS

UNITS OF WEIGHTS AND MEASURES

Throughout history, human beings have sought to establish a unified system of weights and measures. Early decimal systems of measure date back to the Babylonians and the Phoenicians, but as civilizations rose and fell, so too did the various units of measurement. Not until the French Revolution in 1789 did a unified system begin to develop. The French Academy of Sciences proposed an international unit of length against which all other measurements of length would be compared. The academy members chose as this unit one ten-millionth of a meridional distance from the north pole to the equator along a route that passes through Paris. Surveying and calculating this distance took 6 years. The unit became known as the meter, from the Greek word metron (a measure), and now provides the basis for the modern metric system. The metric unit of mass, called the gram, was later defined as the mass of a cube of pure water that is 1/100 of a meter on a side (1 cubic centimeter), measured at the temperature of maximum water density (277.16 K). The unit of fluid capacity was chosen as the volume of a cube 1/10 of a meter on a side, and became known as the liter.

Many scientists immediately welcomed such a standardized system. Its precision and ease of usage represented a substantial improvement over the existing English system of weights and measures. For example, the accepted English unit of the yard was once defined as the distance from the tip of the nose to the outstretched fingers of some forgotten Anglo-Saxon king. As kings come in different sizes, this choice of unit caused immediate problems! Similar problems arose from an early definition of the acre as the area of land that could be plowed by a yoke of oxen in one day, and the definition of the mile as the distance a Roman soldier could travel in one day.

Today, a refined and modernized metric system known as the International System of Units has gained nearly total acceptance throughout the world. The system name is abbreviated SI units from the French "Système International d'Unités." The meter has been redefined as equal to a precise number of wavelengths of Krypton 86, and the unit of mass of 1000 grams is the mass of a particular platinum-iridium alloy kept at the International Bureau of Weights and Measures in France. SI units are widely used in the professional scientific literature, and have been adopted by the U.S. Geological Survey and other federal agencies. In the United States, formal conversion to a metric system of weights and measures was provided by the Metric Conversion Act of 1975.

The foundation of SI units rests on the following seven fundamental units:

Unit	Name	SI Symbol
Length	meter	m
Mass	kilogram	kg
Time	second	s
Electric current	ampere	A
Temperature	kelvin	K
Quantity	mole	mol
Luminous intensity	candela	cd

Two supplementary units for angular measurements can be added to this list. The radian (rad) is for the measurement of plane angles, and the steradian (sr) is for the measurement of solid angles. These units, as well as the seven previously listed, have been carefully and precisely measured. From this brief list, all other SI units are derived. A unique characteristic of SI units is that the product or quotient of any combination of basic units is a single unit of the derived quantity. For example, the SI unit of force is called the newton (N) and is defined[1] as:

$$N = (kg)(m)/s^2 = kg\ m\ s^{-2}$$

that is, one newton (N) is the force that will impart to a mass of 1 kg an acceleration of 1 meter per second per second (1 m s^{-2}). The newton, then, becomes a *derived* unit on which other units of measurement are based. For example, the fundamental unit of energy is the joule (J). It is defined as:

$$J = N\ m = kg\ m^2\ s^{-2}$$

that is, the energy required for a force of one newton to be applied through a distance of one meter. Table A1.1 gives the most common physical quantities used in science, the SI unit for each quantity, and the SI formula for each quantity.

The unit of time, the second, was originally defined according to the earth's rate of rotation, but because that rate changes with time, the unit was redefined to coincide with the duration of time required for a specified large number of radiation cycles associated with the Cesium 133 atom.

The unit of electric current, the ampere (A), is defined as that amount of current required to produce a force of 2×10^{-7} newtons between two parallel wires separated by a distance of 1 meter.

The unit of temperature, the kelvin (K), is defined as 1/273.16, the triple point temperature of H_2O (the unique temperature at which ice, water, and steam coexist in stable equilibrium = 273.16 kelvins).

[1] The scientific notation of superscripts is used throughout. $10^3 = 1000$, $10^2 = 100$, $10^1 = 10$, $10^0 = 1$, $10^{-1} = 1/10 = 0.1$ $10^{-2} = 0.01 = 1/100$, $10^{-3} = 0.001 = 1/1000$, $X^{-2} = 1/X^2$

TABLE A1.1. *Units of Measurement*

Physical Quantity	Name of Unit	SI Symbol for Unit	Formula
Acceleration			m/s^2
Amount of substance	mole	mol	
Angle (plane)	radian	rad	
Angle (solid)	steradian	sr	
Angular acceleration			rad/s^2
Angular momentum			$kg\ m^2/s$
Angular velocity			rad/s
Area			m^2
Bulk modulus			N/m^2
Chemical potential	joule	J	Nm
Compressibility			$1/Pa$
Density			kg/m^3
Electric charge	coulomb	C	A s
Electrical capacitance	farad	F	A s/V
Electrical conductivity	siemens	S	A/V
Electrical current	ampere	A	
Electrical field strength			V/m
Electrical inductance	henry	H	V s/A
Electrical moment			
Electrical resistance	ohm	Ω	V/A
Electromotive force	volt	V	W/A
Energy	joule	J	N m
Enthalpy	joule	J	N m
Entropy			J/K
Force	newton	N	$kg\ m/s^2$
Frequency	hertz	Hz	$(cycle)/s^2$
Gibbs free energy	joule	J	N m
Heat capacity			J/K
Heat flow			$J/s\ m^2$
Helmholtz free energy	joule	J	N m
Illuminance	lux	lx	lm/m^2
Length	meter	m	
Luminance			cd/m^2
Luminous flux	lumen	lm	cd sr
Luminous intensity	candela	cd	
Magnetic field strength			A/m
Magnetic flux	weber	Wb	V s
Magnetic flux density	tesla	T	Wb/m^2
Magnetic permeability			H/m
Magnetic permittivity			F/m
Mass	kilogram	kg	
Power	watt	W	J/s
Pressure	pascal	Pa	N/m^2
Resistivity			Ω m
Shear modulus			N/m^2
Surface tension			N/m
Temperature	kelvin	K	
Thermal conductivity			W/mk
Thermal expansion			1/K
Time	second	s	
Torque			N m
Velocity			m/s
Viscosity (dynamic)			Ns/m^2
Viscosity (kinematic)			m^2/s
Voltage	volt	V	W/A
Volume			m^3
Wavelength			m
Wave number			1/m
Work	joule	J	N/m
Young's modulus			N/m^2

The unit of quantity, the mole (mol), is the amount of substance corresponding to 6.023×10^{23} (Avogadro's number) elementary entities of that substance. In mineralogy, these entities are commonly taken as atoms or ions. One mole of oxygen atoms would therefore consist of 6.023×10^{23} oxygen atoms that would be contained in 0.016 kg ($= 16$ g) of oxygen. Mineralogists also refer to minerals in terms of moles. For example, a single mole of kyanite (Al_2SiO_5) consists of 6.023×10^{23} units of Al_2OSiO_4 contained in 0.16205 kg of Al_2OSiO_4. The latter number is the molecular weight (MW) of kyanite, that is, $2MW(Al) + MW(Si) + 5MW(O) = 162.05$ g.

TABLE A1.2. *Prefixes for Multiples and Submultiples of Units*

Multiple or Submultiple		Prefix	SI Symbol
10^{12}	1000 000 000 000	tera	T
10^{9}	1000 000 000	giga	G
10^{6}	1000 000	mega	M
10^{3}	1000	kilo	k
10^{2}	100	hecto	h
10^{1}	10	deka	da
10^{0}	1		
10^{-1}	0.1	deci	d
10^{-2}	0.01	centi	c
10^{-3}	0.001	milli	m
10^{-6}	0.000 001	micro	μ
10^{-9}	0.000 000 001	nano	n
10^{-12}	0.000 000 000 001	pico	p
10^{-15}		femto	f
10^{-18}		atto	a

TABLE A1.3. *Other Useful Conversion Factors*

Unit	Value
1 Ångstrom	$= 10^{-10}$ m $= 10^{-8}$ cm $= 10^{-7}$ mm $= 10^{-4}$ μm $= 10^{-1}$ nm
1 atmosphere	$= 1.013 \times 10^{5}$ Pa $= 1.01325$ bar
1 bar	$= 10^{5}$ Pa $= 0.9869$ atm $= 14.5038$ p.s.i.
1 calorie (N.B.S.)	$= 4.184$ j $= 3.9657 \times 10^{-3}$ B.T.U.
1.24 electron volts (eV)	$= 28.59$ kcal mol^{-1} ($= 1000$ nm)
1 gigapascal (GPa)	$= 10^{9}$ Pa $= 10^{4}$ bar $= 10$ kbar
1 g/cm^3	$= 10^{3}$ kg m^{-3} $= 1$ Mg/m^3
1 kilobar (kbar)	$= 10^{8}$ Pa $= 986.9$ atm
1 kilopascal (kPa)	$= 1000$ Pa $= 10^{-2}$ bar
1 kilocalorie (N.B.S.)	$= 4.184 \times 10^{10}$ erg $= 4184$ j $= 3.966$ B.T.U.
1 liter	$= 1.0567$ quart (U.S.)
1 megagram m^{-3} (Mg/m^3)	$= 10^{3}$ kg/m^3 $= 1$ g/cm^3
1 meter	$= 3.28084$ feet
1 millimeter	$= 0.0394$ inch
1 micrometer (μm)	$= 10^{4}$ Å $= 10^{-3}$ mm $= 10^{-6}$ m $= 1$ micron (μ)
1 nanometer (nm)	$= 10$ Å $= 10^{-9}$ m
1 ounce (avoir)	$= 28.35$ g
1 point (gem)	$= 10^{-2}$ carat $= 0.04$ pearl grain
1 poise	$= 10^{-1}$ N s m^{-2}
1 pound (avoir)	$= 453.59$ g
1 radian (ra)	$= 57.296$ degrees
1 stoke	$= 10^{-4}$ m^2 s^{-1}

The unit of luminous intensity, the candela (cd), is defined as the intensity of 1.67×10^{-6} square meters of the cone of light emitted from a black body heated to a specified temperature. The unit amount of light is a more familiar but derived SI unit called the lumen (lm). A normal 100-watt light bulb radiates about 1700 lm.

Finally, a radian is defined as that plane angle measured at the center of a circle subtended at its periphery by an arc of a length equal to the circle's radius. The steradian unit (sr) is that solid angle with vertex at the center of a sphere and subtended at the center of the sphere by a specified area on the spherical surface. That area is equal to the square of the radius of the sphere. A radian is equal to 57.296 degrees and underlies the constant π given by

$$\pi = 360/2 \, \text{rad} = 3.14159265$$

Scientists commonly need to describe very large numbers, for example, the mass of a star, and very small numbers, such as the mass of subatomic particles. As the magnitudes of the fundamental SI units are carefully defined, multiples or submultiples of these can be described with appropriate prefixes. These are shown in Table A1.2.

TABLE A1.4.   *Useful Numerical Factors*

Pi (π)	$= 3.141592654$
Natural log	$= e = 2.71828183$
$\log N$	$= 0.43429 \ln N$
$\ln N$	$= 2.303 \log N$
Absolute zero	$= -273.16 \,°C$ exactly
Triple point of H_2O	$= 273.16 \, K$ exactly

TABLE A1.5.   *Physical Constants*

Quantity		Numerical Value (SI)	Numerical Value (cgs)
Acceleration due to gravity at earth's surface	g		
Equator		$9.7805 \, \text{m s}^{-2}$	$9.7805 \times 10^2 \, \text{cm s}^{-2}$
North Pole		$9.8322 \, \text{m s}^{-2}$	$9.8322 \times 10^2 \, \text{cm s}^{-2}$
Acceleration due to gravity at moon's surface		$1.6200 \, \text{m s}^{-2}$	$1.6200 \times 10^2 \, \text{cm s}^{-2}$
Avogadro's constant	N_A	$6.02252 \times 10^{23} \, \text{mol}^{-1}$	$6.02252 \times 10^{23} \, \text{mol}^{-1}$
Boltzmann's constant	k	$1.380622 \times 10^{-23} \, \text{J K}^{-1}$	$1.380622 \times 10^{-16} \, \text{erg K}^{-1}$
Electron charge		$1.6021917 \times 10^{-19} \, \text{C}$	$1.6021917 \times 10^{-20} \, \text{emu}$
Faraday's constant	F	$9.64870 \times 10^4 \, \text{C mol}^{-1}$	$9.64870 \times 10^3 \, \text{emu mol}^{-1}$
Gas constant	R	$8.31432 \, \text{J mol}^{-1} \text{K}^{-1}$	$1.9872 \, \text{cal mol}^{-1} \text{K}^{-1}$
Gravitational constant	G	$6.6732 \times 10^{-11} \, \text{N m}^2 \text{kg}^{-2}$	$6.6732 \times 10^{-8} \, \text{dyne cm}^2 \text{g}^{-2}$
Mass of an electron		$9.109558 \times 10^{-31} \, \text{kg}$	$9.109558 \times 10^{-28} \, \text{g}$
Mass of a neutron		$1.675 \times 10^{-27} \, \text{kg}$	$1.675 \times 10^{-24} \, \text{g}$
Mass of a proton		$1.672614 \times 10^{-27} \, \text{kg}$	$1.672614 \times 10^{-24} \, \text{g}$
Permittivity of a vacuum	ϵ_0	$8.8419413 \times 10^{-12} \, \text{F m}^{-1}$	$1.0 \, \text{dyne cm}^2 \text{statcoul}^{-2}$
Permeability of a vacuum	μ_0	$1.2566371 \times 10^{-6} \, \text{H m}^{-1}$	
Planck's constant	h	$6.626196 \times 10^{-34} \, \text{J s}$	$6.626196 \times 10^{-27} \, \text{erg s}$
Velocity of light in a vacuum	c	$2.99792458 \times 10^8 \, \text{m s}^{-1}$	$2.99792458 \times 10^{10} \, \text{cm s}^{-1}$

Wavelength of $K\alpha$ x-rays	$K\alpha$	$K\alpha_1$	$K\alpha_2$
Cobalt	0.17902 nm	0.178890 nm	0.179279 nm
Copper	0.15418 nm	0.154050 nm	0.154434 nm
Iron	0.19373 nm	0.193597 nm	0.193991 nm
Molybdenum	0.07107 nm	0.070926 nm	0.071354 nm
Chromium	0.22909 nm	0.228962 nm	0.229352 nm

MINERAL HARDNESS AND IDENTIFICATION

TABLE A2.1. *Mineral Hardness and Identification Table*

Hardness	Mineral	Luster	Ref. page	Hardness	Mineral	Luster	Ref. page
1	Talc		322	$2-2\frac{1}{2}$	Kaolinite		321
$1-1\frac{1}{2}$	Molybdenite	M	390	$2-2\frac{1}{2}$	Palygorskite		324
$1-1\frac{1}{2}$	Sulvanite	M	382	$2-2\frac{1}{2}$	Proustite		395
$1-2$	Arsenic	M	376	$2-2\frac{1}{2}$	Sepiolite		324
$1-2$	Calomel		402	$2-3$	Chlorite		323
$1-2$	Erythrite		451	$2-5$	Serpentine		328
$1-2$	Graphite	(M)	377	$2\frac{1}{2}$	Brucite		421
$1-2$	Nitratite		433	$2\frac{1}{2}$	Carnallite		402
$1\frac{1}{2}$	Pyrophyllite		321	$2\frac{1}{2}$	Chalcanthite		442
$1\frac{1}{2}-2$	Covellite	(M)	391	$2\frac{1}{2}$	Chlorargyrite	R	401
$1\frac{1}{2}-2$	Orpiment	R	399	$2\frac{1}{2}$	Cryolite		403
$1\frac{1}{2}-2$	Realgar	R	399	$2\frac{1}{2}$	Halite		401
$1\frac{1}{2}-2\frac{1}{2}$	Sulfur	R	377	$2\frac{1}{2}$	Muscovite		321
2	Carnotite		453	$2\frac{1}{2}$	Phlogopite		322
2	Gypsum		422	$2\frac{1}{2}$	Ulexite		437
2	Niter		433	$2\frac{1}{2}-3$	Anglesite		440
2	Pyrargyrite		395	$2\frac{1}{2}-3$	Biotite		323
2	Stibnite	M	397	$2\frac{1}{2}-3$	Bismuth	M	376
2	Sylvite		401	$2\frac{1}{2}-3$	Boulangerite	M	397
2	Vivianite		451	$2\frac{1}{2}-3$	Chalcocite	M	391
2	Zinc	M	375	$2\frac{1}{2}-3$	Copper	M	375
$2-2\frac{1}{2}$	Argentite	M	391	$2\frac{1}{2}-3$	Crocoite		445
$2-2\frac{1}{2}$	Autunite		453	$2\frac{1}{2}-3$	Gold	M	374
$2-2\frac{1}{2}$	Borax		436	$2\frac{1}{2}-3$	Silver	M	374
$2-2\frac{1}{2}$	Cinnabar		390	$2\frac{1}{2}-3\frac{1}{2}$	Gibbsite		421
$2-2\frac{1}{2}$	Epsomite		442				
$2-2\frac{1}{2}$	Galena	M	387				

NOTE: Minerals that define the Mohs hardness scale are underlined. Page numbers reference appropriate mineral table or discussion in Part II. Symbol M denotes metallic luster; (M) denotes submetallic luster. Symbol R denotes resinous luster. All other minerals have nonmetallic luster.

TABLE A2.1 (*continued*)

Hardness	Mineral	Luster	Ref. page	Hardness	Mineral	Luster	Ref. page
$2\frac{1}{2}$–$6\frac{1}{2}$	Pyrolusite	(M)	409	$4\frac{1}{2}$–5	Apophyllite		325
3	Bornite	M	382	$4\frac{1}{2}$–5	Pectolite		348
3	Calcite		426	$4\frac{1}{2}$–5	Scheelite		445
3	Enargite	M	383	$4\frac{1}{2}$–5	Wollastonite		348
3	Kernite		436	$4\frac{1}{2}$–$5\frac{1}{2}$	Violarite	M	388
3	Vanadinite	R	449	5	Apatite		449
3	Wulfenite		445	5	Augellite		450
3–$3\frac{1}{2}$	Anhydrite		439	5	Columbite	(M)	409
3–$3\frac{1}{2}$	Antimony	M	376	5	Lepidochrosite	(M)	423
3–$3\frac{1}{2}$	Atacamite		402	5	Titanite		366
3–$3\frac{1}{2}$	Barite		439	5–$5\frac{1}{2}$	Brookite		408
3–$3\frac{1}{2}$	Celestite		440	5–$5\frac{1}{2}$	Goethite	(M)	422
3–$3\frac{1}{2}$	Cerussite		432	5–$5\frac{1}{2}$	Hausmanite	(M)	416
$3\frac{1}{2}$	Witherite		430	5–$5\frac{1}{2}$	Monazite		448
3–$3\frac{1}{2}$	Millerite	M	390	5–$5\frac{1}{2}$	Natrolite		317
3–4	Antigorite		322	5–$5\frac{1}{2}$	Nicolite	M	387
3–4	Laumontite		316	5–$5\frac{1}{2}$	Triphylite		448
3–4	Scorodite	R	452	5–$5\frac{1}{2}$	Ullmanite	M	394
3–4	Tetrahedrite	M	397	5–6	Amphiboles		342
3–4	Wavellite		451	5–6	Ilmenite	M	407
$3\frac{1}{2}$	Cubanite	M	383	5–6	Romanechite	(M)	419
$3\frac{1}{2}$	Strontianite		432	5–6	Scapolite		312
$3\frac{1}{2}$–4	Alunite		441	5–6	Uraninite	M	417
$3\frac{1}{2}$–4	Ankerite		429	5–7	Kyanite		363
$3\frac{1}{2}$–4	Antlerite		441	$5\frac{1}{2}$	Analcime		310
$3\frac{1}{2}$–4	Aragonite		430	$5\frac{1}{2}$	Cobaltite	M	394
$3\frac{1}{2}$–4	Azurite		431	$5\frac{1}{2}$	Chromite	M	415
$3\frac{1}{2}$–4	Boehmite		423	$5\frac{1}{2}$	Periclase		406
$3\frac{1}{2}$–4	Chalcopyrite	M	382	$5\frac{1}{2}$	Perovskite	(M)	417
$3\frac{1}{2}$–4	Cuprite	(M)	419	$5\frac{1}{2}$–6	Anatase		408
$3\frac{1}{2}$–4	Dolomite		429	$5\frac{1}{2}$–6	Arsenopyrite	M	396
$3\frac{1}{2}$–4	Heulandite		317	$5\frac{1}{2}$–6	Leucite		310
$3\frac{1}{2}$–4	Malachite		431	$5\frac{1}{2}$–6	Nepheline		310
$3\frac{1}{2}$–4	Pentlandite	M	388	$5\frac{1}{2}$–6	Rhodonite		348
$3\frac{1}{2}$–4	Pyromorphite	R	449	$5\frac{1}{2}$–6	Skutterudite	M	396
$3\frac{1}{2}$–4	Rhodochrosite		427	$5\frac{1}{2}$–6	Sodalite		316
$3\frac{1}{2}$–4	Sphalerite	R	382	$5\frac{1}{2}$–$6\frac{1}{2}$	Franklinite	M	415
$3\frac{1}{2}$–4	Stilbite		317	$5\frac{1}{2}$–$6\frac{1}{2}$	Hematite	M	407
$3\frac{1}{2}$–4	Wurtzite	R	383	$5\frac{1}{2}$–$6\frac{1}{2}$	Magnetite	M	414
$3\frac{1}{2}$–$4\frac{1}{2}$	Margarite		323	$5\frac{1}{2}$–$6\frac{1}{2}$	Pyroxenes		336
$3\frac{1}{2}$–$4\frac{1}{2}$	Pyrrhotite		387	6	Amblygonite		450
4	Fluorite	M	403	6	Lazulite		450
4	Magnesite		426	6	Orthoclase		304
4	Manganite	(M)	421	6	Microcline		304
4	Siderite		426	6	Sanidine		304
4	Variscite	R	452	6	Turquoise	R	452
4–$4\frac{1}{2}$	Colemanite		437	6–$6\frac{1}{2}$	Humite		361
4–$4\frac{1}{2}$	Platinum	M	374	6–$6\frac{1}{2}$	Marcasite	M	396
4–$4\frac{1}{2}$	Smithsonite		427	6–$6\frac{1}{2}$	Plagioclase		305
4–$4\frac{1}{2}$	Wolframite	(M)	446	6–$6\frac{1}{2}$	Prehnite		325
4–$4\frac{1}{2}$	Zincite		406	6–$6\frac{1}{2}$	Pyrite	M	394
4–5	Low Iron	M	375	6–$6\frac{1}{2}$	Rutile		408
4–5	Chabazite		316	6–7	Cassiterite		409
$4\frac{1}{2}$	Xenotime		448	6–7	Sillimanite		363

TABLE A2.1 (*continued*)

Hardness	Mineral	Luster	Ref. page	Hardness	Mineral	Luster	Ref. page
$6\frac{1}{2}$	Chloritoid		366	7	Quartz		298
$6\frac{1}{2}$	Clinozoisite		373	$7-7\frac{1}{2}$	Staurolite		366
$6\frac{1}{2}$	Epidote		373	$7-7\frac{1}{2}$	Tourmaline		354
$6\frac{1}{2}$	Jadeite		337	$7-8$	Lawsonite		371
$6\frac{1}{2}$	Vesuvianite		371	$7\frac{1}{2}$	Andalusite		363
$6\frac{1}{2}-7$	Diaspore		422	$7\frac{1}{2}$	Zircon		369
$6\frac{1}{2}-7$	Olivine		359	$7\frac{1}{2}-8$	Beryl		314
$6\frac{1}{2}-7$	Sinhalite		435	$7\frac{1}{2}-8$	Spinels		414
$6\frac{1}{2}-7$	Spodumene		337	8	Topaz		369
$6\frac{1}{2}-7\frac{1}{2}$	Garnet		356	$8\frac{1}{2}$	Chrysoberyl		416
7	Boracite		435	9	Corundum		407
7	Cordierite		314	10	Diamond		377

CRYSTALLOGRAPHIC TABLES AND CALCULATIONS

TABLE A3.1. *Space Group Symbols and Point Groups*

Lattice	Point Group		Space Group					
	Centric	Acentric	Centric				Acentric	
$\bar{1}$ Triclinic	$\bar{1}$	1	$P\bar{1}$				$P1$	
$2/m$ Monoclinic	$2/m$		$P2/m$	$P2_1/m$	$C2/m$			
			$P2/c$	$P2_1/c$	$C2/c$			
		2					$P2$ $\quad$ $P2_1$ $\quad$ $C2$	
		m					Pm $\quad$ Pc $\quad$ Cm	
							Cc	
$2/m\,2/m\,2/m$ Orthorhombic	$2/m\,2/m\,2/m$		$P2/m2/m2/m$	$P2/n2/n2/n$	$P2/c2/c2/m$			
			$P2/b2/a2/n$	$P2/m2/c2_1/m$	$P2/n2/a2_1/n$			
			$P2/m2/n2_1/a$	$P2/b2/c2_1/b$	$P2_1/b2_1/a2/m$			
			$P2_1/c2_1/c2/n$	$P2_1/b2_1/m2/a$	$P2_1/n2_1/n2/m$			
			$P2_1/m2_1/m2/n$	$P2_1/n2_1/c2/a$	$P2_1/b2_1/c2_1/a$			
			$P2_1/n2_1/m2_1/a$	$I2/m2/m2/m$	$I2/b2/a2/m$			
			$I2_1/b2_1/c2_1/a$	$I2_1/m2_1/m2_1/a$	$C2/m2/m2/m$			
			$C2/c2/c2/m$	$C2/m2/m2/a$	$C2/c2/c2/a$			
			$C2/m2/c2_1/m$	$C2/m2/c2_1/c$	$F2/m2/m2/m$			
			$F2/d2/d2/d$					
		222					$P222$ $\quad$ $P222_1$ $\quad$ $P2_12_12$	
							$P2_12_12_1$ $\quad$ $C222$ $\quad$ $C222_1$	
							$I222$ $\quad$ $I2_12_12_1$ $\quad$ $F222$	
		$m2m$					$Pm2m$ $\quad$ $Pb2b$ $\quad$ $Pm2a$	
							$Pn2b$ $\quad$ $Pc2a$ $\quad$ $Pn2n$	
							$Pb2_1m$ $\quad$ $Pc2_1b$ $\quad$ $Pn2_1m$	
							$Pn2_1a$ $\quad$ $Im2m$ $\quad$ $Ic2a$	
							$Im2a$ $\quad$ $Bm2m$ $\quad$ $Bb2b$	
							$Bb2_1m$ $\quad$ $Am2m$ $\quad$ $Ac2m$	
							$Am2a$ $\quad$ $Ac2a$ $\quad$ $Fm2m$	
							$Fd2d$	

TABLE A3.1 (*continued*)

Lattice	Point Group Centric	Point Group Acentric	Space Group Centric			Space Group Acentric		
$4/m2/m2/m$ Tetragonal	$4/m$		$P4/m$ $P4_2/n$	$P4/n$ $I4/m$	$P4_2/n$ $I4_1/a$			
		4				$P4$ $P4_2$	$P4_1$ $I4$	$P4_3$ $I4_1$
		$\bar{4}$				$P\bar{4}$	$I\bar{4}$	
	$4/m2/m2/m$		$P4/m2/m2/m$ $P4/n2/n2/c$ $P4/n2_1/m2/m$ $P4_2/m2/c2/m$ $P4_2/m2_1/b2/c$ $P4_2/n2_1/c2/m$ $I4_1/a2/m2/d$	$P4/m2/c2/c$ $P4/m2_1/b2/m$ $P4/n2_1/c2/c$ $P4_2/n2/b2/c$ $P4_2/m2_1/n2/m$ $I4/m2/m2/m$ $I4_1/a2/c2/d$	$P4/n2/b2/m$ $P4/m2_1/n2/c$ $P4_2/m2/m2/c$ $P4_2/n2/n2/m$ $P4_2/n2_1/m2/c$ $I4/m2/c2/m$			
		422				$P422$ $P4_222$ $P4_32_12$ $I4_122$	$P4_122$ $P42_12$ $P4_22_12$	$P4_322$ $P4_12_12$ $I422$
		$4mm$				$P4mm$ $P4nc$ $P4_2bc$ $I4cm$	$P4cc$ $P4_2mc$ $P4_2nm$ $I4_1md$	$P4bm$ $P4_2cm$ $I4mm$ $I4_1cd$
		$\bar{4}2m$				$P\bar{4}2m$ $P\bar{4}2_1c$ $P\bar{4}b2$ $I\bar{4}2d$	$P\bar{4}2c$ $P\bar{4}m2$ $P\bar{4}n2$ $I\bar{4}m2$	$P\bar{4}2_1m$ $P\bar{4}c2$ $I\bar{4}2m$ $I\bar{4}c2$
$6/m2/m2/m$ or $\bar{3}2/m$ Trigonal	$\bar{3}$		$P\bar{3}$	$R\bar{3}$				
		3				$P3$ $R3$	$P3_1$	$P3_2$
	$\bar{3}2/m$		$P\bar{3}2/m1$ $P\bar{3}12/c$	$P\bar{3}12/m$ $R\bar{3}2/m$	$P\bar{3}2/c1$ $R\bar{3}2/c$			
		32				$P321$ $P3_112$ $R32$	$P312$ $P3_221$	$P3_121$ $P3_212$
		$3m$				$P3m1$ $P31c$	$P31m$ $R3m$	$P3c1$ $R3c$
$6/m2/m2/m$ Hexagonal	$6/m$		$P6/m$	$P6_3/m$				
		6				$P6$ $P6_2$	$P6_1$ $P6_4$	$P6_5$ $P6_3$
		$\bar{6}$				$P\bar{6}$		
	$6/m2/m2/m$		$P6/m2/m2/m$ $P6_3/m2/c2/m$	$P6/m2/c2/c$	$P6_3/m2/m2/c$			
		622				$P622$ $P6_222$	$P6_122$ $P6_422$	$P6_522$ $P6_322$
		$6mm$				$P6mm$ $P6_3cm$	$P6cc$	$P6_3mc$
		$\bar{6}2m$				$P\bar{6}2m$ $P6c2$	$P\bar{6}2c$	$P\bar{6}m2$

TABLE A3.1 (*continued*)

Lattice	Point Group Centric	Point Group Acentric	Space Group Centric			Space Group Acentric		
$4/m\bar{3}2/m$ Isometric	$2/m\bar{3}$		$P2/m\bar{3}$	$P2/n\bar{3}$	$P2_1/a\bar{3}$			
			$I2/m\bar{3}$	$I2_1/a\bar{3}$	$F2/m\bar{3}$			
			$F2/d\bar{3}$					
		23				$P23$	$P2_13$	$I23$
						$I2_13$	$F23$	
	$4/m\bar{3}2/m$		$P4/m\bar{3}2/m$	$P4/n\bar{3}2/n$	$P4_2/m\bar{3}2/n$			
			$P4_2/n\bar{3}2/m$	$I4/m\bar{3}2/m$	$I4_1/a\bar{3}2/d$			
			$F4/m\bar{3}2/m$	$F4/m\bar{3}2/c$	$F4_1/d\bar{3}2/m$			
			$F4_1/d\bar{3}2/c$					
		432				$P432$	$P4_132$	$P4_332$
						$P4_232$	$I432$	$I4_132$
						$F432$	$F4_132$	
		$\bar{4}3m$				$P\bar{4}3m$	$P\bar{4}3n$	$I\bar{4}3m$
						$I\bar{4}3d$	$F\bar{4}3m$	$F\bar{4}3c$

TABLE A3.2.   *Symmetrically Equivalent Lattice Planes*

		Equivalent Indices of (hkl)	
Operation	Orientation	ORTHOGONAL*	HEXAGONAL
$\bar{1}$	$[uvw]$	$\bar{h}\,\bar{k}\,\bar{l}$	$\bar{h}\,\bar{k}\,\bar{l}$
2	[100]	$h\bar{k}\bar{l}$	$h\;\bar{h}+\bar{k}\;\bar{l}$
	[010]	$\bar{h}k\bar{l}$	$\bar{h}+\bar{k}\;k\;l$
	[001]	$\bar{h}\,\bar{k}l$	
	[110]	$kh\bar{l}$	$kh\bar{l}$
	[210]		$h+k\;\bar{k}\;\bar{l}$
	[120]		$\bar{h}\;h+k\;\bar{l}$
	[$\bar{1}$10]		$\bar{k}\;\bar{h}\;\bar{l}$
	[101]	$l\bar{k}h$	
	[011]	$\bar{h}lk$	
$1/m$	[100]	$\bar{h}kl$	$\bar{h}\;h+k\;l$
	[010]	$h\bar{k}l$	$h+k\;\bar{k}\;l$
	[001]	$hk\bar{l}$	
	[110]	$\bar{k}\;\bar{h}l$	$\bar{k}\;\bar{h}\;l$
	[210]		$\bar{h}+\bar{k}\;k\;l$
	[120]		$h\;\bar{h}+\bar{k}\;l$
	[$\bar{1}$10]		khl
	[101]	$\bar{l}k\bar{h}$	
	[011]	$h\bar{l}\,\bar{k}$	
3	[001]		$k\;\bar{h}+\bar{k}\;l$ $\bar{h}+\bar{k}\;h\;l$
	[111]	klh lhk	
$\bar{3}$	[001]		$k\;\bar{h}+\bar{k}\;l$ $\bar{h}+\bar{k}\;h\;l$ $\bar{h}\,\bar{k}\,\bar{l}$
			$\bar{k}\;h+k\;\bar{l}$ $h+k\;h\;\bar{l}$
	[111]	klh lhk $\bar{h}\,\bar{k}\,\bar{l}$	
		$\bar{k}\,\bar{l}\,\bar{h}$ $\bar{l}\,\bar{h}\,\bar{k}$	
4	[100]	$h\bar{k}\bar{l}$ $h\bar{l}k$ $hl\bar{k}$	
	[010]	$\bar{h}k\bar{l}$ $lk\bar{h}$ $\bar{l}kh$	
	[001]	$\bar{h}\,\bar{k}l$ $k\bar{h}l$ $\bar{k}hl$	
$\bar{4}$	[100]	$h\bar{k}\bar{l}$ $\bar{h}\,\bar{l}k$ $\bar{h}l\bar{k}$	
	[010]	$\bar{h}k\bar{l}$ $l\bar{k}\,\bar{h}$ $\bar{l}\,\bar{k}h$	
	[001]	$\bar{h}\,\bar{k}l$ $k\bar{h}\,\bar{l}$ $\bar{k}h\bar{l}$	
6	[001]		$k\;\bar{h}+\bar{k}\;l$ $\bar{k}\;h+k\;l$ $\bar{h}\,\bar{k}\,l$
			$\bar{h}+\bar{k}\;h\;l$ $h+k\;\bar{h}\;l$
$\bar{6}$	[001]		$k\;\bar{h}+\bar{k}\;l$ $\bar{h}+\bar{k}\;h\;l$ $hk\bar{l}$
			$k\;\bar{h}+\bar{k}\;\bar{l}$ $\bar{h}+\bar{k}\;h\;\bar{l}$

*Includes monoclinic where the symmetry axis is "orthogonal" to the plane of the other axes.

TABLE A3.3. *Symmetrical Equivalence of Atomic Positions*

Operation	Position	ORTHOGONAL*	HEXAGONAL
		Equivalent Position of xyz	

Operation	Position	ORTHOGONAL*	HEXAGONAL
$\bar{1}$	$[uvw]$	$\bar{x}\ \bar{y}\ \bar{z}$	$\bar{x}\ \bar{y}\ \bar{z}$
2	$[100]$	$x\ \bar{y}\ \bar{z}$	$x-y\ \bar{y}\ \bar{z}$
	$[010]$	$\bar{x}\ y\ \bar{z}$	
	$[001]$	$\bar{x}\ \bar{y}\ z$	
	$[110]$	$yx\ \bar{z}$	
	$[101]$	$z\ \bar{y}\ x$	
	$[011]$	$\bar{x}\ zy$	
	$[210]$		$x\ x-y\ \bar{z}$
2_1	$[100]$	$\tfrac{1}{2}+x\ \bar{y}\ \bar{z}$	
	$[010]$	$\bar{x}\ \tfrac{1}{2}+y\ \bar{z}$	
	$[001]$	$\bar{x}\ \bar{y}\ \tfrac{1}{2}+z$	
$1/m$	$[100]$	$\bar{x}\ yz$	$x-y\ y\ z$
	$[010]$	$x\ \bar{y}z$	
	$[001]$	$xy\ \bar{z}$	
	$[110]$	$\bar{y}\ \bar{x}\ z$	
	$[210]$		$\bar{x}\ y-x\ z$
	$[101]$	$\bar{x}\ y\ \bar{z}$	
	$[011]$	$x\ \bar{y}\ \bar{z}$	
$1/a$	$[010]$	$\tfrac{1}{2}+x\ \bar{y}\ z$	
	$[001]$	$\tfrac{1}{2}+x\ y\ \bar{z}$	
$1/b$	$[100]$	$\bar{x}\ \tfrac{1}{2}+y\ z$	
	$[001]$	$x\ \tfrac{1}{2}+y\ \bar{z}$	
$1/c$	$[100]$	$\bar{x}\ y\ \tfrac{1}{2}+z$	$x-y\ y\ \tfrac{1}{2}+z$
	$[010]$	$x\ \bar{y}\ \tfrac{1}{2}+z$	
	$[110]$	$y\ \bar{x}\ \tfrac{1}{2}+z$	
	$[210]$		$\bar{x}\ y-x\ \tfrac{1}{2}+z$
$1/n$	$[100]$	$\bar{x}\ \tfrac{1}{2}+y\ \tfrac{1}{2}+z$	
	$[010]$	$\tfrac{1}{2}+x\ \bar{y}\ \tfrac{1}{2}+z$	
	$[001]$	$\tfrac{1}{2}+x\ \tfrac{1}{2}+y\ \bar{z}$	
	$[110]$	$\tfrac{1}{2}+\bar{y}\ \tfrac{1}{2}+\bar{x}\ z$	
$1/d$	$[100]$	$\bar{x}\ \tfrac{1}{4}+y\ \tfrac{1}{4}+z$	
	$[010]$	$\tfrac{1}{4}+x\ \bar{y}\ \tfrac{1}{4}+z$	
	$[001]$	$\tfrac{1}{4}+x\ \tfrac{1}{4}+y\ \bar{z}$	
	$[110]$	$\tfrac{1}{4}+\bar{y}\ \tfrac{1}{4}+\bar{x}\ z$	
3	$[001]$		$\bar{y}\ x-y\,z\ \ y-x\ \bar{x}\ z$
	$[111]$	$zxy\ \ yzx$	
$\bar{3}$	$[001]$		$\bar{y}\ x-y\,z\ \ y-\bar{x}\ x\,z\ \ \bar{x}\ \bar{y}\ \bar{z}$
			$y\ x-y\ \bar{z}\ \ y-x\ x\,\bar{z}$
	$[111]$	$zxy\ \ yzx\ \ \bar{x}\ \bar{y}\ \bar{z}$	
		$\bar{z}\ \bar{x}\ \bar{y}\ \ \bar{y}\ \bar{z}\ \bar{x}$	
3_1	$[001]$		$\bar{y}\ x-y\ \tfrac{1}{3}+z\ \ y-x\ \bar{x}\ \tfrac{2}{3}+z$
3_2	$[001]$		$\bar{y}\ x-y\ \tfrac{2}{3}+z\ \ y-x\ \bar{x}\ \tfrac{1}{3}+z$
4	$[001]$	$\bar{y}\,xz\ \ \bar{x}\ \bar{y}\,z\ \ y\,\bar{x}\,z$	
$\bar{4}$	$[001]$	$\bar{y}\,x\,\bar{z}\ \ \bar{x}\ \bar{y}\,z\ \ y\,\bar{x}\ \bar{z}$	
4_1	$[001]$	$y\,x\,-+z\ \ x\,y\,-+z\ \ y\,x\,-+z$	
4_2	$[001]$	$\bar{y}\,x\,-+z\ \ \bar{x}\ \bar{y}\,z\ \ y\,\bar{x}\ \tfrac{1}{2}+z$	
4_3	$[001]$	$\bar{y}\,x\,\tfrac{3}{4}+z\ \ \bar{x}\ \bar{y}\ \tfrac{1}{2}+z\ \ y\,\bar{x}\ \tfrac{1}{4}+z$	
6	$[001]$		$x-y\,xz\ \ \bar{y}\ x-yz\ \ \bar{x}\ \bar{y}z$
			$y-x\ \bar{x}\,z\,y\ \ y\ y-x\,z$

*Includes monoclinic where the symmetry axis is orthogonal to the plane of the other axes.

TABLE A3.3 (*continued*)

Operation	Position	Equivalent Position of *xyz* ORTHOGONAL*	HEXAGONAL
$\bar{6}$	[001]		$zxy\ yzx\ xy\bar{z}\ zx\,\bar{y}\ yz\,\bar{x}$
6	[001]		$x-y\,x\,\frac{1}{6}+z\ \ \bar{y}\,x-y\,\frac{1}{3}+z\ \ \bar{x}\ \bar{y}\,\frac{1}{2}+z$
			$y-x\ \bar{x}\,\frac{2}{3}+z\,y\ y-x\,\frac{5}{6}+z$
6	[001]		$x-y\,x\,\frac{1}{3}+z\ \ \bar{y}\,x-y\,\frac{2}{3}+z\ \ \bar{x}\ \bar{y}z$
			$y-x\ \bar{x}\,\frac{1}{3}+z\,y\ y-x\,\frac{2}{3}+z$
6	[001]		$x-y\,x\,\frac{1}{2}+z\ \ \bar{y}\,x-yz\ \bar{x}\ \bar{y}\,\frac{1}{2}+z$
			$y-x\ \bar{x}z\,y\,y\ y-x\,\frac{1}{2}+z$
6	[001]		$x-y\,x\,\frac{2}{3}+z\ \ \bar{y}\,x-y\,\frac{1}{3}+z\ \ \bar{x}\ \bar{y}z$
			$y-x\ \bar{x}\,\frac{2}{3}+z\,y\ y-x\,\frac{1}{3}+z$
6	[001]		$x-y\,x\,\frac{5}{6}+z\ \ \bar{y}\,x-y\,\frac{2}{3}+z\ \ \bar{x}\ \bar{y}\,\frac{1}{2}+z$
			$y-x\ \bar{x}\,\frac{1}{3}+z\,y\,y\ y-x\,\frac{1}{6}+z$

CALCULATION OF UNIT CELL CONTENT, DENSITY, AND MOLAR VOLUME

Unit Cell Content and Density

$$\text{Density} = \frac{\text{mass of unit cell}}{\text{volume of unit cell}} = \frac{MZ}{NV}$$

where

M = in Mg = sum of atomic weights in one formula unit (in grams) $\times 10^{-6}$

N_A = Avogadro's number = 6.02252×10^{23}

or, $1/N_A = 1.66043 \times 10^{-24}$

V = unit cell volume in cubic nanometers

$$V = abc\sqrt{\begin{array}{c}1 - \cos^2\alpha - \cos^2\beta - \cos^2\gamma \\ + 2\cos\alpha\cos\beta\cos\gamma\end{array}}$$

where a, b, c are in nanometers (10^{-9} m) and α, β, γ are in degrees

D = density in Mg/m^3

$$D = \frac{MZ}{NV} = \frac{MZ\,1/N_A}{V}$$

$$= \frac{M(\text{g}) \times 10^{-6} \times 1.660 \times 10^{-24}}{V(\text{nm}^3) \times 10^{-27}}$$

$$D = \frac{MZ\,1.660}{V} \times 10^{-3}$$

$$Z = \frac{DV}{M\,1.660} \times 10^3$$

$$\text{Molar volume} = \frac{\text{formula weight}}{\text{density}} = \frac{M(\text{Mg})}{D(\text{Mg/m}^3)}$$

$$= \frac{M}{D}\ \text{m}^3$$

The following is a molar volume calculation using the mineral beryl.

Beryl, $Al_2Be_3Si_6O_{18}$, space group: $P6/m\,2/c\,2/c$

Equipoints: Al 4, Be 6, Si 12, O(1) 12, O(2) 24

$a_1 = a_2 = 0.923$ and $c = 0.919$ nm

$$Z = 2$$

$$\text{Density} = \frac{MZ\,1.660}{V} \times 10^{-3}$$

$$= \frac{537.5 \times 2 \times 1.660}{0.6780} \times 10^{-3}$$

$$= 2.64\ \text{Mg/m}^3$$

$$M = 2 \times 26.98 + 3 \times 9.012 + 6 \times 28.09$$
$$+ 18 \times 16.00 = 537.5\ \text{Mg}$$

$$V = a_1 \times a_2 \times c\sqrt{1 - \cos^2(120°)}$$

$$= 0.923 \times 0.923 \times 0.919\sqrt{0.75}$$

$$= 0.6780\ \text{nm}^3$$

Note that the theoretical density of beryl, 2.64, is lower than the observed density, 2.75–2.80. The higher observed density is probably due to impurities and interstitial atoms in the beryl structure.

CONVERSION OF RHOMBOHEDRAL TO HEXAGONAL LATTICE

The rhombohedral unit cell is defined by three equal length unit translations, a, and the equal angle between them, α. The rhombohedral lattice parameters can be converted to hexagonal by using the following equations:

$$a_H = 2a_R \sin\alpha/2$$

$$c_H = 3\sqrt{a_R^2 - a_H^2/3}$$

CRYSTAL FORM DIAGRAMS AND TABLES

CRYSTAL FORM DIAGRAMS

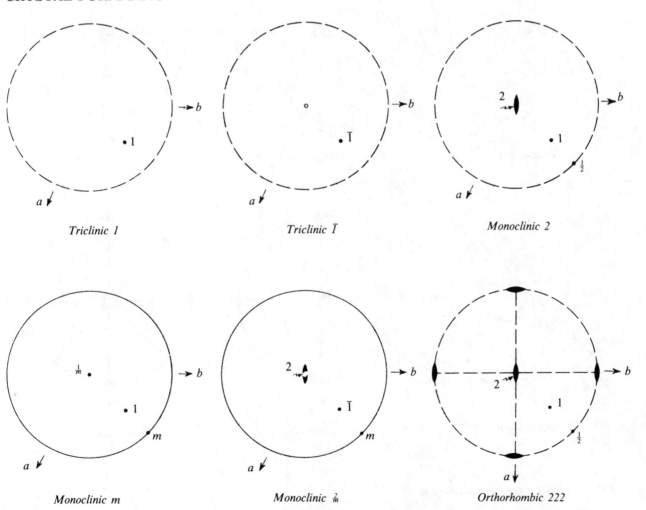

Triclinic 1   Triclinic $\bar{1}$ Monoclinic 2

Monoclinic m   Monoclinic $\frac{2}{m}$ Orthorhombic 222

CRYSTAL FORM DIAGRAMS

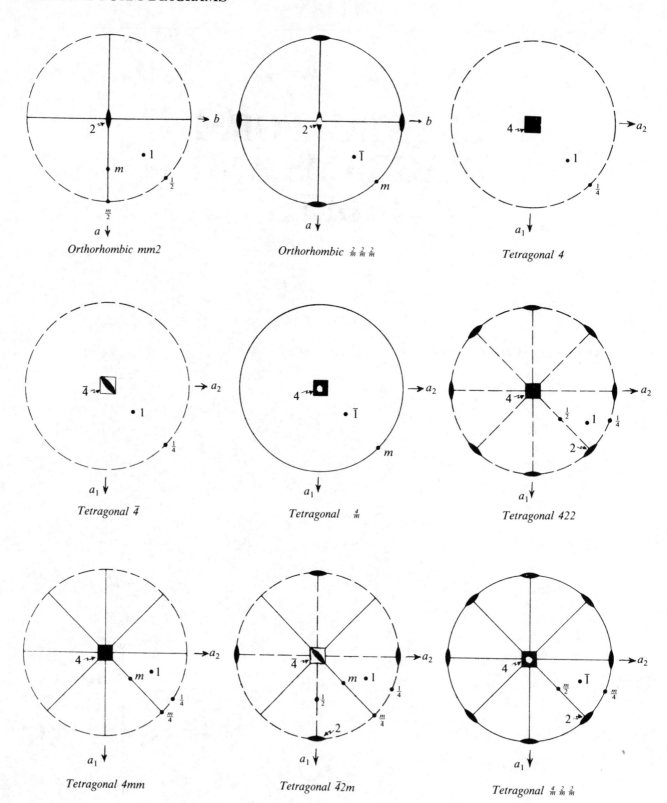

Orthorhombic mm2

Orthorhombic $\frac{2}{m}\frac{2}{m}\frac{2}{m}$

Tetragonal 4

Tetragonal $\bar{4}$

Tetragonal $\frac{4}{m}$

Tetragonal 422

Tetragonal 4mm

Tetragonal $\bar{4}2m$

Tetragonal $\frac{4}{m}\frac{2}{m}\frac{2}{m}$

$\bar{4}m2$ and 312 orientations are not illustrated

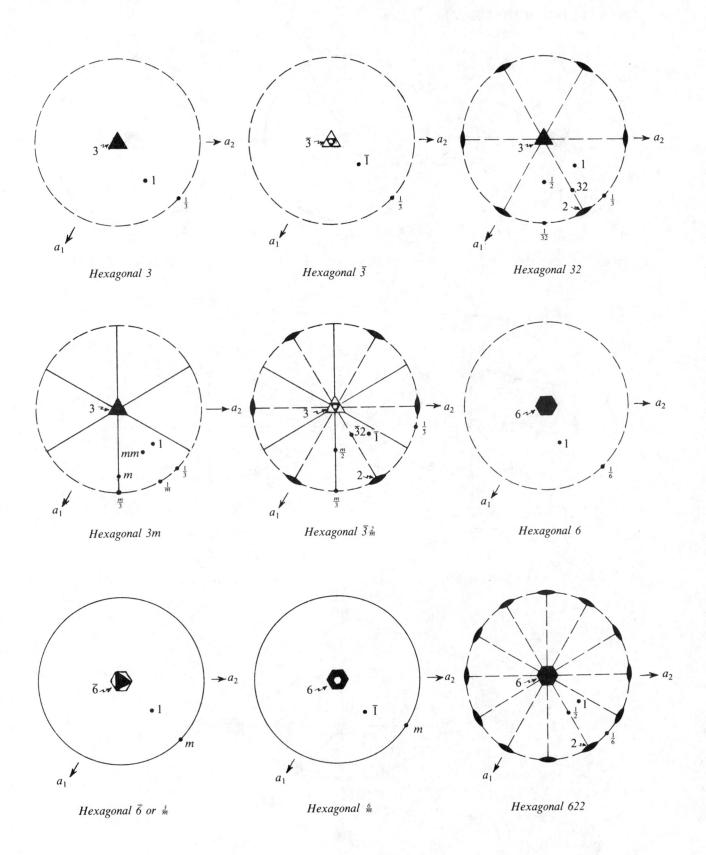

Hexagonal 3

Hexagonal $\bar{3}$

Hexagonal 32

Hexagonal 3m

Hexagonal $\bar{3}\frac{2}{m}$

Hexagonal 6

Hexagonal $\bar{6}$ or $\frac{3}{m}$

Hexagonal $\frac{6}{m}$

Hexagonal 622

CRYSTAL FORM DIAGRAMS

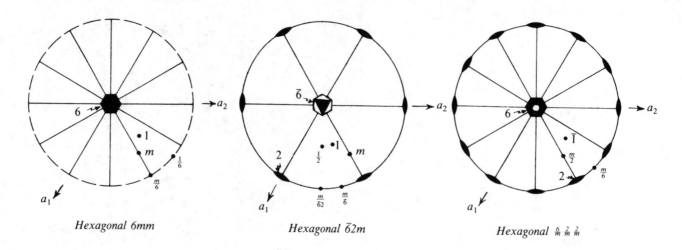

Hexagonal 6mm

Hexagonal 6̄2m

Hexagonal $\frac{6}{m} \frac{2}{m} \frac{2}{m}$

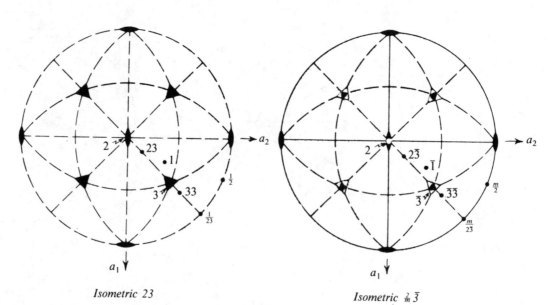

Isometric 23

Isometric $\frac{2}{m} \bar{3}$

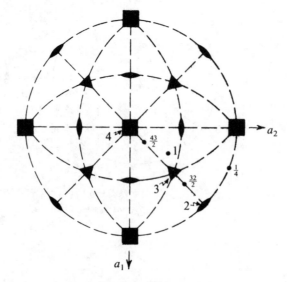

Isometric 432

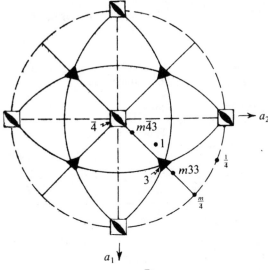

Isometric $\bar{4}3m$

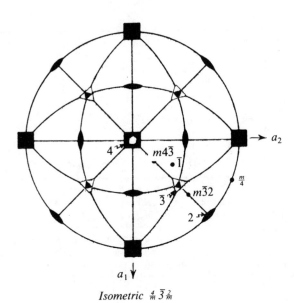

Isometric $\frac{4}{m}\bar{3}\frac{2}{m}$

TABLE A4.1.   *Triclinic*

Crystal Class	Symmetry of Faces	Pole Symmetry	Short Symbol of Form	Classical Name of Crystal Form	Apparent Symmetry	No. of Faces	Indices of a Faces	Other Possible Indices		
$\bar{1}$	1	$\bar{1}$	$\bar{1}$	Pinacoid		2	hkl	$\bar{h}\,\bar{k}\,l$	$h\,\bar{k}\,\bar{l}$	$\bar{h}\,k\,\bar{l}$
1	1	1	1	Pedion		1	hkl	$\bar{h}\,\bar{k}\,l$	$h\,\bar{k}\,\bar{l}$	$\bar{h}\,k\,\bar{l}$
								$\bar{h}\,kl$	$h\,\bar{k}\,l$	$hk\,\bar{l}$
								$\bar{h}\,\bar{k}\,\bar{l}$		

TABLE A4.2. *Monoclinic*

Crystal Class	Symmetry of Faces	Pole Symmetry	Short Symbol of Form	Classical Name of Crystal Form	Apparent Symmetry	No. of Faces	Indices of a Face — 1st	Indices of a Face — 2nd	Other Possible Indices — 1st	Other Possible Indices — 2nd
$2/m$	1	$\bar{1}$	$\bar{1}$	Rhombic prism	$2/m$	4	hkl		$h\bar{k}l$	$hk\bar{l}$
	m	$m/2$	m	Pinacoid		2	$hk0$	$h0l$	$h\bar{k}0$	$h0\bar{l}$
	2	$2/m$	2	Pinacoid		2	001	010		
2	1	1	1	Sphenoid	2	2	hkl		$h\bar{k}l$; $hk\bar{l}$; $h\bar{k}\bar{l}$	$\bar{h}kl$; $\bar{h}\bar{k}l$; $\bar{h}\bar{k}\bar{l}$
	1	$1/2$	$1/2$	Pinacoid		2	$hk0$	$h0l$	$h\bar{k}0$	$\bar{h}0l$
	2	2	2	Pedion		1	001	010	$00\bar{1}$	$0\bar{1}0$
m	1	1	1	Dome	m	2	hkl		$h\bar{k}l$; $\bar{h}kl$; $\bar{h}\bar{k}l$	$\bar{h}kl$; $hk\bar{l}$; $\bar{h}k\bar{l}$
	1	$1/m$	$1/m$	Pinacoid		2	$hk0$	$h0l$	$\bar{h}\bar{k}l$; $\bar{h}k0$; $h\bar{k}0$; $\bar{h}\bar{k}0$	$\bar{h}k\bar{l}$; $\bar{h}0l$; $h0\bar{l}$; $\bar{h}0\bar{l}$
	m	m	m	Pedion		1	001	010		

*Alternative indices are possible as a consequence of different orientations of the *a* and *b*, or *a* and *c* axes, respectively, in first and second settings.

TABLE A4.3. *Orthorhombic*

Crystal Class	Symmetry of Faces	Pole Symmetry	Short Symbol of Form	Classical Name of Crystal Form	Apparent Symmetry	No. of Faces	Indices of a Face	Other Possible Indices	
$2/m\,2/m\,2/m$	1	$\bar{1}$	$\bar{1}$	Rhombic dipyramid	$2/m\,2/m\,2/m$	8	hkl		
	m	$m2\cdot 2/2$	m	Rhombic prism	$2/m\,2/m\,2/m$	4	$hk0$	$h0l$	$0kl$
	$2mm$	$2mm/22m$	2	Pinacoid		2	100	010	001
222	1	1	1	Rhombic disphenoid	222	4	hkl	$hk\bar{l}$	
	1	$2\cdot 2/2$	$1/2$	Rhombic prism	$2/m\,2/m\,2/m$	4	$hk0$	$h0l$	$0kl$
	2	$2/22$	2	Pinacoid		2	100	010	001
$mm2$	1	1	1	Rhombic pyramid	$mm2$	4	hkl	$hk\bar{l}$	$h\bar{k}l$*
	1	$1/2$	$1/2$	Rhombic prism	$2/m\,2/m\,2/m$	4	$hk0$	$h0l$*	
	m	m	m	Dome	$mm2$	2	$h0l$	$0kl$	$hk0$*
	m	$m/2m$	$m/2$	Pinacoid		2	100	010	001*
	$2mm$	$2mm$	2	Pedion		1	001	$00\bar{1}$	010* $0\bar{1}0$*

*In second setting.

TABLE A4.4. *Tetragonal*

Crystal Class	Symmetry of Faces	Pole Symmetry	Short Symbol of Form	Classical Name of Crystal Form	Apparent Symmetry	No. of Faces	Indices of a Face	Other Possible Indices
$4/m\,2/m\,2/m$	1	$\bar{1}$	$\bar{1}$	Ditetragonal dipyramid	$4/m\,2/m\,2/m$	16	hkl	
	m	$m4\cdot2/2$	m/2	Tetragonal dipyramid	$4/m\,2/m\,2/m$	8	hhl	h0l
	m	$m2\cdot2/4$	m/4	Ditetragonal prism	$4/m\,2/m\,2/m$	8	hk0	
	2mm	$2mm/42m$	2	Tetragonal prism	$4/m\,2/m\,2/m$	4	110	100
	4mm	$4mm/22m$	4	Pinacoid		2	001	
4	1	1	1	Tetragonal pyramid	4mm	4	hkl	$hk\bar{l}$
	1	1/4	1/4	Tetragonal prism	$4/m\,2/m\,2/m$	4	hk0	
	4	4	4	Pedion		1	001	$00\bar{1}$
$\bar{4}$	1	1	1	Tetragonal disphenoid	$\bar{4}\,2m$	4	hkl	$hk\bar{l}$
	1	$1/\bar{4}$	$1/\bar{4}$	Tetragonal prism	$4/m\,2/m\,2/m$	4	hk0	
	2	$\bar{4}$	$\bar{4}$	Pinacoid		2	001	$00\bar{1}$
$4/m$	1	$\bar{1}$	$\bar{1}$	Tetragonal dipyramid	$4/m\,2/m\,2/m$	8	hkl	
	m	m/4	m	Tetragonal prism	$4/m\,2/m\,2/m$	4	hk0	
	4	4/m	4	Pinacoid		2	001	
422	1	1	1	Tetragonal trapezohedron	422	8	hkl	$hk\bar{l}$
	1	$4\cdot2/2$	1/2	Tetragonal dipyramid	$4/m\,2/m\,2/m$	8	hhl	h0l
	1	$2\cdot2/4$	1/4	Ditetragonal prism	$4/m\,2/m\,2/m$	8	hk0	
	2	2/42	2	Tetragonal prism	$4/m\,2/m\,2/m$	4	110	100
	4	4/22	4	Pinacoid		2	001	
4mm	1	1	1	Ditetragonal pyramid	4mm	8	hkl	$hk\bar{l}$
	1	1/4	1/4	Ditetragonal prism	$4/m\,2/m\,2/m$	8	hk0	
	m	m	m	Tetragonal pyramid	4mm	4	hhl	$hh\bar{l}$ h0l $h0\bar{l}$
	m	m/4m	m/4	Tetragonal prism	$4/m\,2/m\,2/m$	4	110	100
	4mm	4mm	4	Pedion		1	001	$00\bar{1}$
$\bar{4}\,2m$	1	1	1	Tetragonal scalenohedron	$\bar{4}\,2m$	8	hkl	$hk\bar{l}$
	1	$\bar{4}\cdot2/2$	1/2	Tetragonal dipyramid	$4/m\,2/m\,2/m$	8	h0l	hhl*
	1	$2\cdot(2)/\bar{4}$	$1/\bar{4}$	Ditetragonal prism	$4/m\,2/m\,2/m$	8	hk0	
	m	m	m	Tetragonal disphenoid	$\bar{4}\,2m$	4	hhl	$hh\bar{l}$ h0l* 0kl*
	m	$m2\cdot(2)/\bar{4}m$	$m/\bar{4}$	Tetragonal prism	$4/m\,2/m\,2/m$	4	110	100*
	2	$2/\bar{4}2$	2	Tetragonal prism	$4/m\,2/m\,2/m$	4	100	110*
	2mm	$\bar{4}m/2$	$\bar{4}$	Pinacoid		2	001	

*In $\bar{4}m2$ orientation

TABLE A4.5.   *Trigonal*

Crystal Class	Symmetry of Faces	Pole Symmetry	Short Symbol of Form	Classical Name of Crystal Form	Apparent Symmetry	No. of Faces	Indices of a Face	Other Possible Indices
$\bar{3}\,2/m$	1	$\bar{1}$	$\bar{1}$	Hexagonal scalenohedron	$\bar{3}\,2/m$	12	hkl	$hk\bar{l}$
	1	$3\cdot2$	32	Hexagonal dipyramid	$6/m\,2/m\,2/m$	12	hhl	$h0l*$
	1	$2\cdot(2)/\bar{3}$	$1/\bar{3}$	Dihexagonal prism	$6/m\,2/m\,2/m$	12	$hk0$	
	m	$m/2$	$m/2$	Rhombohedron	$\bar{3}\,2/m$	6	$h0l$	$hhl*$
	m	$m2\cdot(2)/\bar{3}2$	$m/\bar{3}$	Hexagonal prism	$6/m\,2/m\,2/m$	6	100	$110*$
	2	$2/\bar{3}m$	2	Hexagonal prism	$6/m\,2/m\,2/m$	6	110	$100*$
	3m	$\bar{3}m/2$	$\bar{3}$	Pinacoid		2	001	
3	1	1	1	Trigonal pyramid	$3m$	3	hkl	$hk\bar{l}$
	1	$1/3$	$1/3$	Trigonal prism	$\bar{6}\,2m$	3	$hk0$	
	3	3	3	Pedion		1	001	$00\bar{1}$
$\bar{3}$	1	$\bar{1}$	$\bar{1}$	Rhombohedron	$\bar{3}\,2/m$	6	hkl	$hk\bar{l}$
	1	$1/\bar{3}$	$1/\bar{3}$	Hexagonal prism	$6/m\,2/m\,2/m$	6	$hk0$	
	3	$\bar{3}$	$\bar{3}$	Pinacoid		2	001	
32	1	1	1	Trigonal trapezohedron	32	6	hkl	
	1	$3\cdot2$	32	Trigonal dipyramid	$\bar{6}\,2m$	6	hhl	$\bar{h}\,\bar{h}\,l\ h0l*\ 0kl$
	1	$1/2$	$1/2$	Rhombohedron	$\bar{3}\,2/m$	6	$h0l$	$0kl\ hhl*\ \bar{h}\,\bar{h}\,l$
	1	$2\cdot(2)/3$	$1/3$	Ditrigonal prism	$\bar{6}\,2m$	6	$hk0$	
	1	$2\cdot(2)/32$	$1/32$	Hexagonal prism	$6/m\,2/m\,2/m$	6	100	$110*$
	2	$2/3$	2	Trigonal prism	$\bar{6}\,2m$	3	110	$\bar{1}\,\bar{1}\,0\ 100*\ 010$
	3	$3/2$	3	Pinacoid		2	001	
$3m$	1	1	1	Ditrigonal pyramid	$3m$	6	hkl	$hk\bar{l}$
	1	$m\cdot m$	$m\cdot m$	Hexagonal pyramid	$6mm$	6	$h0l$	$h0\bar{l}\ hhl*\ hh\bar{l}*$
	1	$1/3$	$1/3$	Ditrigonal prism	$\bar{6}\,2m$	6	$hk0$	
	1	$1/3m$	$1/m$	Hexagonal prism	$6/m\,2/m\,2/m$	6	100	$110*$
	m	m	m	Trigonal pyramid	$3m$	3	hhl $h0l*$	$hh\bar{l}\ \bar{h}\,\bar{h}\,l\ \bar{h}\,\bar{h}\,\bar{l}$ $0kl*\ h0l*\ 0kl*$
	m	$m/3$	$m/3$	Trigonal prism	$\bar{6}\,2m$	3	110	$\bar{1}\,\bar{1}0\ 100*\ 010*$
	3m	3m	3	Pedion		1	001	$00\bar{1}$

*In $\bar{3}12/m$, 312, and 31m orientations.

TABLE A4.6. *Hexagonal*

Crystal Class	Symmetry of Faces	Pole Symmetry	Short Symbol of Form	Classical Name of Crystal Form	Apparent Symmetry	No. of Faces	Indices of a Face	Other Possible Indices
6/m2/m2/m	1	1̄	1̄	Dihexagonal dipyramid	6/m2/m2/m	24	hkl	
	m	m/2	m/2	Hexagonal dipyramid	6/m2/m2/m	12	hhl	$h0l$
	m	m2·2/6	m/6	Dihexagonal prism	6/m2/m2/m	12	$hk0$	
	2mm	2mm/62m	2	Hexagonal prism	6/m2/m2/m	6	110	100
	6mm	6mm/22m	6	Pinacoid		2	001	
6	1	1	1	Hexagonal pyramid	6mm	6	hkl	$hk\bar{l}$
	1	1/6	1/6	Hexagonal prism	6/m2/m2/m	6	$hk0$	
	6	6	6	Pedion		1	001	00$\bar{1}$
6̄ or 3/m	1	1	1	Trigonal dipyramid	6̄2m	6	hkl	
	m	m/6̄	m	Trigonal prism	6̄2m	3	$hk0$	
	3	6̄/m	6̄	Pinacoid		2	001	
6/m	1	1̄	1̄	Hexagonal dipyramid	6/m2/m2/m	12	hkl	
	m	m/6	m	Hexagonal prism	6/m2/m2/m	6	$hk0$	
	6	6/m	6	Pinacoid		2	001	
622	1	1	1	Hexagonal trapezohedron	622	12	hkl	$hk\bar{l}$
	1	6·2/2	1/2	Hexagonal dipyramid	6/m2/m2/m	12	hhl	$h0l$
	1	2·2/6	1/6	Dihexagonal prism	6/m2/m2/m	12	$hk0$	
	2	2/62	2	Hexagonal prism	6/m2/m2/m	6	110	100
	6	6/22	6	Pinacoid		2	001	
6mm	1	1	1	Dihexagonal pyramid	6mm	12	hkl	$hk\bar{l}$
	1	1/6	1/6	Dihexagonal prism	6/m2/m2/m	12	$hk0$	
	m	m	m	Hexagonal pyramid	6mm	6	hhl	$hh\bar{l}$ $h0l$ $h0\bar{l}$
	m	m/6m	m/6	Hexagonal prism	6/m2/m2/m	6	110	100
	6mm	6mm	6	Pedion		1	001	00$\bar{1}$
6̄2m (or 3/m2m)	1	1	1	Ditrigonal dipyramid	6̄2m	12	hkl	
	1	1/2	1/2	Hexagonal dipyramid	6/m2/m2/m	12	$h0l$	hhl*
	m	m	m	Trigonal dipyramid	6̄2m	6	hhl	$\bar{h}\,\bar{h}\,l$ $h0l$* $0kl$*
	m	m2·(2)/6̄	m/6̄	Ditrigonal prism	6̄2m	6	$hk0$	
	m	m2·(2)/6̄2m	m/6̄2	Hexagonal prism	6/m2/m2/m	6	100	110*
	2mm	2mm/6̄	2	Trigonal prism	6̄2m	3	110	$\bar{1}\,\bar{1}$0 100* 010*
	3m	6̄m/2m	6̄	Pinacoid		2	001	

*In 6̄m2 orientation.

TABLE A4.7.   *Isometric*

Crystal Class	Symmetry of Faces	Pole Symmetry	Short Symbol of Form	Classical Name of Crystal Form	Apparent Symmetry	No. of Faces	Indices of a Face	Other Possible Indices
$4/m\,\overline{3}\,2/m$	1	$\overline{1}$	$\overline{1}$	Hexaoctahedron	$4/m\,\overline{3}\,2/m$	48	*hkl*	
	m	$m\overline{3}\cdot 24/2$	*m*32	Trisoctahedron	$4/m\,\overline{3}\,2/m$	24	*hkl* *h < l*	
	m	$m4\cdot\overline{3}2/2$	$m4\overline{3}$	Trapezohedron	$4/m\,\overline{3}\,2/m$	24	*hkl* *h < l*	
	m	$m4\cdot 2/4$	*m*/4	Tetrahexahedron	$4/m\,\overline{3}\,2/m$	24	*hk*0	
	2*mm*	$2mm/4\overline{3}2m$	2	Rhombic dodecahedron	$4/m\,\overline{3}\,2/m$	12	110	
	3*m*	$\overline{3}m/2$	$\overline{3}$	Octahedron	$4/m\,\overline{3}\,2/m$	8	111	
	4*mm*	$4mm/4\overline{2}m$	4	Cube	$4/m\,\overline{3}\,2/m$	6	100	
23	1	1	1	Tetartoid	23	12	*hkl*	$hk\overline{l}$
	1	$3\cdot(3)2$	33	Deltohedron	$\overline{4}3m$	12	*hhl* *h > l*	$hh\overline{l}$
	1	$2\cdot 3(3)$	23	Tristetrahedron	$\overline{4}3m$	12	*hhl* *h < l*	$hh\overline{l}$
	1	$2\cdot(2)/2$	1/2	Pyritohedron	$2/m\overline{3}$	12	*hk*0	
	1	$2\cdot(2)/23$	1/23	Rhombic dodecahedron	$4/m\,\overline{3}\,2/m$	12	110	
	2	2/2	2	Cube	$4/m\,\overline{3}\,2/m$	6	100	
	3	3	3	Tetrahedron	$\overline{4}3m$	4	111	$11\overline{1}$
$2/m\overline{3}$	1	$\overline{1}$	$\overline{1}$	Diploid	$2/m\overline{3}$	24	*hkl*	
	1	$\overline{3}\cdot(\overline{3})2$	$\overline{3}\,\overline{3}$	Trisoctahedron	$4/m\,\overline{3}\,2/m$	24	*hhl* *h > l*	
	1	$2\cdot\overline{3}(\overline{3})$	$2\overline{3}$	Trapezohedron	$4/m\,\overline{3}\,2/m$	24	*hhl* *h < l*	
	m	$m2\cdot(2)/2$	*m*/2	Pyritohedron	$2/m\overline{3}$	12	*hk*0	
	m	$m2\cdot(2)/2$	$m/2\overline{3}$	Rhombic dodecahedron	$4/m\,\overline{3}\,2/m$	12	110	
	2*mm*	2*m*/2*m*	2	Cube	$4/m\,\overline{3}\,2/m$	6	100	
	3	$\overline{3}$	$\overline{3}$	Octahedron	$4/m\,\overline{3}\,2/m$	8	111	
432	1	1	1	Gyroid	432	24	*hkl*	$hk\overline{l}$
	1	$3\cdot 24/2$	32/2	Trisoctahedron	$4/m\,\overline{3}\,2/m$	24	*hhl* *h > l*	
	1	$4\cdot 32/2$	43/2	Trapezohedron	$4/m\,\overline{3}\,2/m$	24	*hhl* *h < l*	
	1	$4\cdot 2/4$	1/4	Tetrahexahedron	$4/m\,\overline{3}\,2/m$	24	*hk*0	
	2	2/432	2	Rhombic dodecahedron	$4/m\,\overline{3}\,2/m$	12	110	
	3	3/2	3	Octahedron	$4/m\,\overline{3}\,2/m$	8	111	
	4	4/42	4	Cube	$4/m\,\overline{3}\,2/m$	6	100	
$\overline{4}3m$	1	1	1	Hextetrahedron	$\overline{4}3m$	24	*hkl*	$hk\overline{l}$
	1	$\overline{4}\cdot(\overline{4})/\overline{4}$	$1/\overline{4}$	Tetrahexahedron	$4/m\,\overline{3}\,2/m$	24	*hk*0	
	m	$m3\cdot(3)\overline{4}$	*m*33	Deltohedron	$\overline{4}3m$	12	*hhl* *h > l*	$hh\overline{l}$
	m	$m\overline{4}\cdot 3(3)$	$m\overline{4}3$	Tristetrahedron	$\overline{4}3m$	12	*hhl* *h < l*	$hh\overline{l}$
	m	$m\overline{4}\cdot(\overline{4})/\overline{4}3m$	$m/\overline{4}$	Rhombic dodecahedron	$4/m\,\overline{3}\,2/m$	12	110	
	3*m*	3*m*	3	Tetrahedron	$\overline{4}3m$	4	111	$11\overline{1}$
	2*mm*	$\overline{4}m/\overline{4}$	$\overline{4}$	Cube	$4/m\,\overline{3}\,2/m$	6	100	

EFFECTIVE IONIC RADII

TABLE A5.1. *Ionic Radii*

Ions	Coordination Number	1*	2†	Ions	Coordination Number	1*	2†
Ag^{1+}	2	0.067	0.075	Be^{2+}	3	0.017	0.025
	4	0.102	0.110		4	0.027	0.035
	5	0.112	0.120	Bi^{3+}	5	0.099	0.107
	6	0.115	0.123		6	0.102	0.110
	7	0.124	0.132		8	0.111	0.119
	8	0.130	0.138	Bk^{3+}	6	0.096	0.104
Ag^{3+}	4	0.065	0.073	Bk^{5+}	8	0.093	0.101
Al^{3+}	4	0.039	0.047	Br^{1-}	6		0.188
	5	0.048	0.056	Br^{7+}	4	0.026	
	6	0.053	0.061	C^{4+}	3	-0.008	
Am^{3+}	4	0.101	0.108		4		
Am^{4+}	8	0.095	0.103	Ca^{2+}	6	0.100	0.108
As^{5+}	4	0.0335	0.042		7	0.107	0.115
	6	0.050	0.058		8	0.112	0.120
Au^{3+}	4	0.070	0.078		9	0.118	0.126
	6				10	0.128	0.136
B^{3+}	3	0.002	0.010		12	0.135	0.143
	4	0.012	0.020	Cd^{2+}	4	0.080	0.088
Ba^{2+}	6	0.136	0.144		5	0.087	0.095
	7	0.139	0.147		6	0.095	0.103
	8	0.142	0.150		7	0.100	0.108
	9	0.147	0.155		8	0.107	0.115
	10	0.152?	0.160		12	0.131	0.139
	12	0.160	0.168	Ce^{3+}	6	0.101	0.109

NOTE: All radii are given in nanometer units.
SOURCES: *Reprinted courtesy of the International Union of Crystallography from Shannon, R. D., and Prewitt, C. T., 1969. Effective ionic radii in oxides and fluorides. *Acta Crystallographica* 25B:928–929, Table 1. Radii based on an oxygen radius of 0.140 nm (sixfold coordination).

† Reprinted with permission from *Geochimica et Cosmochimica Acta*, 34, by Whittaker, E. J. W., and Muntus, R., Ionic radii for use in geochemistry, pp. 952–953. Copyright © 1970, Pergamon Press, Ltd. Radii based on an oxygen radius of 0.132 nm (sixfold coordination).
**LS and HS refer to low-spin and high-spin radii, respectively.

TABLE A5.1 (*continued*)

Ions	Coordination Number		1*	2†
	8		0.114	0.122
	9		0.115	
	12		0.129	0.137
Ce^{4+}	6		0.080?	0.088
	8		0.097	0.105
Cf^{3+}	6		0.095	0.103
Cl^{1-}	4			0.167
	6			0.172
	8			
Cl^{5+}	3		0.012	0.020
Cl^{7+}	4		0.020	0.028
Cm^{3+}	6		0.098	0.106
Cm^{4+}	8		0.095	0.103
Co^{2+}	4	HS**	0.057	0.065
	5	LS**	0.065	0.073
		HS	0.0745	0.083
Co^{3+}	6	LS	0.0525	0.061
		HS	0.061	0.069
Cr^{2+}	6	LS	0.073	0.081
		HS	0.082	0.091
Cr^{3+}	6		0.0615	0.070
Cr^{4+}	4		0.044	0.052
	6		0.055	0.063
Cr^{5+}	4		0.035	0.043
	8		0.057	
Cr^{6+}	4		0.030	0.038
Cs^{1+}	6		0.170	0.178
	8		0.182?	
	9		0.178	0.186
	10		0.181	0.189
	12		0.188	0.196
Cu^{1+}	2		0.046	0.054
Cu^{2+}	4	(square)	0.062	0.070
	5		0.065	0.073
	6		0.073	0.081
Dy^{3+}	6		0.0912	0.099
	8		0.103	0.111
Er^{3+}	6		0.0890	0.097
	8		0.100	0.108
Eu^{2+}	6		0.117	0.125
	8		0.125	0.133
Eu^{3+}	6		0.0947	0.103
	7		0.103	
	8		0.107	0.115
F^{1-}	2		0.1285	0.121
	3		0.130	0.122
	4		0.131	0.123
	6		0.133	0.125
Fe^{2+}	4	HS	0.063	0.071
	6	LS	0.061	0.069
		HS	0.0780	0.086
Fe^{3+}	4	HS	0.049	0.057
	6	LS	0.055	0.063
		HS	0.0645	0.073
Ga^{3+}	4		0.047	0.055

Ions	Coordination Number		1*	2†
	5		0.055	0.063
	6		0.0620	0.070
Gd^{3+}	6		0.0938	0.102
	7		0.104	
	8		0.106	0.114
Ge^{4+}	4		0.040	0.048
	6		0.0540	0.062
Hf^{4+}	6		0.071	0.079
	8		0.083	0.091
Hg^{1+}	3		0.105	
Hg^{2+}	2		0.069	0.077
	4		0.096	0.104
	6		0.102	0.110
	8		0.114	0.122
Hg^{3+}	6		0.0901	0.098
	8		0.102	0.110
I^{1+}	8			0.197
I^{5+}	6		0.095?	0.103
In^{3+}	6		0.0800	0.088
	8		0.0923	0.100
Ir^{3+}	6		0.073?	0.081
Ir^{4+}	6		0.063	0.071
K^{1+}	6		0.138	0.146
	7		0.146?	0.154
	8		0.151?	0.159
	9		0.155?	0.163
	10		0.159?	0.167
	12		0.160?	0.168
La^{3+}	6		0.1045	0.113
	7		0.110	0.118
	8		0.118	0.126
	9		0.120?	0.128
	10		0.128	0.136
	12		0.132?	0.140
Li^{1+}	4		0.059	0.068
	6		0.074	0.082
Lu^{3+}	6		0.0861	0.094
	8		0.097	0.105
Mg^{2+}	4		0.058	0.066
	5		0.067	
	6		0.0720	0.080
	8		0.089	0.097
Mn^{2+}	6	LS	0.067	0.075
		HS	0.0830	0.091
	8		0.093	0.101
Mn^{3+}	5		0.058	0.066
	6	LS	0.058	0.066
		HS	0.0645	0.073
Mn^{4+}	6		0.054	0.062
Mn^{6+}	4		0.027	0.035
Mn^{7+}	4		0.026	0.034
Mo^{3+}	6		0.067	0.075
Mo^{4+}	6		0.0650	0.073
Mo^{5+}	6		0.063	0.071
Mo^{6+}	4		0.042	0.050

TABLE A5.1 (*continued*)

Ions	Coordination Number		1*	2†	Ions	Coordination Number		1*	2†
	5		0.050	0.058	Pt^{2+}	4		0.060	
	6		0.060	0.068		6			0.068
	7		0.071	0.079	Pt^{4+}	6		0.063	0.071
N^{6+}	3		− 0.012		Pu^{3+}	6		0.101	0.109
Na^{1+}	4		0.099?	0.107	Pu^{4+}	6		0.080?	0.088
	5		0.100?	0.108		8		0.096	0.104
	6		0.102	0.110	Ra^{2+}	8		0.148	
	7		0.113?	0.121		12		0.164	
	8		0.116?	0.124	Rb^{1+}	6		0.149	0.157
	9		0.132?	0.140		7		0.156?	0.164
Nb^{2+}	6		0.071?	0.079		8		0.160	0.168
Nb^{3+}	6		0.070	0.078		12		0.173	0.181
Nb^{4+}	6		0.069	0.077	Re^{4+}	6		0.063	0.071
Nb^{5+}	4		0.032?	0.040	Re^{5+}	6		0.052?	0.060
	6		0.064	0.072	Re^{6+}	6		0.052	0.060
	7		0.066	0.074	Re^{7+}	4		0.040	0.048
Nd^{3+}	6		0.0983	0.106		6		0.057	0.065
	8		0.112	0.120	Rh^{3+}	6		0.0665	0.076
	9		0.109?	0.117	Rh^{4+}	6		0.0615	0.070
Ni^{2+}	6		0.0690	0.077	Ru^{3+}	6		0.068	0.075
Ni^{3+}	6	LS	0.056	0.064	Ru^{4+}	6		0.067	0.071
		HS	0.060	0.068	S^{2-}	6			0.172
Np^{2+}	6		0.110	0.118	S^{6+}	4		0.020	
Np^{3+}	6		0.102	0.110	Sb^{3+}	4		0.077	0.085
Np^{4+}	8		0.098	0.106		5		0.080	0.088
O^{2-}	2		0.135	0.127	Sb^{5+}	6		0.061	0.069
	3		0.136	0.128	Sc^{3+}	6		0.0745	0.083
	4		0.138	0.130		8		0.087	0.095
	6		0.140	0.132	Se^{2-}	6			0.188
	8		0.142	0.134	Se^{6+}	4		0.029	0.037
Os^{4+}	6		0.0630	0.071	Si^{4+}	4		0.026	0.034
P^{5+}	4		0.017	0.025		6		0.0400	0.048
Pa^{4+}	6			0.109	Sm^{3+}	6		0.0958	0.104
	8		0.101			8		0.109	0.117
Pa^{5+}	8		0.091	0.099	Sn^{2+}	8		0.122	0.130
	9		0.095	0.103	Sn^{4+}	6		0.0690	0.077
Pb^{2+}	4		0.091	0.102	Sr^{2+}	6		0.113	0.121
	6		0.118	0.126		7		0.121	0.129
	8		0.129	0.137		8		0.125	0.133
	9		0.133	0.141		10		0.132	0.140
	11		0.139	0.147		12		0.140	0.148
	12		0.149	0.157	Ta^{2+}	6		0.067	0.075
Pb^{4+}	6		0.0775	0.086	Ta^{4+}	6		0.066	0.074
	8		0.094	0.102	Ta^{5+}	6		0.064	0.072
Pd^{1+}	2		0.059	0.067		8		0.069	0.077
Pd^{2+}	4		0.064	0.072	Tb^{3+}	6		0.0923	0.100
	6		0.086	0.094		7		0.102	
Pd^{3+}	6		0.076	0.084		8		0.104	0.112
Pd^{4+}	6		0.062	0.070	Tb^{4+}	6		0.076	0.084
Pm^{3+}	6		0.0997	0.104		8		0.088	0.096
Po^{4+}	8		0.108	0.116	Tc^{4+}	6		0.064	0.072
Pr^{3+}	6		0.1013	0.108	Te^{4+}	6		0.060	
	8		0.114	0.122	Th^{4+}	6		0.100	0.108
Pr^{4+}	6		0.078?	0.086		8		0.104	0.112
	8		0.096	0.107		9		0.109	0.117

TABLE A5.1 (*continued*)

Ions	Coordination Number	1*	2†	Ions	Coordination Number	1*	2†
Ti²⁺	6	0.086	0.094	V²⁺	6	0.079	0.087
Ti³⁺	6	0.067	0.075	V³⁺	6	0.0640	0.072
Ti⁴⁺	5	0.053	0.061	V⁴⁺	6	0.059	0.067
	6	0.0605	0.069	V⁵⁺	4	0.0355	0.044
Tl¹⁺	6	0.150	0.158		5	0.046	0.054
	8	0.160	0.168		6	0.054	0.062
	12	0.176	0.184	W⁴⁺	6	0.0650	0.073
Tl³⁺	6	0.0885	0.097	W⁶⁺	4	0.042	0.050
	8	0.100	0.108		6	0.060	0.068
Tm³⁺	6	0.0880	0.096	Y³⁺	6	0.0900	0.098
	8	0.099	0.107		8	0.1015	0.110
U³⁺	6	0.104	0.112		9	0.110	0.118
U⁴⁺	7	0.098	0.106	Yb³⁺	6	0.0868	0.095
	8	0.100	0.108		8	0.098	0.106
	9	0.105	0.113	Zn²⁺	4	0.060	0.068
U⁵⁺	6	0.076	0.084		5	0.068	0.076
	7	0.096	0.104		6	0.0750	0.083
U⁶⁺	2	0.045	0.053		8	0.090	
	4	0.048	0.056	Zr³⁺	6	0.072	0.080
	6	0.075	0.081		7	0.078	0.086
	7	0.088	0.096		8	0.084	0.092

FIGURE A5.1. *Approximate Covalent Radii of the Elements (Atomic Radii Versus Atomic Number)*

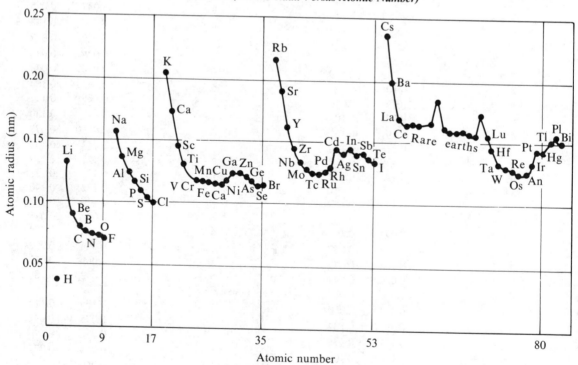

FIGURE A5.2. *Atomic Radii in Metallic Bonds*

Li	Be	B	C						
0.152	0.11–0.114	~0.10	0.071–0.077						
Na	Mg	Al	Si	P					
0.185	0.160	0.143	0.117	0.109					
K	Ca	So	Ti	V	Cr	Mn	Fe	Co	Ni
0.231	0.197	0.160–0.165	0.144–0.147	0.131	0.125	0.123–0.148	0.124	0.125	0.124
Cu	Zn	Ga	Ge	As	Se		Ru	Rh	Pd
0.128	0.133–0.145	0.122–0.140	0.122	0.125–0.157	0.116–0.173		0.132–0.135	0.134	0.137
Rb	Sr	Y	Zr	Nb	Mo	Tc	Os	Ir	Pt
0.246	0.215	0.180–0.183	0.158–0.161	0.143	0.136	0.135–0.136	0.134–0.136	0.135	0.138
Ag	Cd	In	Sn	Sb	Te				
0.144	0.149–0.164	0.162–0.169	0.140–0.159	0.145–0.168	0.143–0.173				
Cs	Ba	La	Hf	Ta	W	Re			
0.263	0.217	0.136–0.187	0.157–0.160	0.143	0.137	0.137–0.138			
Au	Mg	Tl	Pb	Bi	Po				
0.144	0.150	0.170–0.173	0.175	0.155–0.174	0.164–0.167				
			Th	Pa	U				
			0.180	0.160–0.162	0.150				

All ranges of radii are given in nm.

GLOSSARY

absorption edge Characteristic wavelength of an element where its absorbing power drops steeply.

acicular Describing greatly elongated small diameter crystals.

acute bisectrix Line that bisects the acute angle between the optic axes of biaxial minerals. Symbol: BXA.

adamantine Describing luster similar to the brilliance of diamond.

amorphous A substance lacking the development of a regular crystal structure.

analyzer Component of a polarizing microscope producing plane polarized light, usually perpendicular to the polarizer.

angle of incidence Angle the incident beam makes (1) with the diffracting lattice plane (symbol: θ), or (2) with the face normal of the refracting surface of a crystal. Alternate expression: angle of refraction.

Ångstrom Unit of measure, $1\mathring{A} = 10^{-8}$ cm, used in crystal structure and x-ray diffraction studies before the introduction of SI units. *See* **nanometer**.

anhedral crystals Crystal faces, if any, are poorly developed.

anisotropic Optical character of crystals other than isometric.

antiphase boundary Interface between two arrays of atoms out of registry with each other.

apparent symmetry Symmetry revealed by crystal faces; may be higher than the true point group symmetry of the crystal.

asbestiform Crystallization habit characterized by strong and flexible fibers, equivalent to synthetic whiskers of metals and inorganic compounds.

asbestos Commercial term for marketable asbestiform varieties of minerals.

asymmetric unit Basic portion of the unit cell repeated by the equipoint number of the general position.

atomic coordinates Fractional coordinates of the position of atoms in the asymmetric unit of the unit cell. Symbols $x = u/a$, $y = v/b$, and $z = w/c$, where uvw are the actual distances from the origin of the unit cell to the atom along directions a, b, and c.

atomic scattering *See* **scattering of atoms**.

authigenic minerals Minerals formed by diagenesis.

average structure General or ideal structure of a crystal determined from x-ray diffraction data.

axial ratio Ratio of the length of the three (two if $a = b$) unit translations of the crystal. Either the b (triclinic, monoclinic, orthorhombic) or the c (tetragonal, hexagonal) unit translation is designated as unity.

back reflections (or diffractions) Diffractions over 90 degrees of 2θ.

basal cleavage *See* **platy cleavage**.

basic structures Coordination polyhedral frame structures containing a single coordinating polyhedron. *See* **derivative structures**.

basic symmetry operations Symmetry operations that cannot be broken into component operations.

Becke line A band of light concentrated around a mineral in contact with a media of different refractive index when viewed through a slightly unfocused microscope.

biaxial minerals Minerals having two optic axes (orthorhombic, monoclinic, and triclinic).

birefringence The measure of difference (demonstrated by different colors) between two refractive indices in a particular mineral grain or section.

botryoidal Describing a polycrystalline mineral specimen with smooth, curved surfaces.

Bragg equation Expressing conditions of diffraction in terms of radiation wavelength, interplanar spacing, and angle of incidence.

Bravais law Crystal faces are parallel with lattice planes, and the most prominent faces correspond to lattice planes with high reticular density.

Bravais lattices Lattices of the 14 distinct unit cells.

bridged oxygens Oxygens in a crystal structure that are shared by two (or more) equivalent coordinating polyhedra.

bulk composition Chemical composition of a rock.

bundles of fibers Columnar aggregates of (asbestiform) fibers.

calculated intensities The intensities of powder diffraction patterns calculated from structural parameters of a crystal. Contrast: observed intensities.

centric symmetry groups Three-dimensional symmetry (point and space) groups containing a center of inversion.

chain silicates Structures characterized by chains of Si tetrahedra as a primary frame. Equivalent expressions: inosilicates and metasilicates.

chain-width defects Deviations from the regular multiplicity (width) of chains in chain silicates.

characteristic radiation Characteristic narrow wavelength, high-intensity radiations emitted by the target of the x-ray tube.

chemical equilibrium State of one or more substances having equality of temperature, pressure, and chemical potential at every point of the system.

circular sections Sections perpendicular to the optic axes in indicatrices.

clay minerals Minerals with particles of less than 2 micrometers that display plastic properties when hydrated. Also refers to the common hydrous layer silicates that are the major minerals in clays.

cleavage Property that causes a mineral to break along planes of weakest bonding before fracturing along other surfaces.

cleavage energy Bond energy per unit area of a cleavage plane.

closed crystal forms Crystal forms with faces terminating the crystal in all directions.

close-packed structures Similar to closest-packed, but less efficient and having lower coordination number than 12.

closest-packed structures The most efficiently packed structures of like atoms with coordination number 12. Common examples: cubic closest-packed (CCP) and hexagonal closest-packed (HCP).

cockscomb Describing an aggregation of tabular and curved crystals resembling cockscomb. Example: marcasite.

columnar Describing mineral aggregates of parallel prismatic or acicular crystals.

composition plane The common plane of two individual crystals in a twin.

composition space An n-dimensional space defined by n independent chemical units.

compound symmetry operations Symmetry operations composed of two basic operations and repeating a motif after both operations are performed.

conchoidal fracture Describing the curved surfaces of the fracture fragments of a mineral.

congruent symmetry operations Repeating a motif without reversing its sense (right-handedness or left-handedness). Equivalent expressions: operation of the first sort, proper operation. Includes: translation, rotation, and screw axes.

conoscopic illumination Mode of operation of a microscope in which the incident light is focused to a point inside the mineral.

cooperative scattering *See* **diffraction.**

coordination Formation of chemical bonds by atoms or ions with a number of other atoms or ions.

coordination number Number of closest neighbors of an atom or ion with which it forms chemical bonds.

coordination polyhedron Polyhedron formed around an atom or ion by connecting the centers of the coordinated atoms or ions.

Coulomb's law (or force) Two oppositely charged ions will be attracted to each other by a force proportional to the product of the charges divided by the square of the distance between them.

coupled substitution Simultaneous substitution of two or more different ions in a structure in such a way that charge balance is maintained.

covalent bonding Ideal chemical bonding that involves sharing of orbital electrons between elements that have little or no difference in electronegativity.

crystal classes Space point groups.

crystal field Electric field generated by the net negative charge of the anions (ligands) bonded to a cation.

crystal field splitting Separation of the three d levels of a cation because of the geometry of their orbitals in a coordination polyhedron.

crystal structure Spatial arrangement of atoms or ions and their bonds.

crystal systems The six symmetrically distinct crystallographic coordinate systems that correspond to the seven lattice symmetries. The extra lattice symmetry is $\bar{3}2/m$ and is treated as hexagonal. Each system has a crystallographic name, such as triclinic or monoclinic.

crystallization habit Characteristic shape, size, and other properties of minerals or mineral aggregates grown under given conditions.

crystallographic axes Directions of the unit translations a, b, and c that define the lattice and the unit cell of a crystal structure.

crystallographic zone (1) A collection of parallel crystal faces, or (2) faces that meet at a common point.

d **value** *See* **interplanar spacing.**

defects Imperfections, either point or planar, in the crystal structure of a mineral.

dendritic Describing a mineral surface resembling moss or tree branches.

density mass of a unit volume of mineral (expressed in g/cm^3, or in Mg/m^3 in the SI system).

derivative structures Derived from basic structures by substituting lower valent cations in the polyhedral frame and adding extra-frame cations.

diagenesis Rock formation process in the existing chemical-physical environment of a sedimentary basin.

diffraction Cooperative scattering of light from a grating, or scattering of x-rays from a crystal.

diffraction angle Angle between the diffracted beam and diffracting lattice plane of a crystal, equivalent to angle of incidence. Symbol: θ.

diffraction cone Concentric directions of diffractions from a lattice plane of a powdered sample are represented on the cone surface.

diffraction line or arc Recording of a segment of a diffraction cone in a powder camera.

diffraction peak Profile of interception of the diffraction cone by a detector as recorded on chart paper in a diffractometer.

dimensional extent of frames Number of dimensions (0, 1, 2, 3) in which the frame is not terminated.

direct space Space containing the lattice of a crystal, in contrast to reciprocal space containing the reciprocal lattice.

dislocation *See* **edge dislocation** or **screw dislocation**.

disordered structure Random distribution of two or more cations in the same type of coordination polyhedra.

displacive transition *See* **second-order transition**.

distinct unit cells The 14 primitive and multiple unit cells in which the crystallographic axes coincide either with the major rotation axes, or with face normals of mirror planes of the point group symmetry of the crystal, or with both. Equivalent expression: symmetrical unit cells.

divariant equilibrium Equilibrium of *n* phases in an *n* component system with two degrees of freedom.

double refraction Separation of two different velocity rays in a crystal.

double terminated crystals Having faces at both ends of a prismatic or acicular crystal.

drusy Describing a coating layer of small crystals on the surface of another mineral.

ductile Tenacity of a mineral that can be hammered into wire.

edge dislocation Plane of misregistry in a crystal that has one more row of atoms (or lattice points) on one side than on the other.

electron affinity Measure of the energy given up by an atom when it gains an electron and becomes an anion.

electronegativity Measure of the tendency of elements to acquire electrons.

electrostriction effect Property of certain minerals to change shape in response to an electric current.

enantiomorphic symmetry operation Repeating a motif by reversing its sense (right-handedness or left-handedness). Equivalent expressions: operation of the second sort, improper operation. Includes: inversion, reflection, glide, and roto-inversion.

energy dispersive diffraction Application of the Bragg equation at a constant diffraction angle by using an energy dispersive detector.

entropy Measure of the effect of changing temperature on Gibbs free energy at constant pressure.

epitaxial growth (or overgrowth) Growth of one mineral on the surface of another with a plane of registry between the two crystal structures.

equant Describing crystals of approximately equal development in all three dimensions. Equivalent expression: equidimensional.

equipoint number Expresses the multiplicity of a general point in a symmetry group. The corresponding equipoint number of special positions is decreased by a factor expressed by the rank of the (first two) symmetry operations occupied by the point.

euhedral crystals Displaying all or most faces of one or more crystal forms.

exsolution Process by which a homogeneous solid solution separates into two (or more) different minerals.

extinction Blackening of biaxial mineral sections in a polarizing microscope at a given inclination to the vibration directions of the polarizer and the analyzer.

extraordinary ray Ray that vibrates parallel with the optic axis in uniaxial minerals.

face normal Direction perpendicular to a crystal face.

facies diagram Graphical representation of stable mineral assemblages at specified pressure and temperature conditions.

ferrimagnetic Describing a mineral with oppositely aligned atomic dipoles of different magnitude that imparts a net magnetic moment. Example: magnetite.

ferromagnetic Describing a mineral that has a parallel alignment of atomic dipoles. Example: ilmenite.

fiber Crystal that resembles an organic fiber in shape and in apparent properties.

fibril Occasionally used in reference to the smallest component of apparently single (asbestiform) fibers.

fibrous Describing a mineral that appears to be composed of fibers, but is not necessarily separable into smaller fibers.

filiform Describing a wire-shaped mineral.

first-order transition Polymorphic change involving reconstruction of chemical bonds and discontinuous changes in volume and entropy. Example: kyanite and sillimanite. Alternate expression: reconstructive transition (or inversion).

fluoresence Process by which a mineral emits visible light in response to irradiation at a higher energy level (e.g., ultraviolet light).

fractional crystallization Process of rock formation in which the early formed crystals are removed from the melt.

fracture The different patterns and shapes of fragments typically produced by a mineral when crushed. Example: irregular, conchoidal.

Frenkel defect A point defect consisting of a foreign atom occupying an interstitial site in a crystal structure, usually coupled with a Schottky defect.

front reflections Diffractions with lower than 90 degrees of 2θ.

general forms Crystals with face normals in general position.

general positions Positions in a symmetry group not coincident with symmetry operations.

general radiation Mixed wavelength x-rays emitted by a target in an x-ray tube.

Gibbs free energy Measure of the change in energy of a substance brought about by independent changes in pressure and temperature.

Gibbs phase rule The maximum number of variables (pressure, temperature, and chemical elements) that can be changed without altering the number of coexisting phases is the number of components minus the number of phases plus two.

glide plane Compound symmetry operation that repeats a motif after performing a reflection and a translation, τ, located in the plane of reflection. Symbol: g.

gnomonic projection Projection of points located on the surface of the sphere of projection from the center of the sphere of projection onto a flat surface located at the north pole, normal to north-south axis.

grouping of polyhedra Number of identical or similar polyhedral units (of the same dimensional extent) linked together. Example: double chains.

hackly fracture The jagged, rough surfaces of fracture fragments.

hardness Relative resistance of mineral surfaces to scratches.

Haüy's law Crystal faces tend to have Miller index intercepts of the crystallographic axes.

hexagonal sheet The symmetrical sheet of atoms or anions with the plane group symmetry of *p6mm*.

hexagonal system Crystals in system have lattice symmetry of either $\bar{3}\,2/m$ (rhombohedral) or $6/m\,2/m\,2/m$ (hexagonal, or trigonal for point groups containing threefold axes).

high-low transition (or inversion) *See* **second-order transition**.

high-spin state The electron distribution of certain transition metals in the presence of a weak crystal field splitting.

holosymmetric (or normal) crystal class Centric point group in isogonal point groups.

Hund's rule In the orderly structure of atoms, an electron must be placed in each orbital of equal energy before an electron with opposite spin can be placed there.

hybrid orbitals Describing distribution of a shared electron in a covalent bond as a combination of atomic orbitals.

imperfections Discontinuities or interruptions in crystal structure.

improper operation or rotation *See* **enantiomorphic symmetry operation**.

indicatrix Closed geometric body with principal directions representing the vibration directions of refractive indices of a mineral.

integrated intensity Total energy of a diffracted beam as measured by the density of diffraction lines or by the area of diffraction peaks.

interference color Color observed when a mineral is viewed with crossed nicols (polarizer and analyzer). Coloration is related to unequal transmission of component wavelengths of white light.

interference figure Image obtained on the back side of the objective lens of a microscope operated in conoscopic light with crossed nicols.

internal energy Measure of the change in energy of a substance brought about by independent changes in entropy and volume.

interplanar spacing Shortest distance between translationally equivalent lattice planes. Symbol: d.

inversion Basic symmetry operation that repeats a motif by equidistant projection across a point. Symbol: i.

ionic bonding Ideal chemical bonding electrostatic in nature and formed between elements having large differences in electronegativity.

ionic radius Radius of the sphere effectively occupied by an ion in a particular structural environment.

ionization energy Energy required to remove an electron from an atom to create a cation.

island silicates Silicates with single or multiple isolated tetrahedral groups for primary frame. Equivalent expressions: nesosilicates or orthosilicates (single tetrahedra), pyrosilicates (double tetrahedra), sorosilicates (two or more tetrahedra).

isochromatic rings Concentric rings of light representing equal retardation that emanate from a mineral in conoscopic illumination.

isogonal symmetry groups Symmetry groups derivable from the same set of rotation axes and containing the same rotation axis or a compatible set of rotation axes.

isogyre Component of an interference figure that represents coincidence of the principal vibration directions of a mineral with the polarizer and analyzer of a microscope.

isometric (or cubic) system Crystals in system have lattice symmetry of $4/m\,\bar{3}\,2/m$. Each axis is perpendicular to the plane of the other two axes, and all axes have unit translations of equal magnitude.

isomorphism Solid solution series in which the crystalline structure is the same throughout the series.

isostructural Relationship between minerals of different chemical compositions and identical crystal structures.

isotropic Optical character of an isometric mineral.

isotropic section Circular section of the indicatrix that is perpendicular to the optic axis of the mineral.

isotypism Relationship between minerals constructed by different stackings of the same structural unit, but containing different cations in certain structural sites.

K **absorption edge** Absorption edge of the K shell of an element.

K **(α and β) radiations** Characteristic radiations of an element corresponding to the energy release emitted by electron transitions in the K shell.

latent heat of crystallization Measure of energy liberated through the crystallization of a melt or solution at constant temperature.

lattice Three-dimensional representation of translational symmetry of a crystal structure.

lattice parameters The magnitude and angles between unit translations of a lattice.

lattice points Translationally equivalent points in space.

lattice symmetries Centric space point groups compatible with the symmetries of parallelepipeds.

law of crystallography *See* **Haüy's law**.

law of interfacial angles *See* **Steno's law**.

layer (1) In crystal structure, a two-dimensional structural unit such as the tetrahedral layers in micas; (2) in a symmetrically packed structure, the space between two adjacent symmetrical sheets of atoms or anions.

layer silicates Structures with layers of tetrahedra as primary frame. Equivalent expressions: sheet silicates, phyllosilicates, disilicates.

layer unit cell Defined by two unit translations of symmetrical sheets and by the shortest unit translations between them.

ligand refers to the collection of

anions coordinated about a cation.

limiting sphere Portion of the reciprocal lattice that can be diffracted by x-rays of a given wavelength. Radius of the limiting sphere is twice the wavelength of the monochromatic radiation.

liquidus Locus of melt compositions in equilibrium with one or more minerals.

loop Polyhedral ring within a three-dimensional network of polyhedra.

low-spin state Electron distribution of certain transition metals in the presence of strong crystal field splitting effect.

luster Type and nature of reflection of light from the surface of minerals.

Madelung constant A number related to the geometric arrangement of ions in a crystal structure. When multiplied by electrostatic energy of an ion pair, the result is the energy of the structure.

magnetic moment Intensity of the magnetic field produced by orbital electrons about an atom.

magnetic quantum number Number between -1 and $+1$ that specifies the position of an electron in one or more orbitals within each atomic subshell. Symbol: m.

malleable Describing tenacity of a mineral that can be hammered into different shapes without breaking.

mass absorption coefficient Relative absorption power of an element for a given wavelength of x-rays. Coefficient is proportional to atomic number of the element. Symbol: μ.

massive Describing a mineral aggregate composed of small and indistinct crystals.

mesh Two-dimensional equivalent of a three-dimensional lattice.

metallic bonding Chemical bonding between atoms in which the electrons are highly delocalized and free to move from one atom to another.

metamorphism Mineralogical and textural change in a rock induced by changes in temperature or pressure or both.

metasomatism Process by which bulk chemistry of a rock is changed.

metastable equilibrium State of two or more minerals that are at equilibrium but have not achieved the lowest free energy state under conditions of interest.

Miller indices *See* **rational lattice planes**.

mineral groups Groups or series of minerals of similar crystal structure containing different cations in equivalent structural sites. Example: feldspars.

mineral series Groups of minerals of the same crystal structure containing a variable proportion of two or more cations in the same structural site. Example: olivines.

mineral varieties Minerals of the same group with different crystallization habits or different trace element contents resulting in conspicuously dissimilar appearances. Example: crocidolite, asbestiform riebeckite.

mineralogical phase rule Under equilibrium conditions a rock describable by n components will have n or fewer phases (minerals) in each of its possible assemblages.

minerals Naturally occurring, usually inorganic and crystalline substances.

mirror plane A basic symmetry operation across which a mirror image is generated.

miscibility gap Region of composition space with a specific temperature and pressure for which no corresponding minerals exist.

mole Avogadro's number of chemical entities.

molecular crystals Crystals containing discrete molecules bonded together by van der Waals forces.

monochromatic Describing single-wavelength light or x-rays.

monoclinic system Crystals in system have lattice symmetry of $2/m$. Relative magnitude of unit translation not restricted. Twofold axis perpendicular to mirror plane, containing other two axes. (Monoclinic = one axis inclined.)

multiple unit cell These cells contain more than one lattice point and are: side-centered (A, B, C), face-centered (F), body-centered (I), or rhombohedral (R).

multiplicity factor Expresses number of equivalent lattice planes of a given hkl index. Used as correction factor in obtaining true intensity of a lattice plane.

nanometer SI unit of length used in crystal structures and x-ray diffraction. $1 \text{ nm} = 10^{-9}$ m. $1 \text{ nm} = 0.1$ Å.

network silicates Silicates containing three-dimensional networks of tetrahedra in primary frame. Equivalent expression: tectosilicates.

Neumann's principle or Neumann–Curie principle *See* **symmetry principle**.

nicols *See* **polarizer; analyzer**.

obtuse bisectrix Line bisecting the obtuse angle between optic axes of biaxial minerals. Symbol: BXO.

open crystal forms Forms that leave the crystal unterminated in some directions.

open frame structures Polyhedral frame structures of three-dimensional networks containing large openings. These structures may or may not be symmetrically packed structures.

optic angle Acute angle between the optic axes of biaxial minerals. Symbol: $2V$.

optic axial plane Plane containing both optic axes in a biaxial mineral. Symbol: OAP.

optic axis Direction perpendicular to circular sections of uniaxial and biaxial minerals. Symbol: OA.

optic sign For uniaxial minerals, the sign is positive if $n > n$, and negative if the opposite. For biaxial minerals, the optic sign is positive if $Z = $ BXA, and negative if $X = $ BXA.

order of forms Expression ranking identical crystal forms by order of numerical values of the h, k, l indices.

ordered structures Structures with only one kind of cation (or fractions of cations) in a structurally distinct coordination polyhedron. Example: Si/Al in different tetrahedra in microcline.

ordinary ray Direction of ray vibration is perpendicular to optic axis of a uniaxial mineral.

origin of the lattice Lattice point chosen as initial point in description of a lattice. Point can be located anywhere in crystal structure, and need not coincide with an atom.

orthorhombic system Crystal system with lattice symmetry $2/m\,2/m\,2/m$. Relative magnitude of unit translations not restricted, but each axis must be perpendicular to mirrors in the plane of the other two axes. (Orthorhombic = orthogonal axes.)

orthoscopic illumination Parallel bundles of light rays passing through a mineral in polarizing microscope.

parallelepiped Geometric body defined by three pairs of parallel faces.

paramagnetic Describing minerals that have no net magnetic moment due to random orientation of dipole moments of atoms.

parting Property of some minerals to break along crystallographic planes weakened by imperfections or impurities.

Pauli exclusion principle If two electrons have magnetic moments in the same direction, they must occupy different energy levels.

Pauling bond strength Measure of bond strength between a cation and its coordinated anions expressed by the cation's valence divided by coordination number.

Pauling's rules Five rules describing common structural features of ionic crystals.

peak height intensity Intensity of diffraction measured by peak height on diffractometer charts or by maximum density of a diffraction line in films.

periodic bond chain refers to bond between atoms in a chain, which is stronger than bond that would hold the chain to the surface of a crystal. Symbol: PBC.

periodicity of a frame Number of polyhedra in a zero-dimensional extended frame, number of polyhedra in asymmetric unit (repeat unit) in a one-dimensional extended frame, or number of polyhedra in a loop of two-dimensional or three-dimensional extended frames.

phosphorescence Fluorescence in which emission of visible light persists after cessation of irradiation.

piezoelectric effect Production of electric current in acentric minerals in response to pressure.

plane groups Equivalent of space groups in two-dimensional space.

plane polarized light Light constrained to vibrate only in a plane.

plastic flow Process of crystal deformation that does not permanently break chemical bonds.

plastic property Ability of minerals to constitute moldable masses when hydrated. Example: clays.

platy (or basal) cleavage Cleavage pattern of a mineral with only one cleavage plane.

point groups Translation-free equivalents of space groups that describe the symmetries of crystals. There are 10 plane and 32 space groups. Alternate expression: crystal classes.

point symmetry Symmetry of space around a point in plane or space groups.

polar direction A direction in a crystal that has no symmetrically equivalent opposite direction.

polarizer Component of a polarizing microscope that produces plane polarized light, usually in a direction perpendicular to the polarization of the analyzer.

pole symmetry Symmetry of a direction in a space group or space point group.

polyhedral frame One type of coordination polyhedra in a crystal structure that balances half or more than half of the available anion valence.

polyhedral frame structures *See* **primary frame**.

polymorphism Relationship between minerals of the same chemical composition but having different crystal structures.

polysomatic series A relatively large group of minerals (and mineral groups) with a common structural unit in their crystal structures. Example: biopyriboles.

polysynthetic twinning Repeated twinning of a crystal by the same twin law.

polytypism Relationship between minerals of the same chemical composition with different stackings of the same structural unit.

powder pattern or powder x-ray diffraction pattern Collection of diffractions from a powdered crystalline sample.

primary frame Structures characterized by presence of a polyhedral

frame; may or may not be symmetrically packed.

primary voids The regular polyhedra constructed by connecting the centers of the anions surrounding a void. In simple, stuffed, symmetrically packed structures, centers of primary voids are occupied by cations. Example: Na in octahedral voids of Cl in halite.

primitive unit cell Unit cell containing collectively only one lattice point, one eighth at each corner of cell. Symbol: P.

proper operations *See* **congruent symmetry operations**.

pseudohexagonal, pseudoisometric, pseudotetragonal Crystals that have apparent hexagonal, isometric, or tetragonal symmetry but in fact have a lower real symmetry.

pseudomorph One mineral replacing another and adopting its shape.

pyroelectric effect Production of electric current in acentric minerals in response to heat.

radius ratio In crystal-chemical context, the radius of the cation divided by the radius of the anion (taken as unity).

random stacking Absence of stacking order in symmetrically, closest, or close-packed structures.

rank of symmetry operations Number of equivalent positions created by a symmetry operation.

rational features Lattice points and those lattice directions and planes in a lattice containing lattice points.

rational lattice directions Directions in a lattice that contain lattice points, defined by coefficients of the vector sum from the origin to the first lattice point along the direction.

rational lattice planes Lattice planes containing lattice points, defined by reciprocals of the fractions of the unit translations (h, k, l) at intersection of first equivalent plane from origin. Equivalent expression: Miller indices of lattice planes.

ray velocity surface Closed geometric body representing velocity of light transmitted by a crystal in different directions.

real structure Actual small-scale structure of a crystal, containing aberrations from the ideal or average structure.

reciprocal lattice Representation of lattice planes by points defined at a distance reciprocal to the magnitude of the interplanar spacings.

reciprocal lattice plane A plane in the reciprocal lattice containing points representing parallel lattice planes, equivalent to a crystallographic zone.

reciprocal lattice unit Unit of measure used in illustration of the reciprocal lattice, and equal to the reciprocal of the appropriate interplanar spacings multiplied by a factor of enlargement. Symbol: RLU.

reconstructive transition (or inversion) *See* **first-order transitions**.

reduced unit cell Primitive unit cell of the lattice, defined by the three shortest noncoplanar unit translations.

reference sheet The first, arbitrarily chosen sheet in the description of the stacking sequence of a symmetrically packed structure. Reference sheet for hexagonal structures is A, for tetragonal structures S.

reflection plane *See* **mirror plane**.

refractive index A numerical value inversely proportional to the velocity of light through a mineral in a given direction. Isometric minerals have one refractive index (symbol: n), uniaxial minerals have two (symbols: n_ϵ and n_ω) and biaxial minerals have three (symbols: n_α, n_β, and n_γ).

relief The contrast or degree of visibility of transparent minerals relative to their surroundings.

reniform Describing a polycrystalline mineral specimen shaped like a kidney.

repeat unit Number of polyhedra in periodic unit of a polyhedral chain. Symbol: RU.

retardation Measure in nm of the lag of a slower light ray behind a faster ray after passage through a mineral of arbitrary thickness.

reticular density Concentration of lattice points in a rational lattice plane.

reticulated Describing a mineral aggregate of interlaced parallel groups of acicular crystals.

rhombohedral system *See* **hexagonal system**.

ring silicates Silicate structures with isolated rings of tetrahedra for

primary frame. Equivalent expression: cyclosilicates.

saturated solution Solution containing maximum quantity of dissolved phases under given temperature and pressure conditions.

scalar properties Mineral properties without directional character. Example: temperature and density. Alternate expression: zero rank tensor.

scattering of atoms Property of atoms to emit x-rays of the same wavelength when placed in an x-ray beam.

scattering factor Factor expressing the scattering power of an atom for different wavelengths of radiation at different angles of incident beam. Symbol: f with subscript of element.

Schottky defect A point defect; a vacant site in a crystal structure, usually coupled with a Frenkel defect.

Schrödinger equation Probability of finding an electron at any given time and position in terms of electron mass and potential energy at that time.

screw axis Compound symmetry operation repeating a motif after performing a rotation and translation. Symbol: n_m.

screw dislocation Spiral dislocation of structural planes (or lattice planes) around an axis.

second-order transition A continuous polymorphic change, without breaking structural bonds, associated with continuous changes in volume and entropy. Example: high and low quartz. Alternate expressions: high-low transition, displacive transition (or inversion).

secondary frame In crystal structure a frame composed of essentially regular coordination polyhedra, but unlike the primary frame, it balances less than half of the valence of available anions.

secondary voids Faces or edges of primary voids in a symmetrically packed structure. Example: triangular coordination of some Cu in covellite.

sectile Describing tenacity of a mineral that can be cut into shavings with a knife.

semirational lattice planes These

planes are parallel to rational lattice planes. Their interplanar spacings are integer fractions of interplanar spacings of the parallel lattice planes.

sharing coefficient Average number of polyhedra sharing an average polyhedral corner. Symbol: S.

short wave limit Shortest wavelength of the general x-ray radiation produced at a given potential between cathode and anode of an x-ray tube. Symbol: SWL.

simple substitution In a crystal structure, substitution of an ion by a different ion with the same charge.

Snell's law The ratio of the velocity of light in air and in another substance is equal to the angle of incidence divided by the angle of refraction.

solid solution Continuous range of mineral chemistries characterized by the same crystal structure.

solidus Locus of mineral compositions in equilibrum with a melt.

solubility curve Locus of saturated solution chemistries as a function of temperature or other variables.

solvus Locus of mineral chemistries representing the mutual saturation of each phase in the other as a function of temperature or other variables.

space groups Permissible three-dimensional symmetry groups of all types of symmetry operations. Space groups express the symmetries of crystal structures.

special forms Crystal forms with face normals in special positions.

special positions Positions in symmetry groups that coincide with symmetry operations.

specific gravity Density of a mineral with respect to unit density of water.

sphere of projection A sphere marking the intersection points of face normals projected from the center of the sphere.

stacking sequence (or order) Expression of relative orientation of symmetrical sheets in the asymmetric stacking unit of a symmetrically packed structure.

stalactitic Describing mineral samples either formed as stalactites in caves or resembling stalactites.

Steno's law Refers to constant interfacial angles between comparable crystal faces of the same mineral.

stereographic projection Projection of the sphere of projection from the south pole onto a flat surface located at the north pole, perpendicular to north-south axis.

stereoscopic pairs Two slightly rotated images of an object that give a three-dimensional picture of the object when viewed through a stereoscope.

streak Color of a powdered mineral.

streak plate Ceramic plate used for producing a streak of powdered mineral for color determination.

structure factor Entry in the expression of intensity of diffraction that depends on the crystal structure. The structure factor contains the scattering factor of the elements and their atomic coordinates. Symbol: F.

stuffed derivative structures Derived from basic structures by lower valence cation exchange in the polyhedra, and by balancing the negative charge of the frame with additional cations.

stuffed symmetrically packed structures Symmetrically packed structures in which the voids between the anions of the symmetrical sheets are occupied by cations.

sublimation Process of vaporization of a mineral without going through liquid phase.

substructure Small structural unit repeated in various orientations within the unit cell. Reciprocal expression: superstructure.

symmetrical packing index Percentage of space occupied by anions in a stuffed symmetrically packed structure after space between anions (due to larger cations) is deducted. Anion structure is thus contracted into an ideal closest-packed or close-packed structure model. Symbol: SPI.

symmetrical sheet Symmetrical two-dimensional array of atoms or anions. Atoms may or may not be in contact. *See* **hexagonal sheet, tetragonal sheet**.

symmetrically equivalent positions Positions in crystal structure related by symmetry operations.

symmetrically packed structures

Structures constructed by stacking symmetrical sheets. Atoms in adjacent sheets may or may not be in contact. Closest-packed and close-packed structures are also symmetrically packed, and are distinguished (1) by being composed by only one atom or of several structurally equivalent atoms, and (2) by the atoms being in contact between adjacent layers. Symbol: SP.

symmetry axes Rotation, rotoinversion, and screw axes.

symmetry center Inversion center.

symmetry elements Symmetry operations in a symmetry group.

symmetry groups Permissible collections of mutually supportive symmetry operations.

symmetry operations Operations that repeat a motif in a symmetrical pattern.

symmetry planes Mirror planes and glide planes.

symmetry principle The axiomatic relationship between symmetry and properties, stating that the symmetry of an effect is at least as great as the common symmetry of the crystal and the cause (of the effect). Equivalent expressions: Neumann or Neumann–Curie principle.

temperature factor Factor expressing the relative value of vibration of atoms and ions at given temperatures.

tenacity Property of a mineral describing its resistance to breaking, bending, crushing, or cutting. Examples: brittle, flexible.

tensor Mathematical quantity specifying how a physical property varies in one or more directions in a crystal.

tetragonal sheet Symmetrical sheet of atoms or anions with plane group symmetry *p4mm*.

tetragonal system Crystals in system have lattice symmetry $4/m2/m2/m$. All axes are perpendicular to the plane of the other two axes, and magnitude of unit translations of the latter two axes must be equal. (Tetragonal = tetrafold axis.)

toughness Describing the relative difficulty of breaking or crushing a mineral.

translation Basic symmetry opera-

tion that repeats a motif at regular intervals along a direction. Symbol: t, or a, b, c.

triclinic system Crystals in system have lattice symmetry of 1. Neither relative magnitude nor direction of unit translations is restricted. (Triclinic = three axes inclined.)

trigonal system *See* **hexagonal system**.

twin axis Rotational relationship between individual crystals in a twin.

twin laws Definitions of specific twinning patterns.

twin plane Mirror plane expressing the relationship between two individual crystals in a twin.

twinning Development of multiple crystals with common crystal structural planes.

uniaxial Describing minerals with a single optic axis (tetragonal and hexagonal).

unit cell Parallelepiped defined by three noncoplanar unit translations in a lattice.

unit cell content Number of chemical formulas contained in a unit cell. Symbol: Z.

unit translation Vector between two adjacent lattice points.

univariant equilibrium Equilibrium of $n + 1$ phases in an n-component system with one degree of freedom.

vitreous luster Describing glassy appearance of some minerals.

voids In symmetrically closest-packed or close-packed structures, the space between atoms or anions in the symmetrical sheets.

wavelength Distance between wave crests of electromagnetic waves expressed in nm (or in Å).

white radiation *See* **general radiation**.

x-ray fluorescence Method of chemical analysis based on knowledge of the characteristic wavelengths emitted by the elements when radiated by x-rays.

x-rays Electromagnetic radiation produced by sudden deceleration of electrons in an x-ray tube.

AUTHOR INDEX

SUBJECT INDEX

NOTE: Boldface type indicates the name of minerals and the pages on which their property tables and structures appear.

"A" sites in amphiboles, 344, 351
Absolute scale factor (ASF). *See* Powder pattern, indexing
Absorption
  coefficient (μ), 250
  edge, 236–237
  factor, 262
Acanthite, 391
Accessory plate
  gypsum, 285
  quartz, 285–286, 290–291
Acmite (aegerine), 345
Actinolite, 13, 14, 182, **342**, 345
Adamantine luster, 13
Adularia, 196
Agalite, 331. *See also* **Talc**
Akermanite, 223
Alabandite, 387
Alabaster, 442. *See also* **Gypsum**
Alamosite, 55
Albite, 303, **305**, 307
  bond lengths in, 198
  entropy and volume of, 196
  free energy of, 208
  stability diagram of, 227, 307
Albite law, twinning, 64–67, 308
Alkali, feldspars, 111, 198, 303–309
Allanite, 372–373
Almandine, 106, **356**
α plutonium, 134
Aluminosilicates, 362–365
  associations of, 212, 222, 224
  entropy and volume of, 196, 204
  free energy of, 200–202
  stability of, 198, 202, 207, 362
Alunite, 438, **441**
Amblygonite, **450**
Amesite, 328
Amethyst, 301. *See also* **Quartz**
Amosite, 346
Amphiboles, 118–119, 157, 341–347
  associations of, 214, 224–225, 341, 345–346
  biopyriboles, relations to, 352–353
  cation substitutions in, 345
  cleavage in, 13, 288, 340
  color in, 105

compositional zoning in, 181
exsolution in, 194
quadrilateral composition diagram, 341
Amphibolite facies, 220, 223–224
Analbite, 307
Analcime, 122, **309**, **310**, 318
Analyzer, petrographic microscope, 264, 278
Anandite, 328
Anatase, 196, **408**, **411**
Andalusite, 362, **363**, 364–365. *See also* Aluminosilicates
  coordination polyhedra in, 114, 198
  stability of, 202, 207, 362
Andradite, 106, **357**
Anglesite, 438, **440**
  structure. *See* **Anhydrite**
Angular interval, rotational symmetry, 23
Anhydrite, 9, 218, 438, **439**, **443**
Anionic groups, 8, 80
Anisotropic, optical, 265
Ankerite, 425
Annabergite, 451, 455
Annite, 328
Anorthite, 303, **305**, 307
  associations of, 212, 217
  bond lengths in, 198
  stability of, 227, 307
Anorthoclase, 308
Anthophyllite, 341, **342**
Antiferromagnetic, 164
Antigorite, 322
Antimony, **376**
Antiperthite, 308
Antiphase boundary (APB), 165–166
Antiphase domains, 166
Antiprism, coordination polyhedron, 143
Antlerite, 438, **441**, **443**
Apatite, 449
Apophyllite, **325**
Apparent symmetry, 56
Aquamarine, 106, 315. *See also* **Beryl**
Aragonite, 425, **428**, **430**
  entropy and volume of, 196

polymorphism of, 6, 15
twinning, 64
Arborescent habit, 11, 15
Arfvedsonite, 228, 360
Argentite, 391
Argentopyrite, 383
Arsenates. *See* Phosphates, arsenates and vanadates
Arsenic, **376**, **378**
Arsenopyrite, 223, **396**, **398**
Asbestiform habit, 11, 182–183, 331, 346
Asbestos, 182
ASTM cards. *See* Powder diffraction file
Astrophyllite, 118
Asymmetric unit, 32, 153
Atacamite, **402**
Atomic coordinates, 32
Atoms and ions
  Bohr, model of, 71
  shapes and sizes, 80
  wave, model of, 72–73
Attachment energy, crystal growth, 178
Attapulgite, 324. *See also* **Palygorskite**
Attractive energy. *See* Coulomb attractive energy
Augelite, **450**, **454**
Augite, 334, 336
  subcalcic, 334
Autunite, **453**
Auxiliary plates. *See* Accessory plates
Axial glides (a, b, c), 26
  ratio limits. *See* Radius ratio
  ratios, 50
  symmetry, bonds, 93
Axinite, 117
Azimuthal quantum number 1, 74–75
Azurite, 13, 425, **431**

Back reflections, x-ray diffraction, 242
Barite, 10, 438, **439**
  structure. *See* **Anhydrite**
Barite rosette, 217, 443
Barkevikite, 343
Barlow's symmetrical plans of atoms, 126

497

Student Survey

Tibor Zoltai and James H. Stout
MINERALOGY: CONCEPTS AND PRINCIPLES

Students, send us your ideas!

The authors and the publisher want to know how well this book served you and what can be done to improve it for those who will use it in the future. By completing and returning this questionnaire, you can help us develop better textbooks. We value your opinion and want to hear your comments. Thank you.

Your name (optional) _____ School _____

Your mailing address _____

City _____ State _____ ZIP _____

Instructor's name (optional) _____ Course title _____

1. How does this book compare with other texts you have used? (Check one)

   ☐ Superior ☐ Better than most ☐ Comparable ☐ Not as good as most

2. Circle those chapters you especially liked:

   Chapters: *Part I* 1 2 3 4 5 6 7 8 9 10 11 *Part II*

   Comments:

3. Circle those chapters you think could be improved:

   Chapters: *Part I* 1 2 3 4 5 6 7 8 9 10 11 *Part II*

   Comments:

4. Please rate the following. (Check one for each)

	Excellent	Good	Average	Poor
Logical organization	( )	( )	( )	( )
Readability of text material	( )	( )	( )	( )
General layout and design	( )	( )	( )	( )
Match with instructor's course organization	( )	( )	( )	( )
Illustrations that clarify the text	( )	( )	( )	( )
Up-to-date treatment of subject	( )	( )	( )	( )
Explanation of difficult concepts	( )	( )	( )	( )
Selection of topics in the text	( )	( )	( )	( )

OVER, PLEASE

5. List any chapters that your instructor did not assign. _____

_____

6. What additional topics did your instructor discuss that were not covered in the text? _____

_____

7. Did you buy this book new or used?   ☐ New   ☐ Used

    Do you plan to keep the book or sell it?   ☐ Keep it   ☐ Sell it

    Do you think your instructor should continue to assign this book?   ☐ Yes   ☐ No

8. After taking the course, are you interested in taking more courses in this field?   ☐ Yes   ☐ No

    Are you a major in geology?   ☐ Yes   ☐ No

9. Is a lab offered in conjunction with the course?   ☐ Yes   ☐ No

10. Did you purchase the Zoltai–Stout *Problems and Solutions Manual* that accompanies the text?
   ☐ Yes   ☐ No

11. **GENERAL COMMENTS:**

May we quote you in our advertising?   ☐ Yes   ☐ No

Please remove this page and mail to: Mary L. Paulson
                                         Burgess Publishing Company
                                         7108 Ohms Lane
                                         Minneapolis, MN 55435

THANK YOU!